Acta Numerica 2013

Acta Numerica

Volume 22 2013

CAMBRIDGE
UNIVERSITY PRESS

CAMBRIDGE
UNIVERSITY PRESS

Shaftesbury Road, Cambridge CB2 8EA, United Kingdom

One Liberty Plaza, 20th Floor, New York, NY 10006, USA

477 Williamstown Road, Port Melbourne, VIC 3207, Australia

314–321, 3rd Floor, Plot 3, Splendor Forum, Jasola District Centre, New Delhi – 110025, India

103 Penang Road, #05–06/07, Visioncrest Commercial, Singapore 238467

Cambridge University Press is part of Cambridge University Press & Assessment, a department of the University of Cambridge.

We share the University's mission to contribute to society through the pursuit of education, learning and research at the highest international levels of excellence.

www.cambridge.org
Information on this title: www.cambridge.org/9781107043862

First published 2013

A catalogue record for this publication is available from the British Library

ISBN 978-1-107-04386-2 Hardback

Contents

Acta Numerica (2013), pp. 1–131
doi:10.1017/S0962492913000032

Mixed-integer nonlinear optimization*†

Pietro Belotti
Department of Mathematical Sciences,
Clemson University,
Clemson, SC 29634, USA

Christian Kirches
Interdisciplinary Center for Scientific Computing,
Heidelberg University,
69120 Heidelberg, Germany
and
Mathematics and Computer Science Division,
Argonne National Laboratory,
Argonne, IL 60439, USA

Sven Leyffer
Mathematics and Computer Science Division,
Argonne National Laboratory,
Argonne, IL 60439, USA

Jeff Linderoth
Department of Industrial and Systems Engineering,
University of Wisconsin–Madison,
Madison, WI 53706, USA

James Luedtke
Department of Industrial and Systems Engineering,
University of Wisconsin–Madison,
Madison, WI 53706, USA

Ashutosh Mahajan
Industrial Engineering and Operations Research,
Indian Institute of Technology Bombay,
Powai, Mumbai, MH 400076, India

* Colour online for monochrome figures available at journals.cambridge.org/anu.

† This work was supported by the Office of Advanced Scientific Computing Research, Office of Science, US Department of Energy, under Contract DE-AC02-06CH11357.

Many optimal decision problems in scientific, engineering, and public sector applications involve both discrete decisions and nonlinear system dynamics that affect the quality of the final design or plan. These decision problems lead to mixed-integer nonlinear programming (MINLP) problems that combine the combinatorial difficulty of optimizing over discrete variable sets with the challenges of handling nonlinear functions. We review models and applications of MINLP, and survey the state of the art in methods for solving this challenging class of problems.

Most solution methods for MINLP apply some form of tree search. We distinguish two broad classes of methods: single-tree and multitree methods. We discuss these two classes of methods first in the case where the underlying problem functions are convex. Classical single-tree methods include nonlinear branch-and-bound and branch-and-cut methods, while classical multitree methods include outer approximation and Benders decomposition. The most efficient class of methods for convex MINLP are hybrid methods that combine the strengths of both classes of classical techniques.

Non-convex MINLPs pose additional challenges, because they contain non-convex functions in the objective function or the constraints; hence even when the integer variables are relaxed to be continuous, the feasible region is generally non-convex, resulting in many local minima. We discuss a range of approaches for tackling this challenging class of problems, including piecewise linear approximations, generic strategies for obtaining convex relaxations for non-convex functions, spatial branch-and-bound methods, and a small sample of techniques that exploit particular types of non-convex structures to obtain improved convex relaxations.

We finish our survey with a brief discussion of three important aspects of MINLP. First, we review heuristic techniques that can obtain good feasible solution in situations where the search-tree has grown too large or we require real-time solutions. Second, we describe an emerging area of mixed-integer optimal control that adds systems of ordinary differential equations to MINLP. Third, we survey the state of the art in software for MINLP.

CONTENTS

1. Introduction

Many optimal decision problems in scientific, engineering, and public sector applications involve both discrete decisions and nonlinear system dynamics that affect the quality of the final design or plan. Mixed-integer nonlinear programming (MINLP)[1] problems combine the combinatorial difficulty of optimizing over discrete variable sets with the challenges of handling nonlinear functions. MINLP is one of the most general modelling paradigms in optimization and includes both nonlinear programming (NLP) and mixed-integer linear programming (MILP) as subproblems. MINLPs are conveniently expressed as

$$\begin{cases} \underset{x}{\text{minimize}} & f(x), \\ \text{subject to} & c(x) \leq 0, \\ & x \in X, \\ & x_i \in \mathbb{Z}, \ \forall i \in I, \end{cases} \tag{1.1}$$

where $f : \mathbb{R}^n \to \mathbb{R}$ and $c : \mathbb{R}^n \to \mathbb{R}^m$ are twice continuously differentiable functions, $X \subset \mathbb{R}^n$ is a bounded polyhedral set, and $I \subseteq \{1, \ldots, n\}$ is the index set of integer variables. We note that we can readily include maximization and more general constraints, such as equality constraints, or lower and upper bounds $l \leq c(x) \leq u$. More general discrete constraints that are not integers can be modelled by using so-called special-ordered sets of type I (Beale and Tomlin 1970, Beale and Forrest 1976).

Problem (1.1) is an NP-hard combinatorial problem, because it includes MILP (Kannan and Monma 1978), and its solution typically requires searching enormous search trees: see Figure 1.1. Worse, non-convex integer optimization problems are in general undecidable (Jeroslow 1973). Jeroslow provides an example of a quadratically constrained integer program and shows that no computing device exists that can compute the optimum for all problems in this class. In the remainder of this paper, we concentrate on the case where (1.1) is decidable, which we can achieve either by ensuring that X is compact or by assuming that the problem functions are convex.

1.1. MINLP notation and basic definitions

Throughout this paper we use $x^{(k)}$ to indicate iterates of x and $f^{(k)} = f(x^{(k)})$ to denote the evaluation of the objective at $x^{(k)}$. Similar conventions apply to constraints, gradients, or Hessians at $x^{(k)}$; for example, $\nabla f^{(k)} = \nabla f(x^{(k)})$. We use subscripts to denote components; for example, x_i is the ith component of vector x. For a set $J \subset \{1, \ldots, n\}$ we let x_J denote the components

[1] Note that we use 'MINLP' to refer to both 'mixed-integer nonlinear programming' and 'mixed-integer nonlinear program'.

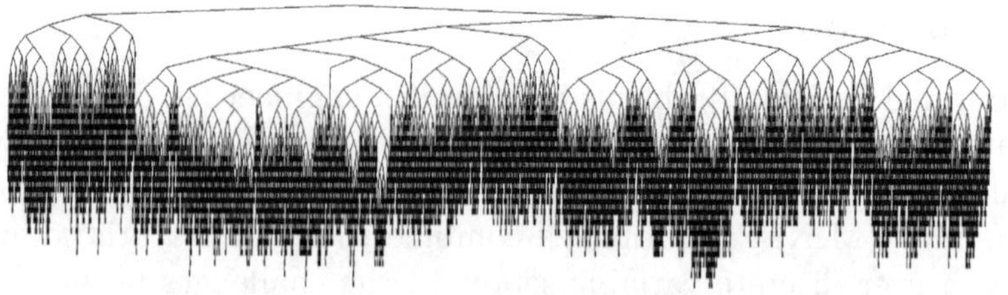

Figure 1.1. A branch-and-bound tree without presolve after 360 s CPU time has more than 10 000 nodes.

of x corresponding to J. In particular, x_I are the integer variables. We also define $C = \{1, \ldots, n\} - I$ and let x_C denote the continuous variables. We denote by p the dimension of the integer space, $p = |I|$. We denote the floor and ceiling operator by $\lfloor x_i \rfloor$ and $\lceil x_i \rceil$, which denote the largest integer smaller than or equal to x_i and the smallest integer larger than or equal to x_i, respectively. Given two $n \times n$ matrices Q and X, $Q \bullet X = \sum_{i=1}^n \sum_{j=1}^n Q_{ij} X_{ij}$ represents their inner product.

In general, the presence of integer variables $x_i \in \mathbb{Z}$ implies that the feasible set of (1.1) is not convex. In a slight abuse of terminology, we distinguish convex from non-convex MINLPs.

Definition 1.1. We say that (1.1) is a *convex MINLP* if the problem functions $f(x)$ and $c(x)$ are convex functions. If either $f(x)$ or any $c_i(x)$ is a non-convex function, then we say that (1.1) is a *non-convex MINLP*.

Throughout this paper, we use the notion of a convex hull of a set S.

Definition 1.2. Given a set S, the *convex hull of* S is denoted by conv(S) and defined as

$$\text{conv}(S) := \{x : x = \lambda x^{(1)} + (1 - \lambda) x^{(0)},\ \forall 0 \le \lambda \le 1,\ \forall x^{(0)}, x^{(1)} \in S\}.$$

If $X = \{x \in \mathbb{Z}^p : l \le x \le u\}$ and $l \in \mathbb{Z}^p$, $u \in \mathbb{Z}^p$, then conv$(X) = [l, u]$ is simply the hypercube. In general, however, even when X itself is polyhedral, it is not easy to find conv(X). The convex hull plays an important role in mixed-integer linear programming. Because an LP obtains a solution at a vertex, we can solve an MILP by solving an LP over its convex hull. Unfortunately, finding the convex hull of an MILP is just as hard as solving the MILP.

The same result does not hold for MINLP, as the following example illustrates:

$$\underset{x}{\text{minimize}} \sum_{i=1}^{n} (x_i - \tfrac{1}{2})^2, \quad \text{subject to } x_i \in \{0, 1\}.$$

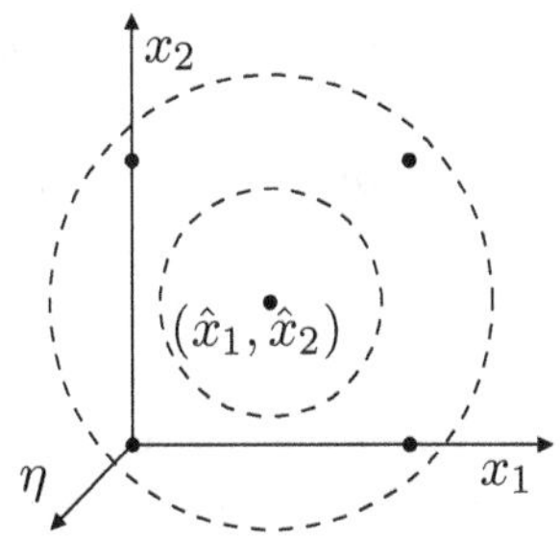

Figure 1.2. Small MINLP to illustrate the need for a linear objective function.

The solution of the continuous relaxation is $x = \left(\frac{1}{2}, \ldots, \frac{1}{2}\right)$, which is not an extreme point of the feasible set and, in fact, lies in the strict interior of the MINLP: see Figure 1.2. Because the continuous minimizer lies in the interior of the convex hull of the integer feasible set, it cannot be separated from the feasible set. However, we can reformulate (1.1) by introducing an objective variable z and a constraint $z \geq f(x)$. We obtain the following equivalent MINLP:

$$\begin{cases} \underset{z,x}{\text{minimize}} & z, \\ \text{subject to} & f(x) \leq z, \\ & c(x) \leq 0, \\ & x \in X, \\ & x_i \in \mathbb{Z}, \ \forall i \in I. \end{cases} \tag{1.2}$$

The optimal solution of (1.2) always lies on the boundary of the convex hull of the feasible set and therefore allows us to use cutting-plane techniques.

1.2. Preview of key building blocks of MINLP algorithms

A wide variety of methods exists for solving MINLP. Here, we briefly introduce the two fundamental concepts underlying these algorithms: *relaxation* and *constraint enforcement*. A relaxation is used to compute a lower bound on the optimal solution of (1.1). A relaxation is obtained by enlarging the feasible set of the MINLP, for example, by ignoring some constraints of the problem. Typically, we are interested in relaxations that are substantially easier to solve than the MINLP itself. Together with upper bounds, which can be obtained from any feasible point, relaxations allow us to terminate the search for a solution whenever the lower bound is larger than the current upper bound. Constraint enforcement refers to procedures used to exclude solutions that are feasible for the relaxation but not to the original MINLP. Constraint enforcement may be accomplished by refining or tightening the relaxation, often by adding valid inequalities, or by *branching*, where the relaxation is divided into two or more separate problems.

In general, upper bounds are obtained from any feasible point. Often, we fix the integer variables at an integral value and solve the resulting NLP to obtain an upper bound (which we set to infinity if the NLP is infeasible).

Relaxations. Formally, an optimization problem $\min\{\xi_R(x) : x \in S_R\}$ is a relaxation of a problem $\min\{\xi(x) : x \in S\}$ if (i) $S_R \supseteq S$ and (ii) $\xi_R(x) \leq \xi(x)$ for each $x \in S$. The feasible set $\mathcal{R}$ of a relaxation of a problem with feasible set $\mathcal{F}$ contains all feasible points of $\mathcal{F}$. The main role of the relaxation is to provide a problem that is easier to solve and for which we can obtain globally optimal solutions that allow us to derive a lower bound. Relaxations that fall into this category are convex NLPs, for which nonlinear optimization solvers will converge to the global minimum, and MILPs, which can often be solved efficiently (for practical purposes) by using a branch-and-cut approach.

Several strategies are used to obtain relaxations of MINLPs.

(1) *Relaxing integrality.* Integrality constraints $x_i \in \mathbb{Z}$ can be relaxed to $x_i \in \mathbb{R}$ for all $i \in I$. This procedure yields a *nonlinear relaxation* of MINLP. This type of relaxation is used in branch-and-bound algorithms (Section 3.1) and is given by

$$\begin{cases} \underset{x}{\text{minimize}} & f(x), \\ \text{subject to} & c(x) \leq 0, \\ & x \in X. \end{cases} \tag{1.3}$$

(2) *Relaxing convex constraints.* Constraints $c(x) \leq 0$ and $f(x) \leq z$ containing convex functions c and f can be relaxed with a set of supporting hyperplanes obtained from first-order Taylor series approximation,

$$z \geq f^{(k)} + \nabla f^{(k)^T}(x - x^{(k)}), \tag{1.4}$$

$$0 \geq c^{(k)} + \nabla c^{(k)^T}(x - x^{(k)}), \tag{1.5}$$

for a set of points $x^{(k)}$, $k = 1, \ldots, K$. When c and f are convex, any collection of such hyperplanes forms a *polyhedral relaxation* of these constraints. This class of relaxations is used in the outer approximation methods discussed in Section 3.2.1.

(3) *Relaxing non-convex constraints.* Constraints $c(x) \leq 0$ and $f(x) \leq z$ containing non-convex functions require more work to be relaxed. One approach is to derive convex underestimators, $\breve{f}(x)$ and $\breve{c}(x)$, which are convex functions that satisfy

$$\breve{f}(x) \leq f(x) \quad \text{and} \quad \breve{c}(x) \leq c(x), \quad \text{for all } x \in \text{conv}(X). \tag{1.6}$$

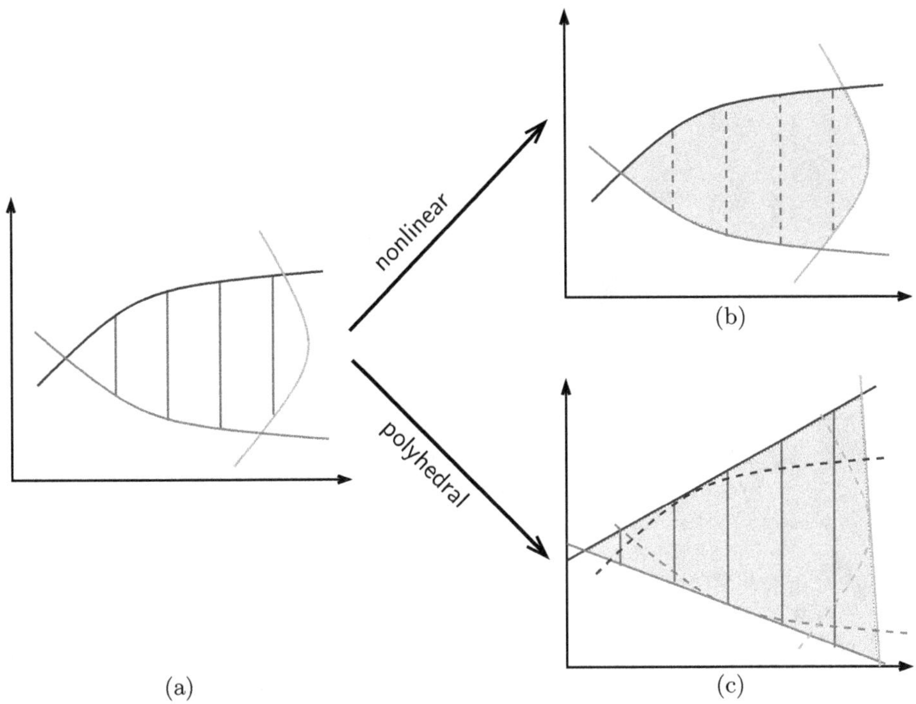

Figure 1.3. Illustration of the two classes of relaxation. (a) The mixed-integer feasible set, (b) the nonlinear relaxation, and (c) the polyhedral relaxation.

Then the constraints $z \geq f(x)$ and $0 \geq c(x)$ are relaxed by replacing them with the constraints

$$z \geq \breve{f}(x) \quad \text{and} \quad 0 \geq \breve{c}(x).$$

In Section 5 we review classes of nonlinear functions for which convex underestimators are known, and we describe a general procedure to derive underestimators for more complex nonlinear functions.

All these relaxations enlarge the feasible set of (1.2), and they can be combined with each other. For example, a convex underestimator of a nonconvex function can be further relaxed by using supporting hyperplanes, yielding a polyhedral relaxation.

Figure 1.3 illustrates the relaxation of integrality constraints and convex nonlinear constraints: (a) the mixed-integer feasible set (the union of the vertical segments), (b) the nonlinear relaxation obtained by relaxing the integrality constraints (the shaded area is the NLP feasible set), and (c) a polyhedral relaxation (the union of the vertical segments) as well as its LP relaxation (the shaded area). We note that an infinite number of

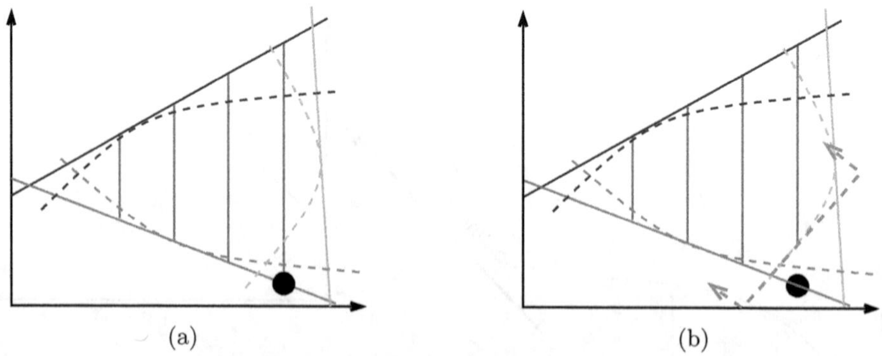

Figure 1.4. Separation of infeasible point (black dot) by adding a separating hyperplane. The dashed line with arrows (b) shows a separating hyperplane with arrows indicating the feasible side.

possible polyhedral relaxations exists, depending on the choice of the points $x^{(k)} \in \text{conv}(X),\ k = 1, \ldots, K$.

If the solution to a relaxation is feasible in (1.2), then it also solves the MINLP. In general, however, the solution is not feasible in (1.2), and we must somehow exclude this solution from the relaxation.

Constraint enforcement. Given a point $\hat{x}$ that is feasible for a relaxation but is not feasible for the MINLP, the goal of constraint enforcement is to exclude this solution, so that the algorithm can eventually converge to a solution that satisfies all the constraints. Two broad classes of constraint enforcement strategies exist: relaxation refinement and branching. Most modern MINLP algorithms use both classes.

The goal of relaxation refinement is to tighten the relaxation in such a way that a solution $\hat{x}$ of the relaxation that is infeasible for the MINLP is removed from the relaxation. Most commonly, this is achieved by adding a new *valid inequality* to the relaxation. A valid inequality is an inequality that is satisfied by all feasible solutions for the MINLP. When a valid inequality successfully excludes a given infeasible solution, it is often called a *cut.* Valid inequalities are usually linear but may be convex. For example, after relaxing a convex constraint with a polyhedral relaxation, a valid inequality can be obtained by linearizing the nonlinear functions about $\hat{x}$. This valid inequality will successfully cut off $\hat{x}$, unless $\hat{x}$ satisfies the nonlinear constraints, $c(\hat{x}) \leq 0$: see Figure 1.4. This class of separation is used in outer approximation, Benders decomposition, and the ECP algorithm discussed in Section 3.2.1. Valid inequalities can also be useful after relaxing integrality. In this case, the goal is to obtain an inequality that is valid because it does not cut off any *integer feasible* solution but will cut off an

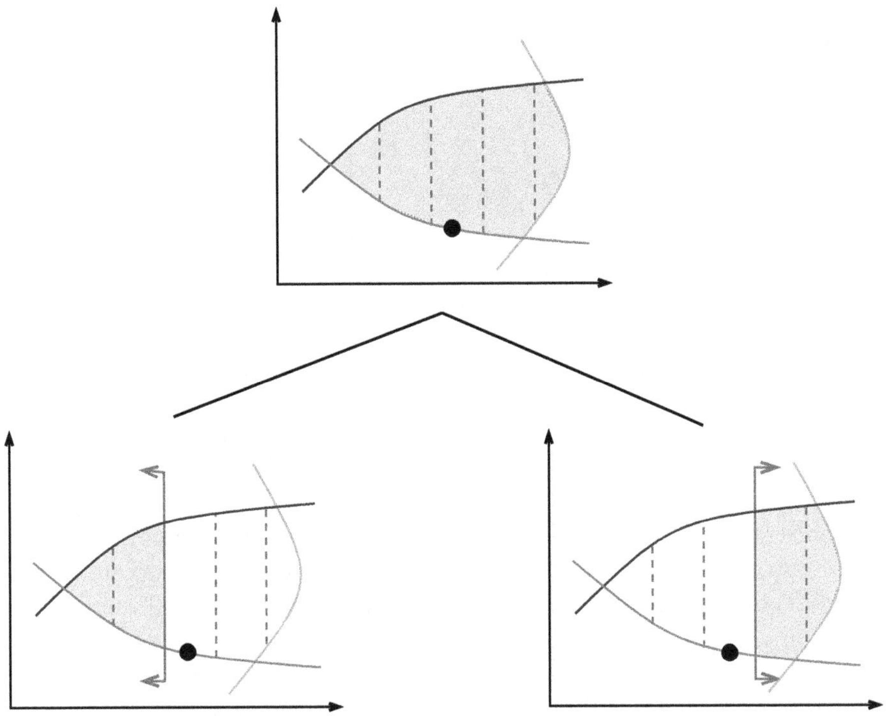

Figure 1.5. Branching on the values of an integer variable creates two new nonlinear subproblems that both exclude the infeasible point, denoted by the black dot.

integer infeasible solution $\hat{x}$. This technique has been critical to the success of algorithms for solving mixed-integer *linear* programs.

The second class of constraint enforcement strategy is branching. Branching consists in dividing the feasible region into subsets such that every solution to MINLP is feasible in one of the subsets. When integrality is relaxed, it can be enforced by branching on an integer variable that takes a fractional value $\hat{x}_i$ for some $i \in I$. Branching creates two new separate relaxations: the constraint $x_i \leq \lfloor \hat{x}_i \rfloor$ is added to the first relaxation, and the constraint $x_i \geq \lceil \hat{x}_i \rceil$ is added to the second relaxation (see Figure 1.5). All solutions of the MINLP now lie in one of these two new relaxations. The resulting subproblems are managed in a search tree that keeps track of all subproblems that remain to be solved. This approach is the basis of the branch-and-bound algorithms described in detail in Section 3.1.

Constraint enforcement for relaxed non-convex constraints involves a combination of branching and relaxation refinement. These techniques are discussed in detail in Section 5, but here we outline the general idea using Figure 1.6. Following the solution of the relaxation (given by the dashed objective function in Figure 1.6(a)), we branch on a *continuous* variable

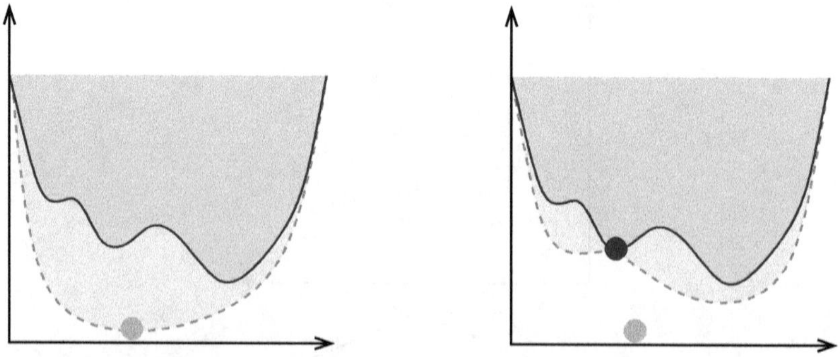

Figure 1.6. Constraint enforcement by using spatial branching for global optimization.

and hence split its domain into two subdomains. We then compute new underestimators for use in (1.6) that are valid on each of the two subdomains (*i.e.*, we refine the relaxation). In the example, these refined underestimators are indicated by the two dashed objective functions in Figure 1.6(b). This approach, which we refer to as *spatial branching*, results in a branch-and-bound algorithm similar to the one for discrete variables. We continue to divide the domain into smaller subdomains until the lower bound on a subdomain is larger than the upper bound, at which point we can exclude this domain from our search. For MINLPs having both integer variables and non-convex constraints, branching may be required on both integer and continuous decision variables.

1.3. Scope and outline

The past 20 years or so have seen a dramatic increase in new mixed-integer nonlinear models and applications, which has motivated the development of a broad range of new techniques to tackle this challenging class of problems. This survey presents a broad overview of *deterministic methodologies for solving mixed-integer nonlinear programs.* In Section 2 we motivate our interest in MINLP methods by presenting some small examples, and we briefly discuss good modelling practices. In Section 3 we present deterministic methods for *convex* MINLPs, including branch-and-bound, outer approximation, and hybrid techniques. We also discuss advanced implementation considerations for these methods. Cutting planes have long played a fundamental role in mixed-integer linear programming, and in Section 4 we discuss their extension to MINLP. We review a range of cutting planes such as conic MIR cuts, disjunctive cuts, and perspective cuts. In Section 5 we outline methods for solving *non-convex* MINLPs. A range of heuristics to obtain good incumbent solutions quickly is discussed in Section 6.

We review two classes of deterministic heuristics: search and improvement heuristics. In Section 7 we discuss an emerging extension of MINLP, the mixed-integer optimal control problem. In Section 8 we review the state of the art in software for MINLP and categorize the different solvers within the context of the previous sections.

Given the wide applicability of MINLP and the ensuing explosion of numerical methods, it would be prohibitive to discuss all methods. Instead, we focus this survey on deterministic methods that tackle both integer variables and nonlinear constraints. In particular, we do not survey the two main subproblems of MINLP, mixed-integer linear programming and nonlinear programming, for which there exist excellent textbooks (Nocedal and Wright 1999, Fletcher 1987, Wolsey 1998, Nemhauser and Wolsey 1988) and recent survey articles (Gould and Leyffer 2003). We also do not cover three other related topics.

(1) *Algorithms that are polynomial when the number of variables is fixed, or which require a polynomial number of calls to an oracle.* Lenstra's algorithm (Lenstra 1983) solves integer linear programming problems in polynomial time when the number of variables in the problem is fixed. Khachiyan and Porkolab (2000) extended this algorithm to integer programs with convex polynomial objective and constraints. Generalizations and improvements in the complexity bound have been made by Heinz (2005), De Loera, Hemmecke, Koppe and Weismantel (2006), Hildebrand and Köppe (2013) and Dadush, Peikert and Vempala (2011*d*). Using ideas from Graver bases, Hemmecke, Onn and Weismantel (2011) derive an algorithm that requires a polynomial number of *augmentation steps* to solve specially structured convex minimization problems over integer sets. Baes *et al.* (2012) investigate the problem of minimizing a strongly convex Lipschitz-continuous function over a set of integer points in a polytope, obtaining an algorithm that provides a solution with a constant approximation factor of the best solution by solving a polynomial number of specially structured quadratic integer programs.

(2) *Properties of closures of convex integer sets.* Recently, Dey and Vielma (2010) and Dadush, Dey and Vielma (2011*a*, 2011*b*, 2011*c*) have been studying *closures* of certain classes of valid inequalities for convex MINLPs. A closure is the set obtained after including all inequalities from a particular class. In particular, for the class of Gomory–Chvátal (G-C) inequalities, inequalities which can be obtained from a simple rounding argument, they have shown that the resulting closure is a rational polyhedron if the original convex set is either strictly convex or compact. This is an interesting result for two reasons: the original convex set may not be described by finitely many linear inequalities,

and also the number of G-C inequalities is not finite. In the same spirit, Dey and Morán (2013) study conditions under which the convex hull of a set of integer points in a convex set is closed, and polyhedral.

(3) *Mixed-integer derivative-free optimization problems.* Such problems arise when the problem functions are not given explicitly and can be evaluated only as the result of a (black-box) simulation $\mathcal{S}(x)$. In this case, derivatives are typically not available, and the simulation is often computationally expensive. One example of this class of problem is the design of landfill soil liners, where integer variables model the choice of material and the permeative and decomposition processes are modelled via simulation (Bartelt-Hunt *et al.* 2006). Other applications include the design of nanophotonic devices (Maria *et al.* 2009, Miller *et al.* 2010) and the determination of loop unroll factors in empirical performance tuning (Balaprakash, Wild and Hovland 2011). Algorithms for derivative-free mixed-integer nonlinear optimization employ either global surrogates (Müller 2012, Müller, Shoemaker and Piché 2013, Davis and Ierapetritou 2009, Hemker 2008, Hemker, Fowler, Farthing and von Stryk 2008, Rashid, Ambani and Cetinkaya 2013), or mesh- and pattern-based techniques (Liuzzi, Lucidi and Rinaldi 2012, Audet and Dennis, Jr 2000, Fowler *et al.* 2008, Abramson, Audet, Chrissis and Walston 2009).

We mention a number of surveys and monographs on MINLP, including a review of MINLP and disjunctive programming (Grossmann 2002), a survey of MINLP algorithms (Grossmann and Kravanja 1997), a survey of algorithms and software for MINLP (Bonami, Kılınç and Linderoth 2012), a comprehensive textbook on MINLP by Floudas (1995), and a collection of articles related to a recent IMA workshop on MINLP (Lee and Leyffer 2012).

2. Nonlinear models with integer variables

In this section we review a small number of models and modelling tricks to motivate the algorithmic developments that are discussed in the subsequent sections. The models are chosen to provide insight into the interactions of the two main modelling paradigms: integer variables and nonlinear equations. We start by presenting a well-known application from chemical engineering that models the design of a multiproduct batch plant (Kocis and Grossmann 1988, Grossmann and Sargent 1979), which can be formulated as a convex MINLP. We then present a non-convex MINLP that arises in the design of water distribution networks (Eiger, Shamir and Ben-Tal 1994, Sherali and Smith 1997, Sherali, Subramanian and Loganathan

2001, Bragalli *et al.* 2006, Bragalli *et al.* 2012). Finally, we present a time- and energy-optimal subway control example (Bock and Longman 1985) that adds time-dependent integer variables and constraints to MINLP.

2.1. Modelling practices for MINLP

Modelling plays a fundamental role in MILP (see the textbook by Williams (1999)) and is arguably more critical in MINLP, because of the additional nonlinear relationships. The nonlinearities often allow for equivalent formulations with more favourable properties. For example, in Section 2.2 we present a model of a multiproduct batch plant, and show that by using a nonlinear transformation we are able to reformulate the non-convex problem as an equivalent convex problem, which is typically easier to solve. Here, we briefly review a number of other modelling tricks that can be useful in formulating easier-to-solve problems.

Convexification of binary quadratic programs. We can always make the quadratic form in a pure binary quadratic program convex, because for any $x_i \in \{0,1\}$ it follows that $x_i^2 = x_i$. For $x \in \{0,1\}^n$ we consider the quadratic form

$$q(x) = x^T Q x + g^T x,$$

and let λ be the smallest eigenvalue of Q. If $\lambda \geq 0$, then $q(x)$ is convex. Otherwise, we define a new Hessian matrix $W := Q - \lambda I$, where I is the identity matrix, and a new gradient $c := g + \lambda e$, where $e = (1, \ldots, 1)$. It follows that

$$q(x) = x^T Q x + g^T x = x^T W x + c^T x,$$

where the second quadratic is now convex.

Exploiting low-rank Hessians. In many applications such as model structure determination and parameter estimation problems (Skrifvars, Leyffer and Westerlund 1998), the Hessian matrix is a large, dense, and low-rank matrix. In this application, the Hessian matrix can be written as $W = Z^T R^{-1} Z$, where $R \in \mathbb{R}^{m \times m}$ and $Z \in \mathbb{R}^{m \times n}$, where $m \ll n$ and Z is sparse. Forming the Hessian matrix and operating with it can be prohibitive. Some nonlinear solvers only require matrix–vector products of the Hessian and can readily exploit this situation. An alternative is to introduce additional variables z and constraints $z = Zx$, and then rewrite $x^T W x = z^T R^{-1} z$, which allows the solver to exploit the sparsity of the constraint matrix Z.

Linearization of constraints. A simple transformation is to rewrite $x_1/x_2 = a$ as $x_1 = a x_2$, where a is a constant. Special-ordered sets provide a systematic way to formulate nonlinear expressions as piecewise linear functions (see Section 5.1), and these can be effective for expressions that involve a small number of variables.

Linearization of x_1x_2, for $x_2 \in \{0,1\}$. We can linearize the expression x_1x_2, when $0 \le x_1 \le u$ and $x_2 \in \{0,1\}$, by observing that the product x_1x_2 is either equal to zero (if $x_2 = 0$) or equal to x_1 (if $x_2 = 1$). By introducing the new variable x_{12} and adding the constraints

$$0 \le x_{12} \le x_2 u \quad \text{and} \quad -u(1-x_2) \le x_1 - x_{12} \le u(1-x_2),$$

we can replace the non-convex term x_1x_2 by x_{12}. This trick readily generalizes to situations where $l \le x_1 \le u$.

Avoiding undefined nonlinear expressions. In many practical instances the MINLP solvers fail, because the nonlinear solver generates an iterate at which the nonlinear functions cannot be evaluated. An example is the constraint

$$c(x_1) = -\ln(\sin(x_1)) \le 0,$$

which cannot be evaluated whenever $\sin(x_1) \le 0$. Because NLP solvers typically remain feasible with respect to simple bounds, we can reformulate this constraint equivalently as

$$\tilde{c}(x_2) = -\ln(x_2) \le 0, \quad x_2 = \sin(x_1), \quad \text{and} \quad x_2 \ge 0.$$

Interior-point methods will never evaluate the constraint $\tilde{c}(x_2)$ at a point $x_2 \le 0$, and even active-set methods can be expected to avoid this region, because the constraint violation becomes unbounded near $x_2 = 0$. An additional benefit of this reformulation is that it reduces the 'degree of nonlinearity' of the resulting constraint set.

A common feature of these reformulation techniques is that convex formulations are typically preferred over non-convex ones and linear formulations over nonlinear formulations. More useful modelling tricks for integer and binary variables are given in the classic textbook by Williams (1999).

2.2. Design of multiproduct batch plants

Multiproduct batch plants produce a number of different products on parallel lines in a multistage batch process. In the following we use subscripts j to index the stages and subscripts i to index the products. A multiproduct batch plant is characterized by the following set of (fixed) parameters:

- M the set of batch processing stages,
- N the set of different products,
- H the time horizon,
- Q_i the required quantity of product $i \in N$,
- t_{ij} processing time, product i, stage $j \in M$, and
- S_{ij} size factor, product $i \in N$, stage $j \in M$.

Our goal is to find the minimum cost design by choosing optimal values for the design (free) variables of the batch process, which are given by:

B_i the batch size of product $i \in N$,
V_j the size of stage $j \in M$, which must satisfy $V_j \geq S_{ij}B_i$ for all i, j,
N_j the number of machines at stage $j \in M$, and
C_i the longest stage time for product $i \in N$, which satisfies $C_i \geq t_{ij}/N_j$ for all i, j.

Given a set of cost parameters, $\alpha_j, \beta_j > 0$, we can formulate the minimum cost design of a multiproduct batch plant as a non-convex MINLP:

$$\begin{cases} \underset{V,C,B,N}{\text{minimize}} & \sum\limits_{j\in M} \alpha_j N_j V_j^{\beta_j}, \\ \text{subject to} & V_j - S_{ij}B_i \geq 0, \qquad \forall i \in N, \forall j \in M, \\ & C_i N_j \geq t_{ij}, \qquad \forall i \in N, \forall j \in M, \\ & \sum\limits_{i\in N} \frac{Q_i}{B_i} C_i \leq H, \\ & V_j \in [V_l, V_u],\ C_i \in [C_l, C_u],\ B_i \in [B_l, B_u], \quad \forall i \in N, \forall j \in M, \\ & N_j \in \{1, 2, \ldots, N_u\}, \qquad \forall j \in M. \end{cases}$$

Unfortunately this model is a non-convex MINLP, because the objective function, the horizon-time constraint, and the constraint defining the longest stage time are non-convex functions. Fortunately, we can apply a variable transformation which convexifies the problem. We introduce new log-transformed variables v_j, n_j, b_i, and c_i defined by

$$v_j = \ln(V_j), \quad n_j = \ln(N_j), \quad b_i = \ln(B_i), \quad c_i = \ln C_i.$$

This transformation provides an equivalent convex reformulation of the multiproduct batch plant design problem:

$$\begin{cases} \underset{v,c,b,n}{\text{minimize}} & \sum\limits_{j\in M} \alpha_j e^{n_j + \beta_j v_j}, \\ \text{subject to} & v_j - \ln(S_{ij}) b_i \geq 0, \qquad \forall i \in N, \forall j \in M, \\ & c_i + n_j \geq \ln(t_{ij}), \qquad \forall i \in N, \forall j \in M, \\ & \sum\limits_{i\in N} Q_i e^{c_i - b_i} \leq H, \\ & v_j \in [v_l, v_u],\ c_i \in [c_l, c_u],\ b_i \in [b_l, b_u], \quad \forall i \in N, \forall j \in M, \\ & n_j \in \{0, \ln(2), \ldots, \ln(N_u)\}, \qquad \forall j \in M, \end{cases}$$

where v_l, v_u, c_l, c_u, b_l, and b_u are suitably transformed bounds on the variables. The last constraint, $n_j \in \{0, \ln(2), \ldots, \ln(N_u)\}$, is difficult to enforce directly, but it can be modelled using a special-ordered set of type I (SOS-1) (Beale and Tomlin 1970). By introducing binary variables $y_{kj} \in \{0, 1\}$ we can replace the last constraint in the convex model by

$$\sum_{k=1}^{K} \ln(k) y_{kj} = n_j \quad \text{and} \quad \sum_{k=1}^{K} y_{kj} = 1, \quad \text{for all } j \in M. \tag{2.1}$$

The resulting model is available as batch on MacMINLP (Leyffer 2003). Similar reformulations have been proposed using $\sqrt{x}$-transforms for models involving bilinear terms (Harjunkoski, Westerlund, Pörn and Skrifvars 1998).

2.3. Design of water distribution networks

Many MINLPs model flows on networks. Examples include the optimal design of water networks (Burgschweiger, Gnädig and Steinbach 2008, Bragalli *et al.* 2006, Bragalli *et al.* 2012), the design of gas networks (Wolf and Smeers 2000) and the optimal regulation of ventilation systems for mines (Wu, Topuz and Karfakis 1991, Maonan and Wenjun 1991). All applications share the same structural flow constraints. Here, we concentrate on the design of water networks. The model is defined by a set on nodes, $i \in \mathcal{N}$, and a set of arcs $(i,j) \in \mathcal{A}$ that define the (fixed) topology of the network. The goal is to determine the pipe diameter for all arcs that minimizes the installation cost, meets the demand at all nodes, and satisfies the hydraulic (flow) constraints along the arcs. The pipe diameters must be chosen from a discrete set of available diameters, giving rise to integrality constraints, while the hydraulic constraints involve nonlinear relationships, giving rise to a non-convex MINLP.

We simplify the model by assuming that certain constants are uniform throughout the network. The sets and (fixed) parameters that define the model are:

- $\mathcal{N}$ the set of nodes in the network,
- $\mathcal{S}$ the set of source nodes in the network, $\mathcal{S} \subset \mathcal{N}$,
- $\mathcal{A}$ the set of arcs in the network, $\mathcal{A} \subset \mathcal{N} \times \mathcal{N}$,
- L_{ij} the length of pipe $(i,j) \in \mathcal{A}$,
- K_{ij} a physical constant that models the roughness of pipe $(i,j) \in \mathcal{A}$,
- D_i the demand at node $i \in \mathcal{N}$,
- $V_{\max}$ the maximum velocity of flow in all pipes,
- $Q_{\max}$ the maximum magnitude of flow in all pipes,
- P_k the pipe diameters available to the network, $k = 1, \ldots, r$,
- H_s the hydraulic head at source node $s \in \mathcal{S}$, and
- H_l, H_u the lower and upper bounds on the hydraulic head.

The variables of the model are:

- q_{ij} the flow in pipe $(i, j) \in \mathcal{A}$ (from node i to node j),
- d_{ij} the diameter of pipe $(i, j) \in \mathcal{A}$, where $d_{ij} \in \{P_1, \ldots, P_r\}$,
- h_i the hydraulic head at node $i \in \mathcal{N}$, where $h_s = H_s$ for all $s \in \mathcal{S}$, and $H_l \leq h_i \leq H_u$ for all $i \in \mathcal{N} - \mathcal{S}$,
- z_{ij} binary variables that model the direction of flow in pipe $(i, j) \in \mathcal{A}$,
- a_{ij} the area of the cross-section of pipe $(i, j) \in \mathcal{A}$, and
- y_{ijk} a set of SOS-1 variables (see (2.1)) that models the pipe type on arc $(i, j) \in \mathcal{A}$.

The conservation of flow at every node gives rise to a linear constraint

$$\sum_{(i,j)\in\mathcal{A}} q_{ij} - \sum_{(j,i)\in\mathcal{A}} q_{ji} = D_i, \quad \text{for all } i \in \mathcal{N} - \mathcal{S}.$$

Because our model contains cross-section variables, $a_{ij} = \pi d_{ij}^2/4$, we can model the bounds on the flow along an arc depending on a linear set of constraints

$$-V_{\max} a_{ij} \leq q_{ij} \leq V_{\max} a_{ij}, \quad \text{for all } (i, j) \in \mathcal{A}.$$

The choice of pipe types, $d_{ij} \in \{P_1, \ldots, P_r\}$, is modelled using the SOS-1 variables y_{ijk}, giving rise to the set of constraints

$$y_{ijk} \in \{0, 1\}, \quad \text{for all } k = 1, \ldots, r, \quad \text{and}$$

$$\sum_{k=1}^{r} y_{ijk} = 1, \quad \text{and} \quad \sum_{k=1}^{r} P_k y_{ijk} = d_{ij}, \quad \text{for all } (i, j) \in \mathcal{A},$$

The MINLP solvers can now either branch on individual y_{ijk} or on SOS-1 sets $(y_{ij1}, \ldots, y_{ijr})$, which is generally more efficient. See Williams (1999) for a discussion on branching on special-ordered sets. We can use the same SOS-1 set to linearize the nonlinear equation, $a_{ij} = \pi d_{ij}^2/4$, by adding the constraints

$$\sum_{k=1}^{r} (\pi P_k^2/4) y_{ijk} = a_{ij}, \quad \text{for all } (i, j) \in \mathcal{A}.$$

The final set of constraints is an empirical model of the pressure loss along the arc $(i, j) \in \mathcal{A}$,

$$h_i - h_j = \frac{\operatorname{sgn}(q_{ij}) |q_{ij}|^{c_1} c_2 L_{ij} K_{ij}^{-c_1}}{d_{ij}^{c_3}}, \quad \text{for all } (i, j) \in \mathcal{A}, \tag{2.2}$$

where $c_1 = 1.852$, $c_2 = 10.7$, and $c_3 = 4.87$ are constants that depend on the medium of the fluid. This last equation appears to be non-smooth because it contains terms of the form $h(x) = \mathrm{sgn}(x)|x|^{c_1}$. However, it is easy to verify that $h(x)$ is continuously differentiable, because $c_1 > 1$. To model $h(x)$, we split the flow into its positive and negative parts by adding binary variables $z_{ij} \in \{0,1\}$ and the following set of constraints,

$$0 \leq q_{ij}^+ \leq Q_{\max} z_{ij}, \quad 0 \leq q_{ij}^- \leq Q_{\max}(1 - z_{ij}), \quad q_{ij} = q_{ij}^+ - q_{ij}^-.$$

An alternative formulation based on complementarity constraints avoids the introduction of the binary variables z_{ij}, and instead models the disjunction as $0 \leq q_{ij}^+ \perp q_{ij}^- \geq 0$, where $\perp$ means that either $q_{ij}^+ = 0$ or $q_{ij}^- = 0$. With these new flow variables, we can now rewrite (2.2) equivalently as

$$h_i - h_j = \frac{\left[(q_{ij}^+)^{c_1} - (q_{ij}^-)^{c_1}\right] c_2 L_{ij} K_{ij}^{-c_1}}{d_{ij}^{c_3}}, \quad \text{for all } (i,j) \in \mathcal{A}.$$

Finally, we can lower the degree of nonlinearity by substituting d_{ij} in this last constraint by $d_{ij} = \sqrt{4a_{ij}/\pi}$. This substitution generally provides better Taylor-series approximations than the 'more nonlinear' version involving d_{ij}, which ensures better practical convergence behaviour of the underlying NLP solvers.

2.4. A dynamic subway operation problem

As an example of a dynamic nonlinear mixed-integer problem, we present a control problem that goes back to work by Bock and Longman (1985) optimizing the subway of the city of New York. The problem has been treated by Sager (2005), for example. We are interested in minimizing the total energy consumption

$$\underset{x(\cdot),u(\cdot),v(\cdot),T}{\text{minimize}} \int_0^T L(x(t), v(t))\,\mathrm{d}t \tag{2.3}$$

of a subway train, subject to a system of ordinary differential equations that models the dynamic behaviour of a subway train's state $x(t)$ comprising position $x_{\mathrm{s}}(t)$ and velocity $x_{\mathrm{v}}(t)$,

$$\dot{x}_{\mathrm{s}}(t) = x_{\mathrm{v}}(t), \qquad t \in [0,T], \tag{2.4}$$

$$\dot{x}_{\mathrm{v}}(t) = f_{\mathrm{v}}(x(t), u(t), v(t)), \quad t \in [0,T], \tag{2.5}$$

on a horizon $[0,T] \subset \mathbb{R}$ with free end time T. The continuous control $0 \leq u(t) \leq u_{\max}$ indicates the braking deceleration. The integer control vector $v(t) \in \{1,2,3,4\}$ indicates the operation of the subway's two electric engines in one of four discrete modes that affect the subway train's acceleration and energy consumption, as follows.

(1) *Serial*, $v(t) = 1$. The energy consumption rate is given by

$$L(x(t), 1) = \begin{cases} e\, p_1 & \text{if } x_\text{v}(t) \le v_1, \\ e\, p_2 & \text{if } v_1 < x_\text{v}(t) \le v_2, \\ e \sum_{i=0}^{5} c_i(1)\big(0.1\gamma x_\text{v}(t)\big)^{-i} & \text{if } v_2 < x_\text{v}(t), \end{cases} \tag{2.6}$$

and the subway train's dynamics are described by

$$W_\text{eff}\, f_\text{v}(t) = \begin{cases} g\, e\, a_1 & \text{if } x_\text{v}(t) \le v_1, \\ g\, e\, a_2 & \text{if } v_1 < x_\text{v}(t) \le v_2, \\ g\, (e\, F(x_\text{v}(t), 1) - R(x_\text{v}(t))) & \text{if } v_2 < x_\text{v}(t). \end{cases} \tag{2.7}$$

(2) *Parallel*, $v(t) = 2$. The energy consumption rate is given by

$$L(x(t), 2) = \begin{cases} 0 & \text{if } x_\text{v}(t) \le v_2, \\ e\, p_3 & \text{if } v_2 < x_\text{v}(t) \le v_3, \\ e \sum_{i=0}^{5} c_i(2)\big(0.1\gamma x_\text{v}(t) - 1\big)^{-i} & \text{if } v_3 < x_\text{v}(t), \end{cases} \tag{2.8}$$

and the subway train's dynamics are described by

$$W_\text{eff}\, f_\text{v}(t) = \begin{cases} 0 & \text{if } x_\text{v}(t) \le v_2, \\ g\, e\, a_3 & \text{if } v_2 < x_\text{v}(t) \le v_3, \\ g\, (e\, F(x_\text{v}(t), 2) - R(x_\text{v}(t))) & \text{if } v_3 < x_\text{v}(t). \end{cases} \tag{2.9}$$

(3) *Coasting*, $v(t) = 3$. The energy consumption rate is zero, $L(x(t), 3) = 0$, and the subway train's dynamics are described by

$$W_\text{eff}\, f_\text{v}(t) = -g\, R(x_\text{v}(t)) - C\, W_\text{eff}. \tag{2.10}$$

(4) *Braking*, $v(t) = 4$. The energy consumption rate is zero, $L(x(t), 4) = 0$, and the subway train's dynamics are described by

$$f_\text{v}(t) = -u(t). \tag{2.11}$$

The forces occurring in these dynamics are given by

$$R(x_\text{v}(t)) = ca\gamma^2 x_\text{v}(t)^2 + bW\gamma x_\text{v}(t) + \frac{1.3}{2000}W + 116, \tag{2.12}$$

$$F(x_\text{v}, 1) = \sum_{i=0}^{5} b_i(1)\big(0.1\gamma x_\text{v}(t) - 0.3\big)^{-i}, \tag{2.13}$$

$$F(x_\text{v}, 2) = \sum_{i=0}^{5} b_i(2)\big(0.1\gamma x_\text{v}(t) - 1\big)^{-i}. \tag{2.14}$$

In addition, the system will satisfy certain path and point constraints as follows. Initial and terminal states for the system trajectory are constrained to

$$x(0) = (0,0)^T, \quad x(T) = (S,0)^T. \tag{2.15}$$

A maximum on the allowable driving time to complete the distance S is imposed,

$$T \le T_{\max}. \tag{2.16}$$

Different scenarios can be defined for this problem by prescribing values for the parameters S and W. In addition, scenarios may include speed limits at certain points or on certain parts of the track. A description of several scenarios, units and numerical values for the model parameters a, a_1, a_2, a_3, b, $b_i(v)$, C, c, $c_i(v)$, e, g, γ, p_1, p_2, p_3, S, $T_{\max}$, $u_{\max}$, v_1, v_2, v_3 , W, W_{eff}, and analytical investigations along with numerical solutions can be found in Bock and Longman (1985) or Sager (2005), for example. This problem shows three challenging features that are typical for real-world dynamic mixed-integer problems: integer variables in time, system-state-dependent switches that need to be modelled appropriately using, for example, additional integer variables, and higher-order polynomial approximations to nonlinear and possibly non-convex system characteristics.

2.5. Summary of MINLP applications

In addition to the models discussed above, MINLPs arise in a broad range of engineering and scientific applications, including chemical, civil, electrical, nuclear, and communication engineering, as well as emerging scientific applications in the design of nanophotonic materials and devices.

Electrical engineering applications of MINLPs include the efficient management of electricity transmission (Bacher 1997, Momoh *et al.* 1997), transmission expansion (Romero, Monticelli, Garcia and Haffner 2002, Garver 1997), transmission switching (Bartholomew, O'Neill and Ferris 2008, Hedman, O'Neill, Fisher and Oren 2008), and contingency analysis, and blackout prevention of electric power systems (Bienstock and Mattia 2007, Donde *et al.* 2005).

MINLPs also arise in many chemical engineering applications, such as the design of water (Bragalli *et al.* 2006, Karuppiah and Grossmann 2006) and gas (Martin, Möller and Moritz 2006) distribution networks, the minimization of the environmental impact of utility plants (Eliceche, Corvalán and Martínez 2007), the integrated design and control of chemical processes (Flores-Tlacuahuac and Biegler 2007), block layout design in the manufacturing and service sectors (Castillo, Westerlund, Emet and Westerlund 2005), and in the optimization of polymerization plants (Prata, Oldenburg, Kroll and Marquardt 2008). A related area is systems biology, where topics such as circadian rhythms (Shaik, Sager, Slaby and Lebiedz 2008) and protein folding (Klepeis and Floudas 2003) are addressed using mixed-integer optimization.

Applications in communication engineering and computer science include the optimal response to a cyber attack (Goldberg, Leyffer and Safro 2012, Altunay, Leyffer, Linderoth and Xie 2011), wireless bandwidth allocation (Bhatia, Segall and Zussman 2006, Sheikh and Ghafoor 2010, Costa-Montenegro, González-Castaño, Rodriguez-Hernández and Burguillo-Rial 2007), selective filtering (Sinha, Yener and Yates 2002, Soleimanipour, Zhuang and Freeman 2002), optical network performance optimization (Elwalid, Mitra and Wang 2006), network design with queuing delay constraints (Boorstyn and Frank 1977), and network design topology (Bertsekas and Gallager 1987, Chi, Jiang, Horiguchi and Guo 2008), multi-vehicle swarm communication network optimization (Abichandani, Benson and Kam 2008), the design of optimal paths (*i.e.*, minimum time) for robotic arms (Gentilini, Margot and Shimada 2013), the synthesis of periodic waveforms by tripolar pulse codes (Callegari, Bizzarri, Rovatti and Setti 2010), and the solution of MILP under uncertainty of the parameters through robust optimization (Ben-Tal and Nemirovski 1995).

Other engineering applications include the operational reloading of nuclear reactors (Quist *et al.* 1998), the optimal response to catastrophic oil spills such as the recent Deepwater oil spill in the Gulf of Mexico (You and Leyffer 2010, 2011), the design of load-bearing thermal insulation system for the Large Hadron Collider (Abramson 2004, Abhishek, Leyffer and Linderoth 2010), concrete structure design (Guerra, Newman and Leyffer 2011), and the design and operation of drinking water networks (Burgschweiger *et al.* 2008), gas networks (Martin *et al.* 2006), electric distribution networks (Lakhera, Shanbhag and McInerney 2011), and mining networks (Pruitt, Leyffer, Newman and Braun 2012). Applications in traffic modelling and optimization are found in Fügenschuh, Herty, Klar and Martin (2006), and Soler, Olivares, Staffetti and Bonami (2011) consider a flight path optimization problem. An important financial application of MINLP arises in portfolio optimization (Bienstock 1996, Jobst, Horniman, Lucas and Mitra 2001).

MINLP models can also be used for stochastic service system design problems (Elhedhli 2006), since performance metrics are often nonlinear in the decision variables.

Another developing application area of both game theory and optimization is resource allocation for homeland security; see, *e.g.*, Bier (2005), Sandler and Arce M (2003) and Zhuang and Bier (2007*b*). Simple versions of these models (involving, for example, a single target and attacker) have closed-form optimal (equilibrium) solutions; see *e.g.*, Powell (2007), Sandler and Siqueira (2006) and Zhuang and Bier (2007*a*). In more realistic models, however, the best defensive strategy must be computed. Such models often have important binary choices, including which targets to assign a high priority in defending (Bier, Nagaraj and Abhichandani 2005, Bier, Oliveros and Samuelson 2007) and which strategies to employ in defending a target (*e.g.*, whether to attempt to deceive attackers about the level of resources invested to defend a given target: see Zhuang and Bier 2010, Zhuang 2008). Moreover, the utility functions of the attackers and defenders in these models are highly nonlinear. Powerful MINLP solvers are expected to provide decision support in this area.

An emerging area with challenging MINLP tasks is human decision analysis and support in complex problem solving (Engelhart, Funke and Sager 2013).

A special domain of MINLP is dynamic problems constrained by ordinary differential equations (ODEs) or differential-algebraic equations (DAEs), often called mixed-integer optimal control problems (MIOCPs). One of the earliest practical problems was the optimal switched operation of the New York subway (Bock and Longman 1985). Recent applications include gear shifts in automotive control (Gerdts 2005, Kirches, Sager, Bock and Schlöder 2010), automated cruise controllers (Terwen, Back and Krebs 2004, Hellström, Ivarsson, Aslund and Nielsen 2009, Kirches 2011), superstructure detection in simulated moving bed processes (Sager *et al.* 2007), and the optimization of batch distillation processes (Oldenburg, Marquardt, Heinz and Leineweber 2003).

3. Deterministic methods for convex MINLP

In general, we resolve the integrality constraints using some form of tree search strategy. MINLPs pose the additional challenge of having nonlinear functions. Consequently, two broad classes of methods for solving (1.1) exist: single-tree methods and multitree methods. In this section we concentrate on methods for convex objective functions and convex constraints. See Section 5 for a discussion of methods for non-convex MINLPs. Throughout this section, we make the following assumptions.

Assumption 3.1. Consider problem (1.1) and assume the following.

A1 The set X is a bounded polyhedral set.

A2 The functions f and c are twice continuously differentiable convex functions.

A3 Problem (1.1) satisfies a constraint qualification for every point in the convex hull of the feasible set of (1.1).

The most restrictive assumption is the convexity assumption A2. Assumption A1 is rather mild, and A3 is technical. We do not specify the

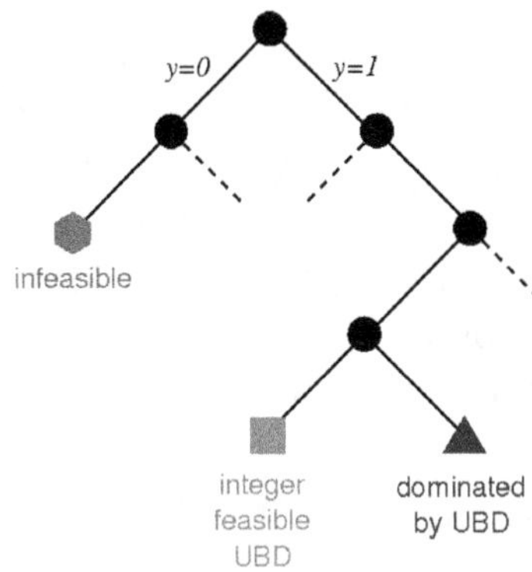

Figure 3.1. Illustration of a nonlinear branch-and-bound algorithm that traverses the tree by solving NLPs at every node of the tree.

particular constraint qualification; it is needed merely to ensure the existence of multipliers and the convergence of the NLP solvers. We note that if we assume the existence of a strictly interior feasible point for (1.1), then A3 follows from A2 as a consequence of Slater's constraint qualification (Griva, Nash and Sofer 2009), which is one of the strongest constraint qualifications.

We start by discussing nonlinear branch-and-bound, whose basic concepts underpin both classes of methods and which is a first example of a single-tree method.

3.1. Nonlinear branch-and-bound

The nonlinear branch-and-bound method for MINLPs dates back to Dakin (1965); see also Gupta and Ravindran (1985). The algorithm starts by solving the NLP relaxation of (1.1), the root node, defined by relaxing the integrality conditions on the integer variables, $x_i, i \in I$. If this relaxation is infeasible, then MINLP is also infeasible. If the solution of the relaxation is integer, then it also solves the MINLP. Otherwise, branch-and-bound searches a tree whose nodes correspond to NLP subproblems, and whose edges correspond to branching decisions. We use both optimality and feasibility of NLP subproblems to prune nodes in the tree. We define a node problem and then define the branching and pruning operations before stating the algorithm, which is also illustrated in Figure 3.1.

A node in the branch-and-bound tree is uniquely defined by a set of bounds, (l, u), on the integer variables and corresponds to the NLP

$$\begin{cases} \underset{x}{\text{minimize}} & f(x), \\ \text{subject to} & c(x) \le 0, \\ & x \in X, \\ & l_i \le x_i \le u_i, \ \forall i \in I. \end{cases} \tag{NLP(l,u)}$$

We note that the root node relaxation corresponds to NLP$(-\infty, \infty)$. Next, we describe the branching and pruning rules for branch-and-bound.

Branching. If the solution x' of (NLP(l,u)) is feasible but not integral, then we branch on any non-integral variable, say x'_i. Branching introduces two new NLP nodes, also referred to as child nodes of (NLP(l,u)). In particular, we initialize bounds for two new problems as $(l^-, u^-) := (l, u)$ and $(l^+, u^+) := (l, u)$ and then modify the bound corresponding to the branching variable

$$u_i^- := \lfloor x'_i \rfloor \quad \text{and} \quad l_i^+ := \lceil x'_i \rceil. \tag{3.1}$$

The new NLP problems are then defined as NLP(l^-, u^-) and NLP(l^+, u^+). In practice, the new problems are stored on a heap $\mathcal{H}$, which is updated with these two new problems.

Pruning rules. The pruning rules for NLP branch-and-bound are based on optimality and feasibility of NLP subproblems. We let U be an upper bound on the optimal value of (1.1) (initialized as $U = \infty$).

- *Infeasible nodes.* If any node (NLP(l,u)) is infeasible, then any problem in the subtree rooted at this node is also infeasible. Thus, we can prune infeasible nodes (indicated by a hexagon in Figure 3.1).
- *Integer feasible nodes.* If the solution $x^{(l,u)}$ of (NLP(l,u)) is integral, then we obtain a new incumbent solution if $f(x^{(l,u)}) < U$, and we set $x^* = x^{(l,u)}$ and $U = f(x^{(l,u)})$. Otherwise, we prune the node because its solution is dominated by the upper bound.
- *Upper bounds on NLP nodes.* If the optimal value of (NLP(l,u)), $f(x^{(l,u)})$ (or in fact a lower bound on the optimal value), is dominated by the upper bound, that is, if $f(x^{(l,u)}) \geq U$, then we can prune this node because there cannot be any better integer solution in the subtree rooted at (NLP(l,u)).

The complete nonlinear branch-and-bound algorithm is described in Algorithm 3.1, and Proposition 3.2 establishes its convergence.

Proposition 3.2. Consider solving (1.1) by nonlinear branch-and-bound. Assume that the problem functions f and c are convex and twice continuously differentiable and that X is a bounded polyhedral set. Then it follows that branch-and-bound terminates at an optimal solution after searching a finite number of nodes or with an indication that (1.1) has no solution.

Proof. The assumptions in Proposition 3.1 ensure that every NLP node can be solved to global optimality. In addition, the boundedness of X ensures that the search tree is finite. The proof now follows similarly to the proof for MILP branch-and-bound; see, for example, Theorem 24.1 of Schrijver (1986). □

Branch-and-bound for MINLP
Choose a tolerance $\epsilon > 0$, set $U = \infty$, and initialize the heap of open problems $\mathcal{H} = \emptyset$.
Add (NLP$(-\infty, \infty)$) to the heap: $\mathcal{H} = \mathcal{H} \cup \{\text{NLP}(-\infty, \infty)\}$.
while $\mathcal{H} \neq \emptyset$ **do**
  Remove a problem (NLP(l, u)) from the heap:
  $\mathcal{H} = \mathcal{H} - \{\text{NLP}(l, u)\}$.
  Solve (NLP(l, u)) and let its solution be $x^{(l,u)}$.
  if (NLP(l, u)) *is infeasible* **then**
    Node can be pruned because it is infeasible.
  else if $f(x^{(l,u)}) > U$ **then**
    Node can be pruned, because it is dominated by upper bound.
  else if $x_I^{(l,u)}$ *integral* **then**
    Update incumbent solution: $U = f(x^{(l,u)})$, $x^* = x^{(l,u)}$.
  else
    BranchOnVariable$(x_i^{(l,u)}, l, u, \mathcal{H})$; see Algorithm 3.2.

Algorithm 3.1. Branch-and-bound for MINLP.

Subroutine: $\mathcal{S} \leftarrow$ **BranchOnVariable** $(x_i^{(l,u)}, l, u, \mathcal{H})$
// *Branch on a fractional* $x_i^{(l,u)}$ *for* $i \in I$
Set $u_i^- = \lfloor x_i^{(l,u)} \rfloor, l^- = l$ and $l_i^+ = \lceil x_i^{(l,u)} \rceil, u^+ = u$.
Add NLP(l^-, u^-) and NLP(l^+, u^+) to the heap:
$\mathcal{H} = \mathcal{H} \cup \{\text{NLP}(l^-, u^-), \text{NLP}(l^+, u^+)\}$.

Algorithm 3.2. Branch on a fractional variable.

We note that the convergence analysis of nonlinear branch-and-bound requires us only to ensure that every node that is pruned is solved to global optimality. The convexity assumptions are one convenient sufficient condition that ensures global optimality, but clearly not the only one!

Our description of branch-and-bound leaves open a number of important questions. In particular, two important strategic decisions concern the selection of branching variable and the next problem to be solved. Details of these important algorithmic decisions can be found in Achterberg, Koch and Martin (2004) for the MILP case, and in Bonami, Lee, Leyffer and Wächter (2011) for the MINLP case. It turns out that most observations generalize from the MILP to the MINLP case. In particular, branching decisions are best based on estimates of the effect of branching on a particular variables.

On the other hand, a depth-first search is typically preferred initially in order to obtain a good incumbent solution, after which most solvers switch to a best-estimate node selection strategy. In the next two sections we discuss the choice of branching variable and search strategy, before discussing other implementation considerations and presenting a generic branch-and-cut approach for convex MINLPs.

3.1.1. Selection of branching variable

Choosing a good branching variable is a crucial component of branch-and-bound. Ideally, we would like to choose the sequence of branching variables that minimizes the size of the tree that we need to search. Doing so, however, is impractical, because the sequence of branching variables is not known *a priori*. A more achievable goal in selecting a branching variable is to choose a variable that maximizes the increase in the lower bound at a node.

We denote by $I_c \subset I$ the set of all candidate branching variables at a particular node in the branch-and-bound tree. For example, I_c could be the index set of all fractional integer variables or a subset chosen according to some user-defined priorities (see, *e.g.*, Williams 1999). A simple branching rule is to select the variable with the largest integer violation for branching, in other words, choose

$$\arg\max_{i \in I_c} \{\min(x_i - \lfloor x_i \rfloor, \lceil x_i \rceil - x_i)\},$$

which is known as *maximum fractional branching*. In practice, however, this branching rule is not efficient: it performs about as well as randomly selecting a branching variable (Achterberg *et al.* 2004).

The most successful branching rules estimate the change in the lower bound after branching. Because we prune a node of the branch-and-bound tree whenever the lower bound for the node is above the current upper bound, we want to increase the lower bound as much as possible. For every integer variable $x_i, i \in I$, we define degradation estimates D_i^+ and D_i^- for the increase in the lower bound value after branching up and down on x_i, respectively. A reasonable choice would be to select the variable for which both D_i^+ and D_i^- are large. We combine the up and down degradations D_i^+ and D_i^- to compute a score for each candidate branching variable; the variable of highest score is selected. A common formula for computing this score is

$$s_i := \mu \min(D_i^+, D_i^-) + (1 - \mu) \max(D_i^+, D_i^-),$$

where $\mu \in [0, 1]$ is a prescribed parameter typically close to 1. We then select the branching variable that maximizes s_i:

$$\arg\max_{i \in I_c} \{s_i\}.$$

We next describe two methods for estimating D_i^+ and D_i^- and show how they can be combined.

Strong branching computes the degradations D_i^+ and D_i^- by solving both child nodes for all branching candidates, $x_i, i \in I_c$, which requires the solution of $2 \times |I_c|$ NLPs. To this end, we let the solution to the current node, (NLP(l,u)), be $f^{(l,u)}$. For every branching candidate $x_i, i \in I_c$ we create two temporary branching problems, NLP$_i(l^-, u^-)$ and NLP$_i(l^+, u^+)$, as in (3.1) and solve them. If both NLPs are infeasible for an index i, then we can prune the parent node (NLP(l,u)); if one of them is infeasible, then we can tighten that integer variable in the parent node and re-solve it; otherwise, we let the solution be f_i^+ and f_i^-, respectively. Given the values f_i^+ and f_i^-, we compute D_i^+ and D_i^- as

$$D_i^+ = f_i^+ - f \quad \text{and} \quad D_i^- = f_i^- - f.$$

Strong branching can significantly reduce the number of nodes in a branch-and-bound tree, but it is often slow overall because of the added computational cost of solving two NLP subproblems for each fractional variable. In order to reduce the computational cost of strong branching, it is often efficient to solve the subproblems only approximately. If the relaxation is an LP, as in the case of LP/NLP-BB (see Section 3.3.1), then one can limit the number of pivots. In the NLP case, we can limit the number of iterations, but that does not reduce the solve time sufficiently. Instead, we can use approximations of the NLP, which can be warm-started much faster than NLPs can. One approach that has been shown to be efficient is to use the basis information from a quadratic program solved at the parent node, and perform a limited number of pivots on this quadratic approximation in order to obtain estimates for f_i^+ and f_i^-, respectively (Bonami *et al.* 2011).

Pseudocost branching keeps a history of the results of past branching decisions for every variable and computes the degradations D_i^+ and D_i^- by averaging the increase in the objective value over the history. For every integer variable, we let n_i^+, n_i^- denote the number of times we have solved the up/down node for variable i. We update the per unit change in the objective when branching on x_i by computing the pseudocost, p_i^+, p_i^-, whenever we solve an up or down child node:

$$\begin{aligned} p_i^+ &= \frac{f_i^+ - f}{\lceil x_i \rceil - x_i} + p_i^+, \quad n_i^+ = n_i^+ + 1 \quad \text{or} \\ p_i^- &= \frac{f_i^- - f}{x_i - \lfloor x_i \rfloor} + p_i^-, \quad n_i^- = n_i^- + 1. \end{aligned} \tag{3.2}$$

The pseudocosts are then used to estimate D_i^+ and D_i^- whenever we need to make a branching decision, *i.e.*,

$$D_i^+ = (\lceil x_i \rceil - x_i)\frac{p_i^+}{n_i^+} \quad \text{and} \quad D_i^- = (x_i - \lfloor x_i \rfloor)\frac{p_i^-}{n_i^-}.$$

Pseudocosts are typically initialized by using strong branching. The update of the pseudocosts is relatively cheap compared with that of strong branching, because the solutions of the parent and child node are available. Statistical experience on MILP problems has shown that pseudocosts are reasonable estimates of the degradation in the objective after branching (Linderoth and Savelsbergh 1999). One difficulty that arises with pseudocosts, however, is how to update the pseudocosts if the NLP is infeasible. Typically the update is skipped in this case, but a fruitful alternative motivated by NLP might be to use the value of the ℓ_1 exact penalty functions (Fletcher 1987, Chapter 12.3), which is readily available at the end of the NLP solve.

Reliability branching is an attractive hybrid between strong branching and pseudocost branching. It performs strong branching on a variable and updates the pseudocosts until n_i^+ or n_i^- are greater than a threshold τ (a typical value for τ is 5). This corresponds to using strong branching early during the tree search, when branching decisions are more critical, and then switching to pseudocost branching once reliable estimates are available.

Branching on general disjunctions is an alternative to branching on a variable. We can branch in many ways. One popular way is to branch on special-ordered sets (Beale and Tomlin 1970). A more general approach is to branch on split disjunctions of the form

$$(a^T x_I \leq b) \vee (a^T x_I \leq b + 1), \tag{3.3}$$

where $a \in \mathbb{Z}^p$ and $b \in \mathbb{Z}$. Split disjunctions have been shown to produce promising results for MILP (Jobst *et al.* 2001) but have not been used in MINLP.

3.1.2. Node selection strategies

The second important strategic decision is which node should be solved next. The goal of this strategy is to find a good feasible solution quickly in order to reduce the upper bound, and to prove optimality of the current incumbent x^* by increasing the lower bound as quickly as possible. We introduce two popular strategies, *depth-first search* and *best-bound search*, and discuss their strengths and weaknesses. We also present two hybrid schemes that aim to overcome the weaknesses of these two strategies.

Depth-first search selects the deepest node in the tree (or the last node that was added to $\mathcal{H}$). One advantage of this strategy is that it keeps the list of open nodes, $\mathcal{H}$, as small as possible. This property made it a popular strategy for the earliest implementations of branch-and-bound (Dakin 1965). Another advantage is that this strategy minimizes the change to subsequent NLP relaxations (NLP(l, u)) that are solved, because only a single bound is changed. This fact allows us to exploit warm-start techniques that re-use the existing basis factors in MILP problems, or make use of a good starting point in MINLP problems. Though some attempts have been made to use warm-starts in MINLP (see Bonami *et al.* 2011 and Mahajan, Leyffer and Kirches 2012), they have generally not been as successful in MINLP, mainly because the Jacobian and Hessian matrices change from one node to another so that factorizations are always outdated. Unfortunately, depth-first search can exhibit extremely poor performance if no upper bound is found, exploring many nodes with a lower bound that is larger than the solution.

Best-bound search selects the node with the best lower bound. Its advantage is that for a fixed sequence of branching decisions it minimizes the number of nodes that are explored, because all nodes that are explored would have been explored independently of the upper bound. On the other hand, the weaknesses of this strategy are that it may require significantly more memory to store the open problems, the sequence of NLP subproblems does not readily allow for warm-starts, and it usually does not find an integer feasible solution before the end of the search. This last point is particularly relevant for very large problems or if the solution time is limited such as in real-time applications, because best-bound search may fail to produce even a feasible point. Like depth-first search, this strategy has also been used since the very first branch-and-bound algorithms (Land and Doig 1960, Lawler and Woods 1966).

Variants of best-bound search. Two variants of best-bound search have been proposed. Both try to estimate the effect of the branching on the bound by using pseudocosts. We let f_p denote the lower bound or estimate of problem p on the heap.

(1) *Best expected bound* selects the node with the best expected bound after branching, which is estimated as

$$b_p^+ = f_p + (\lceil x_i \rceil - x_i)\frac{p_i^+}{n_i^+} \quad \text{and} \quad b_p^- = f_p + (x_i - \lfloor x_i \rfloor)\frac{p_i^-}{n_i^-}.$$

The next node is selected as $\max_p\{\min(b_p^+, b_p^-)\}$.

(2) *Best estimate* chooses the node that is expected to contain the best expected integer solution within its subtree based on pseudocosts. The

best expected solution within a subtree can be estimated as

$$e_p = f_p + \sum_{i: x_i \text{fractional}} \min\left((\lceil x_i \rceil - x_i) \frac{p_i^+}{n_i^+} , \ (x_i - \lfloor x_i \rfloor) \frac{p_i^-}{n_i^-} \right),$$

namely, by adding the pseudocost estimates for *all* non-integral integer variables to the lower bound at that node. The next node to be solved is then chosen as $\max_p \{e_p\}$.

Both strategies have been reasonably successful in the context of MINLP (Goux and Leyffer 2003).

Hybrid search strategies. Good search strategies try to combine depth-first and best-bound search. Two such strategies are the two-phase method and the diving method (Eckstein 1994, Linderoth and Savelsbergh 1999, Achterberg 2005).

(1) *Two-phase methods* start with depth-first search until one or a small number of integer solutions have been found. It then switches to best-bound search in order to prove optimality. If the tree becomes too large during this second phase, then the method switches back to depth-first search in order to keep the number of open nodes manageable.

(2) *Diving methods* are also two-phase methods. The method starts with depth-first search until a leaf node (feasible or infeasible) is found. It then backtracks to the best bound on the tree to start another depth-first dive. Diving methods continue to alternate between this diving step and the backtracking step.

3.1.3. Other implementation considerations

Two other considerations are important when implementing an MINLP branch-and-bound solver: (1) the degree to which inexact subproblem solves can be used during the tree search and (2) the use of heuristics to find good incumbents quickly that will then reduce the size of the search tree.

Inexact NLP subproblem solves. There exists considerable freedom in how exactly NLP subproblems are solved in the tree search. In fact, at any non-leaf node of the tree, we need only to provide a branching variable. We can exploit this freedom to consider inexact solves at NLP subproblems in order to improve the performance of branch-and-bound. The first approach to using inexact NLP solves is due to Borchers and Mitchell (1994). The authors consider a sequential quadratic programming (SQP) approach to solving every NLP subproblem and interrupt SQP after every iteration to check whether SQP appears to converge to a non-integral solution. If convergence to non-integral solution is detected, the authors stop the current NLP solve and proceed to branching on any non-integral integer variable. Note, however, that the inexact solution of an NLP subproblem does not provide a valid lower bound for the node. Hence, the authors propose occasionally solving an augmented Lagrangian dual to obtain a lower bound. This approach can be improved by combining insights from outer approximation to obtain implicit bounds: see Leyffer (2001). An alternative approach is presented by Mahajan, Leyffer and Kirches (2012). The authors search the branch-and-bound tree using a single quadratic program generated at the root node. The advantage of this approach is that quadratic programs, unlike NLPs, can be warm-started efficiently after branching by re-using factors of the basis of the parent node. The authors show how the pruning rules described above can be adapted to ensure that the algorithm guarantees a global solution to a convex MINLP. Numerical experience shows that this approach can reduce the CPU time of branch-and-bound by several orders of magnitude for some problems.

Heuristics for finding good incumbents. The bounding in Algorithm 3.1 shows that it is important to find good incumbents quickly in order to prune more parts of the tree. A range of heuristics exists that can be used at the root node or at intermediate nodes of the tree; these methods are discussed in Section 6.

3.1.4. Cutting planes for nonlinear branch-and-bound

Branch-and-bound algorithms can be enhanced by adding cutting planes, as described by Stubbs and Mehrotra (1999) for convex mixed 0–1 nonlinear programs, based on Balas, Ceria and Cornuéjols (1996) for MILP. The branch-and-cut algorithm extends the branch-and-bound Algorithm 3.1 by an additional step during which one or more cutting planes may be generated and added to a node (NLP(l,u)) in order to cut off a fractional optimal solution $x^{(l,u)}$. A node is branched on only if the relaxed optimal solution remains fractional even after a prescribed number of rounds of cuts have been added or if no suitable cuts could be generated at all. The hope is that adding cutting planes will lead to a significant reduction of the tree size or will ideally remove the need for branching by producing a locally tight description of the convex hull. An outline of the branch-and-cut algorithm for MINLP is given as Algorithm 3.3.

The most significant difference from the branch-and-bound algorithm for MINLP is the generation of additional cutting planes. We rely on a generic subroutine GenerateCuts that produces a new valid inequality for the current tree node. The high-level description of this step is to solve a separation problem. Given a point $x^{(l,u)}$ with $x_j^{(l,u)} \notin \{0,1\}$ and a feasible set $\mathcal{F}(l,u)$ associated with the current node NLP(l,u), find a vector $(\pi_0, \pi^T)^T$ such that the inequality $\pi^T x \leq \pi_0$ holds for all $x \in \mathcal{F}(l,u)$ but cuts off $x^{(l,u)}$ by

Branch-and-cut for MINLP
Choose a tolerance $\epsilon > 0$, and set $U = \infty$.
Initialize the heap of open problems $\mathcal{H} = \emptyset$.
Add (NLP$(-\infty, \infty)$) to the heap: $\mathcal{H} = \mathcal{H} \cup \{\text{NLP}(-\infty, \infty)\}$.
while $\mathcal{H} \neq \emptyset$ **do**
  Remove a problem (NLP(l, u)) from the heap:
  $\mathcal{H} = \mathcal{H} - \{\text{NLP}(l, u)\}$.
  repeat
    Solve (NLP(l, u)) and let its solution be $x^{(l,u)}$.
    if (NLP(l, u)) *is infeasible* **then**
      Node can be pruned because it is infeasible.
    else if $f(x^{(l,u)}) > U$ **then**
      Node can be pruned, because it is dominated by upper bound.
    else if $x_I^{(l,u)}$ *integral* **then**
      Update incumbent solution: $U = f(x^{(l,u)})$, $x^* = x^{(l,u)}$.
    else if *more cuts shall be generated* **then**
      GenerateCuts$(x^{(l,u)}, j)$; see Algorithm 3.4.
  until *no new cuts generated*
  if (NLP(l, u)) *not pruned and not incumbent* **then**
    BranchOnVariable$(x_j^{(l,u)}, l, u, \mathcal{H})$; see Algorithm 3.2.

Algorithm 3.3. Branch-and-cut for MINLP.

Subroutine: GenerateCuts $(x^{(l,u)}, j)$
// *Generate a valid inequality that cuts off* $x_j^{(l,u)} \notin \{0, 1\}$
Solve a separation problem in $x^{(l,u)}$ to obtain an inequality that cuts off $x_j^{(l,u)} \notin \{0, 1\}$ from the feasible set of (NLP(l, u)). Add this inequality to (NLP(l, u)).

Algorithm 3.4. Generate a subgradient cut by solving a separation problem.

satisfying $\pi^T x^{(l,u)} > \pi_0$. Various approaches to this problem are discussed in Section 4.

Like branch-and-bound, a number of implementation issues are open. In their original work Stubbs and Mehrotra (1999) generate cuts only at the root node. Such cuts will then be valid for the entire branch-and-bound tree and are hoped to reduce it in size. In general, cuts generated in tree nodes will be valid only for that particular subtree. Lifting procedures can sometimes be obtained in order to make locally generated cuts also globally valid for the entire branch-and-bound tree. Implementations of branch-and-cut solvers may choose to maintain both a global and a local pool of cuts.

Depending on the type of cut generated, the separation problems may themselves be difficult to solve numerically. The performance of a branch-and-cut scheme may depend crucially on the ability to reliably and precisely solve the arising separation problems. Solver failures may reduce the number of valid cuts obtained. Care must be taken not to cut off the optimal solution because of numerical difficulties.

3.2. Multitree methods for MINLP

One drawback of nonlinear branch-and-bound is that it requires the solution of a large number of NLPs that cannot be easily hot-started (unlike MILP where we can re-use the LP basis factors after branching). This observation led to the development of another class of methods that we term multitree methods because they decompose the MINLP into an alternating sequence of NLP subproblems and MILP relaxations. Here, we review three such methods: outer approximation (Duran and Grossmann 1986, Fletcher and Leyffer 1994, Bonami *et al.* 2008), generalized Benders decomposition (Geoffrion 1972, Van Roy 1983), and the extended cutting-plane method (Westerlund and Pettersson 1995).

3.2.1. Outer approximation

We start by defining the NLP subproblem obtained by fixing the integer variables to $x_I^{(j)}$ in (1.1),

$$\begin{cases} \underset{x}{\text{minimize}} & f(x), \\ \text{subject to} & c(x) \leq 0, \\ & x \in X \quad \text{and} \quad x_I = x_I^{(j)}, \end{cases} \tag{NLP($x_I^{(j)}$)}$$

and we let its solution be $x^{(j)}$. If (NLP($x_I^{(j)}$)) is infeasible, then most NLP solvers will return a solution to a feasibility problem of the form

$$\begin{cases} \underset{x}{\text{minimize}} & \sum_{i \in J^\perp} w_i c_i^+(x), \\ \text{subject to} & c_i(x) \leq 0, \ i \in J, \\ & x \in X \quad \text{and} \quad x_I = x_I^{(j)}, \end{cases} \tag{F($x_I^{(j)}$)}$$

where $w_i > 0$ is a set of weights that can be chosen to reduce to ℓ_1 or ℓ_∞ norm minimization, J is a set of constraints that can be satisfied, and its complement $J^\perp$ is the set of infeasible constraints; see, for example, Fletcher and Leyffer (1994, 2003) and Gould, Leyffer and Toint (2004). An optimal solution of $(\mathrm{F}(x_I^{(j)}))$ with a positive objective is a certificate that the corresponding NLP $(\mathrm{NLP}(x_I^{(j)}))$ is infeasible.

Next we consider (1.2), and observe that the convexity of f and c implies that the linearization about the solution $x^{(j)}$ of $(\mathrm{NLP}(x_I^{(j)}))$ or $(\mathrm{F}(x_I^{(j)}))$, given by

$$\eta \geq f^{(j)} + \nabla f^{(j)^T}(x - x^{(j)}) \quad \text{and} \quad 0 \geq c^{(j)} + \nabla c^{(j)^T}(x - x^{(j)}), \tag{3.4}$$

is an outer approximation of the feasible set of (1.2). We can show that if $(\mathrm{NLP}(x_I^{(j)}))$ is infeasible, then the corresponding set of outer approximations ensures that $x_I = x_I^{(j)}$ violates (3.4).

Lemma 3.3 (Fletcher and Leyffer 1994, Lemma 1). If $(\mathrm{NLP}(x_I^{(j)}))$ is infeasible and $x^{(j)}$ is an optimal solution of $(\mathrm{F}(x_I^{(j)}))$, then $x_I = x_I^{(j)}$ violates (3.4) for all $x \in X$.

Next, we define an index set of all possible feasible integer assignments:

$$\mathcal{X} := \{x^{(j)} \in X : x^{(j)} \text{ solves } (\mathrm{NLP}(x_I^{(j)})) \text{ or } (\mathrm{F}(x_I^{(j)}))\}. \tag{3.5}$$

Because X is bounded, there exists only a finite (albeit exponentially large) number of different integer points $x_I^{(j)}$, and hence $\mathcal{X}$ is a finite set. Now we can construct an MILP that is equivalent to (1.2):

$$\begin{cases} \underset{\eta, x}{\text{minimize}} & \eta, \\ \text{subject to} & \eta \geq f^{(j)} + \nabla f^{(j)^T}(x - x^{(j)}), \ \forall x^{(j)} \in \mathcal{X}, \\ & 0 \geq c^{(j)} + \nabla c^{(j)^T}(x - x^{(j)}), \ \forall x^{(j)} \in \mathcal{X}, \\ & x \in X, \\ & x_i \in \mathbb{Z}, \ \forall i \in I. \end{cases} \tag{3.6}$$

We can show that (3.6) and (1.2) have the same optimal value and that any solution of (1.2) is an optimal solution of (3.6). However, the converse is not true, as the following example from Bonami *et al.* (2008) shows:

$$\underset{x}{\text{minimize}} \ x_3 \quad \text{subject to } (x_1 - \tfrac{1}{2})^2 + x_2^2 + x_3^3 \leq 1, \quad x_1 \in \mathbb{Z} \cap [-1, 2].$$

The MILP created from outer approximations contains no coefficient for x_2, because $x_2 = 0$ is optimal in all NLP subproblems. Hence, any value of x_2 is optimal in the MILP.

Theorem 3.4. Assume that the assumptions in Proposition 3.1 hold, and let x^* solve (1.1). Then it follows that x^* also solves (3.6). Conversely, if (η^*, x^*) solves (3.6), then it follows that the optimal value of (1.1) is η^* and x_I^* is an optimal solution in both problems.

Proof. The proof follows from the results by Bonami *et al.* (2008) and Fletcher and Leyffer (1994). □

Of course, it is not practical to solve (3.6) because by the time it is set up, we already know the solution of (1.1). Instead, Duran and Grossmann (1986) propose a relaxation algorithm that solves an alternating sequence of MILP problems and NLP subproblems. Initially, we solve (NLP($x_I^{(j)}$)) for a given initial point $x^{(j)} = x^{(0)}$ and set up a relaxation of (3.6) in which we replace $\mathcal{X}$ by a subset $\mathcal{X}^k \subset \mathcal{X}$, with $\mathcal{X}^k = \{0\}$. We also add an upper bound on η corresponding to the best solution found so far:

$$\eta < U^k := \min_{j \le k} \{ f^{(j)} \mid (\text{NLP}(x_I^{(j)})) \text{ is feasible} \}.$$

We note, however, that this latter constraint is not enforceable in practice, and is typically replaced by $\eta \le U^k - \epsilon$, where $\epsilon > 0$ is a small tolerance. This upper bound ensures that once we have solved (NLP(l, u)) and added its outer approximations (3.4) to ($M(\mathcal{X}^k)$), $x_i^{(j)}$ is not feasible in ($M(\mathcal{X}^k)$) for $k \ge j$, ensuring that outer approximation terminates finitely. Thus, the MILP master problem solved at iteration k is given by

$$\begin{cases} \underset{\eta, x}{\text{minimize}} & \eta, \\ \text{subject to} & \eta \le U^k - \epsilon, \\ & \eta \ge f^{(j)} + \nabla f^{(j)^T}(x - x^{(j)}), \ \forall x^{(j)} \in \mathcal{X}^k, \\ & 0 \ge c^{(j)} + \nabla c^{(j)^T}(x - x^{(j)}), \ \forall x^{(j)} \in \mathcal{X}^k, \\ & x \in X, \\ & x_i \in \mathbb{Z}, \ \forall i \in I. \end{cases} \qquad (M(\mathcal{X}^k))$$

The outer approximation algorithm is described in Algorithm 3.5, and its convergence result is stated in Theorem 3.5.

The algorithm also detects whether (1.2) is infeasible. If $U^k = \infty$ on exit, then all integer assignments visited by the algorithm are infeasible, and hence (1.2) is infeasible. The use of upper bounds on η and the definition of the set and $\mathcal{X}^k$ ensure that no $x_I^{(j)}$ is replicated by the algorithm. Thus, one can prove that the algorithm terminates after a finite number of steps, provided that there is only a finite number of integer assignments.

Theorem 3.5. If Assumptions 3.1 hold and if the number of integer points in X is finite, then Algorithm 3.5 terminates in a finite number of steps at an optimal solution of (1.1) or with an indication that (1.1) is infeasible.

A proof of Theorem 3.5 was given by Fletcher and Leyffer (1994). The main argument of the proof is that the optimality of $x^{(j)}$ in (NLP($x_I^{(j)}$)) implies that $\eta \ge f^{(j)}$ for any feasible point in ($M(\mathcal{X}^k)$). The upper bound

Outer approximation
Given $x^{(0)}$, choose a tolerance $\epsilon > 0$, set $U^{-1} = \infty$, set $k = 0$, and initialize $\mathcal{X}^{-1} = \emptyset$.
repeat
  Solve (NLP($x_I^{(j)}$)) or (F($x_I^{(j)}$)) and let the solution be $x^{(j)}$.
  if (NLP($x_I^{(j)}$)) *is feasible and* $f^{(j)} < U^{k-1}$ **then**
    Update current best point: $x^* = x^{(j)}$ and $U^k = f^{(j)}$.
  else
    Set $U^k = U^{k-1}$.
  Linearize objective and constraint f and c about $x^{(j)}$ and set $\mathcal{X}^k = \mathcal{X}^{k-1} \cup \{j\}$.
  Solve ($M(\mathcal{X}^k)$), let the solution be $x^{(k+1)}$ and set $k = k + 1$.
until MILP ($M(\mathcal{X}^k)$) *is infeasible.*

Algorithm 3.5. Outer approximation.

$\eta \leq f^{(j)} - \epsilon$ therefore ensures that the choice $x_I = x_I^{(j)}$ in $(M(\mathcal{X}^k))$ is not feasible. Hence, the algorithm is finite. The optimality of the algorithm follows from the convexity of f and c, which ensures that the linearizations are supporting hyperplanes.

Worst-case complexity of outer approximation. In practice, outer approximation often works efficiently. However, Hijazi, Bonami and Ouorou (2010) provide an example where outer approximation takes an exponential number of iterations. In particular, they consider the following MINLP:

$$\underset{x}{\text{minimize}}\ 0 \quad \text{subject to} \quad \sum_{i=1}^{n} \left(x_i - \frac{1}{2} \right)^2 \leq \frac{n-1}{4}\ x \in \{0,1\}^n. \tag{3.7}$$

Geometrically, the problem corresponds to a feasibility problem that seeks a point that lies in the intersection of the n-dimensional ℓ_2 ball with radius $\frac{n-1}{4}$, centred at $\hat{x} = \left(\frac{1}{2}, \ldots, \frac{1}{2}\right)$, with the unit hypercube $\{0,1\}^n$. Since the distance from $\hat{x}$ to any vertex of $\{0,1\}^n$ is $\frac{\sqrt{n}}{2} > \frac{n-1}{4}$, the problem is infeasible. Figure 3.2 shows a 3D plot of this example. The black lines illustrate the unit hypercube, and the spherical surface is the boundary of the nonlinear constraint in (3.7). The authors show that no outer approximation cut can cut more than one vertex. Thus, outer approximation must visit all 2^n vertices of the hypercube. The example generalizes to Benders decomposition and extended cutting-plane methods, because these methods are relaxations of outer approximation.

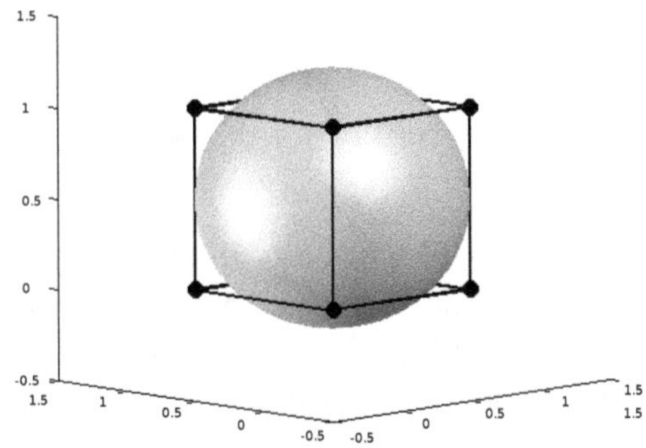

Figure 3.2. 3D plot of worst-case example of outer approximation (3.7).

3.2.2. Generalized Benders decomposition

Generalized Benders decomposition was developed before outer approximation; see Geoffrion (1972). Given outer approximation, however, it is straightforward to develop the Benders cuts and present Benders decomposition. We start by considering the outer approximations (3.4) and assume that the constraints X are inactive. Summing up the outer approximations (3.4) weighted with $(1, \lambda^{(j)})$, where $\lambda^{(j)} \geq 0$ are the multipliers of $(\text{NLP}(x_I^{(j)}))$, we obtain the following valid inequality:

$$\eta \geq \left(f^{(j)} + \lambda^{(j)^T} c^{(j)}\right) + \left(\nabla f^{(j)} + \sum_{i=1}^{m} \lambda_i^{(j)} \nabla c_i^{(j)}\right)^T (x - x^{(j)}). \tag{3.8}$$

We observe that $\lambda^{(j)^T} c^{(j)} = 0$ as a result of complementary slackness and that the continuous variable component of the gradient

$$\nabla_C f^{(j)} + \sum_{i=1}^{m} \lambda_i^{(j)} \nabla_C c_i^{(j)} = 0$$

as a result of the optimality of $x^{(j)}$. Thus, we can rewrite the cut (3.8) in the integer variables only as

$$\eta \geq f^{(j)} + \left(\nabla_I f^{(j)} + \sum_{i=1}^{m} \lambda_i^{(j)} \nabla_I c_i^{(j)}\right)^T (x_I - x_I^{(j)}), \tag{3.9}$$

which is the Benders cut for feasible subproblems. We also observe that the optimality of $(\text{NLP}(x_I^{(j)}))$ implies the existence of multipliers $\mu_I^{(j)}$ of the bounds $x_I = x_I^{(j)}$ and that their value is equal to the gradient in the Benders cut. Thus, we can write the Benders cut compactly as

$$\eta \geq f^{(j)} + \mu_I^{(j)^T} (x_I - x_I^{(j)}). \tag{3.10}$$

With $(\mathrm{F}(x_I^{(j)}))$, a similar derivation shows that the Benders cut for infeasible subproblems is

$$0 \geq \sum_{i \in J^{\perp}} w_i c_i^{+}(x) + \mu_I^{(j)^T}(x_I - x_I^{(j)}), \tag{3.11}$$

where $\mu_I^{(j)}$ are the multipliers of $x_I = x_I^{(j)}$ in $(\mathrm{F}(x_I^{(j)}))$.

The advantage of the Benders cuts is that they involve only the integer variables and one objective variable. A disadvantage is that the Benders cuts are almost always dense. Moreover, the Benders cuts are weaker than the outer approximation cuts from which they are derived.

3.2.3. Extended cutting-plane method

The extended cutting-plane method (Westerlund and Lundqvist 2005, Westerlund and Pettersson 1995) can be viewed as a variant of outer approximation that does not solve NLP subproblems such as $(\mathrm{NLP}(x_I^{(j)}))$ or $(\mathrm{F}(x_I^{(j)}))$. Instead, the extended cutting-plane method linearizes all functions at the solution of the MILP master problem, $x^{(k)}$. If $x^{(k)}$ satisfies all linearizations, then we have solved the MINLP. Otherwise, we choose one (or a small number of) the most violated linearizations and add it to the MILP master problem. The method alternates between the solution of the master problem and the generation of linearizations that are underestimators if the MINLP problem is convex.

Convergence of the extended cutting-plane method follows similarly to that of outer approximation. The convexity of f and c ensures that the linearizations are a separating hyperplane, and the convergence in the continuous space follows from the convergence of Kelley's cutting-plane method (Kelley 1960).

One weakness of the extended cutting-plane method is that it can produce the same integer assignment multiple times, because the cuts are not generated from solutions of $(\mathrm{NLP}(x_I^{(j)}))$ or $(\mathrm{F}(x_I^{(j)}))$. The rate of convergence of Kelley's cutting-plane method is in general linear, and hence it may require a larger number of iterations. In practice, however, the extended cutting-plane method is competitive with outer approximations, and the cutting planes it creates have been used to accelerate outer-approximation-based schemes, such as the LP/NLP-based branch-and-bound method discussed in Section 3.3.1; see for example Abhishek *et al.* (2010).

3.3. Single-tree methods for MINLP

One single-tree method was already described above, namely branch-and-bound. This section shows how we can develop hybrid or integrated approaches that use outer approximation properties but require only a single MILP tree to be searched. The advantages of these approaches are twofold. First, we avoid the need to re-solve related MILP master problems; second, we search a tree whose nodes can be effectively warm-started by re-using basis information from parent nodes.

An alternative motivation for these hybrid techniques is to interpret them as branch-and-cut algorithms for solving the large MILP, (3.6), with the full set of linearizations $\mathcal{X}$ as in (3.5). This problem is clearly intractable, so instead we apply a delayed constraint generation technique of the 'formulation constraints' $\mathcal{X}^k \subset \mathcal{X}$. At integer solutions we can separate cuts by solving the NLP (NLP($x_I^{(j)}$)) or (F($x_I^{(j)}$)) for fixed integer variables.

3.3.1. LP/NLP-based branch-and-bound

Introduced by Quesada and Grossmann (1992), LP/NLP-based branch-and-bound (LP/NLP-BB) methods have emerged as a powerful class of algorithms for solving convex MINLPs. LP/NLP-BB improves outer approximation by avoiding the solution of multiple MILP master problems, which take an increasing amount of computing time. Moreover, since these MILP relaxations are strongly related to one another a considerable amount of information is regenerated each time a relaxation is solved.

Instead, the LP/NLP-BB algorithm solves the continuous relaxation of ($M(\mathcal{X}^k)$) and enforces integrality of the x_I variables by branching. Whenever a new integer solution is found, the tree search is interrupted to solve (NLP($x_I^{(j)}$)), and the master MILP is updated with new outer approximations generated from the solution of the subproblem. Finally the node corresponding to the integer point is re-solved, and the tree search continues.

The previous integer feasible node must be re-solved, because unlike ordinary branch-and-bound, a node cannot be pruned if it produces an integer feasible solution, since the previous solution at this node is cut out by the linearizations added to the master program. Thus, only infeasible nodes can be pruned.

We now formally define the LP/NLP-based branch-and-bound algorithm. It can be viewed as a hybrid algorithm between nonlinear branch-and-bound (Algorithm 3.1) and outer approximation (Algorithm 3.5); see also Bonami *et al.* (2008). We denote by (LP($\mathcal{X}^k, l^i, u^i$)) the LP node (relaxation) of the MILP master problem ($M(\mathcal{X}^k)$) with bounds $l^i \leq x_I \leq u^i$. In particular, (LP($\mathcal{X}^k, -\infty, \infty$)) is the LP root node relaxation of ($M(\mathcal{X}^k)$). The algorithm uses an initial integer point $x^{(0)}$ to set up the initial master problem, but it could also solve the NLP relaxation in order to derive the initial outer approximations.

As in the outer approximation algorithms, the use of an upper bound implies that no integer assignment is generated twice during the tree search. Since both the tree and the set of integer variables are finite, the algorithm eventually encounters only infeasible problems, and the heap is thus emptied so that the procedure stops. This provides a proof of the following corollary to Theorem 3.5.

LP/NLP-based branch-and-bound
Given $x^{(0)}$, choose a tolerance $\epsilon > 0$, set $U^{-1} = \infty$, set $k = 0$, and initialize $\mathcal{X}^{-1} = \emptyset$.
Initialize MILP:
Solve $(\text{NLP}(x_I^{(j)}))$ or $(\text{F}(x_I^{(j)}))$ and let the solution be $x^{(j)}$.
Linearize objective and constraint f and c about $x^{(j)}$ and set $\mathcal{X}^k = \mathcal{X}^{k-1} \cup \{j\}$.
if $(\text{NLP}(x_I^{(j)}))$ *is feasible* **then**
  Update current best point: $x^* = x^{(j)}$ and $U^k = f^{(j)}$.
Initialize MILP Search Tree:
Initialize the heap of open problems $\mathcal{H} = \emptyset$.
Add $(\text{LP}(\mathcal{X}^k, -\infty, \infty))$ to the heap: $\mathcal{H} = \mathcal{H} \cup \{\text{LP}(\mathcal{X}^k, -\infty, \infty)\}$.
while $\mathcal{H} \neq \emptyset$ **do**
  Remove an LP problem from the heap: $\mathcal{H} = \mathcal{H} - \{\text{LP}(\mathcal{X}^k, l, u)\}$.
  Solve $(\text{LP}(\mathcal{X}^k, l, u))$ and let its solution be $x^{(l,u)}$.
  if $(\text{LP}(\mathcal{X}^k, l, u))$ *is infeasible* **then**
    Node can be pruned because $(\text{LP}(\mathcal{X}^k, l, u))$ and hence $(\text{NLP}(x_I^{(j)}))$ are infeasible.
  else if $x_I^{(l,u)}$ *integral* **then**
    Set $x_I^{(j)} = x_I^{(l,u)}$ and solve $(\text{NLP}(x_I^{(j)}))$ or $(\text{F}(x_I^{(j)}))$ and let the solution be $x^{(j)}$.
    Linearize objective and constraint f and c about $x^{(j)}$ and set $\mathcal{X}^k = \mathcal{X}^{k-1} \cup \{j\}$.
    if $(\text{NLP}(x_I^{(j)}))$ *is feasible & $f^{(j)} < U^k$* **then**
      Update current best point: $x^* = x^{(j)}$ and $U^k = f^{(j)}$.
    else
      Set $U^k = U^{k-1}$.
    Add the LP back to the heap: $\mathcal{H} = \mathcal{H} \cup \{\text{LP}(\mathcal{X}^k, l, u)\}$.
    Set $k = k + 1$.
  else
    BranchOnVariable$(x_i^{(l,u)}, l, u, \mathcal{H})$.

Algorithm 3.6. LP/NLP-based branch-and-bound.

Theorem 3.6. If Assumptions 3.1 hold, and if the number of integer points in X is finite, then Algorithm 3.6 terminates in a finite number of steps at a solution of (1.1) or with an indication that (1.1) is infeasible.

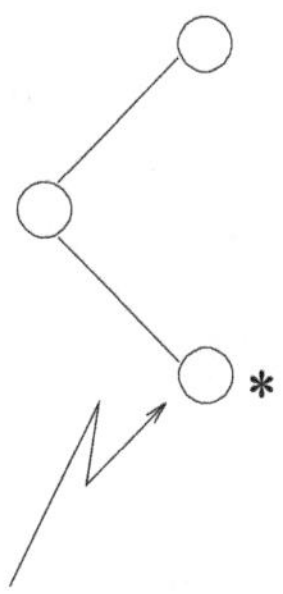

(a) solution not integer feasible: branch

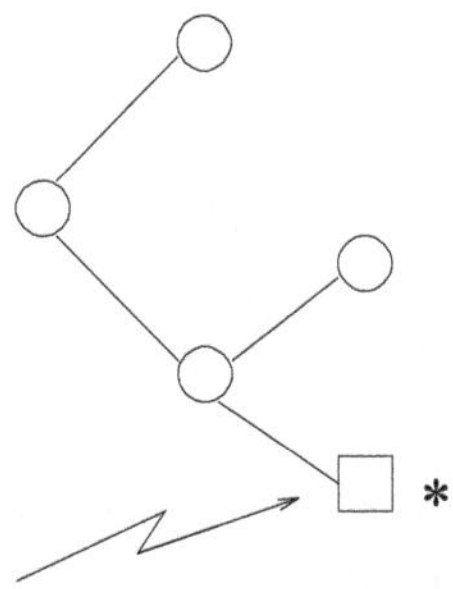

(b) solution integer feasible: solve the NLP subproblem

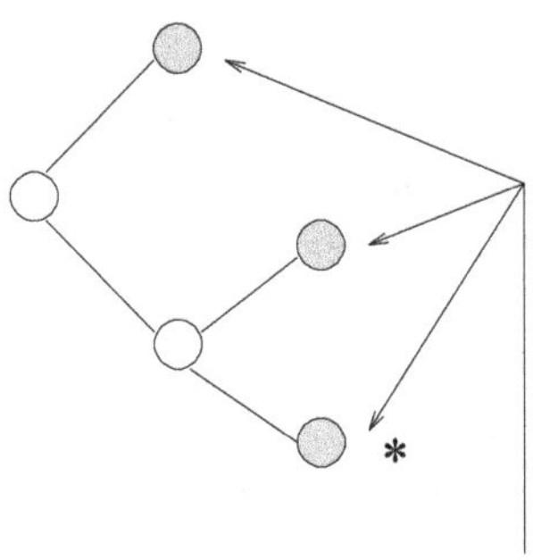

(c) update all problems on the heap

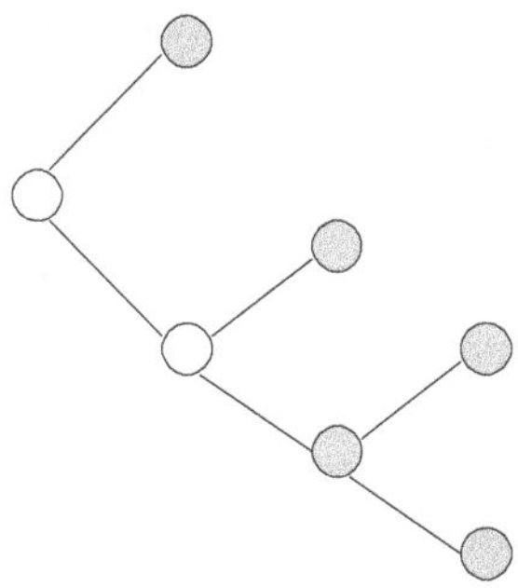

(d) continue the branch-and-bound process

Figure 3.3. Progress of LP/NLP-based branch-and-bound.

Figure 3.3 illustrates the progress of Algorithm 3.6. In Figure 3.3(a) the LP relaxation of the initial MILP has been solved, and two branches have been added to the tree. The LP that is solved next (indicated by *) does not give an integer feasible solution, and two new branches are introduced. The next LP in Figure 3.3(b) produces an integer feasible solution indicated by a box. The corresponding NLP subproblem is solved, and in Figure 3.3(c) all nodes on the heap are updated (indicated by the shaded circles) by adding the linearizations from the NLP subproblem, including the upper bound U^k that cuts out the current assignment x_I. Then, the branch-and-bound process continues on the updated tree by solving the LP marked by *.

Algorithmic refinements of LP/NLP-BB. Abhishek *et al.* (2010) have shown that the branch-and-cut LP/NLP-BB algorithm can be improved significantly by implementing it within a modern MILP solver. Advanced MILP search and cut-management techniques dramatically improve the performance of LP/NLP-BB. It is important to generate cuts at nodes that are not integer feasible, in which case it is advantageous to generate outer approximations around the solution of the LP relaxation, rather than solving an NLP (we term such cuts ECP cuts). We have shown that LP/NLP-BB can be improved dramatically by exploiting MILP domain knowledge, such as strong branching, adaptive node selection, and, most important, cut management. We have also observed that weaker cuts, such as linearizations generated at LP solutions, improve the performance of LP/NLP-BB.

3.3.2. Other single-tree approaches

It is straightforward to develop a single-tree version of generalized Benders decomposition and the extended cutting-plane method. In the case of Benders decomposition, we need only replace the outer approximation cuts (3.4) by the Benders cut (3.10). In fact, the Benders cuts can already be used to condense old outer approximation cuts in order to reduce the size of the LP relaxation. The extended cutting-plane method can be similarly generalized (see Still and Westerlund 2006) by replacing the NLP solver with a simple function and gradient evaluation in order to generate the outer approximations.

The convergence results are readily extended. In the case of Benders decomposition, convergence follows from the convexity and the finiteness of the set of feasible integer variables, because every integer assignment is visited at most once. In the case of the extended cutting-plane method, convergence follows from the finiteness of the integer set and the finite ϵ-convergence of the cutting-plane method.

3.4. Presolve techniques for MINLP

A key component in successful MILP software is an efficient presolve. These techniques were popularized by Savelsbergh (1994), and some of these techniques can be readily extended to MINLP. Here, we briefly review these techniques and demonstrate their importance with two examples: coefficient tightening and constraint disaggregation. The goal of the presolve is to create an equivalent but tighter LP (or NLP) relaxation that will likely result in a significantly smaller search tree.

Presolve techniques fall into two broad categories: basic functions for housekeeping and advanced functions for reformulation. Housekeeping includes checking for duplicate rows (or constraints), tightening of bounds on the variables and constraints, fixing and removing variables, and identifying redundant constraints. Reformulations include improvement of coefficients, disaggregation of constraints, and derivation of implications or conflicts.

3.4.1. Coefficient tightening for MINLP

We observed very large search trees when we tried to solve certain chemical engineering synthesis problems with a range of branch-and-bound solvers.

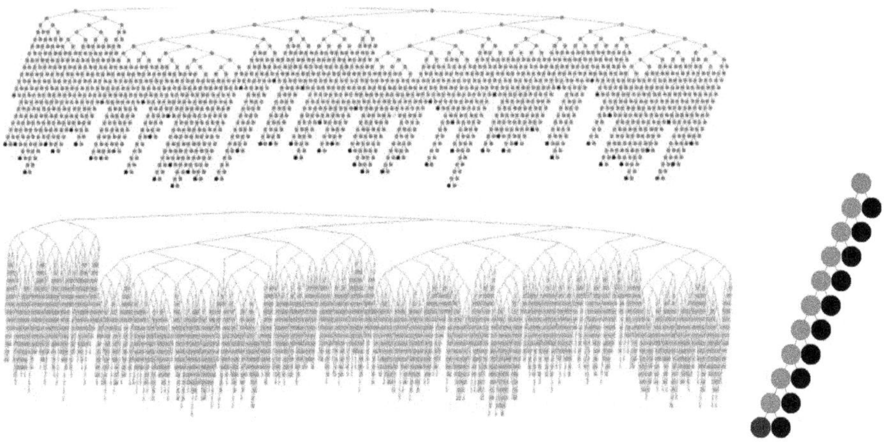
(a) (b)

Figure 3.4. (a) Branch-and-bound tree without presolve after 75 and 200 seconds of CPU time. (b) Complete tree after application of presolve and coefficient tightening.

For example, the search tree generated by MINOTAUR (Mahajan *et al.* 2011) for problem Syn20M04M grows rapidly, as shown in Figure 3.4. Figure 3.4(a) shows the search tree after 75 and 200 seconds of CPU time (the tree after 360 seconds is shown in Figure 1.1). The problem is not solved within 2 hours of CPU time, at which point MINOTAUR has visited 264 000 nodes. Other solvers behave similarly: BONMIN-BB and MINLPBB have searched about 150 000 nodes after 2 hours of CPU time without finding the solution. On the other hand, this problem is solved in 9 seconds by using a hybrid outer approximation branch-and-bound approach (Bonami *et al.* 2008). In this section we show that we can improve the performance of branch-and-bound methods dramatically by extending coefficient tightening to MINLP.

We start by describing the principle of coefficient tightening on a simple MILP, whose feasible set is given by

$$x_1 + 21x_2 \leq 30, \quad 0 \leq x_1 \leq 14, \quad x_2 \in \{0, 1\}. \tag{3.12}$$

The feasible set is the union of the two bold horizontal segments in Figure 3.5. If $x_2 = 1$ then the constraint $x_1 \leq 30 - 21x_2 = 9$ is tight; but if $x_2 = 0$ then $x_1 \leq 30 - 21x_2 = 30$ is loose. The shaded area illustrates the feasible set of the LP relaxation, which is much larger than the convex hull of the integer feasible set. We can tighten the formulation by changing the coefficient of the binary variable x_2. It is easy to see that

$$x_1 + 5x_2 \leq 14, \quad 0 \leq x_1 \leq 14, \quad x_2 \in \{0, 1\}$$

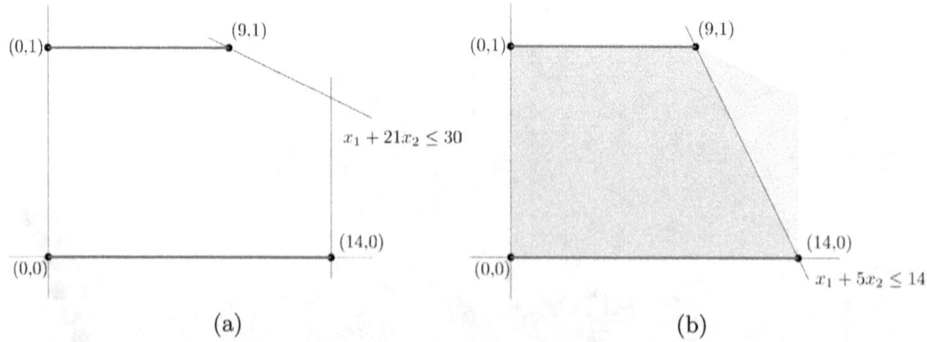

Figure 3.5. (a) Feasible set of MILP example (3.12) and (b) feasible set after coefficient tightening.

is equivalent to (3.12), and corresponds to the convex hull of the two bold horizontal segments (see Figure 3.5(b)).

We can extend this idea to MINLP problems, where we often encounter constraints of the form

$$\begin{aligned} c(x_1, \ldots, x_k) \leq b + M(1-y), \quad y \in \{0,1\}, \quad \text{and} \\ l_i \leq x_i \leq u_i, \quad \text{for all } i = 1, \ldots, k, \end{aligned}$$

where $M > 0$ is a large constant. This form of constraints corresponds to on/off decisions that arise, for example, in process synthesis problems (Türkay and Grossmann 1996).

If the constraint $c(x_1, \ldots, x_k) \leq b + M(1-y)$ is loose for $y = 0$, then we can reduce the value of M by an amount determined by the following optimization problem:

$$\begin{cases} \underset{x}{\text{maximize}} & c(x_1, \ldots, x_k), \\ \text{subject to} & l_i \leq x_i \leq u_i, \ \forall i = 1, \ldots, k. \end{cases} \tag{3.13}$$

Let us denote the optimal value by $c^u := c(x_1^*, \ldots, x_k^*)$. If we have $c^u < b + M$, then we can tighten the coefficient M and arrive at the equivalent formulation

$$\begin{aligned} c(x_1, \ldots, x_k) + (c^u - b)y \leq c^u, \quad y \in \{0,1\}, \quad \text{and} \\ l_i \leq x_i \leq u_i, \quad \text{for all } i = 1, \ldots, k. \end{aligned}$$

Unfortunately, this approach requires solving an NLP for every set of constraints for which we wish to tighten the formulation. Moreover, the optimization problem (3.13) is non-convex, because we maximize a convex function. Thus, we would have to apply global optimization techniques to derive the bound c^u, which would be prohibitive.

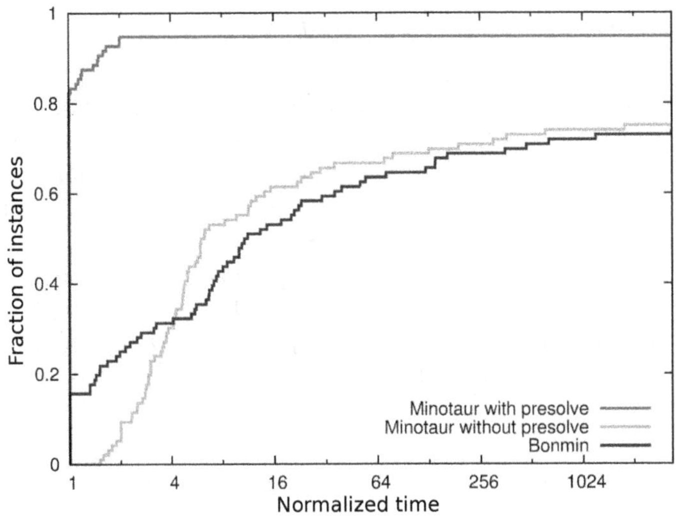

Figure 3.6. Performance profile comparing the effect of presolve on MINLP solver MINOTAUR for Syn* and Rsyn* instances.

We can avoid the solution of this non-convex NLP if the binary variable also appears as an upper bound on the variables. This is indeed the case in many applications, such as the synthesis design problem, where we find the following constraint structure:

$$\begin{aligned} c(x_1, \dots, x_k) \le b + M(1-y), \quad y \in \{0,1\}, \quad \text{and} \\ 0 \le x_i \le u_i y, \quad \text{for all } i = 1, \dots, k, \end{aligned} \tag{3.14}$$

where $y = 0$ now switches the constraints and variables off. In this case, we can simply evaluate the constraint $c^u := c(0, \dots, 0)$ and then derive the following tightened set of constraints (provided that $c^u < b + M$):

$$\begin{aligned} c(x_1, \dots, x_k) + (c^u - b)y \le c^u, \quad y \in \{0,1\}, \quad \text{and} \\ 0 \le x_i \le u_i y, \quad \text{for all } i = 1, \dots, k. \end{aligned} \tag{3.15}$$

The effect of this reformulation is dramatic, as can be seen from Figure 3.4. The tree in Figure 3.4(b) shows the complete search tree from MINOTAUR with presolve, and the MINLP is now solved in 2.3 seconds. Similar improvements are obtained for other synthesis problems. The performance profile (Dolan and Moré 2002) in Figure 3.6 compares the performance of MINOTAUR's presolve on the Syn* and Rsyn* instances from the IBM/CMU library. Clearly, in almost all instances the presolve helps to improve its performance. In addition, the presolve enables MINOTAUR to solve 20% more instances than the version without presolve.

3.4.2. Constraint disaggregation for MINLP

The preceding section provides an example of how problem formulation can have a significant impact on our ability to solve problems. Here we present another idea, known as disaggregation of constraints. A well-known example from MILP is the uncapacitated facility location problem (see, *e.g.*, Wolsey 1998), which can be described as follows. Given a set of customers, $i = 1, \dots, m$, and a set of facilities, $j = 1, \dots, n$, which facilities should we open ($x_j \in \{0, 1\}$, $j = 1, \dots, n$) at cost f_j to serve the customers? The decision that facility j serves customer i is modelled with the binary variable $y_{ij} \in \{0, 1\}$. The constraints that every customer is served by exactly one facility and that only facilities that have customers assigned are open can be modelled as

$$\begin{aligned} \sum_{j=1}^{n} y_{ij} &= 1, \quad \text{for all } i = 1, \dots, m, \quad \text{and} \\ \sum_{i=1}^{m} y_{ij} &\le n x_j, \quad \text{for all } j = 1, \dots, n, \end{aligned} \tag{3.16}$$

respectively. A tighter formulation (in the sense that its LP relaxation is closer to the convex hull of the feasible set) is given by the disaggregated form of the second constraints as

$$\begin{aligned} \sum_{j=1}^{n} y_{ij} &= 1, \quad \text{for all } i = 1, \dots, m, \quad \text{and} \\ y_{ij} &\le x_j, \quad \text{for all } i = 1, \dots, m, \ j = 1, \dots, n. \end{aligned} \tag{3.17}$$

As with (3.15), the difference in solution time can be dramatic. For a small random example with $n = m = 40$, the CPU time is 53 000 seconds for (3.16) versus 2 seconds for (3.17).

Similar disaggregation tricks have been applied to nonlinear functions. Tawarmalani and Sahinidis (2005) consider constraint sets of the form

$$S := \{x \in \mathbb{R}^n : c(x) = h(g(x)) \le 0\}, \tag{3.18}$$

where $g : \mathbb{R}^n \to \mathbb{R}^p$ is a smooth convex function and $h : \mathbb{R}^p \to \mathbb{R}$ is a smooth, convex, and non-decreasing function. These two conditions imply that $c : \mathbb{R}^n \to \mathbb{R}$ is a smooth convex function. We note that this functional form is related to the concept of group partial separability, which is frequently used to compute gradients and Hessians efficiently in NLP (Griewank and Toint 1984, Gay 1991, Bongartz, Conn, Gould and Toint 1995).

We can derive a disaggregated version of the constraint set in (3.18) by introducing new variables $y = g(x) \in \mathbb{R}^p$, which leads to the following convex set:

$$S_d := \{(x, y) \in \mathbb{R}^n \times \mathbb{R}^p : h(y) \le 0, \ y \ge g(x)\}. \tag{3.19}$$

One can show that S is the projection of S_d onto x. This reformulation is of interest because any outer approximation of S_d is stronger than the same outer approximation of S, and hence the formulation (3.19) is preferred in any outer-approximation-based algorithm. In particular, given a set of points $\mathcal{X}^k := \{x^{(1)}, \ldots, x^{(k)}\}$, we construct the outer approximations of S and S_d as

$$S^{oa} := \{x : c^{(l)} + \nabla c^{(l)^T}(x - x^{(l)}) \leq 0, \ \forall x^{(l)} \in \mathcal{X}^k\} \tag{3.20}$$

and

$$\begin{aligned} S_d^{oa} := \{(x, y) :& h^{(l)} + \nabla h^{(l)^T}(y - g(x^{(l)})) \leq 0, \\ & y \geq g^{(l)} + \nabla g^{(l)^T}(x - x^{(l)}), \ \forall x^{(l)} \in \mathcal{X}^k\}, \end{aligned} \tag{3.21}$$

respectively. It can be shown that the projection of S_d^{oa} onto x is contained in S^{oa}. Tawarmalani and Sahinidis (2005) give an example that shows that the outer approximation (3.21) is tighter than (3.20). Moreover, for the two outer approximations to be equivalent, we may require an exponentially larger number of linearization points $x^{(l)}$ in (3.20) compared with (3.21).

Hijazi *et al.* (2010) study a similar structure, namely, the partially separable constraint set:

$$\left\{x : c(x) := \sum_{j=1}^{q} h_j(a_j^T x + b_j) \leq 0\right\}, \tag{3.22}$$

where $h_j : \mathbb{R} \to \mathbb{R}$ are smooth and convex functions, which ensures that this set is convex. We can again derive a disaggregated form of this set by introducing new variables $y \in \mathbb{R}^q$,

$$\left\{(x, y) : \sum_{j=1}^{q} y_j \leq 0, \quad \text{and} \quad y_j \geq h_j(a_j^T x + b_j)\right\}. \tag{3.23}$$

Again, one can show that the outer approximation of (3.23) is tighter than the outer approximation of (3.22). We can apply this technique to the worst-case example, (3.7), choosing two linearization points as $x^{(1)} \in \{0, 1\}^n$ and its complement $x^{(2)} := e - x^{(1)}$, where $e = (1, \ldots, 1)$ is the vector of all ones. The combined outer approximation of (3.23) is then given by

$$\sum_{i=1}^{n} y_i, \quad \text{and} \quad x_i - \tfrac{3}{4} \leq y_i, \quad \text{and} \quad \tfrac{1}{4} - x_i \leq y_i,$$

which together with $x_i \in \{0, 1\}$ implies that $z_i \geq 0$, which leads to $\sum z_i \geq \frac{n}{4} > \frac{n-1}{4}$, which shows that any master problem that includes the linearizations from $x^{(1)}$ and $x^{(2)}$ is infeasible. Hijazi *et al.* (2010) suggest two additional improvements for outer approximation for partially separable constraints. The first improvement is to add additional linearization points for the univariate functions $h_j(t)$ in order to obtain a better description of the nonlinear feasible set. The second improvement is to employ an inner approximation of $h_j(t)$ in order to generate initial feasible points in the case that $a_j^T x + b_j = x_j$.

4. Cutting planes for convex MINLPs

In this section we review different cutting planes for use in a branch-and-cut algorithm solving convex MINLPs. We then review generalized disjunctive cuts and perspective cuts for MINLP. We also cover details of the practical realization of disjunctive cuts in a branch-and-cut framework. The final part of this section addresses the closely related problem class of mixed-integer second-order cone programs (MISOCPs).

4.1. Mixed-integer rounding cuts

The LP/NLP-based branch-and-bound Algorithm 3.1 for MINLP solves MILP relaxations, which we intend to strengthen by iteratively adding cuts to remove fractional solutions from these relaxations as outlined in Algorithm 3.3. We start by considering mixed-integer rounding cuts. These are best introduced for the two-variable set

$$S := \{(x_1, x_2) \in \mathbb{R} \times \mathbb{Z} \mid x_2 \le b + x_1,\ x_1 \ge 0\}, \tag{4.1}$$

where $C = \{1\}$, $I = \{2\}$. Let $f_0 = b - \lfloor b \rfloor$, and observe that the inequality

$$x_2 \le \lfloor b \rfloor + \frac{x_1}{1 - f_0} \tag{4.2}$$

is valid for X by verifying it for the two cases $x_2 \le \lfloor b \rfloor$ and $x_2 \ge \lfloor b \rfloor + 1$. The situation is depicted in Figure 4.1 for the set $S = \{(x_1, x_2) \in \mathbb{R} \times \{0, 1\} \mid x_2 \le \frac{1}{2} + x_1,\ 0 \le x_1 \le 2\}$.

For the general MILP case, it is sufficient to consider the set

$$X := \{(x_C^+, x_C^-, x_I) \in \mathbb{R}^2 \times \mathbb{Z}^p \mid a_I^T x_I + x_C^+ \le b + x_C^-,\ x_C^+ \ge 0, x_C^- \ge 0,\ x_I \ge 0\}. \tag{4.3}$$

This set describes a selected constraint row of an MILP, or a *one-row relaxation* of a subset of constraints aggregated in the vector $a \in \mathbb{R}^n$ and scalar b. Real variables are aggregated in x_C^+ and x_C^- depending on the sign of their coefficient in a_C. The extension from (4.1) is now straightforward by observing the following inequality is valid for X:

$$\sum_{i \in I} \left(\lfloor a_i \rfloor + \frac{\max\{f_i - f_0, 0\}}{1 - f_0} \right) x_i \le \lfloor b \rfloor + \frac{x_C^-}{1 - f_0}, \tag{4.4}$$

where $f_i = a_i - \lfloor a_i \rfloor$ for $i \in I$ and $f_0 = b - \lfloor b \rfloor$ are the fractional parts of a and b.

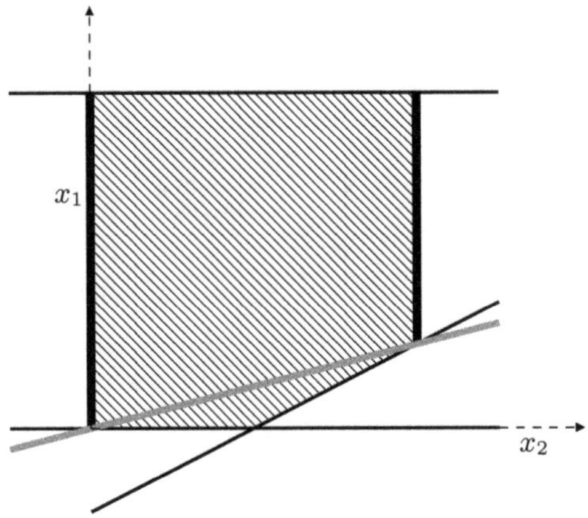

Figure 4.1. Mixed-integer rounding (MIR) cut. Feasible set of the LP relaxation (hatched), integer feasible set (bold black lines), and MIR cut (grey) $x_2 \leq 2x_1$ derived from the inequality $x_2 \leq \frac{1}{2} + x_1$.

Gomory cuts were originally derived by Gomory (1958), Gomory (1960) and Chvátal (1973) for integer linear programs. In the mixed-integer case, a Gomory cut is given by the inequality

$$\sum_{i \in I_1} f_i x_i + \sum_{i \in I_2} \frac{f_0(1-f_i)}{f_i} x_i + x_C^+ + \frac{f_0}{1-f_0} x_C^- \geq f_0, \tag{4.5}$$

where $I_1 = \{i \in I \mid f_i \leq f_0\}$ and $I_2 = I \setminus I_1$. It can be seen to be an instance of a MIR cut by considering the set

$$X = \{(x_C, x_0, x_I) \in \mathbb{R}^2 \times \mathbb{Z} \times \mathbb{Z}^p \mid x_0 + a_I^T x_I + x_C^+ - x_C^- = b,\ x_C \geq 0, x_I \geq 0\}, \tag{4.6}$$

generating a MIR inequality from it, and eliminating the variable x_0.

To apply MIR cuts in MINLP, we can generate cuts from linearized inequality constraints and the linearized objective constraint of the MINLP reformulation (1.2). Hence, it is sufficient to consider cuts for MILPs. Akrotirianakis, Maros and Rustem (2001) report modest performance improvements using this approach for MINLP.

4.2. Perspective cuts for MINLP

Many MINLP problems employ binary variables to indicate the non-positivity of continuous variables. If $x_i, i \in I$ is a binary variable and $x_j, j \in C$ is the associated continuous variable, the relationship can be modelled with what is known as a *variable upper bound* constraint,

$$x_j \leq u_j x_i. \tag{4.7}$$

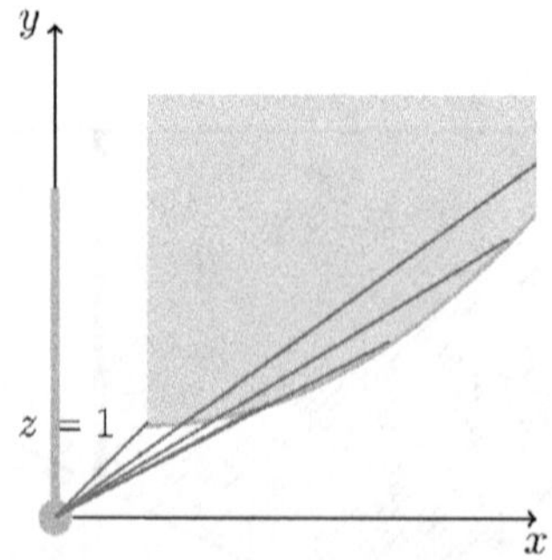

Figure 4.2. The set S.

Note that, since x_i is a $\{0,1\}$ variable, if $x_j > 0$ then $x_i = 1$. If, in addition, the continuous variable x_j appears in a convex, nonlinear constraint, then a reformulation technique called the *perspective reformulation* can lead to significantly improved computational performance. Frangioni and Gentile (2006) pioneered the use of this strengthened relaxation, using cutting planes known as *perspective cuts*. The technique is perhaps best introduced with a simple example. Consider the following mixed-integer set with three variables:

$$S = \{(x_1, x_2, x_3) \in \mathbb{R}^2 \times \{0,1\} : x_2 \geq x_1^2,\ \ ux_3 \geq x \geq 0\}.$$

Figure 4.2 depicts the set S, which is the union of two convex sets $S = S^0 \cup S^1$, where

$$\begin{aligned} S^0 &= \{(0, x_2, 0) \in \mathbb{R}^3 : x_2 \geq 0\}, \\ S^1 &= \{(x_1, x_2, 1) \in \mathbb{R}^3 : x_2 \geq x_1^2,\ u \geq x_1 \geq 0\}. \end{aligned}$$

One can observe from the geometry that the convex hull of S requires the surface defined by the family of line segments connecting the origin in the $x_3 = 0$ plane to the graph of the parabola $x_2 = x_1^2$ in the $x_3 = 1$ plane. Using this geometric intuition, we can define the convex hull of S as

$$\begin{aligned} &\text{conv}(S) = \qquad\qquad (4.8) \\ &\{(x_1, x_2, x_3) \in \mathbb{R}^3 : x_2 x_3 \geq x_1^2,\ ux_3 \geq x_1 \geq 0,\ 1 \geq x_3 \geq 0,\ x_2 \geq 0\}. \end{aligned}$$

The expression $x_2 x_3 \geq x_1^2$ in (4.8) can be explained in terms of the *perspective function* of the left-hand side of the inequality $x_1^2 - x_2 \leq 0$. For a convex function $f : \mathbb{R}^n \to \mathbb{R}$, the *perspective function* $\mathcal{P} : \mathbb{R}^{n+1} \to \mathbb{R}$ of f is

$$\mathcal{P}(x, z) := \begin{cases} 0 & \text{if } z = 0, \\ zf(x/z) & \text{if } z > 0. \end{cases} \qquad (4.9)$$

The epigraph of $\mathcal{P}(x, z)$ is a cone pointed at the origin whose lower shape is $f(x)$. If z_i is an indicator binary variable that forces some variables x to be 0, or else the convex nonlinear constraint $f(x) \le 0$ must hold, then by replacing the constraint $f(x) \le 0$ with

$$z_i f(x/z_i) \le 0 \tag{4.10}$$

results in a convex inequality that describes a significantly tighter relaxation of the feasible region. Günlük and Linderoth (2012) (Lemma 3.2) slightly generalize this construction to the case where the S_0 side of the disjunction is an unbounded ray (as in Figure 4.2).

The general construction is as follows. We consider the problem

$$\min_{(x,z)\in\mathbb{R}^n\times\{0,1\}} \big\{ f(x) + cz \mid Ax \le bz \big\},$$

where (i) $X = \{x \mid Ax \le b\}$ is bounded (also implying $\{x \mid Ax \le 0\} = \{0\}$), (ii) $f(x)$ is a convex function that is finite on X, and (iii) $f(0) = 0$. Under these assumptions, for any $\bar{x} \in X$ and subgradient $s \in \partial f(\bar{x})$, the inequality

$$v \ge f(\bar{x}) + c + s^T(x - \bar{x}) + (c + f(\bar{x}) - s^T\bar{x}))(z - 1) \tag{4.11}$$

is valid for the equivalent mixed-integer program

$$\min_{(x,z,v)\in\mathbb{R}^n\times\{0,1\}\times\mathbb{R}} \big\{ v \mid v \ge f(x) + cz, Ax \le bz \big\}.$$

Inequality (4.11), called the *perspective cut*, was introduced by Frangioni and Gentile (2006) and used dynamically to build a tight formulation. Günlük and Linderoth (2010) show that perspective cuts are indeed outer approximation cuts for the perspective reformulation for this MINLP. Therefore, adding all (infinitely many) perspective cuts has the same strength as the perspective reformulation. It may be computationally more efficient to use linear outer approximation inequalities of the form (1.5) instead of using the nonlinear form (4.10).

4.3. Disjunctive cutting planes for MINLP

Disjunctive cuts for use in a branch-and-cut procedure were first discussed by Stubbs and Mehrotra (1999) for convex mixed 0–1 nonlinear programs, based on Balas *et al.* (1996) for MILP. Simultaneously, Ceria and Soares (1999) derived the disjunctive programming arguments to be presented in a more general setting. We consider the convex mixed 0–1 nonlinear program

$$\begin{cases} \underset{x,\eta}{\text{minimize}} & \eta, \\ \text{subject to} & f(x) \le \eta, \\ & c(x) \le 0, \\ & x \in X, \\ & x_i \in \{0,1\},\ \forall i \in I. \end{cases} \tag{4.12}$$

For the continuous relaxation of this problem, optimal solutions are guaranteed to lie on the boundary of the continuous relaxation $\mathcal{C} = \{x \in X \mid f(x) \leq \eta,\ c(x) \leq 0,\ 0 \leq x_I \leq 1\}$ of the feasible set. We consider a node in a branch-and-bound tree, with optimal solution x', where x'_j is fractional for some $j \in I$. We denote by $I_0 \subseteq I$, $I_1 \subseteq I$ the index sets of integer variables fixed to zero or one, respectively, by the previous branching decisions that led to the tree node under consideration. We denote by F the index set of free real and integer variables.

Instead of separating the fractional solution x' by simply branching to $x'_j = 0$ and $x'_j = 1$, we are interested in adding a valid inequality to the relaxation that cuts off x'. To this end, we consider the disjoint feasible sets obtained by fixing x_j to either choice,

$$\mathcal{C}_j^0 = \{x \in \mathcal{C} \mid x_j = 0,\ 0 \leq x_i \leq 1,\ \forall i \in F, i \neq j\}, \tag{4.13}$$

$$\mathcal{C}_j^1 = \{x \in \mathcal{C} \mid x_j = 1,\ 0 \leq x_i \leq 1,\ \forall i \in F, i \neq j\}. \tag{4.14}$$

We are interested in a description of $\tilde{\mathcal{M}}_j(\mathcal{C}) = \operatorname{conv}(\mathcal{C}_j^0 \cup \mathcal{C}_j^1)$, the convex hull of the continuous relaxation $\mathcal{C}$ of the feasible set with either binary restriction on a single selected x_j. For the set $\tilde{\mathcal{M}}_j(\mathcal{C})$, Stubbs and Mehrotra (1999) give the following description:

$$\tilde{\mathcal{M}}_j(\mathcal{C}) = \tag{4.15}$$
$$\left\{ (x_F, v_0, v_1, \lambda_0, \lambda_1) \;\middle|\; \begin{array}{ll} v_0 + v_1 = x_F, & v_{0j} = 0,\ v_{1j} = \lambda_1, \\ \lambda_0 + \lambda_1 = 1, & \lambda_0, \lambda_1 \geq 0, \\ \mathcal{P}_{c_i}(v_0, \lambda_0) \leq 0, & \mathcal{P}_{c_i}(v_1, \lambda_1) \leq 0,\ 1 \leq i \leq n_c, \end{array} \right\}$$

where $\mathcal{P}_f(v, \lambda)$ is the perspective function (4.9) for the function f. This procedure can be generalized by repetition to describe the lifted convex hull of the disjoint feasible sets obtained for multiple fractional x_j, $j \in F$ all at once. With that description of the convex hull in hand, we can set up and solve a separation NLP to find a point $\hat{x}$ closest to the fractional one x' and in the convex hull:

$$\left\{ \begin{array}{ll} \underset{x, v_0, v_1, \lambda_0, \lambda_1}{\text{minimize}} & \|x - x'\|, \\ \text{subject to} & (x, v_0, v_1, \lambda_0, \lambda_1) \in \tilde{\mathcal{M}}_j(\mathcal{C}), \\ & x_i = 0,\ \forall i \in I_0, \\ & x_i = 1,\ \forall i \in I_1. \end{array} \right. \tag{BC-SEP(x', j)}$$

Let $\hat{x}$ be an optimal solution of (BC-SEP(x', j)), and denote by π_F the Lagrange multipliers for the equality constraint $v_0 + v_1 = x_F$ in (4.15). Then, an inequality that is valid for the current node and cuts off $\bar{x}$ from the feasible set is given by

$$\pi_F^T x_F \leq \pi_F^T \hat{x}_F. \tag{4.16}$$

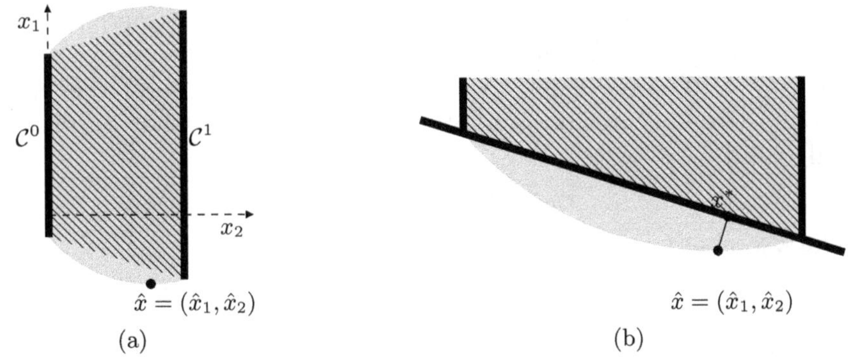

Figure 4.3. (a) NLP relaxation $\mathcal{C}$ (grey), integer feasible convex hull (hatched), and disjoint convex hulls $\mathcal{C}^0$ and $\mathcal{C}^1$ (bold black lines) for the MINLP example (4.17). (b) Solution to the minimum norm problem, and resulting disjunctive cut for the MINLP example (4.17). The next NLP solution including the disjunctive cut will produce the MINLP solution.

As an example of deriving a disjunctive cut, consider the MINLP

$$\begin{cases} \underset{x_1,x_2}{\text{minimize}} & x_1, \\ \text{subject to} & \left(x_1 - \frac{1}{2}\right)^2 + \left(x_2 - \frac{3}{4}\right)^2 \leq 1, \\ & -2 \leq x_1 \leq 2, \\ & x_2 \in \{0,1\}, \end{cases} \tag{4.17}$$

with the relaxed optimal solution $x' = (x_1', x_2') = \left(-\frac{1}{2}, \frac{3}{4}\right)$ shown in Figure 4.3(a). To construct the separation problem for identifying a disjunctive cut that removed this solution from the relaxation, we consider the individual convex hulls $\mathcal{C}^0$ and $\mathcal{C}^1$ for $x_2 = 0$ and $x_2 = 1$, where the limits are found by solving the constraint $\left(x_1 - \frac{1}{2}\right)^2 + \left(x_2 - \frac{3}{4}\right)^2 \leq 1$ for x_1 given a fixed x_2,

$$\begin{aligned} \mathcal{C}^0 &= \left\{(x_1, 0) \in \mathbb{R} \times \{0,1\} \mid 2 - \sqrt{7} \leq 4x_1 \leq 2 + \sqrt{7}\right\}, \\ \mathcal{C}^1 &= \left\{(x_1, 1) \in \mathbb{R} \times \{0,1\} \mid 2 - \sqrt{15} \leq 4x_1 \leq 2 + \sqrt{15}\right\}. \end{aligned} \tag{4.18}$$

With the Euclidean norm $\|\cdot\|_2$, the minimum norm separation problem for x' yields (approximately) the solution $\hat{x} = (\hat{x}_1, \hat{x}_2) = (-0.40, 0.78)$ with steepest descent direction $\pi = (-0.1, -0.03)$ for the norm objective. This identifies the separating hyperplane

$$\begin{pmatrix} -0.1 & -0.03 \end{pmatrix} \begin{pmatrix} x_1 \\ x_2 \end{pmatrix} \leq \begin{pmatrix} -0.1 & -0.03 \end{pmatrix} \begin{pmatrix} -0.40 \\ 0.78 \end{pmatrix} \implies x_1 + 0.3x_2 \geq -0.166, \tag{4.19}$$

shown in Figure 4.3(b).

We are free to choose the norm in (BC-SEP(x', j)), but different choices lead to different reformulations, for example for the 1-norm or the ∞-norm.

Stubbs and Mehrotra (1999) observed the most favourable performance of the generated cuts when using the ∞-norm.

The cut (4.16) is in general valid for the local subtree of the branch-and-bound tree only. Stubbs and Mehrotra (1999) show that a *lifting* to a globally valid cut

$$\pi^T x \leq \pi^T \hat{x} \tag{4.20}$$

may be obtained by assigning

$$\pi_i = \min\{e_i^T H_0^T \mu_0, e_i^T H_1^T \mu_1\}, \quad i \notin F, \tag{4.21}$$

where e_i denotes the ith unit vector; $\mu_0 = (\mu_{0F}, 0)$, $\mu_1 = (\mu_{1F}, 0)$, where μ_{0F}, μ_{1F} are the Lagrange multipliers of the perspective inequalities from (4.15) active in $\hat{x}$; and H_0, H_1 are matrices formed from subgradient rows $\partial_v \mathcal{P}_{c_i}(v_0, \lambda_0)^T$, $\partial_v \mathcal{P}_{c_i}(v_1, \lambda_1)^T$ of those inequalities.

Disjunctive cuts are linear in the lifted variable space. In Stubbs and Mehrotra (2002), a step toward nonlinear cutting planes is taken and convex quadratic inequalities are derived as cuts for convex mixed 0–1 integer programs. Computational evidence so far suggests that such cuts do not benefit branch-and-cut procedures in general, although nonlinear cuts for the mixed-integer conic case show otherwise: see the end of Section 4.5.

4.4. Implementation of disjunctive cuts

The nonlinear separation problem (BC-SEP(x', j)) has twice the number of unknowns as the original problem and is not differentiable everywhere because of the perspective constraints. These drawbacks have hindered the use of disjunctive cutting planes for convex 0–1 MINLP, and Stubbs and Mehrotra (1999) have reported computational results for only four instances with no more than 30 unknowns. Addressing this issue, we mention an LP-based iterative separation procedure by Kılınç, Linderoth and Luedtke (2010) and Kılınç (2011) that replaces (BC-SEP(x', j)), and a nonlinear separation approach by Bonami (2011) that circumvents the difficulties inherent in (BC-SEP(x', j)).

Kılınç *et al.* (2010) and Kılınç (2011) propose an iterative procedure to replace (BC-SEP(x', j)) by a sequence of cut generation LPs. They report both an increased number of solvable problems and a significant reduction in run time for a set of 207 instances. To describe this method, we let $\mathcal{B} \supset \mathcal{C}$ be a relaxation of the original MINLP relaxation, and we consider the sets

$$\mathcal{B}_j^0 = \{x \in \mathcal{B}^0 \mid x_j = 0\}, \quad \mathcal{B}_j^1 = \{x \in \mathcal{B}^0 \mid x_j = 1\}. \tag{4.22}$$

Valid inequalities for $\mathrm{conv}(\mathcal{B}_j^0 \cup \mathcal{B}_j^1)$ are also valid for $\mathrm{conv}(\mathcal{C}_j^0 \cup \mathcal{C}_j^1)$. The separation program becomes a linear one if $\mathcal{B}$ is restricted to be a polyhedron and if an appropriate norm is chosen in (BC-SEP(x', j)). Kılınç *et al.* (2010) and Kılınç (2011) iteratively tighten polyhedral outer approximations of $\mathcal{C}_j^0$

and $\mathcal{C}_j^1$ using inequalities generated as disjunctive cuts. To this end, in iteration t two sets $\mathcal{K}_0^t$, $\mathcal{K}_1^t$ of linearization points are maintained, resulting in the polyhedral approximations

$$\mathcal{F}_t^0 = \left\{ x \in \mathbb{R}^n \mid x_i = 0,\ g(x') + \frac{\partial g(x')}{\partial x}(x - x') \leq 0,\ \forall x \in \mathcal{K}_t^0 \right\}, \quad (4.23)$$

$$\mathcal{F}_t^1 = \left\{ x \in \mathbb{R}^n \mid x_i = 1,\ g(x') + \frac{\partial g(x')}{\partial x}(x - x') \leq 0,\ \forall x \in \mathcal{K}_t^1 \right\}. \quad (4.24)$$

Since the sets $\mathcal{F}_t^0$, $\mathcal{F}_t^1$ are polyhedral relaxations of $\mathcal{C}^0$ and $\mathcal{C}^1$, valid disjunctive cuts can be generated. Initially empty, the sets $\mathcal{K}_t^0$, $\mathcal{K}_t^1$ are augmented by the solutions x_t' of the linear separation problems and with two so-called friendly points y_t and z_t, respectively, that satisfy $x_t' = \lambda y_t + (1 - \lambda) z_t$ for some $\lambda \in [0, 1]$. Kılınç *et al.* (2010) prove this approach to be sufficient to ensure that in the limit the obtained inequality is of the same strength as if the nonlinear separation problem (BC-SEP(x', j)) were solved. The process can be terminated early if it is observed that the cuts are not effective at reducing the solution integrality gap. The procedure is applicable to general convex MINLPs with $x_i \in \mathbb{Z}$ as well.

A similar procedure based on the outer approximating set

$$\mathcal{B} = \left\{ x \in \mathbb{R}^n \mid g(x') + \frac{\partial g}{\partial x}(x - x') \leq 0 \right\}$$

was proposed earlier by Zhu and Kuno (2006), but does not necessarily converge (Kılınç *et al.* 2010, Kılınç 2011).

Bonami (2011) addresses the difficulty of solving (BC-SEP(x', j)) in the special situation when x' with $k < x_j < k + 1$ fractional for some $j \in I$ is to be separated from a relaxation that was obtained by a simple disjunction created from a split relaxation. In this case, the separation problem allows an algebraic reduction to

$$\left\{ \begin{array}{ll} \underset{v_1}{\text{maximize}} & v_{1,j}, \\ \text{subject to} & c_i\left(\frac{v_1}{f_0}\right) \leq f_0 k - v_{1,j}, \\ & c_i\left(\frac{x' - v_1}{1 - f_0}\right) \leq f_0 k - v_{1,j}, \\ & v_1 \in \mathbb{R}^n. \end{array} \right. \quad (4.25)$$

where $f_0 = x_j' - k > 0$ is the fractional part of x_j. Again, the approach is applicable to general convex mixed-integer problems. In (4.25), the perspective is no longer required, the problem size does not grow compared to the MINLP under consideration, and differentiability is maintained. One can show that the optimal objective value is smaller than $f_0 k$ if and only if $x' \notin \mathcal{C}_j^k$. The separating inequality can be computed from an LP model as described for the approach by Kılınç *et al.* (2010) and the particular choice

$$\mathcal{K}_t^0 = \left\{ \frac{x' - v_1^*}{1 - f_0} \right\}, \quad \mathcal{K}_t^1 = \left\{ \frac{v_1^*}{f_0} \right\}, \tag{4.26}$$

if v_1^* denotes the optimal solution of (4.25).

4.5. Mixed-integer second-order cone programs

Both mixed-integer rounding cuts and disjunctive cuts can be generalized from the LP cone $\mathbb{R}^n_+$ to other cones such as second-order cones defined by the set $\{(x_0, x) \in \mathbb{R}\times\mathbb{R}^n \mid x_0 \geq \|x\|_2\}$. We consider the class of mixed-integer linear programs with a second-order cone constraint

$$\begin{cases} \underset{x}{\text{minimize}} & c^T x, \\ \text{subject to} & x \in \mathcal{K}, \\ & x \in X, \\ & x_i \in \mathbb{Z}, \ \forall i \in I. \end{cases} \tag{MISOCP}$$

The conic constraint $x \in \mathcal{K}$ represents the product of $k \geq 1$ smaller cones $\mathcal{K} := \mathcal{K}_1 \times \cdots \times \mathcal{K}_k$, defined by

$$\mathcal{K}_j := \left\{ x_j = (x_{j0}, x_{j1}^T)^T \in \mathbb{R} \times \mathbb{R}^{n_j - 1} : \|x_{j1}\|_2 \leq x_{j0} \right\}, \quad 1 \leq j \leq k, \tag{4.27}$$

and $x = (x_1^T, \ldots, x_k^T)^T$. Convex MINLP solvers are usually not directly applicable to this problem class because the conic constraint is not continuously differentiable. Drewes (2009) and Drewes and Ulbrich (2012) propose a variant of the LP/NLP-based branch-and-bound, Algorithm 3.6, that solves continuous relaxations of (MISOCP) instead of NLPs for a fixed integer assignment x_I^k:

$$\begin{cases} \underset{x}{\text{minimize}} & c^T x, \\ \text{subject to} & x \in \mathcal{K}, \\ & x \in X, \\ & x_I = x_I^k. \end{cases} \tag{SOCP(x_I^k)}$$

The algorithm builds MIP outer approximations of (MISOCP) from subgradient information as follows. Denote by (s, y) the dual variables associated with the conic and linear constraints, and define index sets $J_a(\bar{x})$ and $J_{00}(\bar{x})$, $J_{0+}(\bar{x})$ of active conic constraints differentiable and subdifferentiable, respectively, at $\bar{x}$ by

$$J_a(\bar{x}) := \{j : g_j(\bar{x}) = 0, \ \bar{x} \neq 0\}, \tag{4.28}$$

$$J_{0+}(\bar{x}, \bar{s}) := \{j : \bar{x}_j = 0, \ \bar{s}_{j0} > 0\}, \tag{4.29}$$

$$J_{00}(\bar{x}, \bar{s}) := \{j : \bar{x}_j = 0, \ \bar{s}_{j0} = 0\}, \tag{4.30}$$

where g_i is the conic constraint function. Let $S \subset \mathbb{R}^n$ denote the set of previous primal–dual solutions $(\bar{x}, \bar{s})$ to $(\mathrm{SOCP}(x_I^k))$. Then the linear outer approximation MIP for problem (MISOCP) is

$$\left\{\begin{array}{lll} \underset{x}{\text{minimize}} & c^T x, & \\ \text{subject to} & x \in X, & \\ & c^T x \le c^T \bar{x}, & \bar{x} \in X,\ \bar{x}_I \in \mathbb{Z}^p, \\ & 0 \ge -\|\bar{x}_{j1}\| x_{j0} + \bar{x}_{j1}^T x_{j1}, \quad \forall j \in J_a(\bar{x}), & \bar{x} \in X, \\ & 0 \ge -x_{j0} - \frac{1}{\bar{s}_{j0}} \bar{s}_{j1}^T x_{j1}, \quad \forall j \in J_{0+}(\bar{x}, \bar{s}), & \bar{x} \in X, \\ & 0 \ge -x_{j0}, \quad \forall j \in J_{00}(\bar{x}, \bar{s}), & \bar{x} \in X, \\ & x_i \in \mathbb{Z}, \quad \forall i \in I. & \end{array}\right. \qquad (\mathrm{MIP}(X))$$

For infeasible SOC subproblems, solutions from feasibility problems can be incorporated into $(\mathrm{MIP}(X))$ in a similar way. Convergence under a Slater constraint qualification, or alternatively by using an additional SOCP branching step if this CQ is violated, is shown by Drewes and Ulbrich (2012).

Polyhedral approximations of the second-order cone constraint are discussed for use in an outer approximation algorithm by Vielma, Ahmed and Nemhauser (2008), who use the polynomial-size relaxation introduced by Ben-Tal and Nemirovski (2001), while Krokhmal and Soberanis (2010) generalize this to p-order cones, that is, sets of the form $\{x \in \mathbb{R}^{n+1} : x_{n+1} \ge \|x\|_p\}$. The drawback here is that one has to choose the size of the approximation *a priori*. Consequently, the LP becomes large when the approximation has to be strengthened. An iterative scheme such as the SOCP-based branch-and-cut procedure for strengthening is hence preferable. The use of a polyhedral second-order conic constraint has been generalized by Masihabadi, Sanjeevi and Kianfar (2011).

Two efforts in this rapidly evolving area are the work by Dadush, Dey and Vielma (2011*a*), who prove that conic quadratic inequalities are necessary to represent the split closure of an ellipsoid, and the work by Belotti *et al.* (2012), who study the convex hull of the intersection of a disjunction $\mathcal{A} \cup \mathcal{B}$ and a generic convex set $\mathcal{E}$. In general,

$$\mathcal{A} = \{x \in \mathcal{R}^n : a^T x \le \alpha\}, \quad \mathcal{B} = \{x \in \mathcal{R}^n : b^T x \le \beta\};$$

this is therefore a non-parallel disjunction, and a more general disjunction than discussed in Section 3.1.1 and unlike the previous examples. The authors prove that the convex hull is given by intersecting $\mathcal{E}$ with a cone $\mathcal{K}$ such that $\mathcal{K} \cap \partial\mathcal{A} = \mathcal{E} \cap \partial\mathcal{A}$ and $\mathcal{K} \cap \partial\mathcal{B} = \mathcal{E} \cap \partial\mathcal{B}$, where $\partial\mathcal{S}$ is the frontier of set $\mathcal{S}$, if one such cone exists. The authors then provide an algorithm to find this cone for MISOCP, and they prove that it is a second-order cone. In general, this conic cut, which is shown in two and three dimensions in Figure 4.4, is proved to be more effective than the conic MIR cut presented by Atamtürk and Narayanan (2010), and is discussed below.

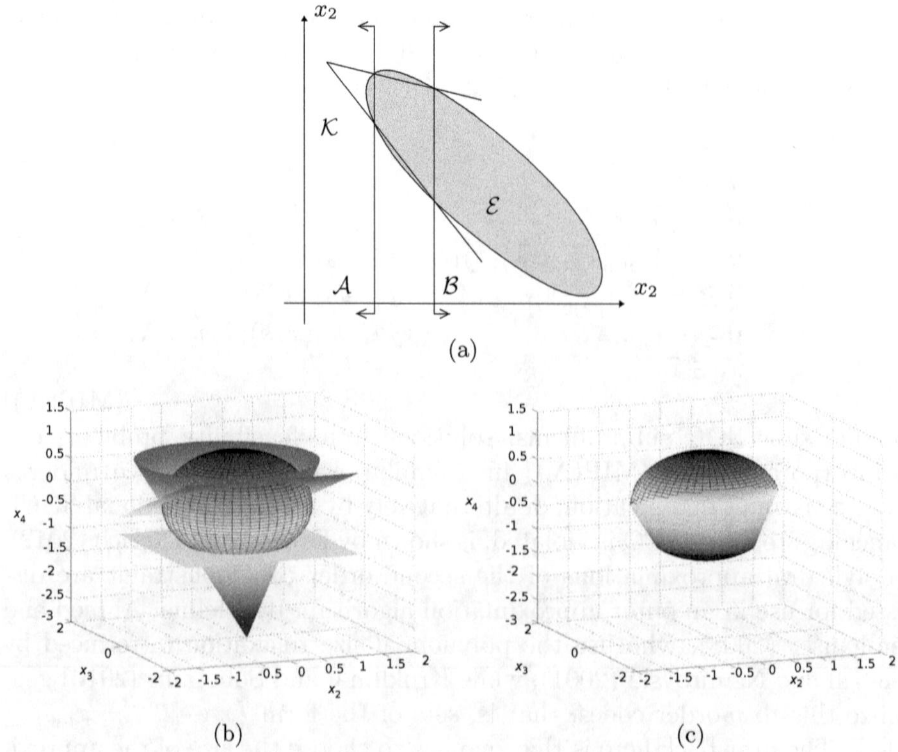

Figure 4.4. Disjunctive conic cuts as generated by Belotti *et al.* (2012). (a) $\mathcal{K}$ is the disjunctive cone generated when intersecting the ellipsoid $\mathcal{E}$ with the intersection $\mathcal{A} \cup \mathcal{B}$. (b) A disjunction described by two half-spaces delimited by non-parallel hyperplanes is intersected with an ellipsoid $\mathcal{E}$, and the intersection with the arising disjunctive cone, also shown in (b), returns a tighter feasible set depicted in (c).

4.5.1. Gomory cuts for MISOCP

Drewes (2009) describes Gomory cuts for (MISOCP) based on the work by Çezik and Iyengar (2005) for pure integer conic programs. We assume here that the bounded polyhedral set X is described by

$$X = \{x \in \mathbb{R}^n \mid Ax = b,\ l \le x \le u\}, \tag{4.31}$$

with $A \in \mathbb{R}^{m \times n}$, $b \in \mathbb{R}^m$. The additional non-negativity requirement $l_i \ge 0$ holds for all $i \in I$. Then, the following theorem describes a Gomory cut for (MISOCP).

Theorem 4.1 (Drewes 2009, Theorem 2.2.6). Assume that the continuous relaxation (SOCP(x_I^k)) and its dual have feasible interior points. Let $\bar{x}$ with $\bar{x}_I \notin \mathbb{Z}^p$ be a solution of (SOCP(x_I^k)), and let $(\bar{s}, \bar{y})$ be the corresponding dual solution. Then the following cut is a valid inequality for (MISOCP),

$$\lceil (A_I^T(\bar{y} - \Delta y)\bar{s}_I \rceil^T s_I \geq \lceil (\bar{y} - \Delta y)^T b \rceil, \tag{4.32}$$

where Δy solves

$$\begin{pmatrix} -A_C \\ A_I \end{pmatrix} \Delta y = \begin{pmatrix} c_C \\ 0 \end{pmatrix}. \tag{4.33}$$

Furthermore, if $(\bar{y} - \Delta y)^T b \notin \mathbb{Z}$, then (4.32) cuts off $\bar{x}$ from the integer feasible set.

Gomory cuts are of restricted applicability because the requirement that $l_I \geq 0$, which turns out to be violated frequently by MISOCP instances of practical interest. Consequently, Gomory cuts were found to be largely ineffective for those MISOCP instances evaluated in the computational studies presented by Drewes (2009).

As an example due to Drewes (2009), we consider the MISOCP

$$\begin{cases} \underset{x}{\text{minimize}} & -x_2, \\ \text{subject to} & -3x_2 + x_3 \leq 0, \\ & 2x_2 + x_3 \leq 3, \\ & 0 \leq x_1, \; x_2 \leq 3, \\ & x_1 \geq \|(x_2, x_3)^T\|_2, \\ & x_1, \; x_2 \in \mathbb{Z}, \end{cases} \tag{4.34}$$

whose SOCP relaxation has the optimal solution $\left(3, \frac{12}{5}, -\frac{9}{5}\right)$. The Gomory cut $x_2 \leq 2$ that can be deduced from this point is shown in Figure 4.5 and cuts off the relaxed solution.

4.5.2. Lift and project cuts for MISOCP

Next, we consider the restricted class of mixed 0–1 second-order cone programs,

$$\begin{cases} \underset{x}{\text{minimize}} & c^T x, \\ \text{subject to} & x \in \mathcal{K}, \\ & Ax = b, \\ & l \leq x \leq u, \\ & x_i \in \{0, 1\}, \; \forall i \in I, \end{cases} \tag{4.35}$$

and follow the notation of the lift and project procedure described for disjunctive cuts in Section 4.3. We are again interested in a finite conic linear description of the convex hull of the feasible set of (4.35), and we investigate the convex hull of the union of the disjoint sets $\mathcal{C}_j^0$ and $\mathcal{C}_j^1$. For MISOCP,

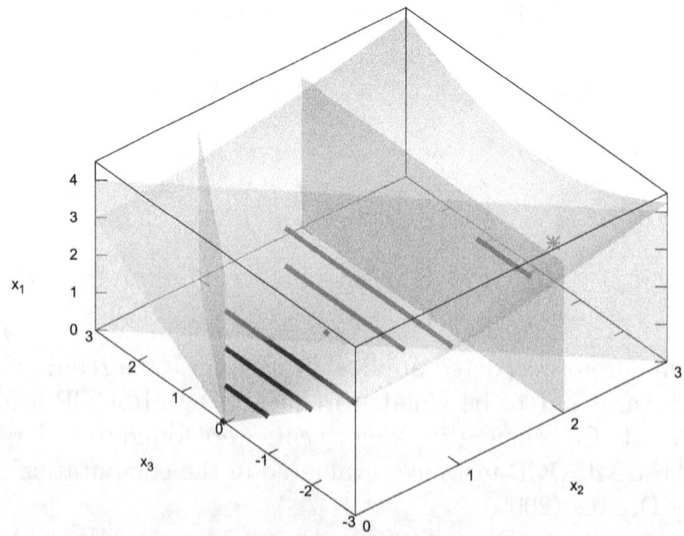

Figure 4.5. MISOCP example (4.34). Feasible cone, linear constraints, integer feasible set (solid lines), relaxed optimal solution (asterisk), and Gomory cut cutting off the relaxed optimal solution (plane at $x_2 = 2$).

the convex hull is described by the set $\tilde{M}_j(\mathcal{C})$:

$$\tilde{M}_j(\mathcal{C}) := \tag{4.36}$$

$$\left\{ (x, v^0, v^1, \lambda^0, \lambda^1) \;\middle|\; \begin{array}{ll} v^0 + v^1 = x, & \\ \lambda^0 + \lambda^1 = 1, & \lambda^0, \lambda^1 \geq 0, \\ Av^0 - \lambda^0 b = 0, & Av^1 - \lambda^1 b = 0, \\ v^0 \in \mathcal{K}, & v^1 \in \mathcal{K}, \\ v^0_j = 0, & v^1_j = \lambda^1, \\ 0 \leq v^0_k \leq \lambda^0, & 0 \leq v^1_k \leq \lambda^1, \; k \in J, k \neq j. \end{array} \right\}$$

If we fix a set $B \subseteq J$, the definition of the convex hull $\tilde{M}_B(\mathcal{C})$ is straightforward by repetition of the lifting process. We are now prepared to state the subgradient cut theorem from Stubbs and Mehrotra (1999) for MISOCP.

Theorem 4.2 (Drewes 2009, Proposition 2.1.6). Fix a set $B \subseteq I$, let $\bar{x} \notin \tilde{M}_B(\mathcal{C})$, and let x^* be the solution of

$$\min_{(x,v^0,v^1,\lambda^0,\lambda^1) \in \tilde{M}_B(\mathcal{C})} \|x - \bar{x}\|_2. \tag{4.37}$$

Then the subgradient cut

$$(x^* - \bar{x})^T x \geq {x^*}^T (x^* - \bar{x}) \tag{SGC}$$

is a valid linear inequality for all $x \in \tilde{M}_B(\mathcal{C})$ that cuts off $\bar{x}$.

Drewes (2009) describes further lift-and-project cuts for MISOCP, based on work by Çezik and Iyengar (2005) and Stubbs and Mehrotra (1999) for integer SOCPs and MILPs. Similar to the procedure described for MINLP disjunctive cuts above, these cuts can be constructed by solving linear, quadratic, and conic auxiliary programs, and have been found to be considerably more efficient than Gomory cuts when used in a MISOCP branch-and-cut procedure.

4.5.3. Mixed-integer rounding cuts for MISOCP

Atamtürk and Narayanan (2010) describe mixed-integer rounding cuts for MISOCPs. To this end, we consider the following formulation of problem (MISOCP):

$$\begin{cases} \underset{x}{\text{minimize}} & c^T x, \\ \text{subject to} & \|A_j x - b_j\|_2 \le d_j^T x - h_{j0},\ 1 \le j \le k, \\ & x \ge 0, \\ & x_i \in \mathbb{Z},\ \forall i \in I, \end{cases} \tag{4.38}$$

where $A_j \in \mathbb{R}^{n_j \times n}$ are matrices, $b_j \in \mathbb{R}^{n_j}$, $d_j \in \mathbb{R}^n$ are column vectors, and $h_{j0} \in \mathbb{R}$ are scalars. Atamtürk and Narayanan (2010) introduce the following *polyhedral second-order conic constraint formulation* that allows the exploitation of polyhedral information on conic constraints of this shape. For simplicity of exposition, we consider the case of $k = 1$ conic constraint of dimension n_1 only. We introduce $t = (t_0, t_1, \ldots, t_{n_1}) \in \mathbb{R}^{1+n_1}$, and we denote by $a_l^T \in \mathbb{R}^n$ the lth row of matrix $A \in \mathbb{R}^{n_1 \times n}$:

$$\begin{cases} \underset{x,t}{\text{minimize}} & c^T x, \\ \text{subject to} & t_l \ge |a_l^T x - b_l|,\ 1 \le l \le n_1, \\ & \|t\|_2 \le t_0 \le d^T x - h_0, \\ & x \ge 0, \\ & x_i \in \mathbb{Z},\ \forall i \in I. \end{cases} \tag{4.39}$$

We denote by S_l the feasible set considering a single component $1 \le l \le n_1$ of the polyhedral conic constraint,

$$S_l := \{x \in \mathbb{R}^n,\ x \ge 0,\ x_i \in \mathbb{Z},\ \forall i \in I,\ t \in \mathbb{R} : t \ge |a_l^T x - b_l|\}. \tag{4.40}$$

A family of valid inequalities for S_l, called conic mixed-integer rounding (MIR) inequalities, is given by the following theorem.

Theorem 4.3 (Atamtürk & Narayanan 2010, Theorem 1). For any $\alpha \neq 0$ the conic mixed-integer rounding (MIR) inequality

$$\sum_{i \in I} \phi_{f_\alpha}(a_i/\alpha) x_i - \phi_{f_\alpha}(b/\alpha) \le (t + x_C^+ + x_C^-)/|\alpha| \tag{4.41}$$

is valid for the set S_l, where x_C^+ and x_C^- aggregate the real variables x_C with positive and negative coefficients a_{lR}, and $\phi_{f_\alpha} : \mathbb{R} \to \mathbb{R}$ is the *conic MIR function* for $0 \leq f_\alpha := b/\alpha - \lfloor b/\alpha \rfloor \leq 1$,

$$\phi_{f_\alpha}(a) := \begin{cases} (1-2f_\alpha)p - (a-p) & \text{if } p \leq a < p + f_\alpha, \\ (1-2f_\alpha)p + (a-p) - 2f_\alpha & \text{if } p + f_\alpha \leq a < p+1, \end{cases} \quad n \in \mathbb{Z}. \tag{4.42}$$

Moreover, the inequalities are shown to be facet-defining under certain conditions. Furthermore, such inequalities can be used efficiently to cut fractional points off the relaxation of S_l.

Theorem 4.4 (Atamtürk & Narayanan 2010, Proposition 4). Conic mixed-integer equalities with $\alpha = a_{lj}$, $j \in I$ are sufficient to cut off all fractional extreme points of the LP relaxation of S_l.

5. Non-convex MINLP

Non-convex MINLPs are especially challenging because they contain non-convex functions in the objective or the constraints; hence even when the integer decision variables are relaxed to be continuous, the feasible region may be non-convex. Therefore, more work needs to be done to obtain an efficiently solvable (convex) relaxation for use in a branch-and-bound framework.

Non-convex MINLP is closely related to global optimization, a topic that also seeks optimal solutions to optimization problems having non-convex functions, although the focus in global optimization has often been on problems with only continuous decision variables. A huge literature on global optimization exists, including several textbooks (Hansen 1992, Horst and Tuy 1993, Horst, Pardalos and Thoai 1995, Floudas 2000). It is beyond the scope of this paper to provide a comprehensive review of global optimization. Our focus will be on describing techniques that either explicitly consider the integer variables appearing in an MINLP or that are essential to many algorithms for solving non-convex MINLPs.

One approach to solving non-convex MINLPs is to replace the non-convex functions with piecewise linear approximations and solve the corresponding approximation by using mixed-integer linear programming solvers. We discuss this approach in Section 5.1. In the remaining sections we discuss components of methods for directly solving non-convex MINLPs. In Section 5.2, we discuss generic strategies for obtaining convex relaxations of non-convex functions. We then describe in Section 5.3 how spatial branching can be used with these relaxations to obtain a convergent algorithm. Section 5.4 provides a sample of some techniques to obtain improved convex relaxations by exploiting particular types of non-convex structures.

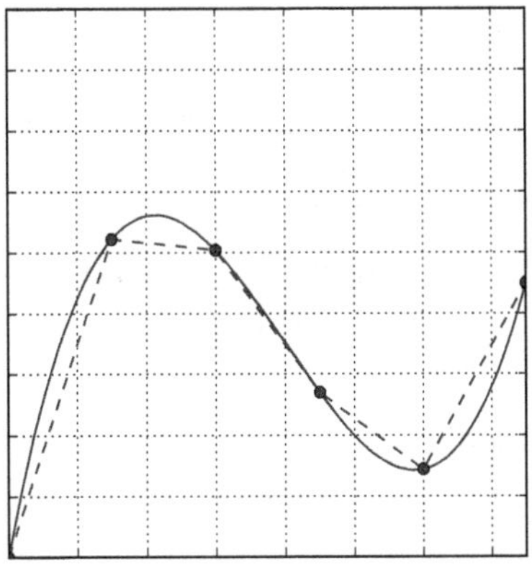

Figure 5.1. Example of a function (solid line) and a piecewise linear approximation to the function (dashed line).

We refer the reader to the works of Tawarmalani and Sahinidis (2002) and Burer and Letchford (2012) for additional surveys focused on non-convex MINLP.

5.1. Piecewise linear modelling

A common approach to the approximate solution of MINLPs with non-convex functions is to replace the nonlinear functions with piecewise linear approximations, leading to an approximation that can be solved by mixed-integer *linear* programming solvers. In fact, the importance of being able to model such functions using binary variables was recognized in some of the earliest work in binary integer programming (Markowitz and Manne 1957, Dantzig 1960). Figure 5.1 shows an example of a non-convex function with a corresponding piecewise linear approximation.

We focus our attention on modelling piecewise linear approximations of a univariate function $f : [l, u] \to \mathbb{R}$, where $l, u \in \mathbb{R}$. A multivariate separable function of the form

$$g(x) = \sum_{i=1}^{K} f_i(x_i)$$

can be approximated by separately obtaining piecewise linear approximations of $f_i(x_i)$. In Section 5.1.3 we briefly introduce extensions of the piecewise linear approximation approach to more general multivariate functions.

Using a piecewise linear modelling technique for solving MINLPs involves two steps: obtaining piecewise linear approximations of the nonlinear functions and modelling the piecewise linear functions in a way that mixed-integer linear programming solvers can handle. We discuss these two steps in Sections 5.1.1 and 5.1.2, respectively. In Section 5.1.3 we provide a brief overview of how these modelling approaches can be extended to multivariate functions.

Our treatment of this approach is necessarily brief. For more details on this topic we refer the reader to Geissler, Martin, Morsi and Schewe (2012), who provide a recent survey of piecewise linear modelling in MINLP, and to Vielma, Ahmed and Nemhauser (2010), who provide a detailed review of methods for modelling piecewise linear functions using binary variables.

5.1.1. Obtaining a piecewise linear approximation

Given a function $f : [l, u] \to \mathbb{R}$, we seek to obtain a piecewise linear function $\hat{f} : [l, u] \to \mathbb{R}$ such that $\hat{f}(x) \approx f(x)$ for all $x \in [l, u]$. If the piecewise linear function $\hat{f}$ consists of d linear segments, then it may be specified by its *break points* $l =: b^0 < b^1 < \cdots < b^d := u$ and the corresponding function values $y^k = \hat{f}(b^k)$, for $k = 0, 1, \ldots, d$. Then the function $\hat{f}$ is given by

$$\hat{f}(x) = y^{k-1} + \left(\frac{y^k - y^{k-1}}{b^k - b^{k-1}}\right)(x - b^{k-1}), \quad x \in [b^{k-1}, b^k], \quad \text{for all } k = 1, \ldots, d. \tag{5.1}$$

Alternatively, if for each $k = 1, \ldots, d$ we let $m_k = (y^k - y^{k-1})/(b^k - b^{k-1})$ be the slope of the line segment in interval k, then $a_k = y^{k-1} - m^k b^{k-1}$ is the y-intercept of the line defining the line segment in interval k. Thus we can equivalently write $\hat{f}$ as

$$\hat{f}(x) = a_k + m_k x, \; x \in [b^{k-1}, b^k], \quad \text{for all } k = 1, \ldots, d. \tag{5.2}$$

Obtaining a piecewise linear approximation involves two competing objectives. The first is to obtain an accurate approximation, where accuracy may be measured in multiple ways. A natural measure of accuracy is the maximum absolute difference between the function and its approximation:

$$\max_{x \in [l,u]} |f(x) - \hat{f}(x)|.$$

The second objective is to obtain an approximation that uses fewer linear segments d. This is important because the time to solve the resulting approximate problem increases with the number of segments used in the approximation, possibly dramatically. Therefore, while one can always obtain an improved approximation by including more segments, this has to be weighed against the computational costs.

The simplest approach to obtaining a piecewise linear approximation is simply to choose a set of break points, for example, uniformly in the interval $[l, u]$, and then let $y^k = f(b^k)$ for each break point b^k. This approach is illustrated in Figure 5.1. For a given number of break points, however, one can obtain significantly better approximations by choosing the location of the break points so that parts of the function that are more nonlinear have more break points. In addition, using a value of $y^k \neq f(b^k)$ may also yield a better approximation.

The sandwich algorithm is another simple approach to obtaining a piecewise linear approximation. This approach begins with a single linear segment approximating the function by using break points $b^0 = l$ and $b^1 = u$ and function values $y^0 = f(b^0)$ and $y^1 = f(b^1)$. At each iteration $k \geq 1$ we have break points $b^0, \ldots, b^k$, and we use $y^i = f(b^i)$ for $i = 0, \ldots, k$. We then select the index $i \in \{1, \ldots, k\}$ such that the error between the function f and the current approximation over the interval $[x_{i-1}, x_i]$ is largest, according to whatever error measure we are interested in. A new break point in the interval (x_{i-1}, x_i) is then selected and added to the set of break points. Many possible rules for selecting the the new break point exist, such as choosing the point where the error is largest or choosing the mid-point of the interval. Rote (1992) analyses these and two other variants of this algorithm when applied to a function f that is either convex or concave, and he shows that using any of these variants the error after k iterations is $O(1/k^2)$.

Piecewise linear approximation methods appear in diverse fields, and so it is beyond the scope of this paper to provide a thorough review of these techniques. We instead provide a sample of references to other approaches. Geoffrion (1977) studies methods for finding optimal piecewise linear approximations of convex or concave functions for a given number of break points. Bellman (1961) introduces a dynamic programming approach. Boyd and Vandenberghe (2004) consider the problem of finding a piecewise linear *convex* function that best approximates a given finite set of $(x, f(x))$ points, where they assume the break points are given and the problem is to find the function values at the break points. Toriello and Vielma (2012) also consider the setting in which a finite set of $(x, f(x))$ data points is given and provide optimization formulations for finding piecewise linear approximations of these points. Notably, the approach of Toriello and Vielma (2012) does not require that the piecewise linear approximations be convex (although this can be enforced if desired) and that one simultaneously choose the break points and the function values of the approximation.

5.1.2. Modelling piecewise linear functions

This section describes techniques for modelling a piecewise linear function $\hat{f} : [l, u] \to \mathbb{R}$, as defined in (5.1). We introduce a variable y that will be used to model the value of this function. That is, we provide formulations that enforce the condition

$$y = \hat{f}(x).$$

We then use y in place of the function $f(x)$ anywhere it appears in the optimization model. In our development, we assume that the piecewise linear function $\hat{f}$ is continuous, although many of these formulations can be extended to lower semicontinuous piecewise linear functions (Vielma *et al.* 2010). In describing the different approaches, we follow the names used by Vielma *et al.* (2010).

The first approach we present is the *multiple choice model*, which uses the representation of $\hat{f}(x)$ given in (5.2). In this model, a set of binary variables z_k, $k = 1, \ldots, d$, is introduced, where $z_k = 1$ indicates that x is in interval k, $[b^{k-1}, b^k]$. In addition, for each interval k a variable w_k is introduced, where w_k will be equal to x if x is in interval k, and $w_k = 0$ otherwise. The model is as follows:

$$\sum_{k=1}^{d} w^k = x, \quad \sum_{k=1}^{d} (m_k w_k + a_k z_k) = y, \quad \sum_{k=1}^{d} z_k = 1, \tag{5.3a}$$

$$b^{k-1} z_k \leq w_k \leq b^k z_k, \quad z_k \in \{0,1\}, \quad k = 1, \ldots, d. \tag{5.3b}$$

This model was introduced by Jeroslow and Lowe (1984) and has been studied by Balakrishnan and Graves (1989) and Croxton, Gendron and Magnanti (2003). In computational experiments conducted by Vielma *et al.* (2010), this model frequently yielded the best computational performance when the number of break points was small or moderate (*e.g.*, $d \leq 16$).

The second model is the *disaggregated convex combination model*, which uses the representation of $\hat{f}(x)$ given in (5.1). In this model, a set of binary variables z_k, $k = 1, \ldots, d$, is introduced, where $z_k = 1$ indicates that $x \in [b^{k-1}, b^k]$. For each interval, this model introduces a pair of continuous variables λ_k and μ_k, which are zero if $z_k = 0$, but otherwise describe x as a convex combination of endpoints of the interval $[b^{k-1}, b^k]$. Consequently, the function value $y = \hat{f}(x)$ can then be described as the same convex combination of y^{k-1} and y^k. The formulation is as follows:

$$\sum_{k=1}^{d} (\lambda_k b^{k-1} + \mu_k b^k) = x, \quad \sum_{k=1}^{d} (\lambda_k y^{k-1} + \mu_k y^k) = y, \tag{5.4a}$$

$$\sum_{k=1}^{d} z_k = 1, \quad \lambda_k + \mu_k = z_k, \quad k = 1, \ldots, d, \tag{5.4b}$$

$$\lambda_k \geq 0, \quad \mu_k \geq 0, \quad z_k \in \{0,1\}, \quad k = 1, \ldots, d. \tag{5.4c}$$

The constraints (5.4b) enforce that exactly one interval has $z_k = 1$, that for this interval the variables λ_k and μ_k sum to one, and that $\lambda_i = \mu_i = 0$ for all other intervals $i \neq k$. Thus, the constraints (5.4a) enforce that x and y be written as a convex combination of the end points of the selected interval, and the function values at those end points, respectively. This approach has been presented by Meyer (1976), Jeroslow and Lowe (1984, 1985), Croxton *et al.* (2003) and Sherali (2001).

The next formulation is the *convex combination model*, also sometimes called the lambda method. In this formulation, a single continuous variable is introduced for each break point. These variables are used to express x and y as a convex combination of the break points, and their function values, respectively. As in the disaggregated convex combination model, binary variables are introduced to determine which interval x lies in. These variables are then used to ensure that only the convex combination variables associated with the end points of this interval are positive. The formulation is as follows:

$$\sum_{k=0}^{d} \lambda_k b^k = x, \quad \sum_{k=0}^{d} \lambda_k y^k = y, \tag{5.5a}$$

$$\sum_{j=k}^{d} \lambda_j \leq \sum_{j=k}^{d} z_j, \quad \sum_{j=0}^{k-1} \lambda_j \leq \sum_{j=1}^{k} z_j, \quad k = 1, \ldots, d, \tag{5.5b}$$

$$\sum_{k=0}^{d} \lambda_k = 1, \quad \sum_{k=1}^{d} z_k = 1, \tag{5.5c}$$

$$\lambda_k \geq 0, \quad k = 0, 1, \ldots, d, \quad z_k \in \{0, 1\}, \quad k = 1, \ldots, d. \tag{5.5d}$$

The constraints (5.5b) enforce that when $z_k = 1$, $\lambda_j = 0$ for all $j \notin \{k - 1, k\}$. In most presentations of the convex combination model (Dantzig 1960, Dantzig 1963, Jeroslow and Lowe 1985, Nemhauser and Wolsey 1988, Wilson 1998, Lee and Wilson 2001, Vielma *et al.* 2010, Geissler *et al.* 2012) this condition is instead formulated by using the following constraints:

$$\lambda_0 \leq z_1, \quad \lambda_d \leq z_d, \quad \lambda_k \leq z_k + z_{k+1}, \quad k = 1, \ldots, d-1. \tag{5.6}$$

However, Padberg (2000) demonstrated that the formulation using (5.6) allows more continuous solutions when the binary constraints on the z_k variables are relaxed, which makes (5.5b) preferable for computational purposes.

The next model we present is the *incremental model*, sometimes referred to as the delta method, which was originally proposed by Markowitz and Manne (1957). In this model, continuous variables δ_k, for each interval $k = 1, \ldots, d$, are introduced to determine what portion of interval k the argument x has 'filled'. Binary variables are introduced to enforce the condition that the intervals are filled in order. This leads to the following formulation:

$$b^0 + \sum_{k=1}^{d} \delta_k (b^k - b^{k-1}) = x, \quad y^0 + \sum_{k=1}^{d} \delta_k (y^k - y^{k-1}) = y, \tag{5.7a}$$

$$\delta_{k+1} \leq z_k \leq \delta_k, \quad k = 1, \ldots, d-1, \tag{5.7b}$$

$$\delta_1 \leq 1, \quad \delta_d \geq 0, \quad z_k \in \{0, 1\}, \quad k = 1, \ldots, d-1. \tag{5.7c}$$

If $\delta_i < 1$ for some i, then constraints (5.7b) combined with the binary restrictions on the z variables enforce that $z_i = 0$ and $\delta_{i+1} = 0$ and then recursively that $z_j = \delta_{j+1} = 0$ for all $j > i$, which is precisely the condition that the intervals should be filled in order. As pointed out by Padberg (2000), the convex combination formulation (5.7) and the incremental formulation (5.7) are related by a simple change of variables (*i.e.*, $\delta_k = \sum_{j=k}^{d} \lambda_j$, $k = 1, \ldots, d$, and similarly for the respective z_k variables).

The fifth model we describe does not use binary variables at all. Instead, it uses the concept of a special-ordered set of variables of type II (SOS2) (Beale and Tomlin 1970, Tomlin 1981, Beale and Forrest 1976, Keha, De Farias Jr and Nemhauser 2006). An ordered set of variables $\lambda = (\lambda_0, \lambda_1, \ldots, \lambda_d)$ is said to be SOS2 if at most two of the variables in the set are non-zero and the non-zero variables are adjacent. This is exactly the condition that the binary variables in the convex combination model (5.5) are used to enforce. Therefore, by instead explicitly stating this condition, the following formulation is obtained:

$$\sum_{k=0}^{d} \lambda_k b^k = x, \quad \sum_{k=0}^{d} \lambda_k y^k = y, \tag{5.8a}$$

$$\sum_{k=0}^{d} \lambda_k = 1, \tag{5.8b}$$

$$\lambda_k \geq 0, \quad k = 0, 1, \ldots, d, \quad (\lambda_0, \lambda_1, \ldots, \lambda_d) \text{ is SOS2.} \tag{5.8c}$$

The condition that an ordered set of variables is SOS2 can be declared in most commercial mixed-integer linear programming solvers (IBM 2009, Gurobi 2012, FICO Xpress 2009), similar to declaring that an individual variable is integer or binary. This condition is relaxed to obtain a linear programming relaxation and is progressively enforced through branching, just as integer restrictions are. The main difference is how the branching is done. In an LP relaxation solution, if the SOS2 condition is violated by the relaxation solution $\hat{\lambda}$, then an index $k \in \{1, \ldots, d\}$ is selected such that there exists an index $j_1 < k$ with $\hat{\lambda}_{j_1} > 0$ and also an index $j_2 > k$ with $\hat{\lambda}_{j_2} > 0$. Then, two branches are created: one that enforces $\lambda_j = 0$ for all $j < k$ and the other that enforces $\lambda_j = 0$ for all $j > k$. Every solution that satisfies the SOS2 condition is feasible for one of the two branching conditions, and hence this branching strategy does not exclude any feasible solutions. In addition, the current relaxation solution, which violates the SOS2 condition, is eliminated from both branches, enabling the relaxation bound to improve and ensuring that the SOS2 condition will be satisfied after a finite number of branches. One also can derive valid inequalities based on the SOS2 condition, analogous to the use of valid inequalities for mixed-integer programming (Keha *et al.* 2006).

Vielma and Nemhauser (2011) have recently developed a novel approach to modelling piecewise linear functions using binary variables. The interesting feature of this approach is that the number of binary variables required is only logarithmic in the number of segments of the linear function. In contrast, the binary formulations we presented here all require one binary variable for each segment. Note, however, that the number of continuous variables required in the formulations of Vielma and Nemhauser (2011) is still linear in the number of segments. The computational results of Vielma and Nemhauser (2011) suggest that this approach is most beneficial when modelling multivariate piecewise linear functions.

5.1.3. Multivariate functions

One approach to using piecewise linear models for problems with multivariate functions is to attempt to reformulate the problem in such a way that only univariate functions appear, and then apply the univariate modelling techniques. We have already mentioned separable functions of the form $g(x) = \sum_{i=1}^{K} f_i(x_i)$ as one example where this approach can work well. More generally, if an algebraic description of a multivariate function g is available, then the techniques described in Section 5.2.1 can be used to first obtain a reformulated problem that contains only univariate and bivariate functions and then construct piecewise linear approximations of these functions. Using further transformations, one may be able to eliminate even the bivariate functions. For example, suppose we have a constraint

$$y = x_1 x_2$$

in our model, where y, x_1, and x_2 are all decision variables, and because of other constraints in the model we know that $x_1 > 0$ and $x_2 > 0$ in any feasible solution. Then, this constraint is equivalent to

$$\ln(y) = \ln(x_1) + \ln(x_2).$$

Each of the functions $\ln(y)$, $\ln(x_1)$, and $\ln(x_2)$ can then be approximated by using univariate piecewise linear models.

The approach of reducing multivariate functions to univariate functions has a few potential disadvantages, which have been pointed out, for example, by Tomlin (1981), Lee and Wilson (2001) and Vielma *et al.* (2010). First, it may not always be easy to find a systematic approach that reduces a given model to one that contains only univariate functions. In our example above, we required $x_1 > 0$ and $x_2 > 0$ in order for the log transformation to be valid. If this were not the case, an alternative would be required. Second, and probably more important, this process of reformulation may allow the errors introduced by the piecewise linear approximations to accumulate and amplify, potentially leading to an approximate model that is either too inaccurate to be useful or requires so many break points in the piecewise linear approximations that it becomes intractable to solve. Finally, in some cases a function is not given analytically. Instead, we may have only an oracle, or 'black box', that allows us to obtain function evaluations at given points, for example, by running a complex simulation or even a physical experiment. In this case, we may wish to obtain a piecewise linear function that approximates the data obtained at a set of trial points.

One can also directly model non-separable multivariate piecewise linear functions using binary variables. We refer the reader to the works of Vielma *et al.* (2010) and Geissler *et al.* (2012) for details of these techniques. In particular, the convex combination method has been extended to multivariate functions by Lee and Wilson (2001), and the incremental method has been extended by Wilson (1998). Tomlin (1981) also extended the notion of special-ordered sets for modelling multivariate piecewise linear functions. D'Ambrosio, Lodi and Martello (2010) provide an interesting alternative for approximating multivariate piecewise linear functions, in which the piecewise linear approximation is not fixed *a priori*, but is instead allowed to be chosen 'optimistically' by the optimization algorithm. This modification enables a relatively compact formulation to be derived. We note, however, that the number of pieces required to obtain an acceptable piecewise linear approximation of a given nonlinear function may grow exponentially in the number of arguments to the function. Hence, all these approaches are, in practice, limited to functions with at most a few arguments.

5.2. Generic relaxation strategies

For general non-convex MINLP problems, methods for finding a relaxation exploit the structure of the problem. For a broad class of MINLP problems, the objective function and the constraints are nonlinear but *factorable*, in other words, they can be expressed as the sum of products of unary functions of a finite set $\mathcal{O}_{\text{unary}} = \{\sin, \cos, \exp, \log, |\cdot|\}$ whose arguments are variables, constants, or other functions, which are in turn factorable. In other words, a factorable function can be written by combining a finite number of elements from a set of operators $\mathcal{O} = \{+, \times, /, \hat{}, \sin, \cos, \exp, \log, |\cdot|\}$. This excludes, among others, integral functions $\int_{x_0}^{x} h(x)\mathrm{d}x$ whose antiderivative is unknown, and *black-box* functions whose value can be computed by running a simulation, for instance. The approach described below is therefore suitable when the symbolic information about the objective function and constraints, that is, their expressions, is known.

Factorable functions can be represented by *expression trees.* These are n-ary arborescences whose leaves are constants or variables and whose non-leaf nodes are, in general, n-ary operators whose children are the arguments of the operator, and which are in turn expression trees (Cohen 2003). The expression tree of the function $f(x_1, x_2) = x_1 \log(x_2) + x_2^3$ is depicted in Figure 5.2.

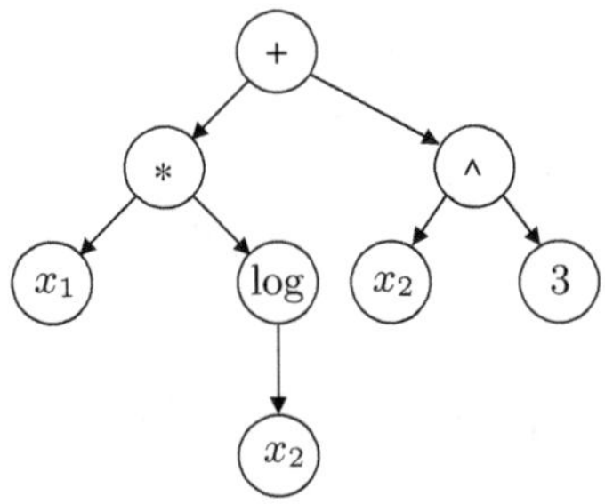

Figure 5.2. The expression tree of $f(x_1, x_2) = x_1 \log(x_2) + x_2^3$. Leaf nodes are for variables x_1 and x_2 and for the constant 3.

5.2.1. Relaxations of factorable functions

If the objective and the constraints of an MINLP of the form (1.1) are factorable, the problem admits a representation where the expression trees of the objective function and all constraints are combined. The root of each expression is one among $c_1(x), c_2(x), \ldots, c_m(x)$, or $f(x)$, and is associated with a lower and an upper bound: $[-\infty, 0]$ for $c_i(x), i = 1, 2, \ldots, m$, and $[-\infty, \bar{\eta}]$ for $f(x)$, where $\bar{\eta}$ is the objective function value of a feasible solution for (1.1), if available, or ∞. The leaf nodes of all expression trees are replaced by a unique set of nodes representing the variables $x_1, x_2, \ldots, x_n$ of the problem. The result is a *directed acyclic graph* (DAG).

Example. Consider the following MINLP:

$$\begin{cases} \text{minimize} & x_1 + x_2^2, \\ \text{subject to} & x_1 + \sin x_2 \le 4, \\ & x_1 x_2 + x_2^3 \le 5, \\ & x_1 \in [-4, 4] \cap \mathbb{Z}, \\ & x_2 \in [0, 10] \cap \mathbb{Z}. \end{cases} \tag{5.9}$$

The DAG of this problem has three nodes without entering arcs (one for the objective function and two for the constraints) and six leaf nodes: two for the variables x_1 and x_2 and four for the constants 2, 3, -4, and -5. It is represented in Figure 5.3.

Factorable problems allow a *reformulation* of problem (1.1) (McCormick 1976, Smith and Pantelides 1997, Tawarmalani and Sahinidis 2002). The reformulation is another MINLP as follows:

$$\begin{cases} \underset{x}{\text{minimize}} & x_{n+q}, & \\ \text{subject to} & x_k = \vartheta_k(x), & k = n+1, n+2, \ldots, n+q, \\ & l_i \le x_i \le u_i, & i = 1, 2, \ldots, n+q, \\ & x \in X, & \\ & x_i \in \mathbb{Z}, \ \forall i \in I, & \end{cases} \tag{5.10}$$

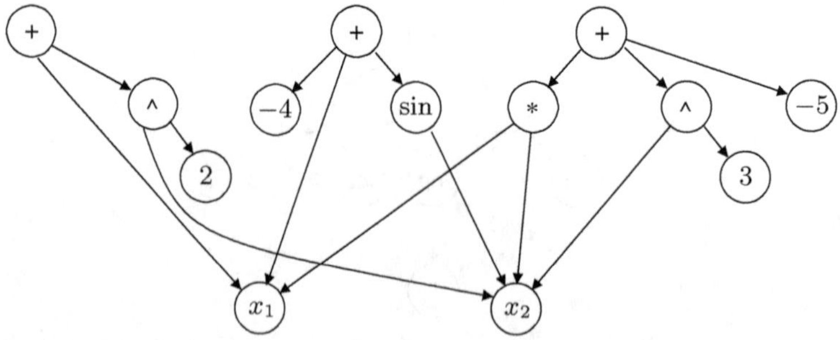

Figure 5.3. The DAG associated with the problem in (5.9). Each node without entering arcs is associated with the root of the expression tree of either a constraint or the objective function. In common with all constraints and the objective are the leaf nodes associated with variables x_1 and x_2.

where ϑ_k is an operator of the set $\mathcal{O}$ introduced above. The bounds on all variables are written explicitly here for the sake of clarity, but they are included in the definition of X; we will use this notation throughout this section. The reformulation contains a set of q new variables known as *auxiliary variables* (or more simply auxiliaries). By convention, the last auxiliary variable replaces the objective function. Each of these variables is constrained to be equal to a function $\vartheta(x)$ such that $\vartheta \in \mathcal{O}$. The lower and upper bounds on each auxiliary variable x_k, as well as its integrality constraint, depend on the operator ϑ_k associated with it and on the bounds on the arguments of ϑ_k (and the integrality of these arguments).

Example. The MINLP shown in (5.9) admits the following reformulation:

$$\left\{\begin{array}{llll}
\multicolumn{4}{l}{\text{minimize } x_9,} \\
\multicolumn{4}{l}{\text{subject to}} \\
\quad x_3 = \sin x_2, & x_7 = x_5 + x_6 - 5, & 0 \le x_2 \le 10, & 0 \le x_6 \le 1000, \\
\quad x_4 = x_1 + x_3 - 4, & x_8 = x_2^2, & -1 \le x_3 \le 1, & -45 \le x_7 \le 0, \\
\quad x_5 = x_1 x_2, & x_9 = x_1 + x_8, & -9 \le x_4 \le 0, & 0 \le x_8 \le 100, \\
\quad x_6 = x_2^3, & -4 \le x_1 \le 4, & -40 \le x_5 \le 40, & -4 \le x_9 \le 104, \\
\quad x_1, x_2, x_5, x_6, x_7, x_8, x_9 \in \mathbb{Z}. & & &
\end{array}\right.$$

Note that x_3, the auxiliary associated with $\sin x_2$, has bounds $[-1, 1]$ because $x_2 \in [0, 10]$, while $x_7 := x_5 + x_6 - 5$, with bounds $[-45, 1035]$, is further constrained by the right-hand side of the constraint $x_1 x_2 + x_2^3 \le 5$. This variable is indeed associated with the root of the expression tree of $c_2(x) = x_1 x_2 + x_2^3 - 5$. The integrality of an auxiliary also depends on whether the function associated with it can return only integer values given the integrality constraints of its arguments. In this case, x_3 and x_4 are not constrained to be integer, while all other variables are because of the integrality constraint on x_1 and x_2.

Problems (5.10) and (1.1) are equivalent in that for any feasible (resp. optimal) solution $\bar{x} \in \mathbb{R}^{n+q}$ of (5.10) one can obtain a feasible (resp. optimal) solution $\tilde{x} \in \mathbb{R}^n$ of (1.1) and *vice versa*. Although still a non-convex MINLP, (5.10) makes it easier to obtain a convex relaxation of (1.1). Consider the non-convex sets

$$\Theta_k = \{x \in \mathbb{R}^{n+q} : x_k = \vartheta_k(x),\ x \in X,\ l \le x \le u,\ x_i \in \mathbb{Z},\ i \in I\},$$

for $k = n+1, n+2, \dots, n+q$. Note that Θ_k are, in general, non-convex because of the equation $x_k = \vartheta_k(x)$ and the integrality constraints. Suppose that a convex set $\breve{\Theta}_k \supseteq \Theta_k$ exists for each $k = n+1, n+2, \dots, n+q$. Then the following is a convex relaxation of (5.10) that intersects $\breve{\Theta}_k$ for $k = n+1, n+2, \dots, n+q$, and hence it is a relaxation of (1.1) (note that the integrality constraints are also relaxed):

$$\begin{cases} \underset{x}{\text{minimize}} & x_{n+q}, \\ \text{subject to} & x \in \breve{\Theta}_k, \qquad k = n+1, n+2, \dots, n+q, \\ & l_i \le x_i \le u_i, \qquad i = 1, 2, \dots, n+q, \\ & x \in X. \end{cases}$$

Convex sets $\breve{\Theta}_k$ are generally polyhedral, that is, they are described by a system of m_k linear inequalities:

$$\breve{\Theta}_k = \left\{x \in \mathbb{R}^{n+q} : a^k x_k + B^k x \ge d^k,\ x \in X,\ l \le x \le u\right\},$$

where $a^k \in \mathbb{R}^{m_k}$, $B^k \in \mathbb{R}^{m_k \times (n+q)}$, and $d^k \in \mathbb{R}^{m_k}$. Hence several practical MINLP solvers for (1.1) use the following linear relaxation to obtain a lower bound:

$$\begin{cases} \underset{x}{\text{minimize}} & x_{n+q}, \\ \text{subject to} & a^k x_k + B^k x \ge d^k, \quad k = n+1, n+2, \dots, n+q, \\ & l_i \le x_i \le u_i, \qquad i = 1, 2, \dots, n+q, \\ & x \in X. \end{cases}$$

Although finding a polyhedral superset of Θ_k is non-trivial, for each operator of a finite set $\mathcal{O}$ we can define a method to find one. We provide two examples. LP relaxations for monomials of odd degree such as $x_k = x_i^{2p+1}$, with $p \in \mathbb{Z}_+$, were proposed by Liberti and Pantelides (2003). For products of two variables $x_k = x_i x_j$, the following four inequalities proposed by McCormick (1976) provide the convex hull of the set

$$\Theta_k = \{(x_i, x_j, x_k) : x_k = x_i x_j, (l_i, l_j, l_k) \le (x_i, x_j, x_k) \le (u_i, u_j, u_k)\},$$

as proved by Al-Khayyal and Falk (1983):

$$\begin{aligned} x_k &\ge l_j x_i + l_i x_j - l_i l_j, & x_k &\le l_j x_i + u_i x_j - u_i l_j, \\ x_k &\ge u_j x_i + u_i x_j - u_i u_j, & x_k &\le u_j x_i + l_i x_j - l_i u_j. \end{aligned} \tag{5.11}$$

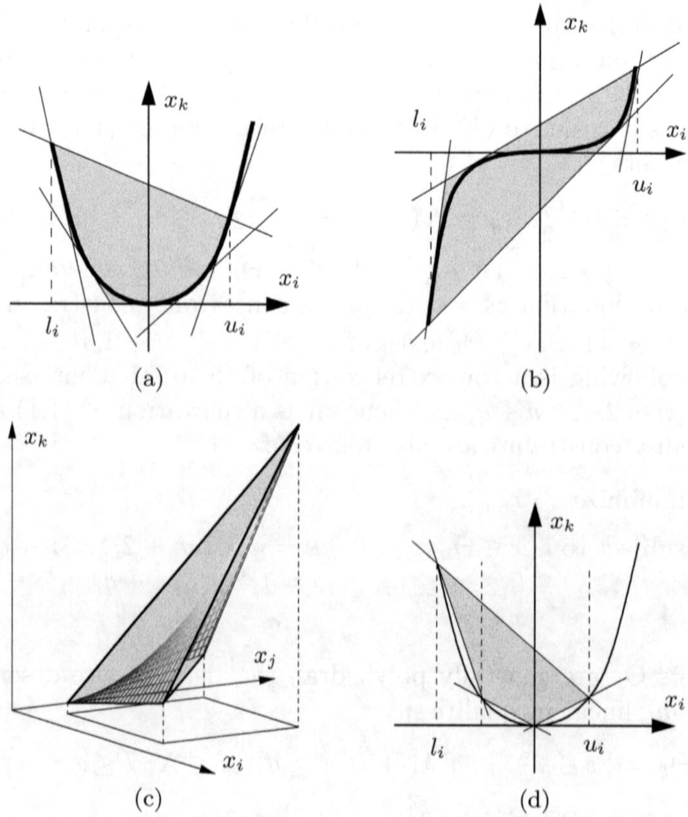

Figure 5.4. Polyhedral relaxations $\breve{\Theta}_k$ for several univariate and bivariate operators: $x_k = x_i^2$, $x_k = x_i^3$, $x_k = x_i x_j$, and $x_k = x_i^2$ with x_i integer. Note that these relaxations are *exact* at the bounds on x_i and x_j.

Figure 5.4 shows polyhedral relaxations for $x_k = x_i^2$, $x_k = x_i^3$, and $x_k = x_i x_j$. If the argument x_i of a function ϑ_k is integer, the linear relaxation can be strengthened. Suppose $x_j = (x_i)^2$ and $x_i \in \mathbb{Z} \cap [-2, 1]$, as in Figure 5.4(d). Then we can add linear inequalities that can be violated by points $(\hat{x}, \hat{x}^2)$ if $\hat{x} \notin \mathbb{Z}$.

Convex relaxations $\breve{\Theta}_k$ of the univariate or multivariate operator ϑ_k are required to be *exact* at the frontier of the bound interval of the arguments of ϑ_k. In particular, let $x_{i_1}, x_{i_2}, \ldots, x_{i_{h(k)}}$ be the arguments of ϑ_k, each of them bounded by the interval $[l_{i_j}, u_{i_j}]$. Define for each $j = 1, 2, \ldots, h(k)$ a value $b_{i_j} \in \{l_{i_j}, u_{i_j}\}$. Then

$$\breve{\Theta}_k \cap \{x \in \mathbb{R}^n : x_{i_j} = b_{i_j}\} = \Theta_k \cap \{x \in \mathbb{R}^n : x_{i_j} = b_{i_j}\}.$$

A direct consequence is the convergence result shown in Section 5.3.

5.2.2. Disjunctive cuts

In the class of convex MINLPs, the only non-convex constraints are represented by integer variables, and these non-convexities are resolved by integer branching, which represents a specific class of *disjunctions*. There are classes of non-convex MINLPs whose nonlinear objective and constraints introduce a different type of disjunction that can be used to create a valid cut. Disjunctive cuts for several classes of convex MINLP problems have been discussed in Section 4.3. For non-convex, factorable MINLP, disjunctions arise from branching rules $x_i \leq b \vee x_i \geq b$ on continuous variables. Other disjunctions that have important applications arise from *complementarity constraints*, that is, constraints of the form $x_i x_j = 0$ that are equivalent to the disjunction $x_i = 0 \vee x_j = 0$. These constraints are the basis of disjunctive cuts developed by Júdice, Sherali, Ribeiro and Faustino (2006).

Disjunctive cuts can be developed by using the branching disjunction $x_i \leq b \vee x_i \geq b$ in MINLPs with factorable functions. Suppose a branching rule is enforced because an auxiliary variable x_k and its related non-convex constraint $x_k = \vartheta_k(x)$ are such that $\hat{x}_k \neq \vartheta_k(\hat{x})$ for a given LP solution $\hat{x}$. A branching rule allows us to refine the LP relaxation of a subproblem as shown in Figure 5.5 (page 79). Figure 5.5(a) depicts the set $\breve{\Theta}_j$ for $x_j = x_i^2$ and shows that $\hat{x}_j \neq \hat{x}_i^2$. The disjunction $x_i \leq b \vee x_i \geq b$ can be used to create two new subproblems, NLP(l^-, u^-) and NLP(l^+, u^+), and consequently two tighter LP relaxations, LP(l^-, u^-) and LP(l^+, u^+), shown in Figure 5.5(b); note that both relaxations exclude $(\hat{x}_i, \hat{x}_j)$. For an example with a different function $x_j = \mathrm{e}^{x_i}$, see Figure 5.5(c). Belotti (2012) creates disjunctive cuts by means of a *cut generating LP* (CGLP) (Balas, Ceria and Cornuéjols 1993), a linear optimization problem used to devise a disjunctive cut that maximizes the violation with respect to $\hat{x}$. The CGLP obtains a disjunctive cut that is valid for both LP(l^-, u^-) and LP(l^+, u^+) and that hence exploits the disjunction to eliminate infeasible solutions without actually creating two subproblems.

This procedure retains the pros and cons of its MILP version: although it allows us to avoid creating two new subproblems and can be effective, each of these cuts is generated by solving very large LPs. Therefore, this procedure can be computationally expensive for large MINLPs.

5.3. Spatial branch-and-bound

The best-known method for solving non-convex MINLP problems is *branch-and-bound* (BB). Most of the global optimization community usually refers to these methods as *spatial BB* (sBB). As outlined in Section 3, a BB method is an implicit enumeration technique that recursively partitions the feasible set. This partitioning yields two or more subproblems whose solution sets are, ideally, *disjoint* from one another, in order to avoid evaluating a feasible solution in more than one subproblem. Most BB implementations for MINLP use the reformulation scheme outlined in Section 5.2.1 to obtain a lower bound (Sahinidis 1996, Ryoo and Sahinidis 1996, Tawarmalani and Sahinidis 2002, Ryoo and Sahinidis 1995, Tawarmalani and Sahinidis 2004, Smith and Pantelides 1997, Belotti *et al.* 2009).

Other approaches employ a relaxation technique called *α-convexification* (Androulakis, Maranas and Floudas 1995). For a non-convex quadratic function $f(x) = x^T Qx + c^T x$ with $x \in [l, u]$, a valid lower bound is provided by

$$\breve{f}(x) = x^T Qx + c^T x + \alpha \sum_{i=1}^{n} (x_i - l_i)(x_i - u_i).$$

Note that $\breve{f}(x) \leq f(x)$ for $x \in [l, u]$ and that $\breve{f}(x)$ can be rewritten as a quadratic function $x^T Px + d^T x$, where $P = Q + \alpha I$, and $\breve{f}(x)$ is convex if $P \succeq 0$. Therefore it suffices to set $\alpha = -\lambda_{\min}(Q)$, namely, the opposite of the minimum eigenvalue of Q, to obtain a convex relaxation. Some implementations of the α-convexification adopt a quadratic approximation of the objective and the constraints and hence guarantee optimality only for quadratic problems (Nowak, Alperin and Vigerske 2003). This method can be extended to non-quadratic functions while preserving the validity of the lower bound. A generalization of this method is at the base of the MINLP solver GloMIQO (Misener and Floudas 2013).

A BB algorithm requires (i) a procedure to compute a lower bound on the optimal objective function value of a subproblem and (ii) a procedure for partitioning the feasible set of a subproblem. In general, the former consists of obtaining a convex relaxation of a subproblem NLP(l, u) and solving it to optimality, while the latter generates two new subproblems, NLP(l^-, u^-) and NLP(l^+, u^+), by appropriately setting new variable bounds. We discuss these techniques in detail below.

The structure of a branch-and-bound algorithm for non-convex MINLP follows the scheme described in Section 3, and we will not repeat it here. The generic subproblem NLP(l, u) of the BB can be defined as a restriction of the original MINLP (1.1) as follows:

$$\begin{cases} \underset{x}{\text{minimize}} & f(x), \\ \text{subject to} & c(x) \leq 0, \\ & x \in X, \\ & l_i \leq x_i \leq u_i, \ \forall i = 1, 2, \ldots, n, \\ & x_i \in \mathbb{Z}, \ \forall i \in I. \end{cases} \tag{5.12}$$

At subproblem NLP(l, u), the BB algorithm seeks a lower bound of the optimal value of $f(x)$ by solving a convex relaxation such as the LP relaxation LP(l, u):

$$\begin{cases} \underset{x}{\text{minimize}} & x_{n+q}, \\ \text{subject to} & a^k x_k + B^k x \geq d^k, \quad k = n+1, n+2, \ldots, n+q, \\ & l_i \leq x_i \leq u_i, \qquad\qquad i = 1, 2, \ldots, n+q, \\ & x \in X. \end{cases} \tag{5.13}$$

Suppose an optimal solution $\hat{x}$ of LP(l, u) is found. If $\hat{x}$ is feasible for (5.12) and hence for (1.1), subproblem NLP(l, u) can be eliminated. If $\hat{x}$ is infeasible for (5.12), then at least one of the following two holds.

(1) $\hat{x}$ is not integer feasible, *i.e.*, there exists $i \in I$ such that $\hat{x}_i \notin \mathbb{Z}$.

(2) At least one of the continuous non-convex constraints of the reformulation is violated, that is,

$$\text{there exists } k \in \{n+1, n+2, \ldots, n+q\} \text{ such that } \hat{x}_k \neq \vartheta_k(\hat{x}).$$

In the first case, one can generate two new subproblems NLP(l^-, u^-) and NLP(l^+, u^+), whose feasible sets $\mathcal{F}(l^-, u^-)$ and $\mathcal{F}(l^+, u^+)$ are amended new bounds on x_i through the branching rule $x_i \leq \lfloor \hat{x}_i \rfloor \vee x_i \geq \lceil \hat{x}_i \rceil$. In the second case, branching may be necessary on a continuous variable. In that case, suppose that x_i is among the arguments of function ϑ_k. Then the branching rule $x_i \leq \hat{x}_i \vee x_i \geq \hat{x}_i$ creates two new subproblems whose feasible sets have a non-empty intersection, where $x_i = \hat{x}_i$. This constitutes a strong point of departure with the subclass of pure integer non-convex MINLPs, where all variables are integer, and with convex MINLP discussed in Section 3. For these subclasses, branching is necessary only on integer variables. Finite bounds on integer variables ensures finite termination of the branch-and-bound algorithm.

Consider the feasible set of a subproblem NLP(l, u) of (1.1):

$$\mathcal{F}(l, u) = \{x \in [l, u] : c_i(x) \leq 0, \ \forall i = 1, 2, \ldots, m, x \in X, x_i \in \mathbb{Z}, i \in I\},$$

whose only difference from the feasible set of (1.1) is new bounds on the variables, as dictated by the branching rules. Consider a branching rule on a continuous variable x_i, subdividing the feasible set of subproblem NLP(l, u) into two subproblems NLP(l^-, u^-) and NLP(l^+, u^+), with feasible sets $\mathcal{F}(l^-, u^-)$ and $\mathcal{F}(l^+, u^+)$. A *bounding operation* yields two subproblems NLP(l^-, u^-) and NLP(l^+, u^+), by applying a branching rule, and lower bounds $\lambda_{\mathcal{F}(l^-,u^-)}$, $\lambda_{\mathcal{F}(l^+,u^+)}$ and upper bounds $\mu_{\mathcal{F}(l^-,u^-)}$, $\mu_{\mathcal{F}(l^+,u^+)}$ for the new subproblems. Such a bounding operation is said to be *consistent* if, at every step, subsets $\mathcal{F}(l^-, u^-)$ and $\mathcal{F}(l^+, u^+)$ are either fathomed or can be further refined in such a way that, for any infinite sequence $\{\mathcal{F}_h\}$ resulting from applying bounding operations, one can guarantee that (Horst and Tuy 1993)

$$\lim_{h \to \infty} \mu_{\mathcal{F}_h} - \lambda_{\mathcal{F}_h} = 0.$$

In addition, a bounding operation is *finitely consistent* if any sequence $\{\mathcal{F}_h\}$ of successively refined partitions of $\mathcal{F}$ is finite. Branching on continuous variables does not imply directly that a finite number of branching rules will be used, yet both in theory and in practice BB algorithms do have finite termination properties, as shown by McCormick (1976) and Horst and Tuy (1993).

Theorem 5.1 (McCormick 1976, Horst & Tuy 1993). If the bounding operation in the BB algorithm is finitely consistent, the algorithm terminates in a finite number of steps.

The value of this result can be made clearer if one considers an MINLP with even just one continuous variable x_1: by branching only on integer variables (in a finite number of BB nodes if all integer variables are bounded), one eventually obtains a possibly non-convex continuous optimization problem. Therefore, branching will become necessary on the continuous variable x_1 as well, although termination is no longer guaranteed by integrality. The result by McCormick (1976) states that convergence is still ensured as long as the bounding operation is finitely consistent.

5.3.1. Spatial branching

Partitioning the feasible set of a subproblem NLP(l, u) yields $h \geq 2$ new subproblems NLP(l', u'), NLP(l'', u''), …, NLP$(l^{(h)}, u^{(h)})$, whose lower bounds $\lambda_{\mathrm{NLP}(l',u')}, \lambda_{\mathrm{NLP}(l'',u'')}, \ldots, \lambda_{\mathrm{NLP}(l^{(h)},u^{(h)})}$ are no smaller than the lower bound of NLP(l, u). We will assume without loss of generality that two new problems NLP(l^-, u^-) and NLP(l^+, u^+) are created. Most practical implementations adopt a *variable branching* $x_i \leq b \vee x_i \geq b$. The performance of the BB algorithm depends strongly on the choice of i and b (Belotti *et al.* 2009, Tawarmalani and Sahinidis 2002). An integer variable is obviously a candidate for selection as a branching variable if its value is fractional in the LP solution. In the remainder of this section, we assume that all integrally constrained variables are integer and hence no branching is possible (integer branching has been discussed in Section 3) and that branching is done because of an auxiliary x_k such that $\hat{x}_k \neq \vartheta_k(\hat{x})$.

An ideal choice of i should balance more than one objective: it should (i) increase both lower bounds $\lambda_{\mathrm{NLP}(l^-,u^-)}$ and $\lambda_{\mathrm{NLP}(l^+,u^+)}$; (ii) shrink both feasible sets $\mathcal{F}(l^-, u^-)$ and $\mathcal{F}(l^+, u^+)$; and (iii) allow for a *balanced* BB tree, among other criteria.

Suppose an optimal solution $\hat{x}$ of the LP relaxation LP(l, u) is found. A continuous variable x_i is a candidate for branching if it is not fixed (*i.e.*, its lower and upper bounds do not coincide), if it is an argument of a function $\vartheta_k(x)$ associated with an auxiliary variable x_k, and $\hat{x}_k \neq \vartheta_k(x)$. For example, if $x_k = \vartheta_k(x) = x_i x_j$, $\hat{x}_k \neq \hat{x}_i \hat{x}_j$, and $l_i < u_i$, then x_i is a candidate for branching.

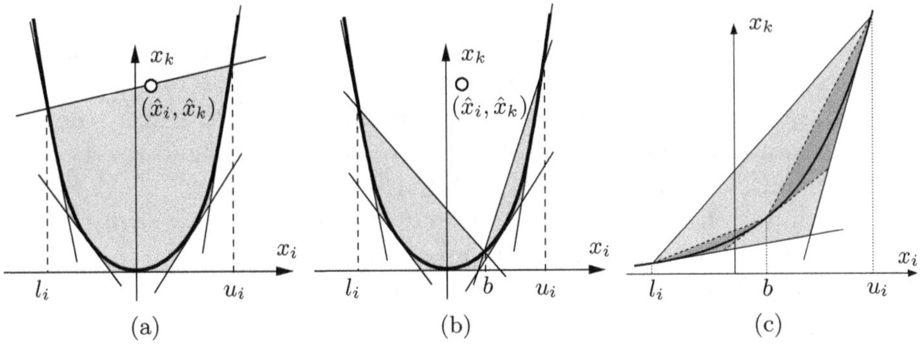

Figure 5.5. Polyhedral relaxations upon branching. (a) The set $\breve{\Theta}_k$ is shown with the components $(\hat{x}_i, \hat{x}_k)$ of the LP solution. Branching on x_i excludes the LP solution: see (b). (c) The LP relaxation before and after branching is shown for $x_k = \mathrm{e}^{x_i}$ in light and dark grey, respectively.

Upon branching, the two generated subproblems will each obtain a lower bound by solving two tighter relaxations than that of their ancestor. Geometrical intuition is provided in Figure 5.5. Suppose the auxiliary x_k is defined as $x_k = \vartheta_k(x_i) = (x_i)^2$ and $x_i \in [l_i, u_i]$ for this subproblem. Because the LP solution $\hat{x}$ is such that $\hat{x}_k \neq (\hat{x}_i)^2$ (see Figure 5.5(a)), one can generate two new subproblems using the branching rule $x_i \leq b \vee x_i \geq b$. The new linear relaxations are the polytopes in Figure 5.5(b), which are disjoint except for the point (b, b^2) and which exclude the point $(\hat{x}_i, \hat{x}_k)$. Figure 5.5(c) provides a similar example for $x_k = \vartheta_k(x_i) = \mathrm{e}^{x_i}$.

Tawarmalani and Sahinidis (2002) introduce *violation transfer* (VT) as a variable selection technique. VT identifies the variable x_i that has the largest impact on the violation of the non-convex constraints $x_k = \vartheta_k(x)$ for all $k = 1, 2, \ldots, n+q$ such that x_i is an argument of ϑ_k. Strong branching, pseudocost branching, and reliability branching, discussed in Section 3, can be applied with little modification to non-convex MINLP and have been implemented in non-convex MINLP solvers (Belotti *et al.* 2009).

The choice of branching point is also crucial, and differs from integer branching in that we have the freedom to choose a branching point for variable x_i that can differ from $\hat{x}_i$. A branching rule should ensure that $\hat{x}$ is infeasible for both $\mathrm{LP}(l^-, u^-)$ and $\mathrm{LP}(l^+, u^+)$; hence the sole branching rule $x_i \leq \hat{x}_i \vee x_i \geq \hat{x}_i$ will not suffice. However, the refined linear relaxations $\mathrm{LP}(l^-, u^-)$ and $\mathrm{LP}(l^+, u^+)$ will be obtained by adding linear inequalities that are violated by $\hat{x}$. While the linear inequalities depend on the new bounds on x_i, setting the branch point to a suitable $b \neq x_i$ does not prevent us from excluding $\hat{x}$.

5.3.2. Bound tightening

Bound tightening (also referred to as bound reduction or domain reduction) is a class of algorithms aiming to reduce the bound intervals on the variables of (1.1). Although these algorithms are optional in a BB solver, they are crucial to obtaining an optimal solution in reasonable time and are therefore implemented in the vast majority of MINLP solvers.

Their importance is directly connected to the LP relaxation (5.13): the tighter the variable bounds, the tighter the linear polyhedra $\breve{\Theta}_k$ for each auxiliary variable x_k and hence the better the lower bound on the objective. Some MINLP solvers use bound reduction as the sole means of obtaining a lower bound on the optimal objective function value of the subproblem (Messine 2004).

The usefulness of bound reduction is not limited to a tighter bound hyperrectangle: as a result of bound reduction, the feasible set might become empty, or the lower bound l_{n+q} on x_{n+q} is above the *cutoff* value of the problem, namely, the objective function value of a feasible solution of (1.1). In these two cases, the procedure proves that the current node is infeasible and can be fathomed without the need to compute a lower bound by solving the convex relaxation.

Consider again the feasible set of an MINLP problem:

$$\mathcal{F} = \{x \in [l,u] : c_i(x) \le 0,\ \forall i = 1,2,\ldots,m, x \in X, x_i \in \mathbb{Z}, i \in I\},$$

and suppose that a feasible solution of (1.1) is known with value $\breve{z}$. For each variable $x_i, i = 1,2,\ldots,n$, valid (and possibly tighter) lower and upper bounds are given by

$$l'_i = \min\{x_i : x \in S, f(x) \le \breve{z}\}, \quad u'_i = \max\{x_i : x \in S, f(x) \le \breve{z}\}. \quad (5.14)$$

Solving the $2n$ optimization problems above would yield tighter bounds, but these problems can be as hard as problem (1.1) itself. The two most important bound reduction techniques are *feasibility-based* (FBBT) and *optimality-based bound tightening* (OBBT). Other commonly used techniques are *probing* and *reduced-cost tightening*; we present these techniques below.

Feasibility-based bound tightening. FBBT has been used in the artificial intelligence literature (Davis 1987) and is a strong component of constraint programming solvers. It is part of nonlinear optimization solvers (Messine 2004) as well as of MILP solvers (Andersen and Andersen 1995, Savelsbergh 1994).

FBBT works by inferring tighter bounds on a variable x_i as a result of a changed bound on one or more other variables x_j that depend, directly or indirectly, on x_i. For example, if $x_j = x_i^3$ and $x_i \in [l_i, u_i]$, then the bound interval of x_j can be tightened to $[l_j, u_j] \cap [l_i^3, u_i^3]$. *Vice versa,* a tightened bound l'_j on x_j implies a possibly tighter bound x_i, *i.e.*, $l'_i = \sqrt[3]{l'_j}$. Another example is given by $x_k = x_i x_j$, with $(1,1,0) \le (x_i, x_j, x_k) \le (5,5,2)$. Lower bounds $l_i = l_j = 1$ imply a tighter lower bound $l_k = l_i l_j = 1 > 0$, while the upper bound $u_k = 2$ implies that $x_i \le \frac{u_k}{l_j}$ and $x_j \le \frac{u_k}{l_i}$, and hence $u'_i = u'_j = 2 < 5$.

Perhaps the best-known example applies to affine functions. Suppose x_k is an auxiliary variable defined as $x_k = a_0 + \sum_{j=1}^{n} a_j x_j$, with $k > n$. Suppose also that

$$J^+ = \{j = 1, 2, \ldots, n : a_j > 0\} \quad \text{and} \quad J^- = \{j = 1, 2, \ldots, n : a_j < 0\}.$$

Then valid bounds on x_k are

$$a_0 + \sum_{j \in J^-} a_j u_j + \sum_{j \in J^+} a_j l_j \le x_k \le a_0 + \sum_{j \in J^-} a_j l_j + \sum_{j \in J^+} a_j u_j.$$

Moreover, explicit bounds $[l_k, u_k]$ on x_k imply new (possibly tighter) bounds l'_j, u'_j on $x_j, j = 1, 2, \ldots, n : a_j \ne 0$:

$$\begin{aligned}
\text{for } a_j > 0, \quad l'_j &= \frac{1}{a_j}\left(l_k - \left(a_0 + \sum_{i \in J^+ \setminus \{j\}} a_i u_i + \sum_{i \in J^-} a_i l_i\right)\right), \\
u'_j &= \frac{1}{a_j}\left(u_k - \left(a_0 + \sum_{i \in J^+ \setminus \{j\}} a_i l_i + \sum_{i \in J^-} a_i u_i\right)\right); \\
\text{for } a_j < 0, \quad l'_j &= \frac{1}{a_j}\left(u_k - \left(a_0 + \sum_{i \in J^+} a_i l_i + \sum_{i \in J^- \setminus \{j\}} a_i u_i\right)\right), \\
u'_j &= \frac{1}{a_j}\left(l_k - \left(a_0 + \sum_{i \in J^+} a_i u_i + \sum_{i \in J^- \setminus \{j\}} a_i l_i\right)\right).
\end{aligned} \tag{5.15}$$

These *implied bounds* are often used as a preprocessing technique (Andersen and Andersen 1995) prior to solving MILP problems. For both MILP and MINLP problems, bound reduction can be obtained by using *pairs* of inequalities (Belotti 2013), specifically through the convex combination of two inequalities $a^T x \ge \alpha$ and $b^T x \ge \beta$ using a parameter $\lambda \in [0,1]$. The resulting inequality $(\lambda a + (1-\lambda) b)^T x \ge \lambda \alpha + (1-\lambda)\beta$ yields bounds on the variables similar to (5.15), but these are a function of λ and can be shown to be tighter than those obtained by single inequalities.

For the general nonlinear case, variable bounds are propagated by using the DAG of the problem. For instance, consider problem (5.9) and bounds $[-4, 4]$ and $[0, 10]$ on x_1 and x_2, respectively. We rewrite the DAG of this problem to reflect these bounds and to show each auxiliary variable next to the root of the expression tree associated to it: see Figure 5.6.

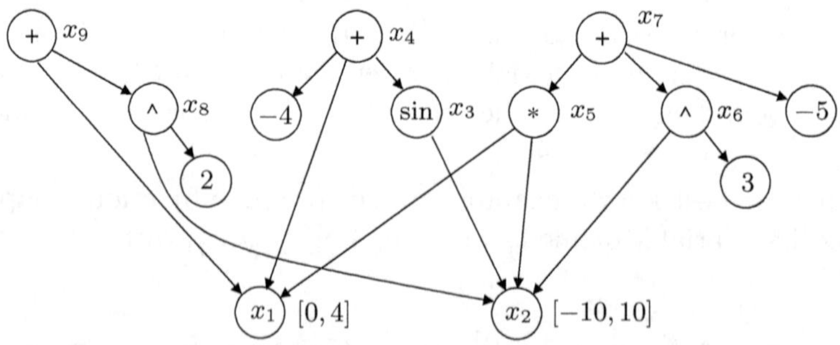

Figure 5.6. Association between auxiliary variables and the nodes of the DAG related to the problem in (5.9).

If a solution $\hat{x}$ is found with $f(\hat{x}) = 10$, then an upper bound on the objective function $x_9 := x_1 + x_8$ and the lower bound on $x_1 \geq -4$ imply that $x_8 \leq 14 < 100$. This in turn is propagated to the expression $x_8 = x_2^2$, which implies that $-\sqrt{14} \leq x_2 \leq \sqrt{14}$, thus tightening x_2. No other variables are tightened because of the new cutoff. In the general case, this procedure *propagates* throughout the DAG of the problem and repeats while there are tightened bounds, terminating when no more bounds are reduced.

FBBT algorithms allow for fast implementation and are commonly used in problems even of very large size. However, they may exhibit convergence issues even at very small scale: consider the trivial problem $\min\{x_1 : x_1 = \alpha x_2, x_2 = \alpha x_1, x_1 \in [-1, 1]\}$, with $\alpha \in \mathbb{R} \setminus \{0, 1\}$. Although by inspection one can see that the only feasible solution $x = (0, 0)$ is also optimal, FBBT will not terminate in a finite number of steps. In fact, a first pass will tighten x_2 to $\left[-\frac{1}{\alpha}, \frac{1}{\alpha}\right]$; this will trigger a reduction of the bound interval of x_1 to $\left[-\frac{1}{\alpha^2}, \frac{1}{\alpha^2}\right]$, which in turn will propagate to yield new bounds $\left[-\frac{1}{\alpha^3}, \frac{1}{\alpha^3}\right]$ on x_2. This procedure does not terminate unless tolerances or iteration limits are imposed, and it does not achieve its *fixed point* in finite time. A linear optimization problem has been proposed for MINLP that achieves the fixed point of a FBBT algorithm applied to the linear relaxation of (1.1) (Belotti, Cafieri, Lee and Liberti 2010).

Optimality-based bound tightening. Solving problems (5.14) is impractical because of the non-convexity of their feasible set, which is the same feasible set of (1.1). A more practical approach considers the feasible set of a convex relaxation of (1.1), such as the one in (5.13):

$$\mathcal{F}(l, u) = \left\{ x \in \mathbb{R}^{n+q} \;\middle|\; \begin{array}{ll} a^k x_k + B^k x \geq d^k, & k = n+1, n+2, \ldots, n+q, \\ l_i \leq x_i \leq u_i, & i = 1, 2, \ldots, n+q, \\ x \in X. & \end{array} \right\}$$

Then the following are valid bounds on variable x_i:

$$l_i' = \min\{x_i : x \in \mathcal{F}(l, u), f(x) \leq \hat{z}\}, \quad u_i' = \max\{x_i : x \in \mathcal{F}(l, u), f(x) \leq \hat{z}\}.$$

Empirical evidence shows that this technique is effective at obtaining tight bounds (Belotti *et al.* 2009), but it requires solution of $2n$ linear programming problems. Thus, its use is limited to the root node or to nodes of small depth.

Probing and reduced-cost bound tightening. Consider the bounds $[l_i, u_i]$ on a variable x_i as defined in (5.13), and set the upper bound to a fictitious value $u'_i < u_i$, regardless of whether u'_i is valid. Then apply a bound-tightening procedure such as FBBT. If the procedure indicates that the tightened bound hyper-rectangle renders the problem infeasible or drives the lower bound l_{n+q} on x_{n+q} above the cutoff value, then we have proof that no optimal solution exists with $x_i \in [l_i, u'_i]$, and the bounds on x_i become $[u'_i, u_i]$.

The same procedure can be repeated by imposing a fictitious lower bound l'_i on x_i and verifying through bound tightening whether tightening x_i to $[l'_i, u_i]$ yields a problem that can be fathomed. Applying this procedure to all variables, possibly in a repeated fashion, can lead to massive reduction in bounds, but it is computationally expensive given that it requires multiple calls to other bound-tightening procedures. Probing is used in MILP (Savelsbergh 1994) and MINLP (Belotti *et al.* 2009, Tawarmalani and Sahinidis 2002) on binary, integer, and continuous variables.

Reduced-cost bound tightening (Ryoo and Sahinidis 1995), akin to the *reduced-cost fixing* technique used in MILP (Nemhauser and Wolsey 1988), uses the solution of an LP relaxation of (1.1) to infer new, and possibly tighter, bounds on the variables. Suppose that an optimal solution $\hat{x}$ of (5.13) has a variable x_i at its lower bound l_i. Suppose also that the optimal solution has an objective function value $z_{\text{LP}} = \hat{x}_{n+q}$ and that a cutoff is known for (1.1) with value $\check{z}$. If the reduced-cost ρ_i of x_i is positive, then increasing x_i by δ yields an increase in the objective function of $\rho_i\delta$. Then a valid upper bound on x_i is $u'_i = l_i + \frac{\check{z}-z_{\text{LP}}}{\rho_i}$, which constitutes a tightening if $u'_i < u_i$. Similarly, if for the optimal solution $x^\star$ we have $x_i = u_i$ and a negative reduced cost ρ_i, then a valid lower bound on x_i is $l'_i = u_i + \frac{\check{z}-z_{\text{LP}}}{\rho_i}$.

5.4. Relaxations of structured non-convex sets

The methods described in Sections 5.2 and 5.3 are broadly applicable. This approach can be used to relax any constraint containing a nonlinear function that can be factored into simpler primitive functions for which we have known relaxations, and then to refine this relaxation after spatial branching. When combined with relaxation and branching on integer variables, this leads to algorithms that can (theoretically) solve almost any MINLP with explicitly given nonlinear constraints. The drawback of this general approach is that the relaxation obtained may be weak compared with the

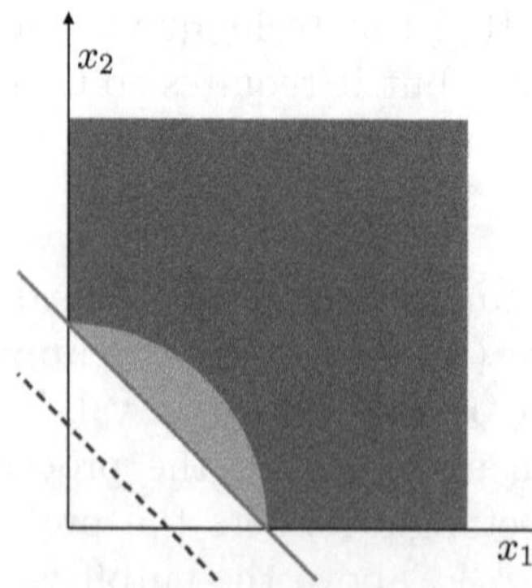

Figure 5.7. The dark shaded area is the feasible region. The convex hull is the entire shaded area, and is defined by the solid line, $x_1 + x_2 \geq 1$. The dashed line corresponds to the weaker inequality $x_1 + x_2 \geq 1/2$.

tightest possible relaxation, the convex hull of feasible solutions, leading to an impractically large branch-and-bound search tree.

As a simple example, consider the non-convex constraint in two variables x_1 and x_2,

$$x_1^2 + x_2^2 \geq 1, \tag{5.16}$$

and suppose $x_1, x_2 \in [0, 2]$. The set defined by these constraints is the dark shaded area in Figure 5.7. One can easily see that the convex hull of this set is given by the bounds on the variables, plus the inequality $x_1 + x_2 \geq 1$ (the solid line). Now consider the relaxation approach of Section 5.2. We first introduce two new decision variables, x_3 and x_4, with $x_3 \leq x_1^2$ and $x_4 \leq x_2^2$, and replace the constraint (5.16) with

$$x_3 + x_4 \geq 1. \tag{5.17}$$

The non-convex constraints $x_3 \leq x_1^2$ and $x_4 \leq x_2^2$ are then relaxed (in the best possible way given the bounds on x_1 and x_2) with $x_3 \leq 2x_1$ and $x_4 \leq 2x_2$. To compare this with the convex hull, we can then eliminate the variables x_3 and x_4 by substituting these inequalities into (5.17), obtaining $2x_1 + 2x_2 \geq x_3 + x_4 \geq 1$, or $x_1 + x_2 \geq 1/2$ (the bold dashed line in the figure). This relaxation is therefore significantly weaker than the convex hull.

Examples like those in the preceding paragraph motivate the study of relaxations that consider more of the problem *simultaneously*, rather than just separately relaxing all components. Of course, considering the entire MINLP feasible region is in general an intractable task. One common strategy, however, is to identify specific structures that may appear in many MINLP problems and study improved relaxations for these structures.

A huge variety of such structures exists, so we cannot provide a complete survey in this paper; but in the following subsections we highlight a couple of examples.

5.4.1. Non-convex quadratic functions

One structure that appears in many MINLP problems is the presence of (non-convex) quadratic or bilinear functions in either the constraints or the objective. These *quadratically constrained quadratic programs* (QCQPs) may also include integer variables and linear constraints. A generic QCQP is a special case of MINLP of the following form:

$$\begin{cases} \underset{x}{\text{minimize}} & x^T Q_0 x + c_0^T x, \\ \text{subject to} & x^T Q_k x + c_k^T x \le b_k, \quad k = 1, \dots, q, \\ & Ax \le b, \\ & 0 \le x \le u, \; x_i \in \mathbb{Z}, \; \forall i \in I, \end{cases} \tag{5.18}$$

where for each $k = 0, 1, \dots, q$, Q_k is an $n \times n$ symmetric matrix, and A is an $m \times n$ matrix. The matrices Q_k are not assumed to be positive semidefinite, so this problem is non-convex even when the integrality constraints are relaxed. It is also possible to have non-zero lower bounds $x \ge \ell$, but we assume here $x \ge 0$ to simplify exposition.

Many relaxation strategies for QCQPs are based on introducing additional variables X_{ij} for all i, j pairs, and reformulating (5.18) as follows:

$$\begin{cases} \underset{x}{\text{minimize}} & Q_0 \bullet X + c_0^T x, \\ \text{subject to} & Q_k \bullet X + c_k^T x \le b_k, \quad k = 1, \dots, q, \\ & Ax \le b, \\ & 0 \le x \le u, \; x_i \in \mathbb{Z}, \; \forall i \in I, \\ & X = xx^T, \end{cases} \tag{5.19}$$

where X is the $n \times n$ matrix containing all the X_{ij} variables, so that the constraint $X = xx^T$ records the non-convex constraints $X_{ij} = x_i x_j$ for all $i, j = 1, \dots, n$. Observe that these constraints are the only nonlinear constraints in this reformulation. Obtaining a relaxation of (5.19) can then be accomplished by relaxing the constraint $X = xx^T$. We discuss two general approaches for relaxing this constraint: the reformulation–linearization technique (RLT) (Adams and Sherali 1986, Sherali and Alameddine 1992, Sherali and Adams 1998) and semidefinite programming. Both approaches have been widely studied, and an exhaustive literature review is beyond the scope of this work. Instead, we introduce the basic idea of each as an example of how *structured* non-convex constraints can be relaxed.

In its most basic form, RLT relaxes the constraint $X_{ij} = x_i x_j$ for any fixed i, j by first deriving the following nonlinear non-convex constraints, based on multiplying pairs of the non-negative quantities $x_i, x_j, u_i - x_i$, and $u_j - x_j$:

$$x_i x_j \geq 0, \quad (u_i - x_i)(u_j - x_j) \geq 0, \quad x_i(u_j - x_j) \geq 0, \quad (u_i - x_i)x_j \geq 0.$$

These inequalities are then *linearized* by replacing the products $x_i x_j$ with the variable X_{ij}, yielding

$$X_{ij} \geq 0, \quad X_{ij} \geq u_i x_j + u_j x_i - u_i u_j, \quad X_{ij} \leq u_j x_i, \quad X_{ij} \leq u_i x_j.$$

Observe that these inequalities are exactly the special case of the inequalities (5.11) where the lower bounds on the variables being multiplied are 0. Using these inequalities in place of $X = xx^T$ yields a polyhedral relaxation of (5.19). Two other techniques are commonly used to further strengthen the RLT relaxation. First, if a decision variable x_i, for $i \in I$, is binary (*i.e.*, $u_i = 1$), then it holds that $x_i^2 = x_i$, and hence the linear constraint $X_{ii} = x_i$ is added to the relaxation. A generalization of this technique to general integer variables, based on Lagrange interpolating polynomials, has been proposed as well (Adams 2011). The second major technique for further improving the RLT relaxation is to multiply linear constraints together to obtain additional quadratic constraints that can then be linearized. For example, multiplying a non-negative decision variable x_i with a linear constraint $b_t - \sum_{j=1}^n a_{tj} x_j \geq 0$ yields the inequality

$$b_t x_i - \sum_{j=1}^{n} a_{tj} x_i x_j \geq 0,$$

which can then be linearized as

$$b_t x_i - \sum_{j=1}^{n} a_{tj} X_{ij} \geq 0.$$

Similar inequalities can be derived by multiplying linear constraints with the non-negative terms $(u_i - x_i)$, and also by multiplying linear constraints with each other, although deriving inequalities from all possible pairs of linear inequalities may yield a very large linear program.

The other general technique for relaxing the constraint $X = xx^T$ in (5.19) is via semidefinite programming. The key observation is that this constraint $X - xx^T$ can be relaxed to the constraint $X - xx^T \succeq 0$, which is equivalent to

$$\begin{bmatrix} 1 & x^T \\ x & X \end{bmatrix} \succeq 0,$$

and hence this yields a semidefinite programming relaxation. Just as in RLT, the corresponding relaxation can be improved by including the constraint $X_{ii} = x_i$ for binary variables x_i. When the QCQP contains linear *equations*, such as $a_t x = b_t$, additional linear constraints for the SDP relaxation can be obtained by 'squaring' the constraints and then linearizing the variables (Anstreicher 2009):

$$a_t^T X a_t = b_t^2.$$

SDP relaxations can be used within a branch-and-bound algorithm to solve QCQPs to optimality (see Burer and Vandenbussche 2009 and Buchheim and Wiegele 2013).

Anstreicher (2009) compares the relaxations obtained using the RLT and SDP approaches, finding that neither strictly dominates the other, and that combining the two approaches can yield a relaxation significantly better than obtained by using either approach individually. Additional linear inequalities related to those obtained for the Boolean Quadric Polytope (Padberg 1989) can also be added to further strengthen the relaxation (Yajima and Fujie 1998, Burer and Letchford 2009). Anstreicher (2012) demonstrated that the resulting relaxations can be exceptionally tight, very often yielding a bound equal to the optimal objective value. Unfortunately, the resulting relaxations, although convex, may be computationally demanding. Consequently, linear cuts have been introduced by Sherali and Fraticelli (2002) and further refined by Qualizza, Belotti and Margot (2012) to provide a polyhedral approximation of the SDP constraint.

For both the RLT and SDP relaxation approaches, improved relaxations can also be obtained by further multiplying linear constraints with each other, obtaining higher-order polynomial constraints. Additional variables can then be introduced, as in (5.19), to linearize these constraints, and then the constraints defining these variables (such as $X_{i,j,k} = x_i x_j x_k$) can be relaxed using an approach similar to that used for (5.19). This approach leads to a hierarchy of relaxations of improving quality (Adams and Sherali 2005, Lasserre 2000, Lasserre 2001). The drawback of this approach is that the size of these formulations grows dramatically, limiting their current practical use. Indeed, even the formulation (5.19) may be significantly larger than the original formulation (5.18), leading Saxena, Bonami and Lee (2011) to study relaxations of QCQPs that do not use the additional variables X_{ij}.

For MINLPs with only a quadratic objective (*i.e.*, problem (5.18) with $q = 0$), Burer (2009) derived another important link with conic optimization. In particular, he showed that an *exact* reformulation of such a problem can be obtained using a reformulation similar to (5.19) with certain enhancements (such as including the constraints $X_{ii} = x_i$ for binary variables) and replacing the constraint $X = xx^T$ with the conic constraint

$$\begin{bmatrix} 1 & x^T \\ x & X \end{bmatrix} \in \mathcal{C},$$

where $\mathcal{C}$ is the *completely positive cone*, the set of matrices Y such that $Y = BB^T$ for some matrix B that is componentwise non-negative.

We close this section by mentioning a few other approaches for solving or relaxing QCQPs. Saxena, Bonami and Lee (2010) study techniques for deriving disjunctive cuts based on the non-convex constraint $xx^T - X \succeq 0$, which is also implied by the constraint $X = xx^T$. Vandenbussche and Nemhauser (2005*b*, 2005*a*) studied polyhedral relaxations of box-constrained QPs, in which a general quadratic function is to be minimized subject to bound constraints on the continuous decision variables. Linderoth (2005) studied a simplicial branch-and-bound algorithm for QCQP. Burer and Letchford (2013) study relaxations of QCQPs with unbounded integer decision variables. The general α-BB relaxation approach (Androulakis *et al.* 1995) has also been used for solving QCQPs, although Anstreicher (2012) showed that the bounds obtained from SDP relaxation are always at least as good as those obtained from the α-BB approach. Misener and Floudas (2012) study the additional use of piecewise linear and edge-concave relaxations of QCQPs. Bao, Sahinidis and Tawarmalani (2009) and Luedtke, Namazifar and Linderoth (2012) consider related approaches for relaxing *multilinear* functions in which the decision variables are bounded, which can also be applied to QCQPs.

5.4.2. Bilinear covering sets

Tawarmalani, Richard and Chung (2010) study a framework for generating valid inequalities for MINLPs that have a certain 'orthogonal disjunction' structure. We do not review this general work but instead highlight the results they obtain by applying this framework to sets they refer to as *bilinear covering sets.*

First consider a pure integer covering set defined by

$$B^{\mathrm{I}} := \left\{ (x, y) \in \mathbb{Z}_+^n \times \mathbb{Z}_+^n \mid \sum_{i=1}^n x_i y_i \geq r \right\},$$

where r is a positive number. Note that, for any i, the convex hull of the two variable integer set

$$B_i^{\mathrm{I}} := \{(x_i, y_i) \in \mathbb{Z}_+ \times \mathbb{Z}_+ \mid x_i y_i \geq r\}$$

is a polyhedron defined by $d \leq \lceil r \rceil + 1$ linear inequalities. Let us denote the inequalities defining the convex hull of B_i^{I} by

$$a^k x_i + b^k y_i \geq 1, \quad k = 1, \ldots, d, \tag{5.20}$$

where we can assume (by scaling) that each inequality has a right-hand side 1. In particular, these inequalities include the constraints $x_i \geq 1$ and $y_i \geq 1$. The remaining inequalities can be computed, for example, by finding all inequalities of the form $ax_i + by_i \geq 1$ that do not cut off any of the points $(x_i^t, y_i^t) = (t, \lceil r/t \rceil)$ for $t = 1, \ldots, \lceil r \rceil$ and that are exactly satisfied by two of these points.

Now, let Π be the collection of all possible mappings of the form $\pi : \{1,\dots,n\} \to \{1,\dots,d\}$. That is, if $\pi \in \Pi$, then for each $i \in 1,\dots,n$, $\pi(i)$ selects an inequality in the description of $\text{conv}(B_i^{\text{I}})$. Then, $\text{conv}(B^{\text{I}})$ is characterized as follows.

Theorem 5.2 (Tawarmalani *et al.* 2010, Proposition 6). The convex hull of B^{I} is given by the set of $x \in \mathbb{R}_+^n, y \in \mathbb{R}_+^n$ that satisfy the inequalities

$$\sum_{i=1}^{n}(a^{\pi(i)}x_i + b^{\pi(i)}y_i) \geq 1, \quad \text{for all } \pi \in \Pi. \tag{5.21}$$

While there are an exponential number of inequalities in (5.21), given a point $(\hat{x}, \hat{y}) \in \mathbb{R}_+^n \times \mathbb{R}_+^n$, separation can be accomplished efficiently by independently considering each $\hat{x}_i, \hat{y}_i$ pair and setting $\pi(i)$ to the index of the most violated constraint in (5.20). Tawarmalani *et al.* (2010) provide a very similar result for the case of a bilinear covering set similar to B^{I}, but where one of the sets of variables is continuous.

We now turn to the case of a continuous bilinear covering set of the form

$$B^{\text{C}} := \left\{(x,y) \in \mathbb{R}_+^n \times \mathbb{R}_+^n \mid \sum_{i=1}^{n}(a_i x_i y_i + b_i x_i + c_i y_i) \geq r\right\},$$

where $r > 0$ and $a_i, b_i, c_i > 0$ for $i = 1,\dots,n$.

Theorem 5.3 (Tawarmalani *et al.* 2010, Proposition 9). The convex hull of B^{C} is given by the set of $x \in \mathbb{R}_+^n, y \in \mathbb{R}_+^n$ that satisfy the inequality

$$\frac{1}{2}\sum_{i=1}^{n}\left(b_i x_i + c_i y_i + \sqrt{(b_i x_i + c_i y_i)^2 + 4a_i r x_i y_i}\right) \geq r.$$

Unfortunately, when the bounds on the variables are also considered (*e.g.*, if x is a set of binary variables), the corresponding covering sets become much more difficult to analyse. In particular, Chung, Richard and Tawarmalani (2011) show that optimizing a linear function over such a set is $\mathcal{NP}$-hard. Chung *et al.* (2011) did, however, study valid inequalities for this set that may be useful in strengthening the continuous relaxation.

6. Heuristics for solving MINLPs

Some real-world applications cannot be solved to global optimality by using the methods described in Sections 3–5, because the problems are too large, generate a huge search tree, or must be solved in real time. In these situations it is more desirable to obtain a good solution quickly than to wait for an optimal solution. In such situations, we may resort to heuristic search techniques that provide a feasible point without any optimality guarantees. Heuristics can also accelerate deterministic techniques by quickly identifying an incumbent with a low value of the objective function. This upper bound can then be used to prune a larger number of the nodes in the branch-and-bound algorithm of Section 3.1 and in the branch-and-cut algorithm of Section 3.1.4. An incumbent solution may additionally be used for 'guided dives' (Danna, Rothberg and LePape 2005), that is, selecting a child node during a dive. Bound tightening, described in Section 5.3.2, can also be applied to the objective function more effectively if a tight upper bound is known.

We distinguish two classes of heuristic search techniques: probabilistic search and deterministic search. Probabilistic search refers to techniques that require at each iteration a random choice of a candidate solution or parameters that determine a solution. Simulated annealing (Kirkpatrick, Gelatt and Vecchi 1983), ant colony optimization (Dorigo, Maniezzo and Colorni 1996), particle-swarm optimization (Kennedy and Eberhart 1995), cross-entropy (Rubinstein and Kroese 2004), tabu search (Glover 1989, Glover 1990) and genetic algorithms (Goldberg 1989) are some methods that fall into this category. Although simple to design and applicable to many combinatorial optimization problems, these methods require implementation and modifications specific to the structure of the problem being solved. We therefore focus only on more general approaches.

We use the term 'deterministic' rather loosely, since the methods in this category may also sometimes need randomization in certain iterations. To begin with, all deterministic techniques discussed in Sections 3–5 can be run as heuristics. For example, we can run branch-and-bound for a fixed time or fixed number of nodes or until it finds its first incumbent. In this section, we discuss more efficient alternatives to such simple heuristics. These heuristics can be classified into two types: search heuristics, which search for a solution without the help of any known solutions, and improvement heuristics, which improve upon a given solution or a set of solutions.

Notation. Throughout this section we use a unified notation to refer to incumbents and solutions of the various steps of the heuristics: x^* refers to the current incumbent, which is feasible in (1.1); x' refers to a (local) solution of the continuous relaxation (1.3); $x^\diamond$ denotes the solution to a polyhedral relaxation of (1.1); and $x^{(j)}$ is a (local) solution of an NLP with fixed integer variables, $(\text{NLP}(x_I^{(j)}))$. Given this notation, we can now describe the deterministic search techniques within a unified terminology.

6.1. Search heuristics

Several heuristics to search for a feasible solution of an MINLP have been proposed recently. They all make clever use of LP, MILP, and NLP solvers to solve problems easier than the MINLP to obtain a feasible point. Some of these heuristics may completely ignore the objective function and focus on finding only a feasible solution. They may use the solution of the relaxation at any node in the branch-and-bound as a starting point and hence try to make up for the lack of focus on the objective function of the MINLP.

6.1.1. MILP-based rounding

Finding a locally optimal solution to the continuous relaxation (1.3) of the MINLP (1.1) is usually easier and computationally faster than solving the MINLP itself. Given a solution x' of the continuous relaxation, one can try rounding fractional values of integer-constrained variables. Unfortunately, such a simple rounding will usually not produce a feasible solution. Nannicini and Belotti (2012) propose overcoming this difficulty by solving an MILP. The constraints of the MILP are linear relaxations of the original MINLP obtained by the methods described in Section 5.2. The objective function is the ℓ_1 norm $\|x - x'\|_1$. The solution $x^\diamond$ of the MILP satisfies the integrality constraints but not necessarily the nonlinear constraints. Another NLP is now solved, this time with all integer variables fixed to the values in $x^\diamond_I$. The process is repeated until we obtain a feasible solution or reach termination criteria (limits on time or iterations).

Nannicini and Belotti (2012) suggest several practical measures for implementing the above general scheme. First, it is not necessary to fully solve the MILP. We can stop as soon as a feasible point of the MILP is found. Second, different initial points x' can be used to initialize the heuristic. In particular, if we are solving the NLP with an interior point method, we can stop the NLP if it finds a feasible point even when the log-barrier coefficient is not close to zero. Third, no-good cuts are added to ensure that the MILP does not include integer solutions ($x^\diamond$) found in previous iterations of the heuristic. A no-good cut is used to model the constraint

$$\sum_{i \in I} |x_i - x^\diamond_i| \geq 1. \tag{6.1}$$

When all integer variables are binary, the constraint (6.1) is simplified to

$$\sum_{i \in I : x^\diamond_i = 0} x_i + \sum_{i \in I : x^\diamond_i = 1} (1 - x_i) \geq 1. \tag{6.2}$$

Auxiliary binary variables may need to be introduced when some variables are general integers instead of binary.

6.1.2. Feasibility pump

The feasibility pump heuristic was introduced by Fischetti, Glover and Lodi (2005) in the context of MILP, and improved by Achterberg and Berthold (2007) and Fischetti and Salvagnin (2009). Bonami, Cornuéjols, Lodi and Margot (2009) have extended this idea to MINLPs. The main idea, like that in MILP-based rounding described above, is that an NLP solver can be used to find a solution that satisfies nonlinear constraints. Integrality is enforced by solving an MILP. An alternating sequence of NLP and MILP is solved that may lead to a solution feasible for the MINLP. The main difference from the rounding approach of Section 6.1.1 is the way MILP is set up. Suppose x' is a locally optimal solution of the NLP relaxation of the MINLP (1.1). Bonami *et al.* (2009) obtain the MILP using the linearization (3.4) used for outer approximation. Thus, if we have a convex MINLP, the MILP is a relaxation. Otherwise, it is only an approximation. The objective function is again the ℓ_1 norm $\|x - x'\|_1$.

If the solution $x^\diamond$ to the MILP satisfies the nonlinear MINLP constraints, $c(x^\diamond) \leq 0$, then we have found a new incumbent. Otherwise, we solve an NLP with all the integer variables fixed to the values of $x^\diamond$. The objective of this NLP is $\|x - x^\diamond\|_2$. Now, x' is updated to the solution of this NLP. If x' does not satisfy integrality constraints, new linearizations are added to the previous MILP.

For a convex MINLP, the more restricted MILP does not contain $x^\diamond$, and hence the no-good cuts are not required. In addition, Bonami *et al.* (2009) add the valid inequality

$$(x' - x^\diamond)(x - x') \geq 0$$

to the MILP. They also prove that in the case of a convex MINLP the heuristic does not cycle, and always terminates in a finite number of iterations. They call their version of the feasibility pump for convex MINLPs the 'enhanced feasibility pump'. A simpler version of the feasibility pump (Bonami and Gonçalves 2012) does not involve solving an MILP. Instead one just rounds x' to the nearest point satisfying integrality constraints. This version requires random changes in x' whenever cycling occurs, similar to the work of Fischetti *et al.* (2005).

D'Ambrosio, Frangioni, Liberti and Lodi (2012) note that feasibility-pump-based heuristics can be viewed as applications of the sequential projection method (SPM) or the alternating projection methods. The SPM has been used extensively for solving convex feasibility problems. We refer to the survey by Bauschke and Borwein (1996) for theory and algorithms. The feasibility of MINLPs is not a convex problem, and so the heuristics along the lines of these algorithms cannot be expected to have a similar performance or even to converge. D'Ambrosio *et al.* (2012) consider two sets related to the feasible region of the MINLP (1.1):

$$A = \{x \mid c(x) \leq 0,\ x \in X\}, \quad \text{and} \tag{6.3}$$
$$B = \{x \mid c_C(x) \leq 0,\ x \in X,\ x_i \in \mathbb{Z},\ \forall\, i \in I\}, \tag{6.4}$$

where c_C refers to the convex constraints in the MINLP. Set A is in general non-convex, while B is a set of feasible points of a convex MINLP. One can now solve optimization problems over A and B repeatedly in order to obtain

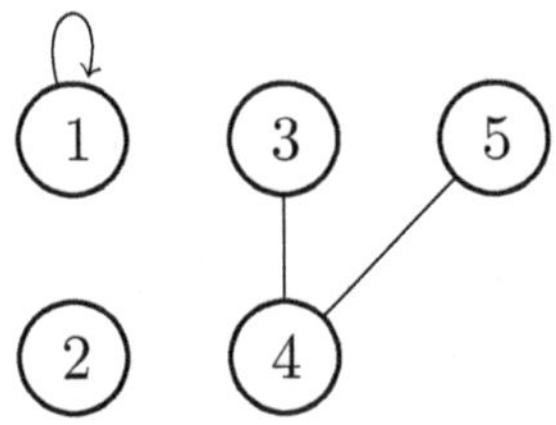

$$\begin{cases} \text{minimize}_x & x_1^3 + x_2 + x_3 x_4 + x_5, \\ \text{subject to} & x_4 x_5, \\ & x_1, x_2, x_3, x_4, x_5 \in [0, 10]. \end{cases}$$

(a) (b)

Figure 6.1. (a) Graph denoting the non-zero structure of the Hessian of the example in (b).

two sequences of solutions $\bar{x}_i$ and $\hat{x}_i$ as follows:

$$\bar{x}_i = \min_{x \in A} \|x - \hat{x}_{i-1}\|,$$
$$\hat{x}_i = \min_{x \in B} \|x - \bar{x}_{i-1}\|.$$

Both the above problems are NP-hard, and D'Ambrosio *et al.* (2012) suggest solving them using well-known heuristics. For instance, a locally optimal solution of the first problem can be obtained by most NLP solvers. The latter problem can be solved as a convex MINLP by methods described in Section 3. By selecting different methods to solve these problems, one can obtain several variants of the feasibility pump.

6.1.3. Undercover

The 'undercover' heuristic (Berthold and Gleixner 2012) is specially designed for non-convex MINLPs. The basic idea is to fix certain variables in the problem to specific values so that the resulting restriction becomes easier to solve. The restriction that Berthold and Gleixner (2012) obtain is an MILP. This MILP can then be solved either exactly or heuristically. Since the MILP is a restriction of the MINLP, any feasible solution of MILP will also satisfy the MINLP.

In order to be successful, the heuristic should fix a minimal number of variables lest the reduction be too restrictive and good solutions be cut off. The sparsity pattern of the Hessian of the Lagrangian tells us which variables appear in nonlinear functions. Berthold and Gleixner create a graph $G(V, E)$ where each vertex $v_i \in V, i = 1, \ldots, n$ denotes a variable of the MINLP. An edge e_{ij} is added to the graph if the Hessian of the Lagrangian of the NLP relaxation has a non-zero entry (i, j). G may also contain loops if a diagonal entry in the Hessian is non-zero. We illustrate a graph for a simple example in Figure 6.1. To make the problem linear, we can fix the variables x_1, x_3, and x_5. However, this choice is not minimal because we can also fix x_1 and x_4 to obtain a linear problem. Berthold and

Gleixner (2012) observe that a minimal vertex-cover of G is also a minimal set of variables that can be fixed to make the problem linear. They solve the minimal vertex-cover problem using an MILP solver that is usually available in an MINLP framework.

The fixed values of variables in a cover are obtained from a solution of the NLP relaxation or an LP relaxation. The variables are fixed sequentially. Each time a variable is fixed, domain propagation is invoked to tighten bounds or to fix other variables. This heuristic works particularly well for problems with quadratic constraints and objective function.

6.1.4. RENS: relaxation enforced neighbourhood search

The RENS (Berthold 2012) heuristic searches for a feasible solution of an MINLP around a point that does not satisfy integrality constraints. Consider an NLP solution x' and suppose $F = \{i \in I \mid x'_i \notin \mathbb{Z}\}$ is the set of all variables that violate integrality constraints at x'. Keeping the values of the variables in the set $I \setminus F$ fixed to those of x', we have $2^{|F|}$ ways of rounding up or down x' so that it satisfies integrality constraints. The RENS heuristic seeks to systematically search through these $2^{|F|}$ combinations to find a feasible solution of the MINLP.

Berthold (2012) creates a possibly much smaller MINLP in order to search for the feasible solution. First, he fixes all variables in the set I that have an integer value in x'. Next, the bounds of each variable $x_i, i \in F$ are changed to $[\lfloor x'_i \rfloor, \lceil x'_i \rceil]$. The resulting MINLP is then solved by using an MINLP solver. If the restricted MINLP is considerably smaller than the original, then the solver can solve it quickly. Additional limits need to be imposed on the solver so that the heuristic does not take excessive time. By means of this heuristic, the authors show that more than half of the test-instances have solutions that could be obtained by rounding.

6.1.5. Diving

The fundamental idea of all diving heuristics is to conduct a depth-first exploration of a possible path from the root node to a leaf node of the branch-and-bound tree, before searching other branches. The hope is that this will lead to a feasible solution and hence an upper bound early in the MINLP solution process.

Bonami and Gonçalves (2012) propose starting the diving process by solving the NLP relaxation of the MINLP to obtain a relaxed solution $x' \in \mathbb{R}^n$. They then fix $x'_i \neq \mathbb{Z}$, $i \in I$ to $\lfloor x'_i \rfloor$ or $\lceil x'_i \rceil$ and re-solve the modified NLP. This process is iterated until all integer variables have been fixed. This heuristic is successful if the obtained leaf NLP is feasible for the MINLP. It reports a failure if the obtained leaf NLP is infeasible, the objective function exceeds the incumbent bound, or an NLP solver termination criterion is met such as a limit on the number of iterations.

Selection of the variable to be fixed and the side to which it is fixed leaves room for some tailoring of diving heuristics. Bonami and Gonçalves (2012) describe *fractional diving*, *vector length diving*, and a modification of these heuristics we refer to as *nonlinear diving*:

- In *fractional diving*, the variable to be rounded is selected from the set of smallest values $|x'_j - [x_j]|$, where the bracket $[\cdot]$ indicates rounding to the nearest integer. The selected variable is fixed to the nearest integer.
- In *vector length diving*, the variable is selected from the set of smallest ratios

$$\frac{(\lceil x'_j \rceil - x'_j)g_j + \varepsilon}{A_j + 1} \text{ if } g_j \geq 0, \quad \text{and} \quad \frac{(\lfloor x'_j \rfloor - x'_j)g_j + \varepsilon}{A_j + 1} \text{ otherwise,}$$

  for $j \in I$. Here, $g_j = \frac{\partial f(x')}{\partial x_j}$, the constant A_j indicates the number of problem functions for which x_j, $j \in I$ has a non-zero coefficient in the linearization, and ε is chosen to be a small positive constant, for example $\varepsilon = 10^{-6}$. The selected variable is rounded up if the gradient with respect to x_j is non-negative, and it is rounded down otherwise. The selection favours rounding of a variable that incurs a small objective function change but affects a large number of problem constraints.

The nonlinear diving heuristic may be applied to either of the above criteria. Here, x_i, $i \in I$ is selected from the subset of nonlinear fractional variables only, with the aim of obtaining an MILP that can then be solved by a (black-box) MILP solver. In a leaf node where all nonlinear integer variables have been fixed and at least one linear integer variable is still fractional, this MILP is obtained by fixing all continuous nonlinear variables to their current value.

Diving heuristics need to take into account the computational effort required for repeated re-solves of the modified NLP. To mitigate this cost, Bonami and Gonçalves (2012) propose to fix $K > 1$ variables at the same time before resolving the modified NLP. Mahajan *et al.* (2012) propose solving QPs instead of NLPs, which speeds up the diving process by exploiting the warm-starting capabilities of active set QP solvers.

6.2. Improvement heuristics

Improvement heuristics start with a given feasible point x^* of the MINLP and try to find a better point. Two well-known heuristics for searching a better solution in the neighbourhood of a known solution have been adapted from MILP to MINLP. We describe them next.

6.2.1. Local branching

Local branching is a heuristic for MINLPs where all integer variables are binary, that is, $x_i \in \{0,1\}$ for all $i \in I$. It was first introduced in the context of MILP by Fischetti and Lodi (2003) and generalizes readily to convex MINLPs. We start by describing local branching for convex MINLPs and then describe an extension to non-convex MINLPs.

The main idea behind local branching is to use a generic MILP solver at a tactical level that is controlled at a strategic level by a simple external branching framework. Assume that we are given a feasible incumbent x^* of (1.1), and consider the following disjunction (generalized branching) for a fixed constant $k \in \mathbb{Z}$:

$$\|x_I - x_I^*\|_1 \le k \text{ (left branch)} \quad \text{or} \quad \|x_I - x_I^*\|_1 \ge k+1 \text{ (right branch)}. \tag{6.5}$$

This disjunction corresponds to the Hamming distance of x_I from x_I^*, and the left branch can also be interpreted as an ℓ_1 trust region around the incumbent. In the case of binary variables, we can rewrite (6.5) as two linear constraints:

$$\begin{aligned} &\sum_{i \in I: x_i^* = 0} x_i + \sum_{i \in I: x_i^* = 1} (1 - x_i) \le k \ \text{(left)} \quad \text{or} \\ &\sum_{i \in I: x_i^* = 0} x_i + \sum_{i \in I: x_i^* = 1} (1 - x_i) \ge k + 1 \ \text{(right)}. \end{aligned} \tag{6.6}$$

The left branch is constructed in such a way that it is much easier to solve than (1.1), typically by choosing $k \in [10, 20]$. We start by solving the left branch using any of the methods introduced in Section 3 and obtain a new incumbent. We can then either solve the right branch or again divide the right branch using a new disjunction (6.6). This creates an outer branch-and-bound tree where each node corresponds to an MINLP. If this local branching tree has been searched to completion, we have solved the MINLP. In general, however, we do not run local branching to completion, because it would be inefficient to regenerate pseudocosts for every MINLP solve, for example. Fischetti and Lodi (2003) propose two enhancements: first they impose a time limit or node limit on each MINLP solve, and second they introduce a diversification mechanism in case the left branch does not improve the solution. If the left branch is not solved completely, one can still obtain a complete search algorithm by modifying the local branching strategy.

Local branching has been extended to non-convex MINLPs (Nannicini, Belotti and Liberti 2008). The extension is based on solving an alternating sequence of (local) NLP relaxations and fixings and (global) MILP outer approximations, and is closely related to the feasibility pump described in Section 6.1.2. The local branching is used only as a no-good cut (6.1) and is not viewed as a strategic (outer) branching technique. A related heuristic is RECIPE (Liberti, Mladenović and Nannicini 2011).

6.2.2. RINS: relaxation-induced neighbourhood search
In the RINS (Danna *et al.* 2005) heuristic, one searches for better solutions in the neighbourhood of an already known solution, much like local branching. However, instead of imposing a distance constraint (6.5) to determine a neighbourhood, variables are fixed to certain values. The variables to be fixed are selected on the basis of the solution of the relaxation x', and the already known incumbent x^*. For all $i \in I$ the variables are fixed to x_i^*, $x_i' = x_i^*$. If the fixing reduces the problem size considerably, then it can be solved by calling the solver again.

Bonami and Gonçalves (2012) extend this idea from MILP to the NLP-based branch-and-bound algorithm of Section 3.1 for convex MINLP. Once they fix the integer variables as above, they solve the smaller MINLP using the LP/NLP-BB algorithm mentioned in Section 3.3.1. They show that the LP/NLP-BB algorithm is much faster on the smaller problems than is the NLP-based branch-and-bound algorithm.

7. Mixed-integer optimal control problems

In this section, we are interested in mixed-integer nonlinear optimization problems constrained by a system of ordinary differential equations, also called mixed-integer optimal control problems (MIOCPs). In principle, these problems can be discretized in time and solved by using any of the techniques of Sections 3–5. Unfortunately, such a discretization destroys the structure of the problem and does not provide successful algorithms. We present an alternative partial outer convexification approach that exploits problem structures and provides an efficient approach to this class of problems.

7.1. Mixed-integer optimal control problem class

We start by defining a class of MIOCPs:

$$\begin{cases} \underset{x(\cdot),u(\cdot),v(\cdot)}{\text{minimize}} & J(x(T)), \\ \text{subject to} & \dot{x}(t) = f_{\text{ODE}}(x(t),u(t),v(t)), \quad \forall t \in [0,T], \\ & x(0) = x_0, \\ & u(t) \in \mathcal{U}, \quad \forall t \in [0,T], \\ & 0 \le c(x(t),u(t)), \quad \forall t \in [0,T], \\ & v(t) \in \Omega, \quad \forall t \in [0,T], \end{cases} \tag{MIOCP}$$

where we strive to determine differential states

$$x : [0,T] \to \mathbb{R}^{n_x},$$

continuous controls

$$u : [0, T] \to \mathcal{U} \subset \mathbb{R}^{n_\mathrm{u}},$$

and integer controls

$$v : [0, T] \to \Omega \subset \mathbb{R}^{n_\mathrm{v}}$$

on the time horizon $[0, T]$ such that a Mayer term $J : \mathbb{R}^{n_\mathrm{x}} \to \mathbb{R}$ evaluated at the end of this horizon is minimized subject to a nonlinear system

$$f_{\mathrm{ODE}} : \mathbb{R}^{n_\mathrm{x}} \times \mathbb{R}^{n_\mathrm{u}} \times \Omega \to \mathbb{R}^{n_\mathrm{x}}$$

of ordinary differential equations (ODEs) with fixed initial values $x_0 \in \mathbb{R}^{n_\mathrm{x}}$. In addition, we consider nonlinear path constraints

$$c : \mathbb{R}^{n_\mathrm{x}} \times \mathcal{U} \to \mathbb{R}^{n_\mathrm{c}}$$

independent of the integer control. The set $\mathcal{U}$ describes a feasible domain for the continuous control $u(\cdot)$, and the set $\Omega := \{v^1, v^2, \ldots, v^{n_\omega}\}$ contains $n_\omega < \infty$ discrete choices $v^i \in \mathbb{R}^{n_\mathrm{v}}$, $i \in I := \{1, \ldots, n_\omega\}$ admissible for the integer control.

Mixed-integer optimal control problems can be discretized in time by using, for example, direct collocation or direct shooting discretizations and can then be solved by any of the MINLP algorithms in Sections 3–5. The discretized MINLPs, however, are typically challenging. Strong nonlinear and transient phenomena inherent in the dynamic model $f_{\mathrm{ODE}}(\cdot)$ may require a fine-grained time discretization, which leads to a large number of integer variables in the discretized MINLP. While this issue is typically successfully addressed by error estimation and adaptivity, incorporating adaptive discretizations into the MINLP model is difficult. Moreover, simply relaxing the integrality in (MIOCP) to $v(t) \in \operatorname{conv} \Omega$ yields a lower bound on the optimal solution that can be obtained by solving a continuous optimal control problem. Unfortunately, this relaxation in general does not give any guarantees on integer feasibility or on the distance to an integer optimal solution. In addition, the ODE constraint poses a particular challenge since the dynamics need not remain well-defined on the continuous relaxation.

To address these issues, we review an NLP-MILP decomposition approach for mixed-integer optimal control, as developed by Sager (2005), Sager, Reinelt and Bock (2009) and Sager, Bock and Diehl (2012). The fundamental insight here is that it is indeed sufficient to compute a relaxed optimal solution for a special convexified reformulation of the mixed-integer optimal control problem. With a direct method for continuous optimal control, this essentially amounts to solving an NLP. From this, we obtain a lower bound on the MIOCP solution, which, unlike in MINLP, can be shown to be tight. An integer feasible solution can be obtained constructively by applying a certain rounding scheme or by solving an MILP.

7.2. Partial outer convexification

To tighten the lower bound obtained from a relaxation and to overcome the problem of ill-defined dynamics, Sager (2005) proposes convexifying the problem with respect to the integer control $v(\cdot)$. In doing so, binary multiplier functions $\omega_i(\cdot) \in \{0, 1\}$ are introduced for each admissible choice of $v(t)$. We represent the integer control as a convex combination,

$$\forall t \in [0, T], \quad v(t) = v^i \;\Leftrightarrow\; \omega_i(t) = 1,\ \omega_j(t) = 0,\ \forall j \in I,\ j \neq i. \tag{7.1}$$

This gives rise to a partially convexified MIOCP:

$$\begin{cases} \underset{x(\cdot),u(\cdot),\omega(\cdot)}{\text{minimize}} & J(x(T)), \\ \text{subject to} & \dot{x}(t) = \sum_{i \in I} \omega_i(t) f_{\text{ODE}}(x(t), u(t), v^i), \quad \forall t \in [0, T], \\ & x(0) = x_0, \\ & u(t) \in \mathcal{U}, \hfill \forall t \in [0, T], \\ & 0 \leq c(x(t), u(t)), \hfill \forall t \in [0, T], \\ & 1 = \sum_{i \in I} \omega_i(t), \hfill \forall t \in [0, T], \\ & \omega(t) \in \{0, 1\}^{n_\omega}, \hfill \forall t \in [0, T]. \end{cases} \tag{MIOCP-BC}$$

This problem is convex only if f_{ODE} and c are convex in x and u. Clearly, the sets of optimal solutions of (MIOCP) and (MIOCP-BC) coincide. Given optimal binary multiplier functions $\omega(\cdot)$, the corresponding original integer control is easily determined. If $v(\cdot)$ is binary and enters linearly, then (MIOCP) is already in the convexified form. Otherwise the special-ordered set of type-1, SOS1, may be used. Introduced by Beale and Tomlin (1970) in the context of MILP, an SOS1 set can be used to model the condition (7.1) by means of the equations

$$\begin{aligned} v(t) &= \sum_{i \in I} v^i \omega_i(t), && \text{for all } t \in [0, T], \\ \sum_{i \in I} \omega_i(t) &= 1, && \text{for all } t \in [0, T]. \end{aligned} \tag{7.2}$$

We are interested in the relaxation of (MIOCP-BC) obtained by replacing $\omega_i(\cdot)$ by relaxed convex multiplier functions $\alpha_i(\cdot) \in [0, 1]$. This relaxation is a purely continuous optimal control problem, denoted by (MIOCP-RC). Direct and all-at-once methods have emerged as the methods of choice to solve it efficiently.

As an illustration, we consider an energy-optimal rocket-car control problem in its integer nonlinear formulation on the left and its partially convexified formulation on the right for $t \in [0, 32]$:

$$\left\{\begin{array}{lll} \underset{x(\cdot),v(\cdot)}{\text{minimize}} & x_2(32), & \\ \text{subject to} & \dot{x}_0(t) = x_1(t), & \forall t, \\ & \dot{x}_1(t) = v(t), & \forall t, \\ & \dot{x}_2(t) = v^2(t), & \forall t, \\ & x(0) = (0,0,0), & \\ & x_0(32) = 300, & \\ & v(t) \in \{-2, 1\}, & \forall t, \end{array}\right. \tag{7.3}$$

$$\Longleftrightarrow \left\{\begin{array}{lll} \underset{x(\cdot),\omega(\cdot)}{\text{minimize}} & x_2(32), & \\ \text{subject to} & \dot{x}_0(t) = x_1(t), & \forall t, \\ & \dot{x}_1(t) = -2\omega_1(t) + \omega_2(t), & \forall t, \\ & \dot{x}_2(t) = 4\omega_1(t) + \omega_2(t), & \forall t, \\ & x(0) = (0,0,0), & \\ & x_0(32) = 300, & \\ & \omega(t) \in \{0, 1\}^2, & \forall t, \\ & \omega_1(t) + \omega_2(t) = 1, & \forall t. \end{array}\right.$$

The multiplier function $\omega_1(t)$ lends itself to elimination by using the SOS1 constraint.

For our example, we use the multiple shooting code MUSCOD-II for direct mixed-integer optimal control (Bock and Plitt 1984, Leineweber *et al.* 2003, Sager 2005, Kirches and Leyffer 2011) available on NEOS (Czyzyk, Mesnier and Moré 1998, Dolan, Fourer, Moré and Munson 2002).

7.3. Relaxation of the partially convexified MIOCP

Optimal solutions to control problems permit a classification into arcs, or sections, of $[0, T]$ on which the solution has a particular structure: see, for example, Bryson and Ho (1975) for an introduction. Of interest for (MIOCP-RC) are arcs on which the optimal solution to the relaxed multipliers $\alpha(\cdot)$ comes to lie on the boundary of the feasible set and is hence feasible also for (MIOCP-BC). One example of such behaviour is *bang-bang* arcs. For particular problem classes, such as time-optimal linear control problems, a bang-bang principle proves the existence of an optimal solution that consists of bang-bang arcs only; see the work of Kirches *et al.* (2010) for a MIOCP example. Solving (MIOCP-RC) then already yields an optimal solution of (MIOCP). In example (7.3), (MIOCP-RC) has a bang-bang arc on $[0, 3.5]$ approximately: see Figure 7.1(e).

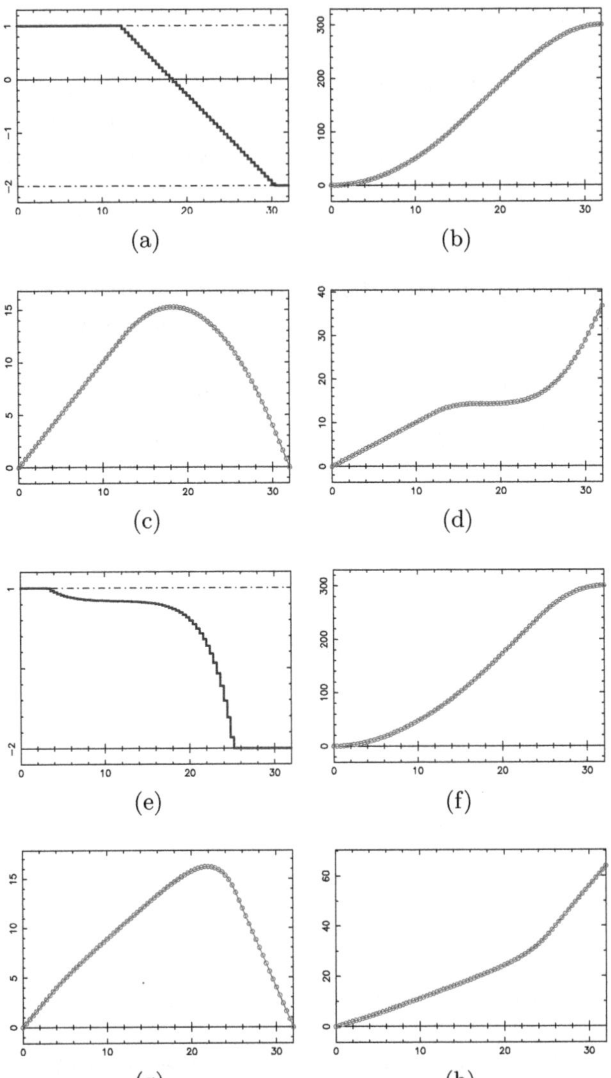

Figure 7.1. The nonlinear relaxed solution to the energy-optimal rocket-car example yields a weak lower bound $x_2(32) \approx 37$ to the MIOCP's optimal objective. The figure shows $v(\cdot) \in [-2, 1]$ relaxed (a), and partial outer convexification $v(\cdot) = -2\alpha_1(t) + 1\alpha_2(t)$ (e), as well as the resulting differential state trajectories $x_0(\cdot)$, $x_1(\cdot)$, $x_2(\cdot)$ (b–d, f–h) for an equidistant piecewise constant control discretization of $N = 80$ intervals.

Arcs on which $\alpha(\cdot)$ is not binary feasible require application of a rounding scheme to construct a binary feasible control $\omega(\cdot)$. We postpone discussion of such rounding schemes to the next section, and first consider the work of Sager *et al.* (2009, 2012), who investigate the difference between the state trajectories generated by relaxed multipliers $\alpha(\cdot)$ and binary ones $\omega(\cdot)$. We consider the ODE right-hand side $\tilde{f}(x,u)$ composed from columns $f_{\text{ODE}}(x,u,v^i)$, $i \in I$, and we fix a measurable control $u^*(\cdot)$.

Theorem 7.1 (Sager *et al.* 2012, Theorem 2 and Corollary 6). Let $x_\alpha(\cdot)$ and $x_\omega(\cdot)$ be the solution of the initial value problems

$$\begin{cases} \dot{x}_\alpha(t) = \tilde{f}(x_\alpha(t), u^*(t)) \cdot \alpha(t), & x_\alpha(0) = x_0, \\ \dot{x}_\omega(t) = \tilde{f}(x_\omega(t), u^*(t)) \cdot \omega(t), & x_\omega(0) = x_0, \end{cases} \tag{7.4}$$

with $t \in [0,T]$, $\alpha, \omega : [0,T] \to [0,1]^{n_\omega}$ measurable functions and $\tilde{f} : \mathbb{R}^{n_x+1} \to \mathbb{R}^{n_x} \times \mathbb{R}^{n_\omega}$ essentially bounded on $[0,T]$ by $M \in \mathbb{R}^+$ and t-differentiable almost everywhere. If there exist C, $L \in \mathbb{R}^+$ such that

$$\begin{aligned} \left\| \frac{\mathrm{d}\tilde{f}}{\mathrm{d}t}(x(t), u^*(t)) \right\| &\le C, \\ \|\tilde{f}(x_\alpha(t), u^*(t)) - \tilde{f}(x_\omega(t), u^*(t))\| &\le L\|x_\alpha(t) - x_\omega(t)\|, \end{aligned} \tag{7.5}$$

for $t \in [0,T]$ almost everywhere, and there exists $\varepsilon \in \mathbb{R}^+$ such that

$$\left\| \int_0^T \alpha(t) - \omega(t)\, \mathrm{d}t \right\| \le \varepsilon, \tag{7.6}$$

then for all $t \in [0,T]$ it holds that

$$\|x_\alpha(t) - x_\omega(t)\| \le (M + Ct)\varepsilon \exp(Lt). \tag{7.7}$$

This theorem differs markedly from the MINLP setting. Given a relaxed optimal solution $(x_\alpha(\cdot), u^*(\cdot), \alpha(\cdot))$, if one can construct a binary solution $\omega(\cdot)$ that satisfies assumption (7.6) on the control deviation bound ε, then the deviation of the resulting state trajectory $x_\omega(\cdot)$ from the relaxed optimal one $x_\alpha(\cdot)$ is bounded linearly in ε.

7.4. Constructing the integer control

A sum-up rounding scheme for construction of $\omega(\cdot)$ from a relaxed optimal solution $\alpha(\cdot)$ is shown to satisfy assumption (7.6) by Sager *et al.* (2012). We first present the case for an affine-linear binary control function $v(\cdot)$, when the SOS1 constraint is absent.

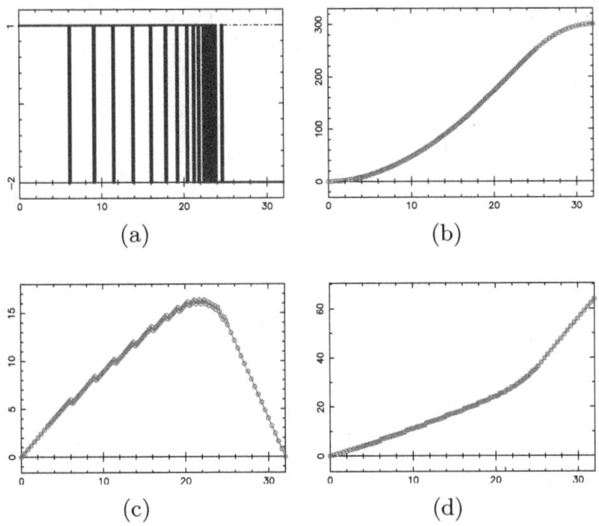

Figure 7.2. Integer feasible sum-up rounding solutions to the energy-optimal rocket-car example. The objective is $x_2(32) = 64$, and the lower bound provided by the convexified relaxed solution is exact in this case (*cf.* Figure 7.1(e–h)). The figure shows the integer control $v(\cdot)$ (a) and the resulting state trajectories $x_0(\cdot)$, $x_1(\cdot)$, $x_2(\cdot)$ (b–d).

Theorem 7.2 (Sager *et al.* 2012, Theorem 3). Let $\alpha\colon [0,T] \to [0,1]^{n_\omega}$ and a time grid $0 = t_0 < t_1 < \cdots < t_N = T$ be given, and define $\omega : [0,T] \to \{0,1\}^{n_\omega}$ by $\omega(t) := p_i$ for $t \in [t_i, t_{i+1})$, $0 \le i < N$, where

$$\begin{aligned} \hat{p}_{i,j} &:= \int_0^{t_i+1} \alpha_j(s)\,\mathrm{d}s - \sum_{k=0}^{i-1} p_{k,j}(t_{k+1} - t_k), \\ p_{i,j} &:= \begin{cases} 1 & \text{if } \hat{p}_{i,j} \ge \frac{1}{2}(t_{i+1} - t_i), \\ 0 & \text{else,} \end{cases} \qquad j \in I. \end{aligned} \tag{7.8}$$

Then it holds that

$$\left\| \int_0^T \alpha(t) - \omega(t)\,\mathrm{d}t \right\| \le \tfrac{1}{2} \max_{0 \le i < N} \{t_{i+1} - t_i\} =: \varepsilon. \tag{7.9}$$

After elimination of $\alpha_1(\cdot)$, example (7.3) is covered by this theorem. Figure 7.2 shows the sum-up rounding solution constructed from the convexified relaxed solution of Figure 7.1(e–h).

When the SOS1 constraint is present in (MIOCP-RC), $\omega(\cdot)$ constructed by the above rounding may violate it. Hence, a more elaborate rounding scheme is required.

Theorem 7.3 (Sager *et al.* 2012, Theorem 5). Let $\alpha\colon [0,T] \to [0,1]^{n_\omega}$ and a time grid $0 = t_0 < t_1 < \cdots < t_N = T$ be given, and define $\omega : [0,T] \to \{0,1\}^{n_\omega}$ by $\omega(t) := p_i$ for $t \in [t_i, t_{i+1})$, $0 \le i < N$, where

$$p_{i,j} := \begin{cases} 1 & \text{if } \forall k \neq j : \hat{p}_{i,j} \ge \hat{p}_{i,k} \text{ and } \forall k \neq j,\ \hat{p}_{i,j} = \hat{p}_{i,k} : j < k, \\ 0 & \text{else,} \end{cases} \qquad j \in I, \tag{7.10}$$

and the $\hat{p}_{i,j}$ are defined as above. Then it holds that

$$\left\| \int_0^T \alpha(t) - \omega(t)\, \mathrm{d}t \right\| \le \gamma(n^\omega) \max_{0 \le i < N} \{t_{i+1} - t_i\} =: \varepsilon. \tag{7.11}$$

For the constant $\gamma(n^\omega)$, the best-known result currently is $\gamma(n^\omega) := n^\omega - 1$ by Sager *et al.* (2012), who conjecture $\gamma(n^\omega) \in O(\log n^\omega)$. The practical implication of these theorems is that the bound ε of Theorem 7.1 can be made arbitrarily small by using a sufficiently fine time discretization, that is, N large enough to allow for $\max_{0 \le i < N}\{t_{i+1} - t_i\}$ to be sufficiently small. As a consequence of Theorems 7.1 and 7.2 or 7.3, the increase in the objective function value and the constraint infeasibility are bounded as well.

Theorem 7.4 (Sager *et al.* 2012, Corollary 8). Assume that J and c are continuous functions, and that the assumptions of Theorem 7.1 hold. Then for all $\delta > 0$ there exists $\Delta t := \max_{0 \le i < N}\{t_{i+1} - t_i\}$ sufficiently small that

$$|J(x_\alpha(T)) - J(x_\omega(T))| \le \delta, \tag{7.12}$$

$$|c_i(x_\alpha(t), u^*(t)) - c_i(x_\omega(t), u^*(t))| \le \delta, \quad 1 \le i \le n_c. \tag{7.13}$$

The binary control $\omega(\cdot)$ obtained from the constructive sum-up rounding process may switch frequently, and the number of switches may increase along with N and need not be bounded, as an example due to Fuller (1963) in Figure 7.3 shows. Sager, Jung and Kirches (2011) hence consider this rounding scheme as a special-case solution to the following switch cost MILP, in which $\sigma_{k,\max}$ is a fixed upper bound on the number of switches allowed in the integer control,

$$\begin{cases} \underset{p}{\text{minimize}}\ \underset{0 \le i < N, k \in I}{\text{maximize}} & \sum_{j=1}^{i} |(\alpha_{jk} - p_{jk})(t_{j+1} - t_j)|, \\ \text{subject to} & \sigma_{k,\max} \ge \sum_{i=0}^{N-1} |p_{ik} - p_{i+1,k}|, \qquad k \in I, \\ & p_i \in \{0,1\}^{n_\omega}, \qquad 0 \le i < N. \end{cases} \tag{7.14}$$

Here, $\omega(\cdot)$ is determined by the solution of this MILP, rather than by Theorem 7.2 or 7.3. If the switch cost constraint is inactive, these theorems give the linear-time solution to this MILP.

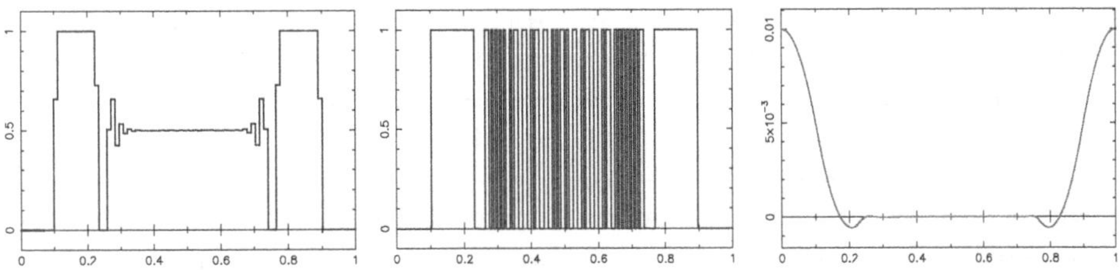

(a) (b) (c)

Figure 7.3. Fuller's problem (Fuller 1963) has an optimal solution that, in the limit $N \to \infty$, switches infinitely often on a finite time horizon. Relaxed optimal control (a), sum-up rounding control with optimized switching times (b), and optimal state trajectory (c) for $N = 80$.

Partial outer convexification, relaxation, and sum-up rounding can hence be seen as an NLP-MILP decomposition approach to solving a time-discretized MIOCP that enjoys ε-optimality and ε-feasibility certificates. The computational advantage of this approach over MINLP branch-and-bound algorithms for MIOCP can hence be arbitrarily large, as demonstrated by Kirches *et al.* (2010) for a benchmark problem due to Gerdts (2005).

7.5. *Extensions to the presented theory*

Problem class (MIOCP) is a special case of a larger class of mixed-integer optimal control problems also covered by the presented theory. An integral Lagrange term in the objective, time-dependent ODE systems, and free continuous model parameters constant in time are easily covered by introducing additional artificial ODE states. Free initial and final times subject to optimization can be realized by introducing two free parameters and an affine-linear transformation of the free horizon. Problem class (MIOCP) is also easily extended to allow for free initial values, for point constraints on the state $x(\cdot)$, including end-point constraints, and for constraints coupled in time, such as periodicity constraints.

Mixed-integer optimal control of systems described by differential-algebraic equations is discussed by Gerdts and Sager (2012). Because of time discretization, the switching times are determined up to the grid granularity only. Sager (2005) uses a switching-time optimization approach to refine the integer solution once the optimal switching structure has been determined and fixed. When (MIOCP) contains integer parameters subject to optimization, a combination of MIOCP techniques and MINLP techniques has to be used.

Constraints that depend on the integer control $v(t)$ are not immediately covered by (MIOCP) and the approximation theory based on Theorem 7.1, because violations of constraints after rounding cannot be ruled out. Kirches

Table 8.1. MINLP solvers (convex). A '—' means that we do not know the language the solver is written in. 'OS' stands for 'open source'.

Name	Algorithm(s)	Interfaces	Language	OS
α-ECP	extended cutting-plane	customized, GAMS	C	no
BONMIN	NLP-BB, LP/NLP-BB, OA, Hybrid	AMPL, C++, GAMS, Matlab	C++	yes
DICOPT	OA	GAMS	—	no
FilMINT	LP/NLP-BB	AMPL	C	no
KNITRO	NLP-BB, LP/NLP-BB	C, C++, Microsoft Excel, AMPL, AIMMS, GAMS, *etc.*	C, C++	no
MILANO	NLP-BB, OA	Matlab	Matlab	yes
MINLPBB	NLP-BB	AMPL, Fortran	Fortran-77	no
MINOPT	OA	customized	C	no
MINOTAUR	NLP-BB, QP-diving	AMPL, C++	C++	yes
SBB	NLP-BB	GAMS	—	no

(2011) discusses this case as well as extensions to mixed-integer nonlinear feedback control. An online benchmark library of mixed-integer optimal control problems is presented by Sager (2012).

8. Software for MINLP

The availability as well as the maturity of software for modelling and solving MINLP has increased significantly in the past fifteen years, and now includes a number of open-source and commercial solvers. We briefly survey the solvers currently available and describe their salient features. Recent surveys of Bussieck and Vigerske (2010) and D'Ambrosio and Lodi (2011) also provide an excellent description of available MINLP solvers. We divide the solvers into those for convex and non-convex MINLPs; little intersection exists between the two categories.

Key characteristics and features of the solvers are summarized in Tables 8.1 and 8.2, where we use the following abbreviations to indicate the type of MINLP method that the solver implements: NLP-BB for nonlinear branch-and-bound (Section 3.1); LP/NLP-BB for LP/NLP-based branch-and-bound (Section 3.3.1); OA for outer approximation (Section 3.2.1); Hybrid is a hybrid between OA and LP/NLP-BB; QP-diving (Section 3.1.3); α-BB for α-branch-and-bound (Section 5.3); LP-BB for LP-based branch-and-bound (Section 5.3).

Table 8.2. MINLP solvers (non-convex). A '—' means that we do not know the language the solver is written in. 'OS' stands for 'open source'.

Name	Algorithm(s)	Interfaces	Language	OS
α-BB	α-BB	customized	—	no
BARON	LP-BB	AIMMS, GAMS	—	no
COCONUT	LP-BB	AMPL, GAMS	C	yes
COUENNE	LP-BB	AMPL, C++, GAMS	C++	yes
GloMIQO	LP-BB	C++, GAMS	C++	no
LGO	sampling, heuristics	AIMMS, AMPL, GAMS, Mathematica	C	no
LindoGlobal	LP-BB	C++, GAMS, Lindo	—	no
SCIP	LP-BB	AMPL, C, GAMS, Matlab, OSiL, ZIMPL	C	yes

We are aware of two software packages, MUSCOD-II and MINOPT, that can solve MIOCPs. While MINOPT relies on MINLP techniques, MUSCOD-II relies on the partial outer convexification approach reviewed in Section 7. We briefly describe it in Section 8.3. Expressing and modelling MINLPs is significantly different from LPs or MILPs because general nonlinear functions are much more difficult to represent with data structures. Hence, good modelling tools are indispensable for MINLPs. In Section 8.4 we describe some tools available for modelling and reformulating MINLPs.

In Appendix A we list the websites of some of the software described below, along with other online resources.

8.1. Convex MINLP solvers

α-ECP is a solver written by Westerlund and Lundqvist (2005) to solve convex MINLPs by using the extended cutting-plane method (Section 3.2.3). It also implements methods for MINLPs with pseudoconvex functions (Westerlund and Pörn 2002). The solver can read MINLPs in an extended LP format and can be called through the GAMS (Brooke, Kendrick, Meeraus and Raman 1992) modelling system. It requires the user to specify an MILP solver for solving the MILP in each iteration. It also provides a graphical user interface for the Microsoft Windows operating system.

BONMIN stands for 'basic open-source nonlinear mixed-integer' optimizer. It is an open-source solver available from the COIN-OR website (Bonami *et al.* 2008). It uses nonlinear branch-and-bound (Section 3.1), LP/NLP-based branch-and-bound (Section 3.3.1), and outer approximation (Section 3.2.1) algorithms. It also implements a hybrid of outer approximation and LP/NLP-based branch-and-bound. It features several primal heuristics, including the feasibility pump, diving, and RINS. It uses the CBC solver to perform all the MILP operations, such as management of cuts and tree search. It can solve NLPs using IPOPT (Wächter and Biegler 2006) or Filter-SQP (Fletcher and Leyffer 1998). The source code and documentation are available on the project website.

DICOPT stands for 'discrete and continuous optimizer'. It implements a variant of outer approximation (Section 3.2.1), which has been generalized to tackle non-convex MINLPs through a penalty function heuristic; see Viswanathan and Grossmann (1990). The problem is input through the GAMS modelling system. The user can specify options to select both the MILP solver and the NLP solver at each iteration of the algorithm.

FilMINT (Abhishek *et al.* 2010) implements the LP/NLP-BB algorithm (Section 3.3.1) with several practical improvements. It exploits the presolve, cutting planes, searching rules, and other MILP tools of the MINTO solver (Nemhauser, Savelsbergh and Sigismondi 1994). Filter-SQP (Fletcher and Leyffer 1998) is used to solve NLPs. It also implements some of the disjunctive cuts described in Section 4.3 and the feasibility pump heuristic (Section 6.1.2).

KNITRO (Byrd, Nocedal and Richard 2006) was initially designed as an NLP solver. Two branch-and-bound-based algorithms (Sections 3.1, 3.3.1) were recently added for solving convex MINLPs (KNITRO 2012).

MILANO is a Matlab-based solver for convex MINLPs. It implements nonlinear branch-and-bound (Section 3.1) and outer approximation (Section 3.2.1). The source code of MILANO is available on the project website. The main focus of this code is to develop efficient warm-starting methods for interior-point methods (Benson 2011, 2012) so as to make them more effective at solving MINLPs.

MINLPBB (Leyffer 1998) is a Fortran-based nonlinear branch-and-bound solver (Section 3.1) for convex MINLPs. Filter-SQP (Fletcher and Leyffer 1998) is used to solve the NLP relaxations. It also has the ability to restart the NLP iterations from different remote points in order to ensure better solutions for non-convex MINLPs, and it provides options for choosing different branching rules and tree search strategies: see Sections 3.1.1 and 3.1.2.

MINOPT is a framework for both modelling and solving MINLPs and MIOCPs developed in 1998 (Schweiger 1999). It uses generalized Benders decomposition (Section 3.2.2) and outer approximation (Section 3.2.1). MINOPT requires linking with an NLP solver and an MILP solver, for which it has built-in routines for different solvers. A license for MINOPT can be obtained by contacting the authors. More information, a reference manual and examples are available on the project website.

MINOTAUR stands for 'mixed-integer nonlinear optimization toolkit: algorithms, underestimators and relaxations'. It is a new open-source toolkit for MINLPs. Currently it only implements nonlinear branch-and-bound (Section 3.1) and QP-diving (Section 3.1.3) for convex MINLPs. It has interfaces to NLP, QP and LP solvers. MINOTAUR has the ability to create and modify computational graphs of nonlinear functions. It can be used to reformulate nonlinear constraints and objective functions. The source code and documentation are available online.

MISQP is a solver designed for practical problems where the nonlinear functions cannot be evaluated when the variables $x_i, i \in I$ are not integers. This solver evaluates the functions and derivates at the integer points only. The algorithm does not guarantee an optimal solution even for convex MINLP. It generalizes the sequential–quadratic programming method with a trust region (Exler and Schittkowski 2007) to MINLPs. Exler, Lehmann and Schittkowski (2012) provide documentation of the solver along with examples. More information is available online.

SBB stands for 'simple branch-and-bound'. Implemented in GAMS, it allows the user to choose an NLP solver for the NLP-BB algorithm (Section 3.1). It can also handle SOS1 and SOS2 constraints (see (7.2) and (5.8)). A user manual is available online.

8.2. Non-convex MINLP solvers

Most modern MINLP solvers designed for non-convex problems utilize a combination of the techniques outlined in the previous sections; in particular, they are branch-and-bound algorithms with at least one rudimentary bound-tightening technique and a lower-bounding procedure. The technique for factorable functions described in Section 5.2.1 is most commonly used.

α-BB is a branch-and-bound solver that uses α-convexification (Adjiman, Androulakis and Floudas 1998) and its variants to obtain quadratic underestimators of non-convex functions in the constraints and objective. A number of special underestimators are available for specific commonly used functions. More information is available on the project website.

BARON is the acronym for 'branch and reduce optimization navigator' (Sahinidis 1996). It implements a branch-and-bound algorithm that computes a lower bound at each subproblem by means of a linear relaxation of (1.1), as discussed in Section 5.2. It includes various bound-tightening techniques (Section 5.3.2) such as probing and the *violation transfer* outlined in Section 5.3.1. It is available through the GAMS and AIMMS modelling systems. More information, examples, and documentation are available on the project website.

COCONUT is an open-source environment for global optimization problems (COCONUT 2004, Schichl 2004). Although it does not solve problems with integer variables, we include it in this section because it uses various techniques common to MINLP solvers: bound tightening (Section 5.3.2), reformulation (Section 5.2), and heuristics. The source code and documentation are available at the project homepage.

COUENNE, or 'convex over- and under-envelopes for nonlinear estimation' (Belotti 2009), is an open-source branch-and-bound algorithm that, similarly to BARON, obtains a lower bound through an LP relaxation using the reformulation technique outlined in Section 5.2.1. It also implements several bound-tightening procedures (Section 5.3.2) as well as a recently introduced feasibility-pump heuristic (Section 6.1.2), a separator of disjunctive cuts (Section 5.2.2), and different branching schemes including strong, pseudocost, and reliability branching (Section 5.3.1). Recently, it also introduced the linear cuts described by Qualizza *et al.* (2012) and briefly discussed in Section 5.4.1. The source code and documentation are available online.

GloMIQO is an evolution of α-BB with additional algorithms based on the work of Misener and Floudas (2013). It dramatically improves the lower-bounding procedure used originally in the α-BB method. It can solve only quadratically constrained quadratic problems. The solver is available via the GAMS modelling system. Related publications and more information are available on the project website.

LaGO is a branch-and-bound algorithm that is guaranteed to return the global optimum for mixed-integer quadratic problems (Nowak *et al.* 2003). It uses α-convexification (see Section 5.3) to obtain a lower bound on each subproblem. Though this technique is guaranteed only to solve quadratic problems to global optimality, LaGO can be used as a heuristic in other cases. The source code and documentation are available online.

LGO, or the 'Lipschitz (continuous) global optimizer', implements a set of heuristics and exact methods for global optimization. A constrained local optimization approach is used to obtain upper bounds on the objective value. Lower bounds are estimated through sampling. LGO assumes that the functions in the objective and the constraints of the problem are Lipschitz-continuous. Thus, it can find a global solution when the Lipschitz constants for all functions in the problem are known. One advantage of LGO is that it does not require gradients of the functions and hence can be applied to problems where the functions are not explicitly known: they could come from a black box or a simulation. A license for LGO may be purchased from the project website.

LindoGlobal is the MINLP solver of the LINGO modelling suite (Lin and Schrage 2009). It is in the same class as BARON and COUENNE in that it employs a lower-bounding technique for factorable functions, as described in Section 5.2.1. This solver may be purchased through LINDO Systems.

SCIP started as an MILP solver (Achterberg 2005) but evolved first into a solver for MINLPs with quadratic objective function and constraints (Berthold *et al.* 2010) and, more recently, into a solver for non-convex MINLP (Berthold *et al.* 2012). Following the basic approach of BARON and COUENNE, it implements a branch-and-bound algorithm (Section 5.2.1) with linear relaxation, various heuristics, and bound-tightening procedures. The source code and documentation are available on the project website.

8.3. An MIOCP solver

MUSCOD-II started as a reference implementation of direct multiple shooting, a direct and all-at-once method for ODE-constrained optimal control problems (Bock and Plitt 1984), and was later extended to DAE-constrained problems (Leineweber *et al.* 2003) and mixed-integer optimal control problems (Sager 2005). Kirches and Leyffer (2011) propose an extension for modelling MIOCPs in the symbolic modelling language AMPL and present an interface to MUSCOD-II. More information about this solver is available from its website on NEOS.

8.4. Modelling languages and online resources

Diverse modelling languages and online resources make it easy to specify and solve MINLP problems without needing to install software, code nonlinear functions, or derivatives. The wide availability of these tools means that MINLP has become accessible to the broader scientific and engineering community. In this section, we briefly summarize these tools.

Modelling languages enable scientists and engineers to express optimization problems in a more natural algebraic form that is close to a mathematical representation of the problem. Most modelling languages include automatic differentiation tools (Griewank 2000) that provide (exact) first and second derivatives of the problem functions and relieve the user of the error-prone tasks of coding derivatives. The most popular modelling languages are AIMMS (Bisschop and Entriken 1993), AMPL (Fourer, Gay and Kernighan 1993), GAMS (Brooke *et al.* 1992), MOSEL (Colombani and Heipcke 2002), TomLab (Holmström and Edvall 2004, Holmström, Göran and Edvall 2010) and YALMIP (Löfberg 2004). TomLab and YALMIP are built on top of Matlab, while the other systems are domain-specific languages that define a syntax for specifying optimization problems that can be parsed by the respective system to provide function and derivative information to the solver through a back-end. YALMIP can also solve small instances of convex and non-convex MINLPs using inbuilt algorithms. Recently, an open-source modelling system, 'Pyomo', has been developed (Hart, Watson and Woodruff 2011), which enables users to express MINLPs using the PYTHON scripting language. Opti Toolbox (Currie and Wilson 2012) is another open-source modelling tool. It enables users to express and solve MINLPs from within the Matlab environment. Currently, users can call BONMIN and SCIP solvers from this toolbox.

Online resources for optimization have grown dramatically over the past 15 years. There exist many libraries of test or benchmark problems in AMPL and GAMS (see Appendix A). Arguably the most important factor in making optimization solvers widely available has been the NEOS server (Czyzyk *et al.* 1998). Many of the solvers described above are now available on NEOS. NEOS provides a collection of state-of-the-art optimization software. Optimization problems are submitted through a web interface (or from within a modelling language session) and solved remotely. The MINLP solvers available on NEOS are α-ECP, BARON, BONMIN, COUENNE, DICOPT, FilMINT, LindoGlobal, MINLPBB, SBB, and SCIP.

A. Online resources

Convex and linear solvers

BONMIN	https://projects.coin-or.org/Bonmin
CBC solver	https://projects.coin-or.org/Cbc
KNITRO	http://www.ziena.com/knitro.htm
MILANO	http://www.pages.drexel.edu/~hvb22/milano
MINOPT	http://titan.princeton.edu/MINOPT
MINOTAUR	http://wiki.mcs.anl.gov/minotaur/
MISQP	http://www.ai7.uni-bayreuth.de/misqp.htm
SBB	http://www.gams.com/dd/docs/solvers/sbb.pdf

Non-convex solvers

α-BB	http://titan.princeton.edu/tools
BARON	http://archimedes.cheme.cmu.edu/?q=baron
COCONUT	http://www.mat.univie.ac.at/~coconut/coconut-environment
COUENNE	https://projects.coin-or.org/Couenne
GloMIQO	http://helios.princeton.edu/GloMIQO/publications.html
LaGO	https://projects.coin-or.org/LaGO
LGO	http://www.pinterconsulting.com
LindoGlobal	http://www.lindo.com
SCIP	http://scip.zib.de
MUSCOD-II	http://www.neos-server.org/neos/solvers/miocp:MUSCOD-II/AMPL.html

Online resources and model libraries

AMPL models	http://wiki.mcs.anl.gov/leyffer/index.php/MacMINLP http://minlp.org/
GAMS models	http://www.gamsworld.org/minlp/ http://minlp.org/
NEOS	http://www.neos-server.org/neos/

B. Optimization subproblems

Acknowledgements

This work was supported by the Office of Advanced Scientific Computing Research, Office of Science, US Department of Energy, under Contract DE-AC02-06CH11357, and by the US Department of Energy through grant DE-FG02-05ER25694. This work was also supported through NSF grant CCF-0830035.

REFERENCES[2]

K. Abhishek, S. Leyffer and J. T. Linderoth (2010), 'FilMINT: An outer-approximation-based solver for nonlinear mixed integer programs', *INFORMS J. Comput.* **22**, 555–567.

P. Abichandani, H. Y. Benson and M. Kam (2008), Multi-vehicle path coordination under communication constraints. In *American Control Conference*, IEEE Conference Publications, pp. 650–656.

M. A. Abramson (2004), 'Mixed variable optimization of a load-bearing thermal insulation system using a filter pattern search algorithm', *Optim. Engng* **5**, 157–177.

[2] The URLs cited in this work were correct at the time of going to press, but the publisher and the authors make no undertaking that the citations remain live or are accurate or appropriate.

M. Abramson, C. Audet, J. Chrissis and J. Walston (2009), 'Mesh adaptive direct search algorithms for mixed variable optimization', *Optim. Lett.* **3**, 35–47.

T. Achterberg (2005), SCIP: A framework to integrate constraint and mixed integer programming. ZIB-Report 04-19, Zuse Institut Berlin.

T. Achterberg and T. Berthold (2007), 'Improving the feasibility pump', *Discrete Optim.* **4**, 77–86.

T. Achterberg, T. Koch and A. Martin (2004), 'Branching rules revisited', *Oper. Res. Lett.* **33**, 42–54.

W. Adams (2011), Use of Lagrange interpolating polynomials in the RLT. In *Wiley Encyclopedia of Operations Research and Management Science*.

W. Adams and H. Sherali (1986), 'A tight linearization and an algorithm for zero-one quadratic programming problems', *Management Sci.* **32**, 1274–1290.

W. Adams and H. Sherali (2005), 'A hierarchy of relaxations leading to the convex hull representation for general discrete optimization problems', *Ann. Oper. Res.* **140**, 21–47.

C. S. Adjiman, I. Androulakis and C. Floudas (1998), 'A global optimization method, αBB, for general twice-differentiable constrained NLPs, II: Implementation and computational results', *Comput. Chem. Engng* **22**, 1159–1179.

I. Akrotirianakis, I. Maros and B. Rustem (2001), 'An outer approximation based branch-and-cut algorithm for convex 0–1 MINLP problems', *Optim. Methods Software* **16**, 21–47.

F. A. Al-Khayyal and J. E. Falk (1983), 'Jointly constrained biconvex programming', *Math. Oper. Res.* **8**, 273–286.

M. Altunay, S. Leyffer, J. T. Linderoth and Z. Xie (2011), 'Optimal security response to attacks on open science grids', *Computer Networks* **55**, 61–73.

E. D. Andersen and K. D. Andersen (1995), 'Presolving in linear programming', *Math. Program.* **71**, 221–245.

I. P. Androulakis, C. D. Maranas and C. A. Floudas (1995), 'αBB : A global optimization method for general constrained nonconvex problems', *J. Global Optim.* **7**, 337–363.

K. Anstreicher (2012), 'On convex relaxations for quadratically constrained quadratic programming', *Math. Program.* **136**, 233–251.

K. M. Anstreicher (2009), 'Semidefinite programming versus the reformulation–linearization technique for nonconvex quadratically constrained quadratic programming', *J. Global Optim.* **43**, 471–484.

A. Atamtürk and V. Narayanan (2010), 'Conic mixed-integer rounding cuts', *Math. Program. A* **122**, 1–20.

C. Audet and J. E. Dennis, Jr (2000), 'Pattern search algorithms for mixed variable programming', *SIAM J. Optim.* **11**, 573–594.

R. Bacher (1997), The Optimal Power Flow (OPF) and its solution by the interior point approach. EES-UETP Madrid, short course.

M. Baes, A. Del Pia, Y. Nesterov, S. Onn and R. Weismantel (2012), 'Minimizing Lipschitz-continuous strongly convex functions over integer points in polytopes', *Math. Program.* **134**, 305–322.

A. Balakrishnan and S. Graves (1989), 'A composite algorithm for a concave-cost network flow problem', *Networks* **19**, 175–202.

P. Balaprakash, S. M. Wild and P. D. Hovland (2011), 'Can search algorithms save large-scale automatic performance tuning?', *Procedia Comput. Sci.* (ICCS 2011) **4**, 2136–2145.

E. Balas, S. Ceria and G. Cornuéjols (1993), 'A lift-and-project cutting plane algorithm for mixed 0–1 programs', *Math. Program.* **58**, 295–324.

E. Balas, S. Ceria and G. Cornuéjols (1996), 'Mixed 0–1 programming by lift-and-project in a branch-and-cut framework', *Management Sci.* **42**, 1229–1246.

X. Bao, N. Sahinidis and M. Tawarmalani (2009), 'Multiterm polyhedral relaxations for nonconvex quadratically constrained quadratic programs', *Optim. Methods Software* **24**, 485–504.

S. Bartelt-Hunt, T. Culver, J. Smith, L. S. Matott and A. Rabideau (2006), 'Optimal design of a compacted soil liner containing sorptive amendments', *J. Environmental Engng* **132**, 769–776.

E. F. Bartholomew, R. P. O'Neill and M. C. Ferris (2008), 'Optimal transmission switching', *IEEE Trans. Power Systems* **23**, 1346–1355.

H. H. Bauschke and J. M. Borwein (1996), 'On projection algorithms for solving convex feasibility problems', *SIAM Rev.* **38**, 367–426.

E. Beale and J. Tomlin (1970), Special facilities in a general mathematical programming system for non-convex problems using ordered sets of variables. In *Proc. 5th International Conference on Operations Research* (J. Lawrence, ed.), Tavistock Publications, pp. 447–454.

E. M. L. Beale and J. J. H. Forrest (1976), 'Global optimization using special ordered sets', *Math. Program.* **10**, 52–69.

R. Bellman (1961), 'On the approximation of curves by line segments using dynamic programming', *Commun. Assoc. Comput. Mach.* **4**, 284.

P. Belotti (2009), COUENNE: A user's manual. Technical report, Lehigh University.

P. Belotti (2012), Disjunctive cuts for non-convex MINLP. In *Mixed Integer Nonlinear Programming*, Vol. 154 of *The IMA Volumes in Mathematics and its Applications*, Springer, pp. 117–144.

P. Belotti (2013), 'Bound reduction using pairs of linear inequalities', *J. Global Optim.*, to appear.

P. Belotti, S. Cafieri, J. Lee and L. Liberti (2010), Feasibility-based bounds tightening via fixed points. In *Combinatorial Optimization and Applications* (W. Wu and O. Daescu, eds), Vol. 6508 of *Lecture Notes in Computer Science*, Springer, pp. 65–76.

P. Belotti, J. Góez, I. Pólik, T. Ralphs and T. Terlaky (2012), A conic representation of the convex hull of disjunctive sets and conic cuts for integer second order cone optimization. Technical report 12T-009, Department of Industrial and Systems Engineering, Lehigh University.
http://www.optimization-online.org/DB_FILE/2012/06/3494.pdf

P. Belotti, J. Lee, L. Liberti, F. Margot and A. Wächter (2009), 'Branching and bounds tightening techniques for non-convex MINLP', *Optim. Methods Software* **24**, 597–634.

A. Ben-Tal and A. Nemirovski (1995), 'Optimal design of engineering structures', *Optima* **47**, 4–8.

A. Ben-Tal and A. Nemirovski (2001), 'On polyhedral approximations of the second-order cone', *Math. Oper. Res.* **26**, 193–205.

H. Y. Benson (2011), 'Mixed integer nonlinear programming using interior point methods', *Optim. Methods Software* **26**, 911–931.

H. Y. Benson (2012), Using interior-point methods within an outer approximation framework for mixed integer nonlinear programming. In *Mixed Integer Nonlinear Programming*, Vol. 154 of *The IMA Volumes in Mathematics and its Applications*, Springer, pp. 225–243.

T. Berthold (2012), RENS: The optimal rounding. ZIB-Report 12-17, Zuse Institut Berlin.

T. Berthold and A. M. Gleixner (2012), Undercover: A primal MINLP heuristic exploring a largest sub-MIP. ZIB-Report 12-07, Zuse Institut Berlin.

T. Berthold, G. Gamrath, A. Gleixner, S. Heinz, T. Koch and Y. Shinano (2012), Solving mixed integer linear and nonlinear problems using the SCIP optimization suite. ZIB-Report 12-27, Zuse Institut Berlin.

T. Berthold, A. Gleixner, S. Heinz and S. Vigerske (2010), Extending SCIP for solving MIQCPs. In *Proc. European Workshop on Mixed Integer Nonlinear Programming*, pp. 181–196.
http://www.lix.polytechnique.fr/~liberti/ewminlp/ewminlp-proceedings.pdf

D. Bertsekas and R. Gallager (1987), *Data Networks*, Prentice Hall.

R. Bhatia, A. Segall and G. Zussman (2006), 'Analysis of bandwidth allocation algorithms for wireless personal area networks', *Wireless Networks* **12**, 589–603.

D. Bienstock (1996), 'Computational study of a family of mixed-integer quadratic programming problems', *Math. Program.* **74**, 121–140.

D. Bienstock and S. Mattia (2007), 'Using mixed-integer programming to solve power grid blackout problems', *Discrete Optim.* **4**, 115–141.

V. M. Bier (2005), Game-theoretic and reliability methods in counterterrorism and security. In *Mathematical and Statistical Methods in Reliability, Series on Quality, Reliability and Engineering Statistics* (A. Wilson, N. Limnios, S. Keller-McNulty and Y. Armijo, eds), World Scientific, pp. 17–28.

V. M. Bier, A. Nagaraj and V. Abhichandani (2005), 'Protection of simple series and parallel systems with components of different values', *Reliability Engineering System Safety* **87**, 315–323.

V. M. Bier, S. Oliveros and L. Samuelson (2007), 'Choosing what to protect', *J. Public Economic Theory* **9**, 563–587.

J. Bisschop and R. Entriken (1993), *AIMMS: The Modeling System*, Paragon Decision Technology.

H. Bock and R. Longman (1985), 'Computation of optimal controls on disjoint control sets for minimum energy subway operation', *Adv. Astronaut. Sci.* **50**, 949–972.

H. Bock and K. Plitt (1984), A multiple shooting algorithm for direct solution of optimal control problems. In *Proc. 9th IFAC World Congress*, Pergamon Press, pp. 242–247.

P. Bonami (2011), Lift-and-project cuts for mixed integer convex programs. In *Integer Programming and Combinatorial Optimization* (O. Günlük and G. Woeginger, eds), Vol. 6655 of *Lecture Notes in Computer Science*, Springer, pp. 52–64.

P. Bonami and J. P. M. Gonçalves (2012), 'Heuristics for convex mixed integer nonlinear programs', *Comput. Optim. Appl.* **51**, 729–747.

P. Bonami, L. Biegler, A. Conn, G. Cornuéjols, I. Grossmann, C. Laird, J. Lee, A. Lodi, F. Margot, N. Sawaya and A. Wächter (2008), 'An algorithmic framework for convex mixed integer nonlinear programs', *Discrete Optim.* **5**, 186–204.

P. Bonami, G. Cornuéjols, A. Lodi and F. Margot (2009), 'A feasibility pump for mixed integer nonlinear programs', *Math. Program.* **119**, 331–352.

P. Bonami, M. Kılınç and J. T. Linderoth (2012), Algorithms and software for convex mixed integer nonlinear programs. In *Mixed Integer Nonlinear Programming*, Vol. 154 of *The IMA Volumes in Mathematics and its Applications*, Springer, pp. 61–92.

P. Bonami, J. Lee, S. Leyffer and A. Wächter (2011), More branch-and-bound experiments in convex nonlinear integer programming. Preprint ANL/MCS-P1949-0911, Mathematics and Computer Science Division, Argonne National Laboratory.

I. Bongartz, A. R. Conn, N. I. M. Gould and P. L. Toint (1995), 'CUTE: Constrained and unconstrained testing environment', *ACM Trans. Math. Software* **21**, 123–160.

R. Boorstyn and H. Frank (1977), 'Large-scale network topological optimization', *IEEE Trans. Communications* **25**, 29–47.

B. Borchers and J. E. Mitchell (1994), 'An improved branch and bound algorithm for mixed integer nonlinear programs', *Comput. Oper. Res.* **21**, 359–368.

S. Boyd and L. Vandenberghe (2004), *Convex Optimization*, Cambridge University Press.

C. Bragalli, C. D'Ambrosio, J. Lee, A. Lodi and P. Toth (2006), An MINLP solution method for a water network problem. In *Algorithms: ESA 2006, 14th Annual European Symposium*, Springer, pp. 696–707.

C. Bragalli, C. D'Ambrosio, J. Lee, A. Lodi and P. Toth (2012), 'On the optimal design of water distribution networks: A practical MINLP approach', *Optim. Engng* **13**, 219–246.

A. Brooke, D. Kendrick, A. Meeraus and R. Raman (1992), *GAMS: A User's Guide*, GAMS Development Corporation.

A. Bryson and Y.-C. Ho (1975), *Applied Optimal Control*, Wiley.

C. Buchheim and A. Wiegele (2013), 'Semidefinite relaxations for non-convex quadratic mixed-integer programming', *Math. Program.*, to appear.

S. Burer (2009), 'On the copositive representation of binary and continuous non-convex quadratic programs', *Math. Program.* **120**, 479–495.

S. Burer and A. Letchford (2009), 'On nonconvex quadratic programming with box constraints', *SIAM J. Optim.* **20**, 1073–89.

S. Burer and A. Letchford (2012), 'Non-convex mixed-integer nonlinear programming: A survey', *Surv. Oper. Res. Management Sci.* **17**, 97–106.

S. Burer and A. Letchford (2013), 'Unbounded convex sets for non-convex mixed-integer quadratic programming', *Math. Program.*, to appear.

S. Burer and D. Vandenbussche (2009), 'Globally solving box-constrained non-convex quadratic programs with semidefinite-based finite branch-and-bound', *Comput. Optim. Appl.* **43**, 181–195.

J. Burgschweiger, B. Gnädig and M. Steinbach (2008), 'Optimization models for operative planning in drinking water networks', *Optim. Engng* **10**, 43–73.

M. R. Bussieck and S. Vigerske (2010), MINLP solver software. In *Wiley Encyclopedia of Operations Research and Management Science* (J. J. Cochran, L. A. Cox, P. Keskinocak, J. P. Kharoufeh, P. Jeffrey and J. C. Smith, eds), Wiley.

R. H. Byrd, J. Nocedal and W. A. Richard (2006), KNITRO: An integrated package for nonlinear optimization. In *Large-Scale Nonlinear Optimization* (G. Pillo and M. Roma, eds), Vol. 83 of *Nonconvex Optimization and its Applications*, Springer, pp. 35–59.

S. Callegari, F. Bizzarri, R. Rovatti and G. Setti (2010), 'On the approximate solution of a class of large discrete quadratic programming problems by $\Delta\Sigma$ modulation: The case of circulant quadratic forms', *IEEE Trans. Signal Process.* **58**, 6126–6139.

I. Castillo, J. Westerlund, S. Emet and T. Westerlund (2005), 'Optimization of block layout design problems with unequal areas: A comparison of MILP and MINLP optimization methods', *Comput. Chem. Engng* **30**, 54–69.

M. Çezik and G. Iyengar (2005), 'Cuts for mixed 0–1 conic programming', *Math. Program. A* **104**, 179–202.

S. Ceria and J. Soares (1999), 'Convex programming for disjunctive optimization', *Math. Program.* **86**, 595–614.

K. Chi, X. Jiang, S. Horiguchi and M. Guo (2008), 'Topology design of network-coding-based multicast networks', *IEEE Trans. Mobile Comput.* **7**, 1–14.

K. Chung, J.-P. Richard and M. Tawarmalani (2011), Lifted inequalities for 0–1 mixed-integer bilinear covering sets.
http://www.optimization-online.org/DB_FILE/2011/03/2949.pdf

V. Chvátal (1973), 'Edmonds polytopes and a hierarchy of combinatorial problems', *Discrete Math.* **4**, 305–337.

COCONUT (2004), The COCONUT benchmark: A benchmark for global optimization and constraint satisfaction.
http://www.mat.univie.ac.at/~neum/glopt/coconut/benchmark.html

J. S. Cohen (2003), *Computer Algebra and Symbolic Computation: Elementary Algorithms*, Universities Press.

Y. Colombani and S. Heipcke (2002), Mosel: An extensible environment for modeling and programming solutions. In *Proc. Fourth International Workshop on Integration of AI and OR Techniques in Constraint Programming for Combinatorial Optimisation Problems: CP-AI-OR'02* (N. Jussien and F. Laburthe, eds), Ecole des Mines, Nantes, pp. 277–290.

E. Costa-Montenegro, F. J. González-Castaño, P. S. Rodriguez-Hernández and J. C. Burguillo-Rial (2007), Nonlinear optimization of IEEE 802.11 mesh networks. In *ICCS 2007, Part IV*, Springer, pp. 466–473.

K. Croxton, B. Gendron and T. Magnanti (2003), 'A comparison of mixed-integer programming models for nonconvex piecewise linear cost minimization problems', *Management Sci.* **49**, 1268–73.

J. Currie and D. I. Wilson (2012), OPTI: Lowering the barrier between open source optimizers and the industrial MATLAB user. In *Foundations of Computer-Aided Process Operations* (N. Sahinidis and J. Pinto, eds).
http://focapo.org

J. Czyzyk, M. Mesnier and J. Moré (1998), 'The NEOS server', *IEEE J. Comput. Sci. Engng* **5**, 68–75.

D. Dadush, S. Dey and J. P. Vielma (2011*a*), 'The split closure of a strictly convex body', *Oper. Res. Lett.* **39**, 121–126.

D. Dadush, S. S. Dey and J. P. Vielma (2011*b*), 'The Chvátal–Gomory closure of a strictly convex body', *Math. Oper. Res.* **36**, 227–239.

D. Dadush, S. S. Dey and J. P. Vielma (2011*c*), On the Chvátal–Gomory closure of a compact convex set. In *Integer Programming and Combinatorial Optimization* (O. Günlük and G. Woeginger, eds), Vol. 6655 of *Lecture Notes in Computer Science*, Springer, pp. 130–142.

D. Dadush, C. Peikert and S. Vempala (2011*d*), Enumerative lattice algorithms in any norm via M-ellipsoid coverings. In *IEEE 52nd Annual Symposium on Foundations of Computer Science: FOCS*, pp. 580–589.

R. J. Dakin (1965), 'A tree search algorithm for mixed programming problems', *Comput. J.* **8**, 250–255.

C. D'Ambrosio and A. Lodi (2011), 'Mixed integer nonlinear programming tools: A practical overview', *4OR* **9**, 329–349.

C. D'Ambrosio, A. Frangioni, L. Liberti and A. Lodi (2012), 'A storm of feasibility pumps for nonconvex MINLP', *Math. Program. B* **136**, 375–402.

C. D'Ambrosio, A. Lodi and S. Martello (2010), 'Piecewise linear approximation of functions of two variables in MILP models', *Oper. Res. Lett.* **38**, 39–46.

E. Danna, E. Rothberg and C. LePape (2005), 'Exploring relaxation induced neighborhoods to improve MIP solutions', *Math. Program.* **102**, 71–90.

G. B. Dantzig (1960), 'On the significance of solving linear programming problems with some integer variables', *Econometrica* **28**, 30–44.

G. B. Dantzig (1963), *Linear Programming and Extensions*, Princeton University Press.

E. Davis (1987), 'Constraint propagation with interval labels', *Artificial Intelligence* **32**, 281–331.

E. Davis and M. Ierapetritou (2009), 'A kriging based method for the solution of mixed-integer nonlinear programs containing black-box functions', *J. Global Optim.* **43**, 191–205.

J. A. De Loera, R. Hemmecke, M. Koppe and R. Weismantel (2006), 'Integer polynomial optimization in fixed dimension', *Math. Oper. Res.* **31**, 147–153.

S. S. Dey and D. A. Morán (2013), 'Some properties of convex hulls of integer points contained in general convex sets', *Math. Program.*, to appear.

S. S. Dey and J. P. Vielma (2010), The Chvátal–Gomory closure of an ellipsoid is a polyhedron. In *Integer Programming and Combinatorial Optimization*, Vol. 6080 of *Lecture Notes in Computer Science*, Springer, pp. 327–340.

E. Dolan and J. Moré (2002), 'Benchmarking optimization software with performance profiles', *Math. Program.* **91**, 201–213.

E. Dolan, R. Fourer, J. Moré and T. Munson (2002), 'Optimization on the NEOS server', *SIAM News* **35**, 8–9.

V. Donde, V. Lopez, B. Lesieutre, A. Pinar, C. Yang and J. Meza (2005), Identification of severe multiple contingencies in electric power networks, in *Proc. 37th North American Power Symposium*, IEEE.

M. Dorigo, V. Maniezzo and A. Colorni (1996), 'The ant system: Optimization by a colony of cooperating agents', *IEEE Trans. Systems, Man and Cybernetics B* **26**, 1–13.

S. Drewes (2009), Mixed integer second order cone programming. PhD thesis, Technische Universität Darmstadt.

S. Drewes and S. Ulbrich (2012), Subgradient based outer approximation for mixed integer second order cone programming. In *Mixed Integer Nonlinear Programming*, Vol. 154 of *The IMA Volumes in Mathematics and its Applications*, Springer, pp. 41–59.

M. A. Duran and I. Grossmann (1986), 'An outer-approximation algorithm for a class of mixed-integer nonlinear programs', *Math. Program.* **36**, 307–339.

J. Eckstein (1994), 'Parallel branch-and-bound algorithms for general mixed integer programming on the CM-5', *SIAM J. Optim.* **4**, 794–814.

G. Eiger, U. Shamir and A. Ben-Tal (1994), 'Optimal design of water distribution networks', *Water Resources Research* **30**, 2637–2646.

S. Elhedhli (2006), 'Service system design with immobile servers, stochastic demand, and congestion', *Manufacturing & Service Operations Management* **8**, 92–97.

A. M. Eliceche, S. M. Corvalán and P. Martínez (2007), 'Environmental life cycle impact as a tool for process optimisation of a utility plant', *Comput. Chem. Engng* **31**, 648–656.

A. Elwalid, D. Mitra and Q. Wang (2006), 'Distributed nonlinear integer optimization for data-optical internetworking', *IEEE J. Selected Areas in Communications* **24**, 1502–1513.

M. Engelhart, J. Funke and S. Sager (2013), 'A decomposition approach for a new test-scenario in complex problem solving', *J. Comput. Sci.*, in press.

O. Exler and K. Schittkowski (2007), 'A trust region SQP algorithm for mixed-integer nonlinear programming', *Optim. Lett.* **1**, 269–280.

O. Exler, T. Lehmann and K. Schittkowski (2012), MISQP: A Fortran subroutine of a trust region SQP algorithm for mixed-integer nonlinear programming, user's guide. Technical report, Department of Computer Science, University of Bayreuth.

FICO Xpress (2009), *FICO Xpress optimization suite: Xpress-BCL reference manual*, Fair Isaac Corporation.

M. Fischetti and A. Lodi (2003), 'Local branching', *Math. Program.* **98**, 23–47.

M. Fischetti and D. Salvagnin (2009), 'Feasibility pump 2.0', *Math. Program. Comput.* **1**, 201–222.

M. Fischetti, F. Glover and A. Lodi (2005), 'The feasibility pump', *Math. Program.* **104**, 91–104.

R. Fletcher (1987), *Practical Methods of Optimization*, Wiley.

R. Fletcher and S. Leyffer (1994), 'Solving mixed integer nonlinear programs by outer approximation', *Math. Program.* **66**, 327–349.

R. Fletcher and S. Leyffer (1998), User manual for filterSQP. University of Dundee Numerical Analysis Report NA-181.

R. Fletcher and S. Leyffer (2003), Filter-type algorithms for solving systems of algebraic equations and inequalities. In *High Performance Algorithms and*

Software for Nonlinear Optimization (G. di Pillo and A. Murli, eds), Kluwer, pp. 259–278.

A. Flores-Tlacuahuac and L. T. Biegler (2007), 'Simultaneous mixed-integer dynamic optimization for integrated design and control', *Comput. Chem. Engng* **31**, 648–656.

C. Floudas (1995), *Nonlinear and Mixed-Integer Optimization*, Topics in Chemical Engineering, Oxford University Press.

C. A. Floudas (2000), *Deterministic Global Optimization: Theory, Algorithms and Applications*, Kluwer.

R. Fourer, D. M. Gay and B. W. Kernighan (1993), *AMPL: A Modeling Language for Mathematical Programming*, The Scientific Press.

K. R. Fowler, J. P. Reese, C. E. Kees, J. E. Dennis, C. T. Kelley, C. T. Miller, C. Audet, A. J. Booker, G. Couture, R. W. Darwin, M. W. Farthing, D. E. Finkel, J. M. Gablonsky, G. A. Gray and T. G. Kolda (2008), 'A comparison of derivative-free optimization methods for water supply and hydraulic capture community problems', *Adv. Water Resources* **31**, 743–757.

A. Frangioni and C. Gentile (2006), 'Perspective cuts for a class of convex 0–1 mixed integer programs', *Math. Program.* **106**, 225–236.

A. Fügenschuh, M. Herty, A. Klar and A. Martin (2006), 'Combinatorial and continuous models for the optimization of traffic flows on networks', *SIAM J. Optim.* **16**, 1155–1176.

A. Fuller (1963), 'Study of an optimum nonlinear control system', *J. Electronics Control* **15**, 63–71.

L. L. Garver (1997), 'Transmission network estimation using linear programming', *IEEE Trans. Power Apparatus Systems* **89**, 1688–1697.

D. M. Gay (1991), Automatic differentiation of nonlinear AMPL models. In *Automatic Differentiation of Algorithms: Theory, Implementation, and Application* (A. Griewank and G. F. Corliss, eds), SIAM, pp. 61–73.

B. Geissler, A. Martin, A. Morsi and L. Schewe (2012), Using piecewise linear functions for solving MINLPs. In *Mixed Integer Nonlinear Programming* (J. Lee and S. Leyffer, eds), Vol. 154 of *The IMA Volumes in Mathematics and its Applications*, Springer, pp. 287–314.

I. Gentilini, F. Margot and K. Shimada (2013), 'The travelling salesman problem with neighbourhoods: MINLP solution', *Optim. Methods Software* **28**, 364–378,

A. M. Geoffrion (1972), 'Generalized Benders decomposition', *J. Optim. Theory Appl.* **10**, 237–260.

A. M. Geoffrion (1977), 'Objective function approximations in mathematical programming', *Math. Program.* **13**, 23–37.

M. Gerdts (2005), 'Solving mixed-integer optimal control problems by branch& bound: A case study from automobile test-driving with gear shift', *Optimal Control Appl. Methods* **26**, 1–18.

M. Gerdts and S. Sager (2012), Mixed-integer DAE optimal control problems: Necessary conditions and bounds. In *Control and Optimization with Differential-Algebraic Constraints* (L. Biegler, S. Campbell and V. Mehrmann, eds), SIAM, pp. 189–212.

F. Glover (1989), 'Tabu search, part I', *ORSA J. Comput.* **1**, 190–206.

F. Glover (1990), 'Tabu search, part II', *ORSA J. Comput.* **2**, 4–32.

D. E. Goldberg (1989), *Genetic Algorithms in Search, Optimization, and Machine Learning*, Addison-Wesley.

N. Goldberg, S. Leyffer and I. Safro (2012), Optimal response to epidemics and cyber attacks in networks. Preprint ANL/MCS-1992-0112, Mathematics and Computer Science Division, Argonne National Laboratory.

R. E. Gomory (1958), 'Outline of an algorithm for integer solutions to linear programs', *Bull. Amer. Math. Monthly* **64**, 275–278.

R. E. Gomory (1960), An algorithm for the mixed integer problem. Technical report RM-2597, The RAND Corporation.

N. I. M. Gould and S. Leyffer (2003), An introduction to algorithms for nonlinear optimization. In *Frontiers in Numerical Analysis* (J. Blowey, A. Craig and T. Shardlow, eds), Springer, pp. 109–197.

N. I. M. Gould, S. Leyffer and P. L. Toint (2004), 'A multidimensional filter algorithm for nonlinear equations and nonlinear least squares', *SIAM J. Optim.* **15**, 17–38.

J.-P. Goux and S. Leyffer (2003), 'Solving large MINLPs on computational grids', *Optim. Engng* **3**, 327–354.

A. Griewank (2000), *Evaluating Derivatives: Principles and Techniques of Algorithmic Differentiation*, Vol. 19 of *Frontiers in Applied Mathematics*, SIAM.

A. Griewank and P. L. Toint (1984), 'On the existence of convex decompositions of partially separable functions', *Math. Program.* **28**, 25–49.

I. Griva, S. G. Nash and A. Sofer (2009), *Linear and Nonlinear Optimization*, second edition, SIAM.

I. E. Grossmann (2002), 'Review of nonlinear mixed-integer and disjunctive programming techniques', *Optim. Engng* **3**, 227–252.

I. E. Grossmann and Z. Kravanja (1997), Mixed-integer nonlinear programming: A survey of algorithms and applications. In *Large-Scale Optimization with Applications, Part II: Optimal Design and Control* (L. T. Biegler, T. F. Coleman, A. R. Conn and F. N. Santosa, eds), Springer.

I. E. Grossmann and R. W. H. Sargent (1979), 'Optimal design of multipurpose batch plants', *Indust. Engng Chem. Process Design and Development* **18**, 343–348.

A. Guerra, A. M. Newman and S. Leyffer (2011), 'Concrete structure design using mixed-integer nonlinear programming with complementarity constraints', *SIAM J. Optim.* **21**, 833–863.

O. Günlük and J. T. Linderoth (2010), 'Perspective relaxation of mixed integer nonlinear programs with indicator variables', *Math. Program. B* **104**, 186–203.

O. Günlük and J. T. Linderoth (2012), Perspective reformulation and applications. In *Mixed Integer Nonlinear Programming*, Vol. 154 of *The IMA Volumes in Mathematics and its Applications*, Springer, pp. 61–92.

O. K. Gupta and A. Ravindran (1985), 'Branch and bound experiments in convex nonlinear integer programming', *Management Sci.* **31**, 1533–1546.

Gurobi (2012), *Gurobi Optimizer Reference Manual, Version 5.0*, Gurobi Optimization.

E. Hansen (1992), *Global Optimization Using Interval Analysis*, Marcel Dekker.

I. Harjunkoski, T. Westerlund, R. Pörn and H. Skrifvars (1998), 'Different transformations for solving non-convex trim loss problems by MINLP', *European J. Oper. Res.* **105**, 594–603.

W. E. Hart, J.-P. Watson and D. L. Woodruff (2011), 'Pyomo: modeling and solving mathematical programs in Python', *Math. Program. Comput.* **3**, 219–260.

K. W. Hedman, R. P. O'Neill, E. B. Fisher and S. S. Oren (2008), 'Optimal transmission switching: Sensitivity analysis and extensions', *IEEE Trans. Power Systems* **23**, 1469–1479.

S. Heinz (2005), 'Complexity of integer quasiconvex polynomial optimization', *J. Complexity* **21**, 543–556.

E. Hellström, M. Ivarsson, J. Aslund and L. Nielsen (2009), 'Look-ahead control for heavy trucks to minimize trip time and fuel consumption', *Control Engng Practice* **17**, 245–254.

T. Hemker (2008), Derivative free surrogate optimization for mixed-integer nonlinear black box problems in engineering. PhD thesis, Technischen Universität Darmstadt, Darmstadt, Germany.

T. Hemker, K. Fowler, M. Farthing and O. von Stryk (2008), 'A mixed-integer simulation-based optimization approach with surrogate functions in water resources management', *Optim. Engng* **9**, 341–360.

R. Hemmecke, S. Onn and R. Weismantel (2011), 'A polynomial oracle-time algorithm for convex integer minimization', *Math. Program.* **126**, 97–117.

H. Hijazi, P. Bonami and A. Ouorou (2010), An outer–inner approximation for separable MINLPs. Technical report, LIF, Faculté des Sciences de Luminy, Université de Marseille.

R. Hildebrand and M. Köppe (2013), 'A new Lenstra-type algorithm for quasiconvex polynomial integer minimization with complexity $2^{O(n \log n)}$', *Discrete Optim.* **10**, 69–84.

K. Holmström and M. Edvall (2004), The TOMLAB optimization environment. In *Modeling Languages in Mathematical Optimization* (J. Kallrath, ed.), Kluwer Academic, pp. 369–378. http://tomopt.com/tomlab/

K. Holmström, A. O. Göran and M. M. Edvall (2010), *User's Guide for TOMLAB 7*, Tomlab Optimization Inc.

H. Horst, P. M. Pardalos and V. Thoai (1995), *Introduction to Global Optimization*, Kluwer.

R. Horst and H. Tuy (1993), *Global Optimization*, Springer.

IBM Ilog CPLEX (2009), *IBM Ilog CPLEX V12.1: User's Manual for CPLEX*, IBM.

R. Jeroslow and J. Lowe (1984), 'Modelling with integer variables', *Math. Program. Studies* **22**, 167–84.

R. Jeroslow and J. Lowe (1985), 'Experimental results on the new techniques for integer programming formulations', *J. Oper. Res. Soc.* **36**, 393–403.

R. G. Jeroslow (1973), 'There cannot be any algorithm for integer programming with quadratic constraints', *Oper. Res.* **21**, 221–224.

N. J. Jobst, M. D. Horniman, C. A. Lucas and G. Mitra (2001), 'Computational aspects of alternative portfolio selection models in the presence of discrete asset choice constraints', *Quant. Finance* **1**, 489–501.

J. J. Júdice, H. D. Sherali, I. M. Ribeiro and A. M. Faustino (2006), 'A complementarity-based partitioning and disjunctive cut algorithm for mathematical programming problems with equilibrium constraints', *J. Global Optim.* **36**, 89–114.

R. Kannan and C. Monma (1978), On the computational complexity of integer programming problems. In *Optimization and Operations Research* (R. Henn, B. Korte and W. Oettli, eds), Vol. 157 of *Lecture Notes in Economics and Mathematical Systems*, Springer, pp. 161–172.

R. Karuppiah and I. E. Grossmann (2006), 'Global optimization for the synthesis of integrated water systems in chemical processes', *Comput. Chem. Engng* **30**, 650–673.

A. B. Keha, I. R. De Farias Jr and G. L. Nemhauser (2006), 'A branch-and-cut algorithm without binary variables for nonconvex piecewise linear optimization', *Oper. Res.* **54**, 847–858.

J. E. Kelley (1960), 'The cutting plane method for solving convex programs', *J. SIAM* **8**, 703712.

J. Kennedy and R. Eberhart (1995), Particle swarm optimization. In *IEEE International Conference on Neural Networks*, Vol. 4, pp. 1942–1948.

L. Khachiyan and L. Porkolab (2000), 'Integer optimization on convex semialgebraic sets', *Discrete Comput. Geom.* **23**, 207–224.

M. Kılınç (2011), Disjunctive cutting planes and algorithms for convex mixed integer nonlinear programming. PhD thesis, Department of Industrial and Systems Engineering, University of Wisconsin–Madison.

M. Kılınç, J. T. Linderoth and J. Luedtke (2010), Effective separation of disjunctive cuts for convex mixed integer nonlinear programs. Technical report 1681, Computer Sciences Department, University of Wisconsin–Madison.

C. Kirches (2011), Fast numerical methods for mixed-integer nonlinear model-predictive control. In *Advances in Numerical Mathematics* (H. Bock, W. Hackbusch, M. Luskin and R. Rannacher, eds), Springer Vieweg. PhD thesis, Ruprecht-Karls-Universität Heidelberg.

C. Kirches and S. Leyffer (2011), TACO: A toolkit for AMPL control optimization. Preprint ANL/MCS-P1948-0911, Mathematics and Computer Science Division, Argonne National Laboratory.

C. Kirches, S. Sager, H. Bock and J. Schlöder (2010), 'Time-optimal control of automobile test drives with gear shifts', *Optimal Control Appl. Methods* **31**, 137–153.

S. Kirkpatrick, C. D. Gelatt and M. P. Vecchi (1983), 'Optimization by simulated annealing', *Science* **220**, 671–680.

J. L. Klepeis and C. A. Floudas (2003), 'ASTRO-FOLD: A combinatorial and global optimization framework for *ab initio* prediction of three-dimensional structures of proteins from the amino acid sequence', *Biophysical J.* **85**, 2119–2146.

KNITRO (2012), *KNITRO Documentation.* Ziena Optimization.

G. R. Kocis and I. E. Grossmann (1988), 'Global optimization of nonconvex mixed-integer nonlinear programming (MINLP) problems in process synthesis', *Indust. Engng Chem. Research* **27**, 1407–1421.

P. A. Krokhmal and P. Soberanis (2010), 'Risk optimization with p-order conic constraints: A linear programming approach', *European J. Oper. Res.* **201**, 653–671.

S. Lakhera, U. V. Shanbhag and M. McInerney (2011), 'Approximating electrical distribution networks via mixed-integer nonlinear programming', *Internat. J. Electric Power and Energy Systems* **33**, 245–257.

A. H. Land and A. G. Doig (1960), 'An automatic method for solving discrete programming problems', *Econometrica* **28**, 497–520.

J. Lasserre (2000), Convergent LMI relaxations for nonconvex quadratic programs. In *Proc. 39th IEEE Conference on Decision and Control*, Vol. 5, IEEE, pp. 5041–5046.

J. Lasserre (2001), An explicit exact SDP relaxation for nonlinear 0–1 programs. In *Integer Programming and Combinatorial Optimization 2001* (K. Aardal and A. Gerards, eds), Vol. 2081 of *Lecture Notes in Computer Science*, Springer, pp. 293–303.

E. L. Lawler and D. E. Woods (1966), 'Branch-and-bound methods: A survey', *Oper. Res.* **14**, 699–719.

J. Lee and S. Leyffer, eds (2012), *Mixed Integer Nonlinear Programming*, Vol. 154 of *The IMA Volumes in Mathematics and its Applications*, Springer.

J. Lee and D. Wilson (2001), 'Polyhedral methods for piecewise-linear functions I: The lambda method', *Discrete Appl. Math.* **108**, 269–285.

D. Leineweber, I. Bauer, A. Schäfer, H. Bock and J. Schlöder (2003), 'An efficient multiple shooting based reduced SQP strategy for large-scale dynamic process optimization (Parts I and II)', *Comput. Chem. Engng* **27**, 157–174.

J. H. W. Lenstra (1983), 'Integer programming with a fixed number of variables', *Math. Oper. Res.* **8**, 538–548.

S. Leyffer (1998), User manual for MINLP-BB. University of Dundee.

S. Leyffer (2001), 'Integrating SQP and branch-and-bound for mixed integer nonlinear programming', *Comput. Optim. Appl.* **18**, 295–309.

S. Leyffer (2003), MacMINLP: Test problems for mixed integer nonlinear programming. http://www.mcs.anl.gov/~leyffer/macminlp

L. Liberti and C. C. Pantelides (2003), 'Convex envelopes of monomials of odd degree', *J. Global Optim.* **25**, 157–168.

L. Liberti, N. Mladenović and G. Nannicini (2011), 'A recipe for finding good solutions to MINLPs', *Math. Program. Comput.* **3**, 349–390.

Y. Lin and L. Schrage (2009), 'The global solver in the LINDO API', *Optim. Methods Software* **24**, 657–668.

J. T. Linderoth (2005), 'A simplicial branch-and-bound algorithm for solving quadratically constrained quadratic programs', *Math. Program. B* **103**, 251–282.

J. T. Linderoth and M. W. P. Savelsbergh (1999), 'A computational study of search strategies in mixed integer programming', *INFORMS J. Comput.* **11**, 173–187.

G. Liuzzi, S. Lucidi and F. Rinaldi (2012), 'Derivative-free methods for bound constrained mixed-integer optimization', *Comput. Optim. Appl.* **53**, 505–526.

J. Löfberg (2004), YALMIP: A toolbox for modeling and optimization in MATLAB. In *IEEE International Symposium on Computer Aided Control Systems Design*, pp. 284–289.

J. Luedtke, M. Namazifar and J. T. Linderoth (2012), 'Some results on the strength of relaxations of multilinear functions', *Math. Program.* **136**, 325–351.

A. Mahajan, S. Leyffer and C. Kirches (2012), Solving mixed-integer nonlinear programs by QP-diving. Preprint ANL/MCS-2071-0312, Mathematics and Computer Science Division, Argonne National Laboratory.

A. Mahajan, S. Leyffer, J. T. Linderoth, J. Luedtke and T. Munson (2011), MINOTAUR: A toolkit for solving mixed-integer nonlinear optimization. Wiki page. http://wiki.mcs.anl.gov/minotaur

L. Maonan and H. Wenjun (1991), The study of choosing optimal plan of air quantities regulation of mine ventilation network. In *Proc. 5th US Mine Ventilation Symposium*, pp. 427–421.

J. Maria, T. T. Truong, J. Yao, T.-W. Lee, R. G. Nuzzo, S. Leyffer, S. K. Gray and J. A. Rogers (2009), 'Optimization of 3D plasmonic crystal structures for refractive index sensing', *J. Phys. Chem. C* **113**, 10493–10499.

H. M. Markowitz and A. S. Manne (1957), 'On the solution of discrete programming problems', *Econometrica* **25**, 84–110.

A. Martin, M. Möller and S. Moritz (2006), 'Mixed integer models for the stationary case of gas network optimization', *Math. Program.* **105**, 563–582.

S. Masihabadi, S. Sanjeevi and K. Kianfar (2011), 'n-step conic mixed integer rounding inequalities', *Optimization Online.* http://www.optimization-online.org/DB_HTML/2011/11/3251.html

G. P. McCormick (1976), 'Computability of global solutions to factorable nonconvex programs I: Convex underestimating problems', *Math. Program.* **10**, 147–175.

F. Messine (2004), 'Deterministic global optimization using interval constraint propagation techniques', *RAIRO-RO* **38**, 277–294.

R. Meyer (1976), 'Mixed integer minimization models for piecewise-linear functions of a single variable', *Discrete Math.* **16**, 163–71.

R. Miller, Z. Xie, S. Leyffer, M. Davis and S. Gray (2010), 'Surrogate-based modeling of the optical response of metallic nanostructures', *J. Phys. Chem. C* **114**, 20741–20748.

R. Misener and C. Floudas (2012), 'Global optimization of mixed-integer quadratically-constrained quadratic programs (MIQCQP) through piecewise-linear and edge-concave relaxations', *Math. Program.* **136**, 155–182.

R. Misener and C. Floudas (2013), 'GloMIQO: Global mixed-integer quadratic optimizer', *J. Global Optim.*, to appear.

J. Momoh, R. Koessler, M. Bond, B. Stott, D. Sun, A. Papalexopoulos and P. Ristanovic (1997), 'Challenges to optimal power flow', *IEEE Trans. Power Systems* **12**, 444–455.

J. Müller (2012), Surrogate model algorithms for computationally expensive black-box global optimization problems. PhD thesis, Tampere University of Technology, Finland.

J. Müller, C. A. Shoemaker and R. Piché (2013), 'SO-MI: A surrogate model algorithm for computationally expensive nonlinear mixed-integer black-box global optimization problems', *Comput. Oper. Res.* **40**, 1383–1400.

G. Nannicini and P. Belotti (2012), 'Rounding-based heuristics for nonconvex MINLPs', *Math. Program. Comput.* **4**, 1–31.

G. Nannicini, P. Belotti and L. Liberti (2008), A local branching heuristic for MINLPs. http://arxiv.org/abs/0812.2188

G. Nemhauser and L. A. Wolsey (1988), *Integer and Combinatorial Optimization*, Wiley.

G. L. Nemhauser, M. W. P. Savelsbergh and G. C. Sigismondi (1994), 'MINTO: A Mixed INTeger Optimizer', *Oper. Res. Lett.* **15**, 47–58.

J. Nocedal and S. Wright (1999), *Numerical Optimization*, Springer.

I. Nowak, H. Alperin and S. Vigerske (2003), LaGO: An object oriented library for solving MINLPs. In *Proc. 1st Global Optimization and Constraint Satisfaction Workshop: COCOS 2002* (C. Bliek, C. Jermann and A. Neumaier, eds), Vol. 2861 of *Lecture Notes in Computer Science*, Springer, pp. 32–42.

J. Oldenburg, W. Marquardt, D. Heinz and D. Leineweber (2003), 'Mixed logic dynamic optimization applied to batch distillation process design', *AIChE J.* **49**, 2900–2917.

M. Padberg (1989), 'The Boolean Quadric Polytope: Some characteristics, facets and relatives', *Math. Program. B* **45**, 139–172.

M. Padberg (2000), 'Approximating separable nonlinear functions via mixed zero-one programs', *Oper. Res. Lett.* **27**, 1–5.

R. Powell (2007), 'Defending against terrorist attacks with limited resources', *Amer. Political Sci. Review* **101**, 527–541.

A. Prata, J. Oldenburg, A. Kroll and W. Marquardt (2008), 'Integrated scheduling and dynamic optimization of grade transitions for a continuous polymerization reactor', *Comput. Chem. Engng* **32**, 463–476.

K. A. Pruitt, S. Leyffer, A. M. Newman and R. Braun (2012), Optimal design and dispatch of distributed generation systems. Preprint ANL/MCS-2004-0112, Mathematics and Computer Science Division, Argonne National Laboratory.

A. Qualizza, P. Belotti and F. Margot (2012), Linear programming relaxations of quadratically constrained quadratic programs. In *Mixed Integer Nonlinear Programming*, Vol. 154 of *The IMA Volumes in Mathematics and its Applications*, Springer, pp. 407–426.

I. Quesada and I. E. Grossmann (1992), 'An LP/NLP based branch-and-bound algorithm for convex MINLP optimization problems', *Comput. Chem. Engng* **16**, 937–947.

A. J. Quist, R. van Gemeert, J. E. Hoogenboom, T. Illes, C. Roos and T. Terlaky (1998), 'Application of nonlinear optimization to reactor core fuel reloading', *Ann. Nuclear Energy* **26**, 423–448.

K. Rashid, S. Ambani and E. Cetinkaya (2013), 'An adaptive multiquadric radial basis function method for expensive black-box mixed-integer nonlinear constrained optimization', *Engng Optim.* **45**, 185–206.

R. Romero, A. Monticelli, A. Garcia and S. Haffner (2002), 'Test systems and mathematical models for transmission network expansion planning', *IEEE Proceedings: Generation, Transmission and Distribution* **149**, 27–36.

G. Rote (1992), 'The convergence rate of the sandwich algorithm for approximating convex functions', *Computing* **48**, 337–61.

R. Y. Rubinstein and D. P. Kroese (2004), *The Cross-Entropy Method: A Unified Approach to Combinatorial Optimization, Monte-Carlo Simulation and Machine Learning*, Springer.

H. S. Ryoo and N. V. Sahinidis (1995), 'Global optimization of nonconvex NLPs and MINLPs with applications in process design', *Comput. Chem. Engng* **19**, 552–566.

H. S. Ryoo and N. V. Sahinidis (1996), 'A branch-and-reduce approach to global optimization', *J. Global Optim.* **8**, 107–139.

S. Sager (2005), *Numerical Methods for Mixed-Integer Optimal Control Problems*, Der Andere Verlag.

S. Sager (2012), A benchmark library of mixed-integer optimal control problems. In *Mixed Integer Nonlinear Programming* (J. Lee and S. Leyffer, eds), Springer, pp. 631–670.

S. Sager, H. Bock and M. Diehl (2012), 'The integer approximation error in mixed-integer optimal control', *Math. Program. A* **133**, 1–23.

S. Sager, M. Diehl, G. Singh, A. Küpper and S. Engell (2007), Determining SMB superstructures by mixed-integer control. In *Proc. OR2006*, Springer, pp. 37–44.

S. Sager, M. Jung and C. Kirches (2011), 'Combinatorial integral approximation', *Math. Methods Oper. Res.* **73**, 363–380.

S. Sager, G. Reinelt and H. Bock (2009), 'Direct methods with maximal lower bound for mixed-integer optimal control problems', *Math. Program.* **118**, 109–149.

N. V. Sahinidis (1996), 'BARON: A general purpose global optimization software package', *J. Global Optim.* **8**, 201–205.

T. Sandler and D. G. Arce M (2003), 'Terrorism and game theory', *Simulation Gaming* **34**, 319–337.

T. Sandler and K. Siqueira (2006), 'Global terrorism: Deterrence versus preemption', *Canad. J. Economics* **39**, 1370–1387.

M. W. P. Savelsbergh (1994), 'Preprocessing and probing techniques for mixed integer programming problems', *ORSA J. Comput.* **6**, 445–454.

A. Saxena, P. Bonami and J. Lee (2010), 'Convex relaxations of non-convex mixed integer quadratically constrained programs: Extended formulations', *Math. Program.* **124**, 383–411.

A. Saxena, P. Bonami and J. Lee (2011), 'Convex relaxations of non-convex mixed integer quadratically constrained programs: Projected formulations', *Math. Program.* **130**, 359–413.

H. Schichl (2004), Global optimization in the COCONUT project. In *Numerical Software with Result Verification*, Springer, pp. 243–249.

A. Schrijver (1986), *Theory of Linear and Integer Programming*, Wiley.

C. A. Schweiger (1999), Process synthesis, design, and control: Optimization with dynamic models and discrete decisions. PhD thesis, Princeton University.

O. Shaik, S. Sager, O. Slaby and D. Lebiedz (2008), 'Phase tracking and restoration of circadian rhythms by model-based optimal control', *IET Systems Biology* **2**, 16–23.

W. Sheikh and A. Ghafoor (2010), 'An optimal bandwidth allocation and data droppage scheme for differentiated services in a wireless network', *Wireless Communications and Mobile Comput.* **10**, 733–747.

H. Sherali and W. Adams (1998), *A Reformulation–Linearization Technique for Solving Discrete and Continuous Nonconvex Problems*, Kluwer.

H. Sherali and A. Alameddine (1992), 'A new reformulation–linearization technique for bilinear programming problems', *J. Global Optim.* **2**, 379–410.

H. Sherali and E. Smith (1997), 'A global optimization approach to a water distribution network design problem', *J. Global Optim.* **11**, 107–132.

H. D. Sherali (2001), 'On mixed-integer zero-one representations for separable lower-semicontinuous piecewise-linear functions', *Oper. Res. Lett.* **28**, 155–160.

H. D. Sherali and B. M. P. Fraticelli (2002), 'Enhancing RLT relaxations via a new class of semidefinite cuts', *J. Global Optim.* **22**, 233–261.

H. D. Sherali, S. Subramanian and G. V. Loganathan (2001), 'Effective relaxations and partitioning schemes for solving water distribution network design problems to global optimality', *J. Global Optim.* **19**, 1–26.

R. Sinha, A. Yener and R. D. Yates (2002), 'Noncoherent multiuser communications: Multistage detection and selective filtering', *EURASIP J. Appl. Signal Processing* **12**, 1415–1426.

H. Skrifvars, S. Leyffer and T. Westerlund (1998), 'Comparison of certain MINLP algorithms when applied to a model structure determination and parameter estimation problem', *Comput. Chem. Engng* **22**, 1829–1835.

E. M. B. Smith and C. C. Pantelides (1997), 'Global optimization of nonconvex MINLPs', *Comput. Chem. Engng* **21**, S791–S796.

M. Soleimanipour, W. Zhuang and G. H. Freeman (2002), 'Optimal resource management in wireless multimedia wideband CDMA systems', *IEEE Trans. Mobile Computing* **1**, 143–160.

M. Soler, A. Olivares, E. Staffetti and P. Bonami (2011), En-route optimal flight planning constrained to pass through waypoints using MINLP. In *Proc. 9th USA/Europe Air Traffic Management Research and Development Seminar*, Berlin.

C. Still and T. Westerlund (2006), 'Solving convex MINLP optimization problems using a sequential cutting plane algorithm', *Comput. Optim. Appl.* **34**, 63–83.

R. Stubbs and S. Mehrotra (1999), 'A branch-and-cut method for 0–1 mixed convex programming', *Math. Program.* **86**, 515–532.

R. Stubbs and S. Mehrotra (2002), 'Generating convex polynomial inequalities for mixed 0–1 programs', *J. Global Optim.* **24**, 311–332.

M. Tawarmalani and N. V. Sahinidis (2002), *Convexification and Global Optimization in Continuous and Mixed-Integer Nonlinear Programming: Theory, Algorithms, Software, and Applications*, Kluwer.

M. Tawarmalani and N. V. Sahinidis (2004), 'Global optimization of mixed integer nonlinear programs: A theoretical and computational study', *Math. Program.* **99**, 563–591.

M. Tawarmalani and N. V. Sahinidis (2005), 'A polyhedral branch-and-cut approach to global optimization', *Math. Program.* **103**, 225–249.

M. Tawarmalani, J.-P. Richard and K. Chung (2010), 'Strong valid inequalities for orthogonal disjunctions and bilinear covering sets', *Math. Program.* **124**, 481–512.

S. Terwen, M. Back and V. Krebs (2004), Predictive powertrain control for heavy duty trucks. In *Proc. IFAC Symposium in Advances in Automotive Control* (G. Rizzo, L. Glielmo, C. Pianese and F. Vasca, eds), Elsevier, pp. 451–457.

J. Tomlin (1981), 'A suggested extension of special ordered sets to non-separable non-convex programming problems', *Ann. Discrete Math.* **11**, 359–370.

A. Toriello and J. P. Vielma (2012), 'Fitting piecewise linear continuous functions', *European J. Oper. Res.* **219**, 86–95.

M. Türkay and I. E. Grossmann (1996), 'Logic-based MINLP algorithms for the optimal synthesis of process networks', *Comput. Chem. Engng* **20**, 959–978.

T. J. Van Roy (1983), 'Cross decomposition for mixed integer programming', *Math. Program.* **25**, 145–163.

D. Vandenbussche and G. L. Nemhauser (2005*a*), 'A branch-and-cut algorithm for nonconvex quadratic programs with box constraints', *Math. Program.* **102**, 559–575.

D. Vandenbussche and G. L. Nemhauser (2005*b*), 'A polyhedral study of nonconvex quadratic programs with box constraints', *Math. Program.* **102**, 531–557.

J. P. Vielma, S. Ahmed and G. Nemhauser (2010), 'Mixed-integer models for non-separable piecewise-linear optimization: Unifying framework and extensions', *Oper. Res.* **58**, 303–315.

J. P. Vielma, S. Ahmed and G. L. Nemhauser (2008), 'A lifted linear programming branch-and-bound algorithm for mixed integer conic quadratic programs', *INFORMS J. Comput.* **20**, 438–450.

J. P. Vielma and G. Nemhauser (2011), 'Modeling disjunctive constraints with a logarithmic number of binary variables and constraints', *Math. Program.* **128**, 49–72.

J. Viswanathan and I. E. Grossmann (1990), 'A combined penalty function and outer-approximation method for MINLP optimization', *Comput. Chem. Engng* **14**, 769–782.

A. Wächter and L. T. Biegler (2006), 'On the implementation of a primal–dual interior point filter line search algorithm for large-scale nonlinear programming', *Math. Program.* **106**, 25–57.

T. Westerlund and K. Lundqvist (2005), Alpha-ECP, version 5.101: An interactive MINLP-solver based on the extended cutting plane method. Technical report 01-178-A, Process Design Laboratory at Åbo University.

T. Westerlund and F. Pettersson (1995), 'A cutting plane method for solving convex MINLP problems', *Comput. Chem. Engng* **19**, s131–s136.

T. Westerlund and R. Pörn (2002), 'Solving pseudo-convex mixed integer optimization problems by cutting plane techniques', *Optim. Engng* **3**, 253–280.

H. P. Williams (1999), *Model Building in Mathematical Programming*, Wiley.

D. L. Wilson (1998), Polyhedral methods for piecewise-linear functions. PhD thesis, University of Kentucky.

D. D. Wolf and Y. Smeers (2000), 'The gas transmission problem solved by an extension of the simplex algorithm', *Management Sci.* **46**, 1454–1465.

L. A. Wolsey (1998), *Integer Programming*, Wiley.

X. Wu, E. Topuz and M. Karfakis (1991), Optimization of ventilation control device locations and sizes in underground mine ventilation systems. In *Proc. 5th US Mine Ventilation Symposium*, Society for Mining, Metallurgy, and Exploration, pp. 391–399.

Y. Yajima and T. Fujie (1998), 'Polyhedral approach for nonconvex quadratic programming problems with box constraints', *J. Global Optim.* **13**, 151–170.

F. You and S. Leyffer (2010), 'Oil spill response planning with MINLP', *SIAG/OPT Views-and-News* **21**, 1–8.

F. You and S. Leyffer (2011), 'Mixed-integer dynamic optimization for oil-spill response planning with integration of a dynamic oil weathering model', *AIChe J.* **57**, 3555–3564.

Y. Zhu and T. Kuno (2006), 'A disjunctive cutting-plane-based branch-and-cut algorithm for 0–1 mixed-integer convex nonlinear programs', *Indust. Engng Chem. Research* **45**, 187–196.

J. Zhuang (2008), Modeling secrecy and deception in homeland security resource allocation. PhD thesis, University of Wisconsin–Madison.

J. Zhuang and V. M. Bier (2007*a*), 'Balancing terrorism and natural disasters: Defensive strategy with endogenous attacker effort', *Oper. Res.* **55**, 976–991.

J. Zhuang and V. M. Bier (2007*b*), 'Investment in security', *Industrial Engineer* **39**, 53–54.

J. Zhuang and V. M. Bier (2010), 'Reasons for secrecy and deception in homeland-security resource allocation', *Risk Analysis* **30**, 1737–1743.

Acta Numerica (2013), pp. 133–288
doi:10.1017/S0962492913000044

High-dimensional integration: The quasi-Monte Carlo way*†

Josef Dick
University of New South Wales,
Sydney, NSW 2052, Australia
E-mail: josef.dick@unsw.edu.au

Frances Y. Kuo
University of New South Wales,
Sydney, NSW 2052, Australia
E-mail: f.kuo@unsw.edu.au

Ian H. Sloan
University of New South Wales,
Sydney, NSW 2052, Australia
E-mail: i.sloan@unsw.edu.au
and
King Fahd University of Petroleum and Minerals,
Dhahran 34463, Saudi Arabia

This paper is a contemporary review of QMC ('quasi-Monte Carlo') methods, that is, equal-weight rules for the approximate evaluation of high-dimensional integrals over the unit cube $[0,1]^s$, where s may be large, or even infinite. After a general introduction, the paper surveys recent developments in lattice methods, digital nets, and related themes. Among those recent developments are methods of construction of both lattices and digital nets, to yield QMC rules that have a prescribed rate of convergence for sufficiently smooth functions, and ideally also guaranteed slow growth (or no growth) of the worst-case error as s increases. A crucial role is played by parameters called 'weights', since a careful use of the weight parameters is needed to ensure that the worst-case errors in an appropriately weighted function space are bounded, or grow only slowly, as the dimension s increases. Important tools for the analysis are weighted function spaces, reproducing kernel Hilbert spaces, and discrepancy, all of which are discussed with an appropriate level of detail.

* Colour online for monochrome figures available at journals.cambridge.org/anu.

† The URLs cited in this work were correct at the time of going to press, but the publisher and the authors make no undertaking that the citations remain live or are accurate or appropriate.

CONTENTS

1. Introduction

Numerical integration in more than one dimension has for a century been a topic that presents many challenges, as witnessed by the classical monograph by Stroud (1971). Among the challenges are those posed by the rich choice of multi-dimensional geometries, whether the cube, ball, sphere, simplex, or something more elaborate. Another challenge is the common occurrence of singularities of the integrand at special points or sub-manifolds of the region. But of all the challenges the one that stands above all others is the difficulty of doing anything effective when the number of integration variables (the 'dimension') passes beyond the number of fingers on two hands. Certainly the simple strategy of using a 'product' of one-dimensional rules becomes infeasible, since even a mere 10-fold product of say two-point Gauss rules requires 2^{10} points. Yet these days problems arise, and are tackled, in which the number of dimensions is in the hundreds, or even the thousands or tens of thousands.

The aim of this survey is to describe and analyse effective methods for numerical integration, with a special emphasis on a high number of dimensions. The price that has to be paid is that we consider only the simplest geometry (namely the unit cube in an arbitrary number s of dimensions), somewhat smooth functions, and only equal-weight integration rules.

Thus the task is to approximate the integral

$$I_s(f) = \int_{[0,1]^s} f(\boldsymbol{x})\,\mathrm{d}\boldsymbol{x}, \tag{1.1}$$

where s is greater than 1 and possibly large, for some integrable function f, by an n-point integration rule of the form

$$Q_{n,s}(f) = \frac{1}{n}\sum_{i=0}^{n-1} f(\boldsymbol{t}_i), \tag{1.2}$$

thus a rule that uses the prescribed sample points $\boldsymbol{t}_0, \ldots, \boldsymbol{t}_{n-1} \in [0,1]^s$. Of course, we shall have much to say about the choice of these sample points. A rule of this equal-weight form is called a *quasi-Monte Carlo* (or QMC) rule.

The name quasi-Monte Carlo comes from a certain analogy with the *Monte Carlo* (MC) method: in its simplest form the MC approximation to the integral (1.1) takes exactly the same form as (1.2), but with one crucial difference, that the sample points are chosen randomly (and independently), from a uniform distribution on the cube $[0,1]^s$. The MC method (Hammersley and Handscomb 1964, Lemieux 2009, Woźniakowski 2013) has some very attractive features, most notably that there exists a probabilistic (root-mean-square) error estimate of the form

$$\frac{\sigma(f)}{\sqrt{n}},$$

where $\sigma^2(f)$ is the variance of f,

$$\sigma^2(f) := I_s(f^2) - (I_s(f))^2,$$

which is finite whenever f is square-integrable. Moreover, $\sigma(f)$ can be estimated by the same MC rule used to estimate $I_s(f)$. But for all the virtues of the MC method, its $O(1/\sqrt{n})$ rate of convergence is often considered to be impossibly slow, especially when f is smooth. For example, one needs four times as many points to reduce the error by a factor of two, or 100 times as many points to reduce the error by an order of magnitude.

The QMC methods were born in the 1950s and 1960s from the (successful) desire to achieve faster convergence than the MC rate of $O(1/\sqrt{n})$. In the next sections we shall meet QMC methods that depend on constructing points sets with small 'discrepancy'. These methods were indeed successful in improving upon the MC convergence rate, achieving a convergence rate of order $O((\log n)^s/n)$ or better if sufficient smoothness of f is assumed. In particular, we will meet the so-called 'lattice rules', which for certain periodic functions (periodic with respect to each component of $\boldsymbol{x}$) could achieve convergence rates of $O((\log n)^s/n^2)$ or faster, and so-called 'higher-order digital nets', which could achieve convergence rates of $O((\log n)^{2s}/n^2)$ or faster even for non-periodic functions.

In the early days of QMC research, the emphasis was mainly on the rate of convergence as n increases, and less attention was paid to what happens as the dimensionality s increases; those early QMC researchers surely never envisaged QMC methods being applied to integrands in hundreds of dimensions. For lattice methods there was much interest in the 1960s in periodization strategies (for converting non-periodic integrands into periodic ones), which can then achieve fast convergence for appropriate QMC rules, but regrettably those periodization strategies often turn easy high-dimensional problems into ones that are impossibly hard. In any case, the rate of convergence is only part of the story: the size of the implied constant is also important.

In the last two decades there has been great research interest in developing QMC methods and theory that can cope with arbitrarily high dimensions, if the circumstances are right. (Note that we do *not* say that all high-dimensional problems can be successfully tackled by QMC methods. Rather, the interest is in recognizing and analysing mathematically the particular features that make *some* high-dimensional problems manageable.)

Applications have also played an important role in the development of QMC methods suitable for high-dimensional problems. In financial mathematics, the numerical experiments of Paskov and Traub (1995), which used low-discrepancy QMC methods in 360 dimensions to value parcels of mortgage-backed obligations, were successful to a degree that caused universal surprise. This led to many theoretical developments, as researchers struggled to understand how such high-dimensionality could be handled successfully (for example Caflisch, Morokoff and Owen 1997, Wang and Fang 2003). Option pricing problems have spurred many developments (for example Acworth, Broadie and Glasserman 1998, L'Ecuyer 2004, Giles, Kuo, Sloan and Waterhouse 2008, Wang and Sloan 2011, Griebel, Kuo and Sloan 2013). A challenging class of problems in multivariate statistics led Kuo *et al.* (2008*a*) to a new understanding of the importance of proper handling of the transformation process from $\mathbb{R}^s$ into $[0,1]^s$ for problems over unbounded regions. The problem of evaluating high-dimensional expected values arising from partial differential equations with random coefficients, typified by the flow of a liquid (oil or water) through a porous material, with the permeability treated as a random field, is the newest driver of innovation (Graham *et al.* 2011, Kuo, Schwab and Sloan 2012). However, in this survey the focus will be on the theoretical innovations rather than on the applications themselves.

The structure of the survey is as follows. In Section 2 we introduce QMC methods, and develop basic definitions and concepts without much theory. In Section 3 we begin to develop key tools (such as reproducing kernel Hilbert spaces) needed for error analysis of QMC methods. In Section 4 we introduce the modern theory of weighted function spaces, in which certain parameters ('weights') are introduced to describe the circumstance that some variables (or groups of variables) are more important than others. In Sections 5 and 6 we introduce more sophisticated material on lattice rules and digital nets. Then in Section 7 we introduce a setting which is currently attracting great interest, namely that of numerical integration in an infinite number of dimensions. Finally in Section 8 we give concluding remarks. At the end of each section we provide notes with references to further related material.

2. Introducing quasi-Monte Carlo methods

2.1. Multivariate numerical integration

Recall from (1.1) that our goal is to approximate a multiple integral over the s-dimensional unit cube $[0,1]^s$,

$$I_s(f) = \int_{[0,1]^s} f(\boldsymbol{x})\,\mathrm{d}\boldsymbol{x} := \int_0^1 \cdots \int_0^1 f(x_1,\ldots,x_s)\,\mathrm{d}x_1 \cdots \mathrm{d}x_s,$$

where s is large, possibly in the hundreds or thousands. In practice most integrals over bounded or unbounded regions can be transformed into this desired form with a suitable change of variables. The transformation plays an important role as it determines the behaviour of the transformed integrand f. We shall defer the choice and strategy of transformation to be discussed later in Section 2.11.

For our analysis in this survey, we will generally assume that f is continuous and has a certain level of smoothness, as will be made precise in the next section.

In the classical theory of numerical integration (Davis and Rabinowitz 1984), an integral in one dimension can be approximated by a *quadrature rule*, as follows:

$$\int_0^1 f(x)\,\mathrm{d}x \approx \sum_{i=0}^{n-1} w_i f(t_i),$$

where $t_0,\ldots,t_{n-1} \in [0,1]$ are the *quadrature points*, and $w_0,\ldots,w_{n-1} \in \mathbb{R}$ are the *quadrature weights* satisfying $\sum_{i=0}^{n-1} w_i = 1$. Well-known examples include the (left) rectangle rule, which uses equally spaced points $t_i = i/n$ and equal weights $w_i = 1/n$ and has the error $f'(\xi)/(2n)$ for some $\xi \in (0,1)$; the trapezoidal rule (which has second-order convergence); Simpson's rule (with fourth-order convergence); and the Gauss rules (which integrate polynomials of degree $2n-1$ exactly).

For integration in s dimensions with $s \geq 2$, we have *cubature rules* (Stroud 1971),

$$\int_{[0,1]^s} f(\boldsymbol{x})\,\mathrm{d}\boldsymbol{x} \approx \sum_{i=0}^{n-1} w_i f(\boldsymbol{t}_i),$$

where $\boldsymbol{t}_0,\ldots,\boldsymbol{t}_{n-1} \in [0,1]^s$ are the *cubature points*, $\boldsymbol{t}_i = (t_{i,1},\ldots,t_{i,s})$, and $w_0,\ldots,w_{n-1} \in \mathbb{R}$ are the *cubature weights* satisfying $\sum_{i=0}^{n-1} w_i = 1$. Explicit cubature rules are available in low dimensions (Cools 1997). An obvious way to approximate a high-dimensional integral is to use a *product rule*,

$$\int_0^1 \cdots \int_0^1 f(x_1,\ldots,x_s)\,\mathrm{d}x_1 \cdots \mathrm{d}x_s \approx \sum_{i_1=0}^{m-1} \cdots \sum_{i_s=0}^{m-1} u_{i_1} \cdots u_{i_s} f(v_{i_1},\ldots,v_{i_s}),$$

in effect applying a one-dimensional quadrature rule $\sum_{i=0}^{m-1} u_i f(v_i)$ to each of the s integrals. The cubature points $\boldsymbol{t}_i$ are given by the s-fold product of the quadrature points v_i, and the cubature weights w_i are the corresponding products of the quadrature weights u_i. The total number of points is $n = m^s$, which is enormous when s is large, while (since f could be a function of only one component, such as $f(x_1, \ldots, x_s) = x_1^2$) the error is only the sth root of the error of the one-dimensional quadrature rule: for example, the product rectangle rule has error of order $O(n^{-1/s})$.

As the title suggests, this survey will focus on *quasi-Monte Carlo methods* for high-dimensional integration. This section is devoted to introducing the basic principles behind these methods. We begin in the next subsection with a brief introduction to the related and widely used classical *Monte Carlo method.*

Another competing technique for high-dimensional integration is based on *sparse grid methods* (Bungartz and Griebel 2004), but we will not discuss these methods in this survey.

2.2. Monte Carlo method

The classical *Monte Carlo (MC) method* is an equal-weight cubature rule of the form

$$Q_{n,s}(f) = \frac{1}{n} \sum_{i=0}^{n-1} f(\boldsymbol{t}_i),$$

where $\boldsymbol{t}_0, \ldots, \boldsymbol{t}_{n-1}$ are i.i.d. (independent and identically distributed) *uniform* random samples from $[0, 1]^s$.

Theorem 2.1 (Monte Carlo root-mean-square error). For all *square-integrable* functions f, we have

$$\sqrt{\mathbb{E}[|I_s(f) - Q_{n,s}(f)|^2]} = \frac{\sigma(f)}{\sqrt{n}},$$

where the expectation is taken with respect to the uniform random samples $\boldsymbol{t}_0, \ldots, \boldsymbol{t}_{n-1}$, and

$$\sigma^2(f) := I_s(f^2) - (I_s(f))^2$$

is the *variance* of f.

Although this result is well known, we give a proof since it will serve as a model for later arguments.

Proof. We have

$$\mathbb{E}[|Q_{n,s}(f) - I_s(f)|^2] = \mathbb{E}[(Q_{n,s}(f))^2] - 2\,\mathbb{E}[Q_{n,s}(f)]I_s(f) + (I_s(f))^2,$$

where

$$\begin{aligned}\mathbb{E}[Q_{n,s}(f)] &= \int_{[0,1]^s} \cdots \int_{[0,1]^s} \left(\frac{1}{n} \sum_{i=0}^{n-1} f(\boldsymbol{t}_i) \right) \mathrm{d}\boldsymbol{t}_0 \ldots \mathrm{d}\boldsymbol{t}_{n-1} \\ &= \frac{1}{n} \sum_{i=0}^{n-1} \int_{[0,1]^s} f(\boldsymbol{t}_i)\, \mathrm{d}\boldsymbol{t}_i \\ &= I_s(f)\end{aligned}$$

and

$$\begin{aligned}\mathbb{E}[(Q_{n,s}(f))^2] &= \int_{[0,1]^s} \cdots \int_{[0,1]^s} \left(\frac{1}{n} \sum_{i=0}^{n-1} f(\boldsymbol{t}_i) \right)^2 \mathrm{d}\boldsymbol{t}_0 \ldots \mathrm{d}\boldsymbol{t}_{n-1} \\ &= \int_{[0,1]^s} \cdots \int_{[0,1]^s} \left(\frac{1}{n^2} \sum_{i=0}^{n-1} \sum_{k=0}^{n-1} f(\boldsymbol{t}_i) f(\boldsymbol{t}_k) \right) \mathrm{d}\boldsymbol{t}_0 \ldots \mathrm{d}\boldsymbol{t}_{n-1} \\ &= \int_{[0,1]^s} \cdots \int_{[0,1]^s} \left(\frac{1}{n^2} \sum_{i=0}^{n-1} f^2(\boldsymbol{t}_i) + \frac{1}{n^2} \sum_{i=0}^{n-1} \sum_{\substack{k=0 \\ k \neq i}}^{n-1} f(\boldsymbol{t}_i) f(\boldsymbol{t}_k) \right) \mathrm{d}\boldsymbol{t}_0 \ldots \mathrm{d}\boldsymbol{t}_{n-1} \\ &= \frac{1}{n^2} \sum_{i=0}^{n-1} \int_{[0,1]^s} f^2(\boldsymbol{t}_i)\, \mathrm{d}\boldsymbol{t}_i + \frac{1}{n^2} \sum_{i=0}^{n-1} \sum_{\substack{k=0 \\ k \neq i}}^{n-1} \int_{[0,1]^s} f(\boldsymbol{t}_i)\, \mathrm{d}\boldsymbol{t}_i \int_{[0,1]^s} f(\boldsymbol{t}_k)\, \mathrm{d}\boldsymbol{t}_k \\ &= \frac{1}{n} I_s(f^2) + \frac{n-1}{n} (I_s(f))^2.\end{aligned}$$

Hence

$$\mathbb{E}[|Q_{n,s}(f) - I_s(f)|^2] = \frac{I_s(f^2) - (I_s(f))^2}{n},$$

as claimed. □

Treating $Q_{n,s}(f)$ as a random variable, we see from the above proof that its *mean* is

$$\mathbb{E}[Q_{n,s}(f)] = I_s(f)$$

(that is to say, the MC method is unbiased), and its *variance* is

$$\mathrm{Var}[Q_{n,s}(f)] = \mathbb{E}[|Q_{n,s}(f) - I_s(f)|^2] = \frac{\sigma^2(f)}{n}.$$

By the central limit theorem, if $0 < \sigma(f) < \infty$ then

$$\lim_{n \to \infty} \mathbb{P}\left(|I_s(f) - Q_{n,s}(f)| \le c \frac{\sigma(f)}{\sqrt{n}} \right) = \frac{1}{\sqrt{2\pi}} \int_{-c}^{c} \mathrm{e}^{-x^2/2}\, \mathrm{d}x.$$

In other words, we have a 'probabilistic' error bound with a convergence rate of $O(n^{-1/2})$. The convergence rate is *independent of the dimension.* Comparing this with, for example, the product rectangle rule whose convergence rate is $O(n^{-1/s})$, we see that the MC method has the faster convergence when $s \geq 3$ even for smooth functions.

Another advantage of the MC method is that it is easy to obtain a practical error estimation. Indeed, an unbiased estimator for $\mathrm{Var}[Q_{n,s}(f)]$ is given by

$$\frac{1}{n(n-1)} \sum_{i=0}^{n-1} (f(\boldsymbol{t}_i) - Q_{n,s}(f))^2 = \frac{1}{n(n-1)} \left(\sum_{i=0}^{n-1} f^2(\boldsymbol{t}_i) - n[Q_{n,s}(f)]^2 \right). \tag{2.1}$$

The square root of this quantity provides an estimate for the root-mean-square MC error.

Although widely applicable, MC methods suffer from the slow $O(n^{-1/2})$ convergence rate. Variance reduction techniques (*e.g.*, importance sampling, stratified sampling, correlated sampling) can be used to improve the efficiency of MC, but in practice MC methods often remain distressingly slow. Bakhvalov (1959) proved that the $O(n^{-1/2})$ rate of convergence cannot be improved for general square-integrable or continuous functions f. For functions with more smoothness, this slow convergence rate is the main motivation for the switch to quasi-Monte Carlo methods.

2.3. Quasi-Monte Carlo methods

Quasi-Monte Carlo (*QMC*) *methods* are equal-weight cubature rules of the form

$$Q_{n,s}(f) = \frac{1}{n} \sum_{i=0}^{n-1} f(\boldsymbol{t}_i),$$

just like MC methods, but now the points $\boldsymbol{t}_0, \ldots, \boldsymbol{t}_{n-1} \in [0,1]^s$ are *chosen deterministically to be better than random*, in the sense that the deterministic nature of QMC leads to guaranteed error bounds, and that the convergence rate may be faster than the MC rate of $O(n^{-1/2})$ for sufficiently smooth functions.

There are two types of QMC methods.

- The 'open' type: this uses the first n points of an *infinite sequence.* Thus to increase n one only needs to evaluate the integrand at the additional cubature points.
- The 'closed' type: this uses a *finite point set* which depends on n. Thus a new value of n means a completely new set of cubature points.

(Here and in the following, a point set is understood to be a multiset, that is, we count points according to their multiplicity.)

The following definition is needed before we provide examples of QMC methods.

Definition 2.2 (radical inverse function). For integers $i \geq 0$ and $b \geq 2$, we define the *radical inverse function* $\phi_b(i)$ as follows:

$$\text{if} \quad i = \sum_{a=1}^{\infty} i_a b^{a-1}, \quad \text{where} \quad i_a \in \{0, 1, \dots, b-1\}, \quad \text{then} \quad \phi_b(i) := \sum_{a=1}^{\infty} \frac{i_a}{b^a}.$$

In other words, if $i = (\cdots i_2 i_1)_b$ denotes the base b representation of i, then $\phi_b(i) := (0.i_1 i_2 \cdots)_b$.

Example 2.3 (van der Corput sequence). The *van der Corput sequence in base* b is the one-dimensional sequence

$$\phi_b(0), \phi_b(1), \phi_b(2), \dots.$$

For example, take $b = 2$. First we write down the natural numbers $0, 1, 2, \dots$ in base 2

$$0, 1_2, 10_2, 11_2, 100_2, 101_2, 110_2, \dots.$$

Then we apply the radical inverse function ϕ_2 to each number, to obtain the sequence

$$0, 0.1_2, 0.01_2, 0.11_2, 0.001_2, 0.101_2, 0.011_2, \dots,$$

which in decimal form is the sequence

$$0, 0.5, 0.25, 0.75, 0.125, 0.625, 0.375, \dots.$$

Example 2.4 (Halton sequence). Let $p_1, p_2, \dots, p_s$ be the first s prime numbers. The *Halton sequence* $\boldsymbol{t}_0, \boldsymbol{t}_1, \dots$ *in* s *dimensions* is given by

$$\boldsymbol{t}_i = \big(\phi_{p_1}(i), \phi_{p_2}(i), \dots, \phi_{p_s}(i)\big), \quad i = 0, 1, \dots,$$

that is, the jth components of points in the Halton sequence form the van der Corput sequence in base p_j, where p_j is the jth prime. The Halton sequence leads to an 'open' QMC method. We have explicitly

$$\begin{aligned}
\boldsymbol{t}_0 &= (0, 0, 0, \dots, 0), \\
\boldsymbol{t}_1 &= (0.1_2, 0.1_3, 0.1_5, \dots, 0.1_{p_s}), \\
\boldsymbol{t}_2 &= (0.01_2, 0.2_3, 0.2_5, \dots, 0.2_{p_s}), \\
\boldsymbol{t}_3 &= (0.11_2, 0.01_3, 0.3_5, \dots, 0.3_{p_s}) \\
&\ \ \vdots
\end{aligned}$$

The Halton sequence satisfies the error bound

$$|I_s(f) - Q_{n,s}(f)| \leq C_s \frac{(\log n)^s}{n} V(f), \quad \text{for all } n \geq 2,$$

where C_s depends only on s, and $V(f)$ is the *variation of f in the sense of Hardy and Krause*; see for instance Kuipers and Niederreiter (1974, p. 147, Definition 5.2). For now we simply observe that, although the convergence rate appears to beat the MC rate of $O(n^{-1/2})$, the MC rate is independent of s, whereas for fixed s, the function $(\log n)^s/n$ increases with increasing n for all $n < \exp(s)$.

Example 2.5 (Hammersley point set). Let $p_1, p_2, \ldots, p_{s-1}$ be the first $s-1$ prime numbers. The *Hammersley point set* $\{\boldsymbol{t}_0, \boldsymbol{t}_1, \ldots, \boldsymbol{t}_{n-1}\}$ *with n points in s dimensions* is given by

$$\boldsymbol{t}_i = \left(\frac{i}{n}, \phi_{p_1}(i), \phi_{p_2}(i), \ldots, \phi_{p_{s-1}}(i)\right), \quad i = 0, 1, \ldots, n-1.$$

The Hammersley point set leads to a 'closed' QMC method. We have explicitly

$$\begin{aligned} \boldsymbol{t}_0 &= (0, 0, \ldots, 0), \\ \boldsymbol{t}_1 &= (\tfrac{1}{n}, 0.1_2, 0.1_3, \ldots, 0.1_{p_{s-1}}), \\ \boldsymbol{t}_2 &= (\tfrac{2}{n}, 0.01_2, 0.2_3, \ldots, 0.2_{p_{s-1}}) \\ &\vdots \\ \boldsymbol{t}_{n-1} &= (\tfrac{n-1}{n}, \ldots). \end{aligned}$$

The QMC method based on the Hammersley point set satisfies the error bound

$$|I_s(f) - Q_{n,s}(f)| \le C_s' \frac{(\log n)^{s-1}}{n} V(f), \quad \text{for all } n \ge 2,$$

where C_s' depends only on s. Note that there is one less power of $\log n$ compared to the error bound for the Halton sequence. Typically the error bounds for QMC methods based on 'closed' point sets are better than those based on 'open' sequences.

Example 2.6 (Kronecker sequence). Let $1, \alpha_1, \alpha_2, \ldots, \alpha_s \in \mathbb{R}$ be linearly independent over $\mathbb{Q}$. The *Kronecker sequence* $\boldsymbol{t}_0, \boldsymbol{t}_1, \ldots$ *in s dimensions* is given by

$$\boldsymbol{t}_i = (\{i\alpha_1\}, \{i\alpha_2\}, \ldots, \{i\alpha_s\}), \quad i = 0, 1, \ldots,$$

where the braces indicate that we take the fractional part of a real number, that is,

$$\{x\} := x - \lfloor x \rfloor. \tag{2.2}$$

For example, a good choice of parameters is $\alpha_j = \sqrt{p_j}$ where p_j is the jth prime number. The error bound for the Kronecker sequence takes the same form as that of the Halton sequence, but with a different multiplying constant, again depending on s.

Two main families of QMC methods are currently under investigation:

- *digital nets* ('closed') and *digital sequences* ('open'),
- *lattice rules* ('closed' and 'open').

We will introduce these two families in the next few subsections.

2.4. Lattice rules

A *lattice* in $\mathbb{R}^s$ is a discrete subset of $\mathbb{R}^s$ which is closed under addition and subtraction. An *integration lattice* in $\mathbb{R}^s$ is a lattice which contains $\mathbb{Z}^s$ as a subset. A *lattice rule* is an equal-weight cubature rule whose cubature points are those points of an integration lattice that lie in the half-open unit cube $[0,1)^s$.

Every lattice point set includes the origin $\mathbf{0}$. The projection of the lattice points onto each axis gives equally spaced points. In a sense the integral in each dimension is approximated by a rectangle rule (or a trapezoidal rule if the integrand is periodic). Every lattice rule can be written as a *multiple sum* involving one or more generating vectors. The minimal number of generating vectors required to generate a lattice rule is known as the *rank* of the rule. Besides rank-one lattice rules involving just one generating vector, there exist lattice rules having rank up to s. Lattice rules were introduced by Korobov (1959). They were originally designed for periodic integrands.

Definition 2.7 (rank-one lattice rule). An n-point *rank-one lattice rule* in s dimensions, also known as the *method of good lattice points*, is a QMC method with cubature points

$$\boldsymbol{t}_i = \left\{\frac{i\boldsymbol{z}}{n}\right\}, \quad i = 0, 1, \ldots, n-1, \tag{2.3}$$

where $\boldsymbol{z} \in \mathbb{Z}^s$, known as the *generating vector*, is an s-dimensional integer vector having no factor in common with n, and the braces around the vector indicate, as in (2.2), that we take the fractional part of each component in the vector.

Example 2.8 (Fibonacci lattice). Let $\boldsymbol{z} = (1, F_k)$ and $n = F_{k+1}$, where F_k and F_{k+1} are consecutive Fibonacci numbers. Then the resulting two-dimensional lattice point set is called a *Fibonacci lattice*; see Figure 2.1 for two examples. Fibonacci lattices in two dimensions have a certain optimality property, but there is no obvious generalization to higher dimensions that retains the optimality property.

Since we are only interested in the fractional part of $i\boldsymbol{z}/n$, the components of $\boldsymbol{z}$ can be restricted to the set

$$\mathbb{Z}_n := \{0, 1, 2, \ldots, n-1\}.$$

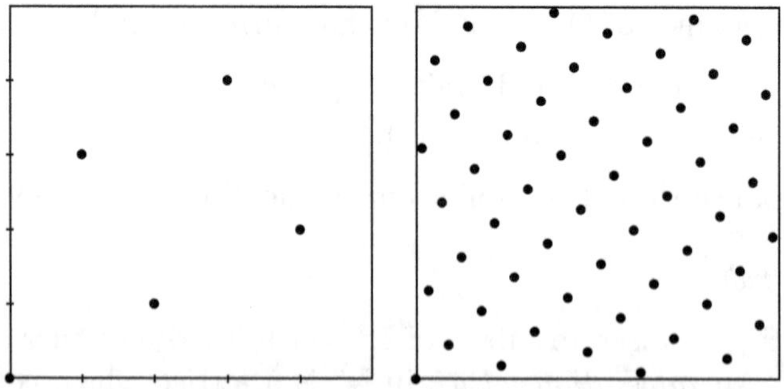

Figure 2.1. The Fibonacci lattices with 5 points and 55 points. The corresponding generating vectors are $(1, 3)$ and $(1, 34)$, respectively.

Moreover, since we would prefer that every one-dimensional projection of the lattice rule has n distinct values, the components of $\boldsymbol{z}$ can be further restricted to the set

$$\mathbb{U}_n := \{z \in \mathbb{Z} : 1 \leq z \leq n-1 \text{ and } \gcd(z, n) = 1\}.$$

The number of elements in the set $\mathbb{U}_n$ is $\varphi(n) := |\mathbb{U}_n|$, the *Euler totient function.* If $n = p_1^{a_1} p_2^{a_2} \cdots p_k^{a_k}$ is the prime factorization of n, then

$$\varphi(n) = (p_1^{a_1} - p_1^{a_1-1})(p_2^{a_2} - p_2^{a_2-1}) \cdots (p_k^{a_k} - p_k^{a_k-1}).$$

Asymptotically, $\varphi(n)$ grows at a rate close to n: $1/\varphi(n) = O((\log\log n)/n)$.

For simplicity, we often assume that n is prime and thus $\varphi(n) = n - 1$. This implies that:

- there are $n - 1$ choices for each component of $\boldsymbol{z}$,
- there are $(n-1)^s$ choices for the generating vector $\boldsymbol{z}$.

For large n and s, an exhaustive search to find a generating vector that minimizes some desired error criterion is practically impossible. We now present two approaches for constructing lattice generating vectors in high dimensions.

Example 2.9 (Korobov construction). Given an integer a satisfying $1 \leq a \leq n-1$ and $\gcd(a, n) = 1$, we define

$$\boldsymbol{z} = \boldsymbol{z}(a) := (1, a, a^2, \ldots, a^{s-1}) \bmod n.$$

There are (at most) $n-1$ choices for the Korobov parameter a, leading to (at most) $n-1$ choices for the generating vector $\boldsymbol{z}$. Thus it is feasible in practice to search through the (at most) $n-1$ choices and take the one that minimizes the desired error criterion.

Example 2.10 (component-by-component (CBC) construction). Given n, we construct a generating vector $\boldsymbol{z} = (z_1, z_2, \ldots)$ as follows.

(1) Set $z_1 = 1$.

(2) With z_1 held fixed, choose z_2 from $\mathbb{U}_n$ to minimize a desired error criterion in 2 dimensions.

(3) With z_1, z_2 held fixed, choose z_3 from $\mathbb{U}_n$ to minimize a desired error criterion in 3 dimensions.

(4) With z_1, z_2, z_3 held fixed, choose z_4 from $\mathbb{U}_n$ to minimize a desired error criterion in 4 dimensions.

(5) ...

The generating vector obtained by the CBC construction is *extensible in s*. In Section 5 we will consider two important aspects of the CBC construction: error analysis and computational cost. We will also discuss how to obtain *extensible lattice sequences* that are extensible in both s and n.

2.5. (t,m,s)-nets and (t,s)-sequences

Nets and sequences provide another method for obtaining well-distributed point sets in the unit cube $[0,1)^s$ that are useful for QMC integration. The concept of (t,m,s)-net is based on subdividing the unit cube into intervals and placing points in the cube such that each interval of a certain size and shape contains the 'correct' number of points.

Definition 2.11 ((t,m,s)-net). Let $t \geq 0$, $m \geq 1$, $s \geq 1$, and $b \geq 2$ be integers with $t \leq m$. A *(t,m,s)-net in base b* is a point set P consisting of b^m points in $[0,1)^s$ such that every *elementary interval* of the form

$$\prod_{j=1}^{s} \left[\frac{a_j}{b^{d_j}}, \frac{a_j+1}{b^{d_j}} \right) \tag{2.4}$$

with each $d_j \geq 0$, $0 \leq a_j < b^{d_j}$, and $d_1 + d_2 + \cdots + d_s = m - t$, contains exactly b^t points of P.

An elementary interval (2.4) has volume $b^{-(d_1+d_2+\cdots+d_s)} = b^{t-m}$, which is precisely the proportion of the points from P that lie in this elementary interval. One would expect such a property to hold if the point set is uniformly distributed. Figure 2.2 provides an illustration of a two-dimensional net with 16 points.

Note that any point set consisting of b^m points in $[0,1)^s$ is trivially an (m,m,s)-net in base b, indicating that the concept of (t,m,s)-net is only useful when $t < m$. For fixed b and m, the (t,m,s)-net condition gets stronger as t gets smaller. Hence the aim is to find (t,m,s)-nets in base

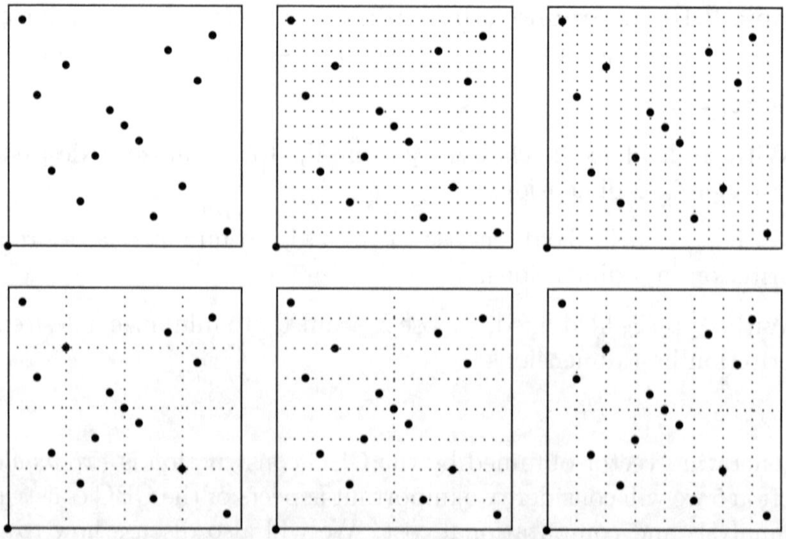

Figure 2.2. Illustration of a $(0,4,2)$-net in base 2: every elementary interval of volume $1/16$ contains exactly one of the 16 points. A point that lies on the dividing line counts toward the interval above or to the right.

b where t is small. A (t,m,s)-net is said to be a *strict* (t,m,s)-net if it is not a $(t-1,m,s)$-net. The 't-value' is often referred to as the *quality parameter*. It is also not useful to choose the base b too large, noting that in the extreme case in which $b = n$ we have $m = 1$, so that even when $t = 0$ the only elementary intervals are the 'boxes' of side length 1 and thickness $1/n$, in which case even the set of points $k(1/n, 1/n, \ldots, 1/n)$ for $k = 0, \ldots, n-1$ lying on the main diagonal is a $(0,m,s)$-net. The relevant quantity for fixed n is the *strength* $k = m - t$ of the (t,m,s)-net. The aim is then to find nets with large strength k.

There is an analogous concept for infinite sequences.

Definition 2.12 ((t,s)-sequences). Let $t \geq 0$ and $s \geq 1$ be integers. A *(t,s)-sequence in base* b is a sequence of points $S = (\boldsymbol{t}_0, \boldsymbol{t}_1, \ldots)$ in $[0,1)^s$ such that for any integers $m > t$ and $\ell \geq 0$, every block of b^m points

$$\boldsymbol{t}_{\ell b^m}, \ldots, \boldsymbol{t}_{(\ell+1)b^m - 1}$$

in the sequence S forms a (t,m,s)-net in base b.

The van der Corput sequence in base b is a $(0,1)$-sequence in base b. Higher-dimensional examples include the *Sobol′ sequence* (Sobol′ 1967), which is a (t,s)-sequence in base 2, where t is a non-decreasing function of s; the *Faure sequence* (Faure 1982); the *Niederreiter sequence* (Niederreiter 1987); and the *Niederreiter–Xing sequence* (Niederreiter and Xing 1995).

More details about these sequences will be given in Section 2.7. These sequences are also commonly referred to as 'low-discrepancy sequences' due to their high uniformity. We will explain the term 'discrepancy' in Section 3.

2.6. Digital construction scheme

We now introduce a method for the construction of (t, m, s)-nets and (t, s)-sequences. This method is based on linear algebra over finite fields.

Let b be prime and let $\mathbb{Z}_b := \{0, 1, \ldots, b-1\}$ denote the equivalence classes of integers modulo b. Using addition $+$ and multiplication $*$ of elements in $\mathbb{Z}_b$ modulo the prime b, we obtain a finite field of order b. In the following we identify the element k in the finite field $\mathbb{Z}_b$ with the integer $0 \le k < b$.

Let $C_1, \ldots, C_s$ be m-by-m matrices with entries in $\mathbb{Z}_b$. Then $t_{i,j}$, the jth component of the ith point in $P = \{\boldsymbol{t}_0, \ldots, \boldsymbol{t}_{b^m-1}\}$, can be constructed as follows.

(1) Write i in its base b representation:

$$i = (i_m \cdots i_2 i_1)_b = i_1 + i_2 b + \cdots + i_m b^{m-1}.$$

(2) Compute

$$\begin{pmatrix} y_1 \\ y_2 \\ \vdots \\ y_m \end{pmatrix} = C_j \begin{pmatrix} i_1 \\ i_2 \\ \vdots \\ i_m \end{pmatrix},$$

where all additions and multiplications are done in the finite field $\mathbb{Z}_b$, *i.e.*, modulo b.

(3) Set

$$t_{i,j} = (0.y_1 y_2 \cdots y_m)_b = \frac{y_1}{b} + \frac{y_2}{b^2} + \cdots + \frac{y_m}{b^m}.$$

The resulting point set $P = \{\boldsymbol{t}_0, \ldots, \boldsymbol{t}_{b^m-1}\}$ is called a *digital net* over $\mathbb{Z}_b$ and the matrices $C_1, \ldots, C_s$ are called the *generating matrices* of the digital net. The following definition connects (t, m, s)-nets in base b with digital nets.

Definition 2.13 (digital (t, m, s)-net). Let b be prime, $t \ge 0$, $m \ge 1$, and $s \ge 1$ be integers. If a digital net P constructed over $\mathbb{Z}_b$ is a (t, m, s)-net in base b, then P is called a *digital (t, m, s)-net over $\mathbb{Z}_b$*.

The following lemma provides the connection between the generating matrices of a digital net and the (t, m, s)-net property.

Lemma 2.14 (t-value for a digital net). A digital net over $\mathbb{Z}_b$ with generating matrices $C_1, \ldots, C_s$ is a digital (t, m, s)-net over $\mathbb{Z}_b$ if and only if, for all $d_1, \ldots, d_s \geq 0$ with $d_1 + \cdots + d_s = m - t$, the set of vectors

$$c_{1,1}, c_{1,2}, \ldots, c_{1,d_1}, c_{2,1}, c_{2,2}, \ldots, c_{2,d_2}, \ldots, c_{s,1}, c_{s,2}, \ldots, c_{s,d_s}$$

is linearly independent over $\mathbb{Z}_b$, where $c_{j,\ell}$ denotes the ℓth row vector of the matrix C_j.

Proof. Let $d_1, \ldots, d_s \geq 0$ such that $d_1 + \cdots + d_s = m - t$, let

$$J = \prod_{j=1}^{s} \left[\frac{a_j}{b^{d_j}}, \frac{a_j + 1}{b^{d_j}} \right)$$

be an elementary interval, and let

$$\frac{a_j}{b^{d_j}} = \frac{u_{j,1}}{b} + \frac{u_{j,2}}{b^2} + \cdots + \frac{u_{j,d_j}}{b^{d_j}}.$$

Thus, for given $d_1, \ldots, d_s$, we can uniquely associate the vector

$$\vec{u} = (u_{1,1}, \ldots, u_{1,d_1}, \ldots, u_{s,1}, \ldots, u_{s,d_s})^\top$$

to the elementary interval J. Let $\{\boldsymbol{t}_0, \boldsymbol{t}_1, \ldots, \boldsymbol{t}_{b^m-1}\}$ be a digital net with generating matrices $C_1, \ldots, C_s \in \mathbb{Z}_b^{m \times m}$. Let $\boldsymbol{t}_i = (t_{i,1}, t_{i,2}, \ldots, t_{i,s})$ with $t_{i,j} = \tau_{i,j,1} b^{-1} + \tau_{i,j,2} b^{-2} + \cdots + \tau_{i,j,m} b^{-m}$. Then $\boldsymbol{t}_i \in J$ if and only if $u_{j,\ell} = \tau_{i,j,\ell}$ for $1 \leq \ell \leq d_j$ and $1 \leq j \leq s$. Let $C_j = (c_{j,1}^\top, \ldots, c_{j,m}^\top)^\top$, that is, $c_{j,\ell}$ is the ℓth row vector of C_j. Let $A = (c_{1,1}^\top, \ldots, c_{1,d_1}^\top, \ldots, c_{s,1}^\top, \ldots, c_{s,d_s}^\top)^\top$. Then the number of points of the digital net which lie in J is given by the number of solutions $(i_1, \ldots, i_m)$ of the equation over $\mathbb{Z}_b$:

$$A \begin{pmatrix} i_1 \\ i_2 \\ \vdots \\ i_m \end{pmatrix} = \vec{u}. \tag{2.5}$$

For each vector $\vec{u}$ the number of solutions of (2.5) is b^{m-t} if and only if the rows of A are linearly independent. Thus the result follows. □

Constructions of good generating matrices for digital nets yielding digital (t, m, s)-nets with small t-value will be discussed in Section 2.7. In the following we give a two-dimensional example of digital $(0, m, 2)$-nets over $\mathbb{Z}_b$ for arbitrary prime b.

Example 2.15. Let b be a prime and let $m \geq 1$ be an integer. Let

$$C_1 = \begin{pmatrix} 1 & 0 & \cdots & \cdots & 0 \\ 0 & 1 & 0 & \cdots & 0 \\ \vdots & \ddots & \ddots & \ddots & \vdots \\ 0 & \cdots & 0 & 1 & 0 \\ 0 & \cdots & \cdots & 0 & 1 \end{pmatrix} \in \mathbb{Z}_b^{m \times m}$$

and

$$C_2 = \begin{pmatrix} 0 & \cdots & \cdots & 0 & 1 \\ 0 & \cdots & 0 & 1 & 0 \\ \vdots & \iddots & \iddots & \iddots & \vdots \\ 0 & 1 & 0 & \cdots & 0 \\ 1 & 0 & \cdots & \cdots & 0 \end{pmatrix} \in \mathbb{Z}_b^{m \times m}.$$

Then Lemma 2.14 implies that C_1 and C_2 are generating matrices of a digital $(0, m, 2)$-net over $\mathbb{Z}_b$.

The concept of digital nets can be extended to infinite sequences $S = (\boldsymbol{t}_0, \boldsymbol{t}_1, \ldots)$ in $[0,1)^s$. To do so, we need infinite matrices $C_1, \ldots, C_s \in \mathbb{Z}^{\mathbb{N}\times\mathbb{N}}$, with $C_j = (c_{j,k,\ell})_{k,\ell\in\mathbb{N}}$ and $c_{j,k,\ell} \in \mathbb{Z}_b$, such that for every ℓ there is a k_ℓ such that $c_{j,k,\ell} = 0$ for all $k > k_\ell$. Then $t_{i,j}$, the jth component of the ith point in the sequence S is constructed in the following way.

(1) Write i in its base b representation:

$$i = (\cdots i_2 i_1)_b = i_1 + i_2 b + \cdots.$$

(Note that only finitely many digits i_k are different from 0.)

(2) Compute

$$\begin{pmatrix} y_1 \\ y_2 \\ \vdots \end{pmatrix} = C_j \begin{pmatrix} i_1 \\ i_2 \\ \vdots \end{pmatrix},$$

where all additions and multiplications are done in the finite field $\mathbb{Z}_b$.

(3) Set

$$t_{i,j} = (0.y_1 y_2 \cdots)_b = \frac{y_1}{b} + \frac{y_2}{b^2} + \cdots.$$

The resulting sequence $S = (\boldsymbol{t}_0, \boldsymbol{t}_1, \ldots)$ is called a *digital sequence* over $\mathbb{Z}_b$ and the matrices $C_1, \ldots, C_s$ are called the *generating matrices* of the digital sequence. The following definition connects (t, s)-sequences in base b with digital sequences.

Definition 2.16 (digital (t, s)-sequences). Let b be prime, and let $t \geq 0$ and $s \geq 1$ be integers. If a digital sequence S constructed over $\mathbb{Z}_b$ is a (t, s)-sequence in base b, then S is called a *digital (t, s)-sequence over $\mathbb{Z}_b$.*

Constructions of good generating matrices for digital sequences yielding digital (t, s)-sequences with small t-value will be discussed in Section 2.7. In the following we give a one-dimensional example of a $(0, 1)$-sequence over $\mathbb{Z}_b$ for arbitrary prime b.

Example 2.17. Let b be a prime. Let $C \in \mathbb{Z}_b^{\mathbb{N}\times\mathbb{N}}$ be the matrix with ones along the main diagonal and zeros everywhere else (*i.e.*, the identity matrix). Then C generates a digital $(0,1)$-sequence over $\mathbb{Z}_b$. In fact, C generates the van der Corput sequence in base b.

2.7. *Constructions of digital nets and sequences*

In this subsection we discuss explicit constructions of digital nets and sequences with small t value.

In the previous subsection we introduced (digital) (t,m,s)-nets and (digital) (t,s)-sequences and we gave explicit constructions for small dimensions. The first constructions in arbitrary dimensions are due to Sobol′ (1967) and Faure (1982), before the general concept of (digital) (t,s)-sequence was introduced in Niederreiter (1987). In the following we discuss these explicit constructions in detail.

We use the following notation. Let $\mathbb{Z}_b[x]$ denote the set of polynomials over the finite field $\mathbb{Z}_b$. The polynomial $p(x) = a_0 + a_1x + \cdots + a_mx^m$ with $a_m \neq 0$ is said to have degree $\deg(p) = m$. For $p = 0$ we set $\deg(p) = -\infty$. A polynomial $p \in \mathbb{Z}_b[x]$ is *irreducible* if it cannot be written as a product of two non-constant polynomials, that is, there are no $q, q' \in \mathbb{Z}_b[x]$ such that $p(x) = q(x)q'(x)$ with $\deg(q), \deg(q') > 0$. A polynomial $p(x) = a_0 + a_1x + \cdots + a_mx^m \in \mathbb{Z}_b[x]$ of degree $m \geq 1$ is *primitive* if and only if $a_m = 1$, $a_0 \neq 0$ and the smallest integer k such that there is a polynomial $q \in \mathbb{Z}_b[x]$ with $p(x)q(x) = x^k - 1 \in \mathbb{Z}_b[x]$ is $k = b^m - 1$. It can be shown that every primitive polynomial is irreducible. Further, if p is primitive, then the polynomial x is a primitive element in $\mathbb{Z}_b[x]/p$, that is, $\{x^k \bmod p : k = 0, 1, \ldots, b^m - 2\} = (\mathbb{Z}_b[x]/p) \setminus \{0\}$.

Example 2.18 (Sobol′ sequence). Sobol′ (1967) was the first to introduce a construction of (t,s)-sequences over $\mathbb{Z}_2$, which are now referred to as *Sobol′ sequences.* These are digital sequences over the finite field $\mathbb{Z}_2$.

(1) Let $p_1, \ldots, p_s \in \mathbb{Z}_2[x]$ be distinct primitive polynomials ordered according to their degree, and let

$$p_j(x) = x^{e_j} + a_{1,j}x^{e_j-1} + a_{2,j}x^{e_j-2} + \cdots + a_{e_j-1,j}x + 1, \quad \text{for } 1 \leq j \leq s,$$

where $a_{j,k} \in \mathbb{Z}_b$. Note that e_j denotes the degree of the polynomial p_j.

(2) Choose odd natural numbers $1 \leq m_{1,j}, \ldots, m_{e_j,j}$ such that $m_{k,j} < 2^k$ for $1 \leq k \leq e_j$, and for all $k > e_j$ define $m_{k,j}$ recursively by

$$m_{k,j} = 2a_{1,j}m_{k-1,j} \oplus \cdots \oplus 2^{e_j-1}a_{e_j-1,j}m_{k-e_j+1,j} \oplus 2^{e_j}m_{k-e_j,j} \oplus m_{k-e_j,j},$$

where $\oplus$ is the bit-by-bit exclusive-or operator.

(3) The so-called *direction numbers* are defined by

$$v_{k,j} := \frac{m_{k,j}}{2^k}, \quad \text{for } k \geq 1.$$

(4) Then for $i \in \mathbb{N}_0$ with dyadic expansion $i = i_0 + 2i_1 + \cdots + 2^{r-1}i_{r-1}$ we define

$$t_{i,j} = i_0 v_{1,j} \oplus i_1 v_{2,j} \oplus \cdots \oplus i_{r-1} v_{r,j}.$$

The Sobol′ sequence is then the sequence of points $\boldsymbol{t}_0, \boldsymbol{t}_1, \ldots,$ where $\boldsymbol{t}_i = (t_{i,1}, \ldots, t_{i,s})$.

Sobol′ sequences are digital (t, s)-sequences with

$$t = \sum_{j=1}^{s} (e_j - 1).$$

This result holds for all choices of direction numbers, and therefore the direction numbers do not influence the overall quality of Sobol′ sequences. Note, however, that the t-value for a Sobol′ net of b^m points for different m can be smaller than the t-value of the full sequence, and hence the choice of direction numbers can affect the quality of a Sobol′ net.

Example 2.19 (Faure sequence). Faure (1982) introduced a construction of $(0, s)$-sequences over prime fields $\mathbb{Z}_b$ with $s \leq b$, which are now called *Faure sequences.* The generating matrices $C_1, \ldots, C_s$ are given by

$$C_j = (P^\top)^{j-1} \pmod b \quad \text{for } 1 \leq j \leq s,$$

where P is the *Pascal matrix* given by

$$P = \begin{pmatrix} \binom{0}{0} & \binom{0}{1} & \binom{0}{2} & \cdots \\ \binom{1}{0} & \binom{1}{1} & \binom{1}{2} & \cdots \\ \binom{2}{0} & \binom{2}{1} & \binom{2}{2} & \cdots \\ \vdots & \vdots & \vdots & \ddots \end{pmatrix},$$

where we set $\binom{k}{\ell} = 0$ for $\ell > k$. A Faure sequence is a digital $(0, s)$-sequence.

Example 2.20 (Niederreiter sequence). Niederreiter (1987) introduced the general concept of digital (t, s)-sequences, which includes both the Sobol′ and Faure constructions as special cases. We describe a special case of this sequence in the following.

Let $\mathbb{Z}_b = \{0, 1, \ldots, b-1\}$ denote the finite field of prime order b with addition and multiplication modulo b. Let $\mathbb{Z}_b[x]$ denote the set of polynomials with coefficients in $\mathbb{Z}_b$. Addition and multiplication of polynomials in $\mathbb{Z}_b$ is defined by using arithmetic in $\mathbb{Z}_b$ (*i.e.*, modulo b). This can also be extended to division of polynomials in $\mathbb{Z}_b$, which gives rise to the set of formal series $\mathbb{Z}_b((x^{-1}))$, which are series of the form

$$\sum_{\ell=w}^{\infty} a_\ell x^{-\ell}$$

for some integer w and coefficients $a_\ell \in \mathbb{Z}_b$. Note that this set contains the set of polynomials $\mathbb{Z}_b[x]$. This set now permits addition, subtraction, multiplication and division (except dividing by 0 of course) of formal series where the coefficients are added, subtracted, multiplied and divided modulo b. Thus $\mathbb{Z}_b((x^{-1}))$ is a field, which is called the field of *formal Laurent series*. The construction of the generating matrices is based on arithmetic in this field.

(1) Let $p_1, p_2, \ldots, p_s \in \mathbb{Z}_b[x]$ be distinct monic irreducible polynomials over $\mathbb{Z}_b$ (monic means that the coefficient of the monomial in p_j with the highest degree is 1). Let $e_j = \deg(p_j)$ for $1 \le j \le s$. The best results are obtained by choosing the degree of the polynomials $p_1, p_2, \ldots, p_s$ as small as possible.

(2) For integers $1 \le j \le s$, $u \ge 1$ and $0 \le k < e_j$, consider the expansions
$$\frac{x^{e_j-k-1}}{p_j(x)^u} = \sum_{\ell=0}^{\infty} a^{(j)}(u,k,\ell) x^{-\ell-1} \tag{2.6}$$
over the field of formal Laurent series $\mathbb{Z}_b((x^{-1}))$.

(3) Then we define the entries in the matrix $C_j = (c_{j,i,\ell})_{i,\ell\in\mathbb{N}}$ in the following way:
$$c_{j,i,\ell} = a^{(j)}(Q+1,k,\ell) \in \mathbb{Z}_b \quad \text{for } 1 \le j \le s, i \ge 1, \ell \ge 0,$$
where $i-1 = Qe_j + k$ with integers $Q = Q(i,j)$ and $k = k(i,j)$, with k satisfying $0 \le k < e_j$.

We briefly describe how to compute the coefficients $a^{(j)}(u,k,\ell)$ recursively. Let u, j be fixed and let $v_\ell = a^{(j)}(u, e_j - 1, \ell)$ for all $\ell \ge 0$. Then $a^{(j)}(u,k,\ell) = v_{\ell+e_j-1-k}$ for $0 \le k < e_j$. Let $p_j(x)^u = x^{ue_j} - b_{ue_j-1}x^{ue_j-1} - \cdots - b_0$, where we assume without loss of generality that $p_j(x)$ is monic (*i.e.*, the coefficient of x^{e_j} is 1). Then (2.6) can be written as
$$1 = (v_0 x^{-1} + v_1 x^{-2} + \cdots)(x^{ue_j} - b_{ue_j-1}x^{ue_j-1} - \cdots - b_0).$$
By comparing coefficients we obtain $v_0 = \cdots = v_{ie_j-2} = 0$, $v_{ie_j-1} = 1$ and
$$v_{ue_j+\ell} = b_{ue_j-1}v_{ue_j+\ell-1} + \cdots + b_0 v_\ell \in \mathbb{Z}_b,$$
for all $\ell \ge 0$.

The Niederreiter sequence is a digital (t,s)-sequence with
$$t = \sum_{j=1}^{s} (e_j - 1).$$

The Faure sequence can be obtained as a special case of the Niederreiter sequence, which corresponds to the case where the base b is a prime number such that $b \geq s$ and $p_j(x) = x - j + 1$ for $1 \leq j \leq s$. The Sobol′ sequence is obtained by choosing $p_1(x) = x$ and $p_2, p_3, \ldots, p_s$ as primitive polynomials. This yields a Sobol′ sequence with a special choice of direction numbers. The Niederreiter sequence can also be generalized, where x^{e_j-k-1} is replaced by a polynomial of degree $e_j - k - 1$. The choice of polynomials then corresponds to choosing different direction numbers. This way one can obtain the generating matrices for a Sobol′ sequence. On the other hand, an algorithm similar to the one introduced above for Sobol′ sequences can also be obtained for Niederreiter sequences, which yields a fast implementation of Niederreiter sequences.

A practical implementation of digital nets and sequences can use a Gray code ordering to improve efficiency.

2.8. Polynomial lattice rule construction

In this subsection we discuss polynomial lattice rules, a concept analogous to lattice rules, but based on linear algebra over finite fields; they form a special class of digital nets.

Let b be prime, let p be a polynomial of degree m with coefficients in $\mathbb{Z}_b$, and let $q_1, \ldots, q_s$ be polynomials of degree at most $m - 1$ with coefficients in $\mathbb{Z}_b$. Then C_j, the jth generating matrix, can be chosen as follows.

(1) Let $u_1, u_2, \ldots \in \mathbb{Z}_b$ be such that

$$\frac{q_j(x)}{p(x)} = \frac{u_1}{x} + \frac{u_2}{x^2} + \cdots .$$

For given $q_j(x)$ and $p(x)$, the values of $u_1, u_2, \ldots$ can be obtained by equating coefficients in $q_j(x) = (u_1/x + u_2/x^2 + \cdots)p(x)$, noting that all additions and multiplications are to be done in $\mathbb{Z}_b$.

(2) Set

$$C_j = \begin{pmatrix} u_1 & u_2 & u_3 & \cdots & u_m \\ u_2 & u_3 & \iddots & \iddots & u_{m+1} \\ u_3 & \iddots & \iddots & \iddots & u_{m+2} \\ \vdots & \iddots & \iddots & \iddots & \vdots \\ u_m & u_{m+1} & u_{m+2} & \cdots & u_{2m-1} \end{pmatrix} \in \mathbb{Z}_b^{m \times m}.$$

The digital net with generating matrices $C_1, \ldots, C_s$ as defined above is called a *polynomial lattice point set*, and a QMC rule using a polynomial lattice point set is called a *polynomial lattice rule*. The polynomial p is referred to as the *modulus*, and the vector of polynomials $(q_1, \ldots, q_s)$ is referred to as the *generating vector*.

There is also a faster method for generating the cubature points of a polynomial lattice rule which does not use the generating matrices. For the following we need to introduce some definitions and some notation. As before, let $\mathbb{Z}_b[x]$ denote the set of polynomials with coefficients in $\mathbb{Z}_b$ and let $\mathbb{Z}_b((x^{-1}))$ denote the set of formal Laurent series

$$\sum_{\ell=w}^{\infty} a_\ell x^{-\ell},$$

where $a_\ell \in \mathbb{Z}_b$. For $m \in \mathbb{N}$ let υ_m be the map from $\mathbb{Z}_b((x^{-1}))$ to the interval $[0,1)$ defined by

$$\upsilon_m\left(\sum_{\ell=w}^{\infty} t_\ell x^{-\ell}\right) = \sum_{\ell=\max(1,w)}^{m} t_\ell b^{-\ell}.$$

We frequently associate a non-negative integer k, with b-adic expansion $k = \kappa_0+\kappa_1 b+\cdots+\kappa_a b^a$, with the polynomial $k(x) = \kappa_0+\kappa_1 x+\cdots+\kappa_a x^a \in \mathbb{Z}_b[x]$ and *vice versa.* Further, for arbitrary $\boldsymbol{k} = (k_1,\ldots,k_s) \in \mathbb{Z}_b[x]^s$ and $\boldsymbol{q} = (q_1,\ldots,q_s) \in \mathbb{Z}_b[x]^s$, we define the 'inner product'

$$\boldsymbol{k}\cdot\boldsymbol{q} = \sum_{i=1}^{s} k_i q_i \in \mathbb{Z}_b[x],$$

and we write $q \equiv 0 \pmod{p}$ if p divides q in $\mathbb{Z}_b[x]$.

With these definitions we can give the following equivalent but simpler form of the construction of $\mathcal{P}(\boldsymbol{q},p)$.

Theorem 2.21. Let b be a prime and let $m, s \in \mathbb{N}$. For $p \in \mathbb{Z}_b[x]$ with $\deg(p) = m$ and $\boldsymbol{q} = (q_1,\ldots,q_s) \in \mathbb{Z}_b[x]^s$, the polynomial lattice point set $\mathcal{P}(\boldsymbol{q},p)$ is the point set consisting of the b^m points

$$\boldsymbol{t}_i = \left(\upsilon_m\left(\frac{i(x)q_1(x)}{p(x)}\right),\ldots,\upsilon_m\left(\frac{i(x)q_s(x)}{p(x)}\right)\right) \in [0,1)^s,$$

for $i \in \mathbb{Z}_b[x]$ with $\deg(i) < m$.

The task of choosing good generating matrices is now replaced by the task of choosing a good generating vector $(q_1,\ldots,q_s)$. The total number of possible choices of generating matrices is reduced from $b^{m^2 s}$ to b^{ms}.

Example 2.22. In this example we show the polynomial lattice rule analogue of Fibonacci lattice rules; they are constructed via the continued fraction expansion of the ratio $q_2(x)/p(x)$. Let b be prime and let $m \geq 1$ be an integer. Set $q_1(x) = 1$. Let $p(x)$ be a polynomial over $\mathbb{Z}_b$ of degree m and let $q_2(x)$ be the polynomial over $\mathbb{Z}_b$ defined by

$$\frac{q_2(x)}{p(x)} = \cfrac{1}{1+x+\cfrac{1}{1+x+\cfrac{1}{1+x+\ddots}}}.$$

The corresponding polynomial lattice point set is a digital $(0, m, 2)$-net over $\mathbb{Z}_b$. For instance, let $b = 2$. Then for $m = 2$ we obtain

$$\frac{q_2(x)}{p(x)} = \frac{1}{1 + x + \frac{1}{1+x}} = \frac{1 + x}{(1 + x)^2 + 1} = \frac{1 + x}{x^2},$$

and thus $p(x) = x^2$ and $q_2(x) = 1 + x$. For $m = 3$ we obtain

$$\frac{q_2(x)}{p(x)} = \frac{1}{1 + x + \frac{1}{1+x+\frac{1}{1+x}}} = \frac{x^2}{(1 + x)x^2 + (1 + x)} = \frac{x^2}{x^3 + x^2 + x + 1},$$

and thus $p(x) = x^3 + x^2 + x + 1$ and $q_2(x) = x^2$.

2.9. Randomization and error estimation: shifted lattice rules

We recall that the mean-square MC error can be estimated in practice using (2.1). In comparison, a fully deterministic QMC method, although having a faster rate of convergence, is biased and lacks a practical error estimate. 'Randomized' QMC methods combine the best of both worlds; their advantages are as follows.

- Randomization yields an unbiased estimator.
- Randomization provides a practical error estimate.
- Randomized QMC methods enjoy faster rates of convergence than the MC method for smooth functions.
- Some randomization technique (see 'scrambling' below) can further improve the QMC rate of convergence by an additional $O(n^{-1/2})$.

Here we discuss the simplest form of randomization called *shifting*, and we explain how an error estimate can be obtained in practice. We begin with a remark that *shifting preserves the lattice structure*, and therefore this randomization technique typically goes with lattice rules, to yield so-called *shifted lattice rules*. Having said that, shifting can be used together with any QMC method to obtain a practical error estimate (however, it need not preserve the original structure of the QMC point set).

The idea behind shifting is to move all the points in the same direction by the same amount. If any point falls outside the unit cube then it is 'wrapped' back into the cube from the opposite side. More precisely, given a vector $\boldsymbol{\Delta} \in [0, 1]^s$, known as the *shift*, the $\boldsymbol{\Delta}$-shift of the QMC points $\boldsymbol{t}_0, \ldots, \boldsymbol{t}_{n-1}$ yields points

$$\{\boldsymbol{t}_i + \boldsymbol{\Delta}\}, \quad i = 0, 1, \ldots, n - 1, \tag{2.7}$$

where, as in (2.2), the braces indicate taking the fractional parts. Figure 2.3 illustrates how shifting is done.

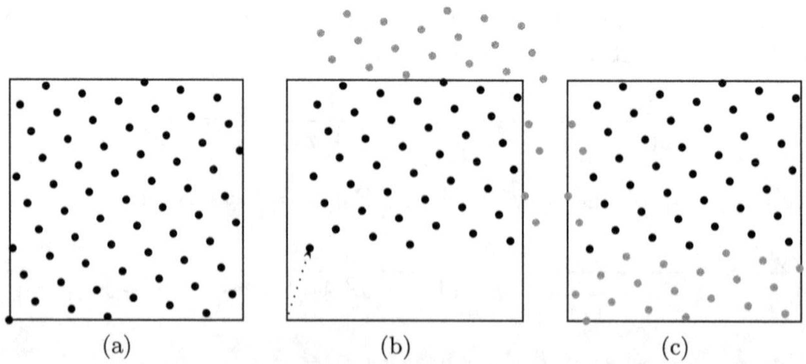

Figure 2.3. Applying a (0.1, 0.3)-shift to a 64-point lattice rule in two dimensions: (a) original lattice rule, (b) moving all points by (0.1, 0.3), (c) wrapping the points back inside the unit cube.

For a random shift $\boldsymbol{\Delta} \in [0,1]^s$, the shifted QMC points $\{\boldsymbol{t}_i + \boldsymbol{\Delta}\}$, $i = 0, 1, \ldots, n-1$ are correlated. Therefore we cannot estimate the variance of the shifted QMC rule using the sample variance as in (2.1). Instead, we need to use a number of independent random shifts as follows.

(1) We generate q independent random shifts $\boldsymbol{\Delta}_0, \boldsymbol{\Delta}_1, \ldots, \boldsymbol{\Delta}_{q-1}$ from the uniform distribution on $[0,1]^s$.

(2) For a given QMC rule, we form the approximations $Q_{n,s}^{(0)}(f)$, $Q_{n,s}^{(1)}(f)$, $\ldots$, $Q_{n,s}^{(q-1)}(f)$, where

$$Q_{n,s}^{(k)}(f) = \frac{1}{n} \sum_{i=0}^{n-1} f(\{\boldsymbol{t}_i + \boldsymbol{\Delta}_k\}), \quad k = 0, 1, \ldots, q-1,$$

is the approximation of the integral using a $\boldsymbol{\Delta}_k$-shift of the original QMC rule.

(3) We take the average

$$\bar{Q}_{n,s,q}(f) = \frac{1}{q} \sum_{k=0}^{q-1} Q_{n,s}^{(k)}(f)$$

as our final approximation to the integral.

(4) An unbiased estimate for the mean-square error of $\bar{Q}_{n,s,q}(f)$ is given by

$$\frac{1}{q(q-1)} \sum_{k=0}^{q-1} (Q_{n,s}^{(k)}(f) - \bar{Q}_{n,s,q}(f))^2.$$

Typically we take n in the thousands or more while keeping q small, say around 10–50. To obtain a fair comparison between the MC method and a randomized QMC method (*e.g.*, for the two methods to have the same number of function evaluations), we should take $n^{\text{MC}} = q \cdot n^{\text{QMC}}$.

Scrambling is a popular but more complicated randomization method. The idea is to randomly and recursively permute the points between elementary intervals so that *the digital net structure is preserved.* More details will be given in the next subsection. For now we just introduce a related randomization technique called *digital shift*, which is similar to shifting. We just need to replace (2.7) by

$$\boldsymbol{t}_i \oplus \boldsymbol{\Delta}, \quad i = 0, 1, \ldots, n-1,$$

where $\oplus$ denotes the digit-wise addition operator in base b defined as follows. If $x, \sigma \in [0,1)$ have base b representations

$$x = (0.x_1x_2\cdots)_b \quad \text{and} \quad \sigma = (0.\sigma_1\sigma_2\cdots)_b,$$

then $y = x \oplus \sigma = (0.y_1y_2\cdots)_b$, where

$$y_i = (x_i + \sigma_i) \bmod b.$$

The procedure for estimating the error is the same as for shifting.

For later use we also define digit-wise subtraction $y = x \ominus \sigma = (0.y_1y_2\cdots)_b$, which is defined by $y_i = (x_i - \sigma_i) \bmod b$.

2.10. Scrambled nets and sequences

Randomization methods are designed to introduce randomness into the point sets but at the same time should preferably keep the relevant structure intact. In the case of (t,m,s)-nets this means that the net should still be a (t,m,s)-net after the randomization (with probability 1). This can be achieved by the following algorithm introduced by Owen (1997*a*).

We first introduce *Owen's scrambling algorithm*, which is most easily described for some generic point $\boldsymbol{x} \in [0,1)^s$, with $\boldsymbol{x} = (x_1, \ldots, x_s)$ and $x_j = x_{j,1}b^{-1} + x_{j,2}b^{-2} + \cdots$. The scrambled point will be denoted by $\boldsymbol{y} \in [0,1)^s$, where $\boldsymbol{y} = (y_1, \ldots, y_s)$ and $y_j = y_{j,1}b^{-1} + y_{j,2}b^{-2} + \cdots$. The point $\boldsymbol{y}$ is obtained by applying permutations to each digit of each coordinate of $\boldsymbol{x}$. The permutation applied to $x_{j,\ell}$ depends on $x_{j,k}$ for $1 \le k < \ell$. Specifically, $y_{j,1} = \pi_j(x_{j,1})$, $y_{j,2} = \pi_{j,x_{j,1}}(x_{j,2})$, $y_{j,3} = \pi_{j,x_{j,1},x_{j,2}}(x_{j,3})$, and in general

$$y_{j,k} = \pi_{j,x_{j,1},\ldots,x_{j,k-1}}(x_{j,k}), \tag{2.8}$$

where $\pi_{j,x_{j,1},\ldots,x_{j,k-1}}$ is a random permutation of $\{0, \ldots, b-1\}$. We assume that permutations with different indices are chosen mutually independent from each other and that each permutation is chosen with the same probability. In this case the scrambled point $\boldsymbol{y}$ is uniformly distributed in $[0,1)^s$. We show this fact in Section 6.

We now introduce some convenient notation to describe Owen's scrambling. For $1 \le j \le s$ let

$$\Pi_j = \{\pi_{j,x_{j,1},\ldots x_{j,k-1}} : k \in \mathbb{N}, x_{j,1},\ldots,x_{j,k-1} \in \{0,\ldots,b-1\}\}$$

(where for $k = 1$ we set $\pi_{j,x_{j,1},\ldots x_{j,k-1}} = \pi_j$) be a given set of permutations, and let $\boldsymbol{\Pi} = (\Pi_1,\ldots,\Pi_s)$. Then, when applying Owen's scrambling using these permutations to some point $\boldsymbol{x} \in [0,1)^s$, we write $\boldsymbol{y} = \boldsymbol{\Pi}(\boldsymbol{x})$, where $\boldsymbol{y}$ is the point obtained by applying Owen's scrambling to $\boldsymbol{x}$ using the permutations $\Pi_1,\ldots,\Pi_s$. For $x \in [0,1)$ we drop the subscript j and just write $y = \Pi(x)$.

To scramble a (t,m,s)-net or a (t,s)-sequence, we choose the permutations in $\boldsymbol{\Pi}$ for each coordinate independently from each other and then apply these permutations to each point of the (t,m,s)-net or (t,s)-sequence as explained above. For instance, for a given (t,m,s)-net $\boldsymbol{t}_0,\boldsymbol{t}_1,\ldots,\boldsymbol{t}_{b^m-1}$, the Owen-scrambled (t,m,s)-net is given by

$$\boldsymbol{\Pi}(\boldsymbol{t}_0),\boldsymbol{\Pi}(\boldsymbol{t}_1),\ldots,\boldsymbol{\Pi}(\boldsymbol{t}_{b^m-1}).$$

In practice, the number of permutations one has to choose is large, but there are simplifications of the randomization procedure which greatly reduce this number. This yields a fast scrambling algorithm. More details are given in Section 6.

In Section 6 we show that scrambling also yields an unbiased estimate of the integral, that is,

$$\mathbb{E}\left(\frac{1}{n}\sum_{i=0}^{n-1} f(\boldsymbol{\Pi}(\boldsymbol{t}_i))\right) = \int_{[0,1]^s} f(\boldsymbol{x})\,\mathrm{d}\boldsymbol{x}.$$

In the same way as for random (digital) shifts, one can also obtain an unbiased estimate of the mean-square error for scrambling. It is sufficient to choose only one set of random permutations $\boldsymbol{\Pi}$. Let

$$Q_{n_1,n_2,s}(f) = \frac{1}{n_2-n_1}\sum_{i=n_1}^{n_2-1} f(\boldsymbol{\Pi}(\boldsymbol{t}_i)).$$

Let $1 \le m' < m$. The mean-square error of the cubature rule applied to f can now be estimated by

$$\frac{1}{b^{m'}(b^{m'}-1)}\sum_{i=0}^{b^{m'}-1}\left(Q_{ib^{m-m'},(i+1)b^{m-m'},s}(f) - Q_{0,b^m,s}(f)\right)^2,$$

where m' can be chosen such that, say, $b^{m'} \approx 30$.

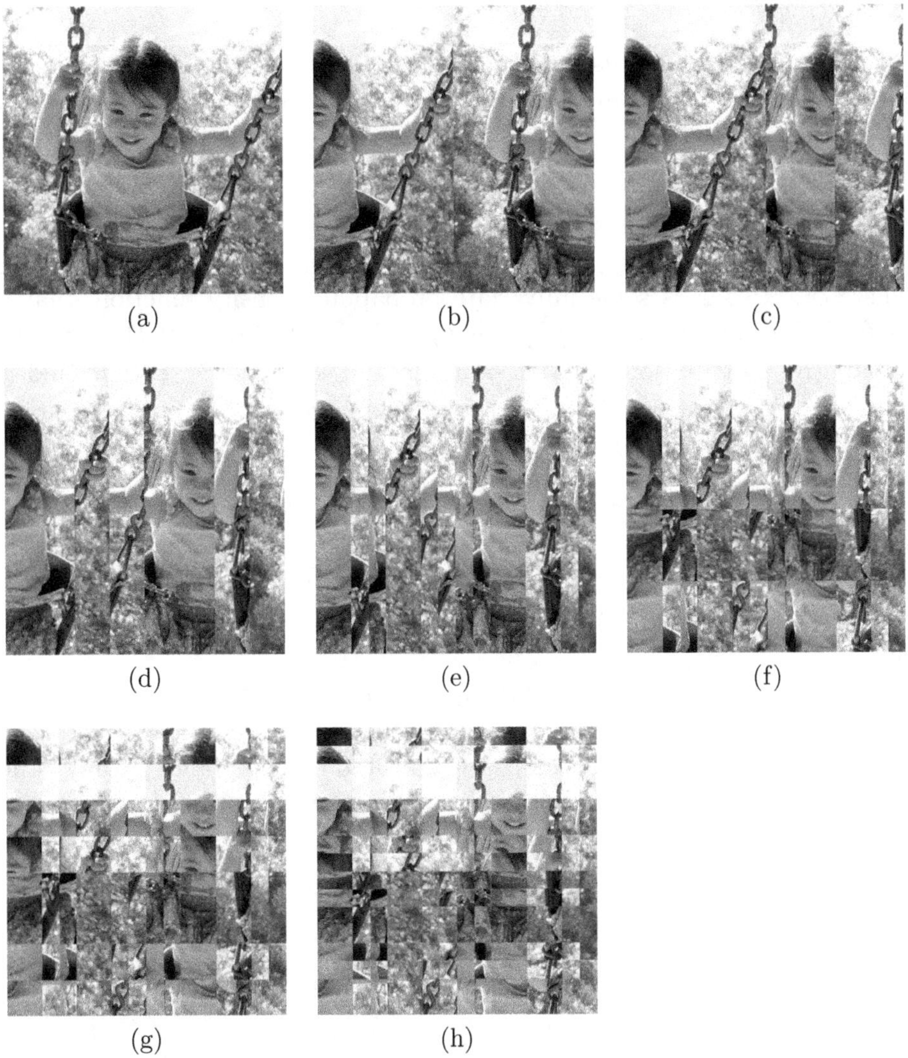

Figure 2.4. Owen's scrambling in base 2: (a) original; (b) swap left and right halves; (c) swap 3rd and 4th vertical quarters; (d) swap 3rd and 4th, 7th and last vertical eighths; (e) swap 3rd and 4th, 7th and 8th, 9th and 10th, 15th and last sixteenths; (f) swap 1st and 2nd horizontal quarters; (g) swap 1st and 2nd, 5th and 6th, 7th and last horizontal eighths; (h) swap 3rd and 4th, 7th and 8th, 9th and 10th, 15th and last horizontal sixteenths.

2.11. *Transformation to the unit cube*

In this subsection we discuss some simple strategies for transforming an integral over $\mathbb{R}^s$ to an integral over $[0,1]^s$. We begin with the univariate case.

Consider an integral

$$\int_{-\infty}^{\infty} g(y)\,\phi(y)\,\mathrm{d}y,$$

where $\phi : \mathbb{R} \to \mathbb{R}$ is some univariate probability density function, that is, $\phi(y) \geq 0$ for all $y \in \mathbb{R}$ and $\int_{-\infty}^{\infty} \phi(y)\,\mathrm{d}y = 1$. Let $\Phi : \mathbb{R} \to [0,1]$ denote the cumulative distribution function of ϕ, that is, $\Phi(y) := \int_{-\infty}^{y} \phi(t)\,\mathrm{d}t$, and let $\Phi^{-1} : [0,1] \to \mathbb{R}$ denote the inverse of the cumulative distribution function. Then we can use the substitution (or change of variables)

$$x = \Phi(y) \iff y = \Phi^{-1}(x),$$

to obtain

$$\int_{-\infty}^{\infty} g(y)\,\phi(y)\,\mathrm{d}y = \int_0^1 g(\Phi^{-1}(x))\,\mathrm{d}x = \int_0^1 f(x)\,\mathrm{d}x,$$

with the transformed integrand defined by $f := g \circ \Phi^{-1}$.

Note that this is not the only way to obtain a transformed integrand. Indeed, we can divide and multiply the original integrand by any other probability density function $\tilde{\phi}$, and then map to $[0,1]$ using its inverse cumulative distribution function $\tilde{\Phi}^{-1}$:

$$\begin{aligned}\int_{-\infty}^{\infty} g(y)\,\phi(y)\,\mathrm{d}y &= \int_{-\infty}^{\infty} \frac{g(y)\,\phi(y)}{\tilde{\phi}(y)}\,\tilde{\phi}(y)\,\mathrm{d}y \\ &= \int_{-\infty}^{\infty} \tilde{g}(y)\,\tilde{\phi}(y)\,\mathrm{d}y = \int_0^1 \tilde{g}(\tilde{\Phi}^{-1}(x))\,\mathrm{d}x = \int_0^1 \tilde{f}(x)\,\mathrm{d}x,\end{aligned}$$

where $\tilde{g}(y) := g(y)\,\phi(y)/\tilde{\phi}(y)$, giving a different transformed integrand $\tilde{f} := \tilde{g} \circ \tilde{\Phi}^{-1}$. Ideally we would like to choose a density function $\tilde{\phi}$ that leads to a 'nice' integrand in the unit cube. This is related to the concept of *importance sampling* for MC methods.

The MC approximation for the integral over $\mathbb{R}$ is

$$\int_{-\infty}^{\infty} g(y)\,\phi(y)\,\mathrm{d}y \approx \frac{1}{n}\sum_{i=0}^{n-1} g(\tau_i),$$

where the MC points $\tau_0, \ldots, \tau_{n-1}$ are randomly sampled following the distribution of ϕ. Depending on the choice of ϕ, there may exist an algorithm to sample from the distribution directly, or it might be necessary to generate random samples from the uniform distribution on $[0,1]$and then map them back to $\mathbb{R}$ using Φ^{-1}. The latter approach is consistent with the MC method over $[0, 1]$,

$$\int_{-\infty}^{\infty} g(y)\,\phi(y)\,\mathrm{d}y = \int_0^1 g(\Phi^{-1}(x))\,\mathrm{d}x \approx \frac{1}{n}\sum_{i=0}^{n-1} g(\Phi^{-1}(t_i)),$$

where $t_0, \ldots, t_{n-1}$ are uniform random samples from $[0, 1]$. If we use a different density function $\tilde{\phi}$ to sample the points from, then we evaluate the points for a different function $\tilde{g}$:

$$\int_{-\infty}^{\infty} g(y)\,\phi(y)\,\mathrm{d}y = \int_{-\infty}^{\infty} \tilde{g}(y)\,\tilde{\phi}(y)\,\mathrm{d}y \approx \frac{1}{n}\sum_{i=0}^{n-1} \tilde{g}(\tau_i).$$

The idea behind importance sampling is to choose a sampling density $\tilde{\phi}$ to minimize the variance of the resulting function $\tilde{g}$.

This transformation strategy can be generalized to s dimensions as follows. If we have a product of univariate densities, then we can apply the mapping Φ^{-1} component-wise,

$$\boldsymbol{y} = \Phi^{-1}(\boldsymbol{x}) := (\Phi^{-1}(x_1), \ldots, \Phi^{-1}(x_s)),$$

to obtain

$$\int_{\mathbb{R}^s} g(\boldsymbol{y}) \prod_{j=1}^{s} \phi(y_j)\,\mathrm{d}\boldsymbol{y} = \int_{[0,1]^s} g(\Phi^{-1}(\boldsymbol{x}))\,\mathrm{d}\boldsymbol{x} = \int_{[0,1]^s} f(\boldsymbol{x})\,\mathrm{d}\boldsymbol{x}.$$

Note that we can always multiply and divide any given integrand by such a product provided that $\phi(y) > 0$ for all $y \in \mathbb{R}$:

$$\int_{\mathbb{R}^s} h(\boldsymbol{y})\,\mathrm{d}\boldsymbol{y} = \int_{\mathbb{R}^s} \frac{h(\boldsymbol{y})}{\prod_{j=1}^{s} \phi(y_j)} \prod_{j=1}^{s} \phi(y_j)\,\mathrm{d}\boldsymbol{y}.$$

Thus, this strategy always works in principle, but the resulting transformed integrand might not be 'nice'.

Many integrals from practical models involve the multivariate normal density. If the multivariate normal density is the dominating part of the entire integrand, then a good strategy is to *factorize the covariance matrix* Σ, that is, to find an $s \times s$ matrix A such that

$$\Sigma = AA^{\mathrm{T}}, \tag{2.9}$$

and then use the substitutions (treating all vectors as column vectors)

$$\boldsymbol{y} = A\boldsymbol{z} \quad \text{followed by} \quad \boldsymbol{z} = \Phi^{-1}(\boldsymbol{x}),$$

to obtain

$$\begin{aligned}\int_{\mathbb{R}^s} g(\boldsymbol{y}) \frac{\exp(-\frac{1}{2}\boldsymbol{y}^{\mathrm{T}}\Sigma^{-1}\boldsymbol{y})}{\sqrt{(2\pi)^s \det(\Sigma)}} \,\mathrm{d}\boldsymbol{y} &= \int_{\mathbb{R}^s} g(A\boldsymbol{z}) \frac{\exp(-\frac{1}{2}\boldsymbol{z}^{\mathrm{T}}\boldsymbol{z})}{\sqrt{(2\pi)^s}} \,\mathrm{d}\boldsymbol{z} \\ &= \int_{\mathbb{R}^s} g(A\boldsymbol{z}) \prod_{j=1}^{s} \frac{\exp(-\frac{1}{2}z_j^2)}{\sqrt{2\pi}} \,\mathrm{d}\boldsymbol{z} \\ &= \int_{[0,1]^s} g(A\Phi^{-1}(\boldsymbol{x})) \,\mathrm{d}\boldsymbol{x} = \int_{[0,1]^s} f(\boldsymbol{x}) \,\mathrm{d}\boldsymbol{x}.\end{aligned}$$

The factorization (2.9) is *not* unique. Two obvious choices are the *Cholesky factorization* with lower triangular matrix A, and the *principal components factorization*, which is given by

$$A = \big[\sqrt{\lambda_1}\boldsymbol{\eta}_1; \cdots ; \sqrt{\lambda_s}\boldsymbol{\eta}_s\big],$$

where $(\lambda_j, \boldsymbol{\eta}_j)_{j=1}^s$ denotes the set of eigenpairs of Σ, with ordered eigenvalues $\lambda_1 \geq \lambda_2 \geq \cdots \geq \lambda_s$ and unit-length column eigenvectors $\boldsymbol{\eta}_1, \ldots, \boldsymbol{\eta}_s$. Other factorizations are possible: for example, in finance applications the covariance matrix arising from the time-discretized Brownian paths can also be factorized using the 'Brownian bridge' technique. In general, factorization of Σ in very high dimensions can be very costly, and the problem can be poorly conditioned.

In some applications the multivariate normal density is *not* the dominating part of the entire integrand. In those cases, other transformation steps (such as recentering and rescaling) might be required to capture the main feature of the integrand: see, for example, Kuo *et al.* (2008*a*).

2.12. Notes

Halton (1960) introduced what is now called the Halton sequence. Discrepancy bounds for Halton sequences where the constant decreases with the dimension have been shown by Atanassov (2004*b*). Kronecker sequences were studied by Niederreiter (1978). Lattice rules were introduced by Korobov (1959) and have also been discovered by Hlawka (1962). For the early history of QMC point sets and related concepts see Niederreiter (1978), Kuipers and Niederreiter (1974), Niederreiter (1992*a*) and Sloan and Joe (1994). More recent results are covered by Dick and Pillichshammer (2010).

For an efficient implementation of Sobol′ sequences see Antonov and Saleev (1979) and Bratley and Fox (1988), and for questions concerning the choice of primitive polynomials and direction numbers, see, for example, Joe and Kuo (2008). The implementation of Faure sequences has been discussed in Atanassov (2004*a*) and Fox (1986). Implementation of the Niederreiter sequence has been discussed in Bratley, Fox and Niederreiter (1992). For digital sequences with improved t-value see Niederreiter and

Xing (1995, 1996*a*, 1996*b*) and Xing and Niederreiter (1995). For an implementation of such sequences see Pirsic (2002).

It has been observed that the quality of QMC point sets usually decreases as the dimension increases. This has led to the suggestion to study 'mixed' point sets, that is, the first few coordinates are QMC point sets with the remaining coordinates filled up by, say, random numbers. This fits in with the structure of integrands whose importance is concentrated in the first few coordinates. The idea of combining the advantages of QMC methods and MC methods was first proposed by Spanier, who applied mixed QMC–MC sequences to particle transport problems. Probabilistic discrepancy bounds for such 'mixed sequences' have been provided in Ökten (1996), Ökten, Tuffin and Burago (2006), Gnewuch (2009), Gnewuch and Roşca (2009) and Aistleitner and Hofer (2012); in these papers it is assumed that the Monte Carlo components of the points consist of idealized random numbers. There have also been studies of mixed point sets whose remaining coordinates are filled up by deterministic pseudo-random numbers or by the components of other deterministic point sets. Deterministic discrepancy bounds for such point sets can be found in Niederreiter (2009, 2010*a*, 2010*b*), Niederreiter and Winterhof (2011), Hofer and Kritzer (2011) and Niederreiter (2012). Further mixed sequences using Halton, Kronecker and digital sequences have been studied by Hofer and Larcher (2010, 2012) and Hofer, Kritzer, Larcher and Pillichshammer (2009). So-called $(t, \boldsymbol{e}, s)$-sequences have been introduced by Tezuka (2013). This concept captures the dependence on the dimension of digital nets more precisely and leads to improvements of discrepancy bounds in terms of their dependence on the dimension. Digital sequences in which the generating matrices have only a finite number of non-zero elements in each row have been studied, for instance, by Hofer and Pirsic (2011) and Hofer and Niederreiter (2013). The latter paper constructs digital (t, s)-sequences with finite-row generating matrices and asymptotically optimal t-values (in the sense of fixed base and dimension s going to ∞). Such sequences with finite-row generating matrices may be advantageous in implementations.

More information on transformations from $\mathbb{R}^s$ to the unit cube $[0, 1]^s$, variance reduction techniques, and other information for the practical use of MC and QMC, can be found in Lemieux (2009). The monographs by Devroye (1986) and Hörmann, Leydold and Derflinger (2010) are primarily concerned with transformations. Integration of functions with singularities has been discussed by Owen (2006). Kuo *et al.* (2008*a*) considered transformation strategies for applying QMC methods to maximum likelihood integrals from statistics. Kuo, Sloan, Wasilkowski and Waterhouse (2010*a*) analysed lattice rules for unbounded integrands arising from the transformation. A construction of digital nets constructed in $\mathbb{R}^s$ has been studied by Dick (2011*b*).

Information relevant to statistical applications can be found in Fang and Wang (1994). Control variates for QMC have been discussed by Hickernell, Lemieux and Owen (2005). QMC methods have been applied to computer graphics by Keller (2006, 2013), to transport problems by Spanier, and to experimental designs by Hickernell (1999).

An introduction to lattice rules can be found in Sloan and Joe (1994); see also Niederreiter (1992*a*, Chapter 5). Dick and Pillichshammer (2010) provide a comprehensive introduction to (digital) (t, m, s)-nets, (digital) (t, s)-sequences and related point sets and sequences. A survey of QMC methods and their randomization strategies is given by Hickernell and Hong (2002).

Sparse grids were discovered independently by Smolyak (1963) and Zenger (1991). See also Wasilkowski and Woźniakowski (1995) and the review article by Bungartz and Griebel (2004).

3. Error, discrepancy, and reproducing kernel

In this section we present the necessary ingredients for the error analysis of QMC methods. To provide an accessible entry point for readers who are new to QMC, we begin by studying QMC in its simplest setting, namely, equal-weight quadrature rules for integration over the unit interval $[0, 1]$.

We will meet the useful notion of reproducing kernel Hilbert spaces, and meet specific examples of function spaces that have proved useful for designing and analysing QMC methods. All of the spaces we deal with in this and the next section are spaces of non-periodic functions. We follow contemporary practice in referring to such spaces as 'Sobolev' spaces, to distinguish them from 'Korobov' spaces of periodic functions, to be considered later in Section 5.8.

3.1. Error analysis in one dimension

We consider real-valued functions f defined on the interval $[0, 1]$. As is common in numerical analysis, we require that the functions have some smoothness and that the fundamental theorem of calculus holds, so that

$$f(x) = f(1) - \int_x^1 f'(y)\,\mathrm{d}y = f(1) - \int_0^1 f'(y) 1_{[0,y]}(x)\,\mathrm{d}y, \tag{3.1}$$

where the indicator function $1_{[0,y]}$ is defined by

$$1_{[0,y]}(x) = \begin{cases} 1 & \text{if } x \in [0, y], \\ 0 & \text{if } x \notin [0, y]. \end{cases}$$

We consider QMC rules of the form

$$Q(f) = \frac{1}{n}\sum_{i=0}^{n-1} f(t_i),$$

and study their integration error given by

$$\mathrm{error}_n(f;Q) = \int_0^1 f(x)\,\mathrm{d}x - \frac{1}{n}\sum_{i=0}^{n-1} f(t_i).$$

Upon substituting (3.1) in the equation above and changing the order of integration, we obtain

$$\mathrm{error}_n(f;Q) = \int_0^1 \left(-\int_0^1 1_{[0,y]}(x)\,\mathrm{d}x + \frac{1}{n}\sum_{i=0}^{n-1} 1_{[0,y]}(t_i) \right) f'(y)\,\mathrm{d}y,$$

or

$$\mathrm{error}_n(f;Q) = \int_0^1 \Delta_P(y) f'(y)\,\mathrm{d}y, \tag{3.2}$$

where $P := \{t_0, \ldots, t_{n-1}\} \subset [0,1]$ denotes the set of quadrature points, and $\Delta_P(y)$ is the *local discrepancy* of the point set P, defined by

$$\Delta_P(y) := \frac{1}{n}\sum_{i=0}^{n-1} 1_{[0,y]}(t_i) - \int_0^1 1_{[0,y]}(x)\,\mathrm{d}x, \quad y \in [0,1]. \tag{3.3}$$

Note that the local discrepancy function Δ_P is just the integration error for the indicator function $1_{[0,y]}$.

Thus from (3.2) the integration error can be viewed as the L_2 inner product of the local discrepancy function Δ_P and the derivative f' of the integrand.

Using Hölder's inequality we obtain from (3.2)

$$\begin{aligned} |\mathrm{error}_n(f;Q)| &\le \left(\int_0^1 |\Delta_P(y)|^p\,\mathrm{d}y\right)^{1/p} \left(\int_0^1 |f'(y)|^q\,\mathrm{d}y\right)^{1/q} \\ &= \|\Delta_P\|_{L_p} \|f'\|_{L_q}, \quad \frac{1}{p} + \frac{1}{q} = 1, \end{aligned} \tag{3.4}$$

where $\|g\|_{L_p} = (\int_0^1 |g(y)|^p\,\mathrm{d}y)^{1/p}$, and we make the usual modification if $p = \infty$. If $p = \infty$ and $q = 1$, this is a special case of *Koksma's inequality* (Koksma's inequality holds when one replaces $\|f'\|_{L_1}$ with the variation of f). The quantity $\|\Delta_P\|_{L_p}$ is referred to as the *discrepancy* of the point set P, and is closely related to the so-called *worst-case error*, which we shall introduce formally in Section 3.3.

Several conclusions can be drawn from this elementary inequality. First, the upper bound (3.4) is best possible: indeed, equality holds if $\Delta_p(y)f'(y) \ge 0$ and $|f'(y)|^q$ is a multiple of $|\Delta_P(y)|^p$. Second, to minimize the integration error for all functions for which $\|f'\|_{L_q} < \infty$, one should choose quadrature points P for which the discrepancy $\|\Delta_P\|_{L_p}$ is, in some sense, small. Third, it is enough to study the integration error of the indicator functions $1_{[0,y]}$ for $y \in [0,1]$, since the local discrepancy function Δ_P evaluated at y is the integration error of the indicator function $1_{[0,y]}$.

The above analysis of the integration error rests entirely on the integral representation of the function f given by

$$f(x) = f(1) - \int_0^1 f'(y) 1_{[0,y]}(x) \, \mathrm{d}y. \tag{3.5}$$

For two functions f, g that permit such a representation, we can define an inner product

$$\langle f, g \rangle := f(1)g(1) + \int_0^1 f'(y) g'(y) \, \mathrm{d}y, \tag{3.6}$$

and a corresponding norm

$$\|f\| = \sqrt{|f(1)|^2 + \int_0^1 |f'(y)|^2 \, \mathrm{d}y}.$$

Then (3.4) holds with $p = q = 2$ for all functions in the function space

$$H := \{f : [0,1] \to \mathbb{R} : f \text{ is absolutely continuous and } \|f\| < \infty\}.$$

3.2. Reproducing kernel Hilbert spaces

We have seen above that the integration error for functions $f \in H$ has the integral representation (3.2) involving the local discrepancy function. This rests in turn on the representation of f given by (3.5). We now express (3.5) in the language of *reproducing kernels*. The idea is to find a function $K : [0,1] \times [0,1] \to \mathbb{R}$ such that

$$\langle f, K(\cdot, x) \rangle = f(x) \quad \text{for all } f \in H \text{ and all } x \in [0,1].$$

For this to hold we must have, from (3.6) and (3.5),

$$f(1)K(1,x) + \int_0^1 f'(y) \frac{\partial K}{\partial y}(x,y) \, \mathrm{d}y = f(1) - \int_0^1 f'(y) 1_{[0,y]}(x) \, \mathrm{d}y,$$

which is clearly satisfied for all $f \in H$ if

$$K(1,x) = 1 \quad \text{and} \quad \frac{\partial K}{\partial y}(x,y) = -1_{[0,y]}(x) \quad \text{for all } x, y \in [0,1].$$

This can be achieved by taking

$$K(x,y) = 2 - \max(x,y). \tag{3.7}$$

The function of two variables $K(x,y)$ given by (3.7) is our first example of a reproducing kernel, and H is our first example of a reproducing kernel Hilbert space.

We now formally define a reproducing kernel Hilbert space of real-valued functions defined on $[0,1]^s$.

Definition 3.1 (reproducing kernel Hilbert space). The Hilbert space $H(K)$ with inner product $\langle\cdot,\cdot\rangle_H$ is a *reproducing kernel Hilbert space* (RKHS) with kernel $K:[0,1]^s\times[0,1]^s\to\mathbb{R}$ if:

- $K(\cdot,\boldsymbol{x})\in H$ for all $\boldsymbol{x}\in[0,1]^s$,
- (the *reproducing property*) $f(\boldsymbol{x})=\langle f,K(\cdot,\boldsymbol{x})\rangle_H$ for all $f\in H$ and all $\boldsymbol{x}\in[0,1]^s$.

For a given RKHS there is a uniquely defined reproducing kernel satisfying the above two properties. In fact, every Hilbert space where *point evaluation is a bounded linear functional* has a reproducing kernel. This follows from the Riesz representation theorem. Reproducing kernels have the additional properties that:

- (*symmetry*) $K(\boldsymbol{x},\boldsymbol{y})=K(\boldsymbol{y},\boldsymbol{x})$ for all $\boldsymbol{x},\boldsymbol{y}\in[0,1]^s$,
- (*positive semidefiniteness*) $\sum_{i=1}^M\sum_{k=1}^M a_ia_kK(\boldsymbol{t}_i,\boldsymbol{t}_k)\geq 0$ for all $M\geq 1$ and all $a_1,\ldots,a_M\in\mathbb{R}$ and $\boldsymbol{t}_1,\ldots,\boldsymbol{t}_M\in[0,1]^s$.

Indeed, any function $K:[0,1]^s\times[0,1]^s\to\mathbb{R}$ which is symmetric and positive semidefinite is a reproducing kernel to which there corresponds a uniquely defined inner product and RKHS. A comprehensive theory on reproducing kernels can be found in Aronszajn (1950).

The thing that makes an RKHS so useful in the problem of numerical integration is that, as we shall demonstrate in Section 3.3, there exists a closed-form expression for the worst-case error. That expression for the worst-case error is expressed in terms of the kernel of the RKHS, and so is particularly useful when there exists a simple closed expression for the kernel K.

We remark that it is not always easy to find the explicit kernel of an RKHS with a given inner product. The particular example above, where the kernel is not only known but is of very simple piecewise linear form, has played an important part, as we shall see, in the recent development of QMC methods. We shall meet other reproducing kernels in Section 3.4.

We now show how to use a reproducing kernel for a space of functions of a single variable as a building block for a multivariate RKHS. First we need the following definition.

Definition 3.2 (tensor product Hilbert space). A Hilbert space H_s of functions on $[0,1]^s$ is a *tensor product* of Hilbert spaces $H_{1,1},H_{1,2},\ldots,H_{1,s}$ of functions on $[0,1]$, written as

$$H_s=H_{1,1}\otimes H_{1,2}\otimes\cdots\otimes H_{1,s},$$

if it is the completion of the span of products $\prod_{j=1}^s f_j(x_j)$, where $f_j \in H_{1,j}$, under the norm in H_s. The H_s norm of the simple product $\prod_{j=1}^s f_j(x_j)$ is just the product of the norms $\|f_j\|_{H_{1,j}}$.

Tensor product spaces provide a popular setting for studying QMC methods, for the very good reason that *the reproducing kernel for a tensor product of reproducing kernel Hilbert spaces is just the product of the kernels.*

Example 3.3 (a key tensor product RKHS). Take each one-dimensional RKHS $H_{1,j}$ to be the space of single-variable absolutely continuous functions with reproducing kernel (3.7). Then the tensor product space H_s is the s-variable RKHS with reproducing kernel

$$K_s(\boldsymbol{x}, \boldsymbol{y}) = \prod_{j=1}^{s} (2 - \max(x_j, y_j)).$$

The inner product in this space is

$$\langle f, g \rangle_{H_s} = \sum_{\mathfrak{u} \subseteq \{1:s\}} \int_{[0,1]^{|\mathfrak{u}|}} \frac{\partial^{|\mathfrak{u}|}}{\partial \boldsymbol{x}_\mathfrak{u}} f(\boldsymbol{x}_\mathfrak{u}; \boldsymbol{1}) \frac{\partial^{|\mathfrak{u}|}}{\partial \boldsymbol{x}_\mathfrak{u}} g(\boldsymbol{x}_\mathfrak{u}; \boldsymbol{1}) \, \mathrm{d}\boldsymbol{x}_\mathfrak{u}, \tag{3.8}$$

and the norm is

$$\|f\|_{H_s} = \left(\sum_{\mathfrak{u} \subseteq \{1:s\}} \int_{[0,1]^{|\mathfrak{u}|}} \left| \frac{\partial^{|\mathfrak{u}|}}{\partial \boldsymbol{x}_\mathfrak{u}} f(\boldsymbol{x}_\mathfrak{u}; \boldsymbol{1}) \right|^2 \mathrm{d}\boldsymbol{x}_\mathfrak{u} \right)^{1/2}.$$

Here $\{1:s\}$ is a shorthand notation for $\{1, 2, \ldots, s\}$, and the sum is over all subsets $\mathfrak{u} \subseteq \{1:s\}$, including the empty set, while for $\boldsymbol{x} \in [0,1]^s$ the symbol $\boldsymbol{x}_\mathfrak{u}$ denotes the set of components x_j of $\boldsymbol{x}$ with $j \in \mathfrak{u}$, and $(\boldsymbol{x}_\mathfrak{u}; \boldsymbol{1})$ indicates that the components of $\boldsymbol{x}$ for $j \notin \mathfrak{u}$ are replaced by 1. The partial derivative $\partial^{|\mathfrak{u}|}/\partial \boldsymbol{x}_\mathfrak{u}$ denotes the mixed first partial derivative with respect to the components $\boldsymbol{x}_\mathfrak{u}$. Which functions lie in this space? The answer is all real-valued functions on $[0,1]^s$ that have square-integrable mixed first derivatives and that are expressible in the form

$$\begin{aligned} f(\boldsymbol{x}) &= \langle f, K(\cdot, \boldsymbol{x}) \rangle_{H_s} \\ &= \sum_{\mathfrak{u} \subseteq \{1:s\}} (-1)^{|\mathfrak{u}|} \int_{[0,1]^{|\mathfrak{u}|}} \frac{\partial^{|\mathfrak{u}|}}{\partial \boldsymbol{y}_\mathfrak{u}} f(\boldsymbol{y}_\mathfrak{u}; \boldsymbol{1}) \, 1_{[0,\boldsymbol{y}_\mathfrak{u}]}(\boldsymbol{x}) \, \mathrm{d}\boldsymbol{y}_\mathfrak{u}, \quad \boldsymbol{x} \in [0,1]^s. \end{aligned}$$

We refer to this space as an *anchored Sobolev space* with the anchor point 1.

3.3. Worst-case error in an RKHS

We have foreshadowed in Section 3.1 that the discrepancy of a point set is closely related to the worst-case error. We now give the formal definition.

Definition 3.4 (worst-case error). The *worst-case error* of a QMC rule $Q_{n,s}$ using the point set $P \subset [0,1]^s$ in a normed space H (not necessarily a Hilbert space) is

$$e_{n,s}(P;H) := \sup_{\|f\|_H \le 1} |I_s(f) - Q_{n,s}(f)|,$$

that is, it is the worst error attained by $Q_{n,s}$ for f in the unit ball of H. The *initial error* is

$$e_{0,s}(H) := \sup_{\|f\|_H \le 1} |I_s(f)|,$$

which is the error obtained with the QMC rule replaced by zero.

Due to linearity, for any function $f \in H$ we have

$$|I_s(f) - Q_{n,s}(f)| \le e_{n,s}(P;H)\, \|f\|_H,$$

which takes the same form as the inequality in (3.4). Later in Section 3.5 we shall see other inequalities of the same form.

Worst-case errors are in general hard to compute, except for the case of an RKHS. In every RKHS of functions defined on the unit cube (not just a tensor product space), the following theorem gives a formula for the worst-case error in terms of the reproducing kernel.

Theorem 3.5 (formula for the squared worst-case error). Let $K : [0,1]^s \times [0,1]^s \to \mathbb{R}$ be a reproducing kernel that satisfies

$$\int_{[0,1]^s} \int_{[0,1]^s} K(\boldsymbol{x},\boldsymbol{y})\,\mathrm{d}\boldsymbol{x}\,\mathrm{d}\boldsymbol{y} < \infty.$$

The *squared worst-case error* and *initial error* for a QMC rule in an RKHS $H_s(K)$ with reproducing kernel K satisfy

$$\begin{aligned} e_{n,s}^2(P;H_s(K)) &= \int_{[0,1]^s} \int_{[0,1]^s} K(\boldsymbol{x},\boldsymbol{y})\,\mathrm{d}\boldsymbol{x}\,\mathrm{d}\boldsymbol{y} - \frac{2}{n}\sum_{i=0}^{n-1} \int_{[0,1]^s} K(\boldsymbol{t}_i,\boldsymbol{y})\,\mathrm{d}\boldsymbol{y} \\ &\quad + \frac{1}{n^2}\sum_{i=0}^{n-1}\sum_{k=0}^{n-1} K(\boldsymbol{t}_i,\boldsymbol{t}_k), \end{aligned} \tag{3.9}$$

and

$$e_{0,s}^2(H_s(K)) = \int_{[0,1]^s} \int_{[0,1]^s} K(\boldsymbol{x},\boldsymbol{y})\,\mathrm{d}\boldsymbol{x}\,\mathrm{d}\boldsymbol{y}. \tag{3.10}$$

Proof. We give here the proof for $s = 1$. The proof for general s can be easily obtained by replacing the unit interval $[0,1]$ in the argument below by the unit cube $[0,1]^s$.

For an $f \in H$ we write the reproducing property $f(x) = \langle f, K(\cdot,x)\rangle_H$ with successive arguments $x = t_1, t_2, \cdots$ and average the results to obtain, for the quadrature sum,

$$\frac{1}{n}\sum_{i=0}^{n-1} f(t_i) = \frac{1}{n}\sum_{i=0}^{n-1}\langle f, K(\cdot,t_i)\rangle_H = \left\langle f, \frac{1}{n}\sum_{i=0}^{n-1} K(\cdot,t_i)\right\rangle_H .$$

In a similar way we find

$$\int_0^1 f(x)\,\mathrm{d}x = \int_0^1 \langle f, K(\cdot,x)\rangle_H \,\mathrm{d}x = \left\langle f, \int_0^1 K(\cdot,x)\,\mathrm{d}x\right\rangle_H ,$$

provided that integration from 0 to 1 is a bounded linear functional in H, or equivalently that $\int_0^1 K(\cdot,x)\,\mathrm{d}x \in H$. In turn this requires

$$\begin{aligned}\left\|\int_0^1 K(\cdot,x)\,\mathrm{d}x\right\|_H^2 &= \int_0^1\int_0^1 \big\langle K(\cdot,x), K(\cdot,y)\big\rangle_H \,\mathrm{d}x\,\mathrm{d}y \\ &= \int_0^1\int_0^1 K(x,y)\,\mathrm{d}x\,\mathrm{d}y < \infty,\end{aligned}$$

which will hold for all the kernels we shall consider. Note that in the last step we used the reproducing property. By subtraction, the integration error is

$$\int_0^1 f(x)\,\mathrm{d}x - \frac{1}{n}\sum_{i=0}^{n-1} f(t_i) = \left\langle f, \int_0^1 K(\cdot,x)\,\mathrm{d}x - \frac{1}{n}\sum_{i=0}^{n-1} K(\cdot,t_i)\right\rangle_H = \langle f, \xi\rangle_H,$$

where

$$\xi(y) := \int_0^1 K(x,y)\,\mathrm{d}x - \frac{1}{n}\sum_{i=0}^{n-1} K(y,t_i), \quad y \in [0,1]. \tag{3.11}$$

We call the function ξ the *representer* of the integration error. Thus

$$e_{n,1}(P;H) = \sup_{\|f\|_H \le 1} |\langle f,\xi\rangle_H| = \|\xi\|_H,$$

since the supremum is attained by the choice $f = \xi/\|\xi\| \in H$. Therefore

$$\begin{aligned} &e_{n,1}^2(P;H) = \langle \xi,\xi\rangle_H \\ &= \left\langle \int_0^1 K(\cdot,x)\,\mathrm{d}x - \frac{1}{n}\sum_{i=0}^{n-1} K(x\cdot,t_i), \int_0^1 K(\cdot,x)\,\mathrm{d}x - \frac{1}{n}\sum_{i=0}^{n-1} K(x,t_i)\right\rangle_H \\ &= \int_0^1\int_0^1 K(x,y)\,\mathrm{d}x\,\mathrm{d}y - \frac{2}{n}\sum_{i=0}^{n-1}\int_0^1 K(x,t_i)\,\mathrm{d}x + \frac{1}{n^2}\sum_{i=0}^{n-1}\sum_{k=0}^{n-1} K(t_i,t_k),\end{aligned}$$

where we again used the reproducing property of the kernel. □

For the particular Hilbert space with reproducing kernel $K(x,y) = 2 - \max(x,y)$, it is easily seen that

$$\int_0^1 K(x,y)\,\mathrm{d}y = \frac{3-x^2}{2} \quad \text{and} \quad \int_0^1\int_0^1 K(x,y)\,\mathrm{d}x\,\mathrm{d}y = \frac{4}{3},$$

so that the worst-case error is

$$e_{n,1}^2(P;H) = \frac{4}{3} - \frac{2}{n}\sum_{i=0}^{n-1}\left(\frac{3-t_i^2}{2}\right) + \frac{1}{n^2}\sum_{i=0}^{n-1}\sum_{k=0}^{n-1}(2-\max(t_i,t_k)).$$

For future purposes we note that the derivative of the error representer ξ is $\xi' = \Delta_P$, the local discrepancy function.

For the corresponding tensor product RKHS the worst-case error is

$$e_{n,s}^2(P;H_s) \tag{3.12}$$
$$= \left(\frac{4}{3}\right)^s - \frac{2}{n}\sum_{i=0}^{n-1}\prod_{j=1}^{s}\left(\frac{3-t_{i,j}^2}{2}\right) + \frac{1}{n^2}\sum_{i=0}^{n-1}\sum_{k=0}^{n-1}\prod_{j=1}^{s}(2-\max(t_{i,j},t_{k,j})),$$

where $t_{i,j}$ denotes the jth component of the ith cubature point $\boldsymbol{t}_i$. This is sometimes called the *squared L_2 discrepancy* of the point set P, for reasons that will soon become clear.

It proves to be very useful (and surprisingly easy) to compute the average of the squared worst-case error over all cubature points,

$$E_{n,s}^2(H_s(K)) := \int_{[0,1]^s}\cdots\int_{[0,1]^s} e_{n,s}^2(\boldsymbol{t}_0,\ldots,\boldsymbol{t}_{n-1};H_s(K))\,\mathrm{d}\boldsymbol{t}_0\cdots\mathrm{d}\boldsymbol{t}_{n-1}.$$

We refer to the quantity $E_{n,s}(K)$ as the *QMC mean* of the worst-case error. (*Why* this is a useful quantity will be explained after the theorem.)

Theorem 3.6 (QMC mean). Let $K : [0,1]^s \times [0,1]^s \to \mathbb{R}$ be a reproducing kernel that satisfies

$$\int_{[0,1]^s} K(\boldsymbol{x},\boldsymbol{x})\,\mathrm{d}\boldsymbol{x} < \infty.$$

The *QMC mean* of the worst-case error in a reproducing kernel Hilbert space $H_s(K)$ of functions on $[0,1]^s$ and with kernel K satisfies

$$E_{n,s}^2(H_s(K)) = \frac{1}{n}\left(\int_{[0,1]^s} K(\boldsymbol{x},\boldsymbol{x})\,\mathrm{d}\boldsymbol{x} - \int_{[0,1]^s}\int_{[0,1]^s} K(\boldsymbol{x},\boldsymbol{y})\,\mathrm{d}\boldsymbol{x}\,\mathrm{d}\boldsymbol{y}\right). \tag{3.13}$$

Proof. Using Theorem 3.5 we have

$$
\begin{aligned}
&E_{n,s}^2(H_s(K)) \\
&= \int_{[0,1]^s} \cdots \int_{[0,1]^s} \Bigg(\int_{[0,1]^s} \int_{[0,1]^s} K(\boldsymbol{x},\boldsymbol{y}) \,\mathrm{d}\boldsymbol{x} \,\mathrm{d}\boldsymbol{y} - \frac{2}{n} \sum_{i=0}^{n-1} \int_{[0,1]^s} K(\boldsymbol{t}_i,\boldsymbol{y}) \,\mathrm{d}\boldsymbol{y} \\
&\qquad + \frac{1}{n^2} \sum_{i=0}^{n-1} K(\boldsymbol{t}_i,\boldsymbol{t}_i) + \frac{1}{n^2} \sum_{i=0}^{n-1} \sum_{\substack{k=0\\k\neq i}}^{n-1} K(\boldsymbol{t}_i,\boldsymbol{t}_k) \Bigg) \mathrm{d}\boldsymbol{t}_0 \cdots \mathrm{d}\boldsymbol{t}_{n-1},
\end{aligned}
$$

noting that in the double sum we separated the diagonal and off-diagonal terms as in the proof of Theorem 2.1. The first term inside the parentheses is independent of $\boldsymbol{t}_0, \ldots, \boldsymbol{t}_{n-1}$, thus these variables can all be integrated out to give the result 1. In the second term, after interchanging the outer integrations and the sum, all variables but $\boldsymbol{t}_i$ can be integrated out. The situation is similar for the third term, while for the fourth term all but $\boldsymbol{t}_i$ and $\boldsymbol{t}_k$ can be integrated out. We are left with

$$
\begin{aligned}
E_{n,s}^2(K) &= \int_{[0,1]^s} \int_{[0,1]^s} K(\boldsymbol{x},\boldsymbol{y}) \,\mathrm{d}\boldsymbol{x} \,\mathrm{d}\boldsymbol{y} - \frac{2}{n} \sum_{i=0}^{n-1} \int_{[0,1]^s} \int_{[0,1]^s} K(\boldsymbol{t}_i,\boldsymbol{y}) \,\mathrm{d}\boldsymbol{t}_i \,\mathrm{d}\boldsymbol{y} \\
&\quad + \frac{1}{n^2} \sum_{i=0}^{n-1} \int_{[0,1]^s} K(\boldsymbol{t}_i,\boldsymbol{t}_i) \,\mathrm{d}\boldsymbol{t}_i + \frac{1}{n^2} \sum_{i=0}^{n-1} \sum_{\substack{k=0\\k\neq i}}^{n-1} \int_{[0,1]^s} \int_{[0,1]^s} K(\boldsymbol{t}_i,\boldsymbol{t}_k) \,\mathrm{d}\boldsymbol{t}_i \,\mathrm{d}\boldsymbol{t}_k \\
&= \int_{[0,1]^s} \int_{[0,1]^s} K(\boldsymbol{x},\boldsymbol{y}) \,\mathrm{d}\boldsymbol{x} \,\mathrm{d}\boldsymbol{y} - 2 \int_{[0,1]^s} \int_{[0,1]^s} K(\boldsymbol{x},\boldsymbol{y}) \,\mathrm{d}\boldsymbol{x} \,\mathrm{d}\boldsymbol{y} \\
&\quad + \frac{1}{n} \int_{[0,1]^s} K(\boldsymbol{x},\boldsymbol{x}) \,\mathrm{d}\boldsymbol{x} + \frac{n-1}{n} \int_{[0,1]^s} \int_{[0,1]^s} K(\boldsymbol{x},\boldsymbol{y}) \,\mathrm{d}\boldsymbol{x} \,\mathrm{d}\boldsymbol{y} \\
&= \frac{1}{n} \int_{[0,1]^s} K(\boldsymbol{x},\boldsymbol{x}) \,\mathrm{d}\boldsymbol{x} - \frac{1}{n} \int_{[0,1]^s} \int_{[0,1]^s} K(\boldsymbol{x},\boldsymbol{y}) \,\mathrm{d}\boldsymbol{x} \,\mathrm{d}\boldsymbol{y}. \qquad \square
\end{aligned}
$$

The importance of Theorem 3.6 is that it gives us an immediate existence result for QMC integration: by applying the principle that *there is always at least one choice as good as the average*, we conclude that there exists a QMC point set P for which

$$
e_{n,s}^2(P; H_s(K)) \le E_{n,s}^2(H_s(K)).
$$

Thus, provided that both integrals on the right-hand side in Theorem 3.6 are finite (as they are in all the cases we will consider), there exists a QMC point set with worst-case error of order $O(n^{-1/2})$, that is, the same rate of convergence as Monte Carlo. Later we will not be satisfied with this result, for two reasons: first, we want not just the Monte Carlo rate of convergence, but rather a rate of convergence close to $O(n^{-1})$, or even better; second, we want not just existence, but also a method of constructing the points.

3.4. Other univariate reproducing kernels

Note that (3.1) can be viewed as a Taylor series with integral remainder, developed at 1. As such it is clear that we could choose an arbitrary anchor point $c \in [0,1]$ instead of 1. Or we can consider function spaces whose members have more than one derivative. Or we can depart from a Taylor series representation of f to obtain other examples of reproducing kernels. All of the following examples can be used as building blocks for multivariate Sobolev spaces.

Example 3.7 (anchored reproducing kernel of smoothness α). If the $(\alpha-1)$th derivative of f is absolutely continuous for some natural number α, then the Taylor series of a univariate function f anchored at 0 is

$$f(x) = \sum_{r=0}^{\alpha-1} \frac{f^{(r)}(0)}{r!} x^r + \int_0^1 f^{(\alpha)}(y) \frac{(x-y)_+^{\alpha-1}}{(\alpha-1)!} \, dy,$$

where $f^{(r)}$ denotes the rth derivative of f and

$$(x-y)_+^{\alpha-1} = \begin{cases} (x-y)^{\alpha-1} & \text{if } x > y, \\ 0 & \text{if } x \le y. \end{cases}$$

For functions f, g permitting such a Taylor series presentation, we can define the inner product

$$\langle f, g \rangle = \sum_{r=0}^{\alpha-1} f^{(r)}(0) g^{(r)}(0) + \int_0^1 f^{(\alpha)}(y) g^{(\alpha)}(y) \, dy.$$

The set of functions whose norm corresponding to this inner product is finite is another RKHS, with kernel given by

$$K_\alpha(x,y) = \sum_{r=0}^{\alpha-1} \frac{x^r}{r!} \frac{y^r}{r!} + \int_0^1 \frac{(x-z)_+^{\alpha-1}}{(\alpha-1)!} \frac{(y-z)_+^{\alpha-1}}{(\alpha-1)!} \, dz.$$

We call the corresponding RKHS the *anchored Sobolev space of smoothness α with anchor* 0.

Example 3.8 (unanchored reproducing kernel of smoothness α). For a univariate function f belonging to the same smoothness class as in the preceding example, consider the representation given by

$$f(x) = \sum_{r=0}^{\alpha} \frac{B_r(x)}{r!} \int_0^1 f^{(r)}(y) \, dy - (-1)^\alpha \int_0^1 \frac{B_\alpha(|x-y|)}{\alpha!} f^{(\alpha)}(y) \, dy.$$

Here, B_r denotes the Bernoulli polynomial of degree r: see Chapter 24 of the *Digital Library of Mathematical Functions* (2012). In this case the inner product is

$$\langle f, g\rangle = \sum_{r=0}^{\alpha-1}\left(\int_0^1 f^{(r)}(y)\,\mathrm{d}y\right)\left(\int_0^1 g^{(r)}(y)\,\mathrm{d}y\right) + \int_0^1 f^{(\alpha)}(y)g^{(\alpha)}(y)\,\mathrm{d}y,$$

and the reproducing kernel is

$$K_\alpha(x,y) = \sum_{r=0}^{\alpha}\frac{B_r(x)}{r!}\frac{B_r(y)}{r!} - (-1)^\alpha\frac{B_{2\alpha}(|x-y|)}{(2\alpha)!}.$$

The corresponding RKHS is called the *unanchored Sobolev space of smoothness* α.

3.5. Geometric discrepancy

We now return to the local discrepancy function

$$\Delta_P(x) = \frac{1}{n}\sum_{i=0}^{n-1} 1_{[0,x]}(t_i) - \int_0^1 1_{[0,x]}(y)\,\mathrm{d}y,$$

which we encountered already in (3.3), and which we also obtained as the derivative ξ' of the representer ξ, where ξ is given by (3.11) with K as in (3.7).

The local discrepancy function has a geometric interpretation: since

$$\sum_{i=0}^{n-1} 1_{[0,x]}(t_i)$$

is the number of points of P lying in the interval $[0,x]$, therefore

$$\frac{1}{n}\sum_{i=0}^{n-1} 1_{[0,x]}(t_i)$$

is the proportion of points of P lying in the interval $[0,x]$; and the local discrepancy is the departure of this proportion from the 'ideal' proportion, which is the length of the interval. By taking the L_p norm of Δ_P we obtain the L_p *discrepancy* of the point set P, given by

$$\|\Delta_P\|_p := \left(\int_0^1 |\Delta_P(x)|^p\,\mathrm{d}x\right)^{1/p}$$

for $1 \le p < \infty$, with the obvious modifications for $p = \infty$. The L_p discrepancy can therefore be understood as a measure for how uniformly the point set P is distributed, that is, it measures the discrepancy between the empirical distribution of the point set P and the uniform distribution.

We consider now a generalization to dimensions $s > 1$. In one dimension, the local discrepancy function can be derived as the derivative of the representer ξ based on the reproducing kernel $K(x,y) = 2-\max(x,y)$. As before we consider the product kernel $K_s(\boldsymbol{x},\boldsymbol{y}) = \prod_{j=1}^s K(x_j,y_j)$. The representer of the error is given by

$$\begin{aligned}\xi(\boldsymbol{x}) &= \int_{[0,1]^s} K_s(\boldsymbol{x},\boldsymbol{y})\,\mathrm{d}\boldsymbol{y} - \frac{1}{n}\sum_{i=0}^{n-1} K_s(\boldsymbol{x},\boldsymbol{t}_i)\\ &= \prod_{j=1}^{s}\left(\frac{3-x_j^2}{2}\right) - \frac{1}{n}\sum_{i=0}^{n-1}\prod_{j=1}^{s}(2-\max(x_j,t_{i,j})).\end{aligned}$$

The mixed first partial derivatives are then given by

$$\frac{\partial^{|\mathfrak{u}|}\xi}{\partial \boldsymbol{x}_{\mathfrak{u}}}(\boldsymbol{x}_{\mathfrak{u}};\boldsymbol{1}) = (-1)^{|\mathfrak{u}|}\left(\prod_{j\in\mathfrak{u}} x_j - \frac{1}{n}\sum_{i=0}^{n-1} 1_{[\boldsymbol{0}_{\mathfrak{u}},\boldsymbol{x}_{\mathfrak{u}}]}(\boldsymbol{t}_{i,\mathfrak{u}})\right),$$

where

$$1_{[\boldsymbol{0}_{\mathfrak{u}},\boldsymbol{x}_{\mathfrak{u}}]}(\boldsymbol{t}_{i,\mathfrak{u}}) = \prod_{j\in\mathfrak{u}} 1_{[0,x_j]}(t_{i,j}).$$

We define the local discrepancy function Δ_P in s dimensions by

$$\Delta_P(\boldsymbol{x}) := \frac{1}{n}\sum_{i=0}^{n-1} 1_{[\boldsymbol{0},\boldsymbol{x}]}(\boldsymbol{t}_i) - \prod_{j=1}^{s} x_j.$$

Then

$$\frac{\partial^{|\mathfrak{u}|}\xi}{\partial \boldsymbol{x}_{\mathfrak{u}}}(\boldsymbol{x}_{\mathfrak{u}};\boldsymbol{1}) = (-1)^{|\mathfrak{u}|+1}\Delta_P(\boldsymbol{x}_{\mathfrak{u}};\boldsymbol{1}).$$

Analogously to the one-dimensional case, we can show that

$$\frac{1}{n}\sum_{i=0}^{n-1} f(\boldsymbol{t}_i) - \int_{[0,1]^s} f(\boldsymbol{x})\,\mathrm{d}\boldsymbol{x} = \sum_{\mathfrak{u}\subseteq\{1:s\}} (-1)^{|\mathfrak{u}|}\int_{[0,1]^{|\mathfrak{u}|}} \frac{\partial^{|\mathfrak{u}|} f}{\partial \boldsymbol{x}_{\mathfrak{u}}}(\boldsymbol{x}_{\mathfrak{u}};\boldsymbol{1})\Delta_P(\boldsymbol{x}_{\mathfrak{u}};\boldsymbol{1})\,\mathrm{d}\boldsymbol{x}_{\mathfrak{u}},$$

where the $\mathfrak{u} = \emptyset$ term is actually zero since $\Delta_P(\boldsymbol{1}) = 0$. This equality is called by different people the *Hlawka identity* or the *Zaremba identity*.

By applying Hölder's inequality for integrals and sums, we then obtain the following inequality.

Theorem 3.9 (Koksma–Hlawka inequality). We have

$$\left|\frac{1}{n}\sum_{i=0}^{n-1} f(\boldsymbol{t}_i) - \int_{[0,1]^s} f(\boldsymbol{x})\,\mathrm{d}\boldsymbol{x}\right| \le \|\Delta_P\|_{p,p'}\|f\|_{q,q'}, \tag{3.14}$$

where $1 \le p, p', q, q' \le \infty$, $\frac{1}{p} + \frac{1}{q} = 1$, $\frac{1}{p'} + \frac{1}{q'} = 1$, and

$$\|\Delta_P\|_{p,p'} = \left(\sum_{\mathfrak{u} \subseteq \{1:s\}} \left(\int_{[0,1]^{|\mathfrak{u}|}} |\Delta_P(\boldsymbol{x}_\mathfrak{u}; \mathbf{1})|^p \, \mathrm{d}\boldsymbol{x}_\mathfrak{u} \right)^{p'/p} \right)^{1/p'}$$

and

$$\|f\|_{q,q'} = \left(\sum_{\mathfrak{u} \subseteq \{1:s\}} \left(\int_{[0,1]^{|\mathfrak{u}|}} \left| \frac{\partial^{|\mathfrak{u}|} f}{\partial \boldsymbol{x}_\mathfrak{u}}(\boldsymbol{x}_\mathfrak{u}; \mathbf{1}) \right|^q \mathrm{d}\boldsymbol{x}_\mathfrak{u} \right)^{q'/q} \right)^{1/q'},$$

with the obvious modifications if one or more of p, p', q, q' are infinite.

In the theorem the norm $\|f\|_{q,q'}$ can obviously be replaced by the corresponding seminorm $|f|_{q,q'}$, defined in exactly the same way but with the $\mathfrak{u} = \emptyset$ term omitted. (This term can be omitted because the QMC rule is exact for constants.)

The inequality (3.14) says that the integration error is bounded by the product of the norm $\|f\|_{q,q'}$ or seminorm $|f|_{q,q'}$ of the integrand and the discrepancy $\|\Delta_P\|_{p,p'}$ of the cubature points. It is often called a Koksma–Hlawka inequality, especially for the case $p = p' = \infty$. If the integrand is given, and cannot be changed, the inequality motivates the search for cubature points with small discrepancy. It also connects QMC with the field of discrepancy theory, since it shows that point sets with low discrepancy are attractive to use as QMC integration points.

The classical Koksma–Hlawka inequality states that

$$\left| \frac{1}{n} \sum_{i=0}^{n-1} f(\boldsymbol{t}_i) - \int_{[0,1]^s} f(\boldsymbol{x}) \, \mathrm{d}\boldsymbol{x} \right| \le D_n^*(P) V(f), \tag{3.15}$$

where $D_n^*(P)$ is the *star discrepancy*

$$D_n^*(P) := \sup_{\boldsymbol{x} \in [0,1]^s} |\Delta_P(\boldsymbol{x})|, \tag{3.16}$$

and $V(f)$ is the *variation of f in the sense of Hardy and Krause.* If f has continuous mixed first partial derivatives, the variation $V(f)$ coincides with

$$|f|_{1,1} = \sum_{\emptyset \ne \mathfrak{u} \subseteq \{1:s\}} \int_{[0,1]^{|\mathfrak{u}|}} \left| \frac{\partial^{|\mathfrak{u}|} f}{\partial \boldsymbol{x}_\mathfrak{u}}(\boldsymbol{x}_\mathfrak{u}; \mathbf{1}) \right| \mathrm{d}\boldsymbol{x}_\mathfrak{u}.$$

3.6. Notes

The classical reference for reproducing kernels is Aronszajn (1950). See also Thomas-Agnan (1996) and Wahba (1990). In the context of QMC, reproducing kernels were introduced by Hickernell (1996*a*); see also Hickernell (1998*a*). For another elementary introduction to RKHSs in the context of QMC integration see Dick and Pillichshammer (2010, Chapter 2).

Theorem 3.5 can be found in Hickernell (1998*a*). Mercer's theorem implies that each reproducing kernel has an expansion in terms of its eigenfunctions

$$K(\boldsymbol{x}, \boldsymbol{y}) = \sum_{r=0}^{\infty} \lambda_r \psi_r(\boldsymbol{x}) \psi_r(\boldsymbol{y}),$$

where $\lambda_r > 0$ are the eigenvalues and ψ_r are the eigenfunctions of the integral operator on $[0,1]^s$ with kernel K. For many reproducing kernels considered in QMC theory these eigenfunctions are explicitly known: see Dick, Nuyens and Pillichshammer (2013), Wasilkowski and Woźniakowski (1999) and Werschulz and Woźniakowski (2009).

The one-dimensional form of (3.15) is due to Koksma (1942/43) and the higher-dimensional result is due to Hlawka (1961).

Geometric discrepancy was studied long before QMC integration. The classical reference to uniform distribution modulo 1 is Weyl (1916). Further monographs dealing with discrepancy theory and various aspects of it are due to Kuipers and Niederreiter (1974), Niederreiter (1992*a*), Drmota and Tichy (1997), Matoušek (1999), Chazelle (2000) and Dick and Pillichshammer (2010).

Novak and Woźniakowski (2009) studied whether general L_2 discrepancies (in Euclidean spaces) are related to the worst-case error of multivariate integration on an RKHS. In particular, they provided a concrete formula for the corresponding reproducing kernel for arbitrary L_2 discrepancies. Gnewuch (2012*a*) extended these results to weighted L_2 discrepancies and weighted RKHS. Here the underlying domain can be more general as in Novak and Woźniakowski (2009). In particular, Gnewuch (2012*a*) covers the relation between infinite-dimensional integration on weighted Hilbert spaces and the corresponding weighted (limiting) L_2 discrepancies.

4. Weighted spaces and tractability

4.1. Meeting the high-dimensional challenge

High-dimensional problems, as noted already in the Introduction, can be very hard. Yet some problems have caused surprise in the opposite direction by turning out to be easier than might have been expected. Especially influential in this respect were certain 360-dimensional calculations carried out in the mid-1990s by Paskov and Traub (1995) at Columbia University. That paper described a problem coming directly from Wall Street, on the valuation of a class of financial derivatives known as mortgage-backed obligations. While the details of the model or the calculation are not important here, broadly the problem was to evaluate a parcel of mortgages held by a bank, where each month borrowers make individual choices as to whether or not to repay the mortgage, and are assumed to make this choice according to some probability distribution. And of course the value of a mortgage held by the bank is affected by whether or not the loan is repaid. Because there are 360 months in the 30-year period of the loan, the problem is to evaluate a 360-dimensional expected value. The standard approach to such a problem is to model the process month by month by a Monte Carlo method. Paskov and Traub (1995), in contrast, treated the problem as one of 360-dimensional integration. To the surprise of most observers, they were able to obtain satisfactory results with a QMC method, at much smaller computational cost than with a Monte Carlo method.

Following the success of those experiments, the idea began to emerge that perhaps problems such as the one described in the previous paragraph are in some hidden sense not really as difficult as might be expected from their nominal high-dimensionality. In this spirit Caflisch, Morokoff and Owen (1997) defined two notions of 'effective dimension' (namely 'truncation dimension' and 'superposition dimension'), with the idea that either or both might be much smaller than the nominal dimension. Sloan and Woźniakowski (1998) sought to build this notion into the mathematics, by introducing function spaces containing 'weights'. It is fair to say that in some form or another 'weighted spaces' have become a standard part of the QMC framework. We will define such spaces formally in the next subsection, but first we describe the underlying idea, and indicate why it has become so popular.

Sloan and Woźniakowski speculated that perhaps the reason that QMC calculations such as those in Paskov and Traub (1995) were successful is that some coordinate directions are more important than others, in the sense of being in some way more difficult. Assuming then that the coordinate directions are ordered in order of difficulty, they sought to quantify this decreasing importance by associating with each coordinate direction a positive number γ_j, where

$$\gamma_1 \geq \gamma_2 \geq \gamma_3 \cdots > 0.$$

By introducing a function space incorporating these weights, they were able to obtain a rather precise result, namely that the worst-case error (defined already in Definition 3.4) is bounded independently of dimension if and only if

$$\sum_{j=0}^{\infty} \gamma_j < \infty.$$

This means that a classical choice of weights (with all weights equal) certainly fails to have worst-case errors bounded independently of the dimension. So too, though only marginally, do weights $\gamma_j = 1/j$. On the other hand weights of the form $\gamma_j = 1/j^a$ lead to uniformly bounded worst-case errors if (and only if) $a > 1$.

The weights described in the preceding paragraph are now called 'product weights', for reasons that will soon become clear. Since that time many other forms of weights have been studied (we shall meet 'general weights', 'finite-order' weights, 'order-dependent' weights, and the most recent addition, 'POD' (for 'product and order-dependent') weights. The driving motivation for this flowering of possibilities has been the desire to describe in a more precise way the influence of particular combinations of the variables. At the (unrealistic) extreme, the greatest freedom attaches to 'general weights', which allow a different weight $\gamma_{\mathfrak{u}}$ for each of the 2^s subsets $\mathfrak{u} \subseteq \{1 : s\}$.

For all these choices of weights the question of 'tractability' arises: loosely, we ask under what conditions on the weights are the worst-case errors uniformly bounded in dimension; or if they are not bounded, then at worst grow only slowly with dimension. We shall discuss tractability in Section 4.5.

4.2. Weighted reproducing kernel Hilbert spaces: anchored Sobolev spaces

In this subsection we first introduce a weighted RKHS in its simplest product form. But we then quickly develop weighted spaces of a more general form.

Arguing as in Example 3.3, we build our first weighted s-variable space out of single-variable weighted spaces with simple kernels. Our one-dimensional space $H_{1,\gamma}$ is almost the same as before (in particular, it is always the space of absolutely continuous functions f defined on $[0,1]$ whose first derivatives are square-integrable), except that we now associate with the space a weight $\gamma > 0$, and in addition we allow the 'anchor' value to be any number $c \in [0,1]$, instead of just 1 as before. Our building block is now the one-dimensional kernel

$$K_{1,\gamma}(x,y) = 1 + \gamma\,\eta(x,y),$$

where

$$\eta(x,y) = \begin{cases} \min(x,y) - c & \text{if } x,y > c, \\ c - \max(x,y) & \text{if } x,y < c, \\ 0 & \text{otherwise.} \end{cases} \tag{4.1}$$

If $c = 1$ and $\gamma = 1$ it is easily seen that this reduces to the reproducing kernel (3.7). For general c and γ it is easy to verify, using no more than the fundamental theorem of calculus, that $K_{1,\gamma}(x,y)$ is the reproducing kernel associated with the inner product

$$\langle f, g\rangle_{1,\gamma} = f(c)g(c) + \frac{1}{\gamma}\int_0^1 f'(x)\,g'(x)\,\mathrm{d}x,$$

which is a generalization of (3.6). The corresponding norm is

$$\|f\|_{1,\gamma} = \langle f, f\rangle_{1,\gamma}^{1/2}.$$

The s-dimensional tensor product RKHS corresponding to the above one-dimensional space is

$$H_{s,\gamma} := H_{1,\gamma_1} \otimes H_{1,\gamma_2} \otimes \cdots \otimes H_{1,\gamma_s},$$

if we allow a different weight γ_j for each component x_j. It has the reproducing kernel

$$K_{s,\gamma}(\boldsymbol{x}, \boldsymbol{y}) = \prod_{j=1}^{s} (1 + \gamma_j \eta(x_j, y_j)).$$

The identity

$$\prod_{j=1}^{s}(1 + a_j) = \sum_{\mathfrak{u}\subseteq\{1:s\}} \prod_{j\in\mathfrak{u}} a_j = 1 + \sum_{\emptyset\neq\mathfrak{u}\subseteq\{1:s\}} \prod_{j\in\mathfrak{u}} a_j \tag{4.2}$$

can now be used to rewrite the product as a sum over all subsets of $\{1:s\}$,

$$K_{s,\gamma}(\boldsymbol{x}, \boldsymbol{y}) = \sum_{\mathfrak{u}\subseteq\{1:s\}} \gamma_{\mathfrak{u}} \prod_{j\in\mathfrak{u}} \eta(x_j, y_j), \tag{4.3}$$

where for this *product weight* case we have

$$\gamma_{\mathfrak{u}} = \prod_{j\in\mathfrak{u}} \gamma_j,$$

with the convention that the empty product has the value 1. The reason for the 'product' tag for these weights now becomes clear: the weight associated with the subset $\mathfrak{u}$ of the variables is the product of the weights for the individual components. One implication of this is that *the scaling of product weights is crucially important.* If, for example, all the weights γ_j are halved then the weights for the subset $\mathfrak{u} = \{1, 3\}$ with two elements are reduced by a factor of 4, and those for the subset $\mathfrak{u} = \{2, 4, 5\}$ by a factor of 8.

Because of the limited flexibility of product weights, other choices of weights have been considered. In principle the formula (4.3) allows a different weight $\gamma_{\mathfrak{u}}$ to be chosen for each subset $\mathfrak{u}$ of the variables, but the cost of even one evaluation of the reproducing kernel would then be prohibitive when s is large. This is the *general weight* case (Dick, Sloan, Wang and Woźniakowski 2006), where we take $\gamma_\emptyset \equiv 1$. There is much interest (as well as some unease (Sloan 2007)) in *finite-order weights* (Sloan, Wang and Woźniakowski 2004, Wasilkowski and Woźniakowski 2004), in which $\gamma_{\mathfrak{u}} = 0$ for all $|\mathfrak{u}|$ greater than some number q^*. *Order-dependent weights,* introduced by Dick *et al.* (2006), take the form

$$\gamma_{\mathfrak{u}} = \Gamma_{|\mathfrak{u}|}$$

for some non-negative numbers $\Gamma_1, \Gamma_2, \ldots$. The most recent addition to the menagerie of weights are the *POD weights* (*i.e., product and order-dependent weights*), in which two sequences $\gamma_1, \gamma_2, \ldots$ and $\Gamma_1, \Gamma_2, \ldots$ of non-negative numbers are defined, and we take

$$\gamma_{\mathfrak{u}} = \Gamma_{|\mathfrak{u}|} \prod_{j \in \mathfrak{u}} \gamma_j. \tag{4.4}$$

Weights of the POD form were found to arise naturally in a recent study of PDEs with random coefficients (Kuo *et al.* 2012).

For general weights $\gamma_{\mathfrak{u}}$ the inner product in our anchored space is

$$\langle f, g \rangle_{s,\gamma} = \sum_{\mathfrak{u} \subseteq \{1:s\}} \gamma_{\mathfrak{u}}^{-1} \int_{[0,1]^{|\mathfrak{u}|}} \frac{\partial^{|\mathfrak{u}|}}{\partial \boldsymbol{x}_{\mathfrak{u}}} f(\boldsymbol{x}_{\mathfrak{u}}; \boldsymbol{c}) \frac{\partial^{|\mathfrak{u}|}}{\partial \boldsymbol{x}_{\mathfrak{u}}} g(\boldsymbol{x}_{\mathfrak{u}}; \boldsymbol{c}) \, \mathrm{d}\boldsymbol{x}_{\mathfrak{u}},$$

where the notation follows that in (3.8), and the term labelled by $\mathfrak{u}$ is to be omitted for $\mathfrak{u} = \emptyset$.

The worst-case errors for spaces with reproducing kernel of the form (4.3) and η given by (4.1) can be found from (3.9). To make use of that expression one needs the single and double integrals of the kernel. For completeness we give the necessary integrals here, as well as the 'diagonal' integral needed for the QMC mean (3.13):

$$\begin{aligned} \int_{[0,1]^s} K_{s,\gamma}(\boldsymbol{x}, \boldsymbol{y}) \, \mathrm{d}\boldsymbol{y} &= \sum_{\mathfrak{u} \subseteq \{1:s\}} \gamma_{\mathfrak{u}} \prod_{j \in \mathfrak{u}} \big(\alpha(x_j) + \beta\big), \\ \int_{[0,1]^s} \int_{[0,1]^s} K_{s,\gamma}(\boldsymbol{x}, \boldsymbol{y}) \, \mathrm{d}\boldsymbol{x} \, \mathrm{d}\boldsymbol{y} &= \sum_{\mathfrak{u} \subseteq \{1:s\}} \gamma_{\mathfrak{u}} \, \beta^{|\mathfrak{u}|}, \\ \int_{[0,1]^s} K_{s,\gamma}(\boldsymbol{x}, \boldsymbol{x}) \, \mathrm{d}\boldsymbol{x} &= \sum_{\mathfrak{u} \subseteq \{1:s\}} \gamma_{\mathfrak{u}} \big(\beta + \tfrac{1}{6}\big)^{|\mathfrak{u}|}, \end{aligned} \tag{4.5}$$

where

$$\alpha(x) := \max(x, c) - \tfrac{x^2}{2} - \tfrac{c^2}{2} - \tfrac{1}{3} \quad \text{and} \quad \beta := c^2 - c + \tfrac{1}{3}. \tag{4.6}$$

When the weights are of the product form, these three integrals simplify (by using the identity (4.2)) to

$$\prod_{j=1}^{s} \big(1 + \gamma_j \big(\alpha(x_j) + \beta\big)\big), \quad \prod_{j=1}^{s} \big(1 + \gamma_j \beta\big), \quad \prod_{j=1}^{s} \big(1 + \gamma_j \big(\beta + \tfrac{1}{6}\big)\big),$$

respectively.

For example, with product weights and $c = 1$, we have from (3.9) and (3.10) that the worst-case error and initial error are given by

$$\begin{aligned} e_{n,s}^2(P; H(K_{s,\gamma})) = \prod_{j=1}^{s} \left(1 + \frac{\gamma_j}{3}\right) - \frac{2}{n} \sum_{i=0}^{n-1} \prod_{j=1}^{s} \left(1 + \frac{\gamma_j}{2}(1 - t_{i,j}^2)\right) \\ + \frac{1}{n^2} \sum_{i=0}^{n-1} \sum_{k=0}^{n-1} \prod_{j=1}^{s} \big(1 + \gamma_j \big[1 - \max(t_{i,j}, t_{k,j})\big]\big), \end{aligned}$$

which reduces to (3.12) if $\gamma_j = 1$ for all j, and

$$e_{0,s}^2(H(K_{s,\gamma})) = \prod_{j=1}^{s}\left(1 + \frac{\gamma_j}{3}\right).$$

Moreover, we have from (3.13) that the QMC mean satisfies

$$E_{n,s}^2(H(K_{s,\gamma})) = \frac{1}{n}\left(\prod_{j=1}^{s}\left(1 + \frac{\gamma_j}{2}\right) - \prod_{j=1}^{s}\left(1 + \frac{\gamma_j}{3}\right)\right).$$

We remark that it is also possible to use a different anchor value c_j for each coordinate direction x_j, thus leading to different η_j, α_j, β_j for each index j. All results can be trivially generalized to that case.

4.3. Unanchored Sobolev spaces

The reproducing kernels considered so far in this section have been of the so-called 'anchored' variety, in which there is a special number $c \in [0, 1]$, called the 'anchor', at which the components of $\boldsymbol{x}$ that are not active (*i.e.*, that are not in the active set $\mathfrak{u}$) are fixed. However, there are other RKHSs with interesting properties (Sloan and Woźniakowski 2002). In particular, the so-called 'unanchored' spaces, in which the inactive variables are integrated over rather than fixed, have some advantages. We give the necessary formulas here; all can be checked by similar arguments to those given previously for the anchored spaces (or see the reference above).

In the one-dimensional weighted case the inner product is now (*cf.* Example 3.8 for $\alpha = 1$)

$$\langle f, g\rangle_{1,\gamma} = \left(\int_0^1 f(x)\,\mathrm{d}\boldsymbol{x}\right)\left(\int_0^1 g(x)\,\mathrm{d}\boldsymbol{x}\right) + \frac{1}{\gamma}\int_0^1 f'(x)\,g'(x)\,\mathrm{d}x,$$

and the corresponding norm is

$$\|f\|_{1,\gamma} = \langle f, f\rangle_{1,\gamma}^{1/2}.$$

The corresponding reproducing kernel is, as before,

$$K_{1,\gamma}(x, y) = 1 + \gamma\,\eta(x, y),$$

but now η is given by

$$\eta(x, y) = \tfrac{1}{2}B_2(|x - y|) + (x - \tfrac{1}{2})(y - \tfrac{1}{2}), \tag{4.7}$$

where $B_2(x) := x^2 - x + \frac{1}{6}$ is the Bernoulli polynomial of degree 2, which has the useful property

$$\int_0^1 B_2(|x - y|)\,\mathrm{d}y = \int_0^1 B_2(y)\,\mathrm{d}y = 0.$$

The corresponding kernel for the s-dimensional general weight case is again given by (4.3), but now with η given by (4.7). The inner product for the s-dimensional general weight unanchored space is

$$\langle f, g\rangle_{s,\gamma} = \sum_{\mathfrak{u}\subseteq\{1:s\}} \gamma_{\mathfrak{u}}^{-1} \int_{[0,1]^{|\mathfrak{u}|}} \left(\int_{[0,1]^{s-|\mathfrak{u}|}} \frac{\partial^{|\mathfrak{u}|}}{\partial \boldsymbol{x}_{\mathfrak{u}}} f(\boldsymbol{x})\,\mathrm{d}\boldsymbol{x}_{-\mathfrak{u}}\right) \times \left(\int_{[0,1]^{s-|\mathfrak{u}|}} \frac{\partial^{|\mathfrak{u}|}}{\partial \boldsymbol{x}_{\mathfrak{u}}} g(\boldsymbol{x})\,\mathrm{d}\boldsymbol{x}_{-\mathfrak{u}}\right) \mathrm{d}\boldsymbol{x}_{\mathfrak{u}},$$

where $\boldsymbol{x}_{-\mathfrak{u}}$ stands for all the components of the s-dimensional vector $\boldsymbol{x}$ that are not included in $\boldsymbol{x}_{\mathfrak{u}}$.

The range of choices for the weights is exactly the same as for the anchored case.

The worst-case error is again given by (3.9). The necessary single and double integrals of the kernel, along with the diagonal integral needed for the QMC mean, now take the simpler forms

$$\int_{[0,1]^s} K_{s,\gamma}(\boldsymbol{x},\boldsymbol{y})\,\mathrm{d}\boldsymbol{y} = 1,$$

$$\int_{[0,1]^s}\int_{[0,1]^s} K_{s,\gamma}(\boldsymbol{x},\boldsymbol{y})\,\mathrm{d}\boldsymbol{x}\,\mathrm{d}\boldsymbol{y} = 1,$$

$$\int_{[0,1]^s} K_{s,\gamma}(\boldsymbol{x},\boldsymbol{x})\,\mathrm{d}\boldsymbol{x} = \sum_{\mathfrak{u}\subseteq\{1:s\}} \gamma_{\mathfrak{u}} \left(\tfrac{1}{6}\right)^{|\mathfrak{u}|},$$

and in the case of product weights the third integral simplifies to

$$\prod_{j=1}^{s}\left(1+\frac{\gamma_j}{6}\right).$$

To avoid having to treat the anchored and unanchored spaces separately, we can write $\eta(x,y)$ from (4.1) and (4.7) in the common form

$$\eta(x,y) = \tfrac{1}{2}B_2(|x-y|) + (x-\tfrac{1}{2})(y-\tfrac{1}{2}) + \alpha(x) + \alpha(y) + \beta, \tag{4.8}$$

where $\alpha(\cdot)$ and β are defined by (4.6) for the anchored case, and $\alpha \equiv 0$ and $\beta = 0$ for the unanchored case.

4.4. Why weighted spaces are interesting

We have defined different kinds of weighted spaces for functions defined on the s-dimensional cube, but we have not so far said why weighted spaces are interesting. The answer is that *the worst-case errors in any of our weighted reproducing kernel Hilbert spaces are bounded independently of the dimension s if (and only if) the weights decay in a suitable way.*

For product weights, the following result, first obtained by Sloan and Woźniakowski (1998) for the anchored case $c = 1$, is particularly striking.

Theorem 4.1. For product weights, and for the weighted anchored or unanchored Sobolev space of Sections 4.2 and 4.3, there exist point sets $P_n \subset [0,1]^s$ for $n = 1, 2, \ldots$ such that the worst-case error $e_{n,s}(P_n; H(K_{s,\gamma}))$ is bounded independently of s if and only if

$$\sum_{j=1}^{\infty} \gamma_j < \infty. \tag{4.9}$$

There are two parts to this theorem: one part is an existence result, stating that 'good' point sets exist if (4.9) holds, without saying how to find them; and the other part is a negative result, saying that if the assumption (4.9) does not hold then, no matter how the points are chosen, the worst-case error is unbounded as $s \to \infty$. We shall address each result separately in the following two lemmas.

Lemma 4.2. For product weights, and for the weighted anchored or unanchored Sobolev space of Sections 4.2 and 4.3, there exist point sets $P_n \subset [0,1]^s$ for $n = 1, 2, \ldots$ such that the worst-case error $e_{n,s}(P_n; H(K_{\gamma,s}))$ satisfies

$$e_{n,s}(P_n; H(K_{s,\gamma})) \le \frac{1}{\sqrt{n}} \exp\left(B \sum_{j=1}^{s} \gamma_j \right) \le \frac{1}{\sqrt{n}} \exp\left(B \sum_{j=1}^{\infty} \gamma_j \right),$$

where $B = (c^2 - c + 1/2)/2$ for the anchored space with anchor value $c \in [0,1]$, and $B = 1/12$ for the unanchored space. (In particular, $B = 1/4$ for the anchor $c = 1$ and $B = 1/8$ for the midpoint anchor $c = 1/2$.)

Thus the worst-case error is bounded independently of s, and has the Monte Carlo rate of convergence, if the infinite sum in (4.9) converges.

Proof. We give the proof for the anchored space with $c = 1$. The other cases follow similarly. It follows from the *averaging argument* (there is always at least one choice as good as the average) that there exists a QMC point set P_n for which the worst-case error is less than or equal to the QMC mean given by (3.13). We produce an upper bound on the latter by omitting the negative term, and then use the third equation in (4.5) to evaluate the integral, obtaining

$$\begin{aligned} e_{n,s}^2(P_n; H(K_{s,\gamma})) \le E_{n,s}^2(H(K_{s,\gamma})) &\le \frac{1}{n} \prod_{j=1}^{s} \left(1 + \frac{\gamma_j}{2} \right) \\ &= \frac{1}{n} \exp\left(\sum_{j=1}^{s} \log\left(1 + \frac{\gamma_j}{2} \right) \right) \le \frac{1}{n} \exp\left(\frac{1}{2} \sum_{j=1}^{s} \gamma_j \right), \end{aligned}$$

where in the last step we used the property that $\log(1+x) \le x$ for all $x > 0$. The rest of the claim in the lemma follows immediately. □

The 'only if' part of Theorem 4.1 for the anchored case comes from the following result established by Sloan and Woźniakowski (1998). The argument there is relatively elementary, starting from (3.9) with the off-diagonal terms of the last term dropped (which is justifiable by the non-negativity assumption).

Lemma 4.3. If the reproducing kernel K_s of some Hilbert space $H(K_s)$ is non-negative, then for any point set $P_n \subset [0,1]^s$ we have

$$e^2_{n,s}(P_n; H(K_s)) \geq e^2_{0,s}(H(K_s))(1 - n\,\kappa_s^2),$$

where

$$\kappa_s := \left(\sup_{\boldsymbol{x} \in [0,1]^s} \frac{a_s(\boldsymbol{x})}{\sqrt{K_s(\boldsymbol{x},\boldsymbol{x})}} \right) \frac{1}{\|a_s\|_s},$$

and

$$a_s(\boldsymbol{x}) := \int_{[0,1]^s} K_s(\boldsymbol{x},\boldsymbol{y})\,\mathrm{d}\boldsymbol{y}.$$

Hence $e_{n,s}(P_n, H(K_s)) \leq \varepsilon e_{0,s}(H(K_s))$ can happen only if

$$n \geq \frac{1-\varepsilon^2}{\kappa_s^2}.$$

This lemma applies to the anchored space of Section 4.2 because the kernel $K_{s,\gamma}$ is in this case manifestly positive. The proof of the second part of Theorem 4.1 then follows in this case by showing that $\kappa_s \to 0$ if the sum in (4.9) diverges. For the unanchored case a different proof is needed, because the kernel $K_{s,\gamma}$ in that case is not necessarily positive. The result for that case is proved in Sloan and Woźniakowski (2002, Theorem 1), but in any case is superseded by the stronger result stated in Theorem 4.5 below.

In the case of our original anchored but unweighted Sobolev space discussed in Example 3.3, it can be shown (Sloan and Woźniakowski 1998) that $\kappa_s \approx 1.055^{-s}$, thus in this case n must be exponentially large to reduce the initial error by some fixed percentage, say 50%.

The upper bound on the worst-case error in Lemma 4.2 is independent of s under the condition (4.9), but the apparent rate of convergence in that lemma is just the Monte Carlo rate of $O(n^{-1/2})$. Fortunately, as we shall see in Sections 5 and 6, a convergence rate arbitrarily close to $O(n^{-1})$ can be attained if the (product) weights satisfy the stronger condition

$$\sum_{j=1}^{\infty} \gamma_j^{1/2} < \infty.$$

4.5. Tractability of multivariate integration

In this subsection we frame the earlier discussion in terms of the valuable notion of tractability. Loosely speaking, the tractability of a problem relates to the question of how quickly the difficulty of a problem increases as the dimension increases. Note that it relates to the difficulty of the *problem*, not to the cost of any particular algorithm.

The standard reference in this area is the recent three-volume work by Novak and Woźniakowski (2008, 2010, 2012). In the present brief discussion we limit ourselves to a very small part of the subject, touching only parts that are relevant to our concerns with multivariate integration over the unit cube.

Thus our problem is that of evaluating the s-dimensional integral $I_s(f)$ defined in (1.1). We will assume that f belongs to a Banach space F_s of real-valued functions defined on $[0,1]^s$. The space F_s could be one of our reproducing kernel spaces H_s, but could also (as in the next subsection) be a non-Hilbert space. The setting is that we are allowed to approximate $I_s(f)$ by any deterministic algorithm $A_{n,s}(f)$ that uses at most n function values of F_s. (In the language of information-based complexity (Traub, Wasilkowski and Woźniakowski 1988), the algorithm is restricted to 'standard information'.) Thus $A_{n,s}$ could be a QMC integration rule, but could also be an integration rule that uses unequal weights, including even negative weights, and $A_{n,s}$ is even allowed to be a nonlinear combination of $f(\boldsymbol{t}_0), \ldots, f(\boldsymbol{t}_{n-1})$.

For a given $s \geq 1$ and a given $\varepsilon \in (0,1]$ we define the nth *minimal number* for this problem by

$$n(\varepsilon, s) := \min\left\{ n \geq 1 : \inf_{A_{n,s}} \sup_{\|f\|_{F_s} \leq 1} |I_s(f) - A_{n,s}(f)| \leq \varepsilon \right\},$$

where the infimum is over all algorithms $A_{n,s}$ that use no more than n function values of f (and no other information about f).

Definition 4.4 (tractability). The integration problem is said to be *polynomially tractable* if there is some $C > 0$ and some $p, q > 0$ and $q \geq 0$ such that

$$n(\varepsilon, s) \leq C \left(\frac{1}{\varepsilon}\right)^p s^q$$

for all $s \in \mathbb{N}$ and $\varepsilon \in (0,1]$. It is *strongly polynomially tractable* if this inequality holds with $q = 0$, that is, if

$$n(\varepsilon, s) \leq C \left(\frac{1}{\varepsilon}\right)^p.$$

Many other variants of tractability are defined in the cited works, but these two will suffice for our present purposes.

The following theorem mimics Theorem 4.1 in the preceding subsection, but there is an important difference: whereas for the 'only if' part of the theorem we previously allowed only QMC integration rules, here *all* algorithms $A_{n,s}$ are allowed, making the statement considerably stronger.

Theorem 4.5 (strong polynomial tractability). For product weights, and for the weighted anchored or unanchored Sobolev space of Sections 4.2 and 4.3, the integration problem is strongly polynomially tractable if and only if

$$\sum_{j=1}^{\infty} \gamma_j < \infty.$$

A necessary and sufficient condition for polynomial tractability is as follows.

Theorem 4.6 (polynomial tractability). For product weights, and for the weighted anchored or unanchored Sobolev space of Sections 4.2 and 4.3, the integration problem is polynomially tractable if and only if

$$\limsup_{s\to\infty} \frac{\sum_{j=1}^{s} \gamma_j}{\log(s+1)} < \infty.$$

The preceding theorem may give the impression that non-trivial weights are needed in order to have (polynomial) tractability, but the remarkable result discussed in the next subsection shows that, while this is true for the Hilbert spaces we have met so far, it is not true for the classical L_1 version of the anchored spaces.

But the Hilbert spaces retain one great advantage, namely that the worst-case error for a given QMC rule is computable, something that turns out (as we shall see in the next two sections) to be helpful in construction. The star discrepancy (whether weighted or not), on the other hand, is notoriously difficult to compute: see Gnewuch, Srivastav and Winzen (2009).

In this subsection we have so far restricted ourselves to product weights. Tractability results for general weights were first considered by Sloan *et al.* (2004). For example, for general weights $\gamma_{\mathfrak{u}}$ and the unanchored Sobolev space it is shown that the integration problem is strongly polynomially tractable if

$$\sum_{|\mathfrak{u}|<\infty} \gamma_{\mathfrak{u}} \left(\tfrac{1}{6}\right)^{|\mathfrak{u}|} < \infty,$$

where now the sum is over all finite subsets of the natural numbers. For other tractability results for general weights and variants of the weights, we refer the reader to Novak and Woźniakowski (2010).

4.6. Tractability of star discrepancy

A surprising result established in Heinrich, Novak, Wasilkowski and Woźniakowski (2001) is that there exists a point set P in $[0,1]^s$ consisting of n points such that its star discrepancy, given by (3.16), satisfies

$$D_n^*(P) \leq C\sqrt{\frac{s}{n}}, \tag{4.10}$$

for some constant $C > 0$ which is independent of n and s. Aistleitner (2011) showed that C can be chosen as 10, for instance. A lower bound by Hinrichs (2004) states that the infimum of $D_n^*(P)$ over all point sets P is bounded by

$$\inf_{P\subset[0,1]^s, |P|=n} D_n^*(P) \geq \min\left(c_0, c\frac{s}{n}\right),$$

where the constant $c > 0$ is independent of s and n and

$$c_0 = \frac{1}{32\mathrm{e}^2},$$

as stated in Gnewuch and Roşca (2009).

In the following we prove a slightly weaker version of (4.10). The proof is based on Hoeffding's inequality, a special case of which we describe in the following. Let $X_0, X_1, \ldots, X_{n-1}$ be independent and identically distributed random variables with mean μ and $X_i \in [-1,1]$ almost surely for $0 \leq i < n$. Let

$$X = \frac{1}{n}\sum_{i=0}^{n-1} X_i.$$

Then

$$\mathbb{P}(|X-\mu| \geq \delta) \leq 2\,\mathrm{e}^{-\delta^2 n/2}, \quad \text{for all } \delta \geq 0.$$

We need an additional lemma.

Lemma 4.7. Let $P = \{\boldsymbol{t}_0, \boldsymbol{t}_1, \ldots, \boldsymbol{t}_{n-1}\} \subset [0,1]^s$ be a point set consisting of n elements. Let $\delta > 0$ and $m = \lceil s/\delta \rceil$. Let Γ_m be the equidistant grid on $[0,1]^s$ with mesh size $1/m$ (and therefore cardinality $(m+1)^s$). Then

$$D_n^*(P) \leq \max_{\boldsymbol{x}\in\Gamma_m} |\Delta_P(\boldsymbol{x})| + \delta.$$

Proof. Let $\eta > 0$. Then there is a vector $\boldsymbol{x}^* = (x_1^*, \ldots, x_s^*) \in [0,1]^s$ such that

$$D_n^*(P) \leq |\Delta_P(\boldsymbol{x}^*)| + \eta.$$

Let $\boldsymbol{y}, \boldsymbol{z} \in \Gamma_m$ be such that $y_j \leq x_j^* \leq z_j$ and $z_j - y_j = m^{-1}$. Then we have

$$\begin{aligned}\prod_{j=1}^{s} z_j - \prod_{j=1}^{s} y_j &= \sum_{r=1}^{s}\left(\left(\prod_{j=r}^{s} z_j\right)\left(\prod_{j=1}^{r-1} y_j\right) - \left(\prod_{j=r+1}^{s} z_j\right)\left(\prod_{j=1}^{r} y_j\right)\right) \\ &= \sum_{r=1}^{s}\left(\prod_{j=r+1}^{s} z_j\right)\left(\prod_{j=1}^{r-1} y_j\right)(z_r - y_r) \leq \frac{s}{m} \leq \delta,\end{aligned}$$

where the empty product is set to 1. Thus we have

$$\begin{aligned}\prod_{j=1}^{s} z_j - \frac{1}{n}\sum_{i=0}^{n-1} 1_{[0,\boldsymbol{z})}(\boldsymbol{t}_i) - \delta &\leq \prod_{j=1}^{s} x_j^* - \frac{1}{n}\sum_{i=0}^{n-1} 1_{[\boldsymbol{0},\boldsymbol{x}^*)}(\boldsymbol{t}_i) \\ &\leq \prod_{j=1}^{s} y_j - \frac{1}{n}\sum_{i=0}^{n-1} 1_{[\boldsymbol{0},\boldsymbol{y})}(\boldsymbol{t}_i) + \delta,\end{aligned}$$

which implies the result. □

We are now ready to prove the following slightly weaker version of (4.10), which is Theorem 1 of Heinrich *et al.* (2001).

Theorem 4.8. For every $n, s \in \mathbb{N}$ there exists a point set P consisting of n points in $[0,1]^s$, whose star discrepancy satisfies

$$D_n^*(P) \leq \frac{2\sqrt{2}}{\sqrt{n}}\sqrt{s\log\left(\left\lceil\frac{s\sqrt{n}}{\sqrt{4\log 2}}\right\rceil + 1\right) + \log 2}.$$

Proof. Let $\boldsymbol{t}_0, \boldsymbol{t}_1, \ldots, \boldsymbol{t}_{n-1}$ be independently and uniformly distributed random variables in $[0,1]^s$. Let $P = \{\boldsymbol{t}_0, \boldsymbol{t}_1, \ldots, \boldsymbol{t}_{n-1}\}$. Then for each $\boldsymbol{x} \in [0,1]^s$

$$\Delta_{\boldsymbol{t}_i}(\boldsymbol{x}) = 1_{[\boldsymbol{0},\boldsymbol{x})}(\boldsymbol{t}_i) - x_1 \cdots x_s$$

is a random variable with mean 0 and $|\Delta_{\boldsymbol{t}_i}(\boldsymbol{x})| \leq 1$ for $0 \leq i < n$. Further, we have

$$\Delta_P(\boldsymbol{x}) = \frac{1}{n}\sum_{i=0}^{n-1} \Delta_{\boldsymbol{t}_i}(\boldsymbol{x}).$$

Thus we can use Hoeffding's inequality, which implies that

$$\begin{aligned}\mathbb{P}\big(D_n^*(P)) \leq 2\delta\big) &\geq \mathbb{P}\Big(\max_{\boldsymbol{x}\in\Gamma_m} |\Delta_P(\boldsymbol{x})| \leq \delta\Big) \\ &\geq 1 - 2(m+1)^s \mathrm{e}^{-\delta^2 n/2}.\end{aligned}$$

We now choose the parameters such that $1 - 2(m+1)^s \mathrm{e}^{-\delta^2 n/2} > 0$, or equivalently

$$\log 2 + s\log(m+1) - \frac{\delta^2 n}{2} < 0.$$

Since $m = \lceil s/\delta \rceil$, this holds for all $\delta > \delta_0 = \delta_0(n, s)$, where δ_0 satisfies

$$\delta_0^2 = 2n^{-1}(s \log(\lceil s/\delta_0 \rceil + 1) + \log 2).$$

This implies that

$$\frac{1}{\delta_0} \le \sqrt{\frac{n}{4 \log 2}},$$

and therefore

$$\delta_0^2 \le 2n^{-1}\biggl(s \log\biggl(\biggl\lceil \frac{s\sqrt{n}}{\sqrt{4 \log 2}} \biggr\rceil + 1\biggr) + \log 2\biggr).$$

Thus for any $\delta > \delta_0$ there exist points $\boldsymbol{t}_0, \boldsymbol{t}_1, \ldots, \boldsymbol{t}_{n-1}$ such that

$$D_n^*(\{\boldsymbol{t}_0, \boldsymbol{t}_1, \ldots, \boldsymbol{t}_{n-1}\}) \le 2\delta.$$

Hence there is a point set $P = \{\boldsymbol{t}_0, \boldsymbol{t}_1, \ldots, \boldsymbol{t}_{n-1}\}$ such that $D_n^*(P) \le 2\delta_0$. $\square$

4.7. Notes

Weighted spaces were first introduced by Sloan and Woźniakowski (1998), and subsequently generalized by Sloan and Woźniakowski (2001, 2002), Dick *et al.* (2006) and Sloan *et al.* (2004). A non-technical introduction to the weighted space setting and lattice rules is given in Kuo and Sloan (2005). For the standard reference on tractability see the recent three-volume work by Novak and Woźniakowski (2008, 2010, 2012).

The 'only if' parts of Theorems 4.5 and 4.6 were proved by Novak and Woźniakowski (2001) for the anchored Sobolev space, and also for the more general class of 'decomposable' kernels. For the unanchored Sobolev space the second parts of the theorems were proved by Sloan and Woźniakowski (2002), using results from Hickernell and Woźniakowski (2001) and in turn Hickernell and Woźniakowski (2000). See also Wasilkowski and Woźniakowski (2004).

A classic paper on numerical integration and discrepancy is by Hickernell (1998*a*), who defined discrepancies which include lower-dimensional projections and introduced the reproducing kernel machinery for numerical integration. Sufficient conditions on the weights for which Sobol′, Halton, and Niederreiter sequences achieve strong tractability have been studied by Wang (2002, 2003). Tractability of so-called 'finite-order weights' has been shown by Sloan *et al.* (2004). Strong tractability of scrambled digital nets and sequences has been studied by Yue and Hickernell (2005, 2006). Tractability questions for the weighted star discrepancy have been considered by Hinrichs, Pillichshammer and Schmid (2008). A strategy for choosing the weights in finance applications has been considered in Wang and Sloan (2006, 2007). The concepts of 'effective dimension', 'truncation dimension' and 'superposition dimension' have been introduced in Caflisch *et al.* (1997). The effective dimension of problems arising from financial applications have been studied by Wang and Fang (2003) and Wang and Sloan (2005) in the context of QMC.

The equidistant grid Γ_m used in Lemma 4.7 is a special δ-cover or (essentially) bracketing cover. The notion of bracketing is well established in empirical process theory (see van der Vart and Wellner 2009, for example), and to achieve better results than in Theorem 4.8 one needs better bracketing covers than Γ_m. In fact, the results from Aistleitner (2011), together with the results on probabilistic discrepancy estimates from Gnewuch and Roşca (2009), Aistleitner and Hofer (2012) and Gnewuch (2009), as well as theoretical arguments from Doerr, Gnewuch and Wahlström (2009) and Doerr, Gnewuch, Kritzer and Pillichshammer (2008) for constructing small discrepancy samples, all rely on the constructive bracketing covers and the induced upper bounds on bracketing numbers from Gnewuch (2008). Aistleitner (2011) and Aistleitner and Hofer (2012) also use Bernstein's inequality and the technique of dyadic chaining.

Attempts have been made to find constructions of point sets whose star discrepancy grows only polynomially with the dimension: see for instance Doerr, Gnewuch and Wahlström (2010) and Doerr *et al.* (2008). An implementation and numerical experiments have been reported in Doerr *et al.* (2009). However, the construction cost of these algorithms depends exponentially on the dimension. An essential problem in these algorithms is the estimation of the star discrepancy of a given point set, which is itself intractable as shown by Gnewuch *et al.* (2009). Algorithms for estimating the star discrepancy have been further investigated by Gnewuch, Wahlström and Winzen (2012).

Let $n(\varepsilon, d)$ be the smallest number necessary to reduce the d-dimensional L_2 discrepancy $\|\Delta_P\|_2$ by a factor of ε. The exponent of discrepancy is then the smallest number p such that $n(\varepsilon, d) \le C\varepsilon^{-p}$ for all $d \ge 1$. It has been shown that $1.0669 \le p \le 1.41274$ for the classical L_2 discrepancy, where the lower bound is by Matoušek (1998*a*) and the upper bound by Wasilkowski and Woźniakowski (2010).

5. Lattice rules

In Section 2 we gave a brief introduction to lattice rules. We defined *rank-one lattice rules* and explained how *randomly shifted lattice rules* can be used for practical error estimation. We also introduced the *component-by-component* (CBC) construction for obtaining good lattice rule generating vectors. Here in this section we provide the theory behind the CBC construction, including error analysis of randomly shifted lattice rules and the fast implementation of the CBC algorithm. We will also discuss *extensible lattice sequences*, which turn lattice rules from 'closed' point sets to 'open' sequences, making them more flexible for practical applications.

For most of this section, we will consider the weighted anchored or unanchored spaces of Section 4, which contain integrands with square-integrable mixed *first* derivatives. The lattice rules constructed for these function spaces can achieve close to $O(n^{-1})$ convergence rate, with implied constants that can be independent of the dimension s if the weights of the function space satisfy a certain condition. We will also discuss the use of the *baker's transformation* to obtain close to $O(n^{-2})$ convergence rate when the integrands have square-integrable mixed *second* derivatives, again with implied constants that can be independent of s. We will also outline other strategies for obtaining even higher orders of convergence using lattice rules, but now the error bounds known so far have exponential growth in s.

5.1. *A brief background on the classical theory of lattice rules*

Lattice rules were originally designed for periodic functions. It is customary to assume that the integrand f has an *absolutely convergent Fourier series*

$$f(\boldsymbol{x}) = \sum_{\boldsymbol{h}\in\mathbb{Z}^s} \widehat{f}(\boldsymbol{h})\,\mathrm{e}^{2\pi\mathrm{i}\boldsymbol{h}\cdot\boldsymbol{x}}, \quad \mathrm{i}^2 = -1, \tag{5.1}$$

with Fourier coefficients

$$\widehat{f}(\boldsymbol{h}) = \int_{[0,1]^s} f(\boldsymbol{x})\,\mathrm{e}^{-2\pi\mathrm{i}\boldsymbol{h}\cdot\boldsymbol{x}}\,\mathrm{d}\boldsymbol{x},$$

where $\boldsymbol{h}\cdot\boldsymbol{x} = h_1x_1 + \cdots + h_sx_s$ denotes the usual vector dot product. It follows from the absolute (and hence uniform) convergence of (5.1) that f is necessarily continuous and also *one-periodic* with respect to each variable, that is, $f(\boldsymbol{x})|_{x_j=0} = f(\boldsymbol{x})|_{x_j=1}$ for all $j = 1,\ldots,s$.

Theorem 5.1 (the lattice rule error). Let $Q_{n,s}$ denote a lattice rule (not necessarily rank-one) and let $\mathcal{L}$ denote the associated integration lattice. If f has an absolutely convergent Fourier series (5.1), then

$$Q_{n,s}(f) - I_s(f) = \sum_{\boldsymbol{h}\in\mathcal{L}^\perp\setminus\{\boldsymbol{0}\}} \widehat{f}(\boldsymbol{h}),$$

where $\mathcal{L}^\perp := \{\boldsymbol{h}\in\mathbb{Z}^s : \boldsymbol{h}\cdot\boldsymbol{x}\in\mathbb{Z} \text{ for all } \boldsymbol{x}\in\mathcal{L}\}$ is the *dual lattice* associated with $\mathcal{L}$.

We will prove this result for rank-one lattice rules (see (2.3)), that is, for QMC rules of the form

$$Q_{n,s}(f) = \frac{1}{n}\sum_{i=0}^{n-1} f\left(\left\{\frac{i\boldsymbol{z}}{n}\right\}\right).$$

In this case, $\mathcal{L}^\perp = \{\boldsymbol{h} \in \mathbb{Z}^s : \boldsymbol{h}\cdot\boldsymbol{z} \equiv 0 \pmod{n}\}$. A proof for general lattice rules is given in Sloan and Joe (1994).

Theorem 5.2 (rank-one lattice rule error). Let $Q_{n,s}$ denote a rank-one lattice rule with generating vector $\boldsymbol{z}$. If f has an absolutely convergent Fourier series (5.1), then

$$Q_{n,s}(f) - I_s(f) = \sum_{\substack{\boldsymbol{h}\in\mathbb{Z}^s\setminus\{\boldsymbol{0}\} \\ \boldsymbol{h}\cdot\boldsymbol{z}\equiv 0 \,(\mathrm{mod}\, n)}} \widehat{f}(\boldsymbol{h}).$$

Proof. Using the Fourier series (5.1), we can write

$$\begin{aligned} Q_{n,s}(f) - I_s(f) &= \frac{1}{n}\sum_{k=0}^{n-1} f\left(\left\{\frac{k\boldsymbol{z}}{n}\right\}\right) - \int_{[0,1]^s} f(\boldsymbol{x})\,\mathrm{d}\boldsymbol{x} \\ &= \sum_{\boldsymbol{h}\in\mathbb{Z}^s} \widehat{f}(\boldsymbol{h})\left(\frac{1}{n}\sum_{k=0}^{n-1} \mathrm{e}^{2\pi\mathrm{i}k\boldsymbol{h}\cdot\boldsymbol{z}/n}\right) - \sum_{\boldsymbol{h}\in\mathbb{Z}^s} \widehat{f}(\boldsymbol{h}) \int_{[0,1]^s} \mathrm{e}^{2\pi\mathrm{i}\boldsymbol{h}\cdot\boldsymbol{x}}\,\mathrm{d}\boldsymbol{x}. \end{aligned}$$

The result then follows from the elementary properties

$$\int_{[0,1]^s} \mathrm{e}^{2\pi\mathrm{i}\boldsymbol{h}\cdot\boldsymbol{x}}\,\mathrm{d}\boldsymbol{x} = \begin{cases} 1 & \text{if } \boldsymbol{h} = \boldsymbol{0}, \\ 0 & \text{otherwise,} \end{cases}$$

and

$$\frac{1}{n}\sum_{k=0}^{n-1} \mathrm{e}^{2\pi\mathrm{i}k\boldsymbol{h}\cdot\boldsymbol{z}/n} = \begin{cases} 1 & \text{if } \boldsymbol{h}\cdot\boldsymbol{z} \equiv 0 \pmod{n}, \\ 0 & \text{otherwise.} \end{cases} \tag{5.2}$$

□

The property (5.2) is sometimes referred to as the 'character property' of lattice rules with respect to the exponential functions. It is one of the key properties we need for the error analysis of lattice rules.

In the classical theory of lattice rules, one considers a class $E_\alpha(c)$ of functions whose Fourier coefficients satisfy, for $\alpha > 1$ and $c > 0$,

$$|\widehat{f}(\boldsymbol{h})| \le \frac{c}{(\overline{h}_1 \cdots \overline{h}_s)^\alpha}, \quad \text{with } \overline{h} := \max(1, |h|).$$

It follows that

$$|Q_{n,s}(f) - I_s(f)| \le c \sum_{\substack{\boldsymbol{h}\in\mathbb{Z}^s\setminus\{\boldsymbol{0}\} \\ \boldsymbol{h}\cdot\boldsymbol{z}\equiv 0 \,(\mathrm{mod}\, n)}} \frac{1}{(\overline{h}_1 \cdots \overline{h}_s)^\alpha}, \quad \text{for } f \in E_\alpha(c).$$

This leads to the definition of a quality measure called P_α,

$$P_\alpha(\boldsymbol{z}, n) := \sum_{\substack{\boldsymbol{h}\in\mathbb{Z}^s\setminus\{\boldsymbol{0}\} \\ \boldsymbol{h}\cdot\boldsymbol{z}\equiv 0\ (\mathrm{mod}\ n)}} \frac{1}{(\overline{h}_1\cdots\overline{h}_s)^\alpha}, \quad \alpha > 1, \tag{5.3}$$

and the goal is to choose $\boldsymbol{z}$ so that $P_\alpha(\boldsymbol{z}, n)$ is as small as possible. By using the character property (5.2) (or by recognizing that P_α is the integration error for a function f with Fourier coefficients $\widehat{f}(\boldsymbol{h}) = 1/(\overline{h}_1\cdots\overline{h}_s)^\alpha$), one can express P_α as

$$\begin{aligned} P_\alpha(\boldsymbol{z}, n) &= -1 + \frac{1}{n}\sum_{k=0}^{n-1}\prod_{j=1}^{s}\left(1 + \sum_{h\in\mathbb{Z}\setminus\{0\}} \frac{\mathrm{e}^{2\pi \mathrm{i} khz_j/n}}{|h|^\alpha}\right) \qquad (5.4) \\ &= -1 + \frac{1}{n}\prod_{j=1}^{s}(1 + 2\zeta(\alpha)) + \frac{1}{n}\sum_{k=1}^{n-1}\prod_{j=1}^{s}\left(1 + \sum_{h\in\mathbb{Z}\setminus\{0\}} \frac{\mathrm{e}^{2\pi \mathrm{i} khz_j/n}}{|h|^\alpha}\right), \end{aligned}$$

where

$$\zeta(x) := \sum_{h=1}^{\infty} \frac{1}{h^x}, \quad x > 1, \tag{5.5}$$

is the Riemann zeta function. For practical computation, α is often taken to be an even integer. This is because the *Bernoulli polynomial of degree* α, with α an even integer, has the Fourier series

$$B_\alpha(x) = \frac{(-1)^{\frac{\alpha}{2}+1}\alpha!}{(2\pi)^\alpha} \sum_{h\in\mathbb{Z}\setminus\{0\}} \frac{\mathrm{e}^{2\pi \mathrm{i} hx}}{h^\alpha}, \quad \text{for } x \in [0,1], \tag{5.6}$$

so that in this case we can write

$$\sum_{h\in\mathbb{Z}\setminus\{0\}} \frac{\mathrm{e}^{2\pi \mathrm{i} khz_j/n}}{|h|^\alpha} = \frac{(-1)^{\frac{\alpha}{2}+1}(2\pi)^\alpha}{\alpha!} B_\alpha\left(\left\{\frac{kz_j}{n}\right\}\right). \tag{5.7}$$

This allows P_α in (5.4) to be computed in $O(n\,s)$ operations. One may therefore use the Korobov construction in Example 2.9 to search for a lattice generating vector $\boldsymbol{z}$, using the criterion that P_α be as small as possible.

It is well known that the rate of decay of the Fourier coefficients of a function is related to the smoothness of the function. For instance, if $\alpha > 1$ is an integer and all partial derivatives

$$\frac{\partial^{q_1+\cdots+q_s} f}{\partial x_1^{q_1}\cdots\partial x_s^{q_s}}, \quad 0 \le q_1, \ldots, q_s \le \alpha,$$

exist and are continuous on $[0,1]^s$, then there exists $c > 0$ for which $f \in E_\alpha(c)$. Thus the class $E_\alpha(c)$ is essentially a class of functions with smoothness determined by α, therefore the parameter α is called the *smoothness parameter.* It is known that a convergence rate of

$$P_\alpha(\boldsymbol{z}, n) = O(n^{-\alpha} (\log n)^{\alpha s})$$

can be achieved. There are also other related quality measures for lattice rules (Niederreiter 1992*a*, Sloan and Joe 1994).

We shall not discuss the classical theory further, because in the classical error analysis all error bounds grow exponentially with dimension. In the following, we will carry out error analysis in the weighted function space setting of Section 4.

5.2. *Shift-averaged worst-case error*

In this subsection we discuss a general strategy for the error analysis of *randomly shifted QMC rules*, as a preparation for our later analysis of randomly shifted lattice rules.

For any QMC point set $P = \{\boldsymbol{t}_0, \ldots, \boldsymbol{t}_{n-1}\}$ and any shift $\boldsymbol{\Delta} \in [0,1]^s$, let

$$P + \boldsymbol{\Delta} = \{\{\boldsymbol{t}_i + \boldsymbol{\Delta}\} : i = 0, 1, \ldots, n-1\}$$

denote the shifted QMC point set, and let $Q_{n,s}(\boldsymbol{\Delta}; f)$ denote the corresponding shifted QMC rule. Then, for any integrand f belonging to some normed space H, it follows from the definition of the worst-case error (see Definition 3.4) that

$$|I_s(f) - Q_{n,s}(\boldsymbol{\Delta}, f)| \le e_{n,s}(P + \boldsymbol{\Delta}; H)\, \|f\|_H.$$

We deduce a bound for the *root-mean-square* error

$$\sqrt{\mathbb{E}\, |I_s(f) - Q_{n,s}(\boldsymbol{\Delta}, f)|^2} \le e^{\mathrm{sh}}_{n,s}(P; H)\, \|f\|_H,$$

where the expectation $\mathbb{E}$ is taken over the random shift $\boldsymbol{\Delta}$ which is uniformly distributed over $[0,1]^s$, and where the quantity

$$e^{\mathrm{sh}}_{n,s}(P; H) := \sqrt{\int_{[0,1]^s} e^2_{n,s}(P + \boldsymbol{\Delta}; H)\, \mathrm{d}\boldsymbol{\Delta}} \tag{5.8}$$

is referred to as the *shift-averaged worst-case error.*

The shift-averaged worst-case error (5.8) will be used as our quality measure for randomly shifted QMC rules. For any given point set P, the averaging argument guarantees the existence of at least one shift $\boldsymbol{\Delta}$ for which

$$e_{n,s}(P + \boldsymbol{\Delta}; H) \le e^{\mathrm{sh}}_{n,s}(P; H).$$

The following theorem gives an explicit formula for the shift-averaged worst-case error when the function space is an RKHS.

Theorem 5.3 (formula for the shift-averaged worst-case error). The *shift-averaged* worst-case error (5.8) for a QMC point set P in an RKHS $H_s(K)$ satisfies

$$[e^{\mathrm{sh}}_{n,s}(P;H_s(K))]^2 = -\int_{[0,1]^s}\int_{[0,1]^s} K(\boldsymbol{x},\boldsymbol{y})\,\mathrm{d}\boldsymbol{x}\,\mathrm{d}\boldsymbol{y} + \frac{1}{n^2}\sum_{i=0}^{n-1}\sum_{k=0}^{n-1} K^{\mathrm{sh}}(\boldsymbol{t}_i,\boldsymbol{t}_k), \tag{5.9}$$

where

$$K^{\mathrm{sh}}(\boldsymbol{x},\boldsymbol{y}) := \int_{[0,1]^s} K(\{\boldsymbol{x}+\boldsymbol{\Delta}\},\{\boldsymbol{y}+\boldsymbol{\Delta}\})\,\mathrm{d}\boldsymbol{\Delta} \quad \text{for all } \boldsymbol{x},\boldsymbol{y}\in[0,1]^s. \tag{5.10}$$

Proof. Using the definition (5.8) and applying the formula (3.9) for the worst-case error $e_{n,s}(P+\boldsymbol{\Delta};H_s(K))$, we obtain

$$\begin{aligned}
&[e^{\mathrm{sh}}_{n,s}(P;H_s(K))]^2\\
&= \int_{[0,1]^s}\int_{[0,1]^s} K(\boldsymbol{x},\boldsymbol{y})\,\mathrm{d}\boldsymbol{x}\,\mathrm{d}\boldsymbol{y} - \frac{2}{n}\sum_{i=0}^{n-1}\int_{[0,1]^s}\int_{[0,1]^s} K(\{\boldsymbol{t}_i+\boldsymbol{\Delta}\},\boldsymbol{y})\,\mathrm{d}\boldsymbol{\Delta}\,\mathrm{d}\boldsymbol{y}\\
&\quad + \frac{1}{n^2}\sum_{i=0}^{n-1}\sum_{k=0}^{n-1}\int_{[0,1]^s} K(\{\boldsymbol{t}_i+\boldsymbol{\Delta}\},\{\boldsymbol{t}_k+\boldsymbol{\Delta}\})\,\mathrm{d}\boldsymbol{\Delta}.
\end{aligned}$$

With a change of variables $\boldsymbol{x} = \{\boldsymbol{t}_i+\boldsymbol{\Delta}\}$, the double integrals in the second term turn into the double integral in the first term. The result then follows from the definition (5.10). □

The function K^{sh} defined by (5.10) is actually a reproducing kernel, with the *shift-invariant property*

$$K^{\mathrm{sh}}(\boldsymbol{x},\boldsymbol{y}) = K^{\mathrm{sh}}(\{\boldsymbol{x}+\boldsymbol{\Delta}\},\{\boldsymbol{y}+\boldsymbol{\Delta}\}) \quad \text{for all } \boldsymbol{x},\boldsymbol{y},\boldsymbol{\Delta}\in[0,1]^s,$$

or equivalently

$$K^{\mathrm{sh}}(\boldsymbol{x},\boldsymbol{y}) = K^{\mathrm{sh}}(\{\boldsymbol{x}-\boldsymbol{y}\},\boldsymbol{0}) \quad \text{for all } \boldsymbol{x},\boldsymbol{y}\in[0,1]^s.$$

Furthermore, it can be verified that the shift-averaged worst-case error $e^{\mathrm{sh}}_{n,s}(P;H_s(K))$ is precisely the worst-case error of the QMC point set P in the RKHS with K^{sh} as the reproducing kernel, that is,

$$\int_{[0,1]^s} e^2_{n,s}(P+\boldsymbol{\Delta};H_s(K))\,\mathrm{d}\boldsymbol{\Delta} = e^2_{n,s}(P;H_s(K^{\mathrm{sh}})).$$

In words, *the average over all possible shifts of the squared worst-case error for a shifted QMC rule in a Hilbert space with reproducing kernel K* is equal to *the squared worst-case error of the original unshifted QMC rule in a Hilbert space with reproducing kernel K^{sh}*.This important connection provides a powerful tool for analysing randomly shifted QMC rules, because it is often easier to work with a shift-invariant kernel. We refer to the kernel $K^{\rm sh}$ as the *shift-invariant kernel associated with* K.

Further, we note that for the reproducing kernels we consider below, the reproducing kernel $K^{\rm sh}$ has the same smoothness properties as K. As the smoothness properties of the reproducing kernel determine the convergence rate of the worst-case error, one obtains the same convergence rate of the worst-case error for $H_s(K^{\rm sh})$ as for $H_s(K)$.

5.3. *Randomly shifted lattice rules in weighted Sobolev spaces*

We are now ready to analyse randomly shifted lattice rules in the weighted anchored or unanchored Sobolev space of Section 4. First we derive the associated shift-invariant kernel needed for the shift-averaged worst-case error.

Lemma 5.4. Let $H(K_{s,\gamma})$ be the weighted anchored or unanchored Sobolev space of Section 4. The shift-invariant kernel associated with $K_{s,\gamma}$ is

$$K^{\rm sh}_{s,\gamma}(\boldsymbol{x},\boldsymbol{y}) = \sum_{\mathrm{u}\subseteq\{1:s\}} \gamma_{\mathrm{u}} \prod_{j\in\mathrm{u}} \big(B_2(|x_j - y_j|) + \beta\big), \tag{5.11}$$

where $\beta = c^2 - c + 1/3$ for the anchored variant with anchor c, and $\beta = 0$ for the unanchored variant.

Proof. From the definitions (5.10) and (4.3), we have

$$K^{\rm sh}_{s,\gamma}(\boldsymbol{x},\boldsymbol{y}) = \sum_{\mathrm{u}\subseteq\{1:s\}} \gamma_{\mathrm{u}} \prod_{j\in\mathrm{u}} \int_0^1 \eta(\{x_j+\Delta_j\},\{y_j+\Delta_j\})\,\mathrm{d}\Delta_j,$$

where the function η can be written in the common form (4.8) for both the anchored and unanchored spaces. Using the symmetry $B_2(x) = B_2(1-x)$ and $\int_0^1 \alpha(x)\,\mathrm{d}x = 0$, we obtain

$$\begin{aligned}
&\int_0^1 \eta(\{x+\Delta\},\{y+\Delta\})\,\mathrm{d}\Delta \\
&= \tfrac{1}{2}B_2(|x-y|) + \int_0^1 \big(\{x+\Delta\}-\tfrac12\big)\big(\{y+\Delta\}-\tfrac12\big)\mathrm{d}\Delta + \beta.
\end{aligned}$$

Therefore it remains to show that

$$\mathcal{I}(x,y) := \int_0^1 \big(\{x+\Delta\}-\tfrac12\big)\big(\{y+\Delta\}-\tfrac12\big)\mathrm{d}\Delta = \tfrac12 B_2(|x-y|).$$

Since $\mathcal{I}(x,y) = \mathcal{I}(y,x)$, without loss of generality we assume that $x \le y$. There are three possible arrangements of the values of $x+\Delta$ and $y+\Delta$

for $x, y, \Delta \in [0,1)$:

$$\begin{aligned} x+\Delta \le y+\Delta < 1 &\Rightarrow \Delta < 1-y, \\ x+\Delta < 1 \le y+\Delta &\Rightarrow 1-y \le \Delta < 1-x, \\ 1 \le x+\Delta \le y+\Delta &\Rightarrow 1-x \le \Delta. \end{aligned}$$

Thus we have

$$\begin{aligned} \mathcal{I}(x,y) &= \int_0^{1-y} \big(x+\Delta-\tfrac{1}{2}\big)\big(y+\Delta-\tfrac{1}{2}\big)\mathrm{d}\Delta \\ &\quad + \int_{1-y}^{1-x} \big(x+\Delta-\tfrac{1}{2}\big)\big(y+\Delta-1-\tfrac{1}{2}\big)\mathrm{d}\Delta \\ &\quad + \int_{1-x}^{1} \big(x+\Delta-1-\tfrac{1}{2}\big)\big(y+\Delta-1-\tfrac{1}{2}\big)\mathrm{d}\Delta \\ &= \tfrac{1}{2}(y-x)^2 - \tfrac{1}{2}(y-x) + \tfrac{1}{12} = \tfrac{1}{2}\big(|x-y|^2 - |x-y| + \tfrac{1}{6}\big). \end{aligned}$$

This completes the proof. □

For simplicity, let

$$e^{\mathrm{sh}}_{n,s}(\boldsymbol{z}) := e^{\mathrm{sh}}_{n,s}(P; H(K_{s,\boldsymbol{\gamma}}))$$

denote the shift-averaged worst-case error for a rank-one lattice rule with point set P and generating vector $\boldsymbol{z}$ in our weighted anchored or unanchored space. The following lemma gives an explicit expression for $[e^{\mathrm{sh}}_{n,s}(\boldsymbol{z})]^2$.

Lemma 5.5 (shift-averaged worst-case error for lattice rules). The shift-averaged worst-case error for a rank-one lattice rule in the weighted anchored or unanchored Sobolev space satisfies

$$[e^{\mathrm{sh}}_{n,s}(\boldsymbol{z})]^2 = \sum_{\emptyset \ne \mathfrak{u} \subseteq \{1:s\}} \gamma_{\mathfrak{u}} \left(\frac{1}{n} \sum_{k=0}^{n-1} \prod_{j \in \mathfrak{u}} \left[B_2\left(\left\{ \frac{k z_j}{n} \right\} \right) + \beta \right] - \beta^{|\mathfrak{u}|} \right), \tag{5.12}$$

where $\beta = c^2 - c + 1/3$ for the anchored variant with anchor c, and $\beta = 0$ for the unanchored variant. For product weights, the expression simplifies to

$$[e^{\mathrm{sh}}_{n,s}(\boldsymbol{z})]^2 = -\prod_{j=1}^{s}(1+\gamma_j\beta) + \frac{1}{n}\sum_{k=0}^{n-1}\prod_{j=1}^{s}\left(1+\gamma_j\left[B_2\left(\left\{\frac{kz_j}{n}\right\}\right)+\beta\right]\right). \tag{5.13}$$

Proof. Substituting $\boldsymbol{t}_i = \{i\boldsymbol{z}/n\}$ into (5.9) and using (4.5) and (5.11), we obtain

$$[e^{\mathrm{sh}}_{n,s}(\boldsymbol{z})]^2 = -\sum_{\mathfrak{u} \subseteq \{1:s\}} \gamma_{\mathfrak{u}} \beta^{|\mathfrak{u}|} + \frac{1}{n^2} \sum_{i=0}^{n-1} \sum_{k=0}^{n-1} \sum_{\mathfrak{u} \subseteq \{1:s\}} \gamma_{\mathfrak{u}} \prod_{j \in \mathfrak{u}} \left[B_2\left(\left\{ \frac{(i-k) z_j}{n} \right\} \right) + \beta \right].$$

As i and k range from 0 to $n-1$, the values of $(i-k) \bmod n$ are just $0, \ldots, n-1$ in some order, with each value occurring n times. Thus we can reduce the double sum in $[e^{\rm sh}_{n,s}(\boldsymbol{z})]^2$ to a single sum. Grouping the sum over $\mathfrak{u}$ and then cancelling the $\mathfrak{u} = \emptyset$ term yields the formula for general weights above. The formula for product weights follows from the identity (4.2). □

5.4. *Component-by-component construction*

Recall that the components of the generating vector $\boldsymbol{z}$ can be restricted to the set

$$\mathbb{U}_n = \{z \in \mathbb{Z} : 1 \leq z \leq n-1 \text{ and } \gcd(z,n) = 1\},$$

whose cardinality is given by the Euler totient function

$$\varphi(n) := |\mathbb{U}_n| = |\{z \in \mathbb{Z} : 1 \leq z \leq n-1 \text{ and } \gcd(z,n) = 1\}|. \tag{5.14}$$

When n is prime $\varphi(n)$ takes its largest value $n-1$, hence there are altogether up to $(n-1)^s$ possible choices for $\boldsymbol{z}$, far too many to allow an exhaustive search for the best one.

The CBC construction below, already foreshadowed in Example 2.10, provides a feasible way to obtain good lattice generating vectors. The first component z_1 is arbitrarily set to 1, since all choices in one dimension lead to the same rectangle rule.

Algorithm 5.6 (CBC construction). Given: n, $s_{\max}$, and weights $\gamma_{\mathfrak{u}}$.

(1) Set $z_1 = 1$.

(2) For $s = 2, 3, \ldots, s_{\max}$, choose z_s in $\mathbb{U}_n$ to minimize $[e^{\rm sh}_{n,s}(z_1, \ldots, z_s)]^2$.

With general weights $\gamma_{\mathfrak{u}}$, the cost of the CBC algorithm is prohibitively expensive, thus when discussing implementation we will always assume there is some special structure, such as product weights, order-dependent weights, finite-order weights, or POD weights. We will discuss the implementation of CBC in later subsections.

In this subsection we focus on error bounds resulting from the CBC construction. The CBC error bounds are proved by induction: under the induction hypothesis that the desired bound holds in $s-1$ dimensions, we show by the averaging argument that the algorithm picks a next component z_s, for which the bound holds in s dimensions.

To illustrate the general argument, we first prove that in the simplest case of product weights and prime n, a generating vector constructed by the CBC algorithm beats the QMC mean (see (3.13) together with (4.5))

$$E^2_{n,s}(K_{s,\gamma}) = \frac{1}{n}\left(\prod_{j=1}^{s}\left(1 + \gamma_j\left(\beta + \frac{1}{6}\right)\right) - \prod_{j=1}^{s}(1 + \gamma_j\beta)\right),$$

which yields a convergence rate of $O(n^{-1/2})$. We will prove a better, if more complicated, result in Theorem 5.8 below.

Theorem 5.7 (CBC error bound beats QMC mean). Consider the weighted anchored or unanchored Sobolev space with product weights $\gamma_1 \geq \gamma_2 \geq \cdots > 0$, and suppose that n is a prime number satisfying

$$n \geq \frac{\gamma_1}{6(1+\gamma_1\beta)}, \tag{5.15}$$

where $\beta = c^2 - c + 1/3$ for the anchored variant with anchor c, and $\beta = 0$ for the unanchored variant. The generating vector $\boldsymbol{z} \in \mathbb{U}_n^s$ constructed by the CBC algorithm, minimizing the squared shift-averaged worst-case error $[e_{n,s}^{\mathrm{sh}}(\boldsymbol{z})]^2$ in the weighted anchored or unanchored Sobolev space in each step, satisfies

$$[e_{n,s}^{\mathrm{sh}}(\boldsymbol{z})]^2 < E_{n,s}^2(K_{s,\boldsymbol{\gamma}}). \tag{5.16}$$

Proof. For the inductive step we need to show, under the assumption $[e_{n,s-1}^{\mathrm{sh}}(z_1,\ldots,z_{s-1})]^2 < E_{n,s-1}^2(K_{s-1,\boldsymbol{\gamma}})$, that $[e_{n,s}^{\mathrm{sh}}(\boldsymbol{z})]^2 < E_{n,s}^2(K_{s,\boldsymbol{\gamma}})$ when $z_s \in \mathbb{U}_n$ is chosen to minimize $[e_{n,s}^{\mathrm{sh}}(\boldsymbol{z})]^2$.

To proceed, we rewrite (5.13) as

$$\begin{aligned}[e_{n,s}^{\mathrm{sh}}(\boldsymbol{z})]^2 &= (1+\gamma_s\beta)\,[e_{n,s-1}^{\mathrm{sh}}(z_1,\ldots,z_{s-1})]^2 + \frac{\gamma_s}{6n}\prod_{j=1}^{s-1}\left(1+\gamma_j\left(\beta+\frac{1}{6}\right)\right) \\ &\quad + \frac{\gamma_s}{n}\sum_{k=1}^{n-1}\left[B_2\left(\left\{\frac{kz_s}{n}\right\}\right)\prod_{j=1}^{s-1}\left(1+\gamma_j\left[B_2\left(\left\{\frac{kz_j}{n}\right\}\right)+\beta\right]\right)\right],\end{aligned}$$

where we separated out the $k=0$ term and used $B_2(0) = 1/6$. Next we average over all possible choices of z_s, forming

$$\mu(z_1,\ldots,z_{s-1}) := \frac{1}{n-1}\sum_{z_s=1}^{n-1}[e_{n,s}^{\mathrm{sh}}(\boldsymbol{z})]^2.$$

Since $B_2(x) = (1/2\pi^2)\sum_{h\in\mathbb{Z}\setminus\{0\}} \mathrm{e}^{2\pi\mathrm{i}hx}/h^2$ for $x \in [0,1]$, for any integer k from 1 to $n-1$ we can write

$$\begin{aligned}\frac{2\pi^2}{n-1}\sum_{z_s=1}^{n-1} B_2\left(\left\{\frac{kz_s}{n}\right\}\right) &= \frac{1}{n-1}\sum_{z=1}^{n-1}\sum_{h\in\mathbb{Z}\setminus\{0\}}\frac{\mathrm{e}^{2\pi\mathrm{i}hkz/n}}{h^2} \\ &= \sum_{\substack{h\in\mathbb{Z}\setminus\{0\}\\ h\equiv 0\,(\mathrm{mod}\,n)}}\frac{1}{h^2} - \frac{1}{n-1}\sum_{\substack{h\in\mathbb{Z}\setminus\{0\}\\ h\not\equiv 0\,(\mathrm{mod}\,n)}}\frac{1}{h^2} \\ &= \sum_{\substack{h\in\mathbb{Z}\setminus\{0\}\\ h\equiv 0\,(\mathrm{mod}\,n)}}\frac{1}{h^2} - \frac{1}{n-1}\left(\sum_{h\in\mathbb{Z}\setminus\{0\}}\frac{1}{h^2} - \sum_{\substack{h\in\mathbb{Z}\setminus\{0\}\\ h\equiv 0\,(\mathrm{mod}\,n)}}\frac{1}{h^2}\right)\end{aligned}$$

$$= \frac{n}{n-1} \sum_{\substack{h \in \mathbb{Z}\setminus\{0\} \\ h \equiv 0 \, (\mathrm{mod}\, n)}} \frac{1}{h^2} - \frac{1}{n-1} \sum_{h \in \mathbb{Z}\setminus\{0\}} \frac{1}{h^2}$$

$$= \frac{2\pi^2 n}{6n^2(n-1)} - \frac{2\pi^2}{6(n-1)} = -\frac{2\pi^2}{6n},$$

where we used $\sum_{h=1}^{\infty} 1/h^2 = \pi^2/6$. Thus we conclude for any integer $1 \le k \le n-1$ that

$$\frac{1}{n-1} \sum_{z_s=1}^{n-1} B_2\left(\left\{\frac{kz_s}{n}\right\}\right) = -\frac{1}{6n}. \tag{5.17}$$

Combining the expressions yields

$$\mu(z_1, \ldots, z_{s-1}) = (1 + \gamma_s \beta)\, [e^{\mathrm{sh}}_{n,s-1}(z_1, \ldots, z_{s-1})]^2$$
$$+ \frac{\gamma_s}{6n} \prod_{j=1}^{s-1} \left(1 + \gamma_j \left(\beta + \frac{1}{6}\right)\right) - \frac{\gamma_s}{6n^2} \sum_{k=1}^{n-1} \prod_{j=1}^{s-1} \left(1 + \gamma_j \left[B_2\left(\left\{\frac{kz_j}{n}\right\}\right) + \beta\right]\right).$$

Rewriting the third expression in terms of $e^2_{n,s-1}(z_1, \ldots, z_{s-1})$ and then collecting similar expressions, we obtain

$$\mu(z_1, \ldots, z_{s-1}) = \left(1 + \gamma_s \beta - \frac{\gamma_s}{6n}\right) [e^{\mathrm{sh}}_{n,s-1}(z_1, \ldots, z_{s-1})]^2$$
$$- \frac{\gamma_s}{6n} \prod_{j=1}^{s-1} (1 + \gamma_j \beta) + \left(\frac{\gamma_s}{6} + \frac{\gamma_s}{6n}\right) \frac{1}{n} \prod_{j=1}^{s-1} \left(1 + \gamma_j \left(\beta + \frac{1}{6}\right)\right).$$

The condition (5.15) ensures that $1 + \gamma_s \beta - \gamma_s/(6n) \ge 0$. Applying the induction hypothesis and dropping a negative term, we finally arrive at $\mu(z_1, \ldots, z_{s-1}) < E^2_{n,s}(K_{s,\gamma})$.

Since $\mu(z_1, \ldots, z_{s-1})$ is the average of $[e^{\mathrm{sh}}_{n,s}(\boldsymbol{z})]^2$ over all $z_s \in \mathbb{U}_n$, the choice of z_s that minimizes $[e^{\mathrm{sh}}_{n,s}(\boldsymbol{z})]^2$ must satisfy

$$[e^{\mathrm{sh}}_{n,s}(\boldsymbol{z})]^2 \le \mu(z_1, \ldots, z_{s-1}) < E^2_{n,s}(K_{s,\gamma}).$$

To complete the proof by induction we note that the error bound (5.16) is easily satisfied for $s = 1$: indeed we may adapt (5.17) to obtain

$$[e^{\mathrm{sh}}_{n,1}(1)]^2 = \frac{\gamma_1}{n} \sum_{k=0}^{n-1} B_2\left(\frac{k}{n}\right) = \frac{\gamma_1}{n}\left(\frac{1}{6} - \frac{n-1}{6n}\right) = \frac{\gamma_1}{6n^2} \le \frac{\gamma_1}{6n}.$$

This completes the proof. □

We now show, through a more complicated averaging argument in the induction proof, that the CBC algorithm yields a convergence rate arbitrarily close to $O(n^{-1})$. The result holds for general n (not necessarily prime), and holds for the unanchored space with general weights (including product weights), but holds for the anchored space only with product weights. For the anchored space with general non-product weights, the same error bound holds, but a modification to the search criterion in the CBC construction is needed. We therefore state the results for the unanchored and anchored spaces separately.

In the following theorems, the choice of z_s at each step is independent of the parameter λ, and the error bound holds for all values of $\lambda \in (1/2, 1]$. The optimal convergence rate close to $O(n^{-1})$ is obtained with $\lambda \to 1/2$, but note that $\lambda = 1/2$ is not permitted because $\zeta(2\lambda) \to \infty$ as $\lambda \to 1/2$. The implied constant in the big-O bound can be independent of s under an appropriate condition on the weights. For example, in the case of product weights, a sufficient condition to obtain close to $O(n^{-1})$ convergence with the implied constant independently of s is $\sum_{j=1}^{\infty} \gamma_j^{1/2} < \infty$.

Theorem 5.8 (optimal CBC error bound: the unanchored case). The generating vector $\boldsymbol{z} \in \mathbb{U}_n^s$ constructed by the CBC algorithm, minimizing the squared shift-averaged worst-case error $[e_{n,s}^{\mathrm{sh}}(\boldsymbol{z})]^2$ for the weighted unanchored Sobolev space in each step, satisfies

$$[e_{n,s}^{\mathrm{sh}}(\boldsymbol{z})]^2 \leq \left(\frac{1}{\varphi(n)} \sum_{\emptyset \neq \mathfrak{u} \subseteq \{1:s\}} \gamma_{\mathfrak{u}}^{\lambda} \left(\frac{2\zeta(2\lambda)}{(2\pi^2)^{\lambda}} \right)^{|\mathfrak{u}|} \right)^{1/\lambda}, \tag{5.18}$$

for all $\lambda \in (1/2, 1]$, where $\zeta(\cdot)$ is the Riemann zeta function (5.5), and $\varphi(n)$ is the Euler totient function (5.14).

Proof. We prove this by induction on s. The base step $s = 1$ is straightforward to verify for all $\lambda \in (1/2, 1]$. Assume now that we have chosen the first $s - 1$ components $z_1, \ldots, z_{s-1}$ and that (5.18) holds with s replaced by $s - 1$. We separate the terms in (5.12) (remembering that $\beta = 0$ for this unanchored space), depending on whether or not the element s is included in the set $\mathfrak{u}$, to obtain the recursive expression

$$[e_{n,s}^{\mathrm{sh}}(z_1, \ldots, z_{s-1}, z_s)]^2 = [e_{n,s-1}^{\mathrm{sh}}(z_1, \ldots, z_{s-1})]^2 + \theta(z_s), \tag{5.19}$$

with (suppressing the dependence of θ on $z_1, \ldots, z_{s-1}$)

$$\begin{aligned}
\theta(z_s) &:= \sum_{s \in \mathfrak{u} \subseteq \{1:s\}} \gamma_{\mathfrak{u}} \left(\frac{1}{n} \sum_{k=0}^{n-1} \prod_{j \in \mathfrak{u}} B_2\left(\left\{ \frac{k z_j}{n} \right\} \right) \right) \\
&= \sum_{s \in \mathfrak{u} \subseteq \{1:s\}} \frac{\gamma_{\mathfrak{u}}}{(2\pi^2)^{|\mathfrak{u}|}} \left(\frac{1}{n} \sum_{k=0}^{n-1} \sum_{\boldsymbol{h}_{\mathfrak{u}} \in (\mathbb{Z} \setminus \{0\})^{|\mathfrak{u}|}} \frac{e^{2\pi i k \boldsymbol{h}_{\mathfrak{u}} \cdot \boldsymbol{z}_{\mathfrak{u}} / n}}{\prod_{j \in \mathfrak{u}} h_j^2} \right)
\end{aligned}$$

$$= \sum_{s\in\mathfrak{u}\subseteq\{1:s\}} \frac{\gamma_{\mathfrak{u}}}{(2\pi^2)^{|\mathfrak{u}|}} \Bigg(\sum_{\substack{\boldsymbol{h}_{\mathfrak{u}}\in(\mathbb{Z}\setminus\{0\})^{|\mathfrak{u}|} \\ \boldsymbol{h}_{\mathfrak{u}}\cdot\boldsymbol{z}_{\mathfrak{u}}\equiv 0 \,(\mathrm{mod}\, n)}} \frac{1}{\prod_{j\in\mathfrak{u}} h_j^2} \Bigg)$$

$$= \sum_{s\in\mathfrak{u}\subseteq\{1:s\}} \frac{\gamma_{\mathfrak{u}}}{(2\pi^2)^{|\mathfrak{u}|}} \Bigg(\sum_{h_s\in\mathbb{Z}\setminus\{0\}} \frac{1}{h_s^2} \sum_{\substack{\boldsymbol{h}_{\mathfrak{u}\setminus\{s\}}\in(\mathbb{Z}\setminus\{0\})^{|\mathfrak{u}|-1} \\ \boldsymbol{h}_{\mathfrak{u}\setminus\{s\}}\cdot\boldsymbol{z}_{\mathfrak{u}\setminus\{s\}}\equiv -h_s z_s \,(\mathrm{mod}\, n)}} \frac{1}{\prod_{j\in\mathfrak{u}\setminus\{s\}} h_j^2} \Bigg),$$

where we made use of the Fourier expansion of B_2 and the character property (5.2).

If z_s^* denotes the value chosen by the CBC algorithm in dimension s, then (since the minimum is always smaller than or equal to the average) we have for all $\lambda \in (0,1]$ that

$$[\theta(z_s^*)]^\lambda \le \frac{1}{\varphi(n)} \sum_{z_s\in\mathbb{U}_n} [\theta(z_s)]^\lambda \le \frac{1}{\varphi(n)} \sum_{z_s\in\mathbb{U}_n} \sum_{s\in\mathfrak{u}\subseteq\{1:s\}} \frac{\gamma_{\mathfrak{u}}^\lambda}{(2\pi^2)^{|\mathfrak{u}|\lambda}}$$
$$\times \Bigg(\sum_{h_s\in\mathbb{Z}\setminus\{0\}} \frac{1}{|h_s|^{2\lambda}} \sum_{\substack{\boldsymbol{h}_{\mathfrak{u}\setminus\{s\}}\in(\mathbb{Z}\setminus\{0\})^{|\mathfrak{u}|-1} \\ \boldsymbol{h}_{\mathfrak{u}\setminus\{s\}}\cdot\boldsymbol{z}_{\mathfrak{u}\setminus\{s\}}\equiv -h_s z_s \,(\mathrm{mod}\, n)}} \frac{1}{\prod_{j\in\mathfrak{u}\setminus\{s\}} |h_j|^{2\lambda}} \Bigg),$$

where we used the inequality (sometimes referred to as Jensen's inequality)

$$\Big(\sum_k a_k\Big)^\lambda \le \sum_k a_k^\lambda, \quad a_k \ge 0,\ \lambda \in (0,1]. \tag{5.20}$$

Next we separate the terms depending on whether or not h_s is a multiple of n, to obtain

$$[\theta(z_s^*)]^\lambda \le \sum_{s\in\mathfrak{u}\subseteq\{1:s\}} \frac{\gamma_{\mathfrak{u}}^\lambda}{(2\pi^2)^{|\mathfrak{u}|\lambda}} \cdot \frac{2\zeta(2\lambda)}{n^{2\lambda}} \Bigg(\sum_{\substack{\boldsymbol{h}_{\mathfrak{u}\setminus\{s\}}\in(\mathbb{Z}\setminus\{0\})^{|\mathfrak{u}|-1} \\ \boldsymbol{h}_{\mathfrak{u}\setminus\{s\}}\cdot\boldsymbol{z}_{\mathfrak{u}\setminus\{s\}}\equiv 0 \,(\mathrm{mod}\, n)}} \frac{1}{\prod_{j\in\mathfrak{u}\setminus\{s\}} |h_j|^{2\lambda}} \Bigg)$$
$$+ \frac{1}{\varphi(n)} \sum_{z_s\in\mathbb{U}_n} \sum_{c=1}^{n-1} \sum_{s\in\mathfrak{u}\subseteq\{1:s\}} \frac{\gamma_{\mathfrak{u}}^\lambda}{(2\pi^2)^{|\mathfrak{u}|\lambda}}$$
$$\times \Bigg(\sum_{\substack{h_s\in\mathbb{Z}\setminus\{0\} \\ h_s\equiv -c z_s^{-1} \,(\mathrm{mod}\, n)}} \frac{1}{|h_s|^{2\lambda}} \sum_{\substack{\boldsymbol{h}_{\mathfrak{u}\setminus\{s\}}\in(\mathbb{Z}\setminus\{0\})^{|\mathfrak{u}|-1} \\ \boldsymbol{h}_{\mathfrak{u}\setminus\{s\}}\cdot\boldsymbol{z}_{\mathfrak{u}\setminus\{s\}}\equiv c \,(\mathrm{mod}\, n)}} \frac{1}{\prod_{j\in\mathfrak{u}\setminus\{s\}} |h_j|^{2\lambda}} \Bigg),$$

where z_s^{-1} denotes the multiplicative inverse of z_s in $\mathbb{U}_n$, that is, $z_s z_s^{-1} \equiv 1 \pmod n$. For fixed c satisfying $1 \le c \le n-1$, we have $\{c z_s^{-1} \bmod n : z_s \in \mathbb{U}_n\} = \{cz \bmod n : z \in \mathbb{U}_n\}$. Let $g = \gcd(c,n)$. Then $\gcd(c/g, n/g) = 1$,

and

$$\sum_{z_s \in \mathbb{U}_n} \sum_{\substack{h_s \in \mathbb{Z}\setminus\{0\} \\ h_s \equiv -c z_s^{-1} \pmod n}} \frac{1}{|h_s|^{2\lambda}} = \sum_{z \in \mathbb{U}_n} \sum_{\substack{h_s \in \mathbb{Z}\setminus\{0\} \\ h_s \equiv -cz \pmod n}} \frac{1}{|h_s|^{2\lambda}}$$

$$= \sum_{z\in\mathbb{U}_n} \sum_{m \in \mathbb{Z}} \frac{1}{|mn - cz|^{2\lambda}} = g^{-2\lambda} \sum_{z\in\mathbb{U}_n} \sum_{m\in\mathbb{Z}} \frac{1}{|m(n/g) - (c/g)z|^{2\lambda}}$$

$$= g^{-2\lambda} \sum_{z \in \mathbb{U}_n} \sum_{\substack{h \in \mathbb{Z}\setminus\{0\} \\ h \equiv -(c/g)z \pmod{n/g}}} \frac{1}{|h|^{2\lambda}} \le g^{-2\lambda} g \sum_{a=1}^{n/g-1} \sum_{\substack{h \in \mathbb{Z}\setminus\{0\} \\ h \equiv a \pmod{n/g}}} \frac{1}{|h|^{2\lambda}}$$

$$= g^{1-2\lambda} \cdot 2\zeta(2\lambda)\big(1 - (n/g)^{-2\lambda}\big) \le 2\zeta(2\lambda),$$

where the last step holds because $g \ge 1$ and $\lambda > 1/2$. (The condition $\lambda > 1/2$ is needed to ensure that $\zeta(2\lambda) < \infty$.) Hence

$$[\theta(z_s^*)]^\lambda \le \sum_{s\in\mathfrak{u}\subseteq\{1:s\}} \frac{\gamma_\mathfrak{u}^\lambda}{(2\pi^2)^{|\mathfrak{u}|\lambda}} \cdot \frac{2\zeta(2\lambda)}{n^{2\lambda}} \left(\sum_{\substack{\boldsymbol{h}_{\mathfrak{u}\setminus\{s\}} \in (\mathbb{Z}\setminus\{0\})^{|\mathfrak{u}|-1} \\ \boldsymbol{h}_{\mathfrak{u}\setminus\{s\}} \cdot \boldsymbol{z}_{\mathfrak{u}\setminus\{s\}} \equiv 0 \pmod n}} \frac{1}{\prod_{j\in\mathfrak{u}\setminus\{s\}} |h_j|^{2\lambda}} \right)$$

$$+ \frac{1}{\varphi(n)} \sum_{s\in\mathfrak{u}\subseteq\{1:s\}} \frac{\gamma_\mathfrak{u}^\lambda}{(2\pi^2)^{|\mathfrak{u}|\lambda}} \cdot 2\zeta(2\lambda) \left(\sum_{\substack{\boldsymbol{h}_{\mathfrak{u}\setminus\{s\}} \in (\mathbb{Z}\setminus\{0\})^{|\mathfrak{u}|-1} \\ \boldsymbol{h}_{\mathfrak{u}\setminus\{s\}} \cdot \boldsymbol{z}_{\mathfrak{u}\setminus\{s\}} \not\equiv 0 \pmod n}} \frac{1}{\prod_{j\in\mathfrak{u}\setminus\{s\}} |h_j|^{2\lambda}} \right)$$

$$\le \frac{1}{\varphi(n)} \sum_{s\in\mathfrak{u}\subseteq\{1:s\}} \gamma_\mathfrak{u}^\lambda \left(\frac{2\zeta(2\lambda)}{(2\pi^2)^\lambda} \right)^{|\mathfrak{u}|}.$$

Now we use (5.19), the induction hypothesis and another use of (5.20) to obtain (5.18). □

Theorem 5.9 (optimal CBC error bound: the anchored case). A generating vector $\boldsymbol{z} \in \mathbb{U}_n^s$ can be constructed by a (modified) CBC algorithm such that the squared shift-averaged worst-case error $e_{n,s}^{\text{sh}}(\boldsymbol{z})$ in the weighted anchored Sobolev space with anchor c satisfies

$$[e_{n,s}^{\text{sh}}(\boldsymbol{z})]^2 \le \left(\frac{1}{\varphi(n)} \sum_{\emptyset\ne\mathfrak{u}\subseteq\{1:s\}} \gamma_\mathfrak{u}^\lambda \left(\frac{2\zeta(2\lambda)}{(2\pi^2)^\lambda} + \beta^\lambda \right)^{|\mathfrak{u}|} \right)^{1/\lambda}$$

for all $\lambda \in (1/2, 1]$, where $\beta = c^2 - c + 1/3$, $\zeta(\cdot)$ is the Riemann zeta function (5.5), and $\varphi(n)$ is the Euler totient function (5.14). For product weights, the algorithm minimizes $[e_{n,s}^{\text{sh}}(\boldsymbol{z})]^2$ step by step as usual. For general non-product weights, the algorithm minimizes an auxiliary quantity $\widetilde{e}_{n,s,d}^{\,2}(\boldsymbol{z})$

depending on s,

$$\widetilde{e}^2_{n,s,d}(\boldsymbol{z}) := \sum_{\emptyset \neq \mathfrak{v} \subseteq \{1:d\}} \widetilde{\gamma}_{s,\mathfrak{v}} \left(\frac{1}{n} \sum_{k=0}^{n-1} \prod_{j \in \mathfrak{v}} B_2\left(\left\{ \frac{kz_j}{n} \right\} \right) \right), \tag{5.21}$$

step by step for each $d = 2, 3, \ldots, s$, with auxiliary weights defined by

$$\widetilde{\gamma}_{s,\mathfrak{v}} := \sum_{\mathfrak{v} \subseteq \mathfrak{u} \subseteq \{1:s\}} \gamma_{\mathfrak{u}}\, \beta^{|\mathfrak{u}|-|\mathfrak{v}|}, \quad \mathfrak{v} \subseteq \{1:s\}. \tag{5.22}$$

Note that the latter algorithm is not extensible in s.

Proof. In the previous proof we made use of the Fourier expansion of B_2, which has no constant term. To allow us to use essentially the same argument here, we start from (5.12), adapting the identity (4.2) and swapping the order of sums, to obtain

$$\begin{aligned}
[e^{\mathrm{sh}}_{n,s}(\boldsymbol{z})]^2 &= \frac{1}{n} \sum_{k=0}^{n-1} \sum_{\mathfrak{u} \subseteq \{1:s\}} \gamma_{\mathfrak{u}} \sum_{\mathfrak{v} \subseteq \mathfrak{u}} \beta^{|\mathfrak{u}|-|\mathfrak{v}|} \prod_{j \in \mathfrak{v}} B_2\left(\left\{ \frac{kz_j}{n} \right\} \right) - \sum_{\mathfrak{u} \subseteq \{1:s\}} \gamma_{\mathfrak{u}} \beta^{|\mathfrak{u}|} \\
&= \frac{1}{n} \sum_{k=0}^{n-1} \sum_{\mathfrak{v} \subseteq \{1:s\}} \sum_{\mathfrak{v} \subseteq \mathfrak{u} \subseteq \{1:s\}} \gamma_{\mathfrak{u}} \beta^{|\mathfrak{u}|-|\mathfrak{v}|} \prod_{j \in \mathfrak{v}} B_2\left(\left\{ \frac{kz_j}{n} \right\} \right) - \sum_{\mathfrak{u} \subseteq \{1:s\}} \gamma_{\mathfrak{u}} \beta^{|\mathfrak{u}|} \\
&= \frac{1}{n} \sum_{k=0}^{n-1} \sum_{\mathfrak{v} \subseteq \{1:s\}} \widetilde{\gamma}_{s,\mathfrak{v}} \prod_{j \in \mathfrak{v}} B_2\left(\left\{ \frac{kz_j}{n} \right\} \right) - \widetilde{\gamma}_{s,\emptyset} \\
&= \sum_{\emptyset \neq \mathfrak{v} \subseteq \{1:s\}} \widetilde{\gamma}_{s,\mathfrak{v}} \left(\frac{1}{n} \sum_{k=0}^{n-1} \prod_{j \in \mathfrak{v}} B_2\left(\left\{ \frac{kz_j}{n} \right\} \right) \right),
\end{aligned} \tag{5.23}$$

where we introduced the auxiliary weights defined in (5.22).

This last expression takes the same form as the squared shift-averaged worst-case error in the unanchored space. However, there is a vital difference that the auxiliary weights $\widetilde{\gamma}_{s,\mathfrak{v}}$ depend on the dimension s. Consequently, a recursive formula of the form (5.19) does not hold.

Case 1. Consider first the case of product weights $\gamma_{\mathfrak{u}} = \prod_{j \in \mathfrak{u}} \gamma_j$. It follows from the definition (5.22) that, for $s \notin \mathfrak{v}$, we have

$$\widetilde{\gamma}_{s,\mathfrak{v}} = \sum_{\mathfrak{v} \subseteq \mathfrak{u} \subseteq \{1:s-1\}} \gamma_{\mathfrak{u}}\, \beta^{|\mathfrak{u}|-|\mathfrak{v}|} + \sum_{\mathfrak{v} \subseteq \mathfrak{u} \subseteq \{1:s-1\}} \gamma_{\mathfrak{u}} \gamma_s\, \beta^{|\mathfrak{u}|-|\mathfrak{v}|+1} = (1 + \gamma_s \beta) \widetilde{\gamma}_{s-1,\mathfrak{v}}.$$

Thus, by separating the terms depending on whether or not the element s is included in the set $\mathfrak{v}$, we obtain

$$[e^{\mathrm{sh}}_{n,s}(\boldsymbol{z})]^2 = (1 + \gamma_s \beta)[e^{\mathrm{sh}}_{n,s-1}(z_1, \ldots, z_{s-1})]^2 + \theta(z_s),$$

where

$$\theta(z_s) := \sum_{s\in\mathfrak{v}\subseteq\{1:s\}} \widetilde{\gamma}_{s,\mathfrak{v}} \left(\frac{1}{n} \sum_{k=0}^{n-1} \prod_{j\in\mathfrak{v}} B_2\left(\left\{\frac{kz_j}{n}\right\}\right)\right).$$

The induction hypothesis is now

$$[e^{\mathrm{sh}}_{n,s-1}(z_1,\ldots,z_{s-1})]^2 \le \left(\frac{1}{\varphi(n)} \sum_{\emptyset\ne\mathfrak{v}\subseteq\{1:s-1\}} \widetilde{\gamma}^{\lambda}_{s-1,\mathfrak{v}} \left(\frac{2\zeta(2\lambda)}{(2\pi^2)^\lambda}\right)^{|\mathfrak{v}|}\right)^{1/\lambda}$$

for all $\lambda \in (1/2, 1]$. For the quantity $\theta(z_s)$ we follow the averaging argument in the previous proof, to obtain

$$[\theta(z_s^*)]^\lambda \le \frac{1}{\varphi(n)} \sum_{s\in\mathfrak{v}\subseteq\{1:s\}} \widetilde{\gamma}^{\lambda}_{s,\mathfrak{v}} \left(\frac{2\zeta(2\lambda)}{(2\pi^2)^\lambda}\right)^{|\mathfrak{v}|}.$$

Combining the estimates, we obtain

$$\begin{aligned}[e^{\mathrm{sh}}_{n,s}(\boldsymbol{z})]^2 \le (1+\gamma_s\beta)&\left(\frac{1}{\varphi(n)} \sum_{\emptyset\ne\mathfrak{v}\subseteq\{1:s-1\}} \widetilde{\gamma}^{\lambda}_{s-1,\mathfrak{v}} \left(\frac{2\zeta(2\lambda)}{(2\pi^2)^\lambda}\right)^{|\mathfrak{v}|}\right)^{1/\lambda} \\ &+ \left(\frac{1}{\varphi(n)} \sum_{s\in\mathfrak{v}\subseteq\{1:s\}} \widetilde{\gamma}^{\lambda}_{s,\mathfrak{v}} \left(\frac{2\zeta(2\lambda)}{(2\pi^2)^\lambda}\right)^{|\mathfrak{v}|}\right)^{1/\lambda}.\end{aligned}$$

Applying (5.20) then yields

$$[e^{\mathrm{sh}}_{n,s}(\boldsymbol{z})]^2 \le \left(\frac{1}{\varphi(n)} \sum_{\emptyset\ne\mathfrak{v}\subseteq\{1:s\}} \widetilde{\gamma}^{\lambda}_{s,\mathfrak{v}} \left(\frac{2\zeta(2\lambda)}{(2\pi^2)^\lambda}\right)^{|\mathfrak{v}|}\right)^{1/\lambda}. \tag{5.24}$$

Finally we express the result in terms of the original weights. Using (5.22) and (5.20), we have

$$\begin{aligned}&\sum_{\emptyset\ne\mathfrak{v}\subseteq\{1:s\}} \widetilde{\gamma}^{\lambda}_{s,\mathfrak{v}} \left(\frac{2\zeta(2\lambda)}{(2\pi^2)^\lambda}\right)^{|\mathfrak{v}|} \\ &\le \sum_{\mathfrak{v}\subseteq\{1:s\}} \sum_{\mathfrak{v}\subseteq\mathfrak{u}\subseteq\{1:s\}} \gamma_{\mathfrak{u}}^{\lambda}\, \beta^{(|\mathfrak{u}|-|\mathfrak{v}|)\lambda} \left(\frac{2\zeta(2\lambda)}{(2\pi^2)^\lambda}\right)^{|\mathfrak{v}|} - \sum_{\mathfrak{u}\subseteq\{1:s\}} \gamma_{\mathfrak{u}}^{\lambda}\, \beta^{|\mathfrak{u}|\lambda} \\ &= \sum_{\mathfrak{u}\subseteq\{1:s\}} \gamma_{\mathfrak{u}}^{\lambda} \sum_{\mathfrak{v}\subseteq\mathfrak{u}} \beta^{(|\mathfrak{u}|-|\mathfrak{v}|)\lambda} \left(\frac{2\zeta(2\lambda)}{(2\pi^2)^\lambda}\right)^{|\mathfrak{v}|} - \sum_{\mathfrak{u}\subseteq\{1:s\}} \gamma_{\mathfrak{u}}^{\lambda}\, \beta^{|\mathfrak{u}|\lambda}\end{aligned}$$

$$= \sum_{\mathfrak{u}\subseteq\{1:s\}} \gamma_{\mathfrak{u}}^{\lambda}\left(\frac{2\zeta(2\lambda)}{(2\pi^2)^{\lambda}} + \beta^{\lambda}\right)^{|\mathfrak{u}|} - \sum_{\mathfrak{u}\subseteq\{1:s\}} \gamma_{\mathfrak{u}}^{\lambda}\,\beta^{|\mathfrak{u}|\lambda}$$

$$\leq \sum_{\emptyset\neq\mathfrak{u}\subseteq\{1:s\}} \gamma_{\mathfrak{u}}^{\lambda}\left(\frac{2\zeta(2\lambda)}{(2\pi^2)^{\lambda}} + \beta^{\lambda}\right)^{|\mathfrak{u}|}. \qquad (5.25)$$

That the bound holds for $s = 1$ follows as before, completing the proof for the case of product weights.

Remark. For the case of product weights, the use of auxiliary weights is only needed in the proof and has no bearing on the actual CBC construction. It is also possible to prove the result by working directly with the simplified worst-case error expression for product weights (see (5.13)) without introducing auxiliary weights. We have chosen to prove the result this way to provide valuable insights into the more complicated case of general non-product weights.

Case 2. Now we consider the case of general non-product weights. Due to the lack of structure in general weights, we are unable to relate $[e_{n,s}^{\mathrm{sh}}(\boldsymbol{z})]^2$ to $[e_{n,s-1}^{\mathrm{sh}}(z_1,\ldots,z_{s-1})]^2$ in a meaningful way. We therefore switch to a modified CBC construction based on the auxiliary quantity (5.21). Note that the auxiliary weights depend only on the final dimension s and not on the induction index d in (5.21). This allows us to write

$$\widetilde{e}_{n,s,d}^{\,2}(z_1,\ldots,z_d) = \widetilde{e}_{n,s,d-1}^{\,2}(z_1,\ldots,z_{d-1}) + \widetilde{\theta}(z_d),$$

where

$$\widetilde{\theta}(z_d) := \sum_{d\in\mathfrak{v}\subseteq\{1:d\}} \widetilde{\gamma}_{s,\mathfrak{v}}\left(\frac{1}{n}\sum_{k=0}^{n-1}\prod_{j\in\mathfrak{v}} B_2\left(\left\{\frac{kz_j}{n}\right\}\right)\right).$$

Note that

$$\widetilde{e}_{n,s,s}^{\,2}(\boldsymbol{z}) = e_{n,s}^2(\boldsymbol{z}), \quad \text{but} \quad \widetilde{e}_{n,s,d}^{\,2}(z_1,\ldots,z_d) \neq e_{n,d}^2(z_1,\ldots,z_d) \quad \text{for } d < s.$$

Hence the CBC construction based on the auxiliary quantity is *not* the same as the original CBC construction. We prove by induction that the CBC construction based on the auxiliary quantity yields, for each $d = 1, 2, \ldots, s$,

$$\widetilde{e}_{n,s,d}^{\,2}(z_1,\ldots,z_d) \leq \left(\frac{1}{\varphi(n)}\sum_{\emptyset\neq\mathfrak{v}\subseteq\{1:d\}} \widetilde{\gamma}_{s,\mathfrak{v}}^{\lambda}\left(\frac{2\zeta(2\lambda)}{(2\pi^2)^{\lambda}}\right)^{|\mathfrak{v}|}\right)^{1/\lambda}$$

for all $\lambda \in (1/2, 1]$. Since the auxiliary weights do not change with the induction index d, the proof is essentially the same as the previous proof for the unanchored space. At the final step $d = s$ we recover the error bound (5.24), which can be expressed in terms of the original weights following (5.25). This completes the proof. □

The following theorem illustrates how we can apply the theory of this section to a given integration problem.

Theorem 5.10 (CBC integration error). Suppose that a given integrand f belongs to either our weighted anchored or unanchored Sobolev space. A lattice rule generating vector $\boldsymbol{z} \in \mathbb{U}_n^s$ can be constructed by a CBC algorithm such that, for all $\lambda \in (1/2, 1]$,

$$\sqrt{\mathbb{E}\,|I_s(f) - Q_{n,s}(\cdot; f)|^2} \leq \left(\frac{1}{\varphi(n)} \sum_{\emptyset \neq \mathfrak{u} \subseteq \{1:s\}} \gamma_{\mathfrak{u}}^{\lambda} \left(\frac{2\zeta(2\lambda)}{(2\pi^2)^{\lambda}} + \beta^{\lambda}\right)^{|\mathfrak{u}|}\right)^{1/(2\lambda)} \|f\|_{s,\boldsymbol{\gamma}},$$

where the expectation is taken with respect to the random shift which is uniformly distributed over $[0,1]^s$, $\beta = c^2 - c + 1/3$ for the anchored variant with anchor c and $\beta = 0$ for the unanchored variant, $\zeta(\cdot)$ is the Riemann zeta function (5.5), and $\varphi(n)$ is the Euler totient function (5.14).

We conclude this subsection with the remark that a different analysis can be used for the anchored Sobolev space if, roughly speaking, the auxiliary weights (5.22) are dominated by the original weights, that is, if

$$\sum_{\substack{|\mathfrak{u}|<\infty \\ \mathfrak{v} \subseteq \mathfrak{u} \subset \mathbb{N}}} \gamma_{\mathfrak{u}} \beta^{|\mathfrak{u}|-|\mathfrak{v}|} \leq C \gamma_{\mathfrak{v}}$$

for all finite subsets $\mathfrak{v} \subset \mathbb{N}$, where $C > 0$ is independent of $\mathfrak{v}$. A similar result holds if the weights are of the following special POD form:

$$\gamma_{\mathfrak{u}} = ((|\mathfrak{u}| + \ell)!)^a \prod_{j \in \mathfrak{u}} \gamma_j,$$

where ℓ is a non-negative integer and a is a positive real number. We explain the alternative analysis for this case below.

Recall that for the anchored space the CBC algorithm must work with auxiliary weights (5.22) which do not preserve the POD structure of the original weights. We can write

$$\begin{aligned}
\widetilde{\gamma}_{s,\mathfrak{v}} &= \sum_{\mathfrak{v} \subseteq \mathfrak{u} \subseteq \{1:s\}} ((|\mathfrak{u}| + \ell)!)^a \prod_{j \in \mathfrak{u}} \gamma_j \\
&= ((|\mathfrak{v}| + \ell)!)^a \left(\prod_{j \in \mathfrak{v}} \gamma_j\right) \sum_{\mathfrak{v} \subseteq \mathfrak{u} \subseteq \{1:s\}} \left(\frac{(|\mathfrak{u}| + \ell)!}{(|\mathfrak{v}| + \ell)!}\right)^a \prod_{j \in \mathfrak{u} \setminus \mathfrak{v}} \gamma_j \\
&= ((|\mathfrak{v}| + \ell)!)^a \left(\prod_{j \in \mathfrak{v}} \gamma_j\right) \sum_{\mathfrak{w} \subseteq \{1:s\} \setminus \mathfrak{v}} \left(\frac{(|\mathfrak{v}| + |\mathfrak{w}| + \ell)!}{(|\mathfrak{v}| + \ell)!}\right)^a \prod_{j \in \mathfrak{w}} \gamma_j.
\end{aligned}$$

Using the simple inequality

$$\frac{(p+q)!}{p!\,q!} = \binom{p+q}{p} \le \sum_{k=0}^{p+q} \binom{p+q}{k} = 2^{p+q},$$

we obtain

$$\begin{aligned}\widetilde{\gamma}_{s,\mathfrak{v}} &\le ((|\mathfrak{v}|+\ell)!)^a \biggl(\prod_{j\in\mathfrak{v}} \gamma_j\biggr) \sum_{\mathfrak{w}\subseteq\{1:s\}\setminus\mathfrak{v}} (|\mathfrak{w}|!\,2^{|\mathfrak{v}|+|\mathfrak{w}|+\ell})^a \prod_{j\in\mathfrak{w}} \gamma_j \\ &\le ((|\mathfrak{v}|+\ell)!)^a \biggl(\prod_{j\in\mathfrak{v}} (2^a\gamma_j)\biggr) 2^{a\ell} \sum_{\mathfrak{w}\subseteq\{1:s\}} (|\mathfrak{w}|!)^a \prod_{j\in\mathfrak{w}} (2^a\gamma_j) = c_{s,\gamma}\,\widetilde{\widetilde{\gamma}}_{\mathfrak{v}},\end{aligned}$$

where

$$\widetilde{\widetilde{\gamma}}_{\mathfrak{v}} := ((|\mathfrak{v}|+\ell)!)^a \prod_{j\in\mathfrak{v}} (2^a\gamma_j), \quad \text{and} \quad c_{s,\gamma} := 2^{a\ell} \sum_{\mathfrak{w}\subseteq\{1:s\}} (|\mathfrak{w}|!)^a \prod_{j\in\mathfrak{w}} (2^a\gamma_j).$$

Thus from (5.23) we can write

$$[e^{\mathrm{sh}}_{n,s}(\boldsymbol{z})]^2 \le c_{s,\gamma} \sum_{\emptyset\neq\mathfrak{v}\subseteq\{1:s\}} \widetilde{\widetilde{\gamma}}_{\mathfrak{v}} \biggl(\frac{1}{n}\sum_{k=0}^{n-1}\prod_{j\in\mathfrak{v}} B_2\biggl(\biggl\{\frac{kz_j}{n}\biggr\}\biggr)\biggr). \tag{5.26}$$

Hence, we can use the expression on the right-hand side of (5.26) (without the factor $c_{s,\gamma}$) as the search criterion for the CBC construction. The benefit is that the new weights $\widetilde{\widetilde{\gamma}}_{\mathfrak{v}}$ are also of the POD form, and they do *not* depend on the dimension s. This means that a fast implementation is possible (see Section 5.6) and that the algorithm is extensible in s. The resulting error bound would be slightly worse (each factor in the product part of the weights would be scaled by 2^a, and there is an additional factor of $c_{s,\gamma}^{1/2}$ in the overall error bound in Theorem 5.10), but the convergence rate and the dependence of the implied constant on the dimension s remain the same as before.

5.5. *Fast CBC construction for product weights*

In this subsection we explain how to implement the CBC construction efficiently for product weights. We begin by presenting a naive implementation of Algorithm 5.6 as a pseudocode (Pseudocode 1, overleaf). Then we discuss the techniques to speed up the computation.

The reader should keep in mind that the general approach described in this subsection applies to all shift-invariant kernels: we can replace $B_2(x)$ by any generic function whose integral over $[0,1]$ is 0.

Recall that

$$\begin{aligned}\mathbb{Z}_n &:= \{z\in\mathbb{Z} : 0\le z\le n-1\},\\ \mathbb{U}_n &:= \{z\in\mathbb{Z}_n : \gcd(z,n)=1\},\end{aligned}$$

Pseudocode 1 (CBC: naive implementation)

for s **from** 1 **to** $s_{\max}$ **do**
  for all $z_s \in \mathbb{U}_n$ **do**

$$e_s^2(z_s) = -\prod_{j=1}^{s}(1+\gamma_j\beta) + \frac{1}{n}\sum_{k=0}^{n-1}\prod_{j=1}^{s}\left(1+\gamma_j\left[B_2\left(\left\{\frac{kz_j}{n}\right\}\right)+\beta\right]\right)$$

  end for
  $z_s = \operatorname{argmin}_{z\in\mathbb{U}_n} e_s^2(z)$
end for

with $\varphi(n) := |\mathbb{U}_n|$. In the pseudocode we used the simplified notation

$$e_s^2(z_s) := [e_{n,s}^{\mathrm{sh}}(z_1,\ldots,z_s)]^2$$

to highlight the fact that z_s is the only component to be determined at step s; all previous components $z_1,\ldots,z_{s-1}$ have already been fixed. To simplify the exposition, we included z_1 in the search as well, even though all choices in the first dimension are equivalent.

The computational cost of this naive implementation is $O(n^2\, s_{\max}^2)$ operations. There are a number of ways to reduce this cost. Firstly, due to the symmetry $B_2(x) = B_2(1-x)$, we have $e_s^2(z_s) = e_s^2(n-z_s)$, thus we only need to search through half of the elements in $\mathbb{U}_n$. Secondly, at step s we can write

$$e_s^2(z_s) = (1+\gamma_s\beta)\, e_{s-1}^2 + \frac{\gamma_s}{n}\sum_{k=0}^{n-1}\Bigg[\underbrace{B_2\left(\left\{\frac{kz_s}{n}\right\}\right)}_{\Omega_n(z_s,k)}\underbrace{\prod_{j=1}^{s-1}\left(1+\gamma_j\left[B_2\left(\left\{\frac{kz_j}{n}\right\}\right)+\beta\right]\right)}_{p_{s-1}(k)}\Bigg]. \tag{5.27}$$

The n values $p_{s-1}(k)$ do not depend on z_s, and can be stored during the search. This reduces the construction cost to $O(n^2\, s_{\max})$ operations, at the expense of $O(n)$ storage.

The values of (5.27) for all $z_s \in \mathbb{U}_n$ can be expressed as a $\varphi(n)\times 1$ vector

$$\boldsymbol{e}_s^2 = [e_s^2(z_s)]_{z_s\in\mathbb{U}_n},$$

which can be computed in terms of a matrix–vector product, with the $\varphi(n)\times n$ matrix

$$\boldsymbol{\Omega}_n = [\Omega_n(z,k)]_{\substack{z\in\mathbb{U}_n\\ k\in\mathbb{Z}_n}} = \left[B_2\left(\frac{kz \bmod n}{n}\right)\right]_{\substack{z\in\mathbb{U}_n\\ k\in\mathbb{Z}_n}}, \tag{5.28}$$

Pseudocode 2 (CBC: matrix–vector form)

$\boldsymbol{p}_0 = \mathbf{1}$
$e_0^2 = 0$
for s **from** 1 **to** $s_{\max}$ **do**
  $\boldsymbol{e}_s^2 = (\mathbf{1} + \gamma_s \boldsymbol{\beta}) e_{s-1}^2 + \dfrac{\gamma_s}{n} \boldsymbol{\Omega}_n \, \boldsymbol{p}_{s-1}$ ▷ compute
  $z_s = \operatorname{argmin}_{z \in \mathbb{U}_n} e_s^2(z)$ ▷ select
  $e_s^2 = e_s^2(z_s)$ ▷ set
  $\boldsymbol{p}_s = \big(\mathbf{1} + \gamma_s(\boldsymbol{\Omega}_n(z_s, :) + \boldsymbol{\beta})\big) .* \, \boldsymbol{p}_{s-1}$ ▷ update
end for

and the $n \times 1$ vector

$$\boldsymbol{p}_{s-1} = [p_{s-1}(k)]_{k \in \mathbb{Z}_n}.$$

This yields our second pseudocode. Here $\boldsymbol{\Omega}_n(z_s, :)$ means taking a particular row of the matrix, while $.*$ means element-wise vector multiplication. Also, $\mathbf{1} = (1, \ldots, 1)$ and $\boldsymbol{\beta} = (\beta, \ldots, \beta)$ are vectors with the appropriate length.

At the update stage, the vector $\boldsymbol{p}_{s-1}$ can be overwritten by the vector $\boldsymbol{p}_s$. Thus the memory requirement is of order $O(n)$.

The trick now is to order the indices $z \in \mathbb{U}_n$ and $k \in \mathbb{Z}_n$ in (5.28) in a clever way in order to allow fast matrix–vector multiplication. Below we explain how this is achieved for the simpler case when n is prime.

When n is prime and using the *Rader factorization*, the matrix $\boldsymbol{\Omega}_n$ can be reordered such that it has a *circulant submatrix* $\boldsymbol{C}_n$ of size $(n-1) \times (n-1)$. This is achieved by taking the indices in the order

$$z = g^i \quad \text{and} \quad k = (g^{-1})^{i'} \quad \text{for } 0 \le i, i' \le n-2,$$

where g is a primitive root of n (*i.e.*, g is a generator for the cyclic group $\mathbb{U}_n$) and g^{-1} denotes its multiplicative inverse. We denote this particular ordering of $\boldsymbol{\Omega}_n$ by $\boldsymbol{\Omega}_n^{\langle g \rangle}$. Thus, for n prime, the element $\Omega_n^{\langle g \rangle}(i, i')$ in the $(i+1)$th row and $(i'+1)$th column of $\boldsymbol{\Omega}_n^{\langle g \rangle}$ is given by

$$\Omega_n^{\langle g \rangle}(i, i') = \begin{cases} \Omega_n(g^i, (g^{-1})^{i'}) & \text{if } 0 \le i, i' \le n-2, \\ 0 & \text{if } i' = n-1, \end{cases}$$

where $g^i, (g^{-1})^{i'} \in \mathbb{U}_n$ for $0 \le i, i' \le n-2$. A graphical illustration of the effect of reordering the indices is given in Figure 5.1. We present a concrete illustration of the reordering in the next example.

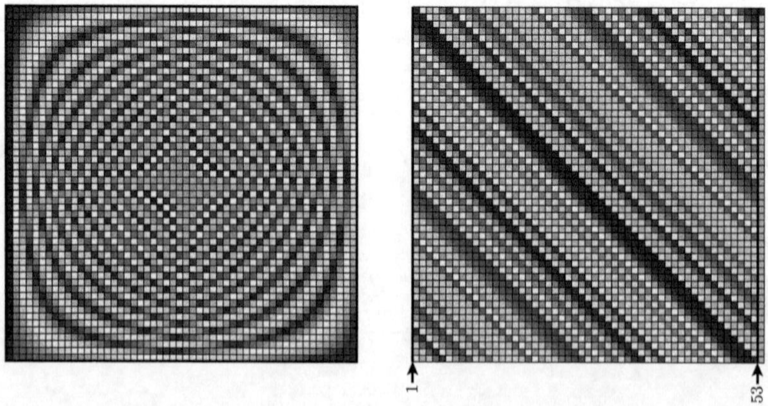

(a) (b)

Figure 5.1. (Image by Dirk Nuyens.) The structure of the matrix Ω_{53} for the fast CBC construction of lattice rules. (a) Original ordering of indices, and (b) after a reordering of indices.

Example 5.11. Take $n = 11$. Then $\mathbb{U}_{11} = \{1, 2, 3, 4, 5, 6, 7, 8, 9, 10\}$. The natural ordering of the indices $z \in \mathbb{U}_{11}$ and $k \in \mathbb{Z}_{11}$ gives the 10×11 matrix

$$\boldsymbol{\Omega}_{11} = B_2\left(\frac{1}{11}\begin{bmatrix} 0 & 1 & 2 & 3 & 4 & 5 & 6 & 7 & 8 & 9 & 10 \\ 0 & 2 & 4 & 6 & 8 & 10 & 1 & 3 & 5 & 7 & 9 \\ 0 & 3 & 6 & 9 & 1 & 4 & 7 & 10 & 2 & 5 & 8 \\ 0 & 4 & 8 & 1 & 5 & 9 & 2 & 6 & 10 & 3 & 7 \\ 0 & 5 & 10 & 4 & 9 & 3 & 8 & 2 & 7 & 1 & 6 \\ 0 & 6 & 1 & 7 & 2 & 8 & 3 & 9 & 4 & 10 & 5 \\ 0 & 7 & 3 & 10 & 6 & 2 & 9 & 5 & 1 & 8 & 4 \\ 0 & 8 & 5 & 2 & 10 & 7 & 4 & 1 & 9 & 6 & 3 \\ 0 & 9 & 7 & 5 & 3 & 1 & 10 & 8 & 6 & 4 & 2 \\ 0 & 10 & 9 & 8 & 7 & 6 & 5 & 4 & 3 & 2 & 1 \end{bmatrix}\right).$$

The generator for the group $\mathbb{U}_{11}$ can be taken to be $g = 2$. The ordering for the z indices is then

$$[g^i : 0 \le i \le 9] = [1, 2, 4, 8, 5, 10, 9, 7, 3, 6].$$

For the ordering of the indices $k \neq 0$ we take the powers of $g^{-1} = 6$,

$$[(g^{-1})^{i'} : 0 \le i' \le 9] = [1, 6, 3, 7, 9, 10, 5, 8, 4, 2],$$

which, leaving aside the index 1, is the reverse order of the z indices. This reordering, together with a column of zeros for $k = 0$, gives the matrix

$$\boldsymbol{\Omega}_{11}^{\langle 2\rangle} = B_2\left(\frac{1}{11}\underbrace{\left[\begin{array}{cccccccccc|c} 1 & 6 & 3 & 7 & 9 & 10 & 5 & 8 & 4 & 2 & 0 \\ 2 & 1 & 6 & 3 & 7 & 9 & 10 & 5 & 8 & 4 & 0 \\ 4 & 2 & 1 & 6 & 3 & 7 & 9 & 10 & 5 & 8 & 0 \\ 8 & 4 & 2 & 1 & 6 & 3 & 7 & 9 & 10 & 5 & 0 \\ 5 & 8 & 4 & 2 & 1 & 6 & 3 & 7 & 9 & 10 & 0 \\ 10 & 5 & 8 & 4 & 2 & 1 & 6 & 3 & 7 & 9 & 0 \\ 9 & 10 & 5 & 8 & 4 & 2 & 1 & 6 & 3 & 7 & 0 \\ 7 & 9 & 10 & 5 & 8 & 4 & 2 & 1 & 6 & 3 & 0 \\ 3 & 7 & 9 & 10 & 5 & 8 & 4 & 2 & 1 & 6 & 0 \\ 6 & 3 & 7 & 9 & 10 & 5 & 8 & 4 & 2 & 1 & 0 \end{array}\right]}_{\boldsymbol{C}_{11}}\right),$$

where the 10×10 submatrix $\boldsymbol{C}_{11}$ is circulant.

We can halve the number of rows and the number of columns of the circulant submatrix $\boldsymbol{C}_{11}$ due to the symmetry $B_2(x) = B_2(1-x)$, to obtain the reduced form (denoted by a $\sim$ instead of $=$)

$$\boldsymbol{\Omega}_{11}^{\langle 2\rangle} \sim B_2\left(\frac{1}{11}\underbrace{\left[\begin{array}{ccccc||c} 1 & 6 & 3 & 7 & 9 & 0 \\ 2 & 1 & 6 & 3 & 7 & 0 \\ 4 & 2 & 1 & 6 & 3 & 0 \\ 8 & 4 & 2 & 1 & 6 & 0 \\ 5 & 8 & 4 & 2 & 1 & 0 \end{array}\right]}_{\widetilde{\boldsymbol{C}}_{11}}\right),$$

where we remember that the full matrix $\boldsymbol{\Omega}_{11}^{\langle 2\rangle}$ can be obtained by doubling the 5×5 circulant submatrix $\widetilde{\boldsymbol{C}}_{11}$ vertically and horizontally up to the double bar. This concludes our simple example.

If we express the reordering of indices induced by g as permutations Π_g^{T} and $\Pi_{g^{-1}}$ on the rows and columns of the matrix $\boldsymbol{\Omega}_n$, then we obtain

$$\boldsymbol{y} = \boldsymbol{\Omega}_n\, \boldsymbol{p}_{s-1} \iff \Pi_g^{\mathrm{T}}\, \boldsymbol{y} = (\Pi_g^{\mathrm{T}}\, \boldsymbol{\Omega}_n\, \Pi_{g^{-1}})\, (\Pi_{g^{-1}}^{\mathrm{T}}\, \boldsymbol{p}_{s-1}) = \boldsymbol{\Omega}_n^{\langle g\rangle}\, \boldsymbol{p}_{s-1}^{\langle g^{-1}\rangle}.$$

Since the initialization of $\boldsymbol{p}_0 = \boldsymbol{1}$ is unaffected by any permutation, we only need to work with the permuted vector throughout. We just need to make sure that the correct index z_s is chosen at the select stage.

A matrix–vector multiplication with a circulant matrix $\boldsymbol{C}_m$ of size $m \times m$, with first column $\boldsymbol{c}$, can be obtained by

$$\begin{aligned}\boldsymbol{C}_m\,\boldsymbol{x} &= \boldsymbol{F}_m^{-1}\,\mathrm{diag}(\boldsymbol{F}_m\,\boldsymbol{c})\;\boldsymbol{F}_m\,\boldsymbol{x}\\ &= \mathrm{IFFT}(\mathrm{FFT}(\boldsymbol{c}).*\;\mathrm{FFT}(\boldsymbol{x})),\end{aligned}$$

where

$$\boldsymbol{F}_m = [\exp(-2\pi\mathrm{i}\,p\,q/m)]_{0\le p,q\le m-1}$$

is the Fourier matrix of size $m \times m$, and

$$\boldsymbol{F}_m^{-1} = \frac{1}{m}[\exp(2\pi\mathrm{i}\,p\,q/m)]_{0\le p,q\le m-1}.$$

Using FFT and IFFT, this matrix–vector multiplication takes $O(m \log m)$ operations instead of the usual $O(m^2)$, and uses $O(m)$ memory.

Using permutation together with FFT, the cost for the CBC construction can be reduced to $O(n \log n\, s_{\max})$ operations, at the expense of $O(n)$ storage. This approach is called the *fast CBC construction*, and is due to Nuyens and Cools (2006*a*, 2006*b*).

For composite n, the general idea is that the complete matrix $\boldsymbol{\Omega}_n$ can be partitioned in blocks which have a circulant or block-circulant structure.

We summarize the fast CBC algorithm in the pseudocode below. We stress again that the key point is to use FFT to compute the matrix–vector product $\boldsymbol{\Omega}_n^{\langle g\rangle}\,\boldsymbol{p}_{s-1}$, taking care to select the correct z_s under the corresponding permuted set of indices.

Pseudocode 3 (Fast CBC: matrix–vector form with permuted matrix)

$\boldsymbol{p}_0 = \boldsymbol{1}$
$e_0^2 = 0$
for s **from** 1 **to** $s_{\max}$ **do**
  $\boldsymbol{e}_s^{2\langle g\rangle} = (\boldsymbol{1} + \gamma_s\boldsymbol{\beta})e_{s-1}^2 + \frac{\gamma_s}{n}\,\boldsymbol{\Omega}_n^{\langle g\rangle}\,\boldsymbol{p}_{s-1}$ ▷ compute – use FFT
  $z_s = \mathrm{argmin}_{z\in\mathbb{U}_n} e_s^2(z)$ ▷ select – pick the correct index
  $e_s^2 = e_s^2(z_s)$ ▷ set
  $\boldsymbol{p}_s = \big(\boldsymbol{1} + \gamma_s(\boldsymbol{\Omega}_n^{\langle g\rangle}(z_s,:) + \boldsymbol{\beta})\big).*\;\boldsymbol{p}_{s-1}$ ▷ update
end for

5.6. *Fast CBC construction for POD weights*

Here we explain how the fast CBC construction can be extended to POD (product and order-dependent) weights (see (4.4)), covering order-dependent weights as a special case. For completely general weights $\gamma_{\mathfrak{u}}$ the cost for evaluating $[e_{n,s}^{\mathrm{sh}}(\boldsymbol{z})]^2$ in (5.12) is prohibitively expensive, whereas for

POD weights their special structure enables us to compute $[e^{\rm sh}_{n,s}(\boldsymbol{z})]^2$ using a recursive argument starting from dimension one.

We restrict ourselves to the unanchored Sobolev space. (Recall that the CBC algorithm for the anchored Sobolev space minimizes an auxiliary quantity that depends on the auxiliary weights (5.22). The POD form of the original weights is not preserved under the auxiliary weights, thus the strategy to be described here is not applicable to the anchored space setting; see also the remark at the end of Section 5.4 for cases where fast CBC for POD weights can be used.) However, the approach can be generalized to other shift-invariant kernels if B_2 is replaced by another function that integrates to zero.

With POD weights $\gamma_{\mathfrak{u}} = \Gamma_{|\mathfrak{u}|}\prod_{j\in\mathfrak{u}|}\gamma_j$, we can write the shift-averaged worst-case error at step s (using again the simplified notation of the previous subsection) as

$$\begin{aligned} e_s^2(z_s) &= \frac{1}{n}\sum_{k=0}^{n-1}\sum_{\emptyset\neq\mathfrak{u}\subseteq\{1:s\}} \Gamma_{|\mathfrak{u}|}\prod_{j\in\mathfrak{u}}\left(\gamma_j\, B_2\left(\left\{\frac{kz_j}{n}\right\}\right)\right) \\ &= \frac{1}{n}\sum_{k=0}^{n-1}\sum_{\ell=1}^{s}\underbrace{\sum_{\substack{\mathfrak{u}\subseteq\{1:s\}\\|\mathfrak{u}|=\ell}} \Gamma_{\ell}\prod_{j\in\mathfrak{u}}\left(\gamma_j\, B_2\left(\left\{\frac{kz_j}{n}\right\}\right)\right)}_{p_{s,\ell}(k)}. \end{aligned} \tag{5.29}$$

We split the sum over $\mathfrak{u}$ into two depending on whether $s\in\mathfrak{u}$, to obtain

$$\begin{aligned} e_s^2(z_s) &= \frac{1}{n}\sum_{k=0}^{n-1}\sum_{\ell=1}^{s}\Bigg[\underbrace{\sum_{\substack{\mathfrak{u}\subseteq\{1:s-1\}\\|\mathfrak{u}|=\ell}} \Gamma_{\ell}\prod_{j\in\mathfrak{u}}\left(\gamma_j\, B_2\left(\left\{\frac{kz_j}{n}\right\}\right)\right)}_{p_{s-1,\ell}(k)} \\ &\quad + \frac{\Gamma_\ell}{\Gamma_{\ell-1}}\gamma_s\underbrace{B_2\left(\left\{\frac{kz_s}{n}\right\}\right)}_{\Omega_n(z_s,k)}\underbrace{\sum_{\substack{\mathfrak{u}\subseteq\{1:s-1\}\\|\mathfrak{u}|=\ell-1}} \Gamma_{\ell-1}\prod_{j\in\mathfrak{u}}\left(\gamma_j\, B_2\left(\left\{\frac{kz_j}{n}\right\}\right)\right)}_{p_{s-1,\ell-1}(k)}\Bigg] \\ &= e_{s-1}^2 + \frac{\gamma_s}{n}\sum_{k=0}^{n-1}\Omega_n(z_s,k)\left(\sum_{\ell=1}^{s}\frac{\Gamma_\ell}{\Gamma_{\ell-1}}p_{s-1,\ell-1}(k)\right). \end{aligned} \tag{5.30}$$

The computation of (5.30) for all $z_s\in\mathbb{U}_n$ can be expressed in matrix–vector form as

$$\boldsymbol{e}_s^2 = e_{s-1}^2\mathbf{1} + \frac{\gamma_s}{n}\boldsymbol{\Omega}_n\left(\sum_{\ell=1}^{s}\frac{\Gamma_\ell}{\Gamma_{\ell-1}}\boldsymbol{p}_{s-1,\ell-1}\right),$$

with the update formula

$$\boldsymbol{p}_{s,\ell} = \boldsymbol{p}_{s-1,\ell} + \frac{\Gamma_\ell}{\Gamma_{\ell-1}} \gamma_s \, \boldsymbol{\Omega}_n(z_s,:).* \, \boldsymbol{p}_{s-1,\ell-1} \quad \text{for } \ell = 1, \dots, s,$$

starting from

$$\boldsymbol{p}_{s,0} = \boldsymbol{1} \quad \text{and} \quad \boldsymbol{p}_{s,\ell} = \boldsymbol{0} \quad \text{for all } s \geq 1 \text{ and } \ell > s.$$

We need to store the vectors $\boldsymbol{p}_{s,\ell}$ for $\ell = 1, \dots, s$, each of length n. These s vectors correspond to the s possible values of $|\mathfrak{u}|$ for $\emptyset \neq \mathfrak{u} \subseteq \{1 : s\}$. These vectors can be overwritten for dimension $s+1$. Thus we require $O(n\, s_{\max})$ storage overall. The update cost for dimension s is $O(n\, s)$ operations.

If we permute the rows and columns of the matrix $\boldsymbol{\Omega}_n$ and use FFT to carry out the matrix–vector multiplication as in the case of product weights, then the search cost for each dimension is $O(n \log n)$ operations. The overall cost, combining search and update in all dimensions up to $s_{\max}$, is

$$O(n \log n \, s_{\max} + n\, s_{\max}^2)$$

operations.

If the weights are of *finite order* q, that is, $\Gamma_\ell = 0$ for all $\ell > q$, then the update cost for every dimension is capped at $O(n\, q)$ operations, with $O(n\, q)$ memory requirement, and the overall cost combining search and update in all dimensions up to $s_{\max}$ is reduced to

$$O(n \log n \, s_{\max} + n\, q\, s_{\max})$$

operations.

The reader may question why we did not leave out the factor Γ_ℓ from the definition of $p_{s,\ell}(k)$ in (5.29). Indeed, the expression (5.30) and the update formula would have been simpler if we had defined $p_{s,\ell}(k)$ that way. The problem is that for certain POD weights we may have Γ_ℓ growing very quickly with increasing ℓ and γ_j decaying with increasing j: for example,

$$\gamma_{\mathfrak{u}} = (|\mathfrak{u}|!)^{4/3} \prod_{j \in \mathfrak{u}} j^{-2.1}.$$

For weights of this kind, the alternative approach would be numerically unstable. The approach we outlined here works with the ratio $\Gamma_\ell/\Gamma_{\ell-1}$, thus avoiding the potential problem of overflow when working with Γ_ℓ.

5.7. *Extensible lattice sequences*

The CBC construction yields a lattice rule which is extensible in dimension, but not extensible in the number of points. Recall that the formula for obtaining the ith point of an n-point lattice rule with generating vector $\boldsymbol{z}$ is

$$\boldsymbol{t}_i = \left\{ \frac{i}{n} \boldsymbol{z} \right\}. \tag{5.31}$$

In an *extensible lattice sequence* we take $\boldsymbol{z}$ to be a vector of b-adic numbers $\sum_{r=0}^{\infty} a_r b^r$, where $a_r \in \{0, 1, \ldots, b-1\}$, and allow n to increase. In an extensible lattice sequence with base $b \geq 2$, the formula (5.31) is changed to

$$\boldsymbol{t}_i = \{\phi_b(i)\, \boldsymbol{z}\}, \tag{5.32}$$

where $\phi_b(\cdot)$ is the radical inverse function in base b (or the *Gray code variant* in which the successive indices differ only in one b-ary digit). The formula (5.32) does not require you to know n in advance, and so in practice you can add more points to your lattice rule approximation until you are satisfied with the error.

When $n = b^m$ for any $m \geq 1$, the formulas (5.31) and (5.32) produce the same set of points, only the ordering of the points is different. Therefore, to obtain the points of an extensible lattice rule only at exact powers of the base, one can avoid the radical inverse function and still use the formula (5.31). For example, if we take $b = 2$ and denote by L_m the lattice point set with 2^m points, then we can illustrate the lattice sequence as follows:

$$\underbrace{\underbrace{\underbrace{\underbrace{\underbrace{\boldsymbol{t}_0 = \boldsymbol{0}}_{L_0},\ \underbrace{\boldsymbol{t}_1 = \tfrac{1}{2}}_{\Delta L_1}}_{L_1},\ \underbrace{\boldsymbol{t}_2, \boldsymbol{t}_3}_{\Delta L_2}}_{L_2},\ \underbrace{\boldsymbol{t}_4, \boldsymbol{t}_5, \boldsymbol{t}_6, \boldsymbol{t}_7}_{\Delta L_3}}_{L_3},\ \underbrace{\boldsymbol{t}_8, \boldsymbol{t}_9, \boldsymbol{t}_{10}, \boldsymbol{t}_{11}, \boldsymbol{t}_{12}, \boldsymbol{t}_{13}, \boldsymbol{t}_{14}, \boldsymbol{t}_{15}}_{\Delta L_4}}_{L_4}, \ \ldots.$$

The number of points in L_m doubles as we move along the sequence, and the set of additional points $\Delta L_m = L_m \setminus L_{m-1}$ is generated by $\{i\boldsymbol{z}/2^m\}$ with i running through all the odd numbers up to $2^m - 1$. In Table 5.1 we show how the points are ordered under radical inverse and the Gray code variant.

Empirical searches for extensible lattice sequences were considered in Hickernell, Hong, L'Ecuyer and Lemieux (2000). Then it was proved by Hickernell and Niederreiter (2003) that good generating vectors exist for extensible lattice sequences, but the proof was non-constructive.

Cools, Kuo and Nuyens (2006) constructed *embedded lattice rules* that are good for n in a large practical range $b^{m_1} \leq n \leq b^{m_2}$. They used a modified CBC algorithm to construct a generating vector $\boldsymbol{z}$, minimizing as much as possible the search criterion

$$X_{m_1,m_2,s}(\boldsymbol{z}) := \max_{m_1 \leq m \leq m_2} \frac{e^{\mathrm{sh}}_{b^m,s}(\boldsymbol{z})}{e^{\mathrm{sh}}_{b^m,s}(\boldsymbol{z}^{(m)})},$$

where $e^{\mathrm{sh}}_{b^m,s}(\boldsymbol{z}^{(m)})$ denotes the shifted-averaged worst-case error for the generating vector obtained by the original CBC construction for b^m points,

Table 5.1. Ordering of lattice points.

	i	Natural order	Radical inverse	Gray code variant
L_0	0	0	0	0
ΔL_1	1	$1/2 = 0.5$	$(0.1)_2 = 0.5$	$(0.1)_2 = 0.5$
ΔL_2	2	$1/4 = 0.25$	$(0.01)_2 = 0.25$	$(0.11)_2 = 0.75$
	3	$3/4 = 0.75$	$(0.11)_2 = 0.75$	$(0.01)_2 = 0.25$
ΔL_3	4	$1/8 = 0.125$	$(0.001)_2 = 0.125$	$(0.011)_2 = 0.375$
	5	$3/8 = 0.375$	$(0.101)_2 = 0.625$	$(0.111)_2 = 0.875$
	6	$5/8 = 0.625$	$(0.011)_2 = 0.375$	$(0.101)_2 = 0.625$
	7	$7/8 = 0.875$	$(0.111)_2 = 0.875$	$(0.001)_2 = 0.125$
ΔL_4	8	$1/16 = 0.0625$	$(0.0001)_2 = 0.0625$	$(0.0011)_2 = 0.1875$
	9	$3/16 = 0.1875$	$(0.1001)_2 = 0.5625$	$(0.1011)_2 = 0.6875$
	10	$5/16 = 0.3125$	$(0.0101)_2 = 0.3125$	$(0.1111)_2 = 0.9375$
	11	$7/16 = 0.4375$	$(0.1101)_2 = 0.8125$	$(0.0111)_2 = 0.4375$
	12	$9/16 = 0.5625$	$(0.0011)_2 = 0.1875$	$(0.0101)_2 = 0.3125$
	13	$11/16 = 0.6875$	$(0.1011)_2 = 0.6875$	$(0.1101)_2 = 0.8125$
	14	$13/16 = 0.8125$	$(0.0111)_2 = 0.4375$	$(0.1001)_2 = 0.5625$
	15	$15/16 = 0.9375$	$(0.1111)_2 = 0.9375$	$(0.0001)_2 = 0.0625$

which is used as the reference error to provide a good normalization. The components of $\boldsymbol{z}$ can take values up to $b^{m_2} - 1$. Due to many structures in the embedding, the computational cost for the modified CBC algorithm is only a factor of $O(m_2)$ more than the cost for constructing a lattice rule with a fixed number of points. Figure 5.2 illustrates the embedding structure.

Numerically, it is found in Cools *et al.* (2006) for $2^{10} \leq n \leq 2^{20}$, s up to 360, and a number of different choices of product weights and order-dependent weights, that the quantity $X_{m_1,m_2,s}(\boldsymbol{z})$ resulting from the modified CBC construction is at most 1.6, indicating that the embedded lattice rules achieve the near-optimal rate of convergence of lattice rules with a fixed number of points, but with an implied constant up to 1.6 times larger.

Dick, Pillichshammer and Waterhouse (2008) introduced a *sieve algorithm* to obtain extensible lattice sequences. The general principle is to begin with a number of generating vectors that are *good* for some value of n, drop those vectors that are not good for a larger value of n, and then repeat this process for larger and larger values of n. Some parameters need to be specified beforehand to ensure that the sieve process leaves a non-empty set of generating vectors. Note that this sieve algorithm yields lattice sequences

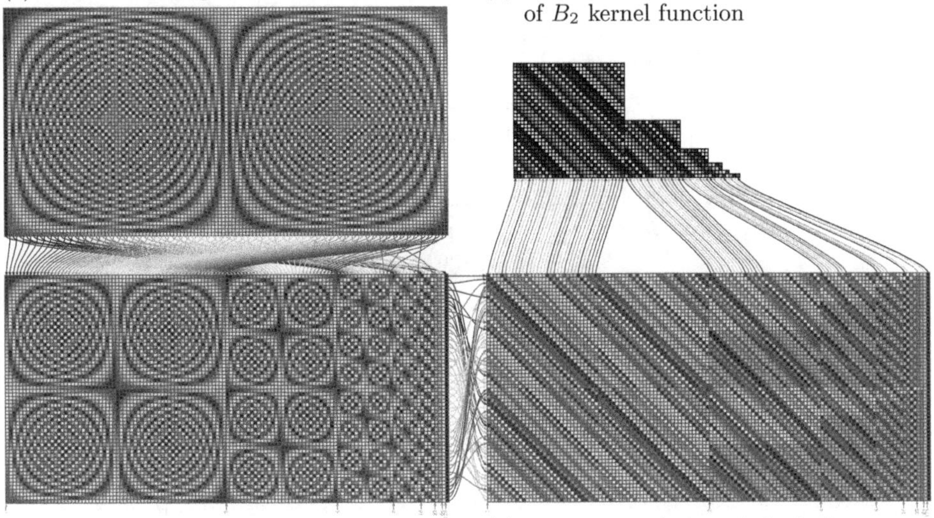

Figure 5.2. (Image by Dirk Nuyens.) The structure of the matrix Ω_{128} for the fast CBC construction of lattice rules, illustrating the reordering of indices and the reduction due to symmetry. The matrices Ω_n for $n = 2^m$ with $m < 7$ are embedded in Ω_{128}.

that are extensible in n but the algorithm is very expensive to implement. Dick *et al.* (2008) also introduced a cheaper version called the *CBC sieve algorithm*, which yields embedded lattice rules that work for a finite set of n and are extensible in s. This theory can also be modified to provide the theoretical justification for the algorithm from Cools *et al.* (2006).

5.8. Lattice rules in weighted Korobov spaces

In the previous subsections we have presented the lattice rule results for weighted anchored or unanchored variants of Sobolev spaces containing functions with square-integrable mixed first derivatives. Analogous results were originally obtained in a different function space setting, the *weighted Korobov spaces* of periodic functions.

We now briefly review the essential ingredients and results for weighted Korobov spaces, with the motive of understanding how one might achieve better than order-one convergence rate for non-periodic integrands. In the Korobov space setting, there is a *smoothness parameter* $\alpha > 1/2$. (The parameter α as used here differs by a factor of two from the usage in Sloan and Woźniakowski 2001.) The Korobov space can be viewed as an L_2 version of the Korobov class $E_\alpha(c)$ mentioned in Section 5.1. It again contains one-periodic functions with absolutely convergent Fourier series (5.1) but with the inner product now given by

$$\langle f, g\rangle_{s,\alpha,\gamma} = \sum_{\boldsymbol{h}\in\mathbb{Z}^s} \frac{\prod_{j\in\mathfrak{u}(\boldsymbol{h})} |h_j|^{2\alpha}}{\gamma_{\mathfrak{u}(\boldsymbol{h})}} \widehat{f}(\boldsymbol{h})\,\overline{\widehat{g}(\boldsymbol{h})},$$

with $\mathfrak{u}(\boldsymbol{h}) := \{j \in \{1:s\} : h_j \neq 0\}$. The reproducing kernel is

$$\begin{aligned} K_{s,\alpha,\gamma}(\boldsymbol{x},\boldsymbol{y}) &= \sum_{\mathfrak{u}\subseteq\{1:s\}} \gamma_{\mathfrak{u}} \prod_{j\in\mathfrak{u}} \Bigg(\sum_{h_j\in\mathbb{Z}\setminus\{0\}} \frac{\mathrm{e}^{2\pi\mathrm{i}h_j(x_j-y_j)}}{|h_j|^{2\alpha}} \Bigg) \\ &= \sum_{\boldsymbol{h}\in\mathbb{Z}^s} \gamma_{\mathfrak{u}(\boldsymbol{h})} \prod_{j\in\mathfrak{u}(\boldsymbol{h})} \frac{\mathrm{e}^{2\pi\mathrm{i}h_j(x_j-y_j)}}{|h_j|^{2\alpha}} \\ &= \sum_{\boldsymbol{h}\in\mathbb{Z}^s} \frac{\gamma_{\mathfrak{u}(\boldsymbol{h})}}{\prod_{j\in\mathfrak{u}(\boldsymbol{h})} |h_j|^{2\alpha}} \mathrm{e}^{2\pi\mathrm{i}\boldsymbol{h}\cdot(\boldsymbol{x}-\boldsymbol{y})}. \end{aligned}$$

From this last expression it is easy to verify the reproducing property.

The parameter α moderates the rate of decay of the Fourier coefficients, and is related to the smoothness of the functions. To illustrate this point, we note that when α is an integer, the norm in one dimension can be written as

$$\|f\|^2_{1,\alpha,\gamma} = \left(\int_0^1 f(t)\,\mathrm{d}t\right)^2 + \frac{1}{(2\pi)^{2\alpha}\gamma} \int_0^1 \big(f^{(\alpha)}(t)\big)^2\,\mathrm{d}t.$$

Thus α is precisely the number of available square-integrable derivatives.

When α is an integer, we can use the property (5.7) to rewrite the kernel as

$$K_{s,\alpha,\gamma}(\boldsymbol{x},\boldsymbol{y}) = \sum_{\mathfrak{u}\subseteq\{1:s\}} \gamma_{\mathfrak{u}} \prod_{j\in\mathfrak{u}} \bigg(\frac{(2\pi)^{2\alpha}}{(-1)^{\alpha+1}(2\alpha)!} B_{2\alpha}\big(|x_j - y_j|\big) \bigg), \qquad (5.33)$$

where $B_{2\alpha}(\cdot)$ is the Bernoulli polynomial of degree 2α. Our earlier results in the Sobolev space setting were originally obtained by observing that the associated shift-invariant kernel (see (5.11)) is in fact the kernel for a Korobov space with $\alpha = 1$ but with some redefined weights.

It is straightforward to deduce, using Theorem 3.5, that the worst-case error for a rank-one lattice rule in a weighted Korobov space satisfies

$$e^2_{n,s}(\boldsymbol{z}) = \frac{1}{n} \sum_{k=0}^{n-1} \sum_{\emptyset\neq\mathfrak{u}\subseteq\{1:s\}} \gamma_{\mathfrak{u}} \prod_{j\in\mathfrak{u}} \sum_{h\in\mathbb{Z}\setminus\{0\}} \frac{\mathrm{e}^{2\pi\mathrm{i}hkz_j/n}}{|h|^{2\alpha}}.$$

For product weights, the expression simplifies to

$$e^2_{n,s}(\boldsymbol{z}) = -1 + \frac{1}{n} \sum_{k=0}^{n-1} \prod_{j=1}^{s} \Bigg(1 + \gamma_j \sum_{h\in\mathbb{Z}\setminus\{0\}} \frac{\mathrm{e}^{2\pi\mathrm{i}hkz_j/n}}{|h|^{2\alpha}} \Bigg),$$

which is precisely the formula for P_α when all $\gamma_j = 1$ (see (5.4)), but with α replaced by 2α.

The following theorem summarizes the error bound for lattice rules constructed by the CBC algorithm in weighted Korobov space. We remark that the result holds for unshifted lattice rules, but since the reproducing kernel is shift-invariant, the result applies also to shifted lattice rules. The best convergence rate is obtained by taking $\lambda \to 1/(2\alpha)$, which yields close to $O(n^{-\alpha})$ convergence.

Theorem 5.12 (optimal CBC error bound in Korobov spaces). The generating vector $\boldsymbol{z}$ constructed by the CBC algorithm in Korobov spaces, minimizing $e_{n,s}^2(\boldsymbol{z})$ in each step, satisfies

$$e_{n,s}^2(\boldsymbol{z}) \le \left(\frac{1}{\varphi(n)} \sum_{\emptyset \ne \mathfrak{u} \subseteq \{1:s\}} \gamma_{\mathfrak{u}}^{\lambda} (2\zeta(2\alpha\lambda))^{|\mathfrak{u}|} \right)^{1/\lambda}$$

for all $\lambda \in (1/(2\alpha), 1]$, where $\zeta(\cdot)$ is the Riemann zeta function (5.5) and $\varphi(n)$ is the Euler totient function (5.14).

Proof. Using the character property (5.2), we can write

$$e_{n,s}^2(\boldsymbol{z}) = \sum_{\emptyset \ne \mathfrak{u} \subseteq \{1:s\}} \gamma_{\mathfrak{u}} \left(\sum_{\substack{\boldsymbol{h}_{\mathfrak{u}} \in (\mathbb{Z}\setminus\{0\})^{|\mathfrak{u}|} \\ \boldsymbol{h}_{\mathfrak{u}} \cdot \boldsymbol{z}_{\mathfrak{u}} \equiv 0 \ (\mathrm{mod}\ n)}} \frac{1}{\prod_{j\in\mathfrak{u}} |h_j|^{2\alpha}} \right).$$

The rest of the proof follows the proof of Theorem 5.8, which is essentially the case $\alpha = 1$, but with weights scaled by $(2\pi)^{|\mathfrak{u}|}$. □

The results for Korobov spaces are not often usable in practice because integrands are typically not fully periodic. However, the analysis in Korobov spaces provides the fundamental framework for the lattice rule analysis in Sobolev spaces. There we achieved order-one convergence via *shifting*: the shift-averaged worst-case error in Sobolev spaces of smoothness one is related to the Korobov space with $\alpha = 1$. Our aim now is to achieve higher convergence rates for non-periodic functions by linking with the results in Korobov space with $\alpha > 1$.

5.9. *The baker's transformation*

Extending Example 3.8 to s dimensions and introducing weights as in Section 4, we obtain the reproducing kernel for the *weighted and unanchored Sobolev space of smoothness* α:

$$K_{s,\alpha,\boldsymbol{\gamma}}(\boldsymbol{x},\boldsymbol{y}) = \sum_{\mathfrak{u} \subseteq \{1:s\}} \gamma_{\mathfrak{u}} \prod_{j\in\mathfrak{u}} \left(\frac{B_{2\alpha}(|x_j - y_j|)}{(-1)^{\alpha+1}(2\alpha)!} + \sum_{r=1}^{\alpha} \frac{B_r(x_j)\, B_r(y_j)}{(r!)^2} \right). \quad (5.34)$$

Comparing with (5.33), we see that this kernel has additional terms involving the lower-degree Bernoulli polynomials $B_1, \ldots, B_\alpha$. We already know that in the case of smoothness one ($\alpha = 1$) the associated shift-invariant kernel does not involve the lower-degree polynomial B_1; in short, *shifting removes* B_1. In this subsection we explain how *shifting followed by folding removes* B_2.

The *baker's transformation* makes use of the function

$$\phi(t) := 1 - |2t - 1| = \begin{cases} 2t & \text{if } t \le \frac{1}{2}, \\ 2 - 2t & \text{if } t > \frac{1}{2}. \end{cases}$$

It is called the baker's transformation since it emulates how a baker stretches and folds bread dough. In one dimension, if we apply the baker's transformation to the n points of a left-rectangle rule $0, 1/n, 2/n, \ldots, (n-1)/n$, where n is even, then we obtain the points

$$0, \frac{2}{n}, \frac{4}{n}, \ldots, 1, \frac{n-2}{n}, \frac{n-4}{n}, \ldots, \frac{2}{n},$$

which are the points for a trapezoidal rule. A rectangle rule has quadrature error $O(n^{-1})$, but a trapezoidal rule has quadrature error $O(n^{-2})$ if the integrand is sufficiently smooth. This is the underlying motivation for considering the baker's transformation. Its application in the context of QMC is due to Hickernell (2002). In the following, we refer to the process of applying the baker's transformation as *folding* and the resulting point set as the *folded* point set.

A shifted and then folded QMC rule takes the form

$$Q^{\phi}_{n,s}(\boldsymbol{\Delta}; f) = \frac{1}{n}\sum_{i=0}^{n-1} f(\phi(\{\boldsymbol{t}_i + \boldsymbol{\Delta}\})).$$

Note that this can also be viewed as applying a shifted lattice rule to a transformed integrand $g = f \circ \phi$. However, this interpretation is not helpful because the transformed integrand g does not have sufficient smoothness to get higher-order convergence.

As in Section 5.2, for a given point set P we let $\phi(P + \boldsymbol{\Delta})$ denote the shifted and then folded QMC point set. Then, for a given shift $\boldsymbol{\Delta}$ we have

$$|I_s(f) - Q^{\phi}_{n,s}(\boldsymbol{\Delta}; f)| \le e_{n,s}(\phi(P + \boldsymbol{\Delta}); H)\, \|f\|_H,$$

and for a randomly shifted and folded rule we have

$$\sqrt{\mathbb{E}|I_s(f) - Q^{\phi}_{n,s}(\boldsymbol{\Delta}; f)|^2} \le e^{\mathrm{sh},\phi}_{n,s}(P; H)\, \|f\|_H,$$

where

$$e_{n,s}^{\mathrm{sh},\phi}(P;H) := \sqrt{\int_{[0,1]^s} e_{n,s}^2(\phi(P+\boldsymbol{\Delta});H)\,\mathrm{d}\boldsymbol{\Delta}}. \tag{5.35}$$

Theorem 5.13 (formula for the shift-averaged and folded worst-case error). The *shift-averaged and folded* worst-case error (5.35) for a QMC point set P in an RKHS $H_s(K)$ satisfies

$$[e_{n,s}^{\mathrm{sh},\phi}(P;H_s(K))]^2 = -\int_{[0,1]^s}\int_{[0,1]^s} K(\boldsymbol{x},\boldsymbol{y})\,\mathrm{d}\boldsymbol{x}\,\mathrm{d}\boldsymbol{y} + \frac{1}{n^2}\sum_{i=0}^{n-1}\sum_{k=0}^{n-1} K^{\mathrm{sh},\phi}(\boldsymbol{t}_i,\boldsymbol{t}_k),$$

where

$$K^{\mathrm{sh},\phi}(\boldsymbol{x},\boldsymbol{y}) := \int_{[0,1]^s} K(\phi\{\boldsymbol{x}+\boldsymbol{\Delta}\},\phi\{\boldsymbol{y}+\boldsymbol{\Delta}\})\,\mathrm{d}\boldsymbol{\Delta} \quad \text{for all } \boldsymbol{x},\boldsymbol{y}\in[0,1]^s. \tag{5.36}$$

Proof. Using the definition (5.35) and applying the formula (3.9) for the worst-case error $e_{n,s}(\phi(P+\boldsymbol{\Delta});H_s(K))$, we obtain

$$\begin{aligned}
&[e_{n,s}^{\mathrm{sh},\phi}(P;H_s(K))]^2\\
&= \int_{[0,1]^s}\int_{[0,1]^s} K(\boldsymbol{x},\boldsymbol{y})\,\mathrm{d}\boldsymbol{x}\,\mathrm{d}\boldsymbol{y} - \frac{2}{n}\sum_{i=0}^{n-1}\int_{[0,1]^s}\int_{[0,1]^s} K(\phi(\{\boldsymbol{t}_i+\boldsymbol{\Delta}\}),\boldsymbol{y})\,\mathrm{d}\boldsymbol{\Delta}\,\mathrm{d}\boldsymbol{y}\\
&\quad + \frac{1}{n^2}\sum_{i=0}^{n-1}\sum_{k=0}^{n-1}\int_{[0,1]^s} K(\phi(\{\boldsymbol{t}_i+\boldsymbol{\Delta}\}),\phi(\{\boldsymbol{t}_k+\boldsymbol{\Delta}\}))\,\mathrm{d}\boldsymbol{\Delta}.
\end{aligned}$$

With a change of variables $\boldsymbol{w} = \{\boldsymbol{t}_i+\boldsymbol{\Delta}\}$, the double integral in the second term becomes $\int_{[0,1]^s}\int_{[0,1]^s} K(\phi(\boldsymbol{w}),\boldsymbol{y})\,\mathrm{d}\boldsymbol{w}\,\mathrm{d}\boldsymbol{y}$, which is easily seen to equal $\int_{[0,1]^s}\int_{[0,1]^s} K(\boldsymbol{x},\boldsymbol{y})\,\mathrm{d}\boldsymbol{x}\,\mathrm{d}\boldsymbol{y}$. The result then follows from the definition (5.36). □

We now demonstrate the effect of shifting followed by folding for the case $\alpha = 2$. To obtain the shifted and folded kernel $K^{\mathrm{sh},\phi}$ associated with (5.34), we need to evaluate the following integrals:

$$\begin{aligned}
\mathcal{I}_1(x,y) &:= \int_0^1 B_1(\phi(\{x+\Delta\}))\,B_1(\phi(\{y+\Delta\}))\,\mathrm{d}\Delta,\\
\mathcal{I}_2(x,y) &:= \int_0^1 B_2(\phi(\{x+\Delta\}))\,B_2(\phi(\{y+\Delta\}))\,\mathrm{d}\Delta,\\
\mathcal{J}_4(x,y) &:= \int_0^1 B_4(|\phi(\{x+\Delta\}) - \phi(\{y+\Delta\}))\,\mathrm{d}\Delta.
\end{aligned}$$

By using the Fourier series expansion (5.6) of the Bernoulli polynomials, it can be shown (Hickernell 2002) that

$$\begin{aligned}
\mathcal{I}_1(x,y) &= -\frac{4}{3}B_4(\{x-y\}) + \frac{4}{3}B_4(\{x-y-\tfrac{1}{2}\}),\\
\mathcal{I}_2(x,y) &= -\frac{4}{3}B_4(\{x-y\}) - \frac{4}{3}B_4(\{x-y-\tfrac{1}{2}\}),\\
\mathcal{J}_4(x,y) &= \frac{32}{3}B_4(\{x-y\}) - \frac{8}{3}B_4(\{x-y-\tfrac{1}{2}\})\\
&\quad + \frac{128}{15}B_6(\{x-y\}) - \frac{128}{15}B_6(\{x-y-\tfrac{1}{2}\}).
\end{aligned}$$

Since the lowest-degree Bernoulli polynomial in the above expressions is B_4, we conclude that the shifted and folded kernel is related to the kernel of the Korobov space with smoothness parameter $\alpha = 2$. This indicates that we get a convergence rate close to $O(n^{-2})$.

In addition, it can be shown that the CBC construction yields lattice rules achieving a convergence rate close to $O(n^{-2})$, and that the implementation can be done in a fast way for product or POD weights. We omit the details.

It is possible to generalize the baker's transformation to obtain even higher rates of convergence, but there is a seemingly inevitable dependence on s, prohibiting the strategy for practical use in high dimensions.

5.10. Periodization

From the results for the Korobov space, we know that lattice rules can yield higher rates of convergence for periodic integrands. This brings us to consider strategies for transforming a non-periodic integrand to a periodic integrand.

We now briefly explain one periodization strategy. Let $\omega : [0,1] \to \mathbb{R}$ be a smooth function, with

$$\omega(0) = \omega(1) = 0, \quad \omega(y) > 0 \quad \text{for all } y \in (0,1), \quad \text{and} \quad \int_0^1 \omega(y)\,\mathrm{d}y = 1.$$

Define

$$\psi(y) := \int_0^y \omega(u)\,\mathrm{d}u.$$

Then ψ is an increasing function satisfying

$$\psi'(y) = \omega(y), \quad \psi(0) = 0 \quad \text{and} \quad \psi(1) = 1.$$

The periodization transformation is essentially a change of variables $\boldsymbol{x} = \psi(\boldsymbol{y}) = (\psi(y_1), \ldots, \psi(y_s))$ so that

$$I_s(f) = \int_{[0,1]^s} f(\boldsymbol{x})\,\mathrm{d}\boldsymbol{x} = \int_{[0,1]^s} F(\boldsymbol{y})\,\mathrm{d}\boldsymbol{y} = I_s(F),$$

with

$$F(\boldsymbol{y}) := f(\psi(\boldsymbol{y})) \prod_{j=1}^{s} \omega(y_j).$$

The function F is periodic since ω vanishes at the end points.

Popular choices of ω include the family of *polynomial transformations* of Korobov (1963) and the family of *trigonometric transformations* of Sidi (1993); see also Laurie (1996). For example, the former family is given by

$$\omega(y) = (2\alpha + 1)\binom{2\alpha}{\alpha} y^{\alpha}(1-y)^{\alpha}$$

for a positive integer α. In particular, we have $\omega^{(r)}(0) = \omega^{(r)}(1)$ for all $r \leq \alpha - 1$. With this choice of ω, and if f is sufficiently smooth, the transformed integrand F belongs to the Korobov space with smoothness parameter α, and the lattice rule convergence rate is therefore close to $O(n^{-\alpha})$.

The problem with the periodization strategy is that the norm of the transformed integrand F can be exponentially large in s, thus is generally only feasible for small to moderate s.

As we explained in the previous subsection, the baker's transformation can also be viewed as a transformation of the integrand. The baker's transformation makes the end points of the function take the same value by reflection. The periodization strategy, on the other hand, makes the function value and perhaps also some derivatives vanish at the end points.

5.11. Notes

Lattice rules date back to the works of number theorists Korobov (1959) and Hlawka (1962). There is an extensive literature up to the late 1990s, which we refer to as the classical results; these are reviewed in detail in the books by Niederreiter (1992*a*) and Sloan and Joe (1994); see also Hickernell (1998*b*). The modern consideration of lattice rules can be said to have begun with Sloan and Woźniakowski (2001), who proved the existence of good lattice rules for periodic integrands in weighted Korobov spaces, and via the shift-invariant kernel, they showed that there exist good shifted lattice rules for non-periodic integrands in weighted Sobolev space of smoothness one. Although the results were non-constructive, they demonstrated that lattice rules have a role to play for truly high-dimensional integration as well as for non-periodic integrands.

The CBC construction was first considered by Korobov (1959) (see also Korobov 1963), who proved error bounds for P_α of order $n^{-\alpha+\delta}$ for any $\delta > 0$ and for arbitrary n. The modern analysis of CBC construction for the lattice rule generating vector began with Sloan and Reztsov (2002) for the classical (unweighted) criterion P_α in the periodic setting. Sloan,

Kuo and Joe (2002*a*) then introduced the CBC algorithm to construct a deterministic shift alongside the generating vector for integration in the non-periodic weighted Sobolev spaces. Subsequently, the CBC algorithm for constructing randomly shifted lattice rules in weighted Sobolev spaces (see Algorithm 5.6) was introduced in Sloan, Kuo and Joe (2002*b*). In these earlier works only $O(n^{-1/2})$ convergence rate was proved. The fact that the CBC algorithm achieves the (optimal) convergence rate close to $O(n^{-\alpha})$ in weighted Korobov and close to $O(n^{-1})$ in weighted Sobolev spaces with smoothness one was later proved by Kuo (2003). Extensions of these results from prime n to composite n were given in Kuo and Joe (2002) with convergence rate of $O(n^{-1/2})$, and Dick (2004) with convergence rate close to $O(n^{-1})$. Other considerations include the CBC construction of intermediate-rank lattice rules (Kuo and Joe 2003), and a modified CBC algorithm making use of the prime factorization of n for the purpose of speeding up the construction (Dick and Kuo 2004*a*, 2004*b*). All of the above results were originally proved in the setting of product weights. Results for general weights in Korobov spaces were obtained in Dick *et al.* (2006), and the corresponding results for weighted Sobolev spaces were obtained in Sloan *et al.* (2004), again via the shift-invariant kernel.

The fast CBC construction was first introduced for prime n in Nuyens and Cools (2006*a*) and later extended to composite n in Nuyens and Cools (2006*b*). The special case of n being a power of 2 is discussed in Cools *et al.* (2006). Fast CBC for POD weights was first considered in Kuo, Schwab and Sloan (2011). Some good generating vectors for lattice rules can be downloaded from http://www.maths.unsw.edu.au/~fkuo/lattice/. Implementation of fast CBC constructions for various forms of weights can be obtained from http://people.cs.kuleuven.be/~dirk.nuyens/fast-cbc/.

Extensible lattice sequences were considered in Maize (1980), Korobov (1982), and Hickernell *et al.* (2000), but the first theoretical result proving that good extensible lattice sequences exist was proved by Hickernell and Niederreiter (2003). Cools *et al.* (2006) introduced a fast CBC construction for embedded lattice rules, while Dick *et al.* (2008) analysed a sieve algorithm and also provided the theoretical justification for the algorithm in Cools *et al.* (2006). An algorithm was introduced in Niederreiter and Pillichshammer (2009) which constructs the lattice rule digit by digit. Hickernell, Kritzer, Kuo and Nuyens (2012) investigated strategies for applying a lattice sequence point by point.

Joe (2004) introduced a CBC construction of lattice rules based on an error criterion called R, which is part of an upper bound to the classical star discrepancy. This was later extended to the weighted setting for product weights in Joe (2006), and to composite n in Sinescu and Joe (2008). General weights were considered by Sinescu and Joe (2007) for prime n, and by Sinescu and L'Ecuyer (2011) for composite n. The analysis for general weights presented difficulties. (Recall that the CBC construction for the anchored space with general non-product weights must work with an auxiliary quantity. There is a similar difficulty here.) Thus, instead of working with weighted-R, the last two papers considered a related quantity which we shall refer to as weighted-$\widetilde{R}$, and which forms part of a different upper bound to the weighted star discrepancy. This argument holds under a restrictive monotonicity assumption on the weights that

$$\gamma_{\mathfrak{v}} \geq \gamma_{\mathfrak{u}} \quad \text{whenever} \quad \mathfrak{v} \subseteq \mathfrak{u}. \tag{5.37}$$

Switching from weighted-R to weighted-$\widetilde{R}$ does not require a redefinition of weights. Thus any structure in the weights, such as the POD form, would be preserved. Note, however, that the given POD weights may not satisfy the monotonicity requirement (5.37) needed to allow for this switch from weighted-R to weighted-$\widetilde{R}$. (We remark that imposing condition (5.37) does not appear to eliminate the need to work with the auxiliary quantity in the anchored space with general non-product weights.)

The baker's transformation (which is called 'tent transformation' in ergodic theory) for shifted lattice rules was first introduced by Hickernell (2002) to gain an extra order of convergence for non-periodic integrands, that is, from nearly $O(n^{-1})$ to nearly $O(n^{-2})$. The recent work by Dick *et al.* (2013) generalizes this strategy to obtain even higher rates of convergence. There is, however, an inevitable dimension dependence. This is similar to the issue with periodization in high dimensions, which was discussed in Kuo, Sloan and Woźniakowski (2007). An interesting discovery by Dick *et al.* (2013) is that the baker's transformation can be applied to an unshifted lattice rule, to achieve close to $O(n^{-1})$ convergence for non-periodic integrands, *without the need to use random shifts*. This is different from most of the results discussed in this section.

Randomly shifted lattice rules for integrands over unbounded domains (including $\mathbb{R}^s$) were considered in Kuo, Wasilkowski and Waterhouse (2006*b*) and Kuo *et al.* (2010*a*). The strategy is to transform the integrals to the unit cube by a change of variables, using the inverse of the cumulative distribution function of some univariate probability density function. The transformed integrands over the unit cube typically do not belong to the standard Sobolev space setting, and this is why new theory is needed. It is proved that good lattice rules can be constructed by a CBC algorithm in a similar way to the standard theory in Sobolev spaces.

Another branch of lattice rule analysis focuses on the trigonometric degree and similar quantities; see for example Cools and Lyness (2001), Lyness and Sørevik (2006), Cools and Nuyens (2008) and the references therein. Recent constructions in terms of trigonometric degrees include the works by Cools, Kuo and Nuyens (2010), Achtsis and Nuyens (2012) and Kämmerer, Kunis and Potts (2012).

Other search criteria for lattice rules were reviewed in L'Ecuyer and Lemieux (2000), Lemieux and L'Ecuyer (2001) and L'Ecuyer and Munger (2012), while L'Ecuyer, Munger and Tuffin (2010) considered the distribution of the integration error with respect to the random shift.

Lattice rules with exponential convergence rates in (modified) weighted Korobov spaces were considered in Dick, Larcher, Pillichshammer and Woźniakowski (2011).

Approximation of functions by lattice rules was considered in Li and Hickernell (2003) and Zeng, Leung and Hickernell (2006). CBC constructions of lattice rules for the approximation of functions were analysed in Kuo, Sloan and Woźniakowski (2006*a*, 2008*b*) and Kuo, Wasilkowski and Woźniakowski (2009).

The choice of weights in relation to the application of lattice rules were considered in Wang and Sloan (2006), Dick (2012) and Kuo *et al.* (2012).

The character property (5.2) is based on the following general algebraic structure: The set $([0,1)^s, \oplus)$ with addition modulo 1 given by $\boldsymbol{x} \oplus \boldsymbol{y} := \{\boldsymbol{x}+\boldsymbol{y}\}$ forms an abelian group, and the set of lattice points $P_{\text{lat}} = \{\boldsymbol{t}_k = \{k\boldsymbol{z}/n\} : 0 \le k < n\}$ is a finite subgroup. The set $\mathcal{K}$ of functions $\chi_{\boldsymbol{h}} : [0,1)^s \to \{z \in \mathbb{C} : |z| = 1\}$, $\boldsymbol{h} \in \mathbb{Z}$, given by

$$\chi_{\boldsymbol{h}}(\boldsymbol{x}) = \mathrm{e}^{2\pi i \boldsymbol{h}\cdot\boldsymbol{x}},$$

constitute a group homomorphism of the group $([0,1)^s, \oplus)$ called characters. They themselves form a group via the multiplication $\chi_{\boldsymbol{h}}\chi_{\boldsymbol{k}} = \chi_{\boldsymbol{h}+\boldsymbol{k}}$ which is isomorphic to the group $(\mathbb{Z}^s, +)$. The character-theoretic dual lattice of the lattice point set P_{lat} is given by the set of characters

$$\mathcal{L}^\perp = \{\chi_{\boldsymbol{h}} : \boldsymbol{h} \in \mathbb{Z}^s, \chi_{\boldsymbol{h}}(\{k\boldsymbol{z}/n\}) = 1 \text{ for all } 0 \le k < n\},$$

which is a subgroup of the set of characters. This can be used to define the quotient group $P_{\text{lat}}^\perp = \mathcal{K}/\mathcal{L}^\perp$, which can be viewed as the dual group of characters of a lattice point set P_{lat}. Since the characters of $([0,1)^s, \oplus)$ are the basis functions of Fourier series, it is natural to study lattice rules for numerical integration of Fourier series. We will use an analogous algebraic structure in the next section.

6. Digital nets and sequences

In the previous section we discussed the modern theory of lattice rules. Many of the ideas there apply also to polynomial lattice rules, or more generally, to nets and sequences. For lattice rules we have seen that Fourier series play an important role. For polynomial lattice rules and nets this role is played by *Walsh series*, which we introduce in the following subsection.

Throughout this section we denote by $\mathbb{N}_0$ the set of non-negative integers and $\mathbb{N}$ the set of positive integers.

6.1. A brief discussion of numerical integration of Walsh series

Walsh (1923) introduced a system of functions denoted by $\{\mathrm{wal}_k : k \in \mathbb{N}_0\}$, which is in some way similar to the trigonometric function system $\{e^{2\pi i k x} : k \in \mathbb{Z}\}$ that is connected to the well-known Fourier theory.

Definition 6.1 (Walsh functions in one dimension). For $b \geq 2$ we denote by $\omega_b := e^{2\pi i/b}$ the primitive bth root of unity. Let $k \in \mathbb{N}_0$ with b-adic expansion $k = \kappa_0 + \kappa_1 b + \cdots + \kappa_{r-1} b^{r-1}$. The *kth b-adic Walsh function* $\mathrm{wal}_k : \mathbb{R} \to \mathbb{C}$, periodic with period one, is defined by

$$\mathrm{wal}_k(x) := \omega_b^{\kappa_0 \xi_1 + \kappa_1 \xi_2 + \cdots + \kappa_{r-1} \xi_r},$$

for $x \in [0,1)$ with b-adic expansion $x = \xi_1 b^{-1} + \xi_2 b^{-2} + \cdots$ (unique in the sense that infinitely many of the digits ξ_i must be different from $b-1$).

In the literature the function system defined above is often called the *generalized Walsh function system* and only in the case $b = 2$ one speaks of Walsh functions.

One of the main differences between Walsh functions and the trigonometric functions is that Walsh functions are only piecewise continuous. This is clear, since Walsh functions are step functions as we now show.

Let $k \in \mathbb{N}$ with b-adic expansion $k = \kappa_0 + \kappa_1 b + \cdots + \kappa_{r-1} b^{r-1}$. Let $J = [a/b^r, (a+1)/b^r)$, with integers $r \geq 1$ and $0 \leq a < b^r$, be a so-called *elementary b-adic interval of order r*. Let a have b-adic expansion of the form $a = \alpha_0 + \alpha_1 b + \cdots + \alpha_{r-1} b^{r-1}$. Then any $x \in J$ has b-adic expansion

$$x = \alpha_{r-1} b^{-1} + \alpha_{r-2} b^{-2} + \cdots + \alpha_0 b^{-r} + \xi_{r+1} b^{-(r+1)} + \cdots$$

for some digits $0 \leq \xi_i \leq b-1$ for $i \geq r+1$, and hence

$$\mathrm{wal}_k(x) = \omega_b^{\kappa_0 \alpha_{r-1} + \kappa_1 \alpha_{r-2} + \cdots + \kappa_{r-1} \alpha_0} = \mathrm{wal}_k(a/b^r).$$

We summarize this result in the following proposition.

Proposition 6.2 (Walsh functions are step functions). Let $k \in \mathbb{N}$ satisfy $b^{r-1} \leq k < b^r$ for some $r \in \mathbb{N}$. Then the kth Walsh function wal_k is constant on elementary b-adic intervals of order r of the form $[a/b^r, (a+1)/b^r)$ with value $\mathrm{wal}_k(a/b^r)$. Further, $\mathrm{wal}_0 \equiv 1$.

Now we generalize the definition of Walsh functions to higher dimensions.

Definition 6.3 (Walsh functions in s dimensions). For dimension $s \geq 2$, and $k_1, \ldots, k_s \in \mathbb{N}_0$ we define the *s-dimensional b-adic Walsh function* $\mathrm{wal}_{k_1,\ldots,k_s} : \mathbb{R}^s \to \mathbb{C}$ by

$$\mathrm{wal}_{k_1,\ldots,k_s}(x_1, \ldots, x_s) := \prod_{j=1}^{s} \mathrm{wal}_{k_j}(x_j).$$

For vectors $\boldsymbol{k} = (k_1, \ldots, k_s) \in \mathbb{N}_0^s$ and $\boldsymbol{x} = (x_1, \ldots, x_s) \in [0,1)^s$ we write, with some abuse of notation,

$$\mathrm{wal}_{\boldsymbol{k}}(\boldsymbol{x}) := \mathrm{wal}_{k_1,\ldots,k_s}(x_1, \ldots, x_s).$$

The system $\{\mathrm{wal}_{\boldsymbol{k}} : \boldsymbol{k} \in \mathbb{N}_0^s\}$ is called the *s-dimensional b-adic Walsh function system.*

Since any s-dimensional Walsh function is a product of one-dimensional Walsh functions, it is clear that s-dimensional Walsh functions are step functions too.

Let $P = \{\boldsymbol{t}_0, \boldsymbol{t}_1, \ldots, \boldsymbol{t}_{b^m-1}\}$ be a digital (t,m,s)-net constructed over $\mathbb{Z}_b$ as introduced in Section 2.6. The set P is an additive group whose dual group of characters is given by the Walsh functions in the same base b, that is, we have the following character property:

$$\frac{1}{b^m}\sum_{i=0}^{b^m-1} \mathrm{wal}_{\boldsymbol{k}}(\boldsymbol{t}_i) = \begin{cases} 1 & \text{if } C_1^\top \vec{k}_1 + \cdots + C_s^\top \vec{k}_s = \vec{0}, \\ 0 & \text{otherwise,} \end{cases} \tag{6.1}$$

where for a vector $k \in \mathbb{N}_0$ with b-adic expansion $k = \kappa_0 + \kappa_1 b + \cdots + \kappa_{m-1} b^{m-1}$ we write $\vec{k} = (\kappa_0, \kappa_1, \ldots, \kappa_{m-1})^\top$. This leads to a similar result to Theorems 5.1 and 5.2 for lattice rules.

Theorem 6.4 (integration error for digital net). Assume we are given a Walsh series

$$f(\boldsymbol{x}) = \sum_{\boldsymbol{k} \in \mathbb{N}_0^s} \widehat{f}(\boldsymbol{k}) \mathrm{wal}_{\boldsymbol{k}}(\boldsymbol{x}),$$

where the Walsh coefficients $\widehat{f}(\boldsymbol{k})$ are given by

$$\widehat{f}(\boldsymbol{k}) = \int_{[0,1]^s} f(\boldsymbol{x}) \overline{\mathrm{wal}_{\boldsymbol{k}}(\boldsymbol{x})} \,\mathrm{d}\boldsymbol{x}.$$

The integration error of approximating the integral of f using a QMC rule $Q_{b^m,s}$ based on a digital net with b^m points is given by

$$Q_{b^m,s}(f) - I_s(f) = \sum_{\boldsymbol{k} \in \mathcal{D} \setminus \{\boldsymbol{0}\}} \widehat{f}(\boldsymbol{k}),$$

where

$$\mathcal{D} := \{\boldsymbol{k} \in \mathbb{N}_0^s : C_1^\top \vec{k}_1 + \cdots + C_s^\top \vec{k}_s = \vec{0}\}. \tag{6.2}$$

is the *dual net* associated with the digital net.

Proof. We have

$$\frac{1}{b^m}\sum_{i=0}^{b^m-1} f(\boldsymbol{t}_i) - \int_{[0,1]^s} f(\boldsymbol{x}) \mathrm{d}\boldsymbol{x} = \sum_{\boldsymbol{k} \in \mathbb{N}_0^s} \widehat{f}(\boldsymbol{k}) \frac{1}{b^m} \sum_{i=0}^{b^m-1} \mathrm{wal}_{\boldsymbol{k}}(\boldsymbol{t}_i) - \widehat{f}(\boldsymbol{0}),$$

The result then follows immediately from the character property (6.1). □

We consider now the case where the Walsh coefficients $\widehat{f}(\boldsymbol{k})$ decay with a certain rate. For $k \in \mathbb{N}_0$ let

$$\mu_1(k) := \begin{cases} 0 & \text{if } k = 0, \\ a & \text{if } k = \kappa_0 + \cdots + \kappa_{a-1}b^{a-1}, \kappa_{a-1} \neq 0. \end{cases} \tag{6.3}$$

(Later in Section 6.6 we generalize the function μ_1 to μ_α.) For vectors $\boldsymbol{k} = (k_1, \ldots, k_s) \in \mathbb{N}_0^s$ let

$$\mu_1(\boldsymbol{k}) := \mu_1(k_1) + \cdots + \mu_1(k_s).$$

Let

$$E_{\vartheta,s}(c) := \{f : [0,1]^s \to \mathbb{R} : |\widehat{f}(\boldsymbol{k})| \leq c\, b^{-\vartheta\mu_1(\boldsymbol{k})} \text{ for all } \boldsymbol{k} \in \mathbb{N}_0^s\},$$

for some $\vartheta > 1$ and $c > 0$. Then we obtain

$$|Q_{b^m,s}(f) - I_s(f)| \leq c \sum_{\boldsymbol{k} \in \mathcal{D} \setminus \{\boldsymbol{0}\}} b^{-\vartheta\mu_1(\boldsymbol{k})} \quad \text{for all } f \in E_{\vartheta,s}(c).$$

The quantity

$$T_\vartheta := \sum_{\boldsymbol{k} \in \mathcal{D} \setminus \{\boldsymbol{0}\}} b^{-\vartheta\mu_1(\boldsymbol{k})}, \quad \vartheta > 1, \tag{6.4}$$

depends only on the function class $E_{\vartheta,s}(1)$ and the digital net and can be understood as the 'worst-case error' for this function class. Thus, for the function class $E_{\vartheta,s}(1)$ one aims at finding digital nets for which T_ϑ is small. Thus the quantity T_ϑ plays a similar role to the quantity P_α for lattice rules (see (5.3)).

We briefly describe the connection between the quantity T_ϑ and the quality parameter t of the digital net. Lemma 2.14 implies that if $\boldsymbol{k} \in \mathcal{D} \setminus \{\boldsymbol{0}\}$, then $\mu_1(\boldsymbol{k}) > m - t$. If the digital net is a strict (t, m, s)-net, then there exists a $\boldsymbol{k} \in \mathcal{D} \setminus \{\boldsymbol{0}\}$ such that $\mu_1(\boldsymbol{k}) = m - t + 1$. Thus, the largest summand in (6.4) is of the form $b^{-\vartheta(m-t+1)}$. On the other hand it can also be shown that the quantity T_ϑ is dominated by its largest term, that is, there is a constant $\tilde{c} > 0$ such that

$$T_\vartheta \leq \tilde{c}\frac{(m-t)^{s-1}}{b^{\vartheta(m-t+1)}}.$$

Thus digital nets for which $m - t$ is large yield a small value of T_ϑ and are therefore useful for numerical integration of Walsh series.

6.2. Digitally shift-averaged worst-case error

In this subsection we discuss the general strategy for the error analysis of randomly digitally shifted QMC rules not specific to digital nets (similar to Section 5.2), and then focus on digital nets in the next subsection.

For any QMC point set $P = \{\boldsymbol{t}_0, \ldots, \boldsymbol{t}_{n-1}\}$ and any digital shift $\boldsymbol{\Delta} \in [0,1]^s$, let

$$P \oplus \boldsymbol{\Delta} = \{\boldsymbol{t}_i \oplus \boldsymbol{\Delta} : i = 0, 1, \ldots, n-1\}$$

denote the digitally shifted QMC point set, and let $Q_{n,s}(\oplus\boldsymbol{\Delta}; f)$ denote the corresponding digitally shifted QMC rule, where $\oplus$ denotes digit-wise addition modulo b already defined in Section 2.9. Then, for any integrand f belonging to some normed space H, it follows from the definition of the worst-case error that

$$|I_s(f) - Q_{n,s}(\oplus\boldsymbol{\Delta}, f)| \le e_{n,s}(P \oplus \boldsymbol{\Delta}; H)\, \|f\|_H.$$

We consider the *root-mean-square* error

$$\sqrt{\mathbb{E}\, |I_s(f) - Q_{n,s}(\oplus\boldsymbol{\Delta}, f)|^2} \le e^{\mathrm{dsh}}_{n,s}(P; H)\, \|f\|_H,$$

where the expectation $\mathbb{E}$ is taken over the random digital shift $\boldsymbol{\Delta}$ which is uniformly distributed over $[0,1]^s$, and where the quantity

$$e^{\mathrm{dsh}}_{n,s}(P; H) := \sqrt{\int_{[0,1]^s} e^2_{n,s}(P \oplus \boldsymbol{\Delta}; H)\, \mathrm{d}\boldsymbol{\Delta}} \tag{6.5}$$

is referred to as the *digitally shift-averaged worst-case error.*

The digitally shift-averaged worst-case error (5.8) will be used as our quality measure for randomly digitally shifted QMC rules. The averaging argument guarantees the existence of at least one digital shift $\boldsymbol{\Delta}$ for which $e_{n,s}(P \oplus \boldsymbol{\Delta}; H) \le e^{\mathrm{dsh}}_{n,s}(P; H)$.

Analogously to Theorem 5.3, the following theorem gives an explicit formula for the digitally shift-averaged worst-case error when the function space is an RKHS.

Theorem 6.5 (formula for the digitally shift-averaged worst-case error). The *digitally shift-averaged* worst-case error (6.5) for a QMC point set P in an RKHS $H_s(K)$ satisfies

$$[e^{\mathrm{dsh}}_{n,s}(P; H_s(K))]^2 = -\int_{[0,1]^s}\int_{[0,1]^s} K(\boldsymbol{x}, \boldsymbol{y})\, \mathrm{d}\boldsymbol{x}\, \mathrm{d}\boldsymbol{y} + \frac{1}{n^2}\sum_{i=0}^{n-1}\sum_{k=0}^{n-1} K^{\mathrm{dsh}}(\boldsymbol{t}_i, \boldsymbol{t}_k),$$

where

$$K^{\mathrm{dsh}}(\boldsymbol{x}, \boldsymbol{y}) := \int_{[0,1]^s} K(\boldsymbol{x} \oplus \boldsymbol{\Delta}, \boldsymbol{y} \oplus \boldsymbol{\Delta})\, \mathrm{d}\boldsymbol{\Delta} \quad \text{for all } \boldsymbol{x}, \boldsymbol{y} \in [0,1]^s. \tag{6.6}$$

The proof follows along the same lines as the proof of Theorem 5.3.

The function K^{dsh} defined by (6.6) is actually a reproducing kernel with the *digitally shift-invariant property*

$$K^{\mathrm{dsh}}(\boldsymbol{x}, \boldsymbol{y}) = K^{\mathrm{dsh}}(\boldsymbol{x} \oplus \boldsymbol{\Delta}, \boldsymbol{y} \oplus \boldsymbol{\Delta}) \quad \text{for all } \boldsymbol{x}, \boldsymbol{y}, \boldsymbol{\Delta} \in [0,1]^s,$$

or equivalently,

$$K^{\mathrm{dsh}}(\boldsymbol{x},\boldsymbol{y}) = K^{\mathrm{dsh}}(\boldsymbol{x}\ominus\boldsymbol{y},\boldsymbol{0}) \quad \text{for all } \boldsymbol{x},\boldsymbol{y}\in[0,1]^s,$$

where $\ominus$ denotes digit-wise subtraction modulo b already defined in Section 2.9. As in Section 5.2, it can be verified that the digitally shift-averaged worst-case error $e^{\mathrm{dsh}}_{n,s}(P;H_s(K))$ is precisely the worst-case error of the QMC point set P in the RKHS with K^{dsh} as the reproducing kernel, that is,

$$\int_{[0,1]^s} e^2_{n,s}(P\oplus\boldsymbol{\Delta};H_s(K))\,\mathrm{d}\boldsymbol{\Delta} = e^2_{n,s}(P;H_s(K^{\mathrm{dsh}})).$$

This important connection provides a powerful tool for analysing randomly digitally shifted QMC rules. We refer to the kernel K^{dsh} as the *digital shift-invariant kernel associated with* K.

6.3. Randomly digitally shifted digital nets in weighted Sobolev spaces

The digital shift-invariant kernel K^{dsh} in the previous subsection has Walsh series expansion

$$K^{\mathrm{dsh}}(\boldsymbol{x},\boldsymbol{y}) = \sum_{\boldsymbol{k}\in\mathbb{N}_0^s} \widehat{K^{\mathrm{dsh}}}(\boldsymbol{k})\mathrm{wal}_{\boldsymbol{k}}(\boldsymbol{x}\ominus\boldsymbol{y})$$

for some coefficients $\widehat{K^{\mathrm{dsh}}}(\boldsymbol{k}) \geq 0$. If K is the unanchored Sobolev space, then it can be shown that the Walsh coefficients of the digital shift-invariant kernel satisfy

$$|\widehat{K^{\mathrm{dsh}}}(\boldsymbol{k})| = b^{-2\mu_1(\boldsymbol{k})}\prod_{j=1}^{s} v_b(k_j),$$

for some function $v_b(k)$ (see (6.9) below). This leads to an expression of the digitally shift-averaged worst-case error for the unanchored Sobolev space using a QMC rule based on a digital net.

Theorem 6.6 (digitally shift-averaged worst-case error for digital nets). The digitally shift-averaged worst-case error for a digital net P in the weighted unanchored Sobolev space $H(K_{s,\gamma})$ satisfies

$$[e^{\mathrm{dsh}}_{b^m,s}(P;H(K_{s,\gamma}))]^2 = \sum_{\emptyset\neq\mathfrak{u}\subseteq\{1:s\}} \gamma_{\mathfrak{u}} \sum_{\boldsymbol{k}_{\mathfrak{u}}\in\mathcal{D}^*_{\mathfrak{u}}} b^{-2\mu_1(\boldsymbol{k}_{\mathfrak{u}})}\prod_{j\in\mathfrak{u}} v_b(k_j), \tag{6.7}$$

where

$$\mathcal{D}^*_{\mathfrak{u}} = \left\{\boldsymbol{k}_{\mathfrak{u}}\in\mathbb{N}^{|\mathfrak{u}|} : \sum_{j\in\mathfrak{u}} C_j^\top\vec{k}_j \equiv \vec{0}\right\}, \tag{6.8}$$

and

$$v_b(k) := \frac{1}{2}\left(\frac{1}{\sin^2(\kappa_{a-1}\pi/b)} - \frac{1}{3}\right), \tag{6.9}$$

where $k = \kappa_{a-1} b^{a-1} + \kappa_{a-2} b^{a-2} + \cdots + \kappa_0$ with $\kappa_{a-1} \in \{1, 2, \ldots, b-1\}$. In particular, if $b = 2$ we have $\kappa_{a-1} = 1$ and $v_b(k)$ is a constant independent of k.

For the weighted anchored Sobolev space $H(K_{s,\gamma})$ with anchor c we have

$$[e^{\mathrm{dsh}}_{b^m,s}(P; H(K_{s,\gamma}))]^2 = \sum_{\emptyset \neq \mathfrak{v} \subseteq \{1:s\}} \widetilde{\gamma}_{s,\mathfrak{v}} \sum_{\boldsymbol{k}_{\mathfrak{v}} \in \mathcal{D}^*_{\mathfrak{v}}} b^{-2\mu_1(\boldsymbol{k}_{\mathfrak{v}})} \prod_{j \in \mathfrak{v}} v_b(k_j), \tag{6.10}$$

where $v_b(k)$ is defined as in (6.9) and $\widetilde{\gamma}_{s,\mathfrak{v}}$ are the auxiliary weights defined in (5.22).

Note that (6.7) and (6.10) are essentially weighted versions of the quantity T_ϑ in (6.4), with some additional factors $v_b(k)$ and a change of weights for the anchored Sobolev space.

The proof of Theorem 6.6 is somewhat intricate and was shown by Dick and Pillichshammer (2005) for the unanchored Sobolev space and by Dick, Kuo, Pillichshammer and Sloan (2005) for the anchored Sobolev space. Instead of giving the proof, we provide an example to show how the decay rate of the Walsh coefficients is connected to the smoothness of the integrand.

Example 6.7. To illustrate how to prove a bound on the decay rate of the Walsh coefficients we consider the simplest possible example. Let $f : [0,1] \to \mathbb{R}$ be absolutely continuous with

$$f(x) = f(1) - \int_0^1 f'(t) 1_{[x,1]}(t) \, \mathrm{d}t.$$

In this example let $b = 2$, and let $k = 2^{a-1} + \kappa_{a-2} 2^{a-2} + \cdots + \kappa_0 \in \mathbb{N}$, for some $a \geq 1$. Then $\mathrm{wal}_k(x) \in \{-1, 1\}$ for all $x \in [0, 1)$. Note that wal_k changes sign on intervals of the form $[\ell 2^{-a+1}, (\ell+1) 2^{-a+1})$ for $0 \leq \ell < 2^{a-1}$, *i.e.*, it has the opposite sign on the interval $[\ell 2^{-a+1}, \ell 2^{-a+1} + 2^{-a})$ compared to the interval $[\ell 2^{-a+1} + 2^{-a}, (\ell+1) 2^{-a+1})$. Thus

$$\begin{aligned}
|\widehat{f}(k)| = \left| \int_0^1 f(x) \overline{\mathrm{wal}_k(x)} \, \mathrm{d}x \right| &\leq \sum_{\ell=0}^{2^{a-1}-1} \int_{\ell 2^{-a+1}}^{\ell 2^{-a+1} + 2^{-a}} |f(x) - f(x + 2^{-a})| \, \mathrm{d}x \\
&\leq \int_0^{1-2^{-a}} |f(x) - f(x + 2^{-a})| \, \mathrm{d}x \\
&\leq \int_0^1 |f'(t)| \int_0^{1-2^{-a}} 1_{[x, x+2^{-a})}(t) \, \mathrm{d}x \, \mathrm{d}t \\
&\leq 2^{-a} \int_0^1 |f'(t)| \, \mathrm{d}t < k^{-1} \left(\int_0^1 |f'(t)|^2 \mathrm{d}t \right)^{1/2}.
\end{aligned}$$

Thus for any $k \in \mathbb{N}_0$ we have

$$|\widehat{f}(k)| \leq \frac{1}{\max(1, k)} \|f\|_H,$$

where $\|\cdot\|_H$ denotes the norm in the unanchored Sobolev space. This can be generalized to arbitrary base b and dimensions $s > 1$.

In particular, if $K : [0,1]^2 \to \mathbb{R}$ is the reproducing kernel for the unanchored Sobolev space, then K has Walsh coefficients $\widehat{K}(k,\ell)$ which satisfy

$$|\widehat{K}(k,\ell)| \le C b^{-\mu_1(k)-\mu_1(\ell)}.$$

For the corresponding digital shift-invariant kernel defined by (6.6), it can be shown that $\widehat{K^{\mathrm{dsh}}}(k) = \widehat{K}(k,k)$. We therefore obtain

$$|\widehat{K^{\mathrm{dsh}}}(k)| = |\widehat{K}(k,k)| \le C b^{-2\mu_1(k)}.$$

6.4. CBC construction of polynomial lattice rules

Analogous to lattice rules, no optimal explicit construction of polynomial lattice rules is known beyond dimension 2. On the other hand, a CBC approach is also possible for polynomial lattice rules. An essential ingredient of the CBC algorithm is an easily computable quality criterion. We now show that the digitally shift-averaged worst-case error (6.7) can be computed efficiently.

Lemma 6.8. The digitally shift-averaged worst-case error for a digital net $P = \{\boldsymbol{t}_0, \ldots, \boldsymbol{t}_{b^m-1}\}$ in the weighted unanchored Sobolev space $H(K_{s,\boldsymbol{\gamma}})$ satisfies

$$[e^{\mathrm{dsh}}_{b^m,s}(P; H(K_{s,\boldsymbol{\gamma}}))]^2 = \frac{1}{b^m} \sum_{i=0}^{b^m-1} \sum_{\emptyset \ne \mathfrak{u} \subseteq \{1:s\}} \gamma_{\mathfrak{u}} \prod_{j\in\mathfrak{u}} \phi_b(t_{i,j}), \tag{6.11}$$

with

$$\phi_b(t) := \begin{cases} 6^{-1} - \frac{\tau_{a_0}(b-\tau_{a_0})}{b^{a_0+1}} & \text{if } 0 < t < 1, \\ 6^{-1} & \text{if } t = 0, \end{cases}$$

where, for $t \in (0,1)$ with b-adic expansion $t = \tau_1 b^{-1} + \tau_2 b^{-2} + \cdots$ (unique in the sense that infinitely many τ_r are different from $b-1$), $a_0 = -\lfloor \log_b t \rfloor$ is the smallest positive integer such that $\tau_1 = \tau_2 = \cdots = \tau_{a_0 - 1} = 0$ and $\tau_{a_0} \ne 0$.

Proof. We can write from (6.7) that

$$\begin{aligned} &[e^{\mathrm{dsh}}_{b^m,s}(P; H(K_{s,\boldsymbol{\gamma}})))]^2 \\ &= \sum_{\emptyset \ne \mathfrak{u} \subseteq \{1:s\}} \gamma_{\mathfrak{u}} b^{-m} \sum_{i=0}^{b^m-1} \sum_{\boldsymbol{k}_{\mathfrak{u}} \in \mathbb{N}^{|\mathfrak{u}|}} b^{-2\mu_1(\boldsymbol{k}_{\mathfrak{u}})} \prod_{j\in\mathfrak{u}} v_b(k_j) \mathrm{wal}_{k_j}(t_{i,j}) \\ &= b^{-m} \sum_{i=0}^{b^m-1} \sum_{\emptyset \ne \mathfrak{u} \subseteq \{1:s\}} \gamma_{\mathfrak{u}} \prod_{j\in\mathfrak{u}} \sum_{k=1}^{\infty} b^{-2\mu_1(k)} v_b(k) \mathrm{wal}_k(t_{i,j}). \end{aligned} \tag{6.12}$$

To simplify the proof we only consider the case $b = 2$ in the following. Then $v_b(k) = 1/3$ for all $k \in \mathbb{N}$. Let $\omega_2 = \mathrm{e}^{2\pi \mathrm{i}/2} = -1$. Then we have

$$\sum_{k=2^{a-1}}^{2^a-1} \mathrm{wal}_k(t) = \prod_{r=1}^{a-1} \sum_{\kappa_{r-1}=0}^{1} \omega_2^{\kappa_{r-1}\tau_r} \sum_{\kappa_{a-1}=1}^{1} \omega_2^{\kappa_{a-1}\tau_a}$$

$$= \begin{cases} 2^{a-1} & \text{if } \tau_1 = \tau_2 = \cdots = \tau_a = 0, \\ -2^{a-1} & \text{if } \tau_1 = \tau_2 = \cdots = \tau_{a-1} = 0, \tau_a \neq 0, \\ 0 & \text{otherwise.} \end{cases}$$

Thus

$$\sum_{k=1}^{\infty} 2^{-2\mu_1(k)} \mathrm{wal}_k(t) = \sum_{a=1}^{\infty} 2^{-2a} \sum_{k=2^{a-1}}^{2^a-1} \mathrm{wal}_k(t)$$

$$= \sum_{a=1}^{a_0-1} 2^{-2a} 2^{a-1} - 2^{-2a_0} 2^{a_0-1} = 2^{-1} - 2^{-a_0}(1 + 2^{-1}).$$

For $t = 0$ we have

$$\sum_{k=1}^{\infty} 2^{-2\mu_1(k)} \mathrm{wal}_k(0) = \sum_{a=1}^{\infty} 2^{-2a} 2^{a-1} = \frac{1}{2}.$$

The result now follows from (6.12). □

The expression (6.11) can be easily evaluated by a computer. Thus we can use the following CBC algorithm to construct polynomial lattice rules. In the following we write

$$[e^{\mathrm{dsh}}_{b^m,s}(P; H(K_{s,\gamma}))]^2 = [e^{\mathrm{dsh}}_{b^m,s}(\boldsymbol{q})]^2 = [e^{\mathrm{dsh}}_{b^m,s}(q_1, \ldots, q_s)]^2$$

if the digital net is a polynomial lattice rule with generating vector $\boldsymbol{q} = (q_1, \ldots, q_s) \in (G^*_{b,m})^s$, where $G^*_{b,m} = \{q \in \mathbb{Z}_b[x] \setminus \{0\} : \deg(q) < m\}$.

Algorithm 6.9 (CBC construction). Given: m, $s_{\max}$, and weights $\gamma_{\mathfrak{u}}$.

(1) Set $q_1 = 1$.

(2) For $s = 2, 3, \ldots, s_{\max}$, choose q_s in $G^*_{b,m}$ to minimize $[e^{\mathrm{dsh}}_{b^m,s}(q_1, \ldots, q_s)]^2$.

The next theorem states that the CBC algorithm yields a convergence rate close to $O(b^{-m})$.

Theorem 6.10 (optimal CBC error bound). The generating vector $\boldsymbol{q}$ constructed by the CBC algorithm, minimizing the shift-averaged worst-case error for the unanchored Sobolev space $e^{\mathrm{dsh}}_{b^m,s}(\boldsymbol{q})$ in each step, satisfies

$$[e^{\mathrm{dsh}}_{b^m,s}(\boldsymbol{q})]^2 \leq \left(\frac{1}{b^m - 1} \sum_{\emptyset \neq \mathfrak{u} \subseteq \{1:s\}} \gamma_{\mathfrak{u}}^{\lambda} (\rho_b(\lambda))^{|\mathfrak{u}|} \right)^{1/\lambda}$$

for any $1/2 < \lambda \le 1$, where

$$\rho_b(\lambda) := \begin{cases} \dfrac{1}{3^\lambda(2^{2\lambda}-2)} & \text{for } b = 2, \\ \dfrac{(b-1)(4b^2-9)^\lambda}{54^\lambda(b^{2\lambda}-b)} & \text{for } b > 2. \end{cases}$$

Proof. Since the algorithm constructs the polynomial lattice rule inductively with respect to the dimension, the proof also uses induction. The argument follows along the same lines as for the lattice rule case. To simplify the proof we only consider $b = 2$ in the following. In this case $v_2(k) = 1/3$ for all $k \in \mathbb{N}$ in (6.7).

First note that we have for $1/2 < \lambda \le 1$ that

$$\sum_{k=1}^{\infty} 2^{-2\lambda\mu_1(k)} = \sum_{a=1}^{\infty} 2^{-2\lambda a} 2^{a-1} = \frac{1}{2^{2\lambda}-2}$$

and

$$\sum_{\substack{k=1\\2^m|k}}^{\infty} 2^{-2\lambda\mu_1(k)} = \sum_{\ell=1}^{\infty} 2^{-2\lambda\mu_1(b^m\ell)} = \sum_{\ell=1}^{\infty} 2^{-2\lambda m} 2^{-2\lambda\mu_1(\ell)} = 2^{-2\lambda m}\frac{1}{2^{2\lambda}-2}.$$

For a polynomial lattice rule with generating vector $(q_1, \ldots, q_s)$ and modulus p, the set $\mathcal{D}_{\mathfrak{u}}^*$ in (6.8) is given by

$$\mathcal{D}_{\mathfrak{u}}^* = \left\{ \boldsymbol{k}_{\mathfrak{u}} \in \mathbb{N}^{|\mathfrak{u}|} : \sum_{j\in\mathfrak{u}} \mathrm{tr}_m(k_j) q_j \equiv 0 \pmod{p} \right\},$$

where for $k_j = \kappa_{j,0} + \kappa_{j,1}2 + \cdots \in \mathbb{N}_0$ we set $\mathrm{tr}_m(k_j) = \kappa_{j,0} + \kappa_{j,1}x + \cdots + \kappa_{j,m-1}x^{m-1}$.

In dimension one we have

$$\begin{aligned}[e^{\mathrm{dsh}}_{2^m,s}(q_1)]^2 &= \gamma_{\{1\}} 3^{-1} \sum_{\substack{k=1\\2^m|k}}^{\infty} 2^{-2\mu_1(k)} \\ &= 2^{-2m}\gamma_{\{1\}} 3^{-1} \sum_{\ell=1}^{\infty} 2^{-2\mu_1(\ell)} = 3^{-1} 2^{-2m-1}\gamma_{\{1\}}.\end{aligned}$$

Thus the result holds for dimension one.

Suppose, for some $1 \le d < s$, we have $(q_1, \ldots, q_d) \in (G^*_{2,m})^d$ and

$$[e^{\mathrm{dsh}}_{2^m,d}(q_1, \ldots, q_d)]^2 \le \left(\frac{1}{2^m-1} \sum_{\emptyset \ne \mathfrak{u} \subseteq \{1:d\}} \gamma_{\mathfrak{u}}^\lambda \left(\frac{1}{3^\lambda(2^{2\lambda}-2)} \right)^{|\mathfrak{u}|} \right)^{1/\lambda}, \quad (6.13)$$

for all $1/2 < \lambda \le 1$. Now we consider $(q_1, \ldots, q_d, q_{d+1})$. We have

$$\begin{aligned}
&[e^{\mathrm{dsh}}_{2^m,d+1}(q_1, \ldots, q_d, q_{d+1})]^2 \\
&= \sum_{\emptyset \neq \mathfrak{u} \subseteq \{1:d\}} \gamma_{\mathfrak{u}} 3^{-|\mathfrak{u}|} \sum_{\boldsymbol{k}_{\mathfrak{u}} \in \mathcal{D}^*_{\mathfrak{u}}} 2^{-2\mu_1(\boldsymbol{k}_{\mathfrak{u}})} + \sum_{\{d+1\} \subseteq \mathfrak{u} \subseteq \{1:d+1\}} \gamma_{\mathfrak{u}} 3^{-|\mathfrak{u}|} \sum_{\boldsymbol{k}_{\mathfrak{u}} \in \mathcal{D}^*_{\mathfrak{u}}} 2^{-2\mu_1(\boldsymbol{k}_{\mathfrak{u}})} \\
&= [e^{\mathrm{dsh}}_{2^m,d}(q_1, \ldots, q_d)]^2 + \theta(q_{d+1}), \qquad (6.14)
\end{aligned}$$

where

$$\theta(q_{d+1}) = \sum_{\{d+1\} \subseteq \mathfrak{u} \subseteq \{1:d+1\}} \gamma_{\mathfrak{u}} \sum_{\boldsymbol{k}_{\mathfrak{u}} \in \mathcal{D}^*_{\mathfrak{u}}} b^{-2\mu_1(\boldsymbol{k}_{\mathfrak{u}})}.$$

According to the algorithm, q_{d+1} is chosen such that the mean-square worst-case error $[e^{\mathrm{dsh}}_{2^m,d+1}(q_1, \ldots, q_d, q_{d+1})]^2$ is minimized. Since the only dependency on q_{d+1} is in $\theta(q_{d+1})$, we have $\theta(q_{d+1}) \le \theta(q)$ for all $q \in G^*_{2,m}$, which implies that for any $1/2 < \lambda \le 1$ we have $\theta(q_{d+1})^\lambda \le \theta(q)^\lambda$ for all $q \in G^*_{2,m}$. This leads to

$$\theta(q_{d+1}) \le \left(\frac{1}{2^m - 1} \sum_{q \in G^*_{2,m}} \theta(q)^\lambda \right)^{1/\lambda}.$$

We obtain a bound on $\theta(q_{d+1})$ through this last inequality.

For λ satisfying $1/2 < \lambda \le 1$ it follows from Jensen's inequality that

$$\theta(q)^\lambda \le \sum_{\{d+1\} \subseteq \mathfrak{u} \subseteq \{1:d+1\}} \gamma_{\mathfrak{u}}^\lambda 3^{-\lambda|\mathfrak{u}|} \sum_{\boldsymbol{k}_{\mathfrak{u}} \in \mathcal{D}^*_{\mathfrak{u}}} 2^{-2\lambda\mu_1(\boldsymbol{k}_{\mathfrak{u}})}.$$

The condition $\boldsymbol{k}_{\mathfrak{u}} \in \mathcal{D}^*_{\mathfrak{u}}$ is equivalent to the equation

$$\mathrm{tr}_m(k_1)q_1 + \cdots + \mathrm{tr}_m(k_d)q_d \equiv -\mathrm{tr}_m(k_{d+1})q \pmod{p}.$$

If k_{d+1} is a multiple of 2^m, then $\mathrm{tr}_m(k_{d+1}) = 0$ and the corresponding term in the sum is independent of q. If k_{d+1} is not a multiple of 2^m, then $\mathrm{tr}_m(k_{d+1})$ can be any polynomial in $G^*_{2,m}$. Moreover, since $q \neq 0$ and p is irreducible, $\mathrm{tr}_m(k_{d+1})q$ is never a multiple of p.

By averaging over all $q \in G^*_{2,m}$, with the above discussion in mind, we obtain

$$\begin{aligned}
&\frac{1}{2^m - 1} \sum_{q \in G^*_{2,m}} \theta(q)^\lambda \\
&\le \sum_{\substack{k_{d+1}=1 \\ 2^m | k_{d+1}}}^\infty 2^{-2\lambda\mu_1(k_{d+1})} \sum_{\emptyset \neq \mathfrak{u} \subseteq \{1:d\}} \gamma^\lambda_{\mathfrak{u} \cup \{d+1\}} 3^{-\lambda(|\mathfrak{u}|+1)} \sum_{\substack{\boldsymbol{k}_{\mathfrak{u}} \in \mathbb{N}^{|\mathfrak{u}|} \\ \mathrm{tr}_m(\boldsymbol{k}_{\mathfrak{u}}) \cdot \boldsymbol{q}_{\mathfrak{u}} \equiv 0 \pmod{p}}} 2^{-2\lambda\mu_1(\boldsymbol{k}_{\mathfrak{u}})} \\
&\quad + \sum_{\substack{k_{d+1}=1 \\ 2^m \nmid k_{d+1}}}^\infty \frac{2^{-2\lambda\mu_1(k_{d+1})}}{2^m - 1} \sum_{\emptyset \neq \mathfrak{u} \subseteq \{1:d\}} \gamma^\lambda_{\mathfrak{u} \cup \{d+1\}} 3^{-\lambda(|\mathfrak{u}|+1)} \sum_{\substack{\boldsymbol{k}_{\mathfrak{u}} \in \mathbb{N}^{|\mathfrak{u}|} \\ \mathrm{tr}_m(\boldsymbol{k}_{\mathfrak{u}}) \cdot \boldsymbol{q}_{\mathfrak{u}} \not\equiv 0 \pmod{p}}} 2^{-2\lambda\mu_1(\boldsymbol{k}_{\mathfrak{u}})}
\end{aligned}$$

$$\le 2^{-2\lambda m}\frac{1}{3^{\lambda(|\mathrm{u}|+1)}(2^{2\lambda}-2)}\sum_{\emptyset\neq\mathrm{u}\subseteq\{1:d\}}\gamma^{\lambda}_{\mathrm{u}\cup\{d+1\}}\sum_{\substack{\boldsymbol{k}_{\mathrm{u}}\in\mathbb{N}^{|\mathrm{u}|}\\ \mathrm{tr}_m(\boldsymbol{k}_{\mathrm{u}})\cdot\boldsymbol{q}_{\mathrm{u}}\equiv 0\,(\mathrm{mod}\,p)}}2^{-2\lambda\mu_1(\boldsymbol{k}_{\mathrm{u}})}$$

$$+\frac{1}{2^m-1}\frac{1-2^{-2\lambda m}}{3^{\lambda(|\mathrm{u}|+1)}(2^{2\lambda}-2)}\sum_{\emptyset\neq\mathrm{u}\subseteq\{1:d\}}\gamma^{\lambda}_{\mathrm{u}\cup\{d+1\}}\sum_{\substack{\boldsymbol{k}_{\mathrm{u}}\in\mathbb{N}^{|\mathrm{u}|}\\ \mathrm{tr}_m(\boldsymbol{k}_{\mathrm{u}})\cdot\boldsymbol{q}_{\mathrm{u}}\not\equiv 0\,(\mathrm{mod}\,p)}}2^{-2\lambda\mu_1(\boldsymbol{k}_{\mathrm{u}})}$$

$$\le\frac{1}{2^m-1}\frac{1}{3^{\lambda(|\mathrm{u}|+1)}(2^{2\lambda}-2)}\sum_{\emptyset\neq\mathrm{u}\subseteq\{1:d\}}\gamma^{\lambda}_{\mathrm{u}\cup\{d+1\}}\sum_{\boldsymbol{k}_{\mathrm{u}}\in\mathbb{N}^{|\mathrm{u}|}}2^{-2\lambda\mu_1(\boldsymbol{k}_{\mathrm{u}})}$$

$$=\frac{1}{2^m-1}\sum_{\{d+1\}\subseteq\mathrm{u}\subseteq\{1:d+1\}}\gamma^{\lambda}_{\mathrm{u}}\left(\frac{1}{3^{\lambda}(2^{2\lambda}-2)}\right)^{|\mathrm{u}|}.$$

Thus we conclude that

$$\theta(q_{d+1})\le\left(\frac{1}{2^m-1}\sum_{\{d+1\}\subseteq\mathrm{u}\subseteq\{1:d+1\}}\gamma^{\lambda}_{\mathrm{u}}\left(\frac{1}{3^{\lambda}(2^{2\lambda}-2)}\right)^{|\mathrm{u}|}\right)^{1/\lambda}$$

which, together with (6.13) and (6.14), yields

$$\begin{aligned}[e^{\mathrm{dsh}}_{2^m,d+1}(q_1,\ldots,q_d,q_{d+1})]^{2\lambda}&=[e^{\mathrm{dsh}}_{b^m,d}(q_1,\ldots,q_d)]^{2\lambda}+\theta(q_{d+1})^{\lambda}\\&\le\frac{1}{2^m-1}\sum_{\emptyset\neq\mathrm{u}\subseteq\{1:d+1\}}\gamma^{\lambda}_{\mathrm{u}}\left(\frac{1}{3^{\lambda}(2^{2\lambda}-2)}\right)^{|\mathrm{u}|}.\end{aligned}$$

Hence the result follows by induction. □

The choice of q_d at each step is independent of λ and therefore the error bound holds for all values of $\lambda\in(1/2,1]$. The optimal convergence rate close to $O(b^{-m})$ is obtained with $\lambda\to 1/2$.

6.5. Fast CBC construction of polynomial lattice rules

In the previous subsection we showed how to construct, for a given modulus p, a generating vector $\boldsymbol{q}$ which yields a polynomial lattice rule that achieves a small integration error using a CBC algorithm. We have seen in Section 5.5 that FFT can be used to reduce the construction cost of the CBC construction of lattice rules. We now show that the same technique can be used for polynomial lattice rules by outlining the key differences from lattice rules. For product weights, it is possible to construct, for a given polynomial p with $\deg(p)=m$, an s-dimensional generating vector $\boldsymbol{q}$ in $O(s\,m\,b^m)$ operations, compared to $O(s^2\,b^{2m})$ operations for a naive implementation of the CBC algorithm.

Let b be a prime. Throughout this subsection we consider the polynomial $p\in\mathbb{Z}_b[x]$ with $\deg(p)=m$ to be irreducible. We restrict ourselves to product weights $\gamma_{\mathfrak{u}} = \prod_{j\in\mathfrak{u}} \gamma_j$, for a sequence of positive real numbers γ_j. The extension to POD weights can be done as in the lattice rule case: see Section 5.6.

Using Algorithm 6.9 we construct, component by component, a generating vector $\boldsymbol{q} = (q_1, \ldots, q_s) \in (G_{b,m}^*)^s$ such that for all $1 \le d \le s$ the quantity $e_d^2(q_d) := [e_{b^m,d}^{\mathrm{dsh}}(q_1, \ldots, q_{d-1}, q_d)]^2$ is minimized with respect to q_d for fixed $q_1, \ldots, q_{d-1}$. Assume that $q_1, \ldots, q_{d-1}$ are already constructed. Then we have to find $q \in G_{b,m}^*$ which minimizes

$$e_d^2(q) = [e_{b^m,d-1}^{\mathrm{dsh}}(q_1, \ldots, q_{d-1})]^2 + \gamma_d \phi_b(0) \frac{\prod_{j=1}^{d-1}(1 + \gamma_j \phi_b(0))}{b^m}$$
$$+ \frac{\gamma_d}{b^m} \sum_{i=1}^{b^m-1} \eta_{d-1}(i) \phi_b\left(v_m\left(\frac{q(x)i(x)}{p(x)} \right)\right),$$

where we used Theorem 2.21, separated out the $i = 0$ term and introduced the column vector $\boldsymbol{\eta}_{d-1} = (\eta_{d-1}(1), \ldots, \eta_{d-1}(b^m - 1))^\top$, where

$$\eta_{d-1}(i) = \prod_{j=1}^{d-1}(1 + \gamma_j \phi_b(t_{i,j})), \quad i = 1, \ldots, b^m - 1.$$

Note that the vector $\boldsymbol{\eta}_{d-1}$ remains fixed for fixed d regardless of the choice of polynomial q. We set $\boldsymbol{\eta}_0 = (1, \ldots, 1) \in \mathbb{R}^{b^m-1}$. The vector $\boldsymbol{\eta}_{d-1}$ can be computed as part of the CBC algorithm with the update formula $\eta_{d-1}(i) = \eta_{d-2}(i)(1 + \gamma_{d-1}\phi_b(t_{i,d-1}))$.

For short we write $\omega := \phi_b \circ v_m$ from now on. We define the $(b^m - 1) \times (b^m - 1)$ matrix

$$\boldsymbol{\Omega}_p := \left(\omega\left(\frac{q(x)i(x)}{p(x)} \right)\right)_{\substack{q=1,\ldots,b^m-1 \\ i=1,\ldots,b^m-1}}.$$

Further, let $\boldsymbol{e}_d^2 = (e_d^2(1), \ldots, e_d^2(b^m - 1))^\top$ be the column vector collecting the quantities $e_d^2(q)$ and let $\boldsymbol{1}$ be the $(b^m - 1)$-dimensional column vector whose components are all 1. Then we have

$$\boldsymbol{e}_d^2 = \boldsymbol{e}_{d-1}^2 + \frac{\prod_{j=1}^{d}(1 + \gamma_j \phi_b(0))}{b^m} \boldsymbol{1} + \frac{\gamma_d}{b^m} \boldsymbol{\Omega}_p \boldsymbol{\eta}_{d-1}.$$

As we are only interested in which q minimizes $e_d^2(q)$, but not the value of $e_d^2(q)$ itself, we only need to compute $\boldsymbol{\Omega}_p \boldsymbol{\eta}_{d-1}$.

The entries of the matrix $\boldsymbol{\Omega}_p$ are of the form $\omega\left(\frac{q(x)i(x)}{p(x)}\right)$, where the product of the non-zero polynomials q and i has to be evaluated in the field $\mathbb{Z}_b[x]/(p)$, and thus modulo p. Since the multiplicative group of every finite field is cyclic, we can find a primitive element g which generates all elements of the multiplicative group $(\mathbb{Z}_b[x]/(p))^*$ $(= (\mathbb{Z}_b[x]/(p)) \setminus \{0\})$. That is, there is a $g \in (\mathbb{Z}_b[x]/(p))^*$ such that $(\mathbb{Z}_b[x]/(p))^* = \{g^0, g^1, g^2, \dots, g^{b^m-1}\}$. Thus, we can write the product of any non-zero polynomials $q, i \in \mathbb{Z}_b[x]/(p)$ as a power of the polynomial g. Nuyens and Cools (2006*a*, 2006*b*) then suggest permuting the rows of $\boldsymbol{\Omega}_p$ by the positive powers of the primitive polynomial g and the columns by the negative powers of the same primitive polynomial. This procedure is often called Rader transform, since it goes back to an idea of Rader (1968).

We now describe the Rader transform in detail. Let $g(x)$ be a primitive element in $(\mathbb{Z}_b[x]/(p))^*$. We define a $(b^m - 1) \times (b^m - 1)$ matrix $\Pi(g) = (\pi_{k,\ell}(g))_{1 \le k,\ell < b^m}$, where

$$\pi_{k,\ell}(g) = \begin{cases} 1 & \text{if } k(x) \equiv g(x)^\ell \pmod{p(x)}, \\ 0 & \text{otherwise.} \end{cases}$$

Here $k(x)$ denotes the polynomial which is associated with the integer k. Since g is a primitive element it follows that each row and each column of $\Pi(g)$ has exactly one entry, which is 1, and the remaining entries are 0. Further, $\Pi(g)\Pi(g)^\top = I$, the identity matrix. In fact, the matrix $\Pi(g)$ is a permutation matrix. That is, for any $(b^m - 1) \times (b^m - 1)$ matrix C, $\Pi(g)C$ just changes the order of the rows of C and $C\Pi(g)$ only changes the order of the columns of C.

Let $C = (c_{k,\ell})_{1 \le k,\ell < b^m}$ and

$$C = (\Pi(g))^\top \boldsymbol{\Omega}_p \Pi(g^{-1}).$$

Then

$$c_{k,\ell} = \sum_{u,v=1}^{b^m-1} \pi_{u,k}(g) \omega\left(\frac{u(x)v(x)}{p(x)}\right) \pi_{v,\ell}(g^{-1}) = \omega\left(\frac{g(x)^k g(x)^{-\ell}}{p(x)}\right).$$

Let

$$c_r = \omega\left(\frac{g(x)^r}{p(x)}\right).$$

Observe that $c_r = c_{r'}$ for all $r, r' \in \mathbb{Z}$ with $r \equiv r' \pmod{b^m - 1}$, since $g(x)^{b^m-1} = 1$. Then we have $c_{k,\ell} = c_{k-\ell}$, and therefore C is circulant (*cf.* Section 5.5).

Note that the circulant matrix C is fully determined by its first column $\boldsymbol{c} = (c_0, c_1, c_2, \dots, c_{b^m-2})^\top$. Such matrices have a similarity transform which has the Fourier matrix as its eigenvectors. For $n \in \mathbb{N}$, let $F_n := (f_{k,\ell})_{k,\ell=0}^{n-1}$ be the Fourier matrix of order n given by $f_{k,\ell} = \omega_n^{-k\ell}$, where $\omega_n = \mathrm{e}^{2\pi\mathrm{i}/n}$. Note that F_n is symmetric and $\frac{1}{n} F_n \overline{F}_n = I$, the identity matrix. Furthermore, let $\mathrm{diag}(a_1, \dots, a_n)$ be the $n \times n$ diagonal matrix $A = (A_{i,j})_{i,j=1}^n$ with $A_{i,i} = a_i$ for $1 \le i \le n$ and $A_{i,j} = 0$ for $i \ne j$. Then we have (*cf.* Section 5.5)

$$\boldsymbol{\Omega}_p = \frac{1}{b^m - 1} \Pi(g) \overline{F}_{b^m-1} D F_{b^m-1} \Pi(g^{-1})^\top,$$

where $\Pi(g), \Pi(g^{-1})^\top$ are permutation matrices, F_{b^m-1} is a Fourier matrix, $\overline{F}_{b^m-1}$ its complex conjugate, and D is the diagonal matrix $D = \mathrm{diag}(\boldsymbol{F}_{b^m-1}\,\boldsymbol{c})$.

For any vector $\boldsymbol{x} = (x_1, \ldots, x_{b^m-1})^\top \in (\mathbb{C}^{b^m-1})^\top$, the matrix–vector multiplications $\Pi(g)\boldsymbol{x}$, $D\boldsymbol{x}$, and $\Pi(g^{-1})^\top\boldsymbol{x}$ can be done in $O(b^m)$ operations (in fact, in the implementation of the algorithm the permutations using $\Pi(g)$ and $\Pi(g^{-1})$ can be avoided altogether: see Section 5.5). Further, $F_{b^m-1}\boldsymbol{x}$ and $\overline{F}_{b^m-1}\boldsymbol{x}$ can be computed in $O(mb^m)$ operations using the fast Fourier transform.

Therefore a fast CBC algorithm requires only $O(s\,m\,b^m)$ operations by using $O(b^m)$ memory space. This is a significant speed-up compared to a straightforward implementation. Only through this reduction of the construction cost does the CBC algorithm become applicable for the generation of polynomial lattice point sets with reasonably large cardinality.

6.6. *Integration of smooth functions*

It is well understood that the convergence rate for numerical integration depends on the smoothness of the integrand. In the following we consider the weighted unanchored Sobolev space of smoothness $\alpha > 1$, $H(K_{s,\gamma,\alpha})$, already encountered in Section 5.9; see also Example 3.8 for the one-dimensional case. The reproducing kernel is given by (5.34), and the inner product is

$$\langle f, g\rangle_{s,\gamma,\alpha} = \sum_{\mathfrak{u}\subseteq\{1:s\}} \gamma_{\mathfrak{u}}^{-1} \sum_{\mathfrak{v}\subseteq\mathfrak{u}} \sum_{\boldsymbol{r}_{\mathfrak{v}}\in\{1:\alpha-1\}^{|\mathfrak{v}|}}$$
$$\int_{[0,1]^{|\mathfrak{u}\setminus\mathfrak{v}|}} \left(\int_{[0,1]^{|\{1:s\}\setminus\mathfrak{u}|}} \int_{[0,1]^{|\mathfrak{v}|}} \frac{\partial^{|\boldsymbol{r}_{\mathfrak{v}}|_1+\alpha|\mathfrak{u}\setminus\mathfrak{v}|} f}{\partial \boldsymbol{x}_{(\boldsymbol{0},\boldsymbol{r},\boldsymbol{\alpha})}} \,\mathrm{d}\boldsymbol{x}_{\mathfrak{v}}\,\mathrm{d}\boldsymbol{x}_{\{1:s\}\setminus\mathfrak{u}}\right)$$
$$\times \left(\int_{[0,1]^{|\{1:s\}\setminus\mathfrak{u}|}} \int_{[0,1]^{|\mathfrak{v}|}} \frac{\partial^{|\boldsymbol{r}_{\mathfrak{v}}|_1+\alpha|\mathfrak{u}\setminus\mathfrak{v}|} g}{\partial \boldsymbol{x}_{(\boldsymbol{0},\boldsymbol{r},\boldsymbol{\alpha})}} \,\mathrm{d}\boldsymbol{x}_{\mathfrak{v}}\,\mathrm{d}\boldsymbol{x}_{\{1:s\}\setminus\mathfrak{u}}\right) \mathrm{d}\boldsymbol{x}_{\mathfrak{u}\setminus\mathfrak{v}},$$

where $|\boldsymbol{r}_{\mathfrak{v}}|_1 := \sum_{j\in\mathfrak{v}} |r_j|$, and

$$\frac{\partial^{|\boldsymbol{r}_{\mathfrak{v}}|_1+\alpha|\mathfrak{u}\setminus\mathfrak{v}|} f}{\partial \boldsymbol{x}_{(\boldsymbol{0},\boldsymbol{r},\boldsymbol{\alpha})}}$$

stands for the partial derivatives of f of order r_j for $j \in \mathfrak{v}$, of order α for $j \in \mathfrak{u}\setminus\mathfrak{v}$ and order 0 otherwise, where $\boldsymbol{r}_{\mathfrak{v}} = (r_j)_{j\in\mathfrak{v}}$. Let the corresponding norm be denoted by $\|f\|_{s,\gamma,\alpha} = \sqrt{\langle f, f\rangle_{s,\gamma,\alpha}}$.

For functions in $H(K_{s,\gamma,\alpha})$ it is known that a convergence rate of order $n^{-\alpha+\delta}$ for any $\delta > 0$ can be achieved with a suitable algorithm. In the previous subsections we have used the decay of the Walsh coefficients for functions in $H(K_{s,\gamma,1})$ to show a convergence rate of order $n^{-1+\delta}$. To generalize this approach to higher order we need to investigate how the

Walsh coefficients of smooth functions decay. Since the proof is involved we describe the main idea, leaving out technical details.

To simplify the notation we restrict ourselves to base $b = 2$. Let $k \in \mathbb{N}$ be given by $k = 2^{a_1-1} + 2^{a_2-1} + \cdots + 2^{a_\nu-1}$ for $a_1 > a_2 > \cdots > a_\nu > 0$. We define for $\alpha \geq 2$ an integer

$$\mu_\alpha(k) := \begin{cases} a_1 + a_2 + \cdots + a_{\min\{\alpha,\nu\}} & \text{if } k \in \mathbb{N}, \\ 0 & \text{if } k = 0. \end{cases}$$

This generalizes the definition of $\mu_1(k)$ given by (6.3). Note that the Walsh functions in base $b = 2$ only take on the values 1 or -1. Thus in the following we write wal_k instead of $\overline{\mathrm{wal}_k}$.

Let $f \in H(K_{s,\gamma,\alpha})$ and $J_k(x) := \int_0^x \mathrm{wal}_k(t)\,\mathrm{d}t$. Then

$$\begin{aligned} \widehat{f}(k) = \int_0^1 f(x)\mathrm{wal}_k(x)\,\mathrm{d}x &= \big[f(x)J_k(x)\big]_{x=0}^1 - \int_0^1 f'(x)J_k(x)\,\mathrm{d}x \\ &= -\int_0^1 f'(x)J_k(x)\,\mathrm{d}x, \end{aligned} \tag{6.15}$$

as $\int_0^1 \mathrm{wal}_k(x)\,\mathrm{d}x = 0$.

We would now like to relate the Walsh coefficient $\widehat{f}(k)$ to some Walsh coefficient of f'. This is done by using the Walsh series expansion of J_k.

Let $k' := k - 2^{a_1-1}$, and hence $0 \leq k' < 2^{a_1-1}$. Then it can be shown that

$$J_k(x) = 2^{-a_1-1}\left(\mathrm{wal}_{k'}(x) - \sum_{c=1}^{\infty} 2^{-c}\mathrm{wal}_{2^{a_1+c-1}+k}(x)\right).$$

Substituting this into (6.15), we obtain approximately

$$\widehat{f}(k) \approx -2^{-a_1}\int_0^1 f'(x)\mathrm{wal}_{k'}(x)\,\mathrm{d}x = -2^{-a_1}\widehat{f'}(k').$$

In fact we obtain an infinite sum on the right-hand side, but the main term is the first one: the remaining terms can be bounded by a suitable expression.

The above step can be repeated. To do so, let $k'' := k' - 2^{a_2-1}$, $k''' := k'' - 2^{a_3-1}$ and so on, where $k^{(\nu)} = 0$ and $k^{(\tau)} = 0$ for $\tau \geq \nu$. We can repeat the last step τ times until either $f^{(\tau)}$ is no longer differentiable, or $k^{(\tau)} = 0$, that is, we can repeat it $\min(\alpha,\nu)$ times. Hence

$$\begin{aligned} \widehat{f}(k) &\approx 2^{-a_1}\widehat{f'}(k^{(1)}) \approx 2^{-a_1-a_2}\widehat{f''}(k^{(2)}) \\ &\;\;\vdots \\ &\approx 2^{-a_1-\cdots-a_{\min(\alpha,\nu)}}\widehat{f^{(\min(\alpha,\nu))}}(k^{(\min(\alpha,\nu))}). \end{aligned}$$

Taking the absolute value and using some estimation we obtain

$$|\widehat{f}(k)| \lessapprox 2^{-a_1-\cdots-a_{\min(\alpha,\nu)}}|\widehat{f}^{(\min(\alpha,\nu))}(k^{(\min(\alpha,\nu))})|$$
$$\leq 2^{-\mu_\alpha(k)}|\widehat{f}^{(\min(\alpha,\nu))}(k^{(\min(\alpha,\nu))})| \leq 2^{-\mu_\alpha(k)} \int_0^1 |f^{(\min(\alpha,\nu))}(x)| \,\mathrm{d}x,$$

where we used

$$\begin{aligned}|\widehat{f}^{(\min(\alpha,\nu))}(k^{(\min(\alpha,\nu))})| &= \left|\int_0^1 f^{(\min(\alpha,\nu))}(x)\overline{\mathrm{wal}_{k^{(\min(\alpha,\nu))}}(x)}\,\mathrm{d}x\right| \\ &\leq \int_0^1 |f^{(\min(\alpha,\nu))}(x)||\overline{\mathrm{wal}_{k^{(\min(\alpha,\nu))}}(x)}|\,\mathrm{d}x \\ &= \int_0^1 |f^{(\min(\alpha,\nu))}(x)|\,\mathrm{d}x.\end{aligned}$$

Thus if f is α-times differentiable, we obtain

$$|\widehat{f}(k)| \lessapprox C_f 2^{-\mu_\alpha(k)},$$

where $C_f > 0$ is a constant which depends only on f and α. By some modification of the above approach it can be shown that the constant C_f can be replaced by a constant which depends only on α (but not on f) and the norm of f, that is, we have

$$|\widehat{f}(k)| \leq C_\alpha \|f\|_{1,(1),\alpha} 2^{-\mu_\alpha(k)}.$$

The same holds for dimension $s > 1$ and integer base $b > 2$, where the constant additionally depends on s and b, that is,

$$|\widehat{f}(\boldsymbol{k})| \leq C_{\alpha,b,s}\|f\|_{s,\mathbf{1},\alpha} b^{-\mu_\alpha(\boldsymbol{k})},$$

where $\mu_\alpha(\boldsymbol{k}) = \mu_\alpha(k_1) + \cdots + \mu_\alpha(k_s)$ for $\boldsymbol{k} = (k_1, \ldots, k_s)$ and $\mathbf{1}$ is the set of weights which are 1 for each projection. For some values of b, this constant $C_{\alpha,b,s}$ goes to 0 exponentially as s increases.

We can now begin to investigate numerical integration of functions in $H(K_{s,\gamma,\alpha})$ using digital nets. From Theorem 6.4 we conclude that the integration error for a digital net $P = \{\boldsymbol{t}_0, \ldots, \boldsymbol{t}_{b^m-1}\}$ satisfies

$$\left|\int_{[0,1]^s} f(\boldsymbol{x})\,\mathrm{d}\boldsymbol{x} - \frac{1}{b^m}\sum_{i=0}^{b^m-1} f(\boldsymbol{t}_i)\right| \leq C_{\alpha,b,s}\|f\|_{s,\mathbf{1},\alpha} \sum_{\boldsymbol{k}\in\mathcal{D}\setminus\{\mathbf{0}\}} b^{-\mu_\alpha(\boldsymbol{k})},$$

where $\mathcal{D}$ is the dual net defined by (6.2).

The last inequality separates the contribution of the function from the contribution of the QMC rule, that is, $\|f\|_{s,\mathbf{1},\alpha}$ depends only on the function f but not on the digital net, whereas $\sum_{\boldsymbol{k}\in\mathcal{D}\setminus\{\mathbf{0}\}} b^{-\mu_\alpha(\boldsymbol{k})}$ depends only on the generating matrices of the digital net and not on the function itself (only on the smoothness of f, *i.e.*, it is the same for all functions that have smoothness α). Therefore, when considering the integration error we can now focus on the term $\sum_{\boldsymbol{k}\in\mathcal{D}\setminus\{\boldsymbol{0}\}} b^{-\mu_\alpha(\boldsymbol{k})}$.

It is possible to also introduce weights in the approach above. A precise result for the unanchored weighted Sobolev space $H(K_{s,\gamma,\alpha})$ for $\alpha > 1$ is the following.

Theorem 6.11. The square worst-case error for multivariate integration in the weighted unanchored Sobolev space $H(K_{s,\gamma,\alpha})$ of smoothness $\alpha > 1$ using a QMC rule based on a digital net P can be bounded by

$$[e(P, H(K_{s,\gamma,\alpha}))]^2 \leq \sum_{\emptyset\neq\mathfrak{u}\subseteq\{1:s\}} \gamma_\mathfrak{u} (D_{\alpha,b})^{|\mathfrak{u}|} \Bigg(\sum_{\boldsymbol{k}_\mathfrak{u}\in\mathcal{D}_\mathfrak{u}^*} b^{-\mu_\alpha(\boldsymbol{k}_\mathfrak{u})}\Bigg)^2,$$

where $\mathcal{D}_\mathfrak{u}^*$ is defined by (6.8) and where

$$D_{\alpha,b} = \max_{1\leq\nu\leq\alpha}\Bigg(\sum_{\tau=\nu}^{\alpha}\Bigg(\bigg(1+\frac{1}{b}+\frac{1}{b(b+1)}\bigg)^{(\tau-2)_+}(2\sin(\pi/b))^{-\tau}\Bigg)^2 b^{-(\tau-\nu)}$$
$$+2\bigg(1+\frac{1}{b}+\frac{1}{b(b+1)}\bigg)^{2\alpha-2}(2\sin(\pi/b))^{-2\alpha}b^{-2(\alpha-\nu)}\Bigg).$$

Baldeaux and Dick (2009) investigated the value of the constant $D_{\alpha,b}$ and showed that for $b = 2$ and 3 and $2 \leq \alpha \leq 8$ the constant $D_{\alpha,b} < 1$.

Theorem 6.11 holds for any digital net, that is, any point set obtained via the digital construction scheme as described in Section 2.6, including the higher-order digital nets to be introduced in the next subsection.

6.7. Higher-order digital nets

The aim is now to find digital nets, that is, generating matrices $C_1, \ldots, C_s \in \mathbb{Z}_b^{w\times m}$ such that

$$\sum_{\boldsymbol{k}\in\mathcal{D}\setminus\{\boldsymbol{0}\}} b^{-\mu_\alpha(\boldsymbol{k})} = O(n^{-\alpha}(\log n)^{\alpha s}),$$

where the number of cubature points is $n = b^m$. Note that we now use generating matrices with w rows, and we usually choose $w \approx \alpha m$ as explained below.

Roughly speaking, the sum $\sum_{\boldsymbol{k}\in\mathcal{D}\setminus\{\boldsymbol{0}\}} b^{-\mu_\alpha(\boldsymbol{k})}$ is dominated by its largest term. To find this largest term, define

$$\mu_\alpha^*(C_1,\ldots,C_s) = \min_{\boldsymbol{k}\in\mathcal{D}\setminus\{\boldsymbol{0}\}} \mu_\alpha(\boldsymbol{k}).$$

The dependence on the generating matrices $C_1, \ldots, C_s$ on the right-hand side of the above equation is via the dual net $\mathcal{D} = \mathcal{D}(C_1, \ldots, C_s)$. The largest term in $\sum_{\boldsymbol{k}\in\mathcal{D}\setminus\{\boldsymbol{0}\}} b^{-\mu_\alpha(\boldsymbol{k})}$ is then $b^{-\mu_\alpha^*(C_1,\ldots,C_s)}$.

In order to achieve a convergence of almost $n^{-\alpha} = b^{-\alpha m}$ we must have that the largest term in $\sum_{\boldsymbol{k}\in\mathcal{D}\setminus\{\boldsymbol{0}\}} b^{-\mu_\alpha(\boldsymbol{k})}$ is also of this order, that is, we must have $\mu_\alpha^*(C_1,\ldots,C_s) \approx \alpha m$ (or say $\mu_\alpha^*(C_1,\ldots,C_s) > \alpha m - t$ for some constant t independent of m). That this condition is also sufficient is quite technical and was shown in Dick (2008, Lemma 5.2).

We can use some analogy to find matrices $C_1,\ldots,C_s \in \mathbb{Z}_b^{w\times m}$ which achieve $\mu_\alpha^*(C_1,\ldots,C_s) \approx \alpha m$. Let $C_j = (\vec{c}_{j,1}^{\top},\ldots,\vec{c}_{j,w}^{\top})^{\top}$, that is, $\vec{c}_{j,\ell} \in \mathbb{Z}_b^m$ is the ℓth row of C_j. Then the matrices $C_1,\ldots,C_s$ generate a classical digital (t,m,s)-net if, for all $i_1,\ldots,i_s \geq 0$ with $i_1 + \cdots + i_s \leq m - t$, the vectors

$$\vec{c}_{1,1},\ldots,\vec{c}_{1,i_1},\ldots,\vec{c}_{s,1},\ldots,\vec{c}_{s,i_s}$$

are linearly independent over $\mathbb{Z}_b$.

Now assume $C_1,\ldots,C_s$ generate a classical digital (t,m,s)-net and that we are given a $\boldsymbol{k} \in \mathbb{N}_0^s \setminus \{\boldsymbol{0}\}$ with $\mu_1(\boldsymbol{k}) \leq m - t$. Let $i_j = \mu_1(k_j)$ for $j = 1,\ldots,s$, then $C_1^{\top}\vec{k}_1 + \cdots + C_s^{\top}\vec{k}_s$ is a linear combination of the vectors $\vec{c}_{1,1},\ldots,\vec{c}_{1,i_1},\ldots,\vec{c}_{s,1},\ldots,\vec{c}_{s,i_s}$. As $\boldsymbol{k} \neq \boldsymbol{0}$ and $i_1 + \cdots + i_s \leq m - t$, which implies that $\vec{c}_{1,1},\ldots,\vec{c}_{1,i_1},\ldots,\vec{c}_{s,1},\ldots,\vec{c}_{s,i_s}$ are linearly independent, it follows that $C_1^{\top}\vec{k}_1 + \cdots + C_s^{\top}\vec{k}_s \neq \boldsymbol{0} \in \mathbb{Z}_b^m$. Thus $\boldsymbol{k} \notin \mathcal{D}$. This shows that if $C_1,\ldots,C_s$ generate a classical digital (t,m,s)-net and $\boldsymbol{k} \in \mathcal{D} \setminus \{\boldsymbol{0}\}$, then $\mu_1(\boldsymbol{k}) > m - t$. This is precisely the type of result described above which we also want to have for $\alpha > 1$.

We now want to generalize this linear independence condition to $\alpha > 1$, that is, we want to have that if $\boldsymbol{k} \in \mathbb{N}_0^s \setminus \{\boldsymbol{0}\}$ with $\mu_\alpha(\boldsymbol{k}) \leq \alpha m - t$, then the generating matrices should have linearly independent rows such that $C_1^{\top}\vec{k}_1 + \cdots + C_s^{\top}\vec{k}_s \neq \boldsymbol{0} \in \mathbb{Z}_b^m$. Let $\boldsymbol{k} = (k_1,\ldots,k_s)$, where $k_j = \kappa_{j,1}b^{a_{j,1}-1} + \cdots + \kappa_{j,\nu_j}b^{a_{j,\nu_j}-1}$, with $a_{j,1} > \cdots > a_{j,\nu_j} > 0$ and $0 < \kappa_{j,1},\ldots,\kappa_{j,\nu_j} < b$. First note that if $w < \alpha m - t$, then $\boldsymbol{k} = (b^w,0,\ldots,0) \in \mathcal{D}$, but $\mu_\alpha(\boldsymbol{k}) = w + 1 \leq \alpha m - t$. In order to avoid this problem we may choose $w = \alpha m$. Hence we may now assume that $a_{j,1} \leq w = \alpha m$ for $j = 1,\ldots,s$, as otherwise $\mu_\alpha(\boldsymbol{k}) > \alpha m$ already and no independence condition on the generating matrices is required in this case.

Now $C_1^{\top}\vec{k}_1 + \cdots + C_s^{\top}\vec{k}_s$ is a linear combination of the rows

$$\vec{c}_{1,a_{1,1}},\ldots,\vec{c}_{1,a_{1,\nu_1}},\ldots,\vec{c}_{s,a_{s,1}},\ldots,\vec{c}_{s,a_{s,\nu_s}}.$$

Thus, if these rows are linearly independent, then

$$C_1^{\top}\vec{k}_1 + \cdots + C_s^{\top}\vec{k}_s \neq \boldsymbol{0} \in \mathbb{Z}_b^m,$$

and therefore $\boldsymbol{k} \notin \mathcal{D}$.

Therefore, if $C_1,\ldots,C_s \in \mathbb{Z}_b^{\alpha m\times m}$ are such that for all choices of $a_{j,1} > \cdots > a_{j,\nu_j} > 0$ for $j = 1,\ldots,s$, with

$$a_{1,1} + \cdots + a_{1,\min(\alpha,\nu_1)} + \cdots + a_{s,1} + \cdots + a_{s,\min(\alpha,\nu_s)} \leq \alpha m - t,$$

the rows

$$\vec{c}_{1,a_{1,1}}, \ldots, \vec{c}_{1,a_{1,\nu_1}}, \ldots, \vec{c}_{s,a_{s,1}}, \ldots, \vec{c}_{s,a_{s,\nu_s}}$$

are linearly independent, then $\boldsymbol{k} \in \mathcal{D} \setminus \{\boldsymbol{0}\}$ implies that $\mu_\alpha(\boldsymbol{k}) > \alpha m - t$. (Note that we also include the case where some $\nu_j = 0$, in which case we just set $a_{j,1} + \cdots + a_{j,\min(\alpha,\nu_j)} = 0$.)

We can now formally define such digital nets for which the generating matrices satisfy such a property. The following definition is a special case of Dick (2008, Definition 4.3).

Definition 6.12 (digital (t, α, m, s)-nets). Let $m, \alpha \geq 1$, and $0 \leq t \leq \alpha m$ be natural numbers. Let $\mathbb{Z}_b$ be the finite field of prime order b and let $C_1, \ldots, C_s \in \mathbb{Z}_b^{\alpha m \times m}$ with $C_j = (\vec{c}_{j,1}^{\top}, \ldots, \vec{c}_{j,\alpha m}^{\top})^{\top}$. If for all $0 < a_{j,\nu_j} < \cdots < a_{j,1}$, where $0 \leq \nu_j$ for all $j = 1, \ldots, s$, with

$$\sum_{j=1}^{s} \sum_{\ell=1}^{\min(\nu_j,\alpha)} a_{j,\ell} \leq \alpha m - t,$$

the vectors

$$\vec{c}_{1,a_{1,\nu_1}}, \ldots, \vec{c}_{1,a_{1,1}}, \ldots, \vec{c}_{s,a_{s,\nu_s}}, \ldots, \vec{c}_{s,a_{s,1}}$$

are linearly independent over $\mathbb{Z}_b$, then the digital net with generating matrices $C_1, \ldots, C_s$ is called a *digital (t, α, m, s)-net over $\mathbb{Z}_b$*.

The need for a more general definition in Dick (2008) arises as we assume therein that the smoothness α of the integrand is not known, so one cannot choose $w = \alpha m$ in this case.

We have the following result.

Theorem 6.13 (worst-case error bound for digital (t, α, m, s)-nets). Let $\alpha \geq 2$ be an integer, and let $b \geq 2$ be a prime number. Then the worst-case error of multivariate integration in the weighted unanchored Sobolev space $H(K_{s,\gamma,\alpha})$ of smoothness $\alpha > 1$, using a QMC rule with a digital (t, α, m, s)-net P over $\mathbb{Z}_b$ as cubature points, is bounded by

$$e(P, H(K_{s,\gamma,\alpha})) \leq b^{-(\alpha m - t)} \left(\sum_{\emptyset \neq \mathfrak{u} \subseteq \{1:s\}} \gamma_{\mathfrak{u}} (D''_{|\mathfrak{u}|,\alpha,b})^2 (\alpha m - t + \alpha + 2)^{2|\mathfrak{u}|\alpha} \right)^{\frac{1}{2}},$$

where $D''_{|\mathfrak{u}|,\alpha,b} = (D_{\alpha,b})^{\frac{|\mathfrak{u}|}{2}} b^{|\mathfrak{u}|\alpha} (b^{-1} + (1 - b^{1/\alpha - 1})^{-|\mathfrak{u}|\alpha})$.

We conclude from this theorem that a digital (t, α, m, s)-net used as cubature points in a QMC rule will yield a convergence of the integration error of order $n^{-\alpha} (\log n)^{\alpha s}$ for integrands with $\|f\|_{s,\gamma,\alpha} < \infty$. The remaining question now is: Do digital (t, α, m, s)-nets exist for all given $\alpha, s \geq 1$ and some fixed t (which may depend on α and s but not on m) for all $m \in \mathbb{N}$? An affirmative answer to this question will be given in the next subsection.

6.8. Construction of higher-order digital nets

We now present explicit constructions of digital (t, α, m, s)-nets. Central to this method is the *digit interlacing function.*

Definition 6.14 (digit interlacing function). The digit interlacing function with interlacing factor $\alpha \in \mathbb{N}$ is given by

$$\mathscr{D}_\alpha : [0,1)^\alpha \to [0,1)$$

$$(x_1, \ldots, x_\alpha) \mapsto \sum_{a=1}^{\infty} \sum_{r=1}^{\alpha} \xi_{r,a} b^{-r-(a-1)\alpha},$$

where $x_r = \xi_{r,1}b^{-1} + \xi_{r,2}b^{-2} + \cdots$ for $1 \le r \le \alpha$. We also define this function for vectors by setting

$$\mathscr{D}_\alpha : [0,1)^{\alpha s} \to [0,1)^s$$

$$(x_1, \ldots, x_{\alpha s}) \mapsto (\mathscr{D}_\alpha(x_1, \ldots, x_\alpha), \ldots, \mathscr{D}_\alpha(x_{(s-1)\alpha+1}, \ldots, x_{s\alpha})).$$

Example 6.15 (construction of higher-order digital nets using point sets). Let $\boldsymbol{t}_0, \boldsymbol{t}_1, \ldots, \boldsymbol{t}_{b^m-1} \in [0,1)^{s\alpha}$ be a digital $(t', m, s\alpha)$-net in base b. Then the point set

$$\mathscr{D}_\alpha(\boldsymbol{t}_0), \mathscr{D}_\alpha(\boldsymbol{t}_1), \ldots, \mathscr{D}_\alpha(\boldsymbol{t}_{b^m-1})$$

is a digital (t, α, m, s)-net in base b, where t is given by Theorem 6.17.

Higher-order digital sequences can be constructed in an analogous way.

To analyse the properties of higher-order digital nets, it is more convenient to describe the above construction using the generating matrices rather than the point set.

Example 6.16 (construction of higher-order digital nets using generating matrices). Let $C_1, \ldots, C_{s\alpha}$ be the generating matrices of a digital $(t', m, s\alpha)$-net. Let $C_j = (\vec{c}_{j,1}^{\top}, \ldots, \vec{c}_{j,m}^{\top})^{\top}$ for $j = 1, \ldots, s\alpha$; that is, $\vec{c}_{j,\ell}$ are the row vectors of C_j. Now let the matrix $C_j^{(\alpha)}$ be made of the first rows of the matrices $C_{(j-1)\alpha+1}, \ldots, C_{j\alpha}$, then the second rows of $C_{(j-1)\alpha+1}, \ldots, C_{j\alpha}$, and so on. The matrix $C_j^{(\alpha)}$ is then an $\alpha m \times m$ matrix, that is,

$$C_j^{(\alpha)} = (\vec{c}_{j,1}^{(\alpha)}, \ldots, \vec{c}_{j,\alpha m}^{(\alpha)})^{\top},$$

where

$$\vec{c}_{j,\ell}^{(\alpha)} = \vec{c}_{u,v},$$

with $\ell = (v-j)\alpha + u$, $1 \le v \le m$, and $(j-1)\alpha < u \le j\alpha$ for $\ell = 1, \ldots, \alpha m$ and $j = 1, \ldots, s$. Then the matrices $C_1^{(\alpha)}, \ldots, C_s^{(\alpha)}$ are the generating matrices of a digital (t, α, m, s)-net over $\mathbb{Z}_b$, where t is given by Theorem 6.17.

As we will see later, the choice of the underlying $(t', m, s\alpha)$-net has a direct impact on the bound on the t-value of the digital (t, α, m, s)-net.

To give the idea why this construction works we may consider the case $s = 1$. Let $\alpha > 1$. To simplify the notation we drop the j (which denotes the coordinate) from the notation for a moment. Let $C^{(\alpha)}$ be constructed from a classical digital (t', m, α)-net with generating matrices $C_1, \ldots, C_\alpha$ as described above. Let $\alpha m \geq a_1 > a_2 > \cdots > a_\nu \geq 1$. Then we need to consider the row vectors $\vec{c}_{a_1}^{(\alpha)}, \ldots, \vec{c}_{a_\nu}^{(\alpha)}$. Now by the construction above, the vector $\vec{c}_{a_1}^{(\alpha)}$ may stem from any of the generating matrices $C_1, \ldots, C_\alpha$. Without loss of generality assume that $\vec{c}_{a_1}^{(\alpha)}$ stems from C_1, *i.e.*, it is the i_1th row of C_1, where $i_1 = \lceil a_1/\alpha \rceil$. Next consider $\vec{c}_{a_2}^{(\alpha)}$. This row vector may again stem from any of the matrices $C_1, \ldots, C_\alpha$. If $\vec{c}_{a_2}^{(\alpha)}$ also stems from C_1, then $\lceil a_2/\alpha \rceil < i_1$. If not, we may assume without loss of generality that it stems from C_2. Indeed, it will be the i_2th row of C_2, where $i_2 = \lceil a_2/\alpha \rceil$. We continue in this fashion and define numbers $i_3, i_4, \ldots, i_\ell$, where $1 \leq \ell \leq \alpha$. Further, we set $i_{\ell+1} = \cdots = i_\alpha = 0$. Then by the (t', m, α)-net property of $C_1, \ldots, C_\alpha$, it follows that $\vec{c}_{a_1}^{(\alpha)}, \ldots, \vec{c}_{a_\nu}^{(\alpha)}$ are linearly independent provided that $i_1 + \cdots + i_\alpha \leq m - t'$. Hence, if we choose t such that $a_1 + \cdots + a_{\min(\alpha,\nu)} \leq \alpha m - t$ implies that $i_1 + \cdots + i_\alpha \leq m - t'$ for all admissible choices of $a_1, \ldots, a_\nu$, then the digital $(t, \alpha, m, 1)$-net property of $C^{(\alpha)}$ follows.

Note that $i_1 = \lceil a_1/\alpha \rceil$ and $i_\ell \leq \lceil a_\ell/\alpha \rceil$ for $\ell = 2, \ldots, \alpha$. Thus

$$\begin{aligned} i_1 + \cdots + i_\alpha &\leq \lceil a_1/\alpha \rceil + \cdots + \lceil a_\alpha/\alpha \rceil \leq (a_1 + \cdots + a_\alpha + \alpha(\alpha - 1))/\alpha \\ &= \frac{a_1 + \cdots + a_\alpha}{\alpha} + \alpha - 1 \leq m - t/\alpha + \alpha - 1. \end{aligned}$$

Thus, if we choose t such that $m - t/\alpha + \alpha - 1 \leq m - t'$, then the result follows. Simple algebra then shows that

$$t = \alpha t' + \alpha(\alpha - 1)$$

will suffice.

A more general and improved result is given in the following.

Theorem 6.17. Let $\alpha \geq 1$ be a natural number and let $C_1, \ldots, C_{s\alpha}$ be the generating matrices of a digital $(t', m, s\alpha)$-net over the finite field $\mathbb{Z}_b$ of prime order b. Let $C_1^{(\alpha)}, \ldots, C_s^{(\alpha)}$ be defined as in Example 6.16. Then the matrices $C_1^{(\alpha)}, \ldots, C_s^{(\alpha)}$ are the generating matrices of a digital (t, α, m, s)-net over $\mathbb{Z}_b$ with

$$t = \alpha \min\left(m, t' + \left\lfloor \frac{s(\alpha - 1)}{2} \right\rfloor\right).$$

This shows that digital (t, α, m, s)-nets can be explicitly constructed for all $\alpha, m, s \geq 1$ with t bounded independently of m. Indeed, the dependence of t on α and s is also known from Dick and Baldeaux (2009): namely, there are constants $c, C > 0$ independent of α and s such that $c\alpha^2 s \leq t \leq C\alpha^2 s$.

Using Theorem 6.17 we therefore obtain explicit constructions of higher-order quasi-Monte Carlo rules which satisfy Theorem 6.13.

6.9. Geometric properties of higher-order digital nets

Figures 6.1–6.4 illustrate the properties of a digital $(2, 2, 4, 2)$-net in base 2. Figure 6.2 shows a partition of the square for which each union of the shaded rectangles contains exactly two points. Figures 6.3 and 6.4 show that also other partitions of the unit square are possible where each union of shaded rectangles contains the 'fair' number of points. Many other partitions of the square are possible where the point set always contains the fair number of points in each union of rectangles (see Dick and Baldeaux 2009), but there are too many to show all of them here. Even in the simple case considered here there are 12 partitions possible, for each of which the point set is fair: this is quite remarkable since the point set itself has only 16 points (we exclude all those partitions for which the fairness would follow already from some other partition, otherwise there would be 34 of them). In the classical case we have 4 such partitions, all of which are shown in Figure 6.1. (The partitions from the classical case are included in the generalized case; so out of the 12 partitions 4 are shown in Figure 6.1, one is shown in Figure 6.2, one is shown in Figure 6.4 and one is indicated in Figure 6.3.)

The subsets of $[0, 1)^s$ which form a partition, each having a fair number of points, are of the form

$$J(\vec{a}_{\vec{\nu}}, \vec{d}_{\vec{\nu}}) = \prod_{j=1}^{s} \bigcup_{\substack{d_{j,\ell}=0 \\ \ell \in \{1,\ldots,\alpha m\} \setminus \{a_{j,1},\ldots,a_{j,\nu_j}\}}}^{b-1} \left[\frac{d_{j,1}}{b} + \cdots + \frac{d_{j,n}}{b^{\alpha m}}, \frac{d_{j,1}}{b} + \cdots + \frac{d_{j,n}}{b^{\alpha m}} + \frac{1}{b^{\alpha m}}\right),$$

where $b \geq 2$ is the base and where $\sum_{j=1}^{s} \sum_{\ell=1}^{\nu_j} a_{j,\ell} \leq \alpha m - t$. For $j = 1, \ldots, s$ we again assume $1 \leq a_{j,\nu_j} < \cdots < a_{j,1} \leq \alpha m$ for $\nu_j > 0$ and $\{a_{j,1}, \ldots, a_{j,\nu_j}\} = \emptyset$ for $\nu_j = 0$. Further, we also use the following notation:

$$\vec{\nu} = (\nu_1, \ldots, \nu_s), \quad |\vec{\nu}|_1 = \sum_{j=1}^{s} \nu_j,$$

$$\vec{a}_{\vec{\nu}} = (a_{1,1}, \ldots, a_{1,\nu_1}, \ldots, a_{s,1}, \ldots, a_{s,\nu_s}),$$

$$\vec{d}_{\vec{\nu}} \in \{0, \ldots, b-1\}^{|\vec{\nu}|_1},$$

$$\vec{d}_{\vec{\nu}} = (d_{1,i_{1,1}}, \ldots, d_{1,i_{1,\nu_1}}, \ldots, d_{s,i_{s,1}}, \ldots, d_{s,i_{s,\nu_s}}),$$

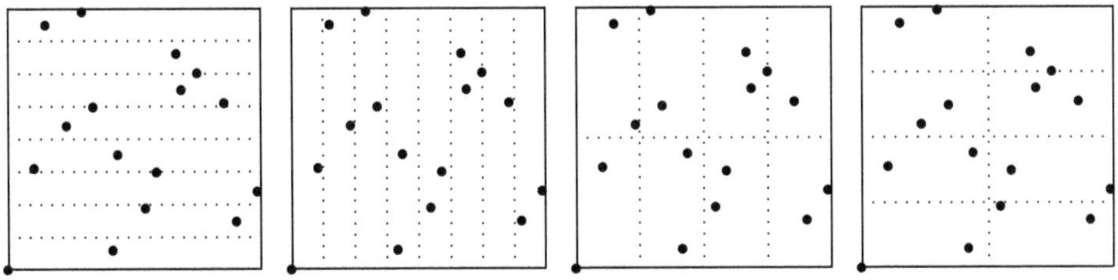

Figure 6.1. Higher-order Sobol′ points: these 16 points are from a Sobol′ point set in dimension four with interlacing factor two (which yields a two-dimensional point set).

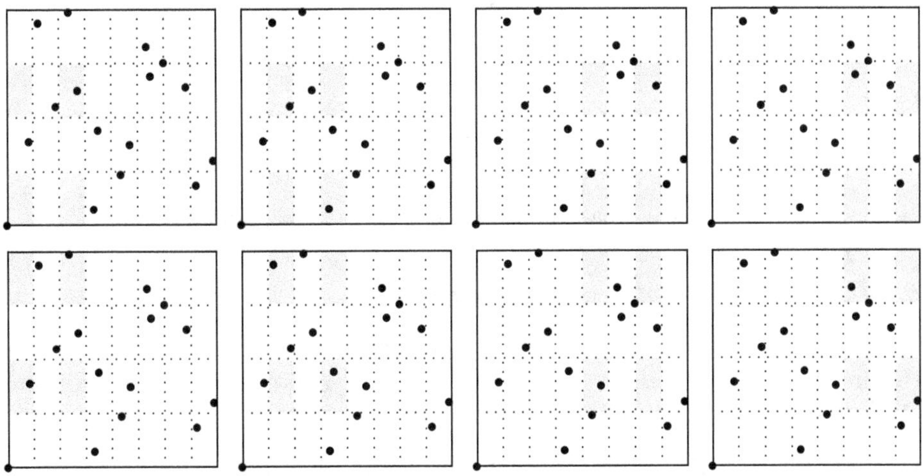

Figure 6.2. Higher-order Sobol′ points: the shaded area in each square contains exactly two points.

where the components $a_{j,\ell}$ and $d_{j,\ell}$, $\ell = 1, \ldots, \nu_j$, do not appear in the vectors $\vec{a}_{\vec{\nu}}$ and $\vec{d}_{\vec{\nu}}$ for $\nu_j = 0$.

Figures 6.1–6.4 give only a few examples of unions of intervals for which each subset of the partition contains the right number of points. As the subsets $J(\vec{a}_{\vec{\nu}}, \vec{d}_{\vec{\nu}})$ for fixed $\vec{\nu}$ and $\vec{a}_{\vec{\nu}}$ (with $\vec{d}_{\vec{\nu}}$ running through all possibilities) form a partition of $[0,1)^s$, it is clear that the right number of points in $J(\vec{a}_{\vec{\nu}}, \vec{d}_{\vec{\nu}})$ has to be $b^m \mathrm{Vol}(J(\vec{a}_{\vec{\nu}}, \vec{d}_{\vec{\nu}}))$. For example, the digital net in Figure 6.2 has 16 points and the partition consists of 8 different subsets $J(\vec{a}_{\vec{\nu}}, \vec{d}_{\vec{\nu}})$, hence each $J(\vec{a}_{\vec{\nu}}, \vec{d}_{\vec{\nu}})$ contains exactly $16/8 = 2$ points. (In general, the volume of $J(\vec{a}_{\vec{\nu}}, \vec{d}_{\vec{\nu}})$ is given by $b^{-|\vec{\nu}|_1}$: see Dick and Baldeaux 2009.)

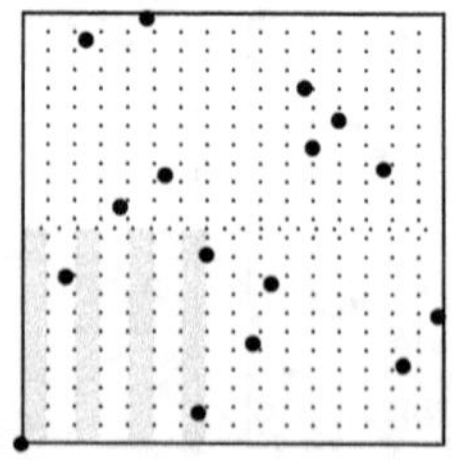

Figure 6.3. Higher-order Sobol′ points: the shaded area contains exactly two points.

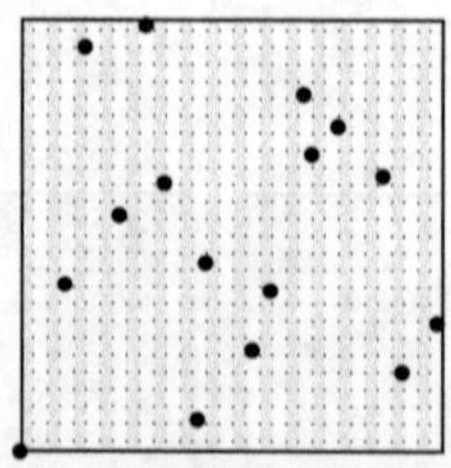

Figure 6.4. Higher-order Sobol′ points: the shaded area contains exactly half of the points.

6.10. Scrambled digital nets and sequences

We now discuss properties of Owen's scrambling algorithm, which was introduced in Section 2.10. First we show that a point $\boldsymbol{x}$ scrambled using Owen's algorithm is uniformly distributed in $[0,1)^s$.

Proposition 6.18. Let $\boldsymbol{x} \in [0,1)^s$ and let $\boldsymbol{\Pi}$ be a uniformly and i.i.d. set of permutations. Then $\boldsymbol{\Pi}(\boldsymbol{x})$ is uniformly distributed in $[0,1)^s$, that is, for any Lebesgue-measurable set $G \subseteq [0,1)^s$, the probability that $\boldsymbol{\Pi}(\boldsymbol{x}) \in G$, denoted by $\mathbb{P}[\boldsymbol{\Pi}(\boldsymbol{x}) \in G]$, satisfies $\mathbb{P}[\boldsymbol{\Pi}(\boldsymbol{x}) \in G] = \lambda_s(G)$, where λ_s denotes the s-dimensional Lebesgue measure.

Proof. We follow Owen (1995, Proof of Proposition 2) in our exposition. We use the notation from above and set $\boldsymbol{y} := \boldsymbol{\Pi}(\boldsymbol{x})$. Consider the case $s = 1$ first and let

$$E = \left[\frac{a}{b^\ell}, \frac{a+1}{b^\ell}\right)$$

be an elementary interval where $\ell \geq 0$ and $0 \leq a < b^\ell$. A technical problem which can arise in the proof below is when y is of the form

$$y_1 = y_{1,1}b^{-1} + \cdots + y_{1,\ell}b^{-\ell} + (b-1)b^{-\ell-1} + (b-1)b^{-\ell-1} + \cdots,$$

since then we have $y_1 = y_{1,1}b^{-1} + \cdots + (y_{1,\ell}+1)b^{-\ell}$. We show that this only happens with probability 0.

The probability that there are $u \geq j_0 \geq 1$ such that $y_{1,j_0} = y_{1,j_0+1} = \cdots = y_{1,u} = b-1$ is given by $((b-1)!)^{-(u-j_0)}$. Hence the probability that $y_{1,j_0} = y_{1,j_0+1} = \cdots = b-1$, that is, all digits of y_1 are $b-1$ from some index j_0 onwards, is 0.

Let $ab^{-\ell} = a_1 b^{-1} + a_2 b^{-2} + \cdots + a_\ell b^{-\ell}$. Then $y_1 \in E$ if and only if $y_1 = a_1, y_2 = a_2, \ldots, y_\ell = a_\ell$. Using (2.8), this is equivalent to

$$\pi_{1,x_{1,1},\ldots,x_{1,k-1}}(x_{1,k}) = a_k, \quad \text{for } 1 \leq k \leq \ell. \tag{6.16}$$

For each $1 \leq k \leq \ell$, the probability that (6.16) holds is b^{-1}. Hence the probability that $y_1 \in E$ is $b^{-\ell}$. The result therefore holds for all elementary intervals of $[0, 1)$.

We now extend the result to the general case. First notice that the result also holds for all subintervals $[ub^{-\ell}, vb^{-\ell})$, where $\ell \geq 0$ and $0 \leq u < v \leq b^{-\ell}$. The endpoints of these intervals are dense in $[0, 1)$. A corollary of Chung (1974, p. 28) extends the result $\mathbb{P}[y_1 \in B] = \lambda_1(B)$ to all Borel-measurable subsets $B \subseteq [0, 1)$. The equality $\mathbb{P}[y_1 \in B] = \lambda_1(B)$ extends to Lebesgue-measurable sets B since subsets of sets of measure zero have probability zero of containing y_1.

Consider now $s > 1$. Let $B_1, \ldots, B_s$ be measurable subsets of $[0, 1)$. Because the components $y_1, \ldots, y_s$ of $\boldsymbol{y}$ are independent, it follows that

$$\mathbb{P}[y_i \in B_i, 1 \leq i \leq s] = \prod_{i=1}^{s} \lambda_1(B_i). \tag{6.17}$$

Finally, λ_s is the unique measure on $[0, 1)^s$ which satisfies (6.17). □

Consider a (t, m, s)-net in base b consisting of points $\boldsymbol{t}_0, \ldots, \boldsymbol{t}_{b^m-1}$, where $\boldsymbol{t}_i = (t_{i,1}, \ldots, t_{i,s})$ and $t_{i,j} = t_{i,j,1} b^{-1} + t_{i,j,2} b^{-2} + \cdots$. We shall denote the scrambled points by $\boldsymbol{y}_0, \ldots, \boldsymbol{y}_{b^m-1}$, where $\boldsymbol{y}_i = (y_{i,1}, \ldots, y_{i,s})$ and $y_{i,j} = y_{i,j,1} b^{-1} + y_{i,j,2} b^{-2} + \cdots$. Specifically, the scrambled points are given by $y_{i,j,k} = \pi_{j,t_{i,j,1},\ldots,t_{i,j,k-1}}(t_{i,j,k})$, for $0 \leq i < b^m$, $1 \leq j \leq s$, and $k \geq 1$.

Similarly, if $(\boldsymbol{t}_0, \boldsymbol{t}_1, \ldots)$ is a (t, s)-sequence, then the scrambled sequence will be denoted by $(\boldsymbol{y}_0, \boldsymbol{y}_1, \ldots)$, where, using the same notation as above, again $y_{i,j,k} = \pi_{j,t_{i,j,1},\ldots,t_{i,j,k-1}}(t_{i,j,k})$, for all $i \geq 0$, $1 \leq j \leq s$, and $k \geq 1$.

We consider now the expected value of $\frac{1}{n}\sum_{i=0}^{n-1} f(\boldsymbol{y}_i)$. For any measurable function f we have

$$\mathbb{E}\left[\frac{1}{n}\sum_{i=0}^{n-1} f(\boldsymbol{y}_i)\right] = \frac{1}{n}\sum_{i=0}^{n-1} \mathbb{E}[f(\boldsymbol{y}_i)] = \int_{[0,1]^s} f(\boldsymbol{y})\mathrm{d}\boldsymbol{y},$$

since each point $\boldsymbol{y}_i$ is uniformly distributed in $[0, 1)^s$ and hence

$$\mathbb{E}[f(\boldsymbol{y}_i)] = \int_{[0,1]^s} f(\boldsymbol{y})\mathrm{d}\boldsymbol{y} \quad \text{for } 0 \leq i < n.$$

In other words, this means that a scrambled point set (note that the above applies even if the underlying point set $\boldsymbol{t}_0, \ldots, \boldsymbol{t}_{n-1}$ is not a digital net) used in a QMC rule yields an unbiased estimator.

The second important property of the randomization algorithm we require is that the (t, m, s)-net structure of the points $\boldsymbol{t}_0, \ldots, \boldsymbol{t}_{b^m-1}$ is retained after applying the scrambling algorithm. For some technical reason this does not quite hold, but it holds with probability one, which is still sufficient. The following proposition was first shown by Owen (1995).

Proposition 6.19. If $\boldsymbol{t}_0, \ldots, \boldsymbol{t}_{b^m-1}$ form a (t, m, s)-net in base b, then $\boldsymbol{y}_0, \ldots, \boldsymbol{y}_{b^m-1}$ is a (t, m, s)-net in base b with probability one. If $\boldsymbol{t}_0, \boldsymbol{t}_1, \ldots$ are obtained from a (t, s)-sequence, then the scrambled points $\boldsymbol{y}_0, \boldsymbol{y}_1, \ldots$ form a (t, s)-sequence with probability one.

Proof. The probability that, for some $0 \le i < b^m$, $1 \le j \le s$, and $\ell \in \mathbb{N}$, all $y_{i,j,k} = b - 1$ for all $k \ge \ell$ is 0. Equivalently, with probability one, infinitely many digits in the b-adic expansion of $y_{i,j}$ are different from $b-1$. Therefore, the probability that $y_{i,j}$ has infinitely many digits in the b-adic expansion of $y_{i,j}$ are different from $b - 1$ for all $0 \le i < b^m$ and $1 \le j \le s$ is 0, since the union of a finite number of zero probability events has probability zero. Hence this holds for each component of each point of a (t, m, s)-net.

For a (t, s)-sequence the same applies, since a countable union of probability zero events itself has probability zero. Hence this also holds for each component of each point of a (t, s)-sequence.

Therefore we may, in the following, assume that infinitely many digits in the b-adic expansion of $y_{i,j}$ differ from $b - 1$ for all $i \in \mathbb{N}_0$ and $1 \le j \le s$.

Assume we are given an elementary interval

$$J = \prod_{j=1}^{s} [a_j b^{-d_j}, (a_j + 1) b^{-d_j}),$$

where $0 \le a_j < b^{d_j}$, $d_j \in \mathbb{N}_0$, and $d_1 + \cdots + d_s \le m - t$. Let $a_j b^{-d_j} = a_{j,1} b^{-1} + a_{j,2} b^{-2} + \cdots + a_{j,d_j} b^{-d_j}$.

Then $\boldsymbol{y}_i \in J$ if and only if $y_{i,j,k} = a_{j,k}$ for all $1 \le k \le d_j$ and all $1 \le j \le s$. Further, $y_{i,j,k} = a_{j,k}$ if and only if $x_{i,j,k} = \pi^{-1}_{j, t_{i,j,1}, \ldots, t_{i,j,k-1}}(a_{j,k})$. Let $a'_{j,k} = \pi^{-1}_{j, t_{i,j,1}, \ldots, t_{i,j,k-1}}(a_{j,k})$. Then $\boldsymbol{y}_i \in J$ if and only if

$$\boldsymbol{t}_i \in J' = \prod_{j=1}^{s} [a'_j b^{-d_j}, (a'_j + 1) b^{-d_j}),$$

where $a'_j b^{-d_j} = a'_{j,1} b^{-1} + \cdots + a'_{j,d_j} b^{-d_j}$. As the points $\boldsymbol{t}_0, \ldots, \boldsymbol{t}_{b^m-1}$ form a (t, m, s)-net, it follows that there are exactly b^{m-t} points of this net in J' and hence there are exactly b^{m-t} points of $\boldsymbol{y}_0, \ldots, \boldsymbol{y}_{b^m-1}$ in J. Thus $\boldsymbol{y}_0, \ldots, \boldsymbol{y}_{b^m-1}$ form a (t, m, s)-net with probability one.

For a (t,s)-sequence $\boldsymbol{t}_0, \boldsymbol{t}_1, \ldots$ for all $k \in \mathbb{N}_0$ and $m \geq t$ the point set consisting of $\boldsymbol{t}_{kb^m}, \ldots, \boldsymbol{t}_{kb^m+b^m-1}$ forms a (t,m,s)-net which is again a (t,m,s)-net after scrambling with probability one. Since the union of countably many zero probability events has probability zero, the result for (t,s)-sequences follows as well. □

6.11. Error analysis of scrambled digital nets

In this subsection we study upper bounds on the variance of scrambled digital nets. To explain the result for scrambled digital nets, we introduce the so-called *nested ANOVA decomposition.*

Theorem 6.20 (nested ANOVA decomposition). Consider a function $f \in L_2([0,1])$ with Walsh series expansion

$$f(x) \sim \sum_{k=0}^{\infty} \widehat{f}(k) \mathrm{wal}_k(x), \tag{6.18}$$

where

$$\widehat{f}(k) = \int_0^1 f(x) \overline{\mathrm{wal}_k(x)} \, \mathrm{d}x.$$

Define $\beta_0 := \int_0^1 f(y) \, \mathrm{d}y$, and for $\ell \geq 1$,

$$\beta_\ell(x) := \sum_{k=b^{\ell-1}}^{b^\ell - 1} \widehat{f}(k) \mathrm{wal}_k(x) \quad \text{and} \quad \sigma_\ell^2(f) := \mathrm{Var}[\beta_\ell].$$

Then we have the *nested ANOVA decomposition* of f

$$\mathrm{Var}[f] = \sum_{\ell=1}^{\infty} \sigma_\ell^2(f).$$

Notice that we do not necessarily have equality in (6.18), since the function f is only assumed to be in $L_2([0,1])$ (and hence may for example be changed arbitrarily on a set of measure zero without changing $\widehat{f}(k)$ for any $k \geq 0$).

Proof. Consider the b^ℓ-term approximation of f given by

$$\sum_{k=0}^{b^\ell - 1} \widehat{f}(k) \mathrm{wal}_k(x) = \sum_{k=0}^{b^\ell - 1} \int_0^1 f(y) \mathrm{wal}_k(x \ominus y) \, \mathrm{d}y = \int_0^1 f(y) D_\ell(x \ominus y) \, \mathrm{d}y,$$

where D_ℓ is the Walsh–Dirichlet kernel given by

$$D_\ell(z) = \sum_{k=0}^{b^\ell - 1} \mathrm{wal}_k(z) = \begin{cases} b^\ell & \text{if } z \in [0, b^{-\ell}), \\ 0 & \text{otherwise.} \end{cases}$$

Hence we have

$$\sum_{k=0}^{b^\ell-1} \widehat{f}(k)\mathrm{wal}_k(x) = b^\ell \int_{\lfloor yb^\ell \rfloor = \lfloor xb^\ell \rfloor} f(y)\,\mathrm{d}y,$$

where the integration is over all y such that $\lfloor yb^\ell \rfloor = \lfloor xb^\ell \rfloor$, that is, the first ℓ digits of x and y coincide. Therefore, for $\ell \geq 1$ we have

$$\beta_\ell(x) = b^\ell \int_{\lfloor yb^\ell \rfloor = \lfloor xb^\ell \rfloor} f(y)\,\mathrm{d}y - b^{\ell-1} \int_{\lfloor yb^{\ell-1} \rfloor = \lfloor xb^{\ell-1} \rfloor} f(y)\,\mathrm{d}y.$$

(Owen 1997*b* defined this function using Haar wavelets.) Notice that β_ℓ is constant on intervals of the form $[ub^{-\ell}, (u+1)b^{-\ell})$, so $\beta_\ell(x) = \beta_\ell(\lfloor b^\ell x \rfloor b^{-\ell})$.

Then, because of the orthogonality of the Walsh functions we obtain

$$\sigma_\ell^2(f) = \int_0^1 |\beta_\ell(x)|^2 \mathrm{d}x = \sum_{k=b^{\ell-1}}^{b^\ell-1} |\widehat{f}(k)|^2$$

and also

$$\int_0^1 \beta_\ell(x)\overline{\beta_{\ell'}(x)}\mathrm{d}x = 0, \quad \text{for } \ell \neq \ell'.$$

Since $f \in L_2([0,1])$ and the Walsh function system is complete, we can use Plancherel's identity to obtain

$$\mathrm{Var}[f] = \int_0^1 |f(y) - \mathbb{E}(f)|^2 \mathrm{d}y = \sum_{k=1}^{\infty} |\widehat{f}(k)|^2 = \sum_{\ell=1}^{\infty} \sigma_\ell^2(f).$$

Therefore we have obtained a decomposition of the variance of f in terms of the variances of β_ℓ. □

Notice that $\sum_{k=1}^\infty |\widehat{f}(k)|^2 = \mathrm{Var}\big[\sum_{k=0}^\infty \widehat{f}(k)\mathrm{wal}_k\big]$. Hence, as a by-product, we obtain that for any $f \in L_2([0,1])$, the variance of f and the variance of its Walsh series coincide, that is,

$$\mathrm{Var}[f] = \mathrm{Var}\left[\sum_{k=0}^{\infty} \widehat{f}(k)\mathrm{wal}_k\right].$$

For functions $f : [0,1]^s \to \mathbb{R}$ the nested ANOVA decomposition can be applied component-wise. Let $f \in L_2([0,1]^s)$ have the following Walsh series expansion:

$$f(\boldsymbol{x}) \sim \sum_{\boldsymbol{k} \in \mathbb{N}_0^s} \widehat{f}(\boldsymbol{k})\mathrm{wal}_{\boldsymbol{k}}(\boldsymbol{x}) =: S(\boldsymbol{x}, f).$$

Let $\boldsymbol{\ell} = (\ell_1, \ldots, \ell_s) \in \mathbb{N}_0^s$ and $L_{\boldsymbol{\ell}} = \{\boldsymbol{k} = (k_1, \ldots, k_s) \in \mathbb{N}_0^s : \lfloor b^{\ell_j - 1} \rfloor \le k_j < b^{\ell_j}$ for $1 \le j \le s\}$. Then let

$$\beta_{\boldsymbol{\ell}}(\boldsymbol{x}) = \sum_{\boldsymbol{k} \in L_{\boldsymbol{\ell}}} \widehat{f}(\boldsymbol{k}) \mathrm{wal}_{\boldsymbol{k}}(\boldsymbol{x})$$

and

$$\sigma_{\boldsymbol{\ell}}^2(f) := \mathrm{Var}[\beta_{\boldsymbol{\ell}}] = \int_{[0,1]^s} |\beta_{\boldsymbol{\ell}}(\boldsymbol{x})|^2 \mathrm{d}\boldsymbol{x} = \sum_{\boldsymbol{k} \in L_{\boldsymbol{\ell}}} |\widehat{f}(\boldsymbol{k})|^2.$$

For $\boldsymbol{\ell} = (\ell_1, \ldots, \ell_s) \in \mathbb{N}_0^s \setminus \{\boldsymbol{0}\}$ let

$$G_{\boldsymbol{\ell}} = \frac{1}{N^2} \sum_{i,i'=0}^{n-1} \prod_{j=1}^{s} \left(\frac{b}{b-1} 1_{\lfloor b^{\ell_j} t_{i,j} \rfloor = \lfloor b^{\ell_j} t_{i',j} \rfloor} - \frac{1}{b-1} 1_{\lfloor b^{\ell_j - 1} t_{i,j} \rfloor = \lfloor b^{\ell_j - 1} t_{i',j} \rfloor} \right),$$

where $\boldsymbol{t}_i = (t_{i,1}, \ldots, t_{i,s})$ for $0 \le i \le n-1$ and where $1_{A=B}$ is 1 if $A = B$ and 0 otherwise.

If $\boldsymbol{k} \in L_{\boldsymbol{\ell}}$, then

$$\frac{1}{n^2} \sum_{i,i'=0}^{n-1} \mathbb{E}\big[\mathrm{wal}_{\boldsymbol{k}}(\boldsymbol{y}_i \ominus \boldsymbol{y}_{i'})\big] = G_{\boldsymbol{\ell}},$$

since the coordinates are randomized independently from each other. Owen (1997*b*) called the numbers $\Gamma_{\boldsymbol{\ell}} := nG_{\boldsymbol{\ell}}$ *gain coefficients*, since, as we see below, they determine how much one gains compared to the classical Monte Carlo algorithm.

We now estimate the variance of the estimator

$$\widehat{Q}_{n,s}(f) = \frac{1}{n} \sum_{i=0}^{n-1} f(\boldsymbol{y}_i) \tag{6.19}$$

for integrands $f \in L_2([0,1]^s)$ when the points $\boldsymbol{y}_0, \ldots, \boldsymbol{y}_{n-1}$ are obtained by applying Owen's scrambling to a (digital) (t, m, s)-net over $\mathbb{Z}_b$. (To emphasize that (6.19) is a random variable, we use the notation $\widehat{Q}_{n,s}$ rather than $Q_{n,s}$.)

For the following results see Owen (1997*a*, 1997*b*, 1998).

Theorem 6.21 (variance of the integral estimator using scrambled nets). Let $f \in L_2([0,1]^s)$ and $\widehat{Q}_{n,s}(f)$ be given by (6.19). Let the point set $\{\boldsymbol{y}_0, \ldots, \boldsymbol{y}_{n-1}\} \subseteq [0,1)^s$ be obtained by applying Owen's scrambling algorithm to the point set $\{\boldsymbol{t}_0, \ldots, \boldsymbol{t}_{n-1}\} \subseteq [0,1)^s$. Then the variance of the estimator $\widehat{Q}_{n,s}(f)$ is given by

$$\mathrm{Var}[\widehat{Q}_{n,s}(f)] = \frac{1}{n} \sum_{\boldsymbol{\ell} \in \mathbb{N}_0^s \setminus \{\boldsymbol{0}\}} \Gamma_{\boldsymbol{\ell}} \sigma_{\boldsymbol{\ell}}^2(f).$$

Theorem 6.22 (a bound on the variance in terms of gain coefficients). Let $f \in L_2([0,1]^s)$ and $\widehat{Q}_{n,s}(f)$ be given by (6.19). Let the points $\{\boldsymbol{t}_0, \dots, \boldsymbol{t}_{b^m-1}\}$ be a digital (t,m,s)-net over $\mathbb{Z}_b$. Then

$$\mathrm{Var}[\widehat{Q}_{n,s}(f)] \le b^{-m+t}\left(\frac{b+1}{b-1}\right)^s \sum_{\substack{\boldsymbol{\ell}\in\mathbb{N}_0^s \\ |\boldsymbol{\ell}|_1 > m-t}} \sigma_{\boldsymbol{\ell}}^2(f).$$

For MC one obtains a variance $\mathrm{Var}[\widehat{Q}_{n,s}(f)] = \frac{1}{n}\sum_{\boldsymbol{\ell}\in\mathbb{N}_0^s\setminus\{\boldsymbol{0}\}} \sigma_{\boldsymbol{\ell}}^2(f)$. Hence the gain of scrambled digital nets lies in the fact that we only sum over $\sigma_{\boldsymbol{\ell}}^2(f)$ for which $|\boldsymbol{\ell}|_1 > m-t$, although one incurs a penalty factor of $b^t((b+1)/(b-1))^s$ using scrambled digital (t,m,s)-nets. Notice that the gain coefficients $\Gamma_{\boldsymbol{\ell}}$ are 0 for $\boldsymbol{\ell} \in \mathbb{N}_0^s$ with $|\boldsymbol{\ell}|_1 \le m-t$ and $\Gamma_{\boldsymbol{\ell}} = b^t((b+1)/(b-1))^s$ for $\boldsymbol{\ell} \in \mathbb{N}_0^s$ with $|\boldsymbol{\ell}|_1 > m-t$.

Theorem 6.22 shows that $\mathrm{Var}[\widehat{Q}_{n,s}(f)]$ for a scrambled $(0,m,s)$-net is always at most by a factor of $((b+1)/(b-1))^s$ larger than the variance for a plain MC algorithm (see also Owen (1997*a*, Theorem 1), who shows that the gain coefficients are always bounded by e in this case). Further, for scrambled (t,m,s)-nets we have

$$b^m\mathrm{Var}[\widehat{Q}_{n,s}(f)] = b^t\left(\frac{b+1}{b-1}\right)^s \sum_{\substack{\boldsymbol{\ell}\in\mathbb{N}_0^s \\ |\boldsymbol{\ell}|_1 > m-t}} \sigma_{\boldsymbol{\ell}}^2(f) \to 0 \quad \text{as } m \to \infty.$$

Hence, scrambled (t,m,s)-nets outperform MC with respect to $\mathrm{Var}[\widehat{Q}_{n,s}(f)]$ asymptotically, as for MC we have $n\mathrm{Var}[\widehat{Q}_{n,s}(f)] = \sum_{\boldsymbol{\ell}\in\mathbb{N}_0^s\setminus\{\boldsymbol{0}\}} \sigma_{\boldsymbol{\ell}}^2(f)$ for all $n \in \mathbb{N}$.

If the integrand f has smoothness $0 < \alpha \le 1$, then one can show that $|\sigma_{\boldsymbol{\ell}}(f)| \le C_f b^{-\alpha\mu_1(\boldsymbol{\ell})}$. Scrambled (t,m,s)-nets can take advantage now, since $\Gamma_{\boldsymbol{\ell}} = 0$ for all $\boldsymbol{\ell} \in \mathbb{N}_0^s$ with $\mu_1(\boldsymbol{\ell}) \le m-t$ and $\Gamma_{\boldsymbol{\ell}}$ is bounded otherwise. Thus, asymptotically one obtains an improved rate of convergence in this case. If the integrand is only in $L_2([0,1]^s)$, then one still gets the MC rate of convergence of $n^{-1/2}$.

Definition 6.23 (generalized Vitali variation). We define the *generalized variation in the sense of Vitali of order* $0 < \alpha \le 1$ by

$$V_\alpha^{(s)}(f) = \sup_{\mathcal{P}} \left(\sum_{J\in\mathcal{P}} \mathrm{Vol}(J) \left| \frac{\Delta(f,J)}{\mathrm{Vol}(J)^\alpha} \right|^2 \right)^{1/2},$$

where the supremum is extended over all partitions $\mathcal{P}$ of $[0,1]^s$ into subintervals and $\mathrm{Vol}(J)$ denotes the volume of the subinterval J.

For $\alpha = 1$ and if the partial derivatives of f are continuous on $[0,1]^s$, we also have the formula

$$V_1^{(s)}(f) = \left(\int_{[0,1]^s} \left| \frac{\partial^s f}{\partial x_1 \cdots \partial x_s}(\boldsymbol{x}) \right|^2 \mathrm{d}\boldsymbol{x} \right)^{1/2}.$$

Indeed we have

$$|\Delta(f, J)| = \left| \int_J \frac{\partial^s f}{\partial x_1 \cdots \partial x_s}(\boldsymbol{x}) \mathrm{d}\boldsymbol{x} \right| = \mathrm{Vol}(J) \left| \frac{\partial^s f}{\partial x_1 \cdots \partial x_s}(\boldsymbol{\zeta}_J) \right|$$

for some $\boldsymbol{\zeta}_J \in \overline{J}$, which follows by applying the mean value theorem to the inequality

$$\begin{aligned} \min_{\boldsymbol{x} \in \overline{J}} \left| \frac{\partial^s f}{\partial x_1 \cdots \partial x_s}(\boldsymbol{x}) \right| &\leq \mathrm{Vol}(J)^{-1} \left| \int_J \frac{\partial^s f}{\partial x_1 \cdots \partial x_s}(\boldsymbol{x}) \mathrm{d}\boldsymbol{x} \right| \\ &\leq \max_{\boldsymbol{x} \in \overline{J}} \left| \frac{\partial^s f}{\partial x_1 \cdots \partial x_s}(\boldsymbol{x}) \right|. \end{aligned}$$

Therefore we have

$$\sum_{J \in \mathcal{P}} \mathrm{Vol}(J) \left| \frac{\Delta(f, J)}{\mathrm{Vol}(J)} \right|^2 = \sum_{J \in \mathcal{P}} \mathrm{Vol}(J) \left| \frac{\partial^s f}{\partial x_1 \cdots \partial x_s}(\boldsymbol{\zeta}_J) \right|^2,$$

which is just a Riemann sum for the integral $\int_{[0,1]^s} \left| \frac{\partial^s f}{\partial x_1 \cdots \partial x_s} \right|^2 \mathrm{d}\boldsymbol{x}$, and thus the equality follows.

So far we have not taken into account projections to lower-dimensional faces.

Definition 6.24 (generalized Hardy and Krause variation). For $\emptyset \neq \mathfrak{u} \subseteq \{1:s\}$, let $V_\alpha^{(|\mathfrak{u}|)}(f_\mathfrak{u}; \mathfrak{u})$ be the generalized Vitali variation with coefficient $0 < \alpha \leq 1$ of the $|\mathfrak{u}|$-dimensional function

$$f_\mathfrak{u}(\boldsymbol{x}_\mathfrak{u}) = \int_{[0,1]^{s-|\mathfrak{u}|}} f(\boldsymbol{x}) \mathrm{d}\boldsymbol{x}_{\{1:s\} \setminus \mathfrak{u}}.$$

For $\mathfrak{u} = \emptyset$ we have $f_\emptyset = \int_{[0,1]^s} f(\boldsymbol{x}) \mathrm{d}\boldsymbol{x}_{\{1:s\}}$ and we define $V_\alpha^{(|\emptyset|)}(f_\emptyset; \emptyset) = |f_\emptyset|$. Then

$$V_\alpha(f) = \left(\sum_{\mathfrak{u} \subseteq \{1:s\}} \left(V_\alpha^{(|\mathfrak{u}|)}(f_\mathfrak{u}; \mathfrak{u}) \right)^2 \right)^{1/2}$$

is called the *generalized Hardy and Krause variation* of f on $[0,1]^s$.

A function f for which $V_\alpha(f) < \infty$ is said to be of *finite variation of order* α.

Theorem 6.25 (a bound on the variance in terms of variation). Let $f : [0,1]^s \to \mathbb{R}$ have bounded variation $V_\alpha(f) < \infty$ of order $0 < \alpha \leq 1$.

Then the variance of the estimator $\mathrm{Var}[\widehat{Q}_{n,s}(f)]$ using a randomly scrambled digital (t, m, s)-net over $\mathbb{Z}_b$ is bounded by

$$\mathrm{Var}[\widehat{Q}_{n,s}(f)] \leq V_\alpha^2(f) b^{-(1+2\alpha)(m-t)} \frac{(b-1)^{(2\alpha-1)_+ + s} b^{2s}}{b^{2\alpha}(b-1)^s} \binom{m-t+s}{s-1},$$

where again $(2\alpha - 1)_+ = \max(2\alpha - 1, 0)$.

6.12. *Simplifications of the scrambling scheme: affine matrix scrambling*

The scrambling algorithm as introduced by Owen requires one to generate and store all the necessary permutations used for the scrambling scheme. Since the number of permutations needed is very large, this is not practical, and therefore simplifications of this algorithm have been found for which the main result still holds.

One version of such a simplified scrambling is the following: Let $x_j = x_{j,1} b^{-1} + x_{j,2} b^{-2} + \cdots$ denote the base b expansion of the jth coordinate of $\boldsymbol{x} = (x_1, \ldots, x_s) \in [0,1)^s$. Let $\boldsymbol{y} \in [0,1)^s$ denote the point which is obtained after scrambling $\boldsymbol{x}$. Assume that the jth coordinate has base b expansion $y_j = y_{j,1} b^{-1} + y_{j,2} b^{-2} + \cdots$. To obtain a simplified scrambling, choose $m_{j,k,\ell}, \sigma_{j,k} \in \mathbb{Z}_b$ for $1 \leq \ell < k$ and $m_{j,k,k} \in \mathbb{Z}_b \setminus \{0\}$ for $1 \leq j \leq s$ independently and uniformly distributed over their ranges. Then we set

$$y_{j,k} = \sum_{\ell=1}^{k} m_{j,k,\ell} x_{j,\ell} + \sigma_{j,k} \pmod{b} \quad \text{for } 1 \leq j \leq s.$$

This can also be written in matrix form. Set $M_j = (m_{j,k,\ell})_{k,\ell}$, where $m_{j,k,\ell} = 0$ for $\ell > k$ and $\vec{\sigma}_j = (\sigma_{j,k})_k^\top$. Note that M_j is a lower triangular matrix. Further, let $\vec{x}_j = (x_{j,1}, x_{j,2}, \ldots)$ and $\vec{y}_j = (y_{j,1}, y_{j,2}, \ldots)^\top$. Then we have

$$\vec{y}_j = M\vec{x}_j + \vec{\sigma}.$$

This method reduces the number of permutations significantly while still preserving the main properties of Owen's scrambling. This scrambling scheme is called *affine matrix scrambling.* The results for Owen's scrambling presented above still hold for the affine matrix scrambling.

6.13. *Higher-order scrambling*

We have seen that scrambling can improve the convergence rate of the variance to $n^{-3/2+\delta}$, for any $\delta > 0$, for instance for functions whose generalized variation is bounded. Since there are higher-order nets which yield deterministic QMC rules with convergence rates of $n^{-\alpha+\delta}$ for arbitrary $\delta > 0$ for functions with smoothness $\alpha \geq 1$, the question also arises of how to generalize Owen's scrambling and its simplifications to achieve higher-order convergence. We briefly describe this in the following.

Let $\boldsymbol{t}_0, \ldots, \boldsymbol{t}_{b^m-1} \in [0,1)^{ds}$ be a randomly scrambled digital (t, m, ds) net over the finite field $\mathbb{Z}_b$ of prime order b (here scrambling can mean Owen's original scrambling or any of the simplifications for which the same results hold as for Owen's scrambling). Then one simply uses the sample points

$$\boldsymbol{y}_i = \mathscr{D}_\alpha(\boldsymbol{x}_i) \in [0,1)^s \quad \text{for } 0 \leq i < b^m,$$

where $\mathscr{D}_\alpha$ is the digit interlacing function of order $\alpha \geq 1$. The integral is then estimated using

$$\widehat{Q}_{n,s}(f) = \frac{1}{b^m} \sum_{n=0}^{b^m-1} f(\boldsymbol{y}_n).$$

This method again yields an unbiased estimator, that is,

$$\mathbb{E}(\widehat{Q}_{n,s}(f)) = \int_{[0,1]^s} f(\boldsymbol{x}) \mathrm{d}\boldsymbol{x}.$$

If the integrand has square-integrable partial mixed derivatives up to order $\alpha \geq 1$ in each variable, then the variance of $\widehat{Q}_{n,s}(f)$ satisfies

$$\mathrm{Var}[\widehat{Q}_{n,s}(f)] = O(n^{-2\min(d,\alpha)-1+\delta})$$

for any $\delta > 0$, where $n = b^m$ is the number of sample points.

6.14. Notes

Section 6.1 is taken partly from Dick and Pillichshammer (2010). Sections 6.4 and 6.5 are adapted from Dick and Pillichshammer (2010, Chapter 10). Section 6.6 is based on Dick and Baldeaux (2009). Sections 6.7 and 6.8 are based on Dick (2009*b*). Sections 6.10 and 6.11 are based on Dick and Pillichshammer (2010, Chapter13).

The monographs by Niederreiter (1992*a*) and by Dick and Pillichshammer (2010) give a comprehensive introduction to (digital) nets and (digital) sequences, discrepancy theory and QMC rules based on such point sets and sequences. More references can be found therein.

Walsh functions were introduced by Walsh (1923); see also Chrestenson (1955) and Fine (1949). For information on Walsh functions in the context of QMC integration, see Dick and Pillichshammer (2010).

Numerical integration of Walsh series using QMC rules based on digital nets was studied, for instance, by Larcher and Traunfellner (1994) and Larcher, Schmid and Wolf (1994, 1996*b*), whereas numerical integration in Haar wavelet spaces was studied by Sobol′ (1969) and Heinrich, Hickernell and Yue (2004). Numerical integration in anchored and unanchored Sobolev spaces using QMC rules based on digital nets was first studied by Dick and Pillichshammer (2005).

Polynomial lattice rules have been introduced by Niederreiter (1992*b*). Shift nets are a subclass of digital nets which were introduced by Schmid (1996) and further studied by Pillichshammer (2002). The so-called Salzburg tables provide a table of digital nets found by computer search with small t-value. These computations are described by Hansen, Mullen and Niederreiter (1993), Larcher, Lauß, Niederreiter and Schmid (1996*a*) and Schmid (2000). An improvement of the method for computing the t-value of digital nets has been studied in Pirsic and Schmid (2001). A table with many of the currently best known t-values for (t, m, s)-nets and (t, s)-sequences for many values of m and s can be found at http://mint.sbg.ac.at/.

The CBC construction of polynomial lattice rules was introduced by Dick *et al.* (2005) for anchored and unanchored Sobolev spaces and by Dick, Kritzer, Leobacher and Pillichshammer (2007*a*) for the weighted star-discrepancy. The case where the modulus is not necessarily irreducible was considered by Kritzer and Pillichshammer (2007).

The fast CBC algorithm was introduced by Nuyens and Cools (2006*a*) using FFT methods for the construction of classical lattice point sets. Due to the similarities between ordinary and polynomial lattice point sets it turned out that their methods can also be carried over to the polynomial case: see Nuyens and Cools (2006*b*). Implementations of the fast algorithm using Matlab can be found in Nuyens and Cools (2006*c*).

Cyclic nets are another subclass of digital nets of size b^m which were introduced by Niederreiter (2004). Hyperplane nets are a subclass of digital nets of size b^{ms} which were introduced by Pirsic, Dick, Pillichshammer (2006). The discrepancy of cyclic and hyperplane nets was considered in Pillichshammer and Pirsic (2009). Extensible polynomial lattice rules were studied by Niederreiter (2003) and a construction algorithm was introduced by Dick (2007*b*). Extensible hyperplane nets were studied by Pirsic and Pillichshammer (2011). Constructions of (digital) nets and (digital) sequences based on existing constructions are called *propagation rules*: see Dick and Pillichshammer (2010, Chapter 9) for a summary.

The basic construction principle of higher-order digital nets and sequences appeared first in Dick (2007*a*) and was slightly modified in Dick (2008). A bound on the decay of the Walsh coefficients has been studied in more detail in Dick (2009*a*). Theorem 6.17 is a special case of Dick (2008, Theorem 4.11), with an improvement for some cases from Dick and Baldeaux (2009) (a proof of this result can be found in Dick and Baldeaux 2009, Dick 2007*a* and Dick and Kritzer 2010). The CBC construction of higher-order polynomial lattice rules was studied by Baldeaux, Dick, Greslehner and Pillichshammer (2011) and Baldeaux *et al.* (2012). A CBC construction of higher-order scrambled polynomial lattice rules was studied by Goda and Dick (2013). Propagation rules for higher-order digital nets were studied by Dick and Kritzer (2010). Tractability of higher-order polynomial lattice rules was studied by Dick and Pillichshammer (2007) and the existence of higher-order polynomial lattice rules with small t-value was studied by Dick, Kritzer, Pillichshammer and Schmid (2007*b*). Recently Matsumoto, Saito and Matoba (2013) and Matsumoto and Yoshiki (2013) constructed QMC rules which achieve higher-order convergence for function classes of very high smoothness.

Scrambling was introduced by Owen (1995) and further studied in Owen (1997*a*, 1997*b*). Scrambled Niederreiter and Xing sequences were studied by Owen (1998). The mean-square discrepancy of scrambled nets was studied by Hickernell and Yue (2000), whereas integration and approximation using scrambled nets were studied by Yue and Hickernell (2001). The gain coefficients of digital nets were further investigated by Yue and Hickernell (2002). Strong tractability of scrambled Niederreiter sequences was studied by Yue and Hickernell (2005), whereas strong tractability in Banach spaces was studied by Yue and Hickernell (2006). Of particular interest in the context of scrambled nets is also the result by Loh (2003), who shows that a central limit theorem holds for the estimate $\widehat{Q}_{n,s}(f)$ for which the cubature points are based on a scrambled $(0, m, s)$-net. This allows one to obtain an approximate confidence interval from the variance estimates of $\widehat{Q}_{n,s}(f)$.

Simplifications of Owen's scrambling algorithm were studied by Hickernell (1996*b*), Matoušek (1998*b*) and Tezuka and Faure (2003). In particular, the affine matrix scrambling is from Matoušek (1998*b*) and is implemented in Matlab. Higher-order scrambling has been studied in Dick (2011*a*). A construction of randomly scrambled polynomial lattice rules which have a better dependence on the dimension has been studied in Baldeaux and Dick (2011) for functions of smoothness at most 1.

7. Infinite-dimensional integration

In this section we consider briefly numerical integration for problems with an infinite number of dimensions. We do not try to be definitive, because the subject is developing and changing rapidly. Rather, we introduce some themes likely to be of continuing importance. We concentrate here on 'anchored' function spaces, because anchoring is a natural idea for functions that depend on an infinite number of variables. The section concludes with a brief discussion of an application that is driving much recent interest in infinite-dimensional problems.

7.1. The infinite-dimensional problem

The problem is to integrate real-valued functions f that have a countably infinite number of variables,

$$f(x_1, x_2, \ldots),$$

with each component lying on the unit interval, $x_j \in [0,1], j \in \mathbb{N}$. Equivalently we write $f(\boldsymbol{x})$ with

$$\boldsymbol{x} = (x_j)_{j\in\mathbb{N}} \in [0,1]^{\mathbb{N}} := \prod_{j=1}^{\infty}[0,1].$$

To avoid formal problems associated with functions of an infinite number of variables, we choose an 'anchor' $c \in [0,1]$, and in our algorithms only allow function evaluation when at most a finite number of components of $\boldsymbol{x}$ have values different from the anchor value c. Often $c = 0$ is the most natural choice for the anchor, but in some applications it is better to take the midpoint $c = 1/2$ as anchor. The components x_j that have the value c are considered to be 'inactive variables' for the particular function evaluation, in contrast to the 'active variables' $x_j \neq c$.

If the set of active variables is $\boldsymbol{x}_{\mathfrak{u}}$, we write the value of f as $f(\boldsymbol{x}_{\mathfrak{u}};\boldsymbol{c})$. In this section we shall assume that $f(\boldsymbol{x}_{\mathfrak{u}};\boldsymbol{c})$ is a continuous function of $\boldsymbol{x}_{\mathfrak{u}}$ on $[0,1]^{|\mathfrak{u}|}$ for every finite subset $\mathfrak{u}$ of the natural numbers. The infinite-dimensional integral may now be defined by

$$I_\infty(f) := \lim_{s\to\infty}\int_{[0,1]^s} f(\boldsymbol{x}_{\{1:s\}};\boldsymbol{c})\,\mathrm{d}\boldsymbol{x}_{\{1:s\}}. \tag{7.1}$$

In Section 7.4 we shall introduce a reproducing kernel Hilbert space setting that ensures that this limit exists, and ensures also that $f(\boldsymbol{x})$ is well defined for all $\boldsymbol{x} \in [0,1]^{\mathbb{N}}$. For the present, however, we concentrate on algorithms and costs.

Our algorithms for approximating this infinite-dimensional integral all have the form

$$Q(f) = \sum_{i=0}^{n-1} w_i f(\boldsymbol{t}^{(i)}_{\mathfrak{u}_i};\boldsymbol{c}), \tag{7.2}$$

where $w_i \in \mathbb{R}$, $\mathfrak{u}_i$ is a finite subset of $\mathbb{N}$, and $\boldsymbol{t}^{(i)}_{\mathfrak{u}_i} \in [0,1]^{|\mathfrak{u}_i|}$. Three kinds of algorithms have been studied in the literature, which we introduce in the following subsection.

7.2. Three kinds of algorithms

In this subsection we describe *single-level* (or *fixed-dimension*), *multi-level* and *changing-dimension* algorithms.

Example 7.1 (single-level algorithm). The *single-level* (SL) or *fixed-dimension* algorithm approximates the infinite-dimensional integral by an s-dimensional QMC rule,

$$Q^{\mathrm{SL}}(f) := \frac{1}{n}\sum_{i=0}^{n-1} f(\boldsymbol{t}^{(i)}_{\{1:s\}};\boldsymbol{c}), \tag{7.3}$$

in which the active variables are the leading variables $x_1, x_2, \ldots, x_s$.

Example 7.2 (multi-level algorithm). A different kind of algorithm comes from adopting a *multi-level* (ML) approach (Heinrich 1998, Giles 2008). We define an infinite sequence of dimensions

$$0 = s_0 < s_1 < s_2 < \cdots < \infty,$$

and write the function as a collapsing sum,

$$f(\boldsymbol{x}) = \sum_{\ell=1}^{\infty} \big(f(\boldsymbol{x}_{\{1:s_\ell\}};\boldsymbol{c}) - f(\boldsymbol{x}_{\{1:s_{\ell-1}\}};\boldsymbol{c})\big) + f(\boldsymbol{c}),$$

and similarly write the integral (7.1) also as a collapsing sum,

$$I_\infty(f) = \sum_{\ell=1}^{\infty} \big(I_{s_\ell}(f) - I_{s_{\ell-1}}(f)\big) + f(\boldsymbol{c}),$$

with $I_0(f) = f(\boldsymbol{c})$, and

$$I_{s_\ell}(f) := \int_{[0,1]^{s_\ell}} f(\boldsymbol{x}_{\{1:s_\ell\}};\boldsymbol{c})\,\mathrm{d}\boldsymbol{x}_{\{1:s_\ell\}}.$$

We then approximate the collapsing sum up to level $L > 0$ by the cubature formula

$$Q^{\mathrm{ML}}(f) := \sum_{\ell=1}^{L} Q_{n_\ell,\ell}(\psi_\ell(f) - \psi_{\ell-1}(f)) + f(\boldsymbol{c}), \tag{7.4}$$

where $\psi_\ell(f) := f(\boldsymbol{x}_{\{1:s_\ell\}};\boldsymbol{c})$, and

$$Q_{n_\ell,\ell}(\psi_\ell(f) - \psi_{\ell-1}(f)) = \frac{1}{n_\ell}\sum_{i=0}^{n_\ell-1}\Big(f(\boldsymbol{t}^{(i)}_{\{1:s_\ell\}};\boldsymbol{c}) - f(\boldsymbol{t}^{(i)}_{\{1:s_{\ell-1}\}};\boldsymbol{c})\Big),$$

and where $\boldsymbol{t}^{(i)}_{\{1:s_\ell\}} \in [0,1]^{s_\ell}$, and $\boldsymbol{t}^{(i)}_{\{1:s_{\ell-1}\}}$ contains the first $s_{\ell-1}$ components of $\boldsymbol{t}^{(i)}_{\{1:s_\ell\}}$. For $n_\ell = 0$ we assume that $Q_{n_\ell,\ell}(\psi_\ell(f) - \psi_{\ell-1}(f)) = 0$. The idea of an ML scheme is that the successive terms in the collapsing sum for the integral should be smaller and smaller, and hence can be well approximated by a smaller number of points n_ℓ as ℓ increases, leaving only relatively low-dimensional integrals needing a large number of points. We assumed in (7.4) that the cubature rules $Q_{n_\ell,\ell}$ are QMC rules, but other choices are clearly possible.

Roughly speaking, ML algorithms are of particular advantage if for a particular f the important subsets $\mathfrak{u}$ involve only a small number of leading variables (that is, in the language of Caflisch *et al.* 1997, if the 'truncation dimension' is small).

Example 7.3 (changing-dimension algorithm). A third kind of algorithm is the *changing-dimension* (CD) algorithm (Kuo, Sloan, Wasilkowski and Woźniakowski 2010*c*). In this approach a different cubature rule is applied to each term $f_{\mathfrak{u}}$ of the so-called 'anchored' decomposition of f. The anchored decomposition has the form

$$f = \sum_{|\mathfrak{u}|<\infty} f_{\mathfrak{u}}, \tag{7.5}$$

where for a finite set $\mathfrak{u}$ the function $f_{\mathfrak{u}}$ depends only on the set $\boldsymbol{x}_{\mathfrak{u}}$ of active variables $x_j, j \in \mathfrak{u}$, and vanishes if any of those active variables has the value c. The anchored decomposition has appeared under many different guises: see for example Sobol′ (1969), Li, Schoendorf, Ho and Rabitz (2004) and Griebel (2006). The terms in the anchored decomposition may be defined recursively by

$$f_{\mathfrak{u}}(\boldsymbol{x}_{\mathfrak{u}}) = f(\boldsymbol{x}_{\mathfrak{u}}; \boldsymbol{c}) - \sum_{\mathfrak{v}\subset\mathfrak{u}} f_{\mathfrak{v}}(\boldsymbol{x}_{\mathfrak{v}}), \tag{7.6}$$

where the sum is over the strict subsets of $\mathfrak{u}$. An explicit formula for $f_{\mathfrak{u}}$ in terms of f is given by

$$f_{\mathfrak{u}}(\boldsymbol{x}_{\mathfrak{u}}) = \sum_{\mathfrak{v}\subseteq\mathfrak{u}} (-1)^{|\mathfrak{u}|-|\mathfrak{v}|} f(\boldsymbol{x}_{\mathfrak{v}}; \boldsymbol{c}), \tag{7.7}$$

where now the sum is over all subsets of $\mathfrak{u}$. The formula is established, for example, in Theorem 1 of Kuo, Sloan, Wasilkowski and Woźniakowski (2010*b*).

The CD algorithms then take the form

$$Q^{\mathrm{CD}}(f) := \sum_{|\mathfrak{u}|<\infty} Q_{n_{\mathfrak{u}},\mathfrak{u}}(f_{\mathfrak{u}}), \tag{7.8}$$

where, for a finite subset $\mathfrak{u}$,

$$Q_{n_{\mathfrak{u}},\mathfrak{u}}(f_{\mathfrak{u}}) = \frac{1}{n_{\mathfrak{u}}} \sum_{i=0}^{n_{\mathfrak{u}}-1} f_{\mathfrak{u}}(\boldsymbol{t}_{\mathfrak{u}}^{(i)}),$$

with $\boldsymbol{t}_{\mathfrak{u}}^{(i)} \in [0,1]^{|\mathfrak{u}|}$, and we use the convention that $Q_{n_{\mathfrak{u}},\mathfrak{u}}(f_{\mathfrak{u}})$ vanishes if $n_{\mathfrak{u}} = 0$. Thus in the CD algorithm a different cubature rule is applied to each component of the anchored decomposition. We assumed for simplicity that the cubature rule $Q_{n_{\mathfrak{u}},\mathfrak{u}}$ is a QMC rule, but other choices are clearly possible. Roughly speaking, this algorithm has a significant advantage if for a particular f the important subsets $\mathfrak{u}$ all have small cardinality (that is, in the language of Caflisch *et al.* 1997, if the 'superposition dimension' is small).

7.3. Cost models and the choice of algorithms

In earlier sections we implicitly assumed that the cost of a single function evaluation is always the same, since in assessing cost we only counted the number of function evaluations. Now that the number of variables is unbounded, it seems reasonable to allow the cost of evaluating $f(\boldsymbol{x}_{\mathfrak{u}};\boldsymbol{c})$ to depend on the set $\mathfrak{u}$. We consider two different cost models.

- Model A: the cost depends on the highest index of the active variables

$$\mathrm{cost}_A(\mathfrak{u}) = \$(\max \mathfrak{u}).$$

- Model B: the cost depends on the number of active variables

$$\mathrm{cost}_B(\mathfrak{u}) = \$(|\mathfrak{u}|).$$

Here the 'dollar' function $\$: \mathbb{N}_0 \to \mathbb{R}_+$ is assumed to be positive and non-decreasing. For example, $\$(k) = [\max(1,k)]^\sigma$ for some $\sigma > 0$. We note that Model A is a simplification of a model introduced in Creutzig, Dereich, Müller-Gronbach and Ritter (2009) and Model B was introduced in Kuo *et al.* (2010*c*).

Whereas in previous sections we assumed that the cost of an algorithm is simply the number of function evaluations, we now assume that the cost of the algorithm given by (7.2) is

$$\mathrm{cost}_X(Q) := \sum_{i=0}^{n-1} \mathrm{cost}_X(\mathfrak{u}_i), \quad X \in \{A, B\}. \tag{7.9}$$

The aim now is to approximate the integral with accuracy $\varepsilon > 0$ by an algorithm with small cost (rather than a small number of function evaluations).

The costs of the SL and ML algorithms under both cost models are, respectively,

$$\mathrm{cost}_X(Q^{\mathrm{SL}}) = n\,\$(s) \quad \text{and} \quad \mathrm{cost}_X(Q^{\mathrm{ML}}) = \sum_{\ell=1}^{L} n_\ell(\$(s_\ell) + \$(s_{\ell-1})),$$

while the cost of the CD algorithm under each cost model is

$$\mathrm{cost}_A(Q^{\mathrm{CD}}) = \sum_{|\mathfrak{u}|<\infty} n_{\mathfrak{u}} \sum_{\mathfrak{v}\subseteq\mathfrak{u}} \$(\max \mathfrak{v}) \le \sum_{|\mathfrak{u}|<\infty} n_{\mathfrak{u}}\,\$(\max \mathfrak{u})\,2^{|\mathfrak{u}|},$$

$$\mathrm{cost}_B(Q^{\mathrm{CD}}) = \sum_{|\mathfrak{u}|<\infty} n_{\mathfrak{u}} \sum_{\mathfrak{v}\subseteq\mathfrak{u}} \$(|\mathfrak{v}|) \le \sum_{|\mathfrak{u}|<\infty} n_{\mathfrak{u}}\,\$(|\mathfrak{u}|)\,2^{|\mathfrak{u}|},$$

where the sum over $\mathfrak{v}$ appears due to the need to compute $f_{\mathfrak{u}}$ using (7.7).

One cost model may be more appropriate than the other, depending on the integrand for the application at hand. The choice of algorithm should therefore also depend on the cost model for the given application.

7.4. A reproducing kernel Hilbert space setting

Now we restrict the function class further. Just as in the finite-dimensional case, we restrict our considerations to functions that lie in a certain reproducing Hilbert space (RKHS), because it is then possible to compute worst-case errors, and to study tractability of the integration problem.

Let $\eta : [0,1]^2 \to \mathbb{R}$ be a reproducing kernel on the interval $[0,1]$ such that there exists a $c \in [0,1]$ with $\eta(c,c) = 0$. Examples of reproducing kernels satisfying this condition are $\min(x,y)$ (see (4.1) with $c=0$) or $1-\max(x,y)$ (see (4.1) with $c=1$). Let the RKHS be denoted by $H(\eta)$, the inner product by $\langle \cdot,\cdot \rangle$ and the corresponding norm by $\|\cdot\|$. Assume that

$$M := \sup_{x\in[0,1]} |\eta(x,x)| < \infty.$$

Then, since

$$|\eta(x,y)| = |\langle \eta(\cdot,x), \eta(\cdot,y)\rangle| \le \|\eta(\cdot,x)\|\|\eta(\cdot,y)\| = \sqrt{\eta(x,x)\eta(y,y)} \qquad (7.10)$$

for all $x,y \in [0,1]$, we have $|\eta(x,c)| \le \sqrt{M\eta(c,c)} = 0$.Thus

$$\eta(x,c) = 0, \quad \text{for all } x \in [0,1], \qquad (7.11)$$

and we have

$$f(c) = \langle f, \eta(\cdot,c)\rangle = \langle f, 0\rangle = 0, \quad \text{for all } f \in H(\eta).$$

Thus all functions $f \in H(\eta)$ vanish at c, in particular, the only constant function in $H(\eta)$ is $f \equiv 0$. This implies that the space of constant functions $H(1)$, defined by the kernel 1, is orthogonal to $H(\eta)$. Moreover, the kernel $1+\eta$ defines an RKHS which is the direct sum of constant functions and functions in $H(\eta)$. This implies that the space of constant functions $H(1)$, defined by the kernel 1, is orthogonal to $H(\eta)$ in the space $H(1+\eta)$.

To define the infinite-dimensional RKHS, for each set $\mathfrak{u} \subset \mathbb{N}$ with $|\mathfrak{u}| < \infty$ let $\gamma_{\mathfrak{u}}$ be a non-negative real number, and for points $\boldsymbol{x}, \boldsymbol{y} \in [0,1]^{\mathbb{N}}$ let

$$K_{\mathfrak{u}}(\boldsymbol{x}_{\mathfrak{u}}, \boldsymbol{y}_{\mathfrak{u}}) := \prod_{j\in\mathfrak{u}} \eta(x_j,y_j) \quad \text{and} \quad K_{\boldsymbol{\gamma}}(\boldsymbol{x},\boldsymbol{y}) := \sum_{|\mathfrak{u}|<\infty} \gamma_{\mathfrak{u}} K_{\mathfrak{u}}(\boldsymbol{x}_{\mathfrak{u}}, \boldsymbol{y}_{\mathfrak{u}}), \qquad (7.12)$$

where again the empty product is defined to be 1. In the following we assume that

$$\sum_{|\mathfrak{u}|<\infty} \gamma_{\mathfrak{u}} M^{|\mathfrak{u}|} < \infty. \qquad (7.13)$$

Condition (7.13) ensures that the series expression for $K_{\boldsymbol{\gamma}}(\boldsymbol{x},\boldsymbol{y})$ converges absolutely and uniformly, since from (7.10) we have $|\eta(x,y)| \le M$. The kernel $K_{\mathfrak{u}}$ defines an RKHS $H(K_{\mathfrak{u}})$ on $[0,1]^{|\mathfrak{u}|}$, and the kernel $K_{\boldsymbol{\gamma}}$ defines an RKHS $H(K_{\boldsymbol{\gamma}})$ on $[0,1]^{\mathbb{N}}$. The inner products in $H(K_{\boldsymbol{\gamma}})$ and $H(K_{\mathfrak{u}})$ will be denoted by $\langle\cdot,\cdot\rangle_{\boldsymbol{\gamma}}$ and $\langle\cdot,\cdot\rangle_{\mathfrak{u}}$ respectively, and the corresponding norms by $\|\cdot\|_{\boldsymbol{\gamma}}$ and $\|\cdot\|_{\mathfrak{u}}$.

Note that a condition significantly weaker than (7.13) would be sufficient to define infinite-dimensional integration, but to avoid some technicalities we shall be content with (7.13). In particular, our assumptions imply that function evaluation is continuous at every point $\boldsymbol{x} \in [0,1]^{\mathbb{N}}$. The idea for weakening the conditions is to impose conditions such that infinite-dimensional integration and function evaluation at finite-dimensional projections are continuous linear functionals. See the work by Gnewuch, Mayer and Ritter (2013) for even weaker conditions.

A function $f \in H(K_\gamma)$ can be decomposed into a sum of functions $f_{\mathfrak{u}} \in H(K_{\mathfrak{u}})$ by setting

$$f_{\mathfrak{u}}(\boldsymbol{x}_{\mathfrak{u}}) := \langle f, \gamma_{\mathfrak{u}} K_{\mathfrak{u}}(\cdot, \boldsymbol{x}_{\mathfrak{u}}) \rangle_\gamma, \tag{7.14}$$

which implies

$$\sum_{|\mathfrak{u}|<\infty} f_{\mathfrak{u}}(\boldsymbol{x}_{\mathfrak{u}}) = \sum_{|\mathfrak{u}|<\infty} \langle f, \gamma_{\mathfrak{u}} K_{\mathfrak{u}}(\cdot, \boldsymbol{x}_{\mathfrak{u}}) \rangle_\gamma = \left\langle f, \sum_{|\mathfrak{u}|<\infty} \gamma_{\mathfrak{u}} K_{\mathfrak{u}}(\cdot, \boldsymbol{x}_{\mathfrak{u}}) \right\rangle_\gamma = f(\boldsymbol{x}). \tag{7.15}$$

The decomposition $f = \sum_{|\mathfrak{u}|<\infty} f_{\mathfrak{u}}$ is orthogonal in $H(K_\gamma)$, since $H(\eta)$ is orthogonal to the space of constant functions and for $\mathfrak{u} \neq \mathfrak{v}$ either there exists $j \in \mathfrak{u}, j \notin \mathfrak{v}$ or there exists $j \in \mathfrak{v}, j \notin \mathfrak{u}$, implying that $\langle K_{\mathfrak{u}}(\cdot, \boldsymbol{x}_{\mathfrak{u}}), K_{\mathfrak{v}}(\cdot, \boldsymbol{y}_{\mathfrak{v}}) \rangle_\gamma = 0$ and hence by (7.14) we have $\langle f_{\mathfrak{u}}, K_{\mathfrak{v}}(\cdot, \boldsymbol{y}_{\mathfrak{v}}) \rangle_\gamma = 0$.

By orthogonality, the norm in $H(K_\gamma)$ is related to the norm of the projections by the formula

$$\|f\|_\gamma^2 = \sum_{|\mathfrak{u}|<\infty} \|f_{\mathfrak{u}}\|_\gamma^2 = \sum_{|\mathfrak{u}|<\infty} \gamma_{\mathfrak{u}}^{-1} \|f_{\mathfrak{u}}\|_{\mathfrak{u}}^2.$$

We now study integration of functions defined on $H(K_\gamma)$. Let

$$h_{\mathfrak{u}}(\boldsymbol{y}_{\mathfrak{u}}) := \int_{[0,1]^{|\mathfrak{u}|}} K_{\mathfrak{u}}(\boldsymbol{y}_{\mathfrak{u}}, \boldsymbol{x}_{\mathfrak{u}}) \, \mathrm{d}\boldsymbol{x}_{\mathfrak{u}} = \prod_{j\in\mathfrak{u}} \int_0^1 \eta(y_j, x_j) \, \mathrm{d}x_j,$$

and

$$h(\boldsymbol{y}) := \sum_{|\mathfrak{u}|<\infty} \gamma_{\mathfrak{u}} h_{\mathfrak{u}}(\boldsymbol{y}_{\mathfrak{u}}).$$

From (7.13) we conclude that this series converges absolutely and uniformly. The functions $h_{\mathfrak{u}}$ are orthogonal in $H(K_\gamma)$ (since $h_{\mathfrak{u}} \in H(K_{\mathfrak{u}})$), from which it follows that

$$\begin{aligned} \|h\|_\gamma^2 &= \sum_{|\mathfrak{u}|<\infty} \gamma_{\mathfrak{u}}^2 \|h_{\mathfrak{u}}\|_\gamma^2 = \sum_{|\mathfrak{u}|<\infty} \gamma_{\mathfrak{u}}^2 \gamma_{\mathfrak{u}}^{-1} \|h_{\mathfrak{u}}\|_{\mathfrak{u}}^2 \qquad (7.16)\\ &= \sum_{|\mathfrak{u}|<\infty} \gamma_{\mathfrak{u}}^2 \gamma_{\mathfrak{u}}^{-1} \int_{[0,1]^{|\mathfrak{u}|}} \int_{[0,1]^{|\mathfrak{u}|}} \langle K_{\mathfrak{u}}(\cdot, \boldsymbol{x}), K_{\mathfrak{u}}(\cdot, \boldsymbol{y}) \rangle_{\mathfrak{u}} \, \mathrm{d}\boldsymbol{x} \, \mathrm{d}\boldsymbol{y} \\ &= \sum_{|\mathfrak{u}|<\infty} \gamma_{\mathfrak{u}} \prod_{j\in\mathfrak{u}} \int_0^1 \int_0^1 \eta(x_j, y_j) \, \mathrm{d}x_j \, \mathrm{d}y_j < \sum_{|\mathfrak{u}|<\infty} \gamma_{\mathfrak{u}} M^{|\mathfrak{u}|} < \infty, \end{aligned}$$

and thus $h \in H(K_\gamma)$. Note that this decomposition of h differs from the decomposition (7.15) (which holds for all functions in $H(K_\gamma)$) by a scaling factor $\gamma_{\mathfrak{u}}$ for each term $h_{\mathfrak{u}}$.

For $f \in H(K_\gamma)$ it follows from (7.1) and (7.12) that

$$\begin{aligned}\int_{[0,1]^{\mathbb{N}}} f(\boldsymbol{x})\,\mathrm{d}\boldsymbol{x} &= \lim_{s\to\infty} \int_{[0,1]^s} f(\boldsymbol{x}_{\{1:s\}};\boldsymbol{c})\,\mathrm{d}\boldsymbol{x} \\ &= \lim_{s\to\infty} \int_{[0,1]^s} \Big\langle f, \sum_{\mathfrak{u}\subseteq\{1:s\}} \gamma_{\mathfrak{u}} K_{\mathfrak{u}}(\boldsymbol{x}_{\mathfrak{u}},\cdot)\Big\rangle_\gamma \,\mathrm{d}\boldsymbol{x} \\ &= \lim_{s\to\infty} \Big\langle f, \sum_{\mathfrak{u}\subseteq\{1:s\}} \gamma_{\mathfrak{u}} h_{\mathfrak{u}}\Big\rangle_\gamma = \langle f, h\rangle_\gamma. \end{aligned} \tag{7.17}$$

Since $\|I_\infty\| = \|h\|_\gamma < \infty$, it follows that I_∞ is a bounded linear functional on $H(K_\gamma)$.

7.5. Error analysis in RKHS

The analysis of the worst-case error is based on the orthogonal decomposition in the space $H(K_\gamma)$. The representer (see (3.11)) of the integration error is given by

$$\xi(\boldsymbol{x}) = \int_{[0,1]^{\mathbb{N}}} K_\gamma(\boldsymbol{y},\boldsymbol{x})\,\mathrm{d}\boldsymbol{y} - Q(K_\gamma(\cdot,\boldsymbol{x})),$$

where, in general, Q is a cubature algorithm of the form (7.2) which is applied to the first variable of K_γ (later we will consider only the SL, ML, CD algorithms). As explained in Section 3.3, the worst-case integration error is given by

$$e(Q;H(K_\gamma)) = \|\xi\|_\gamma.$$

(In this subsection we use the notation $e(Q;H(K_\gamma))$ instead of $e(P;H(K_\gamma))$ as in Section 3.3, since we now allow unequal cubature weights in the algorithm.) Analogously to (7.16), we can obtain an orthogonal decomposition of the worst-case error

$$\begin{aligned} e^2(Q;H(K_\gamma)) = \|\xi\|_\gamma^2 &= \sum_{|\mathfrak{u}|<\infty} \gamma_{\mathfrak{u}}^2 \|\xi_{\mathfrak{u}}\|_\gamma^2 \\ &= \sum_{|\mathfrak{u}|<\infty} \gamma_{\mathfrak{u}} \|\xi_{\mathfrak{u}}\|_{\mathfrak{u}}^2 = \sum_{|\mathfrak{u}|<\infty} \gamma_{\mathfrak{u}}\, e^2(Q_{\mathfrak{u}};H(K_{\mathfrak{u}})), \end{aligned}$$

where $e(Q_{\mathfrak{u}};H(K_{\mathfrak{u}}))$ is the worst-case error in the space $H(K_{\mathfrak{u}})$ using the cubature rule

$$Q_{\mathfrak{u}}(f) = \sum_{i=0}^{n-1} w_i f_{\mathfrak{u}}(\boldsymbol{t}^{(i)}_{\mathfrak{u}_i\cap\mathfrak{u}};\boldsymbol{c}).$$

The relationship between f and its projection $f_{\mathfrak{u}}$ is given by (7.6) and (7.7). Note that if for $\mathfrak{u} \neq \emptyset$ there does not exist at least one $0 \leq i < n$ such that $\mathfrak{u} \subseteq \mathfrak{u}_i$, then $Q_{\mathfrak{u}}(f) = 0$ since $f_{\mathfrak{u}}(\boldsymbol{t}_{\mathfrak{v}}; \boldsymbol{c}) = 0$ for $\mathfrak{v} \subseteq \mathfrak{u}$ and $\mathfrak{v} \neq \mathfrak{u}$. This orthogonal decomposition of the worst-case error is an essential tool in analysing the integration error. We consider now the orthogonal decompositions of the worst-case error for the SL, ML and CD algorithms in more detail.

For the SL algorithm we obtain

$$\begin{aligned} e^2(Q^{\mathrm{SL}}; H(K_\gamma)) &= \sum_{\mathfrak{u} \subseteq \{1:s\}} \gamma_{\mathfrak{u}}\, e^2(Q_{\mathfrak{u}}; H(K_{\mathfrak{u}})) + \sum_{\substack{|\mathfrak{u}|<\infty \\ \mathfrak{u} \not\subseteq \{1:s\}}} \gamma_{\mathfrak{u}}\, e^2(0; H(K_{\mathfrak{u}})) \\ &= e^2(Q^{\mathrm{SL}}; H(K_{s,\gamma})) + \sum_{\substack{|\mathfrak{u}|<\infty \\ \mathfrak{u} \not\subseteq \{1:s\}}} \gamma_{\mathfrak{u}}\, m^{|\mathfrak{u}|}, \end{aligned}$$

where $e^2(0; H(K_{\mathfrak{u}}))$ is the squared initial error (see (3.10))

$$e^2(0; H(K_{\mathfrak{u}})) = m^{|\mathfrak{u}|}, \quad m := \int_0^1 \int_0^1 \eta(x, y)\, \mathrm{d}x\, \mathrm{d}y \leq M,$$

and $e^2(Q^{\mathrm{SL}}; H(K_{s,\gamma}))$ is the squared worst-case error of the cubature rule for the finite-dimensional weighted space $H(K_{s,\gamma})$ with reproducing kernel

$$K_{s,\gamma}(\boldsymbol{x}, \boldsymbol{y}) = \sum_{\mathfrak{u} \subseteq \{1:s\}} \gamma_{\mathfrak{u}} K_{\mathfrak{u}}(\boldsymbol{x}_{\mathfrak{u}}, \boldsymbol{y}_{\mathfrak{u}}), \quad \boldsymbol{x}, \boldsymbol{y} \in [0,1]^s.$$

We can then apply any estimate for $e^2(Q^{\mathrm{SL}}; H(K_{s,\gamma}))$ based on lattice rules or digital nets from the previous sections. We can also use general cubature rules other than QMC rules.

The ML algorithm approximates each level ℓ using a cubature rule with n_ℓ points. To bound the ML cubature error, note that from (7.15) we have

$$f(\boldsymbol{x}_{\{1:s_\ell\}}; \boldsymbol{c}) - f(\boldsymbol{x}_{\{1:s_{\ell-1}\}}; \boldsymbol{c}) = \sum_{\mathfrak{u} \subseteq \{1:s_\ell\}} f_{\mathfrak{u}}(\boldsymbol{x}_{\mathfrak{u}}) - \sum_{\mathfrak{u} \subseteq \{1:s_{\ell-1}\}} f_{\mathfrak{u}}(\boldsymbol{x}_{\mathfrak{u}}).$$

Thus the error for the ML algorithm satisfies

$$\begin{aligned} & e^2(Q^{\mathrm{ML}}; H(K_\gamma)) \\ &= \sum_{\ell=1}^{L} \sum_{\substack{\mathfrak{u} \subseteq \{1:s_\ell\} \\ \mathfrak{u} \not\subseteq \{1:s_{\ell-1}\}}} \gamma_{\mathfrak{u}} e^2(Q_{\mathfrak{u}}^{\mathrm{ML}}; H(K_{\mathfrak{u}})) + \sum_{\substack{|\mathfrak{u}|<\infty \\ \mathfrak{u} \not\subseteq \{1:s_L\}}} \gamma_{\mathfrak{u}}\, e^2(0; H(K_{\mathfrak{u}})) \end{aligned}$$

$$= \sum_{\ell=1}^{L} e^2(Q_{n_\ell,\ell}; H(K_{s_\ell,\gamma} - K_{s_{\ell-1},\gamma})) + \sum_{\substack{|\mathfrak{u}|<\infty \\ \mathfrak{u}\not\subseteq\{1:s_L\}}} \gamma_{\mathfrak{u}}\, m^{|\mathfrak{u}|}$$

$$\le \sum_{\ell=1}^{L} e^2(Q_{n_\ell,\ell}; H(K_{s_\ell,\gamma})) + \sum_{\substack{|\mathfrak{u}|<\infty \\ \mathfrak{u}\not\subseteq\{1:s_L\}}} \gamma_{\mathfrak{u}}\, m^{|\mathfrak{u}|}.$$

We then apply previous results for $e^2(Q_{n_\ell,\ell}; H(K_{s_\ell,\gamma} - K_{s_{\ell-1},\gamma}))$ or simply $e^2(Q_{n_\ell,\ell}; H(K_{s_\ell,\gamma}))$.

For the CD algorithm we have

$$e^2(Q^{\mathrm{CD}}; H(K_\gamma)) = \sum_{|\mathfrak{u}|<\infty} \gamma_{\mathfrak{u}}\, e^2(Q_{n_{\mathfrak{u}},\mathfrak{u}}; H(K_{\mathfrak{u}}))$$

$$= \sum_{\substack{|\mathfrak{u}|<\infty \\ n_{\mathfrak{u}}>0}} \gamma_{\mathfrak{u}}\, e^2(Q_{n_{\mathfrak{u}},\mathfrak{u}}; H(K_{\mathfrak{u}})) + \sum_{\substack{|\mathfrak{u}|<\infty \\ n_{\mathfrak{u}}=0}} \gamma_{\mathfrak{u}}\, m^{|\mathfrak{u}|}.$$

Again, we then apply known estimates for $e^2(Q_{n_{\mathfrak{u}},\mathfrak{u}}; H(K_{\mathfrak{u}}))$.

The idea now is to choose the number of cubature points (n for SL, n_ℓ for ML, $n_{\mathfrak{u}}$ for CD) and the dimensions or sets of variables (s for SL, s_ℓ for ML, active sets $\mathfrak{u}$ for CD), such that the error is small for a given cost, or conversely, the cost is small for a given required level of accuracy. One needs to balance the error obtained by truncating the infinite-dimensional integral against the integration error. This leads to a constrained optimization problem which in some cases can be solved using the technique of Lagrange multipliers. For a given application, one would also need to determine suitable weights $\gamma_{\mathfrak{u}}$ so that the integrand not only belongs to the RKHS setting but also has small norm $\|\cdot\|_\gamma$.

7.6. An infinite-dimensional application

Do infinite-dimensional problems arise in practice? A class of infinite-dimensional problems arises from partial differential equations with random fields as coefficients, with one problem being the flow of a liquid (oil or water) through a porous medium (such as rock). For a given pressure gradient, the rate of flow at a particular location depends on the local 'permeability' of the medium, which can vary rapidly from point to point. Engineers studying the overall flow properties then have two choices: either to model the true permeability as a rapidly varying function over the region, or (as is often done in practice) to model the permeability as a 'random field' whose general characteristics match those of the physical problem. A random field over a two- or three-dimensional region may require a countably infinite number of independent random variables for its complete description, hence the infinite-dimensionality of the problem. Such problems are presently tackled by a variety of methods, beginning with Norbert Wiener's 'polynomial chaos', and continuing to 'generalized polynomial chaos', 'stochastic Galerkin', and 'stochastic collocation' methods. In all those methods the probabilistic aspect of the problem is parametrized by a continuous variable $\boldsymbol{x}$ with a large (possibly infinite) number of components. The solution is approximated by truncation to a manageable number of components of $\boldsymbol{x}$, and an approximate solution depending on both the truncated $\boldsymbol{x}$ and the physical variables is then sought in a finite-dimensional function space. All such methods face great challenges when the effective dimension is large (where the effective dimension may be loosely defined as the number of components of $\boldsymbol{x}$ needed to obtain a reliable approximation). For hard problems of this type MC or QMC methods may be the engineer's only choice (Ghanem and Spanos 1991).

In a recent paper (Kuo *et al.* 2012) a QMC method (randomly shifted lattice rule) was applied to a simple (but infinite-dimensional) model of a partial differential equation with a random coefficient, using the methodology of Section 5, after truncation to a finite dimension s. An interesting feature of this work was the appearance for the first time of POD (product and order-dependent) weights (see (4.4)), obtained by minimizing a certain upper bound on the error of some functional (for example, the overall effective permeability) of the solution. An ML version of the same problem was studied in Kuo, Schwab and Sloan (2013), using an ML scheme more complicated than that above, in that a different spatial discretization was employed at each level ℓ.

7.7. Notes

Early studies of numerical integration for infinite-dimensional problems (and ML methods) include the following: Wasilkowski and Woźniakowski (1995) considered Feynman–Kac path integration, Wasilkowski and Woźniakowski (1996) considered path integration, Heinrich (1998) studied ML in the context of integral equations, Heinrich and Sindambiwe (1999) studied ML in the context of parametric integration, Hickernell and Wang (2002) studied QMC algorithms for infinite-dimensional integration. Giles (2008) considered ML MC schemes for stochastic partial differential equations.

Creutzig *et al.* (2009) introduced the variable subspace sampling model (which generalizes cost model A) and obtained optimality results for ML algorithms for this model, which apply in particular to infinite-dimensional integration of stochastic differential equations. Infinite-dimensional integration using randomized MC algorithms in a hierarchy of finite-dimensional subspaces was studied by Hickernell, Müller-Gronbach, Niu and Ritter (2010). Niu, Hickernell, Müller-Gronbach and Ritter (2011) studied deterministic ML algorithms. Niu and Hickernell (2009) studied MC simulation of infinite-dimensional stochastic integrals arising in financial applications. The CD algorithm combined with QMC methods was introduced in Kuo *et al.* (2010*c*). The above results for ML and CD algorithms were improved by Gnewuch (2012*b*). Plaskota and Wasilkowski (2011) considered the CD algorithm combined with Smolyak algorithms and studied tractability in the worst-case and randomized setting. Higher-order convergence using higher-order polynomial lattice rules in infinite-dimensional anchored spaces was studied by Dick and Gnewuch (2013). Baldeaux (2012) and Baldeaux and Gnewuch (2013) studied randomly scrambled polynomial lattice rules for infinite-dimensional integration, the latter in unanchored Sobolev spaces. Lower bounds were considered by Gnewuch (2013). A probabilistic ML algorithm which yields an unbiased estimator was introduced by Rhee and Glynn (2012).

The mathematics of random fields is well described in the book by Adler (1981). For engineering applications of the MC method to problems of porous flow: see for instance Ghanem and Spanos (1991). The ML MC approach has recently been developed for elliptic problems with random input data in Barth, Schwab and Zollinger (2011), Charrier, Scheichl and Teckentrup (2011), Cliffe, Giles, Scheichl and Teckentrup (2011), Schwab and Gittelson (2011) and Teckentrup, Scheichl, Giles and Ullmann (2012). Application of QMC methods to these PDE problems has been considered in Graham *et al.* (2011), Kuo *et al.* (2012, 2013) and Graham *et al.* (2013).

8. Concluding remarks

In this survey we have tried to capture key concepts and methods in the rapidly developing field of high-dimensional numerical integration. We have concentrated on equal-weight rules, not because they will always be the best rules for a particular problem (indeed, in low dimensions we know that this is certainly not the case), but because they can be used in practice for very high-dimensional problems. We have not covered sparse grid methods, reviewed in a previous *Acta Numerica* article by Bungartz and Griebel (2004), which also provide a well-understood approach to integration in moderately high dimensions, and which are especially attractive for the related topic of approximation of functions.

In this rapidly developing subject it is almost certain that future research will give a different emphasis to the material covered in this survey, but we think it likely that topics such as reproducing kernels, weighted function spaces, component-by-component constructions, and discrepancy in all its forms, will continue to play starring roles in the science of high-dimensional numerical integration.

Acknowledgements

We gratefully acknowledge the the support of the Australian Research Council. J. Dick is supported by an Australian Research Council QEII Fellowship.

We have received valuable comments and corrections from Johann Brauchart, Michael Gnewuch, Mario Hefter, Hernan Leovey, Sebastian Mayer, Harald Niederreiter, Dirk Nuyens, Steffen Omland, Art Owen, Klaus Ritter, Grzegorz Wasilkowski, Markus Weimar, and Henryk Woźniakowski.

REFERENCES

N. Achtsis and D. Nuyens (2012), A component-by-component construction for the trigonometric degree. In *Monte Carlo and Quasi-Monte Carlo Methods 2010* (L. Plaskota and H. Woźniakowski, eds), Springer, pp. 235–253.

P. Acworth, M. Broadie and P. Glasserman (1998), A comparison of some Monte Carlo and quasi-Monte Carlo techniques for option pricing. In *Monte Carlo and Quasi-Monte Carlo Methods 1996* (P. Hellekalek, G. Larcher, H. Niederreiter and P. Zinterhof, eds), Springer, pp. 1–18.

R. J. Adler (1981), *The Geometry of Random Fields*, Wiley.

C. Aistleitner (2011), 'Covering numbers, dyadic chaining and discrepancy', *J. Complexity* **27**, 531–540.

C. Aistleitner and M. Hofer (2012), 'Probabilistic error bounds for the discrepancy of mixed sequences', *Monte Carlo Methods Appl.* **18**, 181–200.

I. A. Antonov and V. M. Saleev (1979), 'An effective method for the computation of λP_τ-sequences', *Ž. Vyčisl. Mat. i Mat. Fiz.* **19**, 243–245, 271.

N. Aronszajn (1950), 'Theory of reproducing kernels', *Trans. Amer. Math. Soc.* **68**, 337–404.

E. I. Atanassov (2004*a*), Efficient CPU-specific algorithm for generating the generalized Faure sequences. In *Large-Scale Scientific Computing*, Vol. 2907 of *Lecture Notes in Computer Science*, Springer, pp. 121–127.

E. I. Atanassov (2004*b*), 'On the discrepancy of Halton sequences', *Math. Balkanica* **18**, 15–32.

N. S. Bakhvalov (1959), 'On approximate computation of integrals' (in Russian), *Vestnik MGU, Ser. Math. Mech. Astron. Phys. Chem.* **4**, 3–18.

J. Baldeaux (2012), Scrambled polynomial lattice rules for infinite-dimensional integration. In *Monte Carlo and Quasi-Monte Carlo Methods 2010* (L. Plaskota and H. Woźniakowski, eds), Springer, pp. 255–263.

J. Baldeaux and J. Dick (2009), 'QMC rules of arbitrary high order: Reproducing kernel Hilbert space approach', *Constr. Approx.* **30**, 495–527.

J. Baldeaux and J. Dick (2011), 'A construction of polynomial lattice rules with small gain coefficients', *Numer. Math.* **119**, 271–297.

J. Baldeaux and M. Gnewuch (2013), Optimal randomized multilevel algorithms for infinite-dimensional integration on function spaces with ANOVA-type decomposition. Submitted.

J. Baldeaux, J. Dick, J. Greslehner and F. Pillichshammer (2011), 'Construction algorithms for higher order polynomial lattice rules', *J. Complexity* **27**, 281–299.

J. Baldeaux, J. Dick, G. Leobacher, D. Nuyens and F. Pillichshammer (2012), 'Efficient calculation of the worst-case error and (fast) component-by-component construction of higher order polynomial lattice rules', *Numer. Algorithms* **59**, 403–431.

A. Barth, C. Schwab and N. Zollinger (2011), 'Multi-level Monte Carlo finite element method for elliptic PDEs with stochastic coefficients', *Numer. Math.* **119**, 123–161.

P. Bratley and B. L. Fox (1988), 'Algorithm 659: Implementing Sobol′s quasirandom sequence generator', *ACM Trans. Math. Softw.* **14**, 88–100.

P. Bratley, B. L. Fox, and H. Niederreiter (1992), 'Implementation and tests of low-discrepancy sequences', *ACM Trans. Model. Comput. Simul.* **2**, 195–213.

H. Bungartz and M. Griebel (2004), Sparse grids. In *Acta Numerica*, Vol. 13, Cambridge University Press, pp. 147–269.

R. E. Caflisch, W. Morokoff and A. Owen (1997), 'Valuation of mortgage backed securities using Brownian bridges to reduce effective dimension', *J. Comput. Finance* **1**, 27–46.

J. Charrier, R. Scheichl and A. L. Teckentrup (2011), Finite element error analysis of elliptic PDEs with random coefficients and its application to multilevel Monte Carlo methods. Preprint 2/11, Bath Institute For Complex Systems.

B. Chazelle (2000), *The Discrepancy Method: Randomness and Complexity*, Cambridge University Press.

H. E. Chrestenson (1955), 'A class of generalized Walsh functions', *Pacific J. Math.* **5**, 17–31.

K. L. Chung (1974), *A Course in Probability Theory*, second edition, Vol. 21 of *Probability and Mathematical Statistics*, Academic Press.

K. A. Cliffe, M. B. Giles, R. Scheichl and A. L. Teckentrup (2011), 'Multilevel Monte Carlo methods and applications to elliptic PDEs with random coefficients', *Comput. Vis. Sci.* **14**, 3–15.

R. Cools (1997), Constructing cubature formulae: The science behind the art. In *Acta Numerica*, Vol. 6, Cambridge University Press, pp. 1–54.

R. Cools and J. N. Lyness (2001), 'Three- and four-dimensional K-optimal lattice rules of moderate trigonometric degree', *Math. Comp.* **70**, 1549–1567.

R. Cools and D. Nuyens (2008), A Belgian view on lattice rules. In *Monte Carlo and Quasi-Monte Carlo Methods 2006* (A. Keller, S. Heinrich and H. Niederreiter, eds), Springer, pp. 3–21.

R. Cools, F. Y. Kuo and D. Nuyens (2006), 'Constructing embedded lattice rules for multivariate integration', *SIAM J. Sci. Comput.* **28**, 2162–2188.

R. Cools, F. Y. Kuo and D. Nuyens (2010), 'Constructing lattice rules based on weighted degree of exactness and worst case error', *Computing* **87**, 63–89.

J. Creutzig, S. Dereich, T. Müller-Gronbach and K. Ritter (2009), 'Infinite-dimensional quadrature and approximation of functions', *Found. Comp. Math.* **9**, 391–429.

P. J. Davis and P. Rabinowitz (1984), *Methods of Numerical Integration*, second edition, Academic.

L. Devroye (1986), *Nonuniform Random Variate Generation*, Springer.

J. Dick (2004), 'On the convergence rate of the component-by-component construction of good lattice rules', *J. Complexity* **20**, 493–522.

J. Dick (2007*a*), 'Explicit constructions of quasi-Monte Carlo rules for the numerical integration of high-dimensional periodic functions', *SIAM J. Numer. Anal.* **45**, 2141–2176.

J. Dick (2007*b*), 'The construction of extensible polynomial lattice rules with small weighted star discrepancy', *Math. Comp.* **76**, 2077–2085.

J. Dick (2008), 'Walsh spaces containing smooth functions and quasi-Monte Carlo rules of arbitrary high order', *SIAM J. Numer. Anal.* **46**, 1519–1553.

J. Dick (2009*a*), 'The decay of the Walsh coefficients of smooth functions', *Bull. Aust. Math. Soc.* **80**, 430–453.

J. Dick (2009*b*), On Quasi-Monte Carlo rules achieving higher order convergence. In *Monte Carlo and Quasi-Monte Carlo Methods 2008* (P. L'Ecuyer and A. B. Owen, eds), Springer, pp. 73–96.

J. Dick (2011*a*), 'Higher order scrambled digital nets achieve the optimal rate of the root mean square error for smooth integrands', *Ann. Statist.* **39**, 1372–1398.

J. Dick (2011*b*), 'Quasi-Monte Carlo numerical integration on $\mathbb{R}^s$: Digital nets and worst-case error', *SIAM J. Numer. Anal.* **49**, 1661–1691.

J. Dick (2012), 'Random weights, robust lattice rules and the geometry of the cbcrc algorithm', *Numer. Math.* **122**, 443–467.

J. Dick and J. Baldeaux (2009), Equidistribution properties of generalized nets and sequences. In *Monte Carlo and Quasi-Monte Carlo Methods 2008* (P. L'Ecuyer and A. B. Owen, eds), Springer, pp. 305–322.

J. Dick and M. Gnewuch (2013), Infinite-dimensional integration in weighted Hilbert spaces: Anchored decompositions, optimal deterministic algorithms, and higher order convergence. Submitted.

J. Dick and P. Kritzer (2010), 'Duality theory and propagation rules for generalized digital nets', *Math. Comp.* **79**, 993–1017.

J. Dick and F. Y. Kuo (2004*a*), 'Reducing the construction cost of the component-by-component construction of good lattice rules', *Math. Comp.* **73**, 1967–1988.

J. Dick and F. Y. Kuo (2004*b*), Constructing good lattice rules with millions of points. In *Monte Carlo and Quasi-Monte Carlo Methods 2002* (H. Niederreiter, ed.), Springer, pp. 181–197.

J. Dick and F. Pillichshammer (2005), 'Multivariate integration in weighted Hilbert spaces based on Walsh functions and weighted Sobolev spaces', *J. Complexity* **21**, 149–195.

J. Dick and F. Pillichshammer (2007), 'Strong tractability of multivariate integration of arbitrary high order using digitally shifted polynomial lattice rules', *J. Complexity* **23**, 436–453.

J. Dick and F. Pillichshammer (2010), *Digital Nets and Sequences*, Cambridge University Press.

J. Dick, P. Kritzer, G. Leobacher and F. Pillichshammer (2007*a*), 'Constructions of general polynomial lattice rules based on the weighted star discrepancy', *Finite Fields Appl.* **13**, 1045–1070.

J. Dick, P. Kritzer, F. Pillichshammer and W. C. Schmid (2007*b*), 'On the existence of higher order polynomial lattices based on a generalized figure of merit', *J. Complexity* **23**, 581–593.

J. Dick, F .Y. Kuo, F. Pillichshammer and I. H. Sloan (2005), 'Construction algorithms for polynomial lattice rules for multivariate integration', *Math. Comp.* **74**, 1895–1921.

J. Dick, G. Larcher, F. Pillichshammer and H. Woźniakowski (2011), 'Exponential convergence and tractability of multivariate integration for Korobov spaces', *Math. Comp.* **80**, 905–930.

J. Dick, D. Nuyens and F. Pillichshammer (2013), Lattice rules for nonperiodic smooth integrands. Submitted.

J. Dick, F. Pillichshammer, and B. J. Waterhouse (2008), 'The construction of good extensible rank-1 lattices', *Math. Comp.* **77**, 2345–2374.

J. Dick, I. H. Sloan, X. Wang and H. Woźniakowski (2006), 'Good lattice rules in weighted Korobov spaces with general weights', *Numer. Math.* **103**, 63–97.

Digital Library of Mathematical Functions, National Institute of Standards and Technology.

B. Doerr, M. Gnewuch, P. Kritzer and F. Pillichshammer (2008), 'Component-by-component construction of low-discrepancy point sets of small size', *Monte Carlo Methods Appl.* **14**, 129–149.

B. Doerr, M. Gnewuch and M. Wahlström (2009), Implementation of a component-by-component algorithm to generate small low-discrepancy samples. In *Monte Carlo and Quasi-Monte Carlo Methods 2008* (P. L'Ecuyer and A. B. Owen, eds), Springer, pp. 323–338.

B. Doerr, M. Gnewuch and M. Wahlström (2010), 'Algorithmic construction of low-discrepancy point sets via dependent randomized rounding', *J. Complexity* **26**, 490–507.

M. Drmota and R. F. Tichy (1997), *Sequences, Discrepancies and Applications*, Vol. 1651 of *Lecture Notes in Mathematics*, Springer.

K. T. Fang and Y. Wang (1994), *Number-Theoretic Methods in Statistics*, Chapman & Hall.

H. Faure (1982), 'Discrépance de suites associées à un système de numération (en dimension s)', *Acta Arith.* **41**, 337–351.

N. J. Fine (1949), 'On the Walsh functions', *Trans. Amer. Math. Soc.* **65**, 372–414.

B. L. Fox (1986), 'Algorithm 647: Implementation and relative efficiency of quasirandom sequence generators', *ACM Trans. Math. Softw.* **12**, 362–376.

R. G. Ghanem and P. D. Spanos (1991), *Stochastic Finite Elements: A Spectral Approach*, Springer.

M. B. Giles (2008), 'Multilevel Monte Carlo path simulation', *Oper. Res.* **56**, 607–617.

M. Giles, F. Y. Kuo, I. H. Sloan and B. J. Waterhouse (2008), 'Quasi-Monte Carlo for finance applications', *ANZIAM J.* **50** (CTAC2008), C308–C323.

M. Gnewuch (2008), 'Bracketing numbers for axis-parallel boxes and applications to geometric discrepancy', *J. Complexity* **24**, 154–172.

M. Gnewuch (2009), 'On probabilistic results for the discrepancy of a hybrid-Monte Carlo sequence', *J. Complexity* **23**, 312–317.

M. Gnewuch (2012*a*), 'Weighted geometric discrepancies and numerical integration on reproducing kernel Hilbert spaces', *J. Complexity* **28**, 2–17.

M. Gnewuch (2012*b*), 'Infinite-dimensional integration on weighted Hilbert spaces', *Math. Comp.* **280**, 2175–2205.

M. Gnewuch (2013), Lower error bounds for randomized multilevel and changing dimension algorithms. In *Monte Carlo and Quasi-Monte Carlo Methods 2012* (J. Dick, F. Y. Kuo, G. W. Peters and I. H. Sloan, eds), Springer, to appear.

M. Gnewuch and A. V. Roşca (2009), 'On G-discrepancy and mixed Monte Carlo and quasi-Monte Carlo sequences', *Acta Univ. Apulensis Math. Inform.* **18**, 97–110.

M. Gnewuch, S. Mayer and K. Ritter (2013), On weighted Hilbert spaces and integration of functions of infinitely many variables. Submitted.

M. Gnewuch, A. Srivastav and C. Winzen (2009), 'Finding optimal volume subintervals with k-points and calculating the star discrepancy are NP-hard problems', *J. Complexity* **25**, 115–127.

M. Gnewuch, M. Wahlström and C. Winzen (2012), 'A new randomized algorithm to approximate the star discrepancy based on threshold accepting', *SIAM J. Numer. Anal.* **50**, 781–807.

T. Goda and J. Dick (2013), Construction of interlaced scrambled polynomial lattice rules of arbitrary high order. Submitted.

I. G. Graham, F. Y. Kuo, J. Nichols, R. Scheichl, C. Schwab and I. H. Sloan (2013), Quasi-Monte Carlo finite element methods for elliptic PDEs with log-normal random coefficients. In preparation.

I. G. Graham, F. Y. Kuo, D. Nuyens, R. Scheichl and I. H. Sloan (2011), 'Quasi-Monte Carlo methods for elliptic PDEs with random coefficients and applications', *J. Comput. Phys.* **230**, 3668–3694.

M. Griebel (2006), Sparse grids and related approximation schemes for higher dimensional problems. In *Foundations of Computational Mathematics, Santander, 2005*, Cambridge University Press, pp. 106–161.

M. Griebel, F. Y. Kuo and I. H. Sloan (2013), 'The smoothing effect of integration in $\mathbb{R}^d$ and the ANOVA decomposition', *Math. Comp.* **82**, 383–400.

J. H. Halton (1960), 'On the efficiency of certain quasi-random sequences of points in evaluating multi-dimensional integrals', *Numer. Math.* **2**, 84–90.

J. M. Hammersley and D. C. Handscomb (1964), *Monte Carlo Methods*, Methuen.

T. Hansen, G. L. Mullen and H. Niederreiter (1993), 'Good parameters for a class of node sets in quasi-Monte Carlo integration', *Math. Comp.* **61**, 225–234.

S. Heinrich (1998), 'Monte Carlo complexity of global solution of integral equations', *J. Complexity* **14**, 151–175.

S. Heinrich and E. Sindambiwe (1999), 'Monte Carlo complexity of parametric integration', *J. Complexity* **15**, 317–341.

S. Heinrich, F. Hickernell and R. X. Yue (2004), 'Optimal quadrature for Haar wavelet spaces', *Math. Comp.* **73**, 259–277.

S. Heinrich, E. Novak, G. W. Wasilkowski, H. Woźniakowski (2001), 'The inverse of the star-discrepancy depends linearly on the dimension', *Acta Arith.* **96**, 279–302.

F. J. Hickernell (1996*a*), 'Quadrature error bounds with applications to lattice rules', *SIAM J. Numer. Anal.* **33**, 1995–2016. Erratum: 'Quadrature error bounds with applications to lattice rules', *SIAM J. Numer. Anal.* **34**, 853–866 (1997).

F. J. Hickernell (1996*b*), 'The mean square discrepancy of randomized nets', *ACM Trans. Modeling Comput. Simul.* **6**, 274–296.

F. J. Hickernell (1998*a*), 'A generalized discrepancy and quadrature error bound', *Math. Comp.* **67**, 299–322.

F. J. Hickernell (1998*b*), Lattice rules: How well do they measure up? In *Random and Quasi-Random Point Sets* (P. Hellekalek and G. Larcher, eds), Springer, pp. 109–166.

F. J. Hickernell (1999), 'Goodness-of-fit statistics, discrepancies and robust designs', *Statist. Probab. Lett.* **44**, 73–78.

F. J. Hickernell (2002), Obtaining $O(N^{-2+\epsilon})$ convergence for lattice quadrature rules. In *Monte Carlo and Quasi-Monte Carlo Methods 2000* (K. T. Fang, F. J. Hickernell and H. Niederreiter, eds), Springer, pp. 274–289.

F. J. Hickernell and H. S. Hong (2002), Quasi-Monte Carlo methods and their randomisations. In *Applied Probability: AMS/IP Studies in Advanced Mathematics*, Vol. 26 (R. Chan, Y.-K. Kwok, D. Yao and Q. Zhang, eds), AMS, pp. 59–77.

F. J. Hickernell and H. Niederreiter (2003), 'The existence of good extensible rank-1 lattices', *J. Complexity* **19**, 286–300.

F. J. Hickernell and X. Wang (2002), 'The error bounds and tractability of quasi-Monte Carlo algorithms in infinite dimension', *Math. Comp.* **71**, 1641–1661.

F. J. Hickernell and H. Woźniakowski (2000), 'Integration and approximation in arbitrary dimensions', *Adv. Comput. Math.* **12**, 25–58.

F. J. Hickernell and H. Woźniakowski (2001), 'Tractability of multivariate integration for periodic functions', *J. Complexity* **17**, 660–682.

F. J. Hickernell and R. X. Yue (2000), 'The mean square discrepancy of scrambled (t,s)-sequences', *SIAM J. Numer. Anal.* **38**, 1089–1112.

F. J. Hickernell, H. S. Hong, P. L'Ecuyer, and C. Lemieux (2000), 'Extensible lattice sequences for quasi-Monte Carlo quadrature', *SIAM J. Sci. Comput.* **22**, 1117–1138.

F. J. Hickernell, P. Kritzer, F. Y. Kuo and D. Nuyens (2012), 'Weighted compound integration rules with higher order convergence for all N', *Numer. Algorithms* **59**, 161–183.

F. J. Hickernell, C. Lemieux and A. B. Owen (2005), 'Control variates for quasi-Monte Carlo', with comments by Pierre L'Ecuyer and Xiao-Li Meng and a rejoinder by the authors. *Statist. Sci.* **20**, 1–31.

F. Hickernell, T. Müller-Gronbach, B. Niu and K. Ritter (2010), 'Monte Carlo algorithms for infinite-dimensional integration on $\mathbb{R}^{\mathbb{N}}$', *J. Complexity* **26**, 229–254.

A. Hinrichs (2004), 'Covering numbers, Vapnik–Červonenkis classes and bounds for the star-discrepancy', *J. Complexity* **20**, 477–483.

A. Hinrichs, F. Pillichshammer and W. C. Schmid (2008), 'Tractability properties of the weighted star discrepancy', *J. Complexity* **24**, 134–143.

E. Hlawka (1961), 'Funktionen von beschränkter Variation in der Theorie der Gleichverteilung', *Ann. Mat. Pura Appl.* **54**, 325–333.

E. Hlawka (1962), 'Zur angenäherten Berechnung mehrfacher Integrale', *Monatsh. Math.* **66**, 140–151.

R. Hofer and P. Kritzer (2011), 'On hybrid sequences built from Niederreiter–Halton sequences and Kronecker sequences', *Bull. Aust. Math. Soc.* **84**, 238–254.

R. Hofer and G. Larcher (2010), 'On existence and discrepancy of certain digital Niederreiter–Halton sequences', *Acta Arith.* **141**, 369–394.

R. Hofer and G. Larcher (2012), 'Metrical results on the discrepancy of Halton–Kronecker sequences', *Math. Z.* **271**, 1–11.

R. Hofer and H. Niederreiter (2013), 'A construction of (t, s)-sequences with finite-row generating matrices using global function fields', *Finite Fields Appl.* **21**, 97–110.

R. Hofer and G. Pirsic (2011), 'An explicit construction of finite-row digital $(0, s)$-sequences', *Unif. Distrib. Theory* **6**, 13–30.

R. Hofer, P. Kritzer, G. Larcher and F. Pillichshammer (2009), 'Distribution properties of generalized van der Corput–Halton sequences and their subsequences', *Int. J. Number Theory* **5**, 719–746.

W. Hörmann, J. Leydold and G. Derflinger (2010), *Automatic Nonuniform Random Variate Generation*, Springer.

S. Joe (2004), Component by component construction of rank-1 lattice rules having $O(n^{-1}(\ln(n))^d)$ star discrepancy. In *Monte Carlo and Quasi-Monte Carlo Methods 2002* (H. Niederreiter, ed.), Springer, pp. 293–298.

S. Joe (2006), Construction of good rank-1 lattice rules based on the weighted star discrepancy. In *Monte Carlo and Quasi-Monte Carlo Methods 2004* (H. Niederreiter and D. Talay, eds), Springer, pp. 181–196.

S. Joe and F. Y. Kuo (2008), 'Constructing Sobol′ sequences with better two-dimensional projection', *SIAM J. Sci. Comput.* **30**, 2635–2654.

L. Kämmerer, S. Kunis and D. Potts (2012), 'Interpolation lattices for hyperbolic cross trigonometric polynomials', *J. Complexity* **28**, 76–92.

A. Keller (2006), Myths of computer graphics. In *Monte Carlo and Quasi-Monte Carlo Methods 2004* (H. Niederreiter and D. Talay, eds), Springer, pp. 217–243.

A. Keller (2013), Quasi-Monte Carlo image synthesis in a nutshell. In *Monte Carlo and Quasi-Monte Carlo Methods 2012* (J. Dick, F. Y. Kuo, G. W. Peters and I. H. Sloan, eds), Springer, to appear.

J. F. Koksma (1942/43), 'Een algemeene stelling uit de theorie der gelijkmatige verdeeling modulo 1', *Mathematica B* (Zutphen) **11**, 7–11.

N. M. Korobov (1959), 'The approximate computation of multiple integrals' (in Russian), *Dokl. Akad. Nauk SSSR* **124**, 1207–1210.

N. M. Korobov (1963), *Number-Theoretic Methods in Approximate Analysis*, Fizmatgiz.

N. M. Korobov (1982), 'On the calculation of optimal coefficients' (in Russian), *Dokl. Akad. Nauk SSSR* **267**, 289–292.

P. Kritzer and F. Pillichshammer (2007), 'Constructions of general polynomial lattices for multivariate integration', *Bull. Austral. Math. Soc.* **76**, 93–110.

L. Kuipers and H. Niederreiter (1974), *Uniform Distribution of Sequences*, Wiley.

F. Y. Kuo (2003), 'Component-by-component constructions achieve the optimal rate of convergence for multivariate integration in weighted Korobov and Sobolev spaces', *J. Complexity* **19**, 301–320.

F. Y. Kuo and S. Joe (2002), 'Component-by-component construction of good lattice rules with a composite number of points', *J. Complexity* **18**, 943–976.

F. Y. Kuo and S. Joe (2003), 'Component-by-component construction of good intermediate-rank lattice rules', *SIAM J. Numer. Anal.* **41**, 1465–1486.

F. Y. Kuo and I. H. Sloan (2005), 'Lifting the curse of dimensionality', *Not. Amer. Math. Soc.* **52**, 1320–1328.

F. Y. Kuo, W. T. M. Dunsmuir, I. H. Sloan, M. P. Wand, R. S. Womersley (2008*a*), 'Quasi-Monte Carlo for highly structured generalised response models', *Method. Comput. Appl. Probab.* **10**, 239–275.

K. Y. Kuo, C. Schwab and I. H. Sloan (2011), 'Quasi-Monte Carlo methods for high-dimensional integration: The standard (weighted Hilbert space) setting and beyond', *ANZIAM J.* **53**, 1–37.

F. Y. Kuo, C. Schwab and I. H. Sloan (2012), Quasi-Monte Carlo finite element methods for a class of elliptic partial differential equations with random coefficients, *SIAM J. Numer. Anal.* **50**, 3351–3374.

F. Y. Kuo, C. Schwab and I. H. Sloan (2013), Multi-level quasi-Monte Carlo finite element methods for a class of elliptic partial differential equations with random coefficients. Submitted.

F. Y. Kuo, I. H. Sloan, G. W. Wasilkowski, and B. J. Waterhouse (2010*a*), 'Randomly shifted lattice rules with the optimal rate of convergence for unbounded integrands', *J. Complexity* **26**, 135–160.

F. Y. Kuo, I. H. Sloan, G. W. Wasilkowski and H. Woźniakowski (2010*b*), 'On decompositions of multivariate functions', *Math. Comp.* **79**, 953–966.

F. Y. Kuo, I. H. Sloan, G. W. Wasilkowski and H. Woźniakowski (2010*c*), 'Liberating the dimension', *J. Complexity* **26**, 422–454.

F. Y. Kuo, I. H. Sloan, and H. Woźniakowski (2006*a*), Lattice rules for multivariate approximation in the worst case setting. In *Monte Carlo and Quasi-Monte Carlo Methods 2004* (H. Niederreiter and D. Talay, eds), Springer, pp. 289–330.

F. Y. Kuo, I. H. Sloan and H. Woźniakowski (2007), 'Periodization strategy may fail in high dimensions', *Numer. Algorithms* **46**, 369–391.

F. Y. Kuo, I. H. Sloan, and H. Woźniakowski (2008*b*), 'Lattice rule algorithms for multivariate approximation in the average case setting', *J. Complexity* **24**, 283–323.

F. Y. Kuo, G. W. Wasilkowski, and B. J. Waterhouse (2006*b*), 'Randomly shifted lattice rules for unbounded integrals', *J. Complexity* **22**, 630–651.

F. Y. Kuo, G. W. Wasilkowski, and H. Woźniakowski (2009), 'Lattice algorithms for multivariate L_∞ approximation in the worst case setting', *Constr. Approx.* **30**, 475–493.

G. Larcher and C. Traunfellner (1994), 'On the numerical integration of Walsh series by number-theoretic methods', *Math. Comp.* **63**, 277–291.

G. Larcher, A. Lauß, H. Niederreiter and W. C. Schmid (1996*a*), 'Optimal polynomials for (t, m, s)-nets and numerical integration of multivariate Walsh series', *SIAM J. Numer. Anal.* **33**, 2239–2253.

G. Larcher, W. C. Schmid and R. Wolf (1994), 'Representation of functions as Walsh series to different bases and an application to the numerical integration of high-dimensional Walsh series', *Math. Comp.* **63**, 701–716.

G. Larcher, W. C. Schmid and R. Wolf (1996*b*), 'Quasi-Monte Carlo methods for the numerical integration of multivariate Walsh series: Monte Carlo and quasi-Monte Carlo methods', *Math. Comput. Modelling* **23**, 55–67.

D. Laurie (1996), 'Periodizing transformations for numerical integration', *J. Comput. Appl. Math.* **66**, 337–344.

P. L'Ecuyer (2004), Quasi-Monte Carlo methods in finance. In *Proc. 2004 Winter Simulation Conference* (R. G. Ingalls, M. D. Rossetti, J. S. Smith and B. A. Peters, eds), IEEE Computer Society Press, pp. 1645–1655.

P. L'Ecuyer and C. Lemieux (2000), 'Variance reduction via lattice rules', *Management Sci.* **46**, 1214–1235.

P. L'Ecuyer and D. Munger (2012), On figures of merit for randomly shifted lattice rules. In *Monte Carlo and Quasi-Monte Carlo Methods 2010* (L. Plaskota and H. Woźniakowski, eds), Springer, pp. 133–159.

P. L'Ecuyer, D. Munger and B. Tuffin (2010), 'On the distribution of integration error by randomly-shifted lattice rules'. *Electron. J. Stat.* **4**, 950–993.

C. Lemieux (2009), *Monte Carlo and Quasi-Monte Carlo Sampling*, Springer.

C. Lemieux and P. L'Ecuyer (2001), 'On selection criteria for lattice rules and other quasi-Monte Carlo point sets', *Math. Comput. Simulation* **55**, 139–148.

D. Li and F. J. Hickernell (2003), Trigonometric spectral collocation methods on lattices. In *Recent Advances in Scientific Computing and Partial Differential Equations* (S. Y. Cheng, C.-W. Shu and T. Tang, eds), Vol. 330 of *AMS Series in Contemporary Mathematics*, AMS, pp. 121–132.

G. Li, J. Schoendorf, T.-S, Ho and H. Rabitz (2004), 'Multicut-HDMR with an application to an ionospheric model', *J. Comput. Chem.* **25**, 1149–1156.

W. L. Loh (2003), 'On the asymptotic distribution of scrambled net quadrature', *Ann. Statist.* **31**, 1282–1324.

J. Lyness and T. Sørevik (2006), 'Five-dimensional k-optimal lattice rules', *Math. Comp.* **75**, 1467–1480.

E. Maize (1980), Contributions to the theory of error reduction in quasi-Monte Carlo methods. PhD thesis, Claremont Graduate School.

J. Matoušek (1998*a*), 'The exponent of discrepancy is at least 1.0669', *J. Complexity* **14**, 448–453.

J. Matoušek (1998*b*), 'On the L_2 discrepancy for anchored boxes', *J. Complexity* **14**, 527–556.

J. Matoušek (1999), *Geometric Discrepancy: An Illustrated Guide*, Vol. 18 of *Algorithms and Combinatorics*, Springer.

M. Matsumoto and T. Yoshiki (2013), Existence of higher order convergent quasi-Monte Carlo rules via Walsh figure of merit. In *Monte Carlo and Quasi-Monte Carlo Methods 2012* (J. Dick, F. Y. Kuo, G. W. Peters and I. H. Sloan, eds), Springer, to appear.

M. Matsumoto, M. Saito, K. Matoba (2013), 'A computable figure of merit for quasi-Monte Carlo point sets', *Math. Comp.*, to appear.

H. Niederreiter (1978), 'Quasi-Monte Carlo methods and pseudo-random numbers', *Bull. Amer. Math. Soc.* **84**, 957–1041.

H. Niederreiter (1987), 'Point sets and sequences with small discrepancy', *Monatsh. Math.* **104**, 273–337.

H. Niederreiter (1988), 'Low-discrepancy and low-dispersion sequences', *J. Number Theory* **30**, 51–70.

H. Niederreiter (1992*a*), *Random Number Generation and Quasi-Monte Carlo Methods*, SIAM.

H. Niederreiter (1992*b*), 'Low-discrepancy point sets obtained by digital constructions over finite fields', *Czechoslovak Math. J.* **42**, 143–166.

H. Niederreiter (2003), 'The existence of good extensible polynomial lattice rules', *Monatsh. Math.* **139**, 295–307.

H. Niederreiter (2004), Digital nets and coding theory. In *Coding, Cryptography and Combinatorics* (K. Q. Feng, H. Niederreiter and C. P. Xing, eds), Birkhäuser, pp. 247–257.

H. Niederreiter (2009), 'On the discrepancy of some hybrid sequences', *Acta Arith.* **138**, 373–398.

H. Niederreiter (2010*a*), 'A discrepancy bound for hybrid sequences involving digital explicit inversive pseudorandom numbers', *Unif. Distrib. Theory* **5**, 53–63.

H. Niederreiter (2010*b*), 'Further discrepancy bounds and an Erdős–Turán–Koksma inequality for hybrid sequences', *Monatsh. Math.* **161**, 193–222.

H. Niederreiter (2012), 'Improved discrepancy bounds for hybrid sequences involving Halton sequences', *Acta Arith.* **155**, 71–84.

H. Niederreiter and F. Pillichshammer (2009), 'Construction algorithms for good extensible lattice rules,' *Constr. Approx.* **30**, 361–393.

H. Niederreiter and A. Winterhof (2011), 'Discrepancy bounds for hybrid sequences involving digital explicit inversive pseudorandom numbers', *Unif. Distrib. Theory* **6**, 33–56.

H. Niederreiter and C. P. Xing (1995), 'Low-discrepancy sequences obtained from algebraic function fields over finite fields', *Acta Arith.* **72**, 281–298.

H. Niederreiter and C. P. Xing (1996*a*), 'Low-discrepancy sequences and global function fields with many rational places', *Finite Fields Appl.* **2**, 241–273.

H. Niederreiter and C. P. Xing (1996*b*), Quasirandom points and global function fields. In *Finite Fields and Applications* Vol. 233 of *London Mathematical Society Lecture Note Series*, Cambridge University Press, pp. 269–296.

B. Niu and F. J. Hickernell (2009), Monte Carlo simulation of stochastic integrals when the cost of function evaluation is dimension dependent. In *Monte Carlo and Quasi-Monte Carlo Methods 2008* (P. L'Ecuyer and A. B. Owen, eds), Springer, pp. 545–560.

B. Niu, F. J. Hickernell, T. Müller-Gronbach and K. Ritter (2011), 'Deterministic multi-level algorithms for infinite-dimensional integration on $\mathbb{R}^{\mathbb{N}}$', *J. Complexity* **27**, 331–351.

E. Novak and H. Woźniakowski (2001), 'Intractability results for integration and discrepancy', *J. Complexity*, **17**, 388–441.

E. Novak and H. Woźniakowski (2008), *Tractability of Multivariate Problems, Vol. I: Linear Information*, EMS.

E. Novak and H. Woźniakowski (2009), L_2 discrepancy and multivariate integration. In *Analytic Number Theory: Essays in Honour of Klaus Roth* (W. W. L. Chen, W. T. Gowers, H. Halberstam, W. M. Schmidt and R. C. Vaughan, eds), Cambridge University Press, pp. 359–388.

E. Novak and H. Woźniakowski (2010), *Tractability of Multivariate Problems, Vol. II: Standard Information for Functionals*, EMS.

E. Novak and H. Woźniakowski (2012), *Tractability of Multivariate Problems, Vol. III: Standard Information for Operators*, EMS.

D. Nuyens and R. Cools (2006*a*), 'Fast algorithms for component-by-component construction of rank-1 lattice rules in shift-invariant reproducing kernel Hilbert spaces', *Math. Comp.* **75**, 903–920.

D. Nuyens and R. Cools (2006*b*), 'Fast component-by-component construction of rank-1 lattice rules with a non-prime number of points', *J. Complexity* **22**, 4–28.

D. Nuyens and R. Cools (2006*c*), Fast component-by-component construction, a reprise for different kernels. In *Monte Carlo and Quasi-Monte Carlo Methods 2004* (H. Niederreiter and D. Talay, eds), Springer, pp. 373–387.

G. Ökten (1996), 'A probabilistic result on the discrepancy of a hybrid Monte Carlo sequence and applications', *Monte Carlo Methods Appl.* **2**, 255–270.

G. Ökten, B. Tuffin and V. Burago (2006), 'A central limit theorem and improved error bounds for a hybrid-Monte Carlo sequence with applications in computational finance', *J. Complexity* **22**, 435–458.

A. B. Owen (1995), Randomly permuted (t, m, s)-nets and (t, s)-sequences. In *Monte Carlo and Quasi-Monte Carlo Methods in Scientific Computing* (Las Vegas, NV, 1994), Vol. 106 of *Lecture Notes in Statistics*, Springer, pp. 299–317.

A. B. Owen (1997*a*), 'Scrambled net variance for integrals of smooth functions', *Ann. Statist.* **25**, 1541–1562.

A. B. Owen (1997*b*), 'Monte Carlo variance of scrambled net quadrature', *SIAM J. Numer. Anal.* **34**, 1884–1910.

A. B. Owen (1998), 'Scrambling Sobol′ and Niederreiter–Xing points', *J. Complexity* **14**, 466–489.

A. B. Owen (2006), Quasi-Monte Carlo for integrands with point singularities at unknown locations. In *Monte Carlo and Quasi-Monte Carlo Methods 2004* (H. Niederreiter and D. Talay, eds), Springer, pp. 403–417.

S. H. Paskov and J. Traub (1995), 'Faster evaluation of financial derivatives', *J. Portfolio Management* **22**, 113–120.

F. Pillichshammer (2002), 'Bounds for the quality parameter of digital shift nets over Z_2', *Finite Fields Appl.* **8**, 444–454.

F. Pillichshammer and G. Pirsic (2009), Discrepancy of hyperplane nets and cyclic nets. In *Monte Carlo and Quasi-Monte Carlo Methods 2008* (P. L'Ecuyer and A. B. Owen, eds), Springer, pp. 573–587.

G. Pirsic (2002), A software implementation of Niederreiter–Xing sequences. In *Monte Carlo and Quasi-Monte Carlo Methods 2000* (K. T. Fang, F. J. Hickernell and H. Niederreiter, eds), Springer, pp. 434–445.

G. Pirsic and F. Pillichshammer (2011), 'Extensible hyperplane nets', *Finite Fields Appl.* **17**, 407–423.

G. Pirsic and W. C. Schmid (2001), 'Calculation of the quality parameter of digital nets and application to their construction', *J. Complexity* **17**, 827–839.

G. Pirsic, J. Dick and F. Pillichshammer (2006), 'Cyclic digital nets, hyperplane nets, and multivariate integration in Sobolev spaces', *SIAM J. Numer. Anal.* **44**, 385–411.

L. Plaskota and G. Wasilkowski (2011), 'Tractability of infinite-dimensional integration in the worst case and randomized setting', *J. Complexity* **27**, 505–518.

L. Plaskota, G. Wasilkowski and H. Woźniakowski (2000), 'A new algorithm and worst case complexity for Feynman–Kac path integration', *J. Comput. Phys.* **164**, 335–353.

C. M. Rader (1968), 'Discrete Fourier transforms when the number of data samples is prime', *Proc. IEEE* **5**, 1107–1108.

C.-H. Rhee and P. W. Glynn (2012), A new approach to unbiased estimation for SDE's. In *Proc. 2012 Winter Simulation Conference* (C. Laroque, J. Himmelspach, R. Pasupathy, O. Rose and A. M. Uhrmacher, eds).

W. C. Schmid (1996), Shift-nets: A new class of binary digital (t, m, s)-nets. In *Monte Carlo and Quasi-Monte Carlo Methods 1996* (H. Niederreiter, P. Hellekalek, G. Larcher and P. Zinterhof, eds), Vol. 127 of *Lecture Notes in Statistics*, Springer, pp. 369–381.

W. C. Schmid (2000), Improvements and extensions of the 'Salzburg Tables' by using irreducible polynomials. In *Monte Carlo and Quasi-Monte Carlo Methods 1998* (H. Niederreiter and J. Spanier, eds), Springer, pp. 436–447.

C. Schwab and C. J. Gittelson (2011), Sparse tensor discretizations of high-dimensional parametric and stochastic PDEs. In *Acta Numerica*, Vol. 20, Cambridge University Press, pp. 291–467.

A. Sidi (1993), A new variable transformation for numerical integration. In *Numerical Integration IV: Oberwolfach, 1992* (H. Brass and G. Hämmerlin, eds), Birkhäuser, pp. 359–373.

V. Sinescu and S. Joe (2007), 'Good lattice rules based on the general weighted star discrepancy', *Math. Comp.* **76**, 989–1004.

V. Sinescu and S. Joe (2008), Good lattice rules with a composite number of points based on the product weighted star discrepancy. In *Monte Carlo and Quasi-Monte Carlo Methods 2006* (A. Keller, S. Heinrich and H. Niederreiter, eds), Springer, pp. 645–658.

V. Sinescu and P. L'Ecuyer (2011), 'Existence and construction of shifted lattice rules with an arbitrary number of points and bounded weighted star discrepancy for general decreasing weights', *J. Complexity* **27**, 449–465.

I. H. Sloan (2007), 'Finite order integration weights can be dangerous', *Comput. Meth. Appl. Math.* **7**, 239–254.

I. H. Sloan and S. Joe (1994), *Lattice Methods for Multiple Integration*, Oxford University Press.

I. H. Sloan and A, V. Reztsov (2002), 'Component-by-component construction of good lattice rules', *Math. Comp.* **71**, 263–273.

I. H. Sloan and H. Woźniakowski (1998), 'When are quasi-Monte Carlo algorithms efficient for high-dimensional integrals?', *J. Complexity* **14**, 1–33.

I. H. Sloan and H. Woźniakowski (2001), 'Tractability of multivariate integration for weighted Korobov classes', *J. Complexity* **17**, 697–721.

I. H. Sloan and H. Woźniakowski (2002), 'Tractability of integration in non-periodic and periodic weighted tensor product Hilbert spaces', *J. Complexity* **18**, 479–499.

I. H. Sloan, F. Y. Kuo and S. Joe (2002*a*) 'On the step-by-step construction of quasi-Monte Carlo integration rules that achieve strong tractability error bounds in weighted Sobolev spaces', *Math. Comp.* **71**, 1609–1640.

I. H. Sloan, F. Y. Kuo and S. Joe (2002*b*) 'Constructing randomly shifted lattice rules in weighted Sobolev spaces', *SIAM J. Numer. Anal.* **40**, 1650–1665.

I. H. Sloan, X. Wang and H. Woźniakowski (2004), 'Finite-order weights imply tractability of multivariate integration', *J. Complexity* **20**, 46–74.

I. M. Sobol′ (1967), 'Distribution of points in a cube and approximate evaluation of integrals' (in Russian), *Ž. Vyčisl. Mat. i Mat. Fiz.* **7**, 784–802.

I. M. Sobol′ (1969), *Multidimensional Quadrature Formulas and Haar Functions* (in Russian), Nauka.

S. Smolyak (1963), 'Quadrature and interpolation formulas for tensor products of certain classes of functions', *Soviet Math. Dokl.* **4**, 240–243. Russian original in *Dokl. Akad. Nauk SSSR* **148** (1963), 1042–1045.

A. H. Stroud (1971), *Approximate Calculation of Multiple Integrals*, Prentice-Hall.

A. L. Teckentrup, R. Scheichl, M. B. Giles and E. Ullmann (2012), Further analysis of multilevel Monte Carlo methods for elliptic PDEs with random coefficient. Technical Report, University of Bath.

S. Tezuka (2013), 'On the discrepancy of generalized Niederreiter sequences, *J. Complexity*, to appear.

S. Tezuka and H. Faure (2003), '*I*-binomial scrambling of digital nets and sequences', *J. Complexity* **19**, 744–757.

C. Thomas-Agnan (1996), 'Computing a family of reproducing kernels for statistical applications', *Numer. Algorithms* **13**, 21–32.

J. F. Traub, G. W. Wasilkowski and H. Woźniakowski (2008), *Information-Based Complexity*, Academic Press.

A. W. van der Vart and J. A. Wellner (2009), *Weak Convergence and Empirical Processes*, Springer Series in Statistics, Springer.

G. Wahba (1990), *Spline Models for Observational Data*, Vol. 59 of *CBMS-NSF Regional Conference Series in Applied Mathematics*, SIAM.

J. L. Walsh (1923), 'A closed set of normal orthogonal functions', *Amer. J. Math.* **45**, 5–24.

X. Wang (2002), 'A constructive approach to strong tractability using quasi-Monte Carlo algorithms', *J. Complexity* **18**, 683–701.

X. Wang (2003), 'Strong tractability of multivariate integration using quasi-Monte Carlo algorithms', *Math. Comp.* **72**, 823–838.

X. Wang and K.-T. Fang (2003), 'Effective dimension and quasi-Monte Carlo integration', *J. Complexity* **19**, 101–124.

X. Wang and I. H. Sloan (2005), 'Why are high-dimensional finance problems often of low effective dimension?', *SIAM J. Sci. Comput.* **27**, 159–183.

X. Wang and I. H. Sloan (2006), 'Efficient weighted lattice rules with applications to finance', *SIAM J. Sci. Comput.* **28**, 728–750.

X. Wang and I. H. Sloan (2007), 'Brownian bridge and principal component analysis: Towards removing the curse of dimensionality', *IMA J. Numer. Anal.* **27**, 631–654.

X. Wang and I. H. Sloan (2011), 'Quasi-Monte Carlo methods in financial engineering: An equivalence principle and dimension reduction', *Oper. Res.* **59**, 80–95.

G. W. Wasilkowski and H. Woźniakowski (1995), 'Explicit cost bounds of algorithms for multivariate tensor product problems', *J. Complexity* **11**, 1–56.

G. W. Wasilkowski and H. Woźniakowski (1996), 'On tractability of path integration', *J. Math.Phys.* **37**, 2071–2088.

G. W. Wasilkowski and H. Woźniakowski (1999), 'Weighted tensor product algorithms for linear multivariate problems', *J. Complexity* **15**, 402–447.

G. W. Wasilkowski and H. Woźniakowski (2004), 'Finite-order weights imply tractability of linear multivariate problems', *J. Approx. Theory* **130**, 57–77.

G. W. Wasilkowski and H. Woźniakowski (2010), 'On the exponent of discrepancy', *Math. Comp.* **79**, 983–992.

A. Werschulz and H. Woźniakowski (2009), 'Tractability of multivariate approximation over a weighted unanchored Sobolev space', *Constr. Approx.* **30**, 395–421.

H. Weyl (1916), 'Über die Gleichverteilung von Zahlen mod. Eins', *Math. Ann.* **77**, 313–352.

H. Woźniakowski (2013), Monte Carlo integration. In *Encyclopedia of Numerical Analysis.*

C. P. Xing and H. Niederreiter (1995), 'A construction of low-discrepancy sequences using global function fields', *Acta Arith.* **73**, 87–102.

R. X. Yue and F. J. Hickernell (2001), 'Integration and approximation based on scramble sampling in arbitrary dimensions', *J. Complexity* **17**, 881–897.

R. X. Yue and F. J. Hickernell (2002), 'The discrepancy and gain coefficients of scrambled digital nets', *J. Complexity* **18**, 135–151.

R. X. Yue and F. J. Hickernell (2005), 'Strong tractability of integration using scrambled Niederreiter points', *Math. Comp.* **74**, 1871–1893.

R. X. Yue and F. J. Hickernell (2006), 'Strong tractability of quasi-Monte Carlo quadrature using nets for certain Banach spaces', *SIAM J. Numer. Anal.* **44**, 2559–2583.

X. Y. Zeng, K. T. Leung, and F. J. Hickernell (2006), Error analysis of splines for periodic problems using lattice designs. In *Monte Carlo and Quasi-Monte Carlo Methods 2004* (H. Niederreiter and D. Talay, eds), Springer, pp. 501–514.

C. Zenger (1991), Sparse grids. In *Parallel Algorithms for Partial Differential Equations* (W. Hackbusch, ed.), Vol. 31 of *Notes on Numerical Fluid Mechanics*, Vieweg.

Acta Numerica (2013), pp. 289–396
doi:10.1017/S0962492913000056

Finite element methods for surface PDEs*

Gerhard Dziuk
Abteilung für Angewandte Mathematik,
Albert-Ludwigs-Universität Freiburg im Breisgau,
Hermann-Herder-Straße 10,
D–79104 Freiburg im Breisgau, Germany
E-mail: gerd@mathematik.uni-freiburg.de

Charles M. Elliott
Mathematics Institute,
University of Warwick,
Coventry CV4 7AL, UK
E-mail: c.m.elliott@warwick.ac.uk

In this article we consider finite element methods for approximating the solution of partial differential equations on surfaces. We focus on surface finite elements on triangulated surfaces, implicit surface methods using level set descriptions of the surface, unfitted finite element methods and diffuse interface methods. In order to formulate the methods we present the necessary geometric analysis and, in the context of evolving surfaces, the necessary transport formulae. A wide variety of equations and applications are covered. Some ideas of the numerical analysis are presented along with illustrative numerical examples.

CONTENTS

* Colour online for monochrome figures available at journals.cambridge.org/anu.

1. Introduction

Surface partial differential equations arise in a wide variety of applications. Further, they are examples of partial differential equations on manifolds. As such they provide a remaining challenge within the general subject of the numerical analysis of partial differential equations. The framework is essentially geometric because the domain in which the equation holds is curved. They are linked naturally to the geometric equations for surfaces such as the minimal surface equation, motion by mean curvature and Willmore flow. In an earlier *Acta Numerica* article (Deckelnick, Dziuk and Elliott 2005) we surveyed numerical methods for geometric partial differential equations and mean curvature flow.

The purpose of this article is to give an account of numerical methods for surface partial differential equations. Our interest and research in this field was stimulated in 2003 during the Isaac Newton Institute programme 'Computational challenges in partial differential equations'. Since this time there has been burgeoning interest in both the numerical analysis of such problems and the application to complex physical models.

The starting point was the use of surface finite elements to compute solutions to the Poisson problem for the Laplace–Beltrami operator on a curved surface proposed and analysed in Dziuk (1988). Here an important concept is the use of triangulated surfaces on which finite element spaces are constructed and then used in variational formulations of surface PDEs using surface gradients. This approach was extended by Dziuk and Elliott (2007*b*) to parabolic (including nonlinear and higher-order) equations on stationary surfaces. The evolving surface finite element method (ESFEM) was introduced by Dziuk and Elliott (2007*a*) in order to treat conservation laws on moving surfaces. The key idea is to use the Leibniz (or transport) formula for the time derivative of integrals over moving surfaces in order to derive weak and variational formulations. An interesting upshot is that the velocity and mean curvature of the surface which appear in certain formulations of the partial differential equation do not appear explicitly in variational formulations. This gives a tremendous advantage to numerical methods that exploit this, such as those of Dziuk and Elliott (2007*a*, 2010). Further numerical analysis of surface finite element methods may be found in Dziuk and Elliott (2012, 2013) and Dziuk, Lubich and Mansour (2012). Applications to complex physical and biological models may be found in Eilks and Elliott (2008), Elliott and Stinner (2010), Barreira, Elliott and Madzvamuse (2011) and Elliott, Stinner and Venkataraman (2012).

Another approach is to use implicit surface methods. The starting point for these methods is the level set method for evolving surfaces (Sethian 1999, Osher and Fedkiw 2003). Here the idea is to use a level set function ϕ to define a degenerate partial differential equation whose solution solves the surface equation on all level sets of ϕ. Such methods are formulated in Bertalmío, Cheng, Osher and Sapiro (2001), Greer, Bertozzi and Sapiro (2006), Burger (2009) and Dziuk and Elliott (2008, 2010).

Further, we describe in some detail numerical schemes based on the use of diffuse interfaces. These arise in phase field approximations of interface problems (Caginalp 1989, Deckelnick *et al.* 2005), and it is natural to exploit the methodology to generate methods for solving partial differential equations on the interfaces (Rätz and Voigt 2006, Elliott, Stinner, Styles and Welford 2011).

An important feature of the methods described in this article is the avoidance of *charts* both in the problem formulation and the numerical methods. The surface finite element method is based simply on triangulated surfaces and requires the geometry solely through knowledge of the vertices of the triangulation. On the other hand, the methods based on implicit surfaces require only the level set function ϕ. All the geometry is then encoded in ϕ.

The layout of the article is as follows. In Section 2 we set basic notation and concepts concerning geometric quantities, surface gradients and integration by parts for parametrized surfaces, using maps X, and hypersurfaces, using level set functions ϕ. Elliptic partial differential equations on surfaces are formulated in Section 3. Surface finite elements on triangulated surfaces are formulated and analysed for elliptic equations in Section 4.

Complex applications involving surfaces and interfaces frequently require the formulation and approximation of parabolic equations on moving surfaces. In Section 5 we formulate a scalar conservation law with a diffusive flux on a moving surface and formulate the evolving surface finite element method. Of particular note is the transport theorem for moving surfaces, which is valid triangle by triangle on an evolving surface. This is exploited together with transport property of the finite element basis functions to obtain a method which does not require explicit knowledge of the curvature or velocity, merely knowledge of the triangle vertices. Some more time-dependent equations are discussed in Section 6.

Implicit surface formulations and their numerical approximations are discussed in Sections 7 and 8. Methods which use finite elements in the higher-dimensional ambient space but with variational forms on surfaces or localized narrow bands are considered in Section 9. Finally, we briefly discuss some applications in Section 10.

2. Parametrized surfaces and hypersurfaces

In this section we introduce the elementary geometric analysis which is necessary to treat partial differential equations on surfaces. It is our opinion that numerical methods have to be intimately related to the analysis of the problems.

We begin in Section 2.1 by recalling some facts from elementary differential geometry concerning parametrized surfaces. We continue in Section 2.2 with hypersurfaces in $\mathbb{R}^{n+1}$ and the basic analysis concepts for such hypersurfaces. We introduce the necessary geometric concepts, for example the notion of curvature. The formula for integration by parts is proved, and we formulate the co-area formula. In Section 2.3 we introduce global coordinates in a neighbourhood of a hypersurface, the Fermi coordinates. These will be quite useful for the numerical analysis of PDEs on surfaces. For theoretical reasons we will introduce the oriented distance function. For the treatment of surface PDEs the Poincaré inequality on surfaces is central; its proof is in Section 2.4.

2.1. Parametrized surfaces

Let $n \in \mathbb{N}$. We call $\Gamma \subset \mathbb{R}^{n+1}$ an *n-dimensional parametrized C^k-surface* ($k \in \mathbb{N} \cup \{\infty\}$) if, for every point $x_0 \in \Gamma$, there exists an open set $U \subset \mathbb{R}^{n+1}$ with $x_0 \in U$, an open connected set $V \subset \mathbb{R}^n$ and a map $X : V \to U \cap \Gamma$ with the properties $X \in C^k(V, \mathbb{R}^{n+1})$, X is bijective and rank $\nabla X = n$ on V.

The map X is called a local *parametrization* of Γ while X^{-1} is called a local *chart*. A collection $(X_i)_{i\in I}, X_i \in C^k(V_i, \mathbb{R}^{n+1})$ of local parametrizations such that $\cup_{i\in I} X_i(V_i) = \Gamma$ is called a *C^k-atlas*. If $X_i(V_i) \cap X_j(V_j) \neq \emptyset$, then the map $X_i^{-1} \circ X_j$ by assumption is a C^k-diffeomorphism.

A function $f : \Gamma \to \mathbb{R}$ is *k-times differentiable* if all the functions $f \circ X_i : V_i \to \mathbb{R}$ are k-times differentiable.

Let $X \in C^2(V, \mathbb{R}^{n+1})$ be a local parametrization of Γ, $\theta \in V$. We define the *first fundamental form* $G(\theta) = (g_{ij}(\theta))_{i,j=1,\dots,n}$, $\theta \in V$ by

$$g_{ij}(\theta) = \frac{\partial X}{\partial \theta_i}(\theta) \cdot \frac{\partial X}{\partial \theta_j}(\theta), \quad i, j = 1, \dots, n.$$

Superscript indices denote the inversion of the matrix G so that

$$(g^{ij})_{i,j=1,\dots,n} = G^{-1},$$

and by $g = \det(G)$ we denote the determinant of the matrix G.

The *Laplace–Beltrami operator* on Γ is defined for a twice differentiable function $f : \Gamma \to \mathbb{R}$ as follows. Let $F(\theta) = f(X(\theta))$, $\theta \in V$. Then

$$(\Delta_\Gamma f)(X(\theta)) = \frac{1}{\sqrt{g(\theta)}} \sum_{i,j=1}^{n} \frac{\partial}{\partial \theta_j}\left(g^{ij}(\theta)\sqrt{g(\theta)}\frac{\partial F}{\partial \theta_i}(\theta)\right). \qquad (2.1)$$

The *tangential gradient* is given by

$$(\nabla_\Gamma f)(X(\theta)) = \sum_{i,j=1}^{n} g^{ij}(\theta)\frac{\partial F}{\partial \theta_j}(\theta)\frac{\partial X}{\partial \theta_i}(\theta). \qquad (2.2)$$

2.2. Hypersurfaces

Definition 2.1. Let $k \in \mathbb{N} \cup \{\infty\}$. $\Gamma \subset \mathbb{R}^{n+1}$ is called a C^k-hypersurface if, for each point $x_0 \in \Gamma$, there exists an open set $U \subset \mathbb{R}^{n+1}$ containing x_0 and a function $\phi \in C^k(U)$ with the property that $\nabla\phi \neq 0$ on $\Gamma \cap U$ and such that

$$U \cap \Gamma = \{x \in U \mid \phi(x) = 0\}. \tag{2.3}$$

The linear space

$$T_x\Gamma = \big\{\tau \in \mathbb{R}^{n+1} \mid \exists \gamma : (-\epsilon, \epsilon) \to \mathbb{R}^{n+1} \text{ differentiable,}$$
$$\gamma((-\epsilon,\epsilon)) \subset \Gamma, \gamma(0) = x \text{ and } \gamma'(0) = \tau\big\}$$

is called the *tangent space* to Γ at $x \in \Gamma$. It is easy to show that $T_x\Gamma = [\nabla\phi(x)]^\perp$, the set of all vectors that are orthogonal to $\nabla\phi(x)$, where ϕ is as in (2.3). In particular, $T_x\Gamma$ is an n-dimensional subspace of $\mathbb{R}^{n+1}$.

A vector $\nu(x) \in \mathbb{R}^{n+1}$ is called a *unit normal vector* at $x \in \Gamma$ if $\nu(x) \perp T_x\Gamma$ and $|\nu(x)| = 1$. In view of the above characterization of $T_x\Gamma$, we then have

$$\nu(x) = \frac{\nabla\phi(x)}{|\nabla\phi(x)|} \quad \text{or} \quad \nu(x) = -\frac{\nabla\phi(x)}{|\nabla\phi(x)|}. \tag{2.4}$$

A C^1-hypersurface is called *orientable* if there exists a continuous vector field $\nu : \Gamma \to \mathbb{R}^{n+1}$ such that $\nu(x)$ is a unit normal vector to Γ for all $x \in \Gamma$.

The connection between the parametrized surfaces of Section 2.1 and hypersurfaces is given by the following well-known little lemma.

Lemma 2.2. Assume that Γ is a C^k-hypersurface in $\mathbb{R}^{n+1}$. Then for every $x \in \Gamma$ there exists an open set $U \subset \mathbb{R}^{n+1}$ with $x \in U$ and a parametrized C^k-surface $X : V \to U \cap \Gamma$ such that X is a bijective map from V onto $U \cap \Gamma$. If $X : V \to U \cap \Gamma$ is a parametrized C^k-surface and $\theta \in V$, then there is an open set $\tilde{V} \subset V$ with $\theta \in \tilde{V}$ such that $X(\tilde{V})$ is a C^k-hypersurface.

This means that locally we can always work with hypersurfaces. And we may use all the definitions from Section 2.1 for hypersurfaces.

Definition 2.3. Let $\Gamma \subset \mathbb{R}^{n+1}$ be a C^1-hypersurface and let $f : \Gamma \to \mathbb{R}$ be differentiable at $x \in \Gamma$. We define the tangential gradient of f at $x \in \Gamma$ by

$$\nabla_\Gamma f(x) = \nabla\overline{f}(x) - \nabla\overline{f}(x) \cdot \nu(x)\, \nu(x) = P(x)\nabla\overline{f}(x),$$

where $P(x)_{ij} = \delta_{ij} - \nu_i(x)\nu_j(x)$ $(i, j = 1, \ldots, n+1)$. Here $\overline{f}$ is a smooth extension of $f : \Gamma \to \mathbb{R}$ to an $(n+1)$-dimensional neighbourhood U of the surface Γ, so that $\overline{f}|_\Gamma = f$. ∇ denotes the gradient in $\mathbb{R}^{n+1}$ and $\nu(x)$ is a unit normal at x.

The *Laplace–Beltrami operator* applied to a twice differentiable function $f \in C^2(\Gamma)$ is given by

$$\Delta_\Gamma f = \nabla_\Gamma \cdot \nabla_\Gamma f = \sum_{i=1}^{n+1} \underline{D}_i \underline{D}_i f. \tag{2.5}$$

See the proof of Theorem 2.10 in Section 2.3 for the construction of an extension $\overline{f}$. We shall use the notation (as in the above definition)

$$\nabla_\Gamma f(x) = (\underline{D}_1 f(x), \ldots, \underline{D}_{n+1} f(x))$$

for the $n+1$ components of the tangential gradient. Note that $\nabla_\Gamma f(x) \cdot \nu(x) = 0$ and hence $\nabla_\Gamma f(x) \in T_x\Gamma$.

Let us show that (2.1) and (2.2) are equivalent to the settings in Definition 2.3. Since the tangential gradient is a tangent vector, $\nabla_\Gamma f \circ X = \sum_{i=1}^n \alpha_i X_{\theta_i}$ with certain scalars α_j. We solve this equation for $\alpha_1, \ldots, \alpha_n$ by multiplying it by X_{θ_k}, to get

$$F_{\theta_k} = \nabla_\Gamma f \circ X \cdot X_{\theta_k} = \sum_{i=1}^{n} \alpha_i X_{\theta_i} \cdot X_{\theta_k} = \sum_{i=1}^{n} \alpha_i g_{ik}. \tag{2.6}$$

For the first equality on the left we have used the fact that, by the chain rule applied to $F(\theta) = f(X(\theta))$, we have $F_{\theta_k} = \sum_{l=1}^n \underline{D}_l f \circ X X_{l\theta_k}$, since X_{θ_k} is a tangent vector. From (2.6) we infer that $\alpha_l = \sum_{k=1}^n F_{\theta_k} g^{kl}$, and this finally gives (2.2). Now it is easy to derive (2.1).

Lemma 2.4. $\nabla_\Gamma f(x)$ only depends on the values of f on $\Gamma \cap U$, where $U \subset \mathbb{R}^{n+1}$ is a neighbourhood of x.

Proof. It is sufficient to show that $f \equiv 0$ on $\Gamma \cap U$ implies that $\nabla_\Gamma f(x) = 0$. Choose $\gamma : (-\epsilon, \epsilon) \to \mathbb{R}^{n+1}$ such that $\gamma(0) = x, \gamma((-\epsilon, \epsilon)) \subset \Gamma \cap U$ and $\gamma'(0) = \nabla_\Gamma f(x)$. Since $\overline{f}(\gamma(t)) = f(\gamma(t)) = 0$ for all $|t| < \epsilon$, we have

$$0 = \nabla \overline{f}(x) \cdot \gamma'(0) = (\nabla_\Gamma f(x) + \nabla \overline{f}(x) \cdot \nu(x)\, \nu(x)) \cdot \nabla_\Gamma f(x) = |\nabla_\Gamma f(x)|^2,$$

which implies the result. □

We denote by $C^1(\Gamma)$ the set of functions $f : \Gamma \to \mathbb{R}$, which are differentiable at every point $x \in \Gamma$ and for which $\underline{D}_j f : \Gamma \to \mathbb{R}, j = 1, \ldots, n+1$ are continuous. Similarly one can define $C^l(\Gamma)$ ($l \in \mathbb{N}$) provided that Γ is a C^k-hypersurface with $k \geq l$.

Definition 2.5. For $\Gamma \in C^2$ we define

$$\mathcal{H}_{ij} = \underline{D}_i \nu_j \quad (i, j = 1, \ldots, n+1). \tag{2.7}$$

It is easily shown that the matrix $\mathcal{H}$ is symmetric and that it possesses an eigenvalue 0 in the normal direction: $\mathcal{H}\nu = 0$. $\mathcal{H}$ is called the *extended*

Weingarten map. The restriction of $\mathcal{H}$ to the tangent space is called the Weingarten map.

For $x \in \Gamma$ the quantity

$$H(x) = \mathrm{trace}\mathcal{H}(x) = \sum_{i=1}^{n+1} \mathcal{H}_{ii}(x) \tag{2.8}$$

is the *mean curvature* of Γ at the point x. It differs from the common definition by a factor of n. We note that the eigenvalues $\kappa_1, \ldots, \kappa_n$ of $\mathcal{H}$ (apart from the trivial eigenvalue 0 in the ν-direction) are the *principal curvatures* of Γ.

Let us have a look at the most simple example. The sphere of radius $R > 0$, $\Gamma = \{x \in \mathbb{R}^{n+1} \mid |x| = R\}$, is given by the level set function $\phi(x) = |x| - R$ for $0 < |x| < \infty$. We may choose

$$\nu = \frac{\nabla \phi}{|\nabla \phi|} = \frac{x}{|x|}$$

and get for $x \in \Gamma$

$$\mathcal{H}_{ij}(x) = \underline{D}_i \nu_j(x) = \underline{D}_i \frac{x_j}{|x|} = \frac{1}{R} \underline{D}_i x_j = \frac{1}{R}(\delta_{ij} - \nu_i \nu_j) = \frac{1}{R}\left(\delta_{ij} - \frac{x_i x_j}{R^2}\right).$$

This matrix has an eigenvalue 0 with eigenvector $\frac{x}{R}$ and n eigenvalues $\kappa_j = \frac{1}{R}$ $(j = 1, \ldots, n)$. The mean curvature of Γ is then given as $H = \frac{n}{R}$.

The following result concerning the exchange of tangential derivatives is easily proved.

Lemma 2.6. For $\Gamma \in C^2$ and $u \in C^2(\Gamma)$ we have

$$\underline{D}_i \underline{D}_j u - \underline{D}_j \underline{D}_i u = (\mathcal{H} \nabla_\Gamma u)_j \nu_i - (\mathcal{H} \nabla_\Gamma u)_i \nu_j. \tag{2.9}$$

for $i, j = 1, \ldots, n+1$.

2.3. Global coordinates

It is quite convenient to use global coordinates in a neighbourhood of a hypersurface, the so-called Fermi coordinates. This avoids working with charts and atlases (see Section 2.1) when proving results and carrying out the numerical analysis. For this one introduces the *oriented distance function* for Γ.

Remark 2.7. In the context of surface finite elements we will use the oriented distance function only for our analysis and numerical analysis. We will not use it in defining the computational methods. We will not need the oriented distance function for the implementation of our algorithms. It may be of use in *implicit surface* methods.

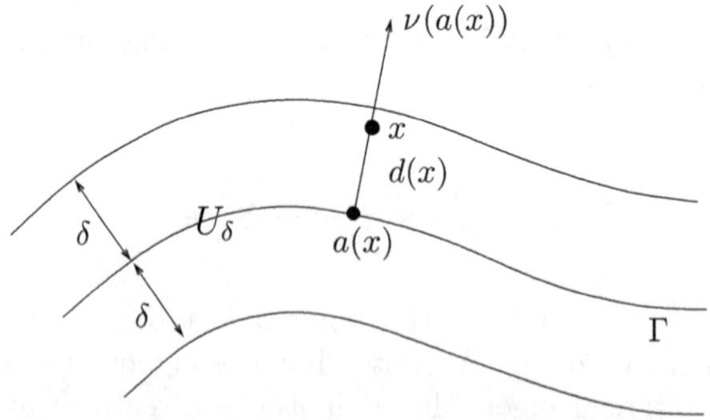

Figure 2.1. Strip U_δ around the hypersurface Γ and normal coordinates $x = a(x) + d(x)\nu(a(x))$.

Assume in the following that $G \subset \mathbb{R}^{n+1}$ is bounded and open with exterior normal ν and assume that $\Gamma = \partial G$ is a C^k-hypersurface ($k \geq 2$). The oriented distance function for Γ is defined by

$$d(x) = \begin{cases} \inf_{y\in\Gamma}|x-y| & x \in \mathbb{R}^{n+1}\setminus\bar{G}, \\ -\inf_{y\in\Gamma}|x-y| & x \in G. \end{cases}$$

One easily verifies that d is globally Lipschitz-continuous with Lipschitz constant 1. Since ∂G is a C^2-hypersurface, it satisfies both a uniform interior and a uniform exterior sphere condition, which means that for each point $x_0 \in \partial\Omega$ there are balls B and B' such that

$$\bar{B} \cap (\mathbb{R}^{n+1}\setminus\Omega) = \{x_0\}, \quad \bar{B}' \cap \bar{\Omega} = \{x_0\},$$

and the radii of B, B' are bounded from below by a positive constant δ uniformly in x_0. With this observation the following lemma is easily proved.

Lemma 2.8. We define

$$U_\delta = \{x \in \mathbb{R}^{n+1} \mid |d(x)| < \delta\}.$$

Then $d \in C^k(U_\delta)$, and for every point $x \in U_\delta$ there exists a unique point $a(x) \in \Gamma$ such that

$$x = a(x) + d(x)\nu(a(x)). \tag{2.10}$$

Moreover, we have that

$$\nabla d(x) = \nu(a(x)), \quad |\nabla d(x)| = 1, \quad \text{for } x \in U_\delta.$$

We also extend the normal constantly in the normal direction: $\nu(x) = \nu(a(x))$ for $x \in U_\delta$. Thus we have introduced a global coordinate system around Γ. Every point $x \in U_\delta$ can be described by its *Fermi coordinates* (normal coordinates) $d(x)$ and $a(x)$ according to (2.10).

The introduction of global coordinates allows us to work with the well-known co-area formula (Evans 1998).

Theorem 2.9. Let $\Gamma(r) = \{x \in \mathbb{R}^{n+1} \mid \mathrm{d}(x) = r\}$ be the parallel surface to $\Gamma = \Gamma(0)$ for $|r| < \delta$. Then

$$\int_{U_\varepsilon} f(x)\,\mathrm{d}x = \int_{-\varepsilon}^{\varepsilon} \int_{\Gamma(r)} f(x)\,\mathrm{d}A(x)\,\mathrm{d}r \tag{2.11}$$

for $f \in C^0(U_\delta)$ and $0 < \varepsilon < \delta$.

We note that this formula changes to

$$\int_{U_\varepsilon} f(x)\,\mathrm{d}x = \int_{-\varepsilon}^{\varepsilon} \int_{\Gamma(r)} f(x)|\nabla\phi(x)|\,\mathrm{d}A(x)\,\mathrm{d}r \tag{2.12}$$

if the surfaces are given by an arbitrary level set function ϕ as in (2.3), $\Gamma(r) = \{x \in \mathbb{R}^{n+1} \mid \phi(x) = r\}$, and the strip around Γ is taken to be $U_\delta = \{x \in \mathbb{R}^{n+1} \mid |\phi(x)| < \delta\}$. In this case one does not work with parallel surfaces to Γ.

With the co-area formula one can prove the formula for *integration by parts* on surfaces Γ.

Theorem 2.10. Assume that Γ is a hypersurface in $\mathbb{R}^{n+1}$ with smooth boundary $\partial\Gamma$ and that $f \in C^1(\overline{\Gamma})$. Then

$$\int_\Gamma \nabla_\Gamma f\,\mathrm{d}A = \int_\Gamma fH\nu\,\mathrm{d}A + \int_{\partial\Gamma} f\mu\,\mathrm{d}A. \tag{2.13}$$

Here, μ denotes the co-normal vector which is normal to $\partial\Gamma$ and tangent to Γ. A compact hypersurface Γ does not have a boundary, $\partial\Gamma = \emptyset$, and the last term on the right-hand side vanishes.

Note that in (2.13) $\mathrm{d}A$ in connection with an integral over Γ denotes the n-dimensional surface measure, while $\mathrm{d}A$ in connection with an integral over $\partial\Gamma$ is the $(n-1)$-dimensional surface measure.

Proof. We extend f to the tubular neighbourhood U_ε of Γ by

$$\overline{f}(x) = f(a(x)) \quad (x \in U_\varepsilon).$$

Then the chain rule gives

$$\frac{\partial\overline{f}}{\partial x_j}(x) = \sum_{k=1}^{n+1} \underline{D}_k f(a(x)) \frac{\partial a_k}{\partial x_j}(x).$$

The tangential derivative $\underline{D}_k f$ appears because we have that

$$\frac{\partial a_k}{\partial x_j}(x) = \frac{\partial}{\partial x_j}(x_k - d(x)\nu_k(x)) = \delta_{jk} - \nu_j(x)\nu_k(x) - d(x)\mathcal{H}_{jk}(x),$$

and the matrix $(\tilde{a}_{kj})_{j,k=1,\dots,n+1}$ $(\tilde{a}_{kj} = a_{k,x_j})$ maps any vector into a tangent vector. Thus we have

$$\nabla\overline{f}(x) = (I - d(x)\mathcal{H}(x))\nabla_\Gamma f(a(x)). \tag{2.14}$$

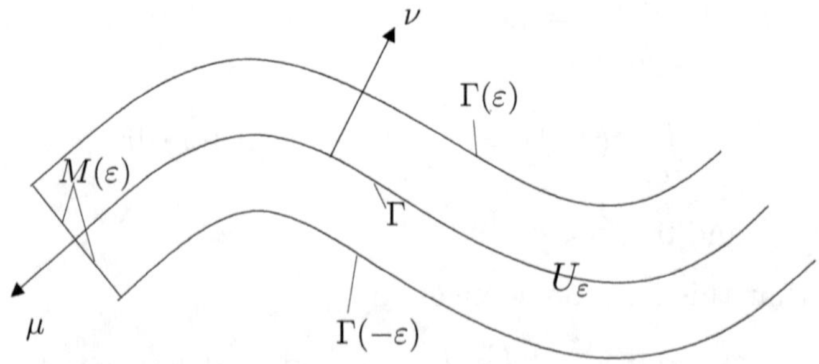

Figure 2.2. Geometric situation around the given surface Γ. Parallel surfaces $\Gamma(\varepsilon), \Gamma(-\varepsilon)$ and normal ν, co-normal μ.

In particular, we obtain $\nabla\overline{f}(x) = \nabla_\Gamma f(x)$ for $x \in \Gamma$. We apply Gauss's theorem to $\overline{f}$ on U_ε and get

$$\int_{U_\varepsilon} \nabla\overline{f}(x)\,\mathrm{d}x = \int_{\partial U_\varepsilon} \overline{f}(x)\nu_{\partial U_\varepsilon}(x)\,\mathrm{d}A(x).$$

We have that $\partial U_\varepsilon = \Gamma(\varepsilon)\cup\Gamma(-\varepsilon)\cup M(\varepsilon)$, where $M(\varepsilon) = \{x + r\nu_\Gamma(x) \mid x \in \partial\Gamma, r \in [-\varepsilon,\varepsilon]\}$ (see Figure 2.2). Thus

$$\begin{aligned}\frac{1}{2\varepsilon}&\int_{U_\varepsilon} (I - d(x)\mathcal{H}(x))\nabla_\Gamma f(a(x))\,\mathrm{d}x \qquad (2.15)\\ &= \frac{1}{2\varepsilon}\bigg(\int_{\Gamma(\varepsilon)} \overline{f}(x)\nu_\Gamma(x)\,\mathrm{d}A(x) - \int_{\Gamma(-\varepsilon)} \overline{f}(x)\nu_\Gamma(x)\,\mathrm{d}A(x)\\ &\qquad + \int_{M(\varepsilon)} \overline{f}(x)\mu_\Gamma(x)\,\mathrm{d}A(x)\bigg),\end{aligned}$$

with the normal ν_Γ and the co-normal μ_Γ to Γ, which do not depend on ε. We take the limit $\varepsilon \to 0$ on both sides of this equation. Obviously, for the left-hand side

$$\lim_{\varepsilon\to 0}\frac{1}{2\varepsilon}\int_{U_\varepsilon} (I - d(x)\mathcal{H}(x))\nabla_\Gamma f(a(x))\,\mathrm{d}x = \int_\Gamma \nabla_\Gamma f(x)\,\mathrm{d}A(x).$$

The limit of the first two terms of the right-hand side of (2.15) is given by

$$\frac{\mathrm{d}}{\mathrm{d}\varepsilon}\bigg|_{\varepsilon=0}\int_{\Gamma(\varepsilon)} \overline{f}(x)\nu_\Gamma(x)\,\mathrm{d}A(x) = \int_\Gamma f(x)H(x)\nu_\Gamma(x)\,\mathrm{d}A(x),$$

the last equality being the transport theorem in Theorem 5.1 with $v = \nu_\Gamma$. In the proof of that theorem we will not use integration by parts.

For the last term on the right-hand side of (2.15) we have that

$$\lim_{\varepsilon\to 0}\frac{1}{2\varepsilon}\int_{M(\varepsilon)}\overline{f}(x)\mu_\Gamma(x)\,\mathrm{d}A(x)=\int_{\partial\Gamma}f(x)\mu_\Gamma(x)\,\mathrm{d}A(x),$$

because the integrand does not depend on ε. □

The formula for integration by parts on Γ leads to the notion of a weak derivative and to the concept of Sobolev spaces on surfaces. Sobolev spaces are the natural spaces for solutions of elliptic partial differential equations. Let $\Gamma\in C^2$ for the following.

For $p\in[1,\infty]$ we let $L^p(\Gamma)$ denote the space of functions $f:\Gamma\to\mathbb{R}$ which are measurable with respect to the surface measure $\mathrm{d}A$ (the n-dimensional Hausdorff measure) and have finite norm, where the norm is defined by

$$\|f\|_{L^p(\Gamma)}=\left(\int_\Gamma |f|^p\,\mathrm{d}A\right)^{\frac{1}{p}}$$

for $p<\infty$, and for $p=\infty$ we mean the essential supremum norm.

$L^p(\Gamma)$ is a Banach space and for $p=2$ a Hilbert space. For $1\le p<\infty$ the spaces $C^0(\Gamma)$ and $C^1(\Gamma)$ are dense in $L^p(\Gamma)$.

Definition 2.11. A function $f\in L^1(\Gamma)$ has the weak derivative $v_i=\underline{D}_i f\in L^1(\Gamma)$ ($i\in\{1,\dots,n+1\}$) if, for every function $\varphi\in C^1(\Gamma)$ with compact support $\overline{\{x\in\Gamma\mid\varphi(x)\neq 0\}}\subset\Gamma$, we have the relation

$$\int_\Gamma f\underline{D}_i\varphi\,\mathrm{d}A=-\int_\Gamma\varphi v_i\,\mathrm{d}A+\int_\Gamma f\varphi H\nu_i\,\mathrm{d}A.$$

The Sobolev space $H^{1,p}(\Gamma)$ is defined by

$$H^{1,p}(\Gamma)=\{f\in L^p(\Gamma)\mid \underline{D}_i f\in L^p(\Gamma),\ i=1,\dots,n+1\}$$

with norm

$$\|f\|_{H^{1,p}(\Gamma)}=\left(\|f\|^p_{L^p(\Gamma)}+\|\nabla_\Gamma f\|^p_{L^p(\Gamma)}\right)^{\frac{1}{p}}.$$

For $k\in\mathbb{N}$ we define

$$H^{k,p}(\Gamma)=\{f\in H^{k-1,p}(\Gamma)\mid \underline{D}_i v^{(k-1)}\in L^p(\Gamma), i=1,\dots,n+1\},$$

where $H^{0,p}(\Gamma)=L^p(\Gamma)$. For $p=2$ we use the notation $H^k(\Gamma)=H^{k,2}(\Gamma)$. We denote by $v^{(l)}$ *all* weak derivatives of order l. Then

$$\|v\|_{H^{k,p}(\Gamma)}=\left(\sum_{l=0}^{k}\|v^{(l)}\|^p_{L^p(\Gamma)}\right)^{\frac{1}{p}}.$$

Note that for the previous definition we have only assumed that $\Gamma\in C^2$. This was done because in the formulation of the weak derivative we used the mean curvature of Γ.

2.4. Poincaré's inequality

For the convenience of the reader we show how the Poincaré inequality for a function with mean value zero on a compact n-dimensional hypersurface can be deduced from the Poincaré inequality in $\mathbb{R}^{n+1}$ with the use of global coordinates.

Theorem 2.12. Assume that $\Gamma \in C^3$ and $1 \leq p < \infty$. Then there is a constant c such that, for every function $f \in H^{1,p}(\Gamma)$ with $\int_\Gamma f \, \mathrm{d}A = 0$, we have the inequality

$$\|f\|_{L^p(\Gamma)} \leq c \|\nabla_\Gamma f\|_{L^p(\Gamma)}. \tag{2.16}$$

Proof. Clearly it is sufficient to prove the inequality for L^1 instead of L^p and it is sufficient to work with smooth functions. Assume that $f \in C^1(\Gamma)$ with $\int_\Gamma f \, \mathrm{d}A = 0$. We extend this function to the tubular neighbourhood U_δ of the surface Γ, δ being sufficiently small, by

$$\overline{f}(x) = f(a(x)), \quad x \in U_\delta.$$

We state the following intermediate lemma.

Lemma 2.13. Let $\Gamma \in C^2$ be a compact hypersurface. Then there is a constant c, such that for every $0 < \varepsilon < \delta$ we have

$$\left| \frac{1}{2\varepsilon} \int_{U_\varepsilon} \overline{f}(x) \, \mathrm{d}x - \int_\Gamma f(x) \, \mathrm{d}A(x) \right| \leq c\varepsilon \int_\Gamma |f(x)| \, \mathrm{d}A(x). \tag{2.17}$$

The proof of this lemma is left to the reader. We note that by the co-area formula from Theorem 2.9 we have

$$\frac{1}{2\varepsilon} \int_{U_\varepsilon} \overline{f}(x) \, \mathrm{d}x = \frac{1}{2\varepsilon} \int_{-\varepsilon}^{\varepsilon} \int_{\Gamma(r)} f(a(x)) \, \mathrm{d}A(x) \, \mathrm{d}r$$

with an integrand which does not depend on ε.

We continue with the proof of Theorem 2.12. From (2.17) we get the inequality

$$\begin{aligned}
(1 - c_1\varepsilon) \int_\Gamma |f(x)| \, \mathrm{d}A(x) &\leq \frac{1}{2\varepsilon} \int_{U_\varepsilon} |\overline{f}(x)| \, \mathrm{d}x \\
&\leq \frac{1}{2\varepsilon} \int_{U_\varepsilon} \left| \overline{f}(x) - \frac{1}{|U_\varepsilon|} \int_{U_\varepsilon} \overline{f}(y) \, \mathrm{d}y \right| \mathrm{d}x + c_2 \left| \frac{1}{|U_\varepsilon|} \int_{U_\varepsilon} \overline{f}(y) \, \mathrm{d}y \right| \\
&\leq c_3(\varepsilon) \int_{U_\varepsilon} |\nabla \overline{f}(x)| \, \mathrm{d}x + c_2 \left| \frac{1}{|U_\varepsilon|} \int_{U_\varepsilon} \overline{f}(y) \, \mathrm{d}y \right|
\end{aligned}$$

by using the Poincaré inequality for $\overline{f}$ on U_ε. We also have that

$$\begin{aligned}\left|\frac{1}{|U_\varepsilon|}\int_{U_\varepsilon}\overline{f}(x)\,\mathrm{d}x\right| &= \frac{2\varepsilon}{|U_\varepsilon|}\left|\frac{1}{2\varepsilon}\int_{U_\varepsilon}\overline{f}(x)\,\mathrm{d}x - \int_\Gamma f(x)\,\mathrm{d}A(x)\right| \\ &\le c_4\varepsilon\int_\Gamma |f(x)|\,\mathrm{d}A(x).\end{aligned}$$

Thus we have the estimate

$$\begin{aligned}(1-c_1\varepsilon-c_4\varepsilon)\int_\Gamma |f(x)|\,\mathrm{d}A(x) &\le c_3(\varepsilon)\int_{U_\varepsilon}|\nabla\overline{f}(x)|\,\mathrm{d}x \\ &\le c_5(\varepsilon)\int_\Gamma |\nabla_\Gamma f(x)|\,\mathrm{d}A(x).\end{aligned}$$

For the last estimate we have used that by definition $\frac{\partial\overline{f}}{\partial\nu}=0$. A suitable choice of $\varepsilon>0$ gives the estimate

$$\int_\Gamma |f(x)|\,\mathrm{d}A(x) \le c\int_\Gamma |\nabla_\Gamma f(x)|\,\mathrm{d}A(x).$$

This is Poincaré's inequality in $L^1(\Gamma)$. For $p>1$ we apply this result to $|f|^p$ instead of f, use that $|\nabla_\Gamma|f|^p| = p|f|^{p-1}|\nabla_\Gamma f|$ and the Hölder inequality, and the theorem is proved. □

The formula for integration by parts on surfaces directly implies Green's formula. From Theorem 2.10, using the summation convention that we sum over doubly appearing indices, we have

$$\begin{aligned}\int_\Gamma \nabla_\Gamma f\cdot\nabla_\Gamma g\,\mathrm{d}A &= \int_\Gamma \underline{D}_i f\underline{D}_i g\,\mathrm{d}A = \int_\Gamma \underline{D}_i(f\underline{D}_i g)\,\mathrm{d}A - \int_\Gamma f\underline{D}_i\underline{D}_i g\,\mathrm{d}A \\ &= \int_\Gamma f\underline{D}_i gH\nu_i\,\mathrm{d}A + \int_{\partial\Gamma} f\underline{D}_i g\mu_i\,\mathrm{d}A - \int_\Gamma f\Delta_\Gamma g\,\mathrm{d}A.\end{aligned}$$

Since $\underline{D}_i g\nu_i = \nabla_\Gamma g\cdot\nu = 0$, we have the following theorem.

Theorem 2.14.

$$\int_\Gamma \nabla_\Gamma f\cdot\nabla_\Gamma g\,\mathrm{d}A = -\int_\Gamma f\Delta_\Gamma g\,\mathrm{d}A + \int_{\partial\Gamma} f\nabla_\Gamma g\cdot\mu\,\mathrm{d}A. \tag{2.18}$$

3. Partial differential equations on surfaces

3.1. Elliptic equations on hypersurfaces

In this section we briefly give the basic ideas for the analysis of elliptic PDEs on hypersurfaces. We assume that Γ is a compact and connected hypersurface.

The Poisson equation

We begin with the model case of the Poisson equation

$$-\Delta_\Gamma u = f \tag{3.1}$$

on a compact hypersurface Γ in $\mathbb{R}^{n+1}$ and thus without boundary. Here f is a given right-hand side or source term which is taken to be from $L^2(\Gamma)$ or more generally from $H^{-1}(\Gamma)$, the dual space of $H^1(\Gamma)$.

A *weak solution* of (3.1) is a function $u \in H^1(\Gamma)$ which satisfies the relation

$$\int_\Gamma \nabla_\Gamma u \cdot \nabla_\Gamma \varphi \, \mathrm{d}A = \int_\Gamma f\varphi \, \mathrm{d}A \tag{3.2}$$

for every test function $\varphi \in H^1(\Gamma)$. Since $\varphi = 1$ is allowed as a test function we have to impose the condition $\int_\Gamma f = 0$ on the right-hand side. If the right-hand side f is a functional only from $H^{-1}(\Gamma)$, then the weak form of the equation reads

$$\int_\Gamma \nabla_\Gamma u \cdot \nabla_\Gamma \varphi \, \mathrm{d}A = \langle f, \varphi \rangle,$$

where the brackets stand for the evaluation of the functional f at the function φ.

Obviously there is no uniqueness of weak solutions in this case, since every constant is a solution. We will fix the free constant by choosing the mean value of u to vanish. The following theorem can easily be proved.

Theorem 3.1. Let $\Gamma \in C^2$ be a compact hypersurface in $\mathbb{R}^{n+1}$ and assume that $f \in H^{-1}(\Gamma)$ with the property $\langle f, 1 \rangle = 0$. Then there exists a unique solution $u \in H^1(\Gamma)$ of (3.2) with $\int_\Gamma u \, \mathrm{d}A = 0$.

The proof is an application of the Lax–Milgram theorem or the Riesz representation theorem. The bilinear form

$$a(u,v) = \int_\Gamma \nabla_\Gamma u \cdot \nabla_\Gamma v \, \mathrm{d}A$$

is a scalar product on the Hilbert space $X = \{u \in H^1(\Gamma) \mid \int_\Gamma u \, \mathrm{d}A = 0\}$ because of Poincaré's inequality (2.16). The right-hand side f was chosen to be in the space $H^{-1}(\Gamma)$ of linear functionals.

Besides the existence of weak solutions the most important ingredient for suitable numerics is the proof of regularity and of *a priori* estimates for solutions of the Poisson equation. We shall show how to prove an *a priori* estimate in the $H^2(\Gamma)$ norm. For this we use the following little lemma, in which we use the usual notational convention for the seminorms $|\cdot|_{H^1(\Gamma)}$ and $|\cdot|_{H^2(\Gamma)}$.

Lemma 3.2. Let $\Gamma \in C^2$ and $u \in H^2(\Gamma)$. Then

$$|u|_{H^2(\Gamma)} \le \|\Delta_\Gamma u\|_{L^2(\Gamma)} + c|u|_{H^1(\Gamma)} \tag{3.3}$$

with the constant $c = \sqrt{\|H\mathcal{H} - 2\mathcal{H}^2\|_{L^\infty(\Gamma)}}$.

Proof. By approximation arguments we can assume that $\Gamma \in C^3$ and $u \in C^3(\Gamma)$. We have

$$|u|^2_{H^2(\Gamma)} = \sum_{i,j=1}^{n+1} \int_\Gamma (\underline{D}_i \underline{D}_j u)^2 \,\mathrm{d}A,$$

and with the formula for integration by parts on Γ (2.13) in combination with Lemma 2.6 we obtain (using the summation convention)

$$\begin{aligned}
\int_\Gamma \underline{D}_i \underline{D}_j u \underline{D}_i \underline{D}_j u \,\mathrm{d}A &= \int_\Gamma \underbrace{\underline{D}_i \underline{D}_j u \underline{D}_j u H \nu_i}_{=0} \,\mathrm{d}A - \int_\Gamma \underline{D}_i \underline{D}_i \underline{D}_j u \underline{D}_j u \,\mathrm{d}A \\
&= -\int_\Gamma \underline{D}_i \big(\underline{D}_j \underline{D}_i u + \underline{D}_k u (\nu_i \mathcal{H}_{jk} - \nu_j \mathcal{H}_{ik})\big) \underline{D}_j u \,\mathrm{d}A \\
&= -\int_\Gamma \underline{D}_i \underline{D}_j \underline{D}_i u \underline{D}_j u \,\mathrm{d}A - \int_\Gamma \underbrace{\underline{D}_i \underline{D}_k u (\nu_i \mathcal{H}_{jk} - \nu_j \mathcal{H}_{ik}) \underline{D}_j u}_{=0} \,\mathrm{d}A \\
&\quad - \int_\Gamma \underline{D}_k u \underline{D}_j u \underline{D}_i (\nu_i \mathcal{H}_{jk} - \nu_j \mathcal{H}_{ik}) \,\mathrm{d}A \\
&= -\int_\Gamma \underline{D}_i \underline{D}_j \underline{D}_i u \underline{D}_j u \,\mathrm{d}A - \int_\Gamma (H\mathcal{H} - \mathcal{H}^2)_{jk} \underline{D}_j u \underline{D}_k u \,\mathrm{d}A.
\end{aligned}$$

For the remaining third-order term we observe that

$$\begin{aligned}
\int_\Gamma \underline{D}_i \underline{D}_j \underline{D}_i u \underline{D}_j u \,\mathrm{d}A &= \int_\Gamma \underline{D}_j \underline{D}_i \underline{D}_i u \underline{D}_j u + \nu_i \mathcal{H}_{jk} \underline{D}_k \underline{D}_i u \underline{D}_j u \,\mathrm{d}A \\
&= \int_\Gamma \underline{D}_j \Delta_\Gamma u \underline{D}_j u \,\mathrm{d}A - \int_\Gamma (\mathcal{H}^2)_{ij} \underline{D}_i u \underline{D}_j u \,\mathrm{d}A \\
&= -\int_\Gamma (\Delta_\Gamma u)^2 \,\mathrm{d}A - \int_\Gamma (\mathcal{H}^2)_{ij} \underline{D}_i u \underline{D}_j u \,\mathrm{d}A.
\end{aligned}$$

Here we have used the fact that

$$\nu_i \underline{D}_k \underline{D}_i u = \underline{D}_k (\nu_i \underline{D}_i u) - \underline{D}_k \nu_i \underline{D}_i u = -\mathcal{H}_{ik} \underline{D}_i u.$$

Altogether we have shown that

$$\begin{aligned}
|u|^2_{H^2(\Gamma)} &= \|\Delta_\Gamma u\|^2_{L^2(\Gamma)} - \int_\Gamma (H\mathcal{H} - 2\mathcal{H}^2) \nabla_\Gamma u \cdot \nabla_\Gamma u \,\mathrm{d}A \\
&\le \|\Delta_\Gamma u\|^2_{L^2(\Gamma)} + \|H\mathcal{H} - 2\mathcal{H}^2\|_{L^\infty(\Gamma)} \|\nabla_\Gamma u\|^2_{L^2(\Gamma)},
\end{aligned}$$

and this finally proves the estimate (3.3). □

With the help of the previous lemma and standard arguments we arrive at the regularity estimate for the solution of the Poisson equation on a compact hypersurface.

Theorem 3.3. Assume that $\Gamma \in C^2$ and that $f \in L^2(\Gamma)$ with $\int_\Gamma f \, \mathrm{d}A = 0$. Then the weak solution from Theorem 3.1 satisfies $u \in H^2(\Gamma)$ and

$$\|u\|_{H^2(\Gamma)} \leq c\|f\|_{L^2(\Gamma)}.$$

For the proof we use the basic estimate (choose $\varphi = u$ in (3.2))

$$|u|_{H^1(\Gamma)} \leq c\|f\|_{L^2(\Gamma)}$$

(for which we use Poincaré's inequality again), together with the PDE pointwise almost everywhere, to obtain

$$|u|_{H^2(\Gamma)} \leq c\|f\|_{L^2(\Gamma)},$$

if the solution has square integrable second derivatives.

The $H^2(\Gamma)$-regularity of u is taken from the theory of linear PDEs on Cartesian domains in $\mathbb{R}^n$. Here the arguments are purely local. For this we parametrize the C^2 surface Γ according to Lemma 2.2 locally by $X \in C^2(\Omega, \Gamma)$, $X = X(\theta)$ with some open domain $\Omega \subset \mathbb{R}^n$. If we set $U(\theta) = u(X(\theta))$, then U is a weak solution of the linear PDE

$$-\big(g^{kj} U_{\theta_j} \sqrt{g}\big)_{\theta_k} = f \circ X \sqrt{g}$$

on Ω. For the notation see Section 2.1. The coefficients of this PDE are in $C^1(\Omega)$ and the right-hand side is in $L^2(\Omega)$ because by assumption $\Gamma \in C^2$. The well-known regularity result from Cartesian PDEs (see for example Gilbarg and Trudinger 1998) then gives $U \in H^2(\Omega')$ for any $\Omega' \subset\subset \Omega$, and this in turn gives $u \in H^2(\Gamma)$.

General elliptic PDEs

In the previous section we have shown how the Poisson equation is solved on a compact surface. The methods are easily extended to general linear elliptic PDEs in divergence form and to boundary value problems (on surfaces with a boundary):

$$-\sum_{i,j=1}^{n+1} \underline{D}_i(a_{ij}\underline{D}_j u) - \sum_{i=1}^{n+1} \underline{D}_i(a_i u) + \sum_{i=1}^{n+1} b_i \underline{D}_i u + cu = f - \sum_{i=1}^{n+1} \underline{D}_i g_i. \quad (3.4)$$

We assume for the given coefficients that

$$a_{ij}, a_i, b_i, c \in L^\infty(\Gamma), g_i \in L^2(\Gamma) \quad (i,j = 1, \ldots, n+1).$$

We also assume that the coefficient vectors $a(x) = (a_1(x), \ldots, a_{n+1}(x))$ and $g(x) = (g_1(x), \ldots, g_{n+1}(x))$ are tangent vectors at $x \in \Gamma$, that is, they lie in $T_x\Gamma$, and that the matrix $\mathcal{A}(x) = (a_{ij}(x))_{i,j=1,\ldots,n+1}$ is symmetric and maps the tangent space $T_x\Gamma$ into itself. We emphasize that the latter condition implies that in general constant coefficients a_{ij} are not admissible. They have to depend on the x-variable. Nevertheless $a_{ij} = \delta_{ij}$ is obviously allowed.

As ellipticity condition we assume a so-called Ladyzhenskaya condition, which says that there exists a number $c_0 > 0$ such that

$$\sum_{i,j=1}^{n+1} a_{ij}\xi_i\xi_j + \sum_{i=1}^{n+1} a_i\xi_0\xi_i + \sum_{i=1}^{n+1} b_i\xi_i\xi_0 + c\xi_0^2 \geq c_0 \sum_{i=1}^{n+1} \xi_i^2,$$

almost everywhere on Γ for all $\xi = (\xi_1, \ldots, \xi_{n+1}) \in \mathbb{R}^{n+1}$ with $\xi \cdot \nu = 0$ and all $\xi_0 \in \mathbb{R}$.

With the PDE (3.4) we associate the bilinear form a,

$$a(u, \varphi) = \int_\Gamma \sum_{i,j=1}^{n+1} a_{ij}\underline{D}_j u \underline{D}_i \varphi + \sum_{i=1}^{n+1} a_i u \underline{D}_i \varphi + \sum_{i=1}^{n+1} b_i \underline{D}_i u \varphi + cu\varphi \, \mathrm{d}A,$$

and the functional F,

$$\langle F, \varphi \rangle = \int_\Gamma f\varphi + \sum_{i=1}^{n+1} g_i \underline{D}_i \varphi \, \mathrm{d}A,$$

for $u, \varphi \in H^1(\Gamma)$, for which we assume $\langle F, 1 \rangle = 0$.

Theorem 3.4. Let $\Gamma \in C^2$ be a compact hypersurface. Assume that the coefficients satisfy the above conditions. Then there exists a unique weak solution of (3.4) with $\int_\Gamma u \, \mathrm{d}A = 0$, that is, there exists a unique $u \in H^1(\Gamma)$ such that

$$a(u, \varphi) = \langle F, \varphi \rangle$$

for every $\varphi \in H^1(\Gamma)$.

Proof. The proof of this theorem is a direct application of the Lax–Milgram theorem. □

4. Triangulated surfaces

In this section we discuss the discretization of surfaces in combination with the discretization of elliptic surface PDEs. The continuous surface Γ is replaced by a piecewise polynomial surface; in the most simple case the approximating surface Γ_h is a polygonal one. This introduces a geometric error between Γ and Γ_h, which we discuss in Section 4.2. In Section 4.4 we introduce finite element spaces on the discrete surface and discretize the PDEs. This is the surface finite element method (SFEM). The error between the continuous solution and the discrete solution is estimated in natural norms in Section 4.5.

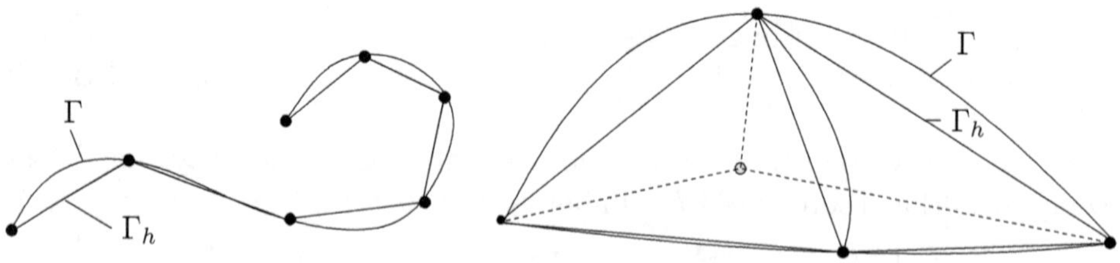

(a) (b)

Figure 4.1. (a) Approximation of a smooth curve by a polygonal curve and (b) approximation of a smooth surface by a polygonal surface.

4.1. Triangulations

The smooth n-dimensional surface Γ $(\partial\Gamma = \emptyset)$ is approximated by a surface Γ_h which lies in the strip U_δ (see Lemma 2.8) and which is a Lipschitz surface. The strip can be chosen with locally varying width: see (4.1). In particular, for $n = 1$, Γ_h is a polygonal curve, and for $n = 2$, it is a triangulated (and hence polyhedral) surface consisting of triangles. A three-dimensional surface in $\mathbb{R}^4$ consists of tetrahedra.

We assume that the discrete triangulated surface Γ_h besides (4.1) has the following properties. Γ_h is the union of finitely many non-degenerate (closed) n-simplices in $\mathbb{R}^{n+1}$. We let $\mathcal{T}_h$ denote the set of these simplices:

$$\Gamma_h = \bigcup_{T \in \mathcal{T}_h} T.$$

The vertices X_j $(j = 1, \ldots, J)$ of the simplices are taken to sit on the smooth surface Γ. For $T, \tilde{T} \in \mathcal{T}_h$, either $T \cap \tilde{T} = \emptyset$ or $T \cap \tilde{T}$ is an $(n-k)$-dimensional side simplex $(k \in \{1, \ldots, n\})$ common to both of the simplices T and $\tilde{T}$. For $T \in \mathcal{T}_h$ we denote by $h(T)$ its diameter and by $\rho(T)$ the in-ball radius. Also,

$$h = \max_{T \in \mathcal{T}_h} h(T), \quad \rho = \min_{T \in \mathcal{T}_h} \rho(T).$$

We assume that the quantity

$$\sigma = \max_{T \in \mathcal{T}_h} \sigma(T), \quad \sigma(T) = \frac{h(T)}{\rho(T)}$$

is uniformly bounded independently of h.

Note that, by Lemma 2.8, for every simplex $T \subset \Gamma_h$ there is a unique curved simplex $\tilde{T} = a(T) \subset \Gamma$. In order to avoid a global double covering (see Figure 4.2), we assume that for each point $a \in \Gamma$ there is at most one point $x \in \Gamma_h$ with $a = a(x)$. This implies that there is a bijective correspondence between the triangles on Γ_h and the induced curvilinear triangles on Γ.

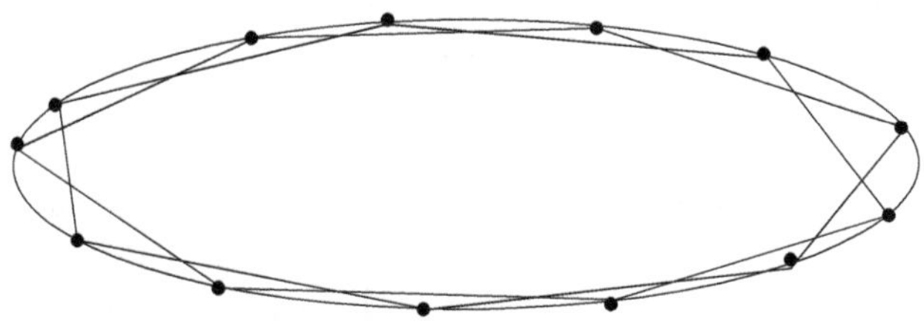

Figure 4.2. Approximation of an ellipse Γ by a polygon Γ_h violating the simple covering condition.

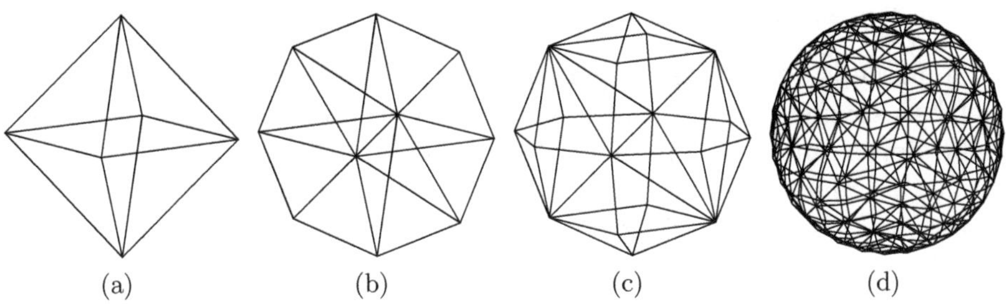

Figure 4.3. Refine and project for the sphere. We start from a macro-triangulation (a) with 6 vertices and 8 triangles, and obtain the first (b), second (c) and sixth (d) refinement. The finest triangulation consists of 258 vertices and 512 triangles. The method of refinement is bisection of triangles.

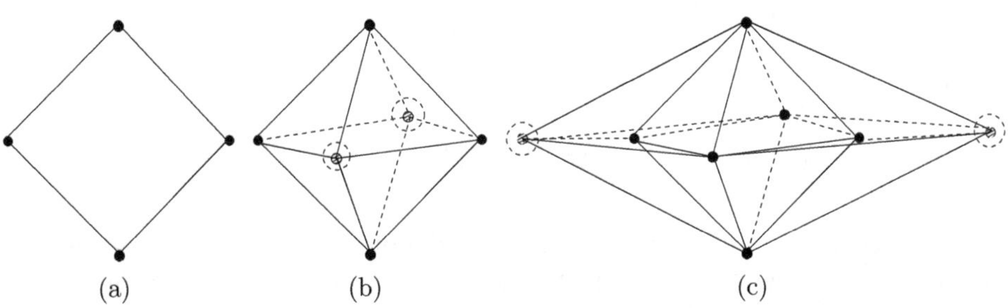

Figure 4.4. Discrete spheres as a macro-triangulation of S^1, S^2 and S^3 (symbolically). We start with the macro-triangulation of S^1 (a), embed it into $\mathbb{R}^3$ and add two new vertices (surrounded by circles) to obtain a macro-triangulation of S^2 (b). We embed it into $\mathbb{R}^4$ and add new vertices to obtain a macro-triangulation of S^3 (c).

For the description of the approximation of surfaces we take the practical point of view. In order to generate triangulated parametric surfaces the two main techniques are as follows.

(1) Construct a macro-triangulation for the given smooth surface in such a way that the coarse discrete surface Γ_h is contained in a strip of unique projection around Γ. Then refine and project the new nodes onto the smooth surface. For an example see Figure 4.3.

(2) Glue together patches (charts). Here one follows the classical differential geometric path. The additional difficulty from the numerical point of view is the joining of two or more different grids.

Both methods can be used to generate a discretization of a standard surface such as a sphere, which can then serve as a parameter domain for the surface to be approximated. One may think of deforming the available discrete surface in order to obtain a new discrete surface. An example of this method is shown in Figure 4.5, where the surface shown is an image of a discretized cylinder.

Refine and project. The following can be understood as setting up a macro-triangulation approximating the smooth surface. For this method we use the description introduced in Section 2.3. We assume that Γ has only one connected component. We start with a discrete surface Γ_h contained in the variable strip U_δ,

$$\Gamma_h \subset U_\delta = \{y + s\nu(y) \mid |s| < \delta(y), y \in \Gamma\}, \tag{4.1}$$

where for a point $y \in \Gamma$ we set

$$\delta(y) = \min\left\{ \left(\max_{i=1,\dots,n} |\kappa_i(y)| \right)^{-1}, \sup_{z\in\Gamma, z\neq y} \frac{L(y,z)}{|y-z|} \right\}$$

with the sectional curvatures κ_i. $L(y,z)$ denotes the geodesic distance between y and z on Γ. Just as in Section 2.3, one can show that each point $x \in U_\delta$ can be uniquely projected onto the smooth surface, yielding the decomposition

$$x = a(x) + d(x)\nu(a(x)), \quad x \in U_\delta,$$

with $a(x)$ the orthogonal projection of x onto Γ and $d(x)$ the oriented distance between x and Γ. Given a refinement of Γ_h, a new refined triangulation of Γ may be obtained by projecting the new nodes on Γ_h onto Γ.

For any function η defined on the discrete surface Γ_h we define an extension or lift onto Γ by

$$\eta^l(a) = \eta(x(a)), \quad a \in \Gamma, \tag{4.2}$$

(a) (b)

Figure 4.5. (a) A complicated hypersurface Γ approximated by a piecewise linear surface. (b) Close-up of the discrete surface Γ_h.

where, by our assumptions, $x(a)$ is defined as the unique solution of

$$x = a + d(x)\nu(a). \tag{4.3}$$

Furthermore, we understand by $\eta^l(x)$ the constant extension from Γ in the normal direction $\nu(a(x))$.

We will use a finite element space

$$S_h = \{\phi_h \in C^0(\Gamma_h) \mid \varphi_h|_T \text{ is linear affine for each } T \in \mathcal{T}_h\}. \tag{4.4}$$

The lifted finite element space is then

$$S_h^l = \{\varphi_h = \phi_h^l \mid \phi_h \in S_h\}. \tag{4.5}$$

4.2. Approximation of geometry

In order to avoid too many technical arguments, we begin with a piecewise linear approximation of a smooth surface $\Gamma \in C^2$ and its geometry. The extension to a higher-order approximation then follows the arguments in the piecewise linear case, and in fact is based upon a piecewise linear approximation.

Piecewise linear approximation of geometry

In this section we collect estimates for the geometric error which is produced by approximating the smooth surface Γ by a discrete surface Γ_h.

The following technical lemma gives more detailed information about the order of approximation of the geometry.

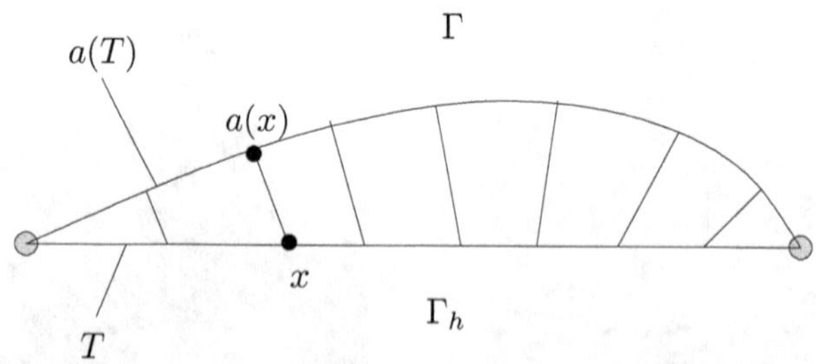

Figure 4.6. A curved 'simplex' $a(T)$ is parametrized over a planar one. Orthogonal projection onto the smooth surface Γ.

Lemma 4.1. Assume Γ and Γ_h are as above. The map $a : \Gamma_h \to \Gamma$ is bijective. For the oriented distance function to Γ we have the estimate

$$\|d\|_{L^\infty(\Gamma_h)} \le ch^2. \tag{4.6}$$

The quotient, δ_h, between the smooth and discrete surface measures $\mathrm{d}A$, and $\mathrm{d}A_h$, defined by $\delta_h \,\mathrm{d}A_h = \mathrm{d}A$, satisfies

$$\|1 - \delta_h\|_{L^\infty(\Gamma_h)} \le ch^2. \tag{4.7}$$

Let P and P_h be the projections onto the tangent planes, $P_{ij} = \delta_{ij} - \nu_i \nu_j$, $P_{h,ij} = \delta_{ij} - \nu_{h,i}\nu_{h,j}$, and let

$$R_h = \frac{1}{\delta_h} P(I - d\mathcal{H})P_h(I - d\mathcal{H}), \tag{4.8}$$

$\mathcal{H}_{ij} = d_{x_i x_j} = \nu_{i_{x_j}}$. Then

$$\|(I - R_h)P\|_{L^\infty(\Gamma_h)} \le ch^2. \tag{4.9}$$

Proof. Let $T \in \mathcal{T}_h$ be a simplex of the discrete surface. By assumption it lies in the strip U_δ (see Lemma 2.8). Without loss of generality we may assume that $T \subset \mathbb{R}^n \times \{0\}$. The corresponding curved triangle $\tilde{T} = a(T)$ is thus parametrized over T. Note that the parametrization is not given as a (vertical) graph. This is crucial for all our arguments! See Figure 4.6 for a sketch of the situation.

We let I_h denote the Lagrange interpolation on T. Since the vertices of T lie on Γ, we have that the interpolant $I_h d$ vanishes identically on T and, with the well-known estimates for the Lagrangian interpolation (see Ciarlet 1978), we obtain

$$\|d\|_{L^\infty(T)} = \|d - I_h d\|_{L^\infty(T)} \le ch^2 |d|_{H^{2,\infty}(T)} \le ch^2 \|d\|_{C^2(U_\delta)}, \tag{4.10}$$

and similarly, for $j = 1, \ldots, n$,

$$\|\nu_j\|_{L^\infty(T)} = \|d_{x_j}\|_{L^\infty(T)} = \|(d - I_h d)_{x_j}\|_{L^\infty(T)} \le ch \|d\|_{C^2(U_\delta)}. \tag{4.11}$$

To derive the estimate of the surface elements (4.7), we observe that

$$\mathrm{d}A = \sqrt{D_1^2 + \cdots + D_{n+1}^2}\,\mathrm{d}x_1 \cdots \mathrm{d}x_n$$

while $\mathrm{d}A_h = \mathrm{d}x_1 \cdots \mathrm{d}x_n$, with

$$D_i = (-1)^{n+1+i} \begin{vmatrix} \frac{\partial a_1}{\partial x_1} & \cdots & \frac{\partial a_1}{\partial x_n} \\ \vdots & \vdots & \vdots \\ \frac{\partial a_{i-1}}{\partial x_1} & \cdots & \frac{\partial a_{i-1}}{\partial x_n} \\ \frac{\partial a_{i+1}}{\partial x_1} & \cdots & \frac{\partial a_{i+1}}{\partial x_n} \\ \vdots & \vdots & \vdots \\ \frac{\partial a_{n+1}}{\partial x_1} & \cdots & \frac{\partial a_{n+1}}{\partial x_n} \end{vmatrix}.$$

Now $a_j(x) = x_j - d(x)\nu_j(x)$, and thus for $j, k = 1, \ldots, n$,

$$\frac{\partial a_j}{\partial x_k} = \delta_{jk} - \nu_k(x)\nu_j(x) - d(x)\mathcal{H}_{jk}(x) = P(x)_{jk} + O(h^2), \tag{4.12}$$

since from (4.10) and (4.11) we know that $|\nu_k \nu_j| \leq ch^2$ for $j, k = 1, \ldots, n$ and $|d\mathcal{H}_{jk}| \leq ch^2$. Similarly, for $j = 1, \ldots, n$,

$$\frac{\partial a_{n+1}}{\partial x_j} = -\nu_{n+1}\nu_j + O(h^2). \tag{4.13}$$

The multilinearity of the determinant, together with (4.12) and (4.13), implies that for $i = 1, \ldots, n$

$$D_i = -\nu_{n+1}\nu_i + O(h^2), \quad D_{n+1} = 1 + O(h^2). \tag{4.14}$$

Then

$$\begin{aligned} \sqrt{D_1^2 + \cdots + D_{n+1}^2} - 1 &= \frac{D_1^2 + \cdots + D_{n+1}^2 - 1}{\sqrt{D_1^2 + \cdots + D_{n+1}^2} + 1} \\ &= \frac{\nu_{n+1}^2(1 - \nu_{n+1}^2) + O(h^2)}{\sqrt{D_1^2 + \cdots + D_{n+1}^2} + 1} = O(h^2), \end{aligned}$$

because $\nu_{n+1}^2 = 1 + O(h^2)$. We have proved (4.7).

The proof of (4.9) follows from the previous estimates when we keep in mind that in our situation $\nu_h = e_{n+1}$. Note that by ν_h we mean the piecewise constant vector defined by the normals to the simplices of Γ_h. We find that

$$(R_h - I)P = PP_hP - P + O(h^2) = O(h^2),$$

since for a unit vector z we have

$$|(PP_hP - P)z| = |z \cdot (\nu_h - (\nu_h \cdot \nu)\nu)(\nu_h - (\nu_h \cdot \nu)\nu)| \leq ch^2,$$

because from (4.11),

$$|\nu_h - (\nu_h \cdot \nu)\nu| = |e_{n+1} - \nu_{n+1}\nu| = \sqrt{1 - \nu_{n+1}^2} = O(h).$$

This proves (4.9). □

In order to compare the norms between functions on Γ_h and their lift (4.2) to Γ we need the following lemma.

Lemma 4.2. Let $\eta : \Gamma_h \to \mathbb{R}$ with lift $\eta^l : \Gamma \to \mathbb{R}$. Then, for the plane $T \subset \Gamma_h$, and smooth curved triangles $\tilde{T} \subset \Gamma$, the following estimates hold if the norms exist. There is a constant $c > 0$ independent of h such that

$$\frac{1}{c}\|\eta\|_{L^2(T)} \leq \|\eta^l\|_{L^2(\tilde{T})} \leq c\|\eta\|_{L^2(T)}, \tag{4.15}$$

$$\frac{1}{c}\|\nabla_{\Gamma_h}\eta\|_{L^2(T)} \leq \|\nabla_\Gamma \eta^l\|_{L^2(\tilde{T})} \leq c\|\nabla_{\Gamma_h}\eta\|_{L^2(T)}, \tag{4.16}$$

$$\|\nabla_{\Gamma_h}^2\eta\|_{L^2(T)} \leq c\|\nabla_\Gamma^2 \eta^l\|_{L^2(\tilde{T})} + ch\|\nabla_\Gamma \eta^l\|_{L^2(\tilde{T})}. \tag{4.17}$$

Proof. The proof is contained in Dziuk (1988). Here we only give the main ideas. In the following let d be the distance function with respect to the smooth surface Γ. By definition (see (4.2))

$$\eta(x) = \eta^l(x - d(x)\nu(x)), \quad x \in \Gamma_h.$$

The chain rule together with the definition of the tangential gradients on smooth and discrete surface, the latter one in a piecewise sense, gives

$$\nabla_{\Gamma_h}\eta(x) = P_h(x)\big(I - d(x)\mathcal{H}(x)\big)\nabla_\Gamma \eta^l(a(x)), \quad x \in \Gamma_h, \tag{4.18}$$

where P_h and $\mathcal{H}$ are as in Lemma 4.1. The results then easily follow from the estimates of that lemma, and in particular the estimate $0 < \frac{1}{c} \leq \delta_h \leq c < \infty$. □

For later use we list interpolation inequalities which are now available. The lemma was proved in Dziuk (1988) for the gradient. It is easily extended to the L^2-estimate.

Lemma 4.3 (interpolation). For $n \leq 3$ and given $\eta \in H^2(\Gamma)$, there exists a unique $I_h\eta \in S_h^l$ such that

$$\|\eta - I_h\eta\|_{L^2(\Gamma)} + h\|\nabla_\Gamma(\eta - I_h\eta)\|_{L^2(\Gamma)} \leq ch^2\big(\|\nabla_\Gamma^2\eta\|_{L^2(\Gamma)} + h\|\nabla_\Gamma\eta\|_{L^2(\Gamma)}\big). \tag{4.19}$$

Proof. The interpolant is constructed in an obvious way. Since $\eta \in H^2(\Gamma)$, by Sobolev's embedding it is in $C^0(\Gamma)$ since Γ is at most three-dimensional. (Compare with Remark 4.10.) Thus the pointwise linear interpolation $\tilde{I}_h\eta \in X_h$ is well defined. The vertices of Γ_h lie on the smooth surface Γ and so the nodal values of η are well defined for this interpolation. We then lift $\tilde{I}_h\eta$ onto Γ by the process $I_h\eta = (\tilde{I}_h\eta)^l$ according to (4.2). □

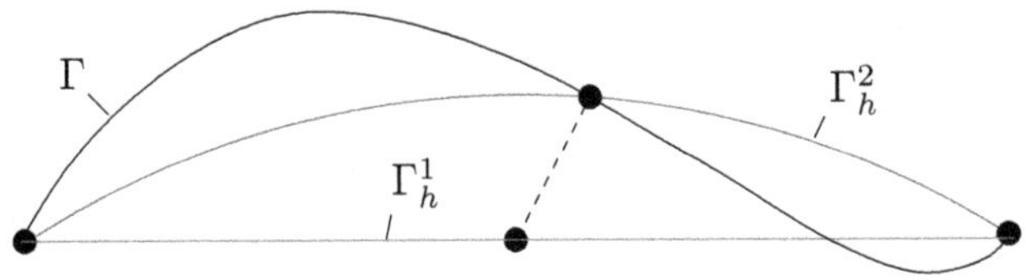

Figure 4.7. Higher-order approximation Γ_h^k to Γ for $k = 2$.

Higher-order approximation of geometry

In the previous subsection we discussed the polygonal approximation of smooth hypersurfaces. The higher-order approximation of the smooth surface Γ is based on the piecewise linear approximation and is related to an isoparametric approximation to boundaries in the usual finite element methods. Let

$$\Gamma_h^1 = \bigcup_{T^1 \in \mathcal{T}_h^1} T^1$$

be the piecewise linear approximation of Γ from the previous subsection. As shown, the map $a : \Gamma_h^1 \to \Gamma$ is bijective. We denote by $a^k = I_h^k a \in \mathbb{P}_k(T^1)$ the kth-order interpolation on the planar simplex T^1 at the Lagrange nodes of T^1. We then define the kth-order discrete surface Γ_h^k by

$$\Gamma_h^k = \bigcup_{T^1 \in \mathcal{T}_h^1} a^k(T^1).$$

Figure 4.7 illustrates this definition. Then the next lemma follows from the proof of Lemma 4.1 in combination with the common estimates for the Lagrange interpolation on (flat) simplices: see Demlow (2009).

Lemma 4.4. Assume that Γ is as in Lemma 4.1 and let Γ_h^k be the kth-order ($k \geq 1$) approximation described above. Then we have the estimates

$$\|d\|_{L^\infty(\Gamma_h^k)} + \|1 - \delta_h^k\|_{L^\infty(\Gamma_h^k)} + \|(I - R_h^k)P\|_{L^\infty(\Gamma_h^k)} \leq ch^{k+1},$$

where δ_h^k is the quotient of the surface measures on Γ and on Γ_h^k, $\mathrm{d}A = \delta_h^k \,\mathrm{d}A_h$, and $\delta_h^k R_h^k = P(I - d\mathcal{H})P_h^k(I - d\mathcal{H})$ with the projection $P_h^k = I - \nu_h^k \otimes \nu_h^k$ onto Γ_h^k with the normal ν_h^k of Γ_h^k.

4.3. Discrete charts

We end this section by adding some information on a discrete differential geometric approach which is consistent with the introduction of parametrized surfaces in Section 2.1. For a complete exposition see Nedelec (1976).

Let Γ be given as in Section 2.1 by local charts $(X_i)_{i\in I}, X_i \in C^k(V_i, \mathbb{R}^{n+1})$. We assume that already

$$\Gamma = \bigcup_{i\in I} X_i(V_{ih}),$$

with $V_{ih} \subset V_i$ and triangulated domains $V_{ih} \subset \mathbb{R}^n$. We interpolate the smooth parametrization $X_{ih} = I_h X_i$, where I_h is the interpolation from Lemma 4.3 on the polygonal domain V_{ih}. We assume that if $U_{ij} = X_{ih}(V_{ih}) \cap X_{jh}(V_{jh}) \neq \emptyset$, then the map $X_{ih}^{-1} \circ X_{jh} : X_{jh}^{-1}(U_{ij}) \to X_{ih}^{-1}(U_{ij})$ is bijective, continuous and piecewise linear. Then the discrete surface is defined by

$$\Gamma_h = \bigcup_{i\in I} X_{ih}(V_{ih}).$$

The advantage of this way to discretize the surface Γ is that it allows the approximation of immersions.

4.4. The surface finite element method (SFEM)

In the previous section we constructed and analytically treated an approximation of the smooth surface Γ by a discrete surface Γ_h. For the following the discrete surface is the union of n-simplices, $\Gamma_h = \cup_{T\in\mathcal{T}_h} T$. The extension to higher-order approximations is then an extension. For a function $g : \Gamma_h \to \mathbb{R}$ we understand the *tangential gradient* on the discrete surface in the sense of Definition 2.3 as

$$\nabla_{\Gamma_h} g = P_h \nabla g$$

in a piecewise sense, on each simplex. Here $(P_h)ij = \delta_{ij} - \nu_{hi}\nu_{hj}$ $(i, j = 1, \ldots, n)$. ∇g denotes the $(n+1)$-dimensional gradient of a continuation of g orthogonal to the simplex.

In the following we will set up a finite element method on this discrete surface in order to solve PDEs. Because this approach is quite transparent, we treat piecewise linear finite elements on the discrete surface first, and we solve the Poisson equation.

We use the finite element space (see (4.4))

$$S_h = \{\phi_h \in C^0(\Gamma_h) \mid \varphi_h|_T \text{ is linear affine for each } T \in \mathcal{T}_h\}. \tag{4.20}$$

This space is spanned by the nodal basis $\chi_1, \ldots, \chi_J$, which is given by

$$\chi_j \in S^h, \quad \chi_j(X_k) = \delta_{jk} \quad (j, k = 1, \ldots, J).$$

Here we denote by $X_j \in \Gamma$ the nodes, *i.e.*, the vertices, of the triangulation $\mathcal{T}_h$. Every function $U_h \in S_h$ has the form

$$U_h(x) = \sum_{j=1}^{J} \alpha_j \chi_j(x) \quad (x \in \Gamma_h)$$

with real constants α_j $(j = 1, \ldots, J)$.

Associated with the finite element space S_h defined on the discrete surface is the lifted finite element space

$$S_h^l = \{\varphi_h = \phi_h^l \mid \phi_h \in S_h\}. \tag{4.21}$$

Note that S_h^l is a subspace of the continuous space $H^1(\Gamma)$. This space will appear in theoretical considerations only.

Intermediate remark. We can define rth-order finite element spaces on kth-order approximations of the smooth surface: see Lemma 4.4. The finite element space of rth order $(r \geq 1)$ on the piecewise linear approximation Γ_h^1 to Γ is given by

$$S_{h,r}^1 = \{\varphi_h \in C^0(\Gamma_h^1) \mid \varphi_h|_T \in \mathbb{P}_r(T), T \in \mathcal{T}_h^1\}.$$

The general isoparametric finite element space on Γ_h^k is then defined by

$$S_{h,r}^k = \{\varphi_h \in C^0(\Gamma_h^k) \mid \varphi_h \circ a^k \in S_{h,r}^1\}.$$

In the following we will continue to work with piecewise linear finite elements on piecewise linear approximations $\Gamma_h = \Gamma_h^1$ and $S_h = S_{h,1}^1$ with S_h from (4.4), to keep the methods transparent. For more information concerning the higher-order case we refer to Demlow (2009).

Definition 4.5. Let $F_h \in L^2(\Gamma_h)$ with the property that $\int_{\Gamma_h} F_h \, \mathrm{d}A_h = 0$. A function $U_h \in S_h$ is a discrete solution of the Poisson equation (3.1) on Γ_h if

$$\int_{\Gamma_h} \nabla_{\Gamma_h} U_h \cdot \nabla_{\Gamma_h} \phi_h \, \mathrm{d}A_h = \int_{\Gamma_h} F_h \phi_h \, \mathrm{d}A_h, \tag{4.22}$$

for every discrete test function $\phi_h \in S_h$.

The discrete Poisson equation is a linear system for the solution $U_h = \sum_{j=1}^{J} \alpha_j \chi_j$. Equation (4.22) is equivalent to

$$\sum_{j=1}^{n} \mathcal{S}_{kj} \alpha_j = F_k \quad (k = 1, \ldots, J),$$

where

$$\mathcal{S}_{kj} = \int_{\Gamma_h} \nabla_{\Gamma_h} \chi_k \cdot \nabla_{\Gamma_h} \chi_j \, \mathrm{d}A_h \quad (k, j = 1, \ldots, J)$$

is the stiffness matrix and $F_k = \int_{\Gamma_h} F_h \chi_k \, \mathrm{d}A_h$ $(k = 1, \ldots, J)$ is the right-hand side. The linear system then reads

$$\mathcal{S}\alpha = F$$

for $\alpha = (\alpha_1, \ldots, \alpha_J)$ and with $F = (F_1, \ldots, F_J)$.

First we note that, for solvability of the discrete system, we need the integral over the right-hand side F_h to vanish, because in (4.22) $\phi_h = 1$ is a possible test function. According to the continuous setting the linear system is not uniquely solvable, but solvable only up to additive constants.

Lemma 4.6. Assume that the situation is as in Definition 4.5. Then there is a unique discrete solution to the Poisson equation with the property that

$$\int_{\Gamma_h} U_h \,\mathrm{d}A_h = 0.$$

Proof. We only have to show uniqueness for a solution. For this we insert $\phi_h = U_h$ as a test function into the homogeneous Poisson equation ((4.22) with $F_h = 0$) and find that $\|\nabla_{\Gamma_h} U_h\|_{L^2(\Gamma_h)} = 0$. This implies that U_h is constant on each simplex of the triangulation separately. Since $U_h \in C^0(\Gamma_h)$, $U_h = c \in \mathbb{R}$ on Γ_h. Since the integral of U_h over Γ_h vanishes, we have that $U_h = 0$ on Γ_h. □

With respect to the implementation of the method we mention that the method is precisely like a Euclidean method except for the fact that the nodes of the triangulation are $(n+1)$-dimensional. Since the stiffness matrix is assembled in a loop over all simplices and set up by calculating element stiffness matrices for $T \in \mathcal{T}_h$,

$$S_{jk}^T = \int_T \nabla_{\Gamma_h} \chi_j^T \cdot \nabla_{\Gamma_h} \chi_k^T \,\mathrm{d}A_h \quad (j,k = 1, \ldots, n+1),$$

the method is quite simple. Note that the element stiffness matrix is a planar integral. By χ_k^T we denote the basis function with respect to the kth vertex of T.

4.5. Error analysis

We begin with an estimate for the geometric error when approximating bilinear forms. The most important bilinear forms are

$$m(u,\varphi) = \int_\Gamma u\varphi \,\mathrm{d}A, \quad a(u,\varphi) = \int_\Gamma \nabla_\Gamma u \cdot \nabla_\Gamma \varphi \,\mathrm{d}A, \tag{4.23}$$

for $u, \varphi \in H^1(\Gamma)$, and the discrete analogues

$$m_h(U_h,\phi_h) = \int_{\Gamma_h} U_h \phi_h \,\mathrm{d}A_h, \quad a_h(U_h,\phi_h) = \int_{\Gamma_h} \nabla_{\Gamma_h} U_h \cdot \nabla_{\Gamma_h} \phi_h \,\mathrm{d}A_h, \tag{4.24}$$

for $U_h, \phi_h \in S_h$.

In the following lemma we bound the geometric perturbation errors in the bilinear forms.

Lemma 4.7. For any $(W_h, \phi_h) \in S_h \times S_h$ with corresponding lifts $(w_h, \varphi_h) \in S_h^l \times S_h^l$, the following bounds hold:

$$|m(w_h, \varphi_h) - m_h(W_h, \phi_h)| \le ch^2 \|w_h\|_{L^2(\Gamma)} \|\varphi_h\|_{L^2(\Gamma)}, \tag{4.25}$$

$$|a(w_h, \varphi_h) - a_h(W_h, \phi_h)| \le ch^2 \|\nabla_\Gamma w_h\|_{L^2(\Gamma)} \|\nabla_\Gamma \varphi_h\|_{L^2(\Gamma)} \tag{4.26}$$

Proof. The bound (4.25) follows by noting that

$$\begin{aligned} &|m(w_h, \varphi_h) - m_h(W_h, \phi_h)| \\ &\quad = \left| \int_\Gamma \left(1 - \frac{1}{\delta_h^l}\right) w_h \varphi_h \right| \le \left\| 1 - \frac{1}{\delta_h} \right\|_{L^\infty(\Gamma_h)} \|w_h\|_{L^2(\Gamma)} \|\varphi_h\|_{L^2(\Gamma)} \end{aligned}$$

and using Lemma 4.1.

In order to prove (4.26) it is convenient to introduce the notation

$$\mathcal{Q}_h = \frac{1}{\delta_h}(I - d\mathcal{H})PP_hP(I - d\mathcal{H}) = \frac{1}{\delta_h}P(I - d\mathcal{H})P_h(I - d\mathcal{H})P$$

on $\Gamma_h(t)$, and its lifted version, $\mathcal{Q}_h^l$ on Γ. Using (4.18) we may write on Γ_h

$$\begin{aligned} &\nabla_{\Gamma_h} W_h \cdot \nabla_{\Gamma_h} \phi_h \\ &\qquad = P_h(I - d\mathcal{H})\nabla_\Gamma w_h(a) \cdot P_h(I - d\mathcal{H})\nabla_\Gamma \varphi_h(a) \\ &\qquad = P_hP(I - d\mathcal{H})\nabla_\Gamma w_h(a) \cdot P_hP(I - d\mathcal{H})\nabla_\Gamma \varphi_h(a) \\ &\qquad = \delta_h \mathcal{Q}_h \nabla_\Gamma w_h(a) \cdot \nabla_\Gamma \varphi_h(a). \end{aligned}$$

We use the geometry estimate (4.8) from Lemma 4.1 and get the estimate

$$|P - \mathcal{Q}_h| \le ch^2.$$

Hence the bound (4.26) follows from

$$\begin{aligned} &a(w_h, \varphi_h) - a_h(W_h, \phi_h) \\ &\quad = \int_\Gamma \nabla_\Gamma w_h \cdot \nabla_\Gamma \varphi_h \, \mathrm{d}A - \int_{\Gamma_h} \delta_h \mathcal{Q}_h \nabla_\Gamma w_h \circ a \cdot \nabla_\Gamma \varphi_h \circ a \, \mathrm{d}A_h \\ &\quad = \int_\Gamma (P - \mathcal{Q}_h) \nabla_\Gamma w_h \cdot \nabla_\Gamma \varphi_h \, \mathrm{d}A. \end{aligned}$$

Thus we have proved Lemma 4.7. □

Example 4.8. As an example of the solution of the Poisson equation on a surface without boundary we take the surface from Figure 4.8. The hypersurface was constructed by mapping a discretization of the unit sphere S^2 onto the surface Γ by

$$F(y_1, y_2, y_3) = \left(2y_1, y_2, \frac{1}{2}y_3\left(1 + \frac{1}{2}\sin(2\pi y_1)\right)\right), \quad y = (y_1, y_2, y_3) \in S^2.$$

Table 4.1. Errors and experimental orders of convergence (EOC) for Example 4.8 in the $L^\infty(\Gamma)$, $L^2(\Gamma)$ and $H^1(\Gamma)$ norms.

h	$E_{L^\infty(\Gamma)}(h)$	EOC	$E_{L^2(\Gamma)}(h)$	EOC	$E_{H^1(\Gamma)}(h)$	EOC
2.0615	0.62256	–	0.85015	–	0.42326	–
1.4577	0.47755	0.76	0.78648	0.22	0.17300	2.58
1.0126	0.17590	2.74	0.20326	3.71	4.3717×10^{-2}	3.77
0.60406	4.4493×10^{-2}	2.66	4.5902×10^{-2}	2.87	3.3671×10^{-2}	0.50
0.33968	1.1708×10^{-2}	2.31	1.0577×10^{-2}	2.54	1.3849×10^{-2}	1.54
0.17613	2.9819×10^{-3}	2.08	2.6590×10^{-3}	2.10	7.9195×10^{-3}	0.85
8.8805×10^{-2}	7.5899×10^{-4}	1.99	6.6591×10^{-4}	2.02	3.7204×10^{-3}	1.10
4.4476×10^{-2}	1.8974×10^{-4}	2.00	1.6651×10^{-4}	2.00	1.9332×10^{-3}	0.94
2.2242×10^{-2}	4.7476×10^{-5}	1.99	4.1628×10^{-5}	2.00	9.5198×10^{-4}	1.02

The grid is shown in Figure 4.9. The representation of $\Gamma = F(S^2)$ as a hypersurface $\{x \in \mathbb{R}^3 \mid \phi(x) = 0\}$ follows from $y_1^2 + y_2^2 + y_3^2 = 1$ with the level set function

$$\phi(x) = \frac{1}{4}x_1^2 + x_2^2 + \frac{4x_3^2}{(1 + \frac{1}{2}\sin(\pi x_1))^2} - 1.$$

From ϕ we calculate the normal $\nu = \frac{\nabla\phi}{|\nabla\phi|}$ of Γ. As exact solution we choose $u(x) = x_1 x_2$ $(x \in \Gamma)$. We then calculate the right-hand side f from u as $f(x) = -\Delta_\Gamma u(x)$ $(x \in \Gamma)$, and after some calculations we get

$$f(x) = 2\nu_1(x)\nu_2(x) + H(x)(x_2\nu_1(x) + x_1\nu_2(x)), \quad x \in \Gamma, \tag{4.27}$$

with the mean curvature H of Γ. According to (2.8) the mean curvature can be calculated from the formula

$$H = \nabla \cdot \frac{\nabla\phi}{|\nabla\phi|} = \frac{1}{|\nabla\phi|}\sum_{j,k=1}^{3}\left(\delta_{jk} - \frac{\phi_{x_j}\phi_{x_k}}{|\nabla\phi|^2}\right)\phi_{x_j x_k}.$$

Figure 4.10 shows the numerical solution with piecewise linear finite elements to the Poisson equation for the right-hand side f and with mean value equal to zero on Γ.

In Table 4.1 we show the errors in the norms $E_{L^\infty(\Gamma)}(h) = \|u - u_h\|_{L^\infty(\Gamma)}$, $E_{L^2(\Gamma)}(h) = \|u - u_h\|_{L^2(\Gamma)}$ and $E_{H^1(\Gamma)}(h) = \|\nabla_\Gamma(u - u_h)\|_{L^2(\Gamma)}$. For errors $E(h_1)$ and $E(h_2)$ for the grid sizes h_1 and h_2, the experimental order of convergence is defined by

$$\mathrm{EOC}(h_1, h_2) = \log\frac{E(h_1)}{E(h_2)}\left(\log\frac{h_1}{h_2}\right)^{-1}.$$

The results confirm the theoretical results from Theorem 4.9.

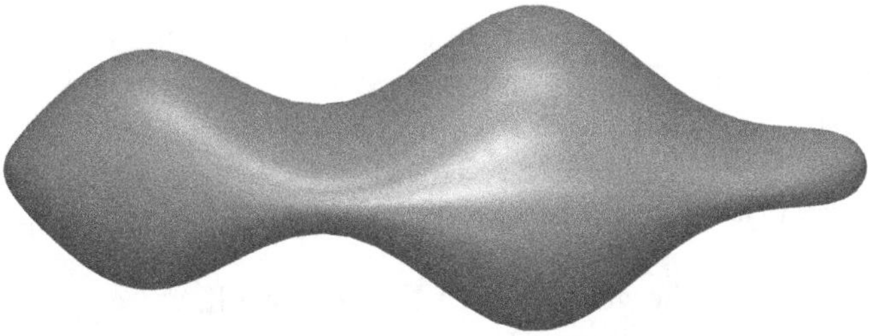

Figure 4.8. The hypersurface for Example 4.8.

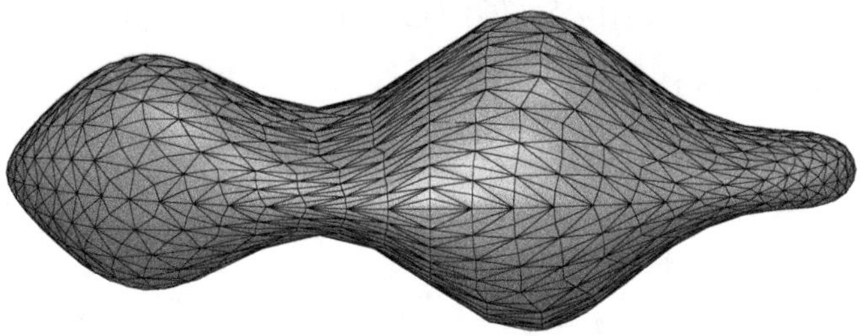

Figure 4.9. Computational grid on the surface from Figure 4.8 obtained after eight bisections of the macro-triangulation.

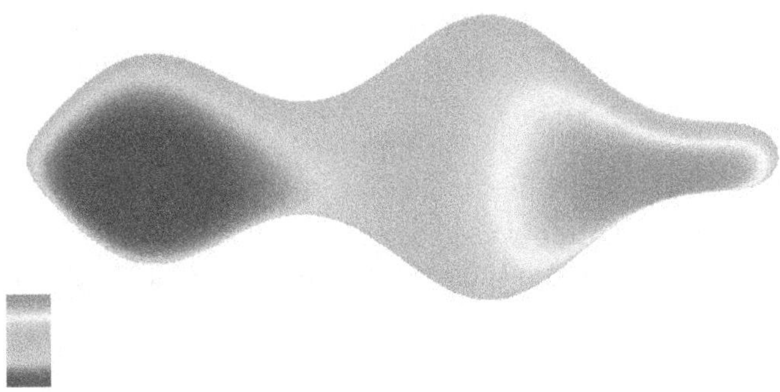

Figure 4.10. Solution u to Example 4.8 on Γ. Values between -1 (blue) and 1 (red).

We now prove the following result on the numerical solution of the Poisson equation by piecewise linear finite elements on a compact connected surface.

Theorem 4.9. Assume that Γ and Γ_h are as above with $n \leq 3$. Let $F_h \in L^2(\Gamma_h)$ with $\int_{\Gamma_h} F_h = 0$. Let S_h be the space of piecewise linear finite elements on the discrete surface Γ_h as in (4.4) and S_h^l its lifted version as in (4.21). Then there is a discrete solution of (3.1), $U_h \in S_h$, which is uniquely determined up to a constant,

$$\int_{\Gamma_h} \nabla_{\Gamma_h} U_h \cdot \nabla_{\Gamma_h} \phi_h \, \mathrm{d}A = \int_{\Gamma_h} F_h \phi_h \, \mathrm{d}A \tag{4.28}$$

for all $\phi_h \in S_h$. Fix the free constant by the requirement $\int_{\Gamma_h} U_h = 0$.

The error in the natural norm between the continuous solution u with $\int_\Gamma u = 0$ for $f \in L^2(\Gamma)$, $\int_\Gamma f \, \mathrm{d}A = 0$, from Theorem 3.1 and the lifted discrete solution $u_h = U_h^l$ can be estimated for $h \le h_0$ as follows:

$$\|\nabla_\Gamma(u - u_h)\|_{L^2(\Gamma)} \le ch\|f\|_{L^2(\Gamma)} + c\|f - f_h\|_{L^2(\Gamma)}, \tag{4.29}$$

with c depending on the geometry of Γ. Here we have set $f_h = F_h^l$.

For the $L^2(\Gamma)$-error we have the estimate

$$\|u - u_h\|_{L^2(\Gamma)} \le ch^2\|f\|_{L^2(\Gamma)} + c\|f - f_h\|_{L^2(\Gamma)}. \tag{4.30}$$

If we choose F_h so that

$$\|f - f_h\|_{L^2(\Gamma)} \le c_f h^2,$$

then

$$\|u - u_h\|_{L^2(\Gamma)} \le ch^2, \quad \|\nabla_\Gamma(u - u_h)\|_{L^2(\Gamma)} \le ch. \tag{4.31}$$

Remark 4.10. The assumption that the dimension n of the surface Γ is less than or equal to 3 is necessary only for the use of the interpolation inequalities in Lemma 4.3. The error estimates of the theorem are valid for any dimension n if it is guaranteed that the estimates (4.19) hold.

Note that for the $H^1(\Gamma)$ error estimate it is sufficient to require first-order approximation of the right-hand side f by f_h.

The proof of this theorem was given in Dziuk (1988). For the convenience of the reader and because the theorem concerns a model problem, we present the proof.

Proof. For sufficiently small grid size $h \le h_0$ the discrete bilinear form a_h is coercive. From Lemma 4.7 we get for $\phi_h \in S_h$ and $\varphi_h = \phi_h^l$ with the coercivity of the bilinear form a,

$$\begin{aligned} a_h(\phi_h, \phi_h) &\ge a(\varphi_h, \varphi_h) - |a(\varphi_h, \varphi_h) - a_h(\phi_h, \phi_h)| \\ &\ge \|\nabla_\Gamma \varphi_h\|^2_{L^2(\Gamma)} - ch^2\|\nabla_\Gamma \varphi_h\|^2_{L^2(\Gamma)} && (4.32) \\ &\ge \frac{1}{2}\|\nabla_\Gamma \varphi_h\|^2_{L^2(\Gamma)} && (4.33) \end{aligned}$$

for $h \le h_0$.

From the continuous equation (3.2) and the discrete equation (4.28) we have

$$a(u, \varphi) = m(f, \varphi), \quad a_h(U_h, \phi_h) = m_h(F_h, \phi_h) \tag{4.34}$$

for all $\varphi \in H^1(\Gamma)$ and all $\phi_h \in S_h$.

Following the common arguments of the first Strang lemma, we infer from (4.33) and (4.34) for arbitrary $\phi_h \in S_h$, $\varphi_h = \phi_h^l$ that

$$\begin{aligned}
\frac{1}{2}\|\nabla_\Gamma(u_h-\varphi_h)\|^2_{L^2(\Gamma)} &\le a_h(U_h-\phi_h, U_h-\phi_h)\\
&= a(u-\varphi_h, u_h-\varphi_h)\\
&\quad + a(\varphi_h, u_h-\varphi_h) - a_h(\phi_h, U_h-\phi_h)\\
&\quad - (m(f, u_h-\varphi_h-c) - m_h(F_h, U_h-\phi_h-c))\\
&= a(u-\varphi_h, u_h-\varphi_h)\\
&\quad + a(\varphi_h, u_h-\varphi_h) - a_h(\phi_h, U_h-\phi_h)\\
&\quad - (m(f, u_h-\varphi_h-c) - m(f_h, u_h-\varphi_h-c))\\
&\quad - (m(f_h, u_h-\varphi_h-c) - m_h(F_h, U_h-\phi_h-c)).
\end{aligned}$$

Because the mean value of f over Γ and the mean value of F_h over Γ_h vanish, we were able to smuggle in an arbitrary real constant c. We have set $f_h = F_h^l$. So,

$$\begin{aligned}
&\frac{1}{2}\|\nabla_\Gamma(u_h-\varphi_h)\|^2_{L^2(\Gamma)}\\
&\quad\le \|\nabla_\Gamma(u-\varphi_h)\|_{L^2(\Gamma)}\|\nabla_\Gamma(u_h-\varphi_h)\|_{L^2(\Gamma)}\\
&\qquad + |a(\varphi_h, u_h-\varphi_h) - a_h(\phi_h, U_h-\phi_h)|\\
&\qquad + \|f-f_h\|_{L^2(\Gamma)}\|u_h-\varphi_h-c\|_{L^2(\Gamma)}\\
&\qquad + |m(f_h, u_h-\varphi_h-c) - m_h(F_h, U_h-\phi_h-c)|.
\end{aligned}$$

We continue the estimate with the results from Lemma 4.7 and get

$$\begin{aligned}
&\frac{1}{2}\|\nabla_\Gamma(u_h-\varphi_h)\|^2_{L^2(\Gamma)}\\
&\quad\le \|\nabla_\Gamma(u-\varphi_h)\|_{L^2(\Gamma)}\|\nabla_\Gamma(u_h-\varphi_h)\|_{L^2(\Gamma)}\\
&\qquad + ch^2\|\nabla_\Gamma\varphi_h\|_{L^2(\Gamma)}\|\nabla_\Gamma(u_h-\varphi_h)\|_{L^2(\Gamma)}\\
&\qquad + \|f-f_h\|_{L^2(\Gamma)}\|u_h-\varphi_h-c\|_{L^2(\Gamma)}\\
&\qquad + ch^2\|f_h\|_{L^2(\Gamma)}\|u_h-\varphi_h-c\|_{L^2(\Gamma)}.
\end{aligned}$$

The choice $c = \frac{1}{|\Gamma|}\int_\Gamma u_h - \varphi_h \,\mathrm{d}A$ allows the application of Poincaré's inequality from Theorem 2.16,

$$\|u_h-\varphi_h-c\|_{L^2(\Gamma)} \le \tilde{c}\|\nabla_\Gamma(u_h-\varphi_h)\|_{L^2(\Gamma)}.$$

We finally arrive at the estimate

$$\begin{aligned}
&\frac{1}{2}\|\nabla_\Gamma(u_h-\varphi_h)\|_{L^2(\Gamma)}\\
&\quad\le \|\nabla_\Gamma(u-\varphi_h)\|_{L^2(\Gamma)} + ch^2\|\nabla_\Gamma\varphi_h\|_{L^2(\Gamma)}\\
&\qquad + c\|f-f_h\|_{L^2(\Gamma)} + ch^2\|f\|_{L^2(\Gamma)}.
\end{aligned}$$

We choose $\varphi_h = I_h u$ and use the interpolation estimates from Lemma 4.3:

$$\begin{aligned}
&\frac{1}{2}\|\nabla_\Gamma(u_h - I_h u)\|_{L^2(\Gamma)} \\
&\qquad \le (1+ch^2)\|\nabla_\Gamma(u - I_h u)\|_{L^2(\Gamma)} + ch^2\|\nabla_\Gamma u\|_{L^2(\Gamma)} \\
&\qquad\qquad + c\|f - f_h\|_{L^2(\Gamma)} + ch^2\|f\|_{L^2(\Gamma)} \\
&\qquad \le ch\|u\|_{H^2(\Gamma)} + c\|f - f_h\|_{L^2(\Gamma)} + ch^2\|f\|_{L^2(\Gamma)}.
\end{aligned}$$

The *a priori* estimate from Theorem 3.3 now finally gives

$$\|\nabla_\Gamma(u - u_h)\|_{L^2(\Gamma)} \le ch\|f\|_{L^2(\Gamma)} + c\|f - f_h\|_{L^2(\Gamma)},$$

and the estimate (4.29) of the theorem is proved.

The $L^2(\Gamma)$ error estimate follows with the usual Aubin–Nitsche trick. Let $v \in H^2(\Gamma)$ be the solution of the problem

$$-\Delta_\Gamma v = u - u_h - \frac{1}{|\Gamma|}\int_\Gamma u - u_h \,\mathrm{d}A$$

on Γ with $\int_\Gamma v \,\mathrm{d}A = 0$. It follows from Theorem 3.3 that

$$\|v\|_{H^2(\Gamma)} \le c\|u - u_h\|_{L^2(\Gamma)}.$$

The PDE for v gives

$$\|u - u_h\|^2_{L^2(\Gamma)} - \frac{1}{|\Gamma|}\left(\int_\Gamma u - u_h \,\mathrm{d}A\right)^2 = -\int_\Gamma (u - u_h)\Delta_\Gamma v \,\mathrm{d}A. \tag{4.35}$$

The right-hand side of this equation is equal to $a(u-u_h, v)$ by Theorem 2.14. Using the results from Lemma 4.7 we have, for $\varphi_h = I_h v \in S_h^l$,

$$\begin{aligned}
&a(u - u_h, v) \\
&\qquad = a(u - u_h, v - \varphi_h) + a(u - u_h, \varphi_h) \\
&\qquad = a(u - u_h, v - \varphi_h) + m(f, \varphi_h) \\
&\qquad\qquad - m_h(F_h, \phi_h) - (a(u_h, \varphi_h) - a_h(U_h, \phi_h)) \\
&\qquad \le \|\nabla_\Gamma(u - u_h)\|_{L^2(\Gamma)}\|\nabla_\Gamma(v - \varphi_h)\|_{L^2(\Gamma)} \\
&\qquad\qquad + c(\|f - f_h\|_{L^2(\Gamma)} + h^2\|f_h\|_{L^2(\Gamma)})\|\nabla_\Gamma \varphi_h\|_{L^2(\Gamma)} \\
&\qquad\qquad + ch^2\|\nabla_\Gamma u_h\|_{L^2(\Gamma)}\|\nabla_\Gamma \varphi_h\|_{L^2(\Gamma)} \\
&\qquad \le c\big(h^2\|f\|_{L^2(\Gamma)} + \|f - f_h\|_{L^2(\Gamma)}\big)\|v\|_{H^2(\Gamma)} \\
&\qquad \le c\big(h^2\|f\|_{L^2(\Gamma)} + \|f - f_h\|_{L^2(\Gamma)}\big)\|u - u_h\|_{L^2(\Gamma)}.
\end{aligned}$$

For the left-hand side of (4.35) we observe the following:

$$\begin{aligned}
\frac{1}{|\Gamma|}\left(\int_\Gamma u - u_h \,\mathrm{d}A\right)^2 &= cm(u - u_h, 1)^2 = cm(u_h, 1)^2 \\
&= c(m(u_h, 1) - m_h(U_h, 1))^2 \le ch^4\big(\|u\|^2_{L^2(\Gamma)} + \|u - u_h\|^2_{L^2(\Gamma)}\big).
\end{aligned}$$

But then

$$\begin{aligned}(1-ch^4)\|u-u_h\|^2_{L^2(\Gamma)} \\ \leq ch^4\|f\|^2_{L^2(\Gamma)} + c\big(h^2\|f\|_{L^2(\Gamma)} + \|f-f_h\|_{L^2(\Gamma)}\big)\|u-u_h\|_{L^2(\Gamma)}.\end{aligned}$$

This finally gives the estimate (4.30) for $h \leq h_0$. □

4.6. Other methods on discrete surfaces

First note that the use of surface finite elements for solving parabolic equations on stationary surfaces follows rather naturally from the developments in this section (Dziuk and Elliott 2007*b*). Higher-order finite element spaces for elliptic equations were analysed in Demlow (2009), an adaptive finite element method for stationary surfaces was considered in Demlow and Dziuk (2007), and coupling of surface and bulk elliptic equations was analysed in Elliott and Ranner (2013).

A discontinuous Galerkin surface finite element method was considered by Dedner, Madhavan and Stinner (2013). An extension of the idea of the surface finite element method (SFEM) is to use surface finite volumes. An analysis of elliptic equations using general meshes is given in Ju and Du (2009) and Ju, Tian and Wang (2009). A method for parabolic equations on stationary surfaces using logically Cartesian grids is presented in Calhoun and Helzel (2009). See also Berger, Calhoun, Helzel and Leveque (2009) for conservation laws on the sphere.

5. Partial differential equations on moving surfaces

Quite often we have to solve PDEs which live on a moving surface or interface. We refer to examples in Section 10. In this chapter we will treat the most basic linear PDE on an evolving surface. The motion of the surface will be prescribed. The geometry will be described in Section 5.1. It will be important to use the space–time structure of the given geometry. We will describe in Section 5.2 how the standard conservation law can be derived. In Section 5.3 we will work with moving triangulated surfaces and use them to discretize the heat equation on a moving surface.

5.1. The geometry of moving surfaces

For each $t \in [0,T]$, let $\Gamma(t)$ be a compact hypersurface oriented by the normal vector field $\nu(\cdot,t)$ and $\Gamma_0 = \Gamma(0)$. We assume that there exists a map $G(\cdot,t) : \Gamma_0 \to \Gamma(t)$, $G \in C^1([0,T], C^2(\Gamma_0))$, such that $G(\cdot,t)$ is a diffeomorphism from Γ_0 to $\Gamma(t)$, and we define the velocity of $\Gamma(t)$ by

$$v(G(\cdot,t),t) = \frac{\partial G}{\partial t}(\cdot,t),$$

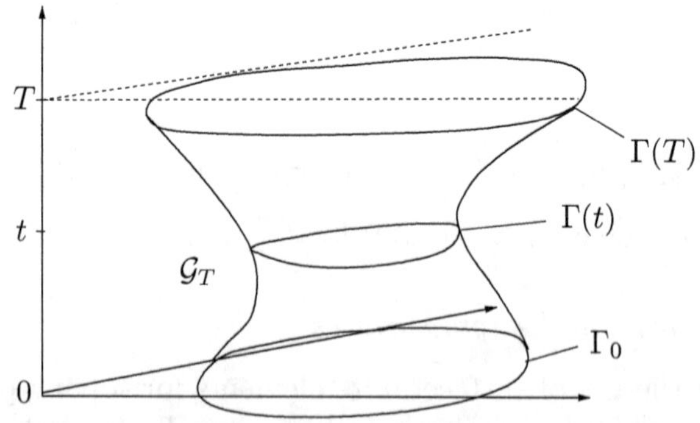

Figure 5.1. Space–time surface $\mathcal{G}_T$ for the dimension $n = 1$. Here each $\Gamma(t)$ is a curve.

$G(\cdot, 0) = \mathrm{Id}$. We assume that $v(\cdot, t) \in C^2(\Gamma(t))$. The normal velocity of Γ is then defined by $v_\nu = v \cdot \nu\nu$.

We use the appropriate time derivative, that is,

$$\partial^\bullet f = \frac{\partial f}{\partial t} + v \cdot \nabla f. \tag{5.1}$$

Obviously this derivative only depends on values of the function f on $\mathcal{G}_T$. It is quite often convenient to work with the space–time surface (see Figure 5.1)

$$\mathcal{G}_T = \bigcup_{t \in [0,T]} \Gamma(t) \times \{t\}. \tag{5.2}$$

Note that for a function $f : \mathcal{G}_T \to \mathbb{R}$ the time derivative $\frac{\partial f}{\partial t}$ and the spatial derivatives ∇f do not make sense separately.

Leibniz formulae, transport theorems

The following formulae for the differentiation of time-dependent surface integrals are called *transport formulae*, and are proved in Dziuk and Elliott (2007*a*, 2012).

Theorem 5.1. Let $\mathcal{M}(t)$ be an evolving surface with normal velocity v_ν. Let v_τ be a tangential velocity field on $\mathcal{M}(t)$. Let the boundary $\partial\mathcal{M}(t)$ evolve with the velocity $v = v_\nu + v_\tau$. Assume that f is a function such that all the following quantities exist. Then

$$\frac{\mathrm{d}}{\mathrm{d}t} \int_{\mathcal{M}(t)} f \,\mathrm{d}A = \int_{\mathcal{M}(t)} \partial^\bullet f + f \,\nabla_\Gamma \cdot v \,\mathrm{d}A. \tag{5.3}$$

Proof. Let $\Omega \subset \mathbb{R}^n$ be open and let $X = X(\theta, t)$, $\theta \in \Omega$, $X(\cdot, t) : \Omega \to U \cap \Gamma$ be a local regular parametrization of the open portion $U \cap \Gamma$ of the surface Γ which evolves so that $X_t = v(X(\theta, t), t)$. The induced metric $(g_{ij})_{i,j=1,\dots,n}$ is given by $g_{ij} = X_{\theta_i} \cdot X_{\theta_j}$ with determinant $g = \det(g_{ij})$. Let $(g^{ij}) = (g_{ij})^{-1}$.

See also Section 2.1. Define

$$F(\theta,t) = f(X(\theta,t),t) \quad \text{and} \quad \mathcal{V}(\theta,t) = v(X(\theta,t),t).$$

Then, with the Euler relation for the derivative of the determinant,

$$\frac{\partial}{\partial t}\sqrt{g} = \sqrt{g}\sum_{i,j=1}^{n} g^{ij} X_{\theta_i}\cdot \mathcal{V}_{\theta_j},$$

we have the following proof of (5.3):

$$\begin{aligned}
\frac{\mathrm{d}}{\mathrm{d}t}\int_{\Gamma\cap U} f\,\mathrm{d}A &= \frac{\mathrm{d}}{\mathrm{d}t}\int_{\Omega} F\sqrt{g}\,\mathrm{d}\theta = \int_{\Omega}\frac{\partial F}{\partial t}\sqrt{g} + F\frac{\partial\sqrt{g}}{\partial t}\,\mathrm{d}\theta \\
&= \int_{\Omega}\left(\frac{\partial f}{\partial t}(X,\cdot) + \nabla f(X,\cdot)\cdot X_t\right)\sqrt{g} + f(X,\cdot)\sqrt{g}\sum_{i,j=1}^{n} g^{ij}X_{\theta_i}\cdot\mathcal{V}_{\theta_j}\,\mathrm{d}\theta \\
&= \int_{\Gamma\cap U}\dot{f} + f\nabla_\Gamma\cdot v\,\mathrm{d}A,
\end{aligned}$$

where in the last step we used that $\mathcal{V} = X_t$ and that the tangential divergence of v is given by

$$(\nabla_\Gamma\cdot v)(X,\cdot) = \sum_{i,j=1}^{n} g^{ij}X_{\theta_i}\cdot\mathcal{V}_{\theta_j}.$$

The theorem is proved. $\square$

We give transport formulae for the time derivative of the most important bilinear forms m, a and g given by

$$\begin{aligned}
m(\phi(\cdot,t),\psi(\cdot,t)) &= \int_{\Gamma(t)}\phi(x,t)\psi(x,t)\,\mathrm{d}A(x), \\
a(\phi(\cdot,t),\psi(\cdot,t)) &= \int_{\Gamma(t)}\mathcal{A}(\cdot,t)\nabla_{\Gamma(t)}\phi(\cdot,t)\cdot\nabla_{\Gamma(t)}\psi(\cdot,t)\,\mathrm{d}A(x), \\
g(v;\phi(\cdot,t),\psi(\cdot,t)) &= \int_{\Gamma(t)}\phi(x,t)\psi(x,t)\nabla_\Gamma\cdot v(x,t)\,\mathrm{d}A(x).
\end{aligned}$$

Note that these bilinear forms now explicitly depend on time too. But instead of writing, say, $m(t,\phi(\cdot,t),\psi(\cdot,t))$, we suppress the explicit dependence on t. It will always be clear from the arguments ϕ and ψ at which time the bilinear form has to be evaluated.

Lemma 5.2. For $\varphi,\psi\in H^1(\mathcal{G}_T)$, we have

$$\frac{\mathrm{d}}{\mathrm{d}t}m(\varphi,\psi) = m(\partial^\bullet\varphi,\psi) + m(\varphi,\partial^\bullet\psi) + g(v;\varphi,\psi) \tag{5.4}$$

and

$$\frac{\mathrm{d}}{\mathrm{d}t}a(\varphi,\psi) = a(\partial^\bullet\varphi,\psi) + a(\varphi,\partial^\bullet\psi) + b(v;\varphi,\psi), \tag{5.5}$$

with the bilinear form

$$b(v;\varphi,\psi) = \int_\Gamma \mathcal{B}(v)\nabla_\Gamma\varphi\cdot\nabla_\Gamma\psi\,\mathrm{d}A. \tag{5.6}$$

With the deformation tensor

$$D(v)_{ij} = \frac{1}{2}\sum_{k=1}^{n+1}(\mathcal{A}_{ik}(\nabla_\Gamma)_k v_j + \mathcal{A}_{jk}(\nabla_\Gamma)_k v_i) \quad (i,j=1,\ldots,n+1)$$

and the tensor

$$\mathcal{B}(v) = \partial^\bullet\mathcal{A} + \nabla_\Gamma\cdot v\,\mathcal{A} - 2D(v), \tag{5.7}$$

we have the formula

$$\frac{\mathrm{d}}{\mathrm{d}t}\int_{\mathcal{M}(t)}\mathcal{A}\nabla_\Gamma f\cdot\nabla_\Gamma g\,\mathrm{d}A = \tag{5.8}$$
$$\int_{\mathcal{M}(t)}\mathcal{A}\nabla_\Gamma\partial^\bullet f\cdot\nabla_\Gamma g + \mathcal{A}\nabla_\Gamma f\cdot\nabla_\Gamma\partial^\bullet g\,\mathrm{d}A + \int_{\mathcal{M}(t)}\mathcal{B}(v)\nabla_\Gamma f\cdot\nabla_\Gamma g\,\mathrm{d}A.$$

For the convenience of the reader we derive the transport formula for Dirichlet's integral,

$$\int_\Gamma |\nabla_\Gamma f|^2\,\mathrm{d}A,$$

for a time-dependent surface. We continue to use the notation of the previous proof. The generalization to the more general case in Lemma 5.2 then follows easily. We first observe that we have

$$|(\nabla_\Gamma f)(X,\cdot)|^2 = \sum_{i,j=1}^{n} g^{ij}F_{\theta_i}F_{\theta_j}, \tag{5.9}$$

so that

$$\frac{1}{2}\frac{\mathrm{d}}{\mathrm{d}t}\int_{\Gamma\cap U}|\nabla_\Gamma f|^2\,\mathrm{d}A = \int_\Omega\sqrt{g}\sum_{i,j=1}^{n} g^{ij}F_{\theta_i}F_{\theta_j t}\,\mathrm{d}\theta$$
$$+\frac{1}{2}\int_\Omega\sqrt{g}\sum_{i,j=1}^{n} g_t^{ij}F_{\theta_i}F_{\theta_j}\,\mathrm{d}\theta + \frac{1}{2}\int_\Omega\sqrt{g}\sum_{i,j,k,l=1}^{n} g^{ij}g^{kl}X_{\theta_k}\cdot\mathcal{V}_{\theta_l}F_{\theta_i}F_{\theta_j}\,\mathrm{d}\theta.$$

An easy calculation shows that

$$\begin{aligned} g_t^{ij} &= -\sum_{k,l=1}^{n} g^{ik} g^{jl} g_{kl,t} = -\sum_{k,l=1}^{n} g^{ik} g^{jl} (X_{\theta_k} \cdot X_{\theta_l})_t \\ &= -\sum_{k,l=1}^{n} g^{ik} g^{jl} (\mathcal{V}_{\theta_k} \cdot X_{\theta_l} + X_{\theta_k} \cdot \mathcal{V}_{\theta_l}), \end{aligned}$$

and we arrive at

$$\begin{aligned} \frac{1}{2}\frac{\mathrm{d}}{\mathrm{d}t} \int_{\Gamma\cap U} |\nabla_\Gamma f|^2 \,\mathrm{d}A &= \int_{\Gamma\cap U} \nabla_\Gamma f \cdot \nabla_\Gamma \partial^\bullet f \,\mathrm{d}A \\ &- \int_{\Gamma\cap U} \sum_{i,j=1}^{n} \underline{D}_i v_j \underline{D}_i f \underline{D}_j f \,\mathrm{d}A + \frac{1}{2} \int_{\Gamma\cap U} |\nabla_\Gamma f|^2 \nabla_\Gamma \cdot v \,\mathrm{d}A. \end{aligned}$$

The formula (5.8) for $\mathcal{A} = I$ then follows by polarization.

5.2. Conservation and diffusion on moving surfaces

Conservation law

Let u be the density of a scalar quantity on $\Gamma(t)$ (for example mass per unit area $n = 2$ or mass per unit length $n = 1$). We suppose there is a surface flux q. The basic conservation law we wish to consider can be formulated for an arbitrary portion $\mathcal{M}(t)$ of $\Gamma(t)$, which is the image of a portion $\mathcal{M}(0)$ of $\Gamma(0)$ evolving with the prescribed velocity $v = v_\nu$. In the following we write $\partial^\circ u$ for the material time derivative of u with respect to this purely normal velocity:

$$\partial^\circ u = u_t + v_\nu \cdot \nabla u.$$

$\partial^\circ u$ is sometimes known as the *normal time derivative* (Cermelli, Fried and Gurtin 2005).

The conservation law is that for every $\mathcal{M}(t)$

$$\frac{\mathrm{d}}{\mathrm{d}t} \int_{\mathcal{M}(t)} u \,\mathrm{d}A = -\int_{\partial\mathcal{M}(t)} q \cdot \mu \,\mathrm{d}A, \tag{5.10}$$

where $\partial\mathcal{M}(t)$ is the boundary of $\mathcal{M}(t)$ (a curve if $n = 2$ and the end points of a curve if $n = 1$) and μ is the co-normal on $\partial\mathcal{M}(t)$. Thus μ is the unit normal to $\partial\mathcal{M}(t)$ pointing out of $\mathcal{M}(t)$ and tangential to $\Gamma(t)$. The surface flux is denoted by q. Observe that components of q normal to $\mathcal{M}$ do not contribute to the flux, so we may assume that q is a tangent vector.

With the use of integration by parts, (2.13), we obtain

$$\int_{\partial\mathcal{M}(t)} q \cdot \mu \,\mathrm{d}A = \int_{\mathcal{M}(t)} \nabla_\Gamma \cdot q \,\mathrm{d}A - \int_{\mathcal{M}(t)} q \cdot \nu H \,\mathrm{d}A = \int_{\mathcal{M}(t)} \nabla_\Gamma \cdot q \,\mathrm{d}A.$$

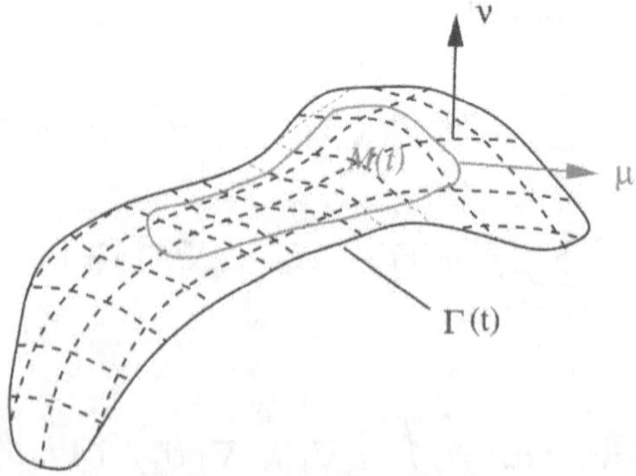

Figure 5.2. Conservation on a moving surface. Moving surface $\Gamma(t)$ and subset $M(t)$ with co-normal μ.

On the other hand, by the transport formula (5.3) we have

$$\frac{\mathrm{d}}{\mathrm{d}t}\int_{\mathcal{M}(t)} u\,\mathrm{d}A = \int_{\mathcal{M}(t)} \partial^\circ u + u\nabla_\Gamma \cdot v_\nu\,\mathrm{d}A,$$

so that

$$\int_{\mathcal{M}(t)} \partial^\circ u + u\,\nabla_\Gamma \cdot v_\nu + \nabla_\Gamma \cdot q\,\mathrm{d}A = 0,$$

which implies the pointwise conservation law

$$\partial^\circ u + u\,\nabla_\Gamma \cdot v_\nu + \nabla_\Gamma \cdot q = 0. \tag{5.11}$$

This may also be written as

$$u_t + V\frac{\partial u}{\partial \nu} + HVu + \nabla_\Gamma \cdot q = 0, \tag{5.12}$$

where $V = v_\nu \cdot \nu$; see also Stone (1990).

We wish to consider a diffusive flux $q_d = -\mathcal{A}\nabla_\Gamma u$ and an advective flux $q_a = uv_\tau$, where v_τ is an advective tangential velocity field, that is, $v_\tau \cdot \nu = 0$, so that

$$q = q_d + q_a = -\mathcal{A}\nabla_\Gamma u + uv_\tau.$$

Then we arrive at the PDE

$$\partial^\bullet u + u\,\nabla_\Gamma \cdot v - \nabla_\Gamma \cdot (\mathcal{A}\nabla_\Gamma u) = 0. \tag{5.13}$$

In the following we assume that $\mathcal{A}$ is a sufficiently smooth symmetric $(n+1)\times(n+1)$ matrix which maps the tangent space of Γ at each point into itself and is positive definite on the tangent space, that is,

$$\mathcal{A}\xi \cdot \xi \geq c_0|\xi|^2 \quad \text{for all } \xi \in \mathbb{R}^{m+1},\ \xi \cdot \nu = 0, \tag{5.14}$$

with some constant $c_0 > 0$. For the definition of a solution we assume that the elements of $\mathcal{A}$ belong to $L^\infty(\mathcal{G}_T)$. A weak solution of the PDE (5.13) is a function $u \in H^1(\mathcal{G}_T)$ that satisfies the equation

$$\frac{\mathrm{d}}{\mathrm{d}t}\int_{\Gamma(t)} u\varphi \,\mathrm{d}A + \int_{\Gamma(t)} \mathcal{A}\nabla_\Gamma u \cdot \nabla_\Gamma \varphi \,\mathrm{d}A = \int_{\Gamma(t)} u\partial^\bullet \varphi \,\mathrm{d}A \tag{5.15}$$

almost everywhere on $(0,T)$, where φ is an arbitrary test function defined on the space–time surface $\mathcal{G}_T$.

In Dziuk and Elliott (2007*a*) we proved the existence of a weak solution.

Theorem 5.3. Assume that the initial data $u_0 \in H^1(\Gamma_0)$, where $\Gamma_0 = \Gamma(0)$. Then there exists a unique weak solution $u \in H^1(\mathcal{G}_T)$ of the PDE (5.13), that is, equation (5.15) is satisfied for almost every $t \in (0,T)$, which satisfies the initial condition $u(\cdot,0) = u_0$ on Γ_0. Furthermore, if $\mathcal{A}$ and $v \in C^1(\mathcal{G}_T)$, the solution satisfies the energy estimates

$$\sup_{(0,T)} \|u\|^2_{L^2(\Gamma)} + \int_0^T \|\nabla_\Gamma u\|^2_{L^2(\Gamma)} \,\mathrm{d}t \le c\|u_0\|^2_{L^2(\Gamma_0)}, \tag{5.16}$$

$$\int_0^T \|\partial^\bullet u\|^2_{L^2(\Gamma)} \,\mathrm{d}t + \sup_{(0,T)} \|\nabla_\Gamma u\|^2_{L^2(\Gamma)} \le c\|u_0\|^2_{H^1(\Gamma_0)}, \tag{5.17}$$

where $c = c(\mathcal{A}, v, \mathcal{G}_T, T)$.

Proof. For the convenience of the reader we include a proof of the *a priori* estimates. For (5.16) we set $\varphi = u$ in (5.15) and, using Lemma 5.2, we get

$$\frac{\mathrm{d}}{\mathrm{d}t}m(u,u) + a(u,u) = m(u,\partial^\bullet u) = \frac{1}{2}\frac{\mathrm{d}}{\mathrm{d}t}m(u,u) - \frac{1}{2}g(v;u,u).$$

This gives

$$\frac{1}{2}\frac{\mathrm{d}}{\mathrm{d}t}m(u,u) + a(u,u) + \frac{1}{2}g(v;u,u) = 0,$$

and with a Gronwall argument this implies (5.16).

For (5.17) we use (5.4). The weak equation (5.15) implies

$$m(\partial^\bullet u,\varphi) + g(v;u,\varphi) + a(u,\varphi) = 0.$$

We insert $\varphi = \partial^\bullet u$ and get from (5.5)

$$m(\partial^\bullet u,\partial^\bullet u) + g(v;u,\partial^\bullet u) + \frac{1}{2}\frac{\mathrm{d}}{\mathrm{d}t}a(u,u) - \frac{1}{2}b(v;u,u) = 0.$$

Standard arguments then lead to (5.17). □

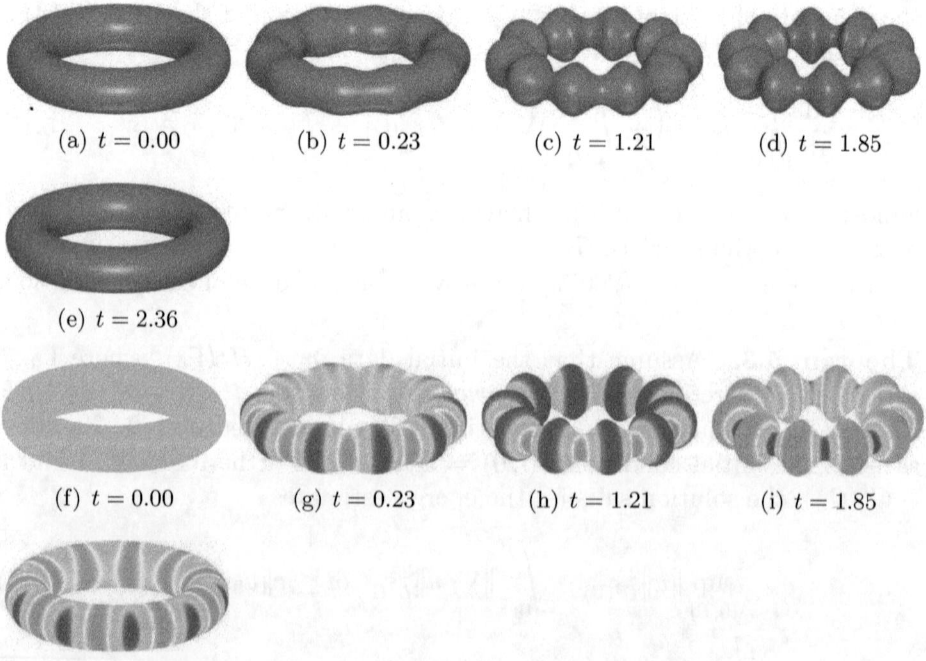

(a) $t = 0.00$ (b) $t = 0.23$ (c) $t = 1.21$ (d) $t = 1.85$

(e) $t = 2.36$

(f) $t = 0.00$ (g) $t = 0.23$ (h) $t = 1.21$ (i) $t = 1.85$

(j) $t = 2.36$

Figure 5.3. (a–e) Deformation of a torus and (f–j) solution of equation (5.13). Colours indicate the magnitude of the solution. There was no source term. The surface is deformed and reaches the initial form of a round torus again at time $t = 2.36$. The initial value was constant: $u_0 = 10.0$ (green at $t = 0.00$). The solution at final time $t = 2.36$ is due purely to geometric motion.

For $\varphi, \psi \in H^1(\Gamma)$ we use the bilinear forms

$$a(\varphi(\cdot,t),\psi(\cdot,t)) = \int_{\Gamma(t)} \mathcal{A}(\cdot,t)\nabla_\Gamma\varphi(\cdot,t)\cdot\nabla_\Gamma\psi(\cdot,t)\,\mathrm{d}A, \tag{5.18}$$

$$m(\varphi(\cdot,t),\psi(\cdot,t)) = \int_{\Gamma(t)} \varphi(\cdot,t)\psi(\cdot,t)\,\mathrm{d}A, \tag{5.19}$$

$$g(v(\cdot,t);\varphi(\cdot,t),\psi(\cdot,t)) = \int_{\Gamma(t)} \varphi(\cdot,t)\psi(\cdot,t)\nabla_\Gamma\cdot v(\cdot,t)\,\mathrm{d}A. \tag{5.20}$$

Using this notation, the weak form (5.15) of the PDE (5.13) becomes

$$\frac{\mathrm{d}}{\mathrm{d}t}m(u,\varphi) + a(u,\varphi) = m(u,\partial^\bullet\varphi). \tag{5.21}$$

Of course, the variational problem may be posed on the initial surface Γ_0. This would lead to non-constant coefficients even with $\mathcal{A} = \mathcal{I}$. It would also require knowledge of the map G in Section 5.1 which we choose to avoid.

5.3. Discretization

Evolving triangulated surfaces

The smooth evolving surface $\Gamma(t)$ $(\partial\Gamma(t)=\emptyset)$ is approximated by an evolving triangulated surface

$$\Gamma_h(t) \subset U_\delta(t) \quad (\partial\Gamma_h(t)=\emptyset),$$

which for each t is polygonal and is smooth in time. $U_\delta(t)$ is as in Lemma 2.8. Let us suppose for simplicity that δ does not depend on time. We assume that every surface $\Gamma(t)$ is approximated as described in Section 4.2. So, $\Gamma_h(t)$ is homeomorphic to $\Gamma(t)$ for every $t\in[0,T]$. $U_\delta(t)$ is a neighbourhood in $\mathbb{R}^{n+1}$ of $\Gamma(t)$ such that for each $x\in U_\delta(t)$ there is a unique $a(x,t)\in\Gamma(t)$ which is the normal projection of x onto $\Gamma(t)$, and $x=a(x,t)+d(x,t)\nu(a(x,t),t)$, where $d(x,t)$ is the oriented distance function with respect to $\Gamma(t)$. We have that

$$\Gamma_h(t) = \bigcup_{E(t)\in\mathcal{T}_h(t)} E(t)$$

with the admissible triangulation $\mathcal{T}_h(t)$ as in Section 4.1. We suppose that the maximum diameter of the simplices in $\mathcal{T}_h(t)$ is bounded uniformly in time by h. Note that for each $E(t)\subset\Gamma_h(t)$ there is a unique $\tilde{E}(t)\subset\Gamma(t)$, $\tilde{E}(t)=a(E(t),t)$, whose edges are the unique projections of the side simplices of $E(t)$ onto $\Gamma(t)$. This induces an exact 'triangulation' of $\Gamma(t)$ with curved simplices.

As in Section 4.2, we consider triangulated surfaces for which the vertices $X_j(t)$ $(j=1,\ldots,J)$ of the simplices sit on $\Gamma(t)$ so that $\Gamma_h(t)$ is an interpolation. Furthermore, we advect the nodes in the tangential direction with the advective velocity v_τ, as well as keeping them on the surface using the normal velocity v_ν, so that with $v=v_\nu+v_\tau$

$$\frac{\mathrm{d}X_j}{\mathrm{d}t}(t) = v(X_j(t),t) \quad (j=1,\ldots,J). \tag{5.22}$$

As discrete analogues of the bilinear forms (5.18), (5.19) and (5.20), we define for $\phi_h(\cdot,t), W_h(\cdot,t)\in S_h(t)$

$$a_h(\phi_h(\cdot,t),W_h(\cdot,t)) = \sum_{E(t)\in\mathcal{T}_h(t)} \int_{E(t)} \mathcal{A}^{-l}(\cdot,t)\nabla_{\Gamma_h}\phi_h(\cdot,t)\cdot\nabla_{\Gamma_h}W_h(\cdot,t)\,\mathrm{d}A_h,$$

$$m_h(\phi_h(\cdot,t),W_h(\cdot,t)) = \int_{\Gamma_h(t)} \phi_h(\cdot,t)W_h(\cdot,t)\,\mathrm{d}A_h,$$

$$g_h(V_h(\cdot,t);\phi_h(\cdot,t),W_h(\cdot,t)) = \int_{\Gamma_h(t)} \phi_h(\cdot,t)W_h(\cdot,t)\nabla_{\Gamma_h}\cdot V_h(\cdot,t)\,\mathrm{d}A_h.$$

We keep in mind that the forms explicitly depend on t.

Evolving finite element spaces

As in the stationary case in Section 4.4, we use piecewise linear finite elements. But now they live on the evolving discrete surface $\Gamma_h(t)$. We use the lift of functions from (4.2). This has two purposes. First we use the lift from $\Gamma_h(t)$ to $\Gamma(t)$ in order to define extensions of our finite element space, which allows an error analysis of the discretization on the smooth surface $\Gamma(t)$. Second, since the numerical method is based upon integration of finite element functions upon $\Gamma_h(t)$, we define approximations of the data, $\mathcal{A}$, given on $\Gamma(t)$ using $\mathcal{A}^{-l}$.

Definition 5.4. For each t and $t^n = n\tau$, $\tau > 0$, we define the finite element spaces

$$S_h(t) = \{\phi_h(\cdot,t) \in C^0(\Gamma_h(t)) \mid \phi_h(\cdot,t)|_{E(t)} \text{ is linear affine for each } E(t) \in \mathcal{T}_h(t)\},$$

$$S_h^l(t) = \{\varphi_h(\cdot,t) = \phi_h(\cdot,t)^l \mid \phi_h(\cdot,t) \in S_h(t)\},$$

$$S_h^n = S_h(t^n), \quad S_h^{n,l} = S_h^l(t^n).$$

For each $\varphi_h \in S_h^l\,(\varphi_h^n \in S_h^{n,l})$ there is a unique $\phi_h \in S_h\,(\phi_h^n \in S_h^n)$ such that

$$\varphi_h = \phi_h^l\,(\varphi_h^n = \phi_h^{n,l}).$$

By $\chi_1, \ldots, \chi_J$ we denote the nodal basis of S_h.

5.4. *Evolving surface finite element method* (ESFEM)

There is an astonishingly simple extension of the Leibniz formulae from Theorem 5.1 and Lemma 5.2 to the case of discrete surfaces. We formulate this as follows.

A discrete material velocity, V_h, for $x = X(t) \in \Gamma_h(t)$ on the surface $\Gamma_h(t)$ is defined by

$$\dot{X}(t) = V_h(X(t),t), \quad V_h(x,t) = I_h v(x,t) := \sum_{j=1}^{N} \dot{X}_j(t)\chi_j(x,t), \tag{5.23}$$

and an associated discrete material velocity, v_h, for $Y(t) = a(X(t),t)$ on $\Gamma(t)$ is defined by

$$\dot{Y}(t) = v_h(Y(t),t) = \frac{\partial a}{\partial t}(X(t),t) + V_h(X(t),t) \cdot \nabla a(X(t),t). \tag{5.24}$$

Note that (i) edges of $\tilde{E}(t)$ (which are the projections onto $\Gamma(t)$ of $E(t) \subset \Gamma_h(t)$) evolve with the material velocity $v_h(\cdot,t)$, and (ii) the discrete material velocity v_h is *not* the interpolation of v in $S_h^l(t)$.

In analogy to (5.1), we define the *discrete material derivatives* on $\Gamma_h(t)$ and $\Gamma(t)$ element-by-element via the equations

$$\partial_h^\bullet \phi_h|_{E(t)} := (\phi_{ht} + V_h \cdot \nabla \phi_h)|_{E(t)}, \tag{5.25}$$

$$\partial_h^\bullet \varphi_h|_{e(t)} := (\varphi_{ht} + v_h \cdot \nabla \varphi_h)|_{e(t)}. \tag{5.26}$$

With these discrete material derivatives we can formulate one of the main properties of ESFEM.

Lemma 5.5 (transport property of the basis functions). The basis functions satisfy the *transport property* that

$$\partial_h^\bullet \chi_j = 0, \quad \partial_h^\bullet \chi_j^l = 0. \tag{5.27}$$

Proof. We prove the property for the discrete material derivative of the basis functions on the discrete surface which follows from the definition of the nodal basis. Take a simplex $E(t) \in \mathcal{T}_h(t)$ with $X_k(t) \in \partial E(t)$. We have that $\chi_j(X_k(t), t) = \delta_{jk}$. The time derivative of this equation gives

$$\frac{\partial \chi_j}{\partial t}(X_k(t), t) + \nabla \chi_j(X_k(t), t) \cdot \frac{\mathrm{d}X_k}{\mathrm{d}t}(t) = 0$$

where the gradient is taken on $E(t)$. Since by definition

$$V_h(X_k(t), t) = \frac{\mathrm{d}X_k}{\mathrm{d}t}(t),$$

it follows that $\partial_h^\bullet \chi_j(X_k(t), t) = 0$. Since χ_j (and $\partial_h^\bullet \chi_j$ too) is a linear polynomial on $E(t)$ this then implies $\partial_h^\bullet \chi_j = 0$. Using $\chi_j(x, t) = \chi_j^l(a(x, t), t)$ and the definition of the projection $a(\cdot, t)$, we find that

$$\begin{aligned} 0 = \partial_h^\bullet \chi_j &= (\chi_{jt} + V_h \cdot \nabla \chi_j) \\ &= (\chi_{jt}^l + (a_t + (V_h \cdot \nabla) a) \cdot \nabla \chi_j^l)(a, \cdot) = \partial_h^\bullet \chi_j^l(a, \cdot). \end{aligned}$$ □

We can now formulate the transport theorem on the discrete evolving surface. This is possible because in the proof, which is similar to the proofs of Theorem 5.1 and Lemma 5.2, we do not use integration by parts.

Lemma 5.6. Let $\Gamma_h(t)$ be an evolving admissible triangulation with material velocity V_h. Then

$$\frac{\mathrm{d}}{\mathrm{d}t} \int_{\Gamma_h(t)} f \,\mathrm{d}A_h = \int_{\Gamma_h(t)} \partial_h^\bullet f + f \,\nabla_{\Gamma_h} \cdot V_h \,\mathrm{d}A_h. \tag{5.28}$$

For $\phi \in S_h(t), W_h \in S_h(t)$,

$$\frac{\mathrm{d}}{\mathrm{d}t} m_h(\phi, W_h) = m_h(\partial_h^\bullet \phi, W_h) + m_h(\phi, \partial_h^\bullet W_h) + g_h(V_h; \phi, W_h),$$

$$\frac{\mathrm{d}}{\mathrm{d}t} a_h(\phi, W_h) = a_h(\partial_h^\bullet \phi, W_h) + a_h(\phi, \partial_h^\bullet W_h) + b_h(V_h; \phi, W_h),$$

with the bilinear form

$$b_h(V_h;\phi,W_h)=\sum_{E(t)\in\mathcal{T}_h(t)}\int_{E(t)}\mathcal{B}_h(V_h)\nabla_{\Gamma_h}\phi\cdot\nabla_{\Gamma_h}W_h\,\mathrm{d}A_h,\tag{5.29}$$

where

$$\mathcal{B}_h(V_h)=\partial_h^\bullet\mathcal{A}^{-l}+\nabla_{\Gamma_h}\cdot V_h\,\mathcal{A}^{-l}-2D_h(V_h),$$

$$D_h(V_h)_{ij}=\frac{1}{2}\sum_{k=1}^{n+1}(\mathcal{A}_{ik}^{-l}(\nabla_{\Gamma_h})_kV_{hj}+\mathcal{A}_{jk}^{-l}(\nabla_{\Gamma_h})_kV_{hi}),\quad i,j=1,\ldots,n+1.$$

Let $\Gamma(t)$ be an evolving surface decomposed into curved elements $\tilde{E}(t)$ whose edges move with velocity v_h. Then

$$\frac{\mathrm{d}}{\mathrm{d}t}\int_{\Gamma(t)}f\,\mathrm{d}A=\int_{\Gamma(t)}\partial_h^\bullet f+f\,\nabla_{\Gamma_h}\cdot v_h\,\mathrm{d}A.\tag{5.30}$$

For $\varphi(\cdot,t),w(\cdot,t),\partial_h^\bullet\varphi(\cdot,t),\partial_h^\bullet w(\cdot,t)\in H^1(\Gamma(t))$,

$$\frac{\mathrm{d}}{\mathrm{d}t}m(\varphi,w)=m(\partial_h^\bullet\varphi,w)+m(\varphi,\partial_h^\bullet w)+g(v_h;\varphi,w),\tag{5.31}$$

$$\frac{\mathrm{d}}{\mathrm{d}t}a(\varphi,w)=a(\partial_h^\bullet\varphi,w)+a(\varphi,\partial_h^\bullet w)+b(v_h;\varphi,w).\tag{5.32}$$

Remark 5.7. The continuous surface moves with the smooth velocity v; the discrete surface moves with the piecewise linear velocity V_h. If we project the discrete surface onto the continuous one by $a(\cdot,t)$, then this induces another velocity on (of) the smooth surface, which we call v_h.

From the smoothness of Γ and $\mathcal{A}$ and the fact that V_h is the interpolant of the smooth velocity v, we have that

$$\|\nabla_{\Gamma_h}V_h\|_{L^\infty(\Gamma_h)}+\|\mathcal{B}_h(V_h)\|_{L^\infty(\Gamma_h)}\leq c$$

uniformly in time.

We are now in a position to cleanly formulate a discretization of the continuous PDE (5.15).

Definition 5.8 (ESFEM). Given $U_{h0}\in S_h(0)$, determine $U_h\in S_h^T$,

$$S_h^T=\{\phi_h\in C^0(\mathcal{G}_T^h)\mid\partial_h^\bullet\phi_h\in C^0(\mathcal{G}_T^h)\text{ and }\phi_h(\cdot,t)\in S_h(t)\;\forall t\in[0,T]\},$$

such that for all $\phi_h\in S_h^T$ and all $t\in(0,T]$,

$$\frac{\mathrm{d}}{\mathrm{d}t}m_h(U_h,\phi_h)+a_h(U_h,\phi_h)=m_h(U_h,\partial_h^\bullet\phi_h),\;U_h(\cdot,0)=U_{h0}.\tag{5.33}$$

U_h is then called a discrete solution of (5.15) with initial value U_{h0}.

Using the transport property Lemma 5.5, it follows that this definition is equivalent to

$$\frac{\mathrm{d}}{\mathrm{d}t} m_h(U_h, \chi_j) + a_h(U_h, \chi_j) = 0, \quad U_h(\cdot, 0) = U_{h0}, \tag{5.34}$$

for all $j = 1, \ldots, J$.

Setting $\mathcal{M}(t)$ to be the evolving mass matrix

$$\mathcal{M}(t)_{jk} = \int_{\Gamma_h(t)} \chi_j \chi_k \, \mathrm{d}A_h,$$

$\mathcal{S}(t)$ to be the evolving stiffness matrix

$$\mathcal{S}(t)_{jk} = \int_{\Gamma_h(t)} \mathcal{A}^{-l} \nabla_{\Gamma_h} \chi_j \cdot \nabla_{\Gamma_h} \chi_k \, \mathrm{d}A_h,$$

and $U_h = \sum_{j=1}^{J} \alpha_j \chi_j$, $\alpha = (\alpha_1, \ldots, \alpha_J)$, we arrive at the following simple version of the finite element approximation:

$$\frac{\mathrm{d}}{\mathrm{d}t} (\mathcal{M}(t)\alpha) + \mathcal{S}(t)\alpha = 0, \tag{5.35}$$

which does not explicitly involve the velocity of the surface.

Since the mass matrix $\mathcal{M}(t)$ is uniformly positive definite for $t \in [0, T]$ and the stiffness matrix $\mathcal{S}(t)$ is positive semidefinite, we get existence and uniqueness of the semidiscrete finite element solution.

Observe that our method and analysis includes the case of advection–diffusion on a stationary surface in which $v_\nu = 0$ and the vertices are moved with the tangential velocity $v_\mathcal{T}$. Moreover, note that the numerical method simply requires knowledge of the location of the vertices of the triangulation and avoids knowledge of the map G in Section 5.1. This is of particular advantage in applications where an approximate triangulated surface is often calculated as part of the solution process.

Lemma 5.9. There exists a unique solution of (5.33). The lifted discrete solution $u_h = U_h^l$, $u_{h0} = U_{h0}^l$, satisfies the *a priori* bounds

$$\sup_{(0,T)} \|u_h\|^2_{L^2(\Gamma)} + \int_0^T \|\nabla_\Gamma u_h\|^2_{L^2(\Gamma)} \, dt \leq c \|u_{h0}\|^2_{L^2(\Gamma(0))}, \tag{5.36}$$

$$\int_0^T \|\partial_h^\bullet u_h\|^2_{L^2(\Gamma)} \, \mathrm{d}t + \sup_{(0,T)} \|\nabla_\Gamma u_h\|^2_{L^2(\Gamma)} \leq c \|u_{h0}\|^2_{H^1(\Gamma(0))}. \tag{5.37}$$

The proof of this lemma is done similarly to the proof of Theorem 5.3 but now with the use of Lemma 5.6.

5.5. *Error analysis of ESFEM*

With the help of the formulae from Lemma 5.6 we have proved the following error estimates for the evolving surface finite element method in Dziuk and Elliott (2007*a*, 2013).

Theorem 5.10 (convergence). Let u be a sufficiently smooth solution of (5.15) satisfying

$$\int_0^T \|u\|^2_{H^2(\Gamma)} + \|\partial^\bullet u\|^2_{H^2(\Gamma)}\,\mathrm{d}t < \infty, \tag{5.38}$$

and let $u_h(,t) = U_h^l(\cdot,t)$, $t \in [0,T]$, be the spatially discrete solution from Lemma 5.9 with initial data $u_{h0} = U_{h0}^l$ satisfying

$$\|u(\cdot,0) - u_{h0}\|_{L^2(\Gamma(0))} \leq ch^2.$$

Then the error estimates

$$\sup_{t\in(0,T)} \|u(\cdot,t) - u_h(\cdot,t)\|_{L^2(\Gamma(t))} \leq ch^2 \tag{5.39}$$

and

$$\int_0^T \|u(\cdot,t) - u_h(\cdot,t)\|^2_{H^1(\Gamma(t))}\,\mathrm{d}t \leq ch^2$$

hold for $h \leq h_0$ with a constant c independent of h but depending on the norms (5.38) and on the geometry of $\mathcal{G}_T$.

Under suitable assumptions on $\mathcal{G}_T$, $\mathcal{A}$, v and u_0 it can be shown that (5.38) holds: see Dziuk and Elliott (2007*a*).

5.6. *Time discretization*

Let N be a positive integer and set $\tau = T/N$. For each $n \in \{0,\ldots,N\}$ set $t^n = n\tau$. For a discrete time sequence f^n, $n \in \{0,\ldots,N\}$, we use the notation

$$\partial_\tau f^n = \frac{1}{\tau}(f^{n+1} - f^n).$$

For the time-discrete case we introduce a suitable material derivative. Assume that $\phi_h^n \in S_h^n$, $\phi_h^n = \sum_{j=1}^J \phi_j^n \chi_j^n$ with the lift

$$\varphi_h^n = \sum_{j=1}^J \phi_j^n \chi_j^{n,l} \in S_h^{n,l}.$$

Then we define

$$\partial_h^\bullet \phi_h^n = \sum_{j=1}^J \partial_\tau \phi_j^n \chi_j^n \in S_h^n, \quad \partial_h^\bullet \varphi_h^n = \sum_{j=1}^J \partial_\tau \phi_j^n \chi_j^{n,l} \in S_h^{n,l}. \tag{5.40}$$

We write the fully discrete scheme in the following form.

Algorithm 5.11. Given $U_h^0 \in S_h^0$, find

$$U_h^n \in S_h^n, \quad n \in \{1, \dots, N\}$$

such that for all $\phi_h^n \in S_h^n$ and $\phi_h^{n+1} \in S_h^{n+1}$ and $n \in \{0, \dots, N-1\}$

$$\partial_\tau m_h(U_h^n, \phi_h^n) + a_h(U_h^{n+1}, \phi_h^{n+1}) = m_h(U_h^n, \partial_h^\bullet \phi_h^n). \tag{5.41}$$

Let us discuss the matrix–vector form of this fully discrete scheme. Choosing $\phi_h^n = \chi_i^n$ in (5.41), it follows that this definition is *equivalent* to

$$\partial_\tau m_h(U_h^n, \chi_i^n) + a_h(U_h^{n+1}, \chi_i^{n+1}) = 0 \tag{5.42}$$

for all $i = 1, \dots, J$.

Setting $\mathcal{M}(t), \mathcal{M}^n$ to be the time-dependent mass matrices

$$\mathcal{M}(t)_{jk} = \int_{\Gamma_h(t)} \chi_j(\cdot, t)\chi_k(\cdot, t)\, \mathrm{d}A_h, \quad \mathcal{M}^n = \mathcal{M}(t^n),$$

$(j, k = 1, \dots, J)$, and $\mathcal{S}(t), \mathcal{S}^n$ to be the time-dependent stiffness matrices

$$\mathcal{S}(t)_{jk} = \int_{\Gamma_h(t)} \mathcal{A}^{-l}(\cdot, t)\nabla_{\Gamma_h}\chi_j(\cdot, t) \cdot \nabla_{\Gamma_h}\chi_k(\cdot, t)\, \mathrm{d}A_h, \quad \mathcal{S}^n = \mathcal{S}(t^n),$$

we arrive at the following simple version of the fully discrete finite element approximation:

$$\partial_\tau(\mathcal{M}^n \alpha^n) + \mathcal{S}^{n+1}\alpha^{n+1} = 0.$$

Equivalently,

$$(\mathcal{M}^{n+1} + \tau\mathcal{S}^{n+1})\alpha^{n+1} = \mathcal{M}^n\alpha^n, \tag{5.43}$$

where

$$U_h^{n+1} = \sum_{j=1}^{N} \alpha_j^{n+1}\chi_j^{n+1}$$

has to be determined as the solution of the sparse system of linear equations (5.43). Since for each n the mass matrix $\mathcal{M}^n$ is uniformly positive definite and the stiffness matrix $\mathcal{S}^n$ is positive semidefinite, we get existence and uniqueness of the discrete finite element solution.

We show how to prove stability for the fully discrete scheme. This is the fully discrete analogue of the continuous estimates (5.16) and (5.17). In the lemma below we use the notation U_h^L and u_h^L to denote linear interpolations in time of the time level values U_h^n and u_h^n, $n = 0, 1, 2, \dots, N$. Note that the material derivatives are piecewise constant in time and we indicate in the natural way which constants are to be used in the sums.

Lemma 5.12. The fully discrete solution U_h^k with lift

$$u_h^k = (U_h^k)^l \quad (k = 0, \ldots, N)$$

satisfies the following *a priori* bounds for $\tau \le \tau_0$:

$$|U_h^n|^2_{L^2(\Gamma_h(t^n))} + \tau \sum_{k=1}^{n} |\nabla_{\Gamma_h} U_h^k|^2_{L^2(\Gamma_h(t^k))} \le c|U_h^0|^2_{L^2(\Gamma_h(0))},$$

$$|u_h^n|^2_{L^2(\Gamma(t^n))} + \tau \sum_{k=1}^{n} |\nabla_{\Gamma} u_h^k|^2_{L^2(\Gamma(t^k))} \le c|u_h^0|^2_{L^2(\Gamma(0))},$$

$$\tau \sum_{k=0}^{n-1} |\partial_h^\bullet U_h^L(\cdot, t^{k+1} - 0)|^2_{h,k} + |\nabla_{\Gamma_h} U_h^n|^2_{L^2(\Gamma_h(t^n))}$$
$$\le c\big(|U_h^0|^2_{L^2(\Gamma_h(0))} + |\nabla_{\Gamma_h} U_h^0|^2_{L^2(\Gamma_h(0))}\big),$$

$$\tau \sum_{k=0}^{n-1} |\partial_h^\bullet u_h^L(\cdot, t^{k+1} - 0)|^2_{k} + |\nabla_{\Gamma} u_h^n|^2_{L(\Gamma(t^n))}$$
$$\le c\big(|u_h^0|^2_{L^2(\Gamma(0))} + |\nabla_{\Gamma} u_h^0|^2_{L^2(\Gamma(0))}\big).$$

The constants depend on the data of the problem including the final time T.

In Dziuk and Elliott (2012) we proved the following error estimate for the fully discrete scheme.

Theorem 5.13. Let u be a sufficiently smooth solution of (5.13), and assume that

$$\|u_0 - u_h^0\|_{L^2(\Gamma_0)} \le ch^2. \tag{5.44}$$

Let $u_h^k = (U_h^k)^l$ $(k = 0, \ldots, N)$ be the lift of the solution of the fully discrete scheme (5.41). Then for $0 < \tau \le \tau_0$ and $0 < h \le h_0$ we have the error bounds

$$\|u(\cdot, t^n) - u_h^n\|^2_{L^2(\Gamma(t^n))} + h^2 \tau \sum_{k=1}^{n} \|\nabla_\Gamma u(\cdot, t^k) - \nabla_\Gamma u_h^k\|^2_{L^2(\Gamma(t^k))} \le c(\tau^2 + h^4) \tag{5.45}$$

for all $n \in \{0, \ldots, N\}$, with a constant c which is independent of τ and h. Note that c, h_0 and τ_0 depend on the data of the problem.

We close this section with an example from Dziuk and Elliott (2007*a*).

Example 5.14. The example confirms the theoretical results. We solve the PDE

$$\partial^\bullet u + u \nabla_\Gamma \cdot v - \Delta_\Gamma u = f \tag{5.46}$$

Table 5.1. Errors and experimental orders of convergence (EOC) for the example (5.46), (5.47).

$h(T)$	$L^\infty(L^\infty)$	EOC	$L^\infty(L^2)$	EOC	$L^2(H^1)$	EOC
0.82737	0.095488	–	0.15424	–	0.29287	–
0.43422	0.057944	0.77	0.097788	0.71	0.17507	0.80
0.21939	0.018764	1.65	0.033083	1.59	0.074327	1.26
0.10994	0.0050819	1.89	0.0089784	1.89	0.033367	1.16
0.055007	0.0013038	1.97	0.0022950	1.97	0.016053	1.06

on the moving surface

$$\Gamma(t) = \left\{ x \in \mathbb{R}^3 \mid \frac{x_1^2}{1 + 0.25 \sin t} + x_2^2 + x_3^2 = 1 \right\}. \tag{5.47}$$

As the exact solution of the PDE we have chosen the function $u(x,t) = \mathrm{e}^{-6t} x_1 x_2$ and calculated the right-hand side f from the PDE. We then calculated the following errors from the exact solution and the computed solution for the time interval $[0,T]$ with $T = 2$:

$$L^\infty(L^\infty) = \sup_{(0,T)} \|u - u_h\|_{L^2(\Gamma)}, \quad L^\infty(L^2) = \sup_{(0,T)} \|u - u_h\|_{L^2(\Gamma)},$$

$$L^2(H^1) = \left(\int_0^T \|\nabla_\Gamma(u - u_h)\|^2_{L^2(\Gamma)} \,\mathrm{d}t \right)^{\frac{1}{2}}.$$

In our computations we have chosen $\tau = h^2$ in order to reveal the quadratic convergence in the L^2 norm.

For a more appealing computational example see Figure 5.3.

5.7. ALE ESFEM

Another definition of an evolving hypersurface $\Gamma(t)$ is the zero level set of a time-dependent level set function $\phi : \mathcal{U} \times (0,T) \to \mathbb{R}$, where $\mathcal{U}$ is an open set in $\mathbb{R}^3$ so that

$$\Gamma(t) = \{x \in \mathcal{U} \mid \phi(x,t) = 0\}$$

and $\nabla\phi(x,t) \neq 0,\ x \in \Gamma(t)$. For such a hypersurface we define the oriented normal ν and the normal velocity V by

$$\nu(x,t) = \frac{\nabla\phi(x,t)}{|\nabla\phi(x,t)|}, \quad V(x,t) = -\frac{\phi_t(x,t)}{|\nabla\phi(x,t)|}.$$

See Section 7.5 for a complete discussion. Thus moving a point $P_0 \in \Gamma(0)$ by the velocity

$$\dot{P} = V(P,\cdot)\nu(P,\cdot) + v_\tau^a(P,\cdot), \quad P(0) = P_0,$$

where v_τ^a is an arbitrary tangential velocity field satisfying $v_\tau^a(x,t)\cdot\nu(x,t) = 0$, keeps $P(t)$ on the surface $\Gamma(t)$ because

$$\frac{\mathrm{d}}{\mathrm{d}t}\phi(P(t),t) = \nabla\phi(P(t),t)\cdot\dot{P}(t) + \phi_t(P(t),t) = 0.$$

Thus we may recover the description in Section 5.1 of a moving hypersurface: $G : \Gamma(0)\times(0,T)\to\Gamma(t)$ so that

$$\Gamma(t) = \{x = G(P_0,t) \mid P_0 \in \Gamma(0)\} \quad \text{and} \quad G_t\cdot\nu = V.$$

Choices for a tangential velocity v_τ^a include the following.

- Frequently in mathematical models there is a material velocity $v = v_\nu + v_\tau$. For example, this might arise when the hypersurface is a fluid–material interface, and choosing $P_t(t) = v(P(t),t)$, $P(0) = P_0 \in \Gamma_0$ implies that points $P(t)$ evolve as material points. In this case there is a natural physical tangential velocity and one may wish to use the material velocity to define the map. For example, this is the case when using ESFEM for the advection–diffusion equation.

- In applications it may be that the hypersurface is determined by a geometric evolution law for which there is a natural partial differential equation which evolves a parametrization. For example, the solution of the equation

$$X_t - \frac{1}{|X_\theta|}\left(\frac{X_\theta}{|X_\theta|}\right)_\theta = 0$$

  for the parametrization $X = X(\theta,t)$ of a closed curve ($X(\theta+2\pi,t) = X(\theta,t)$, $\theta\in[0,2\pi]$) defines motion by curvature, called curve shortening flow, for which the solution $\Gamma(t) = X([0,2\pi],t)$ moves in a normal direction. On the other hand the equation

$$X_t - \frac{X_{\theta\theta}}{|X_\theta|^2} = 0$$

  evolves a closed planar curve in the normal direction with velocity given by the curvature, but has a tangential velocity defined by the equation.

- It may be that we wish to choose a tangential velocity in such a way as to yield a *nice* map G and thus a nice grid.

As an example, consider the triangulated surfaces in Figure 5.4. These are interpolations of a hypersurface defined as the zero level set of (*cf.* Barreira *et al.* 2011, Elliott and Styles 2012)

$$\phi(x,t) = x_1^2 + x_2^2 + a(t)^2 G(x_3^2/L(t)^2) - a(t)^2,$$

where $G(s) = 200s(s - 199/200)$, $a(t) = 0.1 - 0.04\sin(2\pi t)$ and $L(t) = 1 + 0.5\sin(\pi t)$. The triangulated surfaces in plots (b), (d) and (f) are obtained using pure normal motion for the position of the triangle vertices, while those in plots (a), (c) and (e) have a non-zero tangential velocity. Plots (a,b) show the initial triangulated surface, which is the same in each case.

From this figure we see that although both sets of surfaces are interpolations which evolve from the same mesh, the difference between the two meshes is quite pronounced. In particular, the nodes of the triangulated surface on the left, where tangential motion is present, are quite uniformly distributed over the surface, while the nodes of the triangulated surface on the left are very coarsely separated over some parts of the surface.

This is an example which suggests that when evolving triangulations it may be useful to use an arbitrary tangential velocity rather than using a material velocity. This arbitrary tangential velocity may be used to generate a *good* triangulation of the surface. This motivates the arbitrary Lagrangian Eulerian evolving surface finite element method (ALE ESFEM). The goal is to solve the advection–diffusion equation (5.13) numerically. In this approach the surface $\Gamma_h(t)$ interpolates $\Gamma(t)$ in such a way that the velocity of the vertices may not be the material velocity associated with the equation (5.13), so that the nodes move with a velocity $V_h^M \neq V_h = I_h v$ (see also Section 5.4). It follows that since

$$\partial_M^\bullet \chi_i = \frac{\partial \chi_i}{\partial t} + V_h^M \cdot \nabla \chi_i = 0, \quad \text{for all } \ i \in \{1, \ldots, J\},$$

the nodal basis functions, $\chi_i(\cdot, t)$ $(i = 1 \ldots J)$, of $S^h(\Gamma_h(t))$ satisfy the transport property

$$\partial_h^\bullet \chi_i = (V_h - V_h^M) \cdot \nabla \chi_i.$$

This leads to the following discretization of (5.13) with $\mathcal{A} = I$:

$$\begin{aligned}\frac{\mathrm{d}}{\mathrm{d}t} \int_{\Gamma_h(t)} U_h \chi_i \,\mathrm{d}A_h &+ \int_{\Gamma_h(t)} \nabla_{\Gamma_h} U_h \cdot \nabla_{\Gamma_h} \chi_i \,\mathrm{d}A_h \\ &= \int_{\Gamma_h(t)} U_h (V_h - V_h^M) \cdot \nabla_{\Gamma_h} \chi_i \,\mathrm{d}A_h\end{aligned}$$

for all $i = 1, \ldots, J$.

For a fully discrete approximation one can use a fully implicit time discretization.

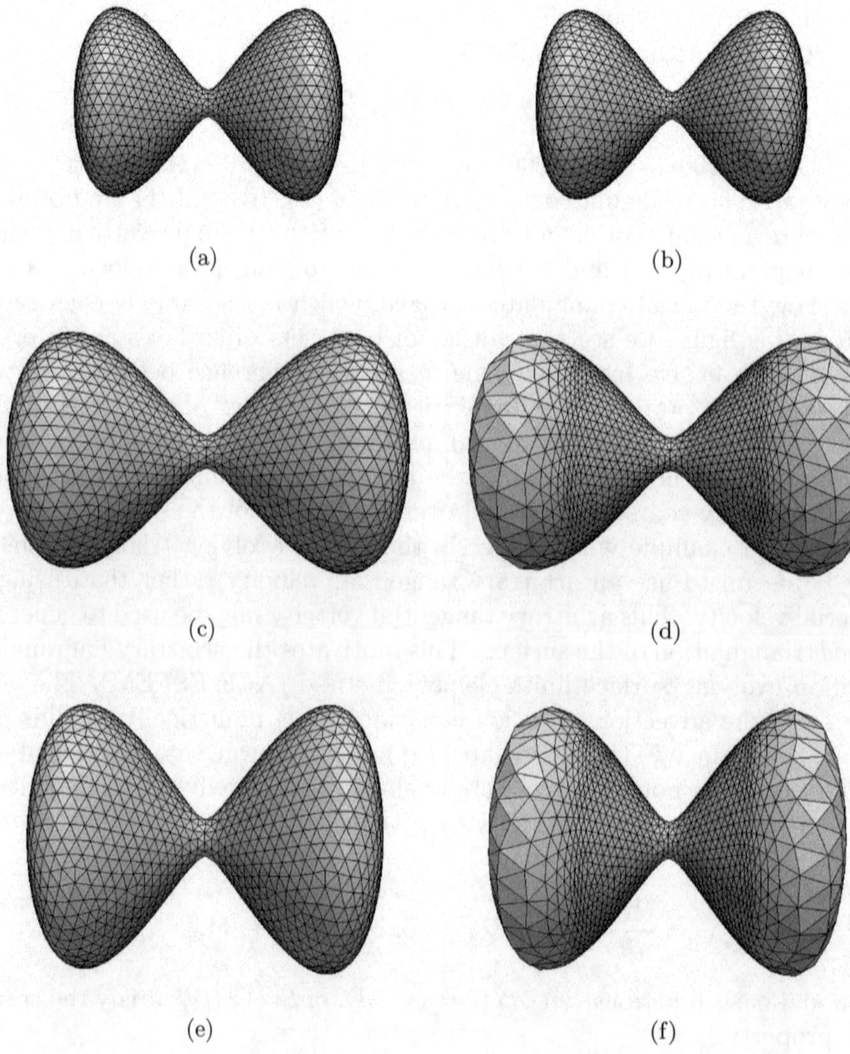

Figure 5.4. Interpolated triangulated surfaces evolving from the same mesh. Vertices (a), (c) and (e) have a tangential and normal velocity whereas vertices (b), (d) and (f) evolve with only a normal velocity.

Algorithm 5.15. Given Γ_h^{m-1}, Γ_h^m and $U_h^{m-1} \in S_h^{m-1}$, find $U_h^m \in S_h^m$ such that

$$\frac{1}{\tau}\int_{\Gamma_h^m} U_h^m \chi_h^m \,\mathrm{d}A_h - \frac{1}{\tau}\int_{\Gamma_h^{m-1}} U_h^{m-1}\chi_h^{m-1}\,\mathrm{d}A_h + \int_{\Gamma_h^m} \nabla_{\Gamma_h^m} U_h^m \cdot \nabla_{\Gamma_h^m}\chi_h^m \,\mathrm{d}A_h$$
$$+ \int_{\Gamma_h^m} (V_h^{M,m} - V_h^m) U_h^m \cdot \nabla_{\Gamma_h^m}\chi_h^m \,\mathrm{d}A_h = 0, \quad \text{for all } \chi_h^m \in S_h^m.$$

Here $V_h^{M,m} = \sum_{i=1}^J \dot{X}_i^m \chi_i^m$ and $V_h^m = V_h(\cdot, m\tau)$.

The key difference between the two methods is that in the ESFEM method the normal velocity of the surface and the advective velocity do not explicitly appear in the discretization. The numerical method only requires knowledge of the position of the vertices and there is no advective term to consider in the discretization. On the other hand, in the ALE ESFEM an advective velocity term appears explicitly in the discretization. If $\dot{X}_i^m = v(X_i^m, m\tau)$ then the ALE ESFEM reduces to ESFEM.

We refer to Elliott and Styles (2012) for examples suggesting the possible accuracy advantages of using the ALE version of ESFEM. It may be of particular value when the surface evolution is coupled to the equation on the surface and the surface has to be approximated in some way. The numerical methods of Barrett, Garcke and Nürnberg (2007, 2008*a*, 2008*b*) for geometric surface evolution yield arbitrary tangential velocities which often yield good mesh properties. This has been exploited for various applications by Elliott and Stinner (2012), Elliott *et al.* (2012) and Elliott and Styles (2012).

6. More surface PDEs

6.1. Nonlinear conservation and diffusion on a stationary surface

Let us turn to more realistic equations for conservation and diffusion on a surface. We follow Dziuk and Elliott (2007*b*) and begin with the derivation from Section 5.2 for a stationary surface Γ.

Conservation. $u = u(x,t)$ ($x \in \Gamma$, $t \in [0,T]$) is the density of a scalar quantity on Γ. The basic conservation law we wish to consider can be formulated for an arbitrary portion $\mathcal{M}$ of Γ using a surface flux q. The law is that, for every $\mathcal{M}$,

$$\frac{\mathrm{d}}{\mathrm{d}t}\int_{\mathcal{M}} u \,\mathrm{d}A = -\int_{\partial\mathcal{M}} q\cdot\mu \,\mathrm{d}A, \tag{6.1}$$

where $\partial\mathcal{M}$ is the boundary of $\mathcal{M}$ and μ is the co-normal on $\partial\mathcal{M}$. Using the divergence theorem Theorem 2.10 and the fact that without loss of generality q is a tangential vector, we obtain as in Section 5.2

$$u_t + \nabla_\Gamma \cdot q = 0 \quad \text{on } \Gamma. \tag{6.2}$$

We take q to be the diffusive flux with respect to a scalar function w on Γ,

$$q = -\mathcal{A}\nabla_\Gamma w, \tag{6.3}$$

where $\mathcal{A}$ is a positive semidefinite symmetric mobility tensor with the property that it maps the tangent space into itself at every point of Γ; see also (5.14). This leads to the equation

$$u_t - \nabla_\Gamma \cdot (\mathcal{A}\nabla_\Gamma w) = 0 \quad \text{on } \Gamma. \tag{6.4}$$

If the surface has no boundary, $\partial\Gamma = \emptyset$, then there is no need for boundary conditions. This would be the case if Γ is the bounding surface of a domain. On the other hand, if $\partial\Gamma$ is non-empty then we may impose boundary conditions similar to the flat case. For example, we may impose the homogeneous Dirichlet boundary condition

$$u = 0 \quad \text{on } \partial\Gamma.$$

or the no-flux condition

$$\mathcal{A}\nabla_\Gamma w \cdot \mu = 0 \quad \text{on } \partial\Gamma.$$

The variational form is obtained in the standard way by multiplying equation (6.4) by an arbitrary test function $\varphi \in H^1(\Gamma)$ and integrating over Γ. Using (2.10) we obtain

$$\int_\Gamma u_t\varphi \,\mathrm{d}A + \int_\Gamma \mathcal{A}\nabla_\Gamma w \cdot \nabla_\Gamma\varphi \,\mathrm{d}A = 0. \tag{6.5}$$

Of course, for parabolic equations we have to impose an initial condition $u(\cdot, 0) = u_0$ on Γ with given u_0. In the following we will not mention this condition when it is obviously required. Note that for arbitrary tensor $\mathcal{A}$ this variational equation implies

$$u_t - \nabla_\Gamma \cdot (P\mathcal{A}P\nabla_\Gamma w) = 0 \quad \text{on } \Gamma \tag{6.6}$$

with the projection P onto Γ. Note that, in general, constant coefficient tensors $\mathcal{A}$ will not satisfy the assumption that $\mathcal{A}$ maps the tangent space into itself and P will not be the constant coefficient. Finally we note that this is indeed a conservation equation by taking $\varphi = 1$ in (6.5), which yields the conservation equation

$$\frac{\mathrm{d}}{\mathrm{d}t}\int_\Gamma u \,\mathrm{d}A = 0.$$

Linear diffusion. The heat equation on surfaces is obtained by setting $w = u$ and $\mathcal{A} = I$, where I is the identity tensor,

$$u_t - \Delta_\Gamma u = 0. \tag{6.7}$$

Figure 6.1. Heating up a torus. Solution of the inhomogeneous heat equation (6.8) with right-hand side (6.10) at three successive times. The colouring indicates the magnitude of the solution. The colours range between blue (0.0) and red (5.0).

This can be generalized in obvious ways. For example, an inhomogeneous variable coefficient parabolic equation is

$$u_t - \sum_{i,j=1}^{n+1} \underline{D}_i(a_{ij}\underline{D}_j u) = f, \tag{6.8}$$

where $\mathcal{A} = (a_{ij}(x,t))_{i,j=1,\ldots,n+1}$ with a symmetric matrix $\mathcal{A}$ which satisfies (5.14).

We show an example of the inhomogeneous heat equation from Dziuk and Elliott (2007*b*).

Example 6.1. In Figure 6.1 we display the solution at three successive times of (6.8) with $\mathcal{A} = \mathcal{I}$ on the torus,

$$\Gamma = \left\{ x \in \mathbb{R}^3 \mid \left(\sqrt{x_1^2 + x_2^2} - 1\right)^2 + x_3^2 = \frac{1}{16} \right\}, \tag{6.9}$$

with the right-hand side being a regularized version of the characteristic function

$$f(x,t) = 100\, \chi_G(x), \quad x \in \Gamma, \tag{6.10}$$

with $G = \{x \in \Gamma \mid |x - (0,1,0)| < 0.25\}$ and with initial value $u_0 = 0$.

Nonlinear diffusion. We find the nonlinear diffusion equation

$$u_t - \nabla_\Gamma \cdot (g(u)\nabla_\Gamma u) = 0$$

by setting

$$w = f(u) \quad \text{and} \quad \mathcal{A} = m(u)I$$

for given continuous functions f and m, where

$$g(u) = m(u)f'(u)$$

and g is positive if f is monotone increasing and m is positive. Clearly, by using suitable choices one recovers linear diffusion and the porous medium equation on a stationary surface Γ.

Parabolic surface p-Laplace equation. Setting $w = u$ and, for $1 < p < \infty$, $\mathcal{A} = |\nabla_\Gamma u|^{p-2} I$ yields the parabolic surface p-Laplace equation

$$u_t - \nabla_\Gamma \cdot (|\nabla_\Gamma u|^{p-2} \nabla_\Gamma u) = 0.$$

This is $L^2(\Gamma)$-gradient flow for the energy

$$E_p(u) = \frac{1}{p} \int_\Gamma |\nabla_\Gamma u|^p \, \mathrm{d}A.$$

Total variation flow. We obtain a singular degenerate equation by setting $w = u$ and taking $\mathcal{A} = |\nabla_\Gamma u|^{-1} I$ which leads (formally) to the surface total variation flow

$$u_t - \nabla_\Gamma \cdot \frac{\nabla_\Gamma u}{|\nabla_\Gamma u|} = 0.$$

Variants of this equation may be useful in processing images on curved surfaces.

Fourth-order linear diffusion. The choice $w = -\Delta_\Gamma u$ leads to the fourth-order linear diffusion equation

$$u_t + \nabla_\Gamma \cdot (\mathcal{A} \nabla_\Gamma \Delta_\Gamma u) = 0.$$

An error analysis of the surface finite element approximation of this equation using the splitting into two second-order equations (Elliott, French and Milner 1989) was carried out by Dziuk and Elliott (2007*b*).

Surface Cahn–Hilliard equation. Setting

$$w = -\epsilon \Delta_\Gamma u + \frac{1}{\epsilon} \psi'(u),$$

where $\psi : \mathbb{R} \to \mathbb{R}$ is typically a double-well potential, for example the classical quartic potential

$$\psi(u) = \frac{1}{4}(1 - u^2)^2$$

leads to the fourth-order Cahn–Hilliard equation

$$u_t + \nabla_\Gamma \cdot \left(\mathcal{A} \nabla_\Gamma \left(\epsilon \Delta_\Gamma u - \frac{1}{\epsilon} \psi'(u) \right) \right) = 0.$$

Surface Allen–Cahn equation. This is an example of an equation not in conservation form. $L^2(\Gamma)$-gradient flow for the energy functional

$$E(v) = \int_\Gamma \frac{\epsilon}{2} |\nabla_\Gamma v|^2 + \frac{\psi(v)}{\epsilon} \, \mathrm{d}A \tag{6.11}$$

($\epsilon > 0$) leads to

$$\epsilon u_t - \epsilon \Delta_\Gamma u - \frac{1}{\epsilon} \psi'(u) = 0. \tag{6.12}$$

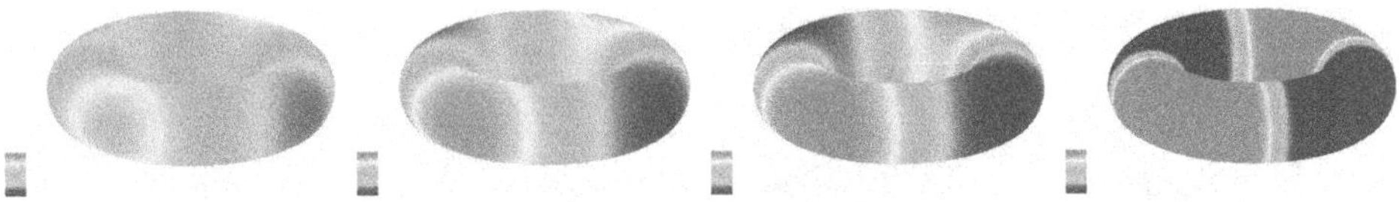

(a) $t = 0.0$ (b) $t = 0.0045$ (c) $t = 0.009$ (d) $t = 0.36$

Figure 6.2. Solution of the surface Allen–Cahn equation on a thick torus.

Here the double-well potential ψ gives the classical Allen–Cahn equation on a surface Γ. Note that the Cahn–Hilliard equation may be interpreted as an $H^{-1}(\Gamma)$-gradient flow for this energy functional.

In Figure 6.2 we show the results of the numerical solution of the surface Allen–Cahn equation on a torus with initial data

$$u_0(x) = \frac{x_1 x_2}{\sqrt{x_1^2 + x_2^2 + x_3^2}}.$$

We observe the evolution to a pattern on the torus which consists of regions where the solution is close to -1 (blue) and regions where the solution is close to 1 (red) with a transition region of width ε. We have chosen $\varepsilon = 0.1$.

6.2. Level set equations on surfaces

We have in mind the evolution, on the fixed surface Γ in $\mathbb{R}^3$, of a closed curve $\mathcal{C}(t)$ which is evolving in the *intrinsic* normal direction ν_g with a velocity V_g. The curve $\mathcal{C}(t)$ is given by the zero level set on Γ of $u(\cdot, t)$ and

$$\nu_g = \frac{\nabla_\Gamma u}{|\nabla_\Gamma u|}, \quad \kappa_g = \nabla_\Gamma \cdot \frac{\nabla_\Gamma u}{|\nabla_\Gamma u|}, \quad V_g = -\frac{u_t}{|\nabla_\Gamma u|}.$$

Using surface gradient notation, it is straightforward to define analogues of various geometric level set equations in the flat case to level set equations on surfaces.

Level set geodesic mean curvature flow. In the case that the the velocity V_g is given by minus its geodesic curvature κ_g, we formulate the level set equation

$$u_t - |\nabla_\Gamma u| \, \nabla_\Gamma \cdot \frac{\nabla_\Gamma u}{|\nabla_\Gamma u|} = 0.$$

Figure 6.3 shows how circles on a dumbbell-shaped surface move under geodesic mean curvature flow. The surface Γ is given as the image of the sphere, $\Gamma = F(S^2)$, under the map $F(x) = (x_1, \eta(x)x_2, \eta(x)x_3)$ with

$$\eta(x) = \sqrt{1 - x_1^2}\sqrt{1 - 0.8(1 - x_1^2)^2} / \sqrt{x_2^2 + x_3^2}$$

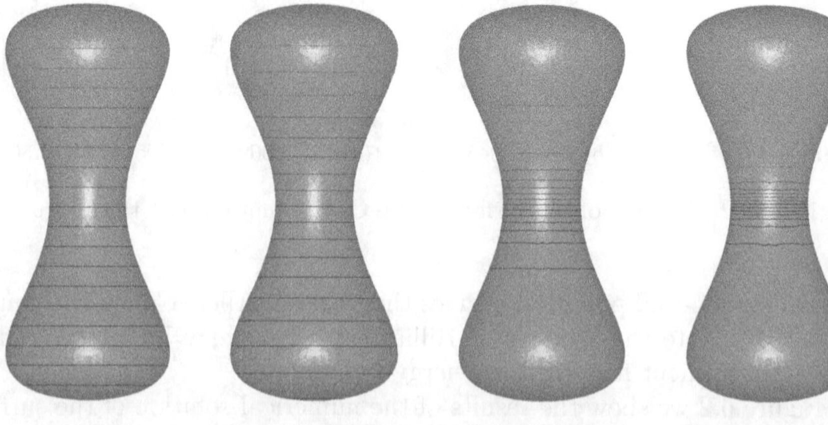

Figure 6.3. Geodesic curve shortening flow on a dumbbell. Initial circles either shrink to a point or to a geodesic during the evolution.

for $x \in S^2$. The initial function is $u_0(x) = x_1 - 0.25$. In Figure 6.3 we display all the level lines $\{x \in \Gamma \mid u(x,t) = c\}$ for c between -1.05 and 1.25 with intervals of 0.2. This example is taken from Dziuk and Elliott (2007*b*).

We observe that circles shrink and either move to the centre of the dumbbell's neck or shrink to round points at the extreme ends of the dumbbell. In this computation of geodesic curve shortening flow we regularized the equation by replacing $|\nabla_\Gamma u|$ by $\sqrt{\varepsilon^2 + |\nabla_\Gamma u|^2}$, and we have taken the parameter ε proportional to the grid size h.

Level set surface active contours. We formulate the level set equation

$$u_t - |\nabla_\Gamma u| \nabla_\Gamma \cdot \left(f \frac{\nabla_\Gamma u}{|\nabla_\Gamma u|} \right) = 0, \tag{6.13}$$

where $f = (1 + |\nabla_\Gamma I_\sigma|^2)^{-1}$. Here typically I_σ is a smoothing of an image which is essentially a characteristic function with sharp edges. The evolution of the zero level set curve $\mathcal{C}(t)$ is designed to detect the edge.

Anisotropic geodesic level set mean curvature flow. Anisotropic geodesic mean curvature flow on the given surface Γ may be formulated by the level set equation

$$\mu(\nabla_\Gamma u) u_t - |\nabla_\Gamma u| \nabla_\Gamma \cdot (D\gamma(\nabla_\Gamma u)) = 0,$$

where $\gamma : \mathbb{R}^{n+1} \setminus \{0\} \to (0, \infty)$, $\gamma(0) = 0$, is an anisotropy function, smooth and positively homogeneous of degree one. Here $D\gamma$ denotes the gradient of γ. μ is a positive and 0-homogeneous function.

Level set geodesic surface diffusion. The evolution, on the fixed surface Γ in $\mathbb{R}^3$, of a closed curve $\mathcal{C}(t)$ which is evolving in the 'intrinsic' normal direction ν_g with a velocity V given by the geodesic Laplacian of the geodesic curvature κ_g may be formulated as the level set equation

$$u_t - \nabla_\Gamma \cdot (|\nabla_\Gamma u|(\mathcal{I} - \nu_g \otimes \nu_g)\nabla_\Gamma w) = 0, \quad w = \nabla_\Gamma \cdot \frac{\nabla_\Gamma u}{|\nabla_\Gamma u|}.$$

Level set geodesic Willmore flow. Now suppose the zero level set of u is the curve $\mathcal{C}(t)$ constrained to lie on Γ, which evolves according to L^2 gradient flow for the energy $\mathcal{E}_\mathcal{C} = \frac{1}{2}\int_\mathcal{C} \kappa_g^2$. The level set equation is then

$$\begin{aligned} u_t + |\nabla_\Gamma u|\nabla_\Gamma \cdot \left(\frac{1}{|\nabla_\Gamma u|}(I - \nu_g \otimes \nu_g)\nabla_\Gamma w\right) & \\ +\frac{1}{2}|\nabla_\Gamma u|\nabla_\Gamma \cdot \left(\frac{w^2}{|\nabla_\Gamma u|^2}\frac{\nabla_\Gamma u}{|\nabla_\Gamma u|}\right) &= 0, \\ w + |\nabla_\Gamma u|\, \nabla_\Gamma \cdot \frac{\nabla_\Gamma u}{|\nabla_\Gamma u|} &= 0. \end{aligned}$$

Note that in these last two examples the equations are fourth-order in space but have been written as two coupled second-order equations, and that the definition of w is different in each case.

6.3. The Jenner equation

This PDE was derived by the authors during a stay at Jenner (California, USA). It is a linear wave equation on a moving surface. Starting with a given moving surface $\Gamma(t)$, $t \in [0, T]$, as in Section 5.1, we define the action integral

$$E(u) = \frac{1}{2}\int_0^T \int_{\Gamma(t)} (\partial^\bullet u)^2 - |\nabla_\Gamma u|^2 \,\mathrm{d}A\,\mathrm{d}t,$$

where we have used the material derivative from (5.1). We use the principle of stationary action and, for the first variation in the direction of $\varphi : \mathcal{G}_T \to \mathbb{R}$, vanishing in a neighbourhood of $\partial\mathcal{G}_T$, find that

$$0 = \langle E'(u), \varphi\rangle = \frac{\mathrm{d}}{\mathrm{d}\varepsilon}\bigg|_{\varepsilon=0} E(u + \varepsilon\varphi) = \int_0^T \int_{\Gamma(t)} \partial^\bullet\varphi\partial^\bullet u - \nabla_\Gamma\varphi \cdot \nabla_\Gamma u \,\mathrm{d}A\,\mathrm{d}t.$$

Let us derive the PDE in classical form. For this we use the Leibniz formula from Theorem 5.1 together with integration by parts on Γ from Theorem 2.10, and obtain

$$\begin{aligned} 0 &= \int_0^T \int_{\Gamma(t)} \partial^\bullet\varphi\partial^\bullet u - \nabla_\Gamma\varphi \cdot \nabla_\Gamma u \,\mathrm{d}A\,\mathrm{d}t \\ &= \int_0^T \int_{\Gamma(t)} \partial^\bullet(\varphi\partial^\bullet u)\,\mathrm{d}A\,\mathrm{d}t - \int_0^T \int_{\Gamma(t)} \varphi(\partial^\bullet\partial^\bullet u - \Delta_\Gamma u)\,\mathrm{d}A\,\mathrm{d}t \end{aligned}$$

$$= \int_0^T \frac{\mathrm{d}}{\mathrm{d}t} \int_{\Gamma(t)} \varphi \partial^\bullet u \,\mathrm{d}A\,\mathrm{d}t - \int_0^T \int_{\Gamma(t)} \varphi \partial^\bullet u \nabla_\Gamma \cdot v \,\mathrm{d}A\,\mathrm{d}t$$
$$- \int_0^T \int_{\Gamma(t)} \varphi(\partial^\bullet \partial^\bullet u - \Delta_\Gamma u)\,\mathrm{d}A\,\mathrm{d}t$$
$$= - \int_0^T \int_{\Gamma(t)} \varphi(\partial^\bullet \partial^\bullet u + \partial^\bullet u \nabla_\Gamma \cdot v - \Delta_\Gamma u)\,\mathrm{d}A\,\mathrm{d}t.$$

The outcome is a linear wave equation on the moving surface Γ, the Jenner equation

$$\partial^\bullet \partial^\bullet u + \partial^\bullet u \nabla_\Gamma \cdot v - \Delta_\Gamma u = 0$$

on $\mathcal{G}_T$, and one has to add suitable initial conditions

$$u(\cdot, 0) = u_0, \quad \partial^\bullet u(\cdot, 0) = u_1 \quad \text{on } \Gamma(0).$$

Discretizations of this PDE in space and time have been studied in Lubich and Mansour (2012).

6.4. *First-order conservation laws on moving surfaces*

One parametrizes the flux q in the conservation law

$$\partial^\bullet u + u\nabla_\Gamma \cdot v + \nabla_\Gamma \cdot q = 0$$

by some nonlinear vector-valued function

$$q = f(x, u), \quad f = (f_1, \ldots, f_{n+1}),$$

where $f(x, s) \cdot \nu(x) = 0$ for all $x \in \Gamma(t)$ and all $s \in \mathbb{R}$ (see also (5.11)). Thus the nonlinearity f *has* to depend on the spatial variable x. In order to obtain bounded weak solutions to the conservation law, one has to assume that f is divergence-free with respect to x, $\nabla_\Gamma \cdot f(\cdot, s) = 0$ for all $s \in \mathbb{R}$. This then leads to the scalar conservation law

$$\partial^\bullet u + u\nabla_\Gamma \cdot v + \sum_{j=1}^{n+1} \frac{\partial f_j}{\partial u}(\cdot, u)\underline{D}_j u = 0 \tag{6.14}$$

on $\mathcal{G}_T$, and one has to impose the initial condition $u(\cdot, 0) = u_0$ on $\Gamma(0)$.

Using the parabolic regularization

$$\partial^\bullet u^\varepsilon + u^\varepsilon \nabla_\Gamma \cdot v + \sum_{j=1}^{n+1} \frac{\partial f_j}{\partial u}(\cdot, u^\varepsilon)\underline{D}_j u^\varepsilon - \varepsilon \nabla_\Gamma \cdot (\mathcal{A}\nabla_\Gamma u^\varepsilon) = 0,$$

Dziuk, Kröner and Müller (2012) prove existence and uniqueness for an entropy solution of (6.14). The PDE is discretized using the finite volume method.

6.5. Coupling of surface and bulk equations

In applications (see Section 10) one frequently encounters the coupling of surface and bulk processes. A model problem concerns the coupling of diffusion within a bulk domain to diffusion on the surface. Let Γ be the boundary of a bounded open domain in $\mathbb{R}^{n+1}$. Consider the coupled linear system to find $u\colon \Omega \times [0,T] \to \mathbb{R}$ and $v\colon \Gamma \times [0,T] \to \mathbb{R}$ such that

$$\begin{aligned} u_t - \Delta u + u &= f \quad \text{in } \Omega, \\ (\alpha u - \beta v) + \frac{\partial u}{\partial \nu} &= 0 \quad \text{on } \Gamma, \\ v_t - \Delta_\Gamma v + v + \frac{\partial u}{\partial \nu} &= g \quad \text{on } \Gamma. \end{aligned}$$

Here we assume that α and β are given positive constants and f and g are known functions on Ω and Γ respectively.

In applications it may be that one has nonlinear equations or a nonlinear coupling equation or time-dependent domains $\Omega(t)$ and $\Gamma(t)$. Such equations arise in the modelling of surfactants on fluid interfaces (see references cited in Section 10.2) and also in the study of diffusion within biological cells (Novak *et al.* 2007).

7. PDEs on implicit surfaces

The idea of this section is based on formulating a partial differential equation on a domain in the ambient bulk space which is equivalent to solving the surface equation on all level set hypersurfaces of a prescribed function ϕ: see Bertalmío *et al.* (2001). In order to do this we define ϕ-surface gradients by using a projection of the gradient in $\mathbb{R}^{n+1}$ onto the level surfaces of ϕ. These ϕ-surface gradients are used to define weak forms of surface elliptic operators and so generate weak formulations of surface elliptic and parabolic equations. The resulting elliptic operators are degenerate because the surface gradient is always tangential to the level surfaces and thus orthogonal to the normal. In Section 8 we describe approximation of the resulting degenerate equations by the finite element method.

7.1. ϕ-surface gradients, calculus and function spaces

We follow the development in Dziuk and Elliott (2008), Burger (2009) and Deckelnick, Dziuk, Elliott and Heine (2010). Suppose that $\Omega \subset \mathbb{R}^{n+1}$ is a bounded domain with a Lipschitz boundary and let $\phi \in C^2(\bar{\Omega})$ satisfy $\nabla\phi(x) \neq 0$, for $x \in \bar{\Omega}$. We say that ϕ is a non-degenerate level set function. For $r \in \mathbb{R}$ we denote by $\Gamma(r)$ the r-level set of ϕ given by

$$\Gamma(r) = \{x \in \Omega \mid \phi(x) = r\}.$$

It is a C^2-hypersurface if $\Gamma(r) \neq \emptyset$: see Section 2.2. We recall that $\nu : \bar{\Omega} \to \mathbb{R}^{n+1}$ and $H_\phi : \bar{\Omega} \to \mathbb{R}$, given by

$$\nu(x) = \frac{\nabla\phi(x)}{|\nabla\phi(x)|}, \quad H_\phi(x) = \sum_{j=1}^{n+1} \nu_{j,x_j}(x), \quad x \in \Omega,$$

when restricted to $\Gamma(r)$, are the unit normal field and the mean curvature of $\Gamma(r)$ respectively. Finally, we associate a ϕ-gradient with a differentiable function $f : \Omega \to \mathbb{R}$:

$$\nabla_\phi f(x) = \nabla f(x) - (\nabla f(x) \cdot \nu(x))\, \nu(x), \quad x \in \Omega.$$

Note that this is simply the tangential gradient (Definition 2.3) on level set surfaces since

$$(\nabla_\phi f)|_{\Gamma(r)} = \nabla_{\Gamma(r)} f|_{\Gamma(r)}$$

if $\Gamma(r) \neq \emptyset$. We may rewrite the ϕ-projected gradient in the following way:

$$\nabla_\phi f = P_\phi \nabla f, \quad P_\phi = I - \nu \otimes \nu.$$

It may be used to construct degenerate elliptic operators. For example, the ϕ-Laplacian applied to f is given by

$$\Delta_\phi f = \nabla_\phi \cdot \nabla_\phi f.$$

Degeneracy of this PDE is a consequence of $P_\phi \nu = 0$.

Lemma 7.1 (ϕ-integration by parts). Let Ω, ϕ be as above and let $f \in H^{1,1}(\Omega)$. Then

$$\int_\Omega \nabla_\phi f |\nabla\phi| \, \mathrm{d}x = \int_\Omega f H_\phi \nu |\nabla\phi| \, \mathrm{d}x + \int_{\partial\Omega} f(\nu_{\partial\Omega} - (\nu \cdot \nu_{\partial\Omega})\nu)|\nabla\phi| \, \mathrm{d}A,$$

where $\nu_{\partial\Omega}$ denotes the unit outer normal to $\partial\Omega$.

Let $f \in H^{1,2}(\Omega), g \in H^{1,2}(\Omega)^{n+1}$. Then

$$\int_\Omega \nabla_\phi \cdot (fg)|\nabla\phi| \, \mathrm{d}x = \int_\Omega f H_\phi g \cdot \nu |\nabla\phi| \, \mathrm{d}x + \int_{\partial\Omega} fg \cdot (\nu_{\partial\Omega} - (\nu \cdot \nu_{\partial\Omega})\nu)|\nabla\phi| \, \mathrm{d}A. \tag{7.1}$$

The boundary terms in the integration by parts formulae above disappear when $\nu = \nu_{\partial\Omega}$.

For a proof see Dziuk and Elliott (2008). The following lemma is proved in Deckelnick *et al.* (2010).

Lemma 7.2. Let $g : \Omega \to \mathbb{R}^{n+1}$ be a differentiable vector field with $g \cdot \nu = 0$ in Ω. Then

$$\nabla_\phi \cdot g = \frac{1}{|\nabla\phi|} \nabla \cdot (g\, |\nabla\phi|) \quad \text{in } \Omega.$$

In order to formulate variational problems we introduce some function spaces. To begin we define weak surface derivatives. For a function

$$f \in L^1_{\text{loc}}(\Omega),\ i \in \{1, \dots, n+1\},$$

we say that $g = \nabla_\phi f$ weakly if $g \in L^1_{\text{loc}}(\Omega, \mathbb{R}^{n+1})$, and

$$\int_\Omega f\, \nabla_\phi \zeta |\nabla\phi| \,\mathrm{d}x = -\int_\Omega g\,\zeta\, |\nabla\phi| \,\mathrm{d}x - \int_\Omega f\,\zeta\, H_\phi \nu\, |\nabla\phi| \,\mathrm{d}x \tag{7.2}$$

for all $\zeta \in C_0^\infty(\Omega)$. It is not difficult to see that (7.2) is equivalent to

$$\int_\Omega f\,(\nabla_\phi \zeta)_i \,\mathrm{d}x = -\int_\Omega (\nabla_\phi f)_i\, \zeta \,\mathrm{d}x + \int_\Omega f\,\zeta (h_i - H_\phi \nu_i) \,\mathrm{d}x \tag{7.3}$$

for all $\zeta \in C_0^\infty(\Omega)$, where $h_i = \nu_{i,x_j}\nu_j$ $(i = 1, \dots, n+1)$. For $1 \le p \le \infty$ we define

$$H^{1,p}_\phi(\Omega) = \{f \in L^p(\Omega) \mid \nabla_\phi f \in L^p(\Omega)^{n+1}\},$$

which we equip with the norm

$$\|f\|_{H^{1,p}_\phi(\Omega)} = \left(\int_\Omega |f|^p + |\nabla_\phi f|^p \,\mathrm{d}x\right)^{\frac{1}{p}}$$

for $p < \infty$ with the usual modification in the case $p = \infty$. Similarly we define the spaces $H^{k,p}_\phi(\Omega)$ for $k \in \mathbb{N}, k \ge 2$. Further, let $H^{0,p}_\phi(\Omega) = L^p(\Omega)$. We note that the spaces $H^{k,2}_\phi(\Omega)$ are Hilbert spaces.

In implicit approaches to surface equations a possible choice for the level set function ϕ is the signed distance function d to Γ (if it is available) and in that case $|\nabla\phi| = |\nabla d| = 1$ on Ω.

7.2. ϕ-elliptic equation

In this section we prove existence and regularity for solutions of a model equation. We consider the implicit surface equation

$$-\Delta_\phi u + cu = f \quad \text{in } \Omega.$$

It is convenient to consider domains Ω with the special form

$$\Omega = \{x \in \mathbb{R}^{n+1} \mid \alpha < \phi(x) < \beta\}, \quad -\infty < \alpha < \beta < \infty \tag{7.4}$$

and recall that $\Gamma(r) = \{x \in \Omega \mid \phi(x) = r\}$ is the r-level set of ϕ. We suppose that $\Gamma(\alpha)$ and $\Gamma(\beta)$ are closed hypersurfaces so that Ω is an annular domain.

Let Ω, ϕ be as above and assume that $f \in L^2(\Omega), c \in L^\infty(\Omega)$ with $c \ge \bar{c}$ a.e. in Ω where $\bar{c}$ is a positive constant.

Using Lemma 7.2 we may rewrite the equation in the form

$$-\nabla \cdot (P_\phi \nabla u\, |\nabla\phi|) + cu\, |\nabla\phi| = f\, |\nabla\phi| \quad \text{in } \Omega. \tag{7.5}$$

Multiplying by a test function, integrating over Ω and using integration by parts leads to the following variational problem: find $u \in H^{1,2}_\phi(\Omega)$ such that

$$\int_\Omega \nabla_\phi u \cdot \nabla_\phi \varphi \, |\nabla \phi| \, dx + \int_\Omega cu\varphi |\nabla\phi| \, dx = \int_\Omega f\varphi |\nabla\phi| \, dx \tag{7.6}$$

for all $\varphi \in H^{1,2}_\phi(\Omega)$. Note that the boundary term vanishes because the unit outer normal to $\partial\Omega$ points in the direction of $\nabla\phi$. It is clear from the developments above (see also Burger 2009) that (7.6) has a unique solution $u \in H^{1,2}_\phi(\Omega)$. The next result gives a characterization of functions in $H^{1,p}_\phi(\Omega)$ in terms of the spaces $H^{1,p}(\Gamma(r))$ (see Definition 2.11). Again from Deckelnick *et al.* (2010) we have the following result.

Lemma 7.3. Let $1 < p < \infty$, $u \in L^p(\Omega)$ and let Ω be of the form (7.4). Then $u \in H^{1,p}_\phi(\Omega)$ if and only if $u|_{\Gamma(r)} \in H^{1,p}(\Gamma(r))$ for almost all $r \in (\alpha, \beta)$ and the map $r \mapsto \|u\|_{H^{1,p}(\Gamma(r))}$ $(r \in (\alpha,\beta))$ is in $L^p(\alpha,\beta)$.

Using the co-area formula, (2.12), one can show that $u|_{\Gamma(r)}$ is the weak solution of

$$-\Delta_\Gamma u + cu = f \quad \text{on } \Gamma(r)$$

for almost all $r \in (\alpha,\beta)$. Since $f \in L^2(\Gamma(r))$ for almost all $r \in (\alpha,\beta)$, the regularity theory for elliptic partial differential equations on surfaces (see Theorem 3.3) implies that $u \in H^{2,2}(\Gamma(r))$ for almost all $r \in (\alpha,\beta)$ and

$$\|u\|_{H^{2,2}(\Gamma(r))} \le c\|f\|_{L^2(\Gamma(r))}.$$

Hence,

$$\int_\alpha^\beta \|u\|^2_{H^{2,2}(\Gamma(r))} \, dr \le c \int_\alpha^\beta \|f\|^2_{L^2(\Gamma(r))} \, dr = c \int_\Omega |f|^2 |\nabla\phi| \, dx < \infty,$$

so that Lemma 7.3 implies that $u \in H^{2,2}_\phi(\Omega)$ with

$$\|u\|_{H^{2,2}_\phi(\Omega)} \le c\|f\|_{L^2(\Omega)},$$

and we have proved the following theorem.

Theorem 7.4. Let $\Omega \subset \mathbb{R}^{n+1}$ satisfy (7.4) with $\phi \in C^2(\overline{\Omega})$, $\nabla\phi \neq 0$ on $\overline{\Omega}$. Then there exists a unique solution $u \in H^{2,2}_\phi(\Omega)$ of equation (7.6) and

$$\|u\|_{H^{2,2}_\phi(\Omega)} \le c\|f\|_{L^2(\Omega)}.$$

Further regularity results may be obtained. The following example comes from Deckelnick *et al.* (2010).

Theorem 7.5. In addition to the assumptions of Theorem 7.4, suppose that $\phi \in C^3(\overline{\Omega})$ and that the coefficients satisfy

$$\frac{\partial c}{\partial \nu} \in L^\infty(\Omega) \quad \text{and} \quad \frac{\partial f}{\partial \nu} \in L^2(\Omega).$$

Then $\frac{\partial u}{\partial \nu} \in H^{2,2}_\phi(\Omega')$ for all $\Omega' \subset\subset \Omega$, and

$$\left\|\frac{\partial u}{\partial \nu}\right\|_{H^{2,2}_\phi(\Omega')} \leq c\left(\|f\|_{L^2(\Omega)} + \left\|\frac{\partial f}{\partial \nu}\right\|_{L^2(\Omega)}\right).$$

The constant c depends on $\Omega, \Omega', \phi, \bar{c}$ and c.

Note that for a smooth function $g : \mathbb{R} \to \mathbb{R}$ we have that $u = g(\phi)$ is ϕ-harmonic, that is, $\Delta_\phi u = 0$.

7.3. ϕ-parabolic equation

Conservation and diffusion. We begin with describing a model for conservation and diffusion on level surfaces (see Dziuk and Elliott 2008). This is exactly analogous to the case of a single hypersurface (see Sections 5.2 and 6.1) and we proceed accordingly.

Let $\phi : \Omega \to \mathbb{R}$ be a prescribed non-degenerate level set function and let $\mathcal{Q} : \Omega \to \mathbb{R}^{n+1}$ be a given flux. Then the conservation law we consider is

$$\frac{\mathrm{d}}{\mathrm{d}t}\int_R |\nabla\phi| u \,\mathrm{d}x = -\int_{\partial R} \mathcal{Q}\cdot \nu_{\partial R}\,\mathrm{d}A$$

for each subdomain R of Ω where ν_R is the outward unit normal to ∂R. In particular we consider a flux of the form

$$\mathcal{Q} = |\nabla\phi|\; q_\phi,$$

where $q_\phi : \Omega \to \mathbb{R}^{n+1}$ is a flux satisfying

$$q_\phi \cdot \nu = 0. \tag{7.7}$$

Since

$$\frac{\mathrm{d}}{\mathrm{d}t}\int_R |\nabla\phi| u \,\mathrm{d}x = \int_R u_t |\nabla\phi| \,\mathrm{d}x$$

and

$$\int_{\partial R} q_\phi \cdot \nu_{\partial R} |\nabla\phi| \,\mathrm{d}x = \int_R \nabla_\phi \cdot q_\phi |\nabla\phi| \,\mathrm{d}x, \tag{7.8}$$

where we have used ϕ-integration by parts ((7.1), (7.7)), it follows that

$$\int_R (u_t + \nabla_\phi \cdot q_\phi)|\nabla\phi| \,\mathrm{d}x = 0$$

for every subdomain R, which implies the partial differential equation

$$u_t + \nabla_\phi \cdot q_\phi = 0 \quad \text{in } \Omega.$$

For the constitutive law, in a similar way to Section 6.1 we take q_ϕ to be a diffusive flux given by

$$q_\phi = -\mathcal{A}\nabla_\phi w, \tag{7.9}$$

where w is a field variable which will be defined in terms of u by a another constitutive relation. Again it is natural to take $\mathcal{A}$ to be a symmetric diffusion tensor with the property $\mathcal{A}\nu^\perp \cdot \nu = 0$ for every tangent vector $\nu^\perp$, that is, $\nu^\perp \cdot \nu = 0$, and for which there exists a $c_0 > 0$ such that

$$z \cdot \mathcal{A}z \geq c_0 z \cdot z, \quad \text{for all } z \in \mathbb{R}^{n+1},\ z \cdot \nu = 0.$$

Thus we obtain the diffusion equation

$$u_t - \nabla_\phi \cdot (\mathcal{A}\nabla_\phi w) = 0 \quad \text{on } \Omega. \tag{7.10}$$

Throughout the following we assume the initial condition $u(\cdot, 0) = u_0(\cdot)$.

Observe that (7.10) can be written as

$$u_t - P_\phi \nabla \cdot (\mathcal{A}P_\phi \nabla w) = 0,$$

which can be seen (for $w = u$) to be a degenerate parabolic equation because P_ϕ has a zero eigenvalue in the normal direction ν.

The variational form is then obtained by multiplying equation (7.10) by a test function η and also by $|\nabla\phi|$, and then integrating to obtain

$$\int_\Omega (u_t - \nabla_\phi \cdot (\mathcal{A}\nabla_\phi w))\, \eta\, |\nabla\phi|\, \mathrm{d}x = 0.$$

Integration by parts (7.1), together with the observation that $\mathcal{A}\nabla_\phi w \cdot \nu = 0$, gives

$$\int_\Omega \mathcal{A}\nabla_\phi w \cdot \nabla_\phi \eta |\nabla\phi|\, \mathrm{d}x = -\int_\Omega \nabla_\phi \cdot \mathcal{A}\nabla_\phi w \eta |\nabla\phi|\, \mathrm{d}x + \int_{\partial\Omega} \mathcal{A}\nabla_\phi w \cdot \nu_{\partial\Omega} \eta |\nabla\phi|\, \mathrm{d}A.$$

In order to proceed we need a boundary condition for w on $\partial\Omega$. It is natural to impose the zero flux condition

$$|\nabla\phi| \mathcal{A}\nabla_\phi w \cdot \nu_{\partial\Omega} = 0 \quad \text{on } \partial\Omega$$

and obtain the equivalent variational equation

$$\int_\Omega u_t \eta |\nabla\phi|\, \mathrm{d}x + \int_\Omega \mathcal{A}\nabla_\phi w \cdot \nabla_\phi \eta |\nabla\phi|\, \mathrm{d}x = 0, \quad \text{for all } \eta \in H^{1,2}_\phi(\Omega). \tag{7.11}$$

The solution of the equation satisfies conservation on each level surface. Let $\xi : \mathbb{R} \to \mathbb{R}$ be an arbitrary smooth function and set $\eta = \xi(\phi)$. Since $\nabla_\phi \eta = \xi'(\phi)\nabla_\phi \phi = 0$, we find the conservation equation

$$\frac{\mathrm{d}}{\mathrm{d}t} \int_\Omega u\xi(\phi)|\nabla\phi|\, \mathrm{d}x = 0.$$

It follows from the co-area formula that

$$\int_{\inf_\Omega \phi}^{\sup_\Omega \phi} \xi(r)\left(\int_{\Gamma(r)} u \, \mathrm{d}A\right) \mathrm{d}r = \int_{\inf_\Omega \phi}^{\sup_\Omega \phi} \xi(r)\left(\int_{\Gamma(r)} u_0 \, \mathrm{d}A\right) \mathrm{d}r$$

from which we infer that on each level surface $\Gamma(r) = \{x \mid \phi(x) = r\}$ of ϕ we have conservation,

$$\int_{\Gamma(r)} u \, \mathrm{d}A = \int_{\Gamma(r)} u_0 \, \mathrm{d}A.$$

Diffusion in a layered medium. Observing that

$$\nabla_\phi \cdot \tau = \frac{1}{|\nabla \phi|} \nabla \cdot (|\nabla \phi| \tau), \quad \text{for all } \tau \cdot \nu = 0,$$

we can rewrite the diffusion equation (7.10) as

$$|\nabla \phi| u_t = \nabla \cdot (\mathcal{A}_\phi \nabla w),$$

where $\mathcal{A}_\phi = |\nabla \phi| \mathcal{A} P_\phi$. Thus we may view (7.10) as the usual diffusion equation in $\mathbb{R}^{n+1}$ with very special forms of the diffusivity tensor and mass density. Since there is no diffusion in directions normal to the level surfaces of ϕ, we might interpret this as a diffusion equation for a stratified material whose layers are infinitesimally thin, tangential to the level surfaces of ϕ, and insulated from each other.

ϕ-heat equation. Setting $w = u$ and $\mathcal{A} = I$, we find the ϕ-heat equation on all level surfaces of ϕ:

$$u_t - \Delta_\phi u = 0.$$

The initial value problem for (7.11) becomes

$$\int_\Omega u_t \eta |\nabla \phi| \, \mathrm{d}x + \int_\Omega \nabla_\phi u \cdot \nabla_\phi \eta |\nabla \phi| \, \mathrm{d}x = 0 \quad \text{for all } \eta \in H^{1,2}_\phi(\Omega), \tag{7.12}$$

$$u(\cdot, 0) = u_0. \tag{7.13}$$

Setting $g(r) = \frac{1}{|\Gamma(r)|} \int_{\Gamma(r)} u_0 \, \mathrm{d}x$, we have that in the case of no-flux boundary conditions the long-time steady-state solution is $u_\infty = g(\phi)$.

Let u be a weak solution of (7.12) and (7.13). Then, choosing $\eta = u$ leads to

$$\frac{1}{2} \frac{\mathrm{d}}{\mathrm{d}t} \int_\Omega u^2 |\nabla \phi| \, \mathrm{d}x + \int_\Omega |\nabla_\phi u|^2 |\nabla \phi| \, \mathrm{d}x = 0$$

and choosing $\eta = u_t$ leads to

$$\int_\Omega u_t^2 |\nabla \phi| \, \mathrm{d}x + \frac{1}{2} \frac{\mathrm{d}}{\mathrm{d}t} \int_\Omega |\nabla_\phi u|^2 |\nabla \phi| \, \mathrm{d}x = 0.$$

7.4. *Examples of ϕ-equations*

It is straightforward to generate analogues of well-known equations.

Fourth-order linear diffusion. Setting $w = -\Delta_\phi u$ leads to the fourth-order linear diffusion equation

$$u_t + \nabla_\phi \cdot (\mathcal{A}\nabla_\phi \Delta_\phi u) = 0.$$

Nonlinear diffusion. Setting $w = f(u)$ and $\mathcal{A} = m(u)I$, we find the nonlinear diffusion equation

$$u_t - \nabla_\phi \cdot (K(u)\nabla_\phi u) = 0,$$

where $K(u) = m(u)f'(u)$. Linear diffusion and the porous medium equation on level sets are recovered by suitable choices of f and m.

Parabolic surface p-Laplacian equation. Setting $w = u$ and $\mathcal{A} = |\nabla_\phi u|^{p-2}I$ for $p > 1$ yields the parabolic surface p-Laplacian equation

$$u_t - \nabla_\phi \cdot (|\nabla_\phi u|^{p-2}\nabla_\phi u) = 0,$$

which is gradient flow for the energy

$$E_p(u) = \frac{1}{p}\int_\Omega |\nabla_\phi u|^p |\nabla\phi| \, \mathrm{d}x.$$

ϕ-Cahn–Hilliard equation. Setting

$$w = -\epsilon\Delta_\phi u + \frac{1}{\epsilon}\psi'(u),$$

where ψ is a double-well potential (*e.g.*, $\psi(u) = \frac{1}{4}(u^2 - 1)^2$), leads to the fourth-order Cahn–Hilliard equation on level sets:

$$u_t + \nabla_\phi \cdot \mathcal{A}\nabla_\phi\left(\epsilon\Delta_\phi u - \frac{1}{\epsilon}\psi'(u)\right) = 0.$$

ϕ-Allen–Cahn equation. Consideration of the L^2 gradient flow for the gradient energy functional,

$$E(v) = \int_\Omega \left\{\frac{\epsilon}{2}|\nabla_\phi v|^2 + \frac{1}{\epsilon}\psi(v)\right\}|\nabla\phi| \, \mathrm{d}x,$$

leads to an Allen–Cahn equation:

$$\epsilon u_t = \epsilon\Delta_\phi u - \frac{1}{\epsilon}\psi'(u).$$

7.5. *Eulerian approach to parabolic equations on moving surfaces*

We follow the development in Dziuk and Elliott (2010).

Notation
It is straightforward to extend the previous notation for fixed implicit surfaces to moving surfaces by using a time-dependent smooth level set function $\phi = \phi(x,t)$ $x \in \mathbb{R}^{n+1}$, $t \in [0,T]$. In particular, we suppose that for some $k \geq 3$ and some $0 < \alpha < 1$, $\phi \in C^1([0,T], C^{k,\alpha}(\overline{\Omega}))$. Thus, for each $t \in [0,T]$, $T > 0$, we let $\Gamma(t)$ be a compact smooth orientable hypersurface (without boundary) in $\mathbb{R}^{n+1}$ so that

$$\Gamma(t) = \{x \in \Omega \mid \phi(x,t) = 0\},$$

where Ω is a bounded domain in $\mathbb{R}^{n+1}$ with Lipschitz boundary $\partial\Omega$. In what follows we assume that ϕ satisfies the non-degeneracy condition $\nabla\phi \neq 0$ on $\Omega \times (0,T)$. We assume that $\partial\Omega \bigcap \Gamma(t)$ is empty and set $\nu_{\partial\Omega}$ to be the unit outward pointing normal to $\partial\Omega$. We set $\Omega_T = \Omega \times (0,T)$.

The orientation of $\Gamma(t)$ is set by taking the normal ν to Γ to be in the direction of increasing ϕ, yielding the normal vector field

$$\nu(x,t) = \frac{\nabla\phi(x,t)}{|\nabla\phi(x,t)|},$$

so that the normal ν_Γ to $\Gamma(t)$ is equal to $\nu|_{\Gamma(t)}$ and the normal velocity V of Γ is given by

$$V(x,t) = -\frac{\phi_t(x,t)}{|\nabla\phi(x,t)|}.$$

The material derivative and Leibniz formulae
Let $v : \Omega_T \to \mathbb{R}^{n+1}$ be a prescribed velocity field which has the decomposition

$$v = V\nu + v_\tau$$

into a normal velocity field $V = v \cdot \nu$ and a tangential velocity field v_τ orthogonal to ν and thus tangential to all level surfaces of ϕ. We use the standard notation for the material derivative of a scalar function $f = f(x,t)$ defined on Ω_T:

$$\partial^\bullet f = \frac{\partial f}{\partial t} + v \cdot \nabla f.$$

In particular, we note that $\partial^\bullet f$ restricted to a given level surface

$$\Gamma(r) = \{(x,t) \mid x \in \Omega, t \in (0,T), \phi(x,t) = r\}$$

depends only on the values of f on that level surface in space–time.

It is convenient to note that

$$\nabla_\phi \cdot v = VH_\phi + \nabla_\phi \cdot v_\tau$$

and

$$\nabla_\phi \cdot v = \operatorname{trace}(P_\phi \nabla v).$$

We have the following version of the transport formula, Theorem 5.1.

Lemma 7.6 (implicit surface transport formula). Let ϕ be a level set function and let f be an arbitrary function defined on Ω_T such that the following quantities exist. Then

$$\frac{\mathrm{d}}{\mathrm{d}t}\int_\Omega f|\nabla\phi|\,\mathrm{d}x = \int_\Omega (\partial^\bullet f + f\nabla_\phi\cdot v)|\nabla\phi|\,\mathrm{d}x - \int_{\partial\Omega} f\,v\cdot\nu_{\partial\Omega}|\nabla\phi|\,\mathrm{d}A. \tag{7.14}$$

Proof. Since $\partial_t|\nabla\phi| = \nu\cdot\nabla\phi_t$, a straightforward calculation yields

$$\frac{\mathrm{d}}{\mathrm{d}t}\int_\Omega f|\nabla\phi|\,\mathrm{d}x = \int_\Omega f_t|\nabla\phi| + f\nu\cdot\nabla\phi_t\,\mathrm{d}x.$$

Integration by parts on the second term in the integrand above gives

$$\frac{\mathrm{d}}{\mathrm{d}t}\int_\Omega f|\nabla\phi|\,\mathrm{d}x =$$
$$\int_\Omega |\nabla\phi|(f_t + V\nabla f\cdot\nu + fv\cdot\nu H_\phi)\,\mathrm{d}x - \int_{\partial\Omega} f\nu\cdot\nu_{\partial\Omega}V|\nabla\phi|\,\mathrm{d}A.$$

The ϕ-divergence theorem (7.1) gives

$$\int_\Omega fv\cdot\nu H_\phi|\nabla\phi|\,\mathrm{d}x =$$
$$\int_\Omega \nabla_\phi\cdot(fv)|\nabla\phi|\,\mathrm{d}x - \int_{\partial\Omega} fv\cdot(\nu_{\partial\Omega} - \nu\cdot\nu_{\partial\Omega}\nu)|\nabla\phi|\,\mathrm{d}A,$$

and observing that

$$V\nabla f\cdot\nu + \nabla_\phi\cdot(fv) = v\cdot\nabla f + f\nabla_\phi\cdot v$$

yields the desired result. □

Eulerian conservation and diffusion on moving surfaces

Let $\phi:\Omega_T\to\mathbb{R}$ be a prescribed non-degenerate level set function. Let $Q:\Omega_T\to\mathbb{R}^{n+1}$ be a given flux. Then the Eulerian conservation law we consider is

$$\frac{\mathrm{d}}{\mathrm{d}t}\int_R u|\nabla\phi|\,\mathrm{d}x = -\int_{\partial R}(Q + |\nabla\phi|uv)\cdot\nu_{\partial R}\,\mathrm{d}A$$

for each subdomain R of Ω, where $\nu_{\partial R}$ is the outward unit normal to ∂R. In particular, we consider a flux of the form

$$Q = |\nabla\phi|q_\phi,$$

where $q_\phi:\Omega_T\to\mathbb{R}^{n+1}$ is a flux satisfying

$$q_\phi\cdot\nu = 0. \tag{7.15}$$

It follows by the implicit surface transport formula (7.14) that

$$\frac{\mathrm{d}}{\mathrm{d}t}\int_R u|\nabla\phi|\,\mathrm{d}x = \int_R (\partial^\bullet u + u\nabla_\phi\cdot v)|\nabla\phi|\,\mathrm{d}x - \int_{\partial R} uv\cdot\nu_{\partial R}|\nabla\phi|\,\mathrm{d}A$$

and by ϕ-integration by parts and (7.15) that

$$\int_{\partial R} q_\phi\cdot\nu_{\partial R}|\nabla\phi|\,\mathrm{d}A = \int_R \nabla_\phi\cdot q_\phi|\nabla\phi|\,\mathrm{d}x.$$

It follows that

$$\int_R |\nabla\phi|(\partial^\bullet u + u\nabla_\phi\cdot v + \nabla_\phi\cdot q_\phi)\,\mathrm{d}x = 0$$

for every subdomain R, which implies the partial differential equation

$$\partial^\bullet u + u\nabla_\phi\cdot v + \nabla_\phi\cdot q_\phi = 0 \quad \text{in } \Omega_T.$$

Taking q_ϕ to be the diffusive flux $q_\phi = -\mathcal{A}\nabla_\phi u$ leads to the diffusion equation

$$\partial^\bullet u + u\nabla_\phi\cdot v - \nabla_\phi\cdot(\mathcal{A}\nabla_\phi u) = 0. \tag{7.16}$$

Here $\mathcal{A}\geq 0$ again is a symmetric mobility tensor with the property that it maps the tangent space $\mathcal{T} = \{\nu^\perp\in\mathbb{R}^{n+1} \mid \nu\cdot\nu^\perp = 0\}$ into itself. Observe that (7.16) is a linear degenerate parabolic equation because P_ϕ has a zero eigenvalue in the normal direction ν.

Another form of this PDE is

$$u_t + V\frac{\partial u}{\partial\nu} + \nabla_\phi\cdot(uv_\tau) + VH_\phi u - \nabla_\phi\cdot(\mathcal{A}\nabla_\phi u) = 0. \tag{7.17}$$

A variational form of (7.16) is obtained in the standard way. In order to proceed we need a boundary condition for u on $\partial\Omega$. For convenience here, we impose the zero flux condition

$$|\nabla\phi|\mathcal{A}\nabla_\phi u\cdot\nu_{\partial\Omega} = 0 \quad \text{on } \partial\Omega\times(0,T) \tag{7.18}$$

and assume that

$$v\cdot\nu_{\partial\Omega} = 0 \quad \text{on } \partial\Omega\times(0,T). \tag{7.19}$$

For each level surface of ϕ we multiply equation (7.16) by a test function η and by $|\nabla\phi|$ to give

$$\int_\Omega (\partial^\bullet u + u\nabla_\phi\cdot v - \nabla_\phi\cdot(\mathcal{A}\nabla_\phi u))\eta|\nabla\phi|\,\mathrm{d}x = 0.$$

The transport formula (7.14) gives

$$\frac{\mathrm{d}}{\mathrm{d}t}\int_\Omega u\eta|\nabla\phi|\,\mathrm{d}x =$$
$$\int_\Omega (\partial^\bullet u + u\nabla_\phi\cdot v)\eta|\nabla\phi|\,\mathrm{d}x + \int_\Omega u\partial^\bullet\eta|\nabla\phi|\,\mathrm{d}x - \int_{\partial\Omega} u\eta v\cdot\nu_{\partial\Omega}|\nabla\phi|\,\mathrm{d}A,$$

and because of $\mathcal{A}\nabla_\phi u \cdot \nu = 0$, integration by parts gives

$$\int_\Omega \mathcal{A}\nabla_\phi u \cdot \nabla_\phi \eta |\nabla\phi| \,\mathrm{d}x = \\ -\int_\Omega \eta \nabla_\phi \cdot (\mathcal{A}\nabla_\phi u)|\nabla\phi| \,\mathrm{d}x + \int_{\partial\Omega} \mathcal{A}\nabla_\phi u \cdot \nu_{\partial\Omega} \eta |\nabla\phi| \,\mathrm{d}A.$$

Finally we obtain the variational equation

$$\frac{\mathrm{d}}{\mathrm{d}t}\int_\Omega u\eta|\nabla\phi| \,\mathrm{d}x + \int_\Omega \mathcal{A}\nabla_\phi u \cdot \nabla_\phi \eta |\nabla\phi| \,\mathrm{d}x = \int_\Omega u\partial^\bullet \eta |\nabla\phi| \,\mathrm{d}x \tag{7.20}$$

for each sufficiently smooth test function.

The conservation equation

$$\frac{\mathrm{d}}{\mathrm{d}t}\int_\Omega u|\nabla\phi| \,\mathrm{d}x = 0$$

is obtained by taking $\eta = 1$.

Weak solution and energy estimate

The variational form (7.20) allows us to define a notion of weak solution of the degenerate Eulerian parabolic equation (7.16). To do this we introduce the normed linear spaces

$$L^2_\phi(\Omega_T) = \{\eta \mid \eta \text{ is measurable}, \|\eta\|_{L^2_\phi(\Omega_T)} < \infty\},$$
$$\|\eta\|^2_{L^2_\phi(\Omega_T)} = \int_0^T \int_\Omega \eta^2 |\nabla\phi| \,\mathrm{d}x \,\mathrm{d}t$$

and

$$H^1_\phi(\Omega_T) = \{\eta \in L^2_\phi(\Omega_T) \mid \partial^\bullet\eta, \nabla_\phi\eta \in L^2_\phi(\Omega_T)\},$$
$$\|\eta\|^2_{H^1_\phi(\Omega_T)} = \|\eta\|^2_{L^2_\phi(\Omega_T)} + \|\partial^\bullet\eta\|^2_{L^2_\phi(\Omega_T)} + \|\nabla_\phi\eta\|^2_{L^2_\phi(\Omega_T)}.$$

Definition 7.7. A function $u \in H^1_\phi(\Omega_T)$ is said to be a weak solution of (7.16) and (7.18) if, for almost every $t \in (0, T)$,

$$\frac{\mathrm{d}}{\mathrm{d}t}\int_\Omega u\eta|\nabla\phi| \,\mathrm{d}x + \int_\Omega \mathcal{A}\nabla_\phi u \cdot \nabla_\phi \eta |\nabla\phi| \,\mathrm{d}x = \int_\Omega u\partial^\bullet \eta |\nabla\phi| \,\mathrm{d}x \tag{7.21}$$

for every $\eta \in H^1_\phi(\Omega_T)$.

Weak solutions satisfy the following basic energy estimate, whose discrete counterpart implies stability for the numerical scheme. Throughout we will assume the initial condition $u(\cdot, 0) = u_0$ on Ω.

Lemma 7.8. Let u satisfy (7.21). Then

$$\frac{1}{2}\frac{\mathrm{d}}{\mathrm{d}t}\int_\Omega u^2|\nabla\phi| \,\mathrm{d}x + \int_\Omega \mathcal{A}\nabla_\phi u \cdot \nabla_\phi u |\nabla\phi| \,\mathrm{d}x + \frac{1}{2}\int_\Omega u^2 \,\nabla_\phi \cdot v|\nabla\phi| \,\mathrm{d}x = 0.$$

Proof. We choose $\eta = u$ in (7.21) and obtain

$$\begin{aligned}\frac{\mathrm{d}}{\mathrm{d}t}\int_\Omega u^2|\nabla\phi|\,\mathrm{d}x + \int_\Omega \mathcal{A}\nabla_\phi u\cdot\nabla_\phi u|\nabla\phi|\,\mathrm{d}x &= \int_\Omega u\partial^\bullet u|\nabla\phi|\,\mathrm{d}x\\ &= \frac{1}{2}\int_\Omega \partial^\bullet(u^2)\,|\nabla\phi|\,\mathrm{d}x\\ &= \frac{1}{2}\frac{\mathrm{d}}{\mathrm{d}t}\int_\Omega u^2|\nabla\phi|\,\mathrm{d}x - \frac{1}{2}\int_\Omega u^2\,\nabla_\phi\cdot v|\nabla\phi|\,\mathrm{d}x + \frac{1}{2}\int_{\partial\Omega} u^2 v\cdot\nu_{\partial\Omega}|\nabla\phi|\,\mathrm{d}A,\end{aligned}$$

where the last term vanishes because of (7.19), and this was the claim. □

8. Implicit surface finite element method

In this section we describe the numerical approximation of the implicit surface ϕ-equations of Section 7. For computational purposes it is natural to seek to exploit these formulations by using a bulk triangulation and finite element space independent of the level set function. Of course, we may be interested only in the solution on just one level set of ϕ which, for convenience, we choose to be the zero level set and label it as Γ. Using this approach avoids the necessity of constructing a surface mesh, and has the potential advantage of using a bulk finite element mesh unaligned to the surface Γ. However, the resulting equation is then to be solved in a space one dimension higher. In this section we describe a naive approach in which the solution to a surface PDE is computed on all level surfaces of the prescribed level set function, ϕ, in a bulk domain Ω using a finite element space defined on a triangulation of Ω. In Section 9.1 we will localize to a narrow band.

8.1. Finite element scheme

We suppose that the level set function satisfies $\nabla\phi \neq 0$ in Ω. For simplicity we consider the variational model equations

$$\int_\Omega \nabla_\phi u\cdot\nabla_\phi\eta|\nabla\phi|\,\mathrm{d}x + \int_\Omega cu\eta|\nabla\phi|\,\mathrm{d}x = \int_\Omega f\eta|\nabla\phi|\,\mathrm{d}x, \tag{8.1}$$

for all $\eta \in H^{1,2}_\phi(\Omega)$, and

$$\int_\Omega u_t\eta|\nabla\phi|\,\mathrm{d}x + \int_\Omega \nabla_\phi u\cdot\nabla_\phi\eta|\nabla\phi|\,\mathrm{d}x = 0, \quad \text{for all } \eta\in H^{1,2}_\phi(\Omega),$$
$$u(\cdot,0) = u_0.$$

Here the data c, f and u_0 are defined on a domain $\Omega\subset\mathbb{R}^{n+1}$ containing the zero level set Γ of ϕ. The data may be interpreted as suitable extensions of data given on Γ. Note that implicitly we have assumed the zero flux boundary condition

$$|\nabla\phi|\mathcal{A}\nabla_\phi u\cdot\nu_{\partial\Omega} = 0 \quad \text{on } \partial\Omega. \tag{8.2}$$

If Ω satisfies (7.4) then this is automatic, and all level surfaces of ϕ are closed in Ω. Otherwise it is possible that level surfaces of ϕ intersect $\partial\Omega$ and in that case a no-flux boundary condition has been imposed.

Let $\mathcal{T}_h$ be a triangulation of Ω and let $h = \max_{T\in\mathcal{T}_h} \operatorname{diam}(T)$ be the maximum mesh size. For ease of exposition we suppose that elements adjacent to the boundary $\partial\Omega$ have a curved edge so the triangulation of Ω is exact. Let S_h be the space of continuous piecewise linear finite element functions on Ω,

$$S_h = \{\eta_h \in C^0(\overline{\Omega}) \mid \eta_h|_T \text{ is a linear polynomial, } T \in \mathcal{T}_h\},$$

generated by the nodal basis functions, that is, $S_h = \operatorname{span}\{\chi_1, \dots, \chi_J\}$.

The natural finite element method for the elliptic equation (8.1) is to find $U_h \in S_h$ such that

$$\int_\Omega \nabla_\phi U_h \cdot \nabla_\phi \eta_h |\nabla\phi| \, \mathrm{d}x + \int_\Omega c U_h \eta_h |\nabla\phi| \, \mathrm{d}x = \int_\Omega f \eta_h |\nabla\phi| \, \mathrm{d}x,$$

for all $\eta_h \in S_h$. Setting $U_h = \sum_{j=1}^J \alpha_j \chi_j$, we find

$$(\mathcal{S} + \mathcal{C})\alpha = \mathcal{F},$$

where the weighted stiffness matrix $\mathcal{S}$ and the matrix $\mathcal{C}$ are given by

$$\mathcal{S}_{jk} = \int_\Omega \nabla_\phi \chi_j \cdot \nabla_\phi \chi_k |\nabla\phi| \, \mathrm{d}x, \quad \mathcal{C}_{jk} = \int_\Omega c \chi_j \chi_k |\nabla\phi| \, \mathrm{d}x$$

for $j, k = 1, \dots, J$, and $\mathcal{F}$ is given by $\mathcal{F} = (\mathcal{F}_1, \dots, \mathcal{F}_J)$,

$$\mathcal{F}_j = \int_\Omega f \chi_j |\nabla\phi| \, \mathrm{d}x \quad (j = 1, \dots, J).$$

Similarly, a semidiscrete Eulerian method for the parabolic equation is to find $U_h(\cdot, t) \in S_h$ such that

$$\int_\Omega U_{ht} \eta_h |\nabla\phi| \, \mathrm{d}x + \int_\Omega \nabla_\phi U_h \cdot \nabla_\phi \eta_h |\nabla\phi| \, \mathrm{d}x = 0, \quad \text{for all } \eta_h \in S_h.$$

Setting $U_h(\cdot, t) = \sum_{j=1}^J \alpha_j(t) \chi_j$ and $\alpha = (\alpha_1, \dots, \alpha_J)$ yields

$$\int_\Omega \sum_{j=1}^J \alpha_{j,t} \chi_j \eta_h |\nabla\phi| \, \mathrm{d}x + \int_\Omega \sum_{j=1}^J \alpha_j(t) \nabla_\phi \chi_j \cdot \nabla_\phi \eta_h |\nabla\phi| \, \mathrm{d}x = 0,$$

for all $\eta_h \in S_h$, and taking $\eta_h = \chi_k$ for $k = 1, \dots, J$, we obtain

$$\mathcal{M}\dot{\alpha} + \mathcal{S}\alpha = 0,$$

where $\mathcal{M}$ is the weighted mass matrix

$$\mathcal{M}_{jk} = \int_\Omega \chi_j \chi_k |\nabla\phi| \, \mathrm{d}x, \quad j, k = 1, \dots, J.$$

A time discretization for this spatially discrete parabolic equation may be carried out by standard methods. As an example consider the implicit Euler method. Let $\tau > 0$ be a time step size, $m_T\tau = T$, and let us use upper indices for the time levels so that U_h^m denotes $U_h(\cdot, m\tau)$. With this notation we propose the following method.

Algorithm 8.1. Given $U_h^0 \in S_h$, for $m = 0, \ldots, m_T$, solve the linear system

$$\frac{1}{\tau}\int_\Omega U_h^{m+1}\eta_h|\nabla\phi|\,\mathrm{d}x + \int_\Omega \nabla_\phi U_h^{m+1}\cdot\nabla_\phi\eta_h|\nabla\phi|\,\mathrm{d}x = \frac{1}{\tau}\int_\Omega U_h^m\eta_h|\nabla\phi|\,\mathrm{d}x,$$

for all $\eta_h \in S_h$.

This leads to the linear algebraic system

$$(\mathcal{M} + \tau\mathcal{S})\alpha^{m+1} = \mathcal{M}\alpha^m. \tag{8.3}$$

Because of the assumption on ϕ the matrices $\mathcal{C}$ and $\mathcal{M}(t)$ are uniformly positive definite, so that, in each case, we get existence and uniqueness of the finite element solutions. A significant feature of our approach is the fact that the matrices $\mathcal{M}$, $\mathcal{C}$ and $\mathcal{S}$ depend only on the evaluation of the gradient of level set function ϕ and not on explicit numerical evaluation of surface quantities.

Numerical experiments concerning parabolic equations may be found in Dziuk and Elliott (2008). We refer to Bertalmío *et al.* (2001), Greer *et al.* (2006), Greer (2006) and Burger (2009) for further descriptions, applications and extensions of these methods.

8.2. Eulerian scheme for conservation and diffusion on moving surfaces

Using implicit descriptions of evolving surfaces, we formulate an Eulerian ESFEM which is based on the weak form (7.20) of the ϕ-diffusion equation and uses fixed-in-time finite element spaces as in Section 8.1. The test functions will now satisfy $\partial^\bullet\eta_h = v\cdot\nabla\eta_h$.

Definition 8.2 (semidiscretization in space). Find $U_h(\cdot,t) \in S_h$ such that

$$\begin{aligned}\frac{\mathrm{d}}{\mathrm{d}t}\int_\Omega U_h\eta_h|\nabla\phi|\,\mathrm{d}x &+ \int_\Omega \mathcal{A}\nabla_\phi U_h\cdot\nabla_\phi\eta_h|\nabla\phi|\,\mathrm{d}x\\ &= \int_\Omega U_h v\cdot\nabla\eta_h|\nabla\phi|\,\mathrm{d}x, \quad \text{for all } \eta_h \in S_h.\end{aligned}$$

Using the Leibniz formula it is easily seen that an equivalent formulation is

$$\int_\Omega \partial^\bullet U_h\eta_h|\nabla\phi|\,\mathrm{d}x + \int_\Omega U_h\eta_h\nabla_\phi\cdot v|\nabla\phi|\,\mathrm{d}x + \int_\Omega \mathcal{A}\nabla_\phi U_h\cdot\nabla_\phi\eta_h|\nabla\phi|\,\mathrm{d}x = 0$$

for all discrete test functions $\eta_h \in S_h$. Setting $U_h(\cdot, t) = \sum_{j=1}^{J} \alpha_j(t)\chi_j$, we find that

$$\frac{\mathrm{d}}{\mathrm{d}t}\left(\sum_{j=1}^{J} \alpha_j \int_\Omega \chi_j \eta_h |\nabla\phi| \,\mathrm{d}x\right) + \sum_{j=1}^{J} \alpha_j \int_\Omega \mathcal{A}\nabla_\phi \chi_j \cdot \nabla_\phi \eta_h |\nabla\phi| \,\mathrm{d}x$$
$$= \sum_{j=1}^{J} \alpha_j \int_\Omega \chi_j v \cdot \nabla \eta_h |\nabla\phi| \,\mathrm{d}x$$

for all $\eta_h \in S_h$, and taking $\eta_h = \chi_k$, $k = 1, \ldots, J$, we obtain

$$\frac{\mathrm{d}}{\mathrm{d}t}(\mathcal{M}(t)\alpha) + \mathcal{S}(t)\alpha = \mathcal{C}(t)\alpha.$$

In the above $\mathcal{M}(t)$ is the evolving mass matrix,

$$\mathcal{M}(t)_{jk} = \int_\Omega \chi_j \chi_k |\nabla\phi| \,\mathrm{d}x,$$

$\mathcal{C}(t)$ is a transport matrix,

$$\mathcal{C}(t)_{jk} = \int_\Omega \chi_j v \cdot \nabla \chi_k |\nabla\phi| \,\mathrm{d}x,$$

and $\mathcal{S}(t)$ is the evolving stiffness matrix,

$$\mathcal{S}(t)_{jk} = \int_\Omega \mathcal{A}\nabla_\phi \chi_j \cdot \nabla_\phi \chi_k |\nabla\phi| \,\mathrm{d}x$$

$(j, k = 1, \ldots, J)$. Existence and uniqueness of the semidiscrete finite element solution follows easily since the mass matrix $\mathcal{M}(t)$ is uniformly positive definite on $[0, T]$ and the other matrices are bounded.

Again a significant feature of our approach is the fact that the matrices $\mathcal{M}(t)$, $\mathcal{C}(t)$ and $\mathcal{S}(t)$ depend only on the evaluation of the gradient of the level set function ϕ and the velocity field v (in the case of $\mathcal{C}$). The method does not require a numerical evaluation of the curvature: see (7.17).

Stability

The basic stability result for our spatially discrete scheme from Definition 8.2 is given in the following lemma, which follows analogously to the continuous case of Lemma 7.8 from the transport formula in Lemma 7.6 and a standard Gronwall argument.

Lemma 8.3 (stability). Let U_h be a solution of the semidiscrete scheme as in Definition 8.2 with initial value $U_h(\cdot, 0) = U_{h0}$. Assume that

$$\int_0^T \|\nabla_\phi \cdot v\|_{L^\infty(\Omega)} \,\mathrm{d}t < \infty,$$

and that $\mathcal{A}$ satisfies the ellipticity condition (5.14). Then the following stability estimate holds:

$$\sup_{(0,T)} \|U_h\|^2_{L^2_\phi(\Omega)} + \int_0^T \|\nabla_\phi U_h\|^2_{L^2_\phi(\Omega)} \,\mathrm{d}t \leq c\|U_{h0}\|^2_{L^2_\phi(\Omega)}. \tag{8.4}$$

Numerical examples illustrating the scheme may be found in Dziuk and Elliott (2010). Level set methods for approaching this problem which do not use this variational approach and which use finite difference methods on the equation (7.17) were formulated in Adalsteinsson and Sethian (2003) and Xu and Zhao (2003).

8.3. Examples of higher-order equations

We formulate mixed finite element schemes based on the splitting into second-order elliptic operators: see Elliott *et al.* (1989) and Copetti and Elliott (1992).

Cahn–Hilliard equation. Find $(U_h(\cdot,t), W_h(\cdot,t)) \in (S_h)^2$ such that

$$\int_\Omega U_{ht}\eta_h|\nabla\phi| \,\mathrm{d}x + \int_\Omega \mathcal{A}\nabla_\phi W_h \cdot \nabla_\phi \eta_h |\nabla\phi| \,\mathrm{d}x = 0,$$

$$\varepsilon\int_\Omega \nabla_\phi U_h \cdot \nabla_\phi \eta_h|\nabla\phi| \,\mathrm{d}x + \int_\Omega \frac{1}{\varepsilon} I_h(\psi'(U_h))\eta_h|\nabla\phi| \,\mathrm{d}x - \int_\Omega W_h \eta_h |\nabla\phi| \,\mathrm{d}x = 0,$$

for every $\eta_h \in S_h$. Here we use I_h to denote the interpolation operator for S_h. Setting

$$U_h(\cdot,t) = \sum_{j=1}^J \alpha_j(t)\chi_j(\cdot), \quad W_h(\cdot,t) = \sum_{j=1}^J \beta_j(t)\chi_j(\cdot), \tag{8.5}$$

we find that, for all $\eta \in S_h$,

$$\int_\Omega \sum_{j=1}^J \dot\alpha_j \chi_j \eta_h |\nabla\phi| \,\mathrm{d}x + \int_\Omega \mathcal{A}\sum_{j=1}^J \beta_j \nabla_\phi\chi_j \cdot \nabla_\phi\eta_h|\nabla\phi| \,\mathrm{d}x = 0,$$

$$\varepsilon\int_\Omega \sum_{j=1}^J \alpha_j \nabla_\phi\chi_j \cdot \nabla_\phi\eta_h|\nabla\phi| \,\mathrm{d}x + \frac{1}{\varepsilon}\int_\Omega \sum_{j=1}^J \psi'(\alpha_j)\chi_j\eta_h|\nabla\phi| \,\mathrm{d}x - \int_\Omega \sum_{j=1}^J \beta_j\chi_j\eta_h|\nabla\phi| \,\mathrm{d}x = 0,$$

and taking $\eta_h = \chi_k$ for $k = 1, \dots, J$ we obtain

$$\mathcal{M}\dot\alpha + \mathcal{S}\beta = 0, \quad \varepsilon\mathcal{S}^0\alpha + \frac{1}{\varepsilon}\mathcal{M}\psi'(\alpha) - \mathcal{M}\beta = 0,$$

where $\psi'(\alpha) = (\psi'(\alpha_1), \dots, \psi'(\alpha_J))$, which yields

$$\mathcal{M}\dot{\alpha} + \varepsilon \mathcal{S}\mathcal{M}^{-1}\mathcal{S}^0\alpha + \frac{1}{\varepsilon}\mathcal{S}\psi'(\alpha) = 0.$$

Of course, the method for fourth-order linear diffusion is recovered by taking ψ to vanish.

9. Unfitted bulk finite element method

In some applications it may be convenient to avoid triangulating a surface. This may be of particular value when the surface partial differential equation is coupled to a bulk system of equations and when the surface is described as a level set function or approximated by the zero level set of a phase field function and/or may be evolving. Further, it may be appropriate when one wishes to take advantage of existing bulk finite element codes. A common feature of a number of approaches is to use a bulk triangulation of a domain in the higher-dimensional ambient space on which bulk finite element spaces are used. The surface quantity is then approximated using the bulk finite element space and an appropriate discretization of the surface partial differential equation. Of course, the Eulerian approach described in Section 8 is of this form, but there the computational approach is to approximate the equation on all level sets of a prescribed level set function. From the practical perspective the solution to a surface PDE will often be required for just one surface. Thus it is desirable to localize the implicit surface approach.

There are a number of variants which we discuss below. In Section 9.1 we describe the method of Deckelnick *et al.* (2010) for a surface elliptic equation which uses a narrow computational band around the given surface in which the ϕ-elliptic equation is approximated. It is natural to consider varying the width of the narrow band in the above approach, and it transpires that one obtains a new method by taking the width to zero: see Section 9.2. This leads to the interesting method proposed by Olshanskii, Reusken and Grande (2009); see also Olshanskii and Reusken (2010) and Demlow and Olshanskii (2012). In this approach bulk finite element spaces are used in a weak form of the surface equation based on tangential gradients and integration on an approximation of the surface. An extension of this method (Deckelnick, Elliott and Ranner 2013) uses the full gradient in the weak formulation. This is also described in Section 9.2.

In Figure 9.1(a) the surface is a smooth curve that is the zero level set of a function ϕ. The triangulation is completely independent of this curve. Another piecewise linear curve approximating the desired surface is shown, which passes through black points. This is the computational surface: it is the zero level set of the bulk finite element piecewise linear interpolation of ϕ. The dark triangles are those triangles which are intersected by the

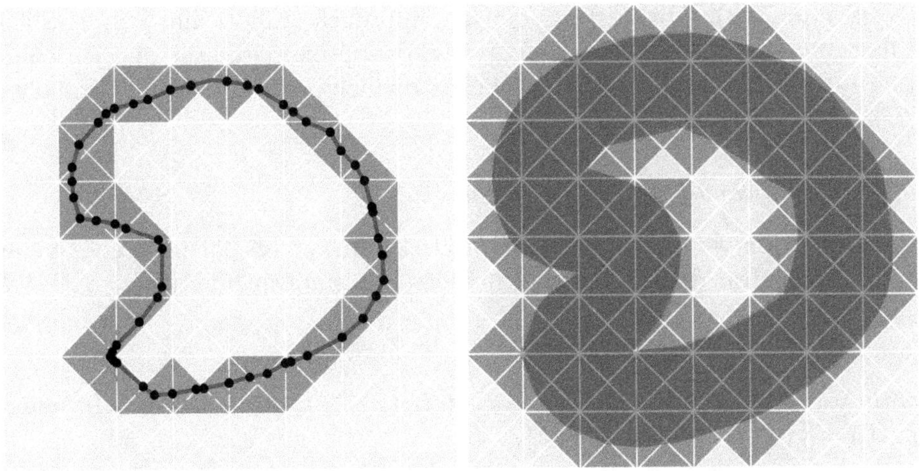

(a) (b)

Figure 9.1. Unfitted bulk finite element meshes.

approximate surface. It is on these triangles that a bulk finite element space is employed to find an approximation to the solution of the surface equation using a variational formulation which involves integration on the computational surface. In Figure 9.1(b) we see a set-up related to the narrow band method described in Section 9.1. Again the dark triangles are associated with the bulk finite element solution space, but now the domain of integration for the variational formulation is the narrow band defined by $\pm\delta$ level sets of the interpolation of the level set function describing the exact surface.

Meshes that form the computational domain but which are not fitted to the domain in which the PDE holds give rise to *unfitted finite element methods*, introduced in Barrett and Elliott (1982, 1984) for elliptic equations in curved domains. The motivation for using finite element spaces on meshes not fitting to the domain came from the desire to solve free or moving boundary problems; see also Barrett and Elliott (1985, 1987*a*, 1987*b*, 1988), Hansbo and Hansbo (2002), Bastian and Engwer (2009) and Engwer and Heimann (2012). In this setting we are concerned with bulk meshes independent of the surface.

Another approach involving bulk unfitted meshes arises from phase field methodology. Phase field models (see Caginalp 1989 and Deckelnick *et al.* 2005, for example) lead to an approximation of the interface by a *diffuse transition layer*, usually of width ϵ, across which an order or phase field parameter ranges from -1 to 1. In this setting it is natural to consider formulating a method for solving surface equations in the diffuse layer (see Section 9.3). This *diffuse interface* approach (Elliott and Stinner 2009, Elliott *et al.* 2011, Rätz and Voigt 2006) approximates the surface equation by a bulk equation with coefficients which are zero or small outside a transition layer.

9.1. Narrow band bulk finite element scheme

In this section we follow the work of Deckelnick *et al.* (2010). Let Ω be an open polyhedral domain for which there is a smooth function $\phi : \overline{\Omega} \to \mathbb{R}$ such that

$$\Gamma = \{x \in \Omega \mid \phi(x) = 0\}$$

with $\nabla\phi \neq 0$ on $\overline{\Omega}$, and assume that $U_\delta(\Gamma) \subset \Omega$ for some $\delta > 0$, where $U_\delta(\Gamma) = \{x \in \Omega \mid \operatorname{dist}(x, \Gamma) < \delta\}$.

Variational form. We consider the linear PDE

$$-\Delta_\Gamma u + u = f \tag{9.1}$$

on a smooth compact surface Γ, and seek to solve the weak form of (9.1): find $u \in H^1(\Gamma)$ such that

$$\int_\Gamma \nabla_\Gamma u \cdot \nabla_\Gamma \eta + u\eta \,\mathrm{d}A = \int_\Gamma f\eta \,\mathrm{d}A, \quad \text{for all } \eta \in H^1(\Gamma), \tag{9.2}$$

with $f \in L^2(\Gamma)$ given.

For a function $v : \Gamma \to \mathbb{R}$, we define the unique closest point extension of v to U_δ by $v^e : U_\delta \to \mathbb{R}$ given by $v^e(x) = v(a(x))$, where $a(x) \in \Gamma$ is the closest point on Γ to $x \in U_\delta$ and δ is chosen sufficiently small. For the notation see Lemma 2.8. Let $\mathcal{T}_h$ be a quasi-uniform triangulation of Ω with maximum mesh size $h = \max_{T\in\mathcal{T}_h} \operatorname{diam}(T)$. The nodes of the triangulation and corresponding linear basis functions are denoted by $X_1, \ldots, X_N$ and by $\chi_1, \ldots, \chi_N$, that is, $\chi_i \in C^0(\bar{\Omega}), \chi_i|_T \in \mathbb{P}_1(T)$ for all $T \in \mathcal{T}_h$ satisfying $\chi_i(X_j) = \delta_{ij}$ for all $i, j = 1, \ldots, J$.

Let

$$\phi_h = I_h\phi = \sum_{i=1}^N \phi(X_i)\chi_i$$

be the usual Lagrange interpolation of a function ϕ. In view of the smoothness of ϕ we have

$$\|\phi - \phi_h\|_{L^\infty(\Omega)} + h\|\nabla\phi - \nabla\phi_h\|_{L^\infty(\Omega)} \leq Ch^2,$$

from which we infer in particular that $|\nabla\phi_h| \geq c_0 > 0$ in $\bar{\Omega}$ for $0 < h \leq h_0$. Hence we can use, element by element, the piecewise constant projection $P_h = I - \nu_h \otimes \nu_h$, where

$$\nu_h = \frac{\nabla\phi_h}{|\nabla\phi_h|}$$

is the element-by-element piecewise constant normal.

Given $\gamma > 0$, the computational domain is taken as a narrow band defined by the γh levels of the interpolant of ϕ, that is,

$$D_h = \{x \in \Omega \,|\, |\phi_h(x)| < \gamma h\}.$$

Let $\mathcal{T}_h^C = \{T \in \mathcal{T}_h \mid T \cap D_h \neq \emptyset\}$ denote the computational elements, and consider the nodes of all simplices T belonging to $\mathcal{T}_h^C$. After relabelling we may assume that these nodes are given by $X_1, \ldots, X_J$. Let $S_h = \operatorname{span}\{\chi_1, \ldots, \chi_J\}$. The discrete problem now reads: find $U_h \in S_h$ such that

$$\int_{D_h} P_h \nabla U_h \cdot \nabla \eta_h \, |\nabla \phi_h| \, \mathrm{d}x + \int_{D_h} U_h \, \eta_h \, |\nabla \phi_h| \, \mathrm{d}x = \int_{D_h} f^e \, \eta_h \, |\nabla \phi_h| \, \mathrm{d}x \quad (9.3)$$

for every discrete test function $\eta_h \in S_h$.

Narrow band approximation assumption. We require the narrow band approximation assumption

$$\Gamma \subset \bigcup_{\Gamma \cap T \neq \emptyset} T \subseteq \bar{D}_h, \quad 0 < h \leq h_1, \quad (9.4)$$

and

$$|D_h| \leq Ch, \quad (9.5)$$

where $|D_h|$ denotes the volume of D_h.

The condition (9.4) can be satisfied by choosing

$$\gamma \geq \max_{x \in \bar{\Omega}} |\nabla \phi(x)|.$$

This can be seen by noting that, if $x \in \Gamma \cap T, y \in T$, then, because ϕ_h is piecewise linear, we have

$$|\phi_h(y)| \leq |\phi_h(X)| = |\phi(X)| = |\phi(X) - \phi(x)| \leq h \, \max_{\bar{\Omega}} |\nabla \phi| = \gamma h,$$

for any vertex X of T. When ϕ is a signed distance function to Γ, we can choose $\gamma = 1$.

The bound on the measure of D_h, (9.5), follows from the observation that for h sufficiently small $|\phi| < 2\gamma h$ on D_h and since $|\nabla \phi| \geq c_\phi > 0$ on Ω, using the co-area formula we have

$$|D_h| \leq \int_{-2\gamma h}^{2\gamma h} \int_{\Gamma(r)} \frac{1}{|\nabla \phi|} \, \mathrm{d}A \, \mathrm{d}r \leq \frac{4\gamma |\Gamma_*|}{c_\phi} h,$$

where $|\Gamma_*|$ is a bound for the measure of the level sets of ϕ in Ω.

Error bound. The main numerical analysis result of Deckelnick *et al.* (2010) is the following error bound.

Theorem 9.1. Assume that the solution u of (9.1) belongs to $H^{2,\infty}(\Omega)$ and that $f^e \in H^{1,\infty}(\Omega)$. Let u_h be the trace of the solution U_h on Γ of the finite element scheme (9.3) satisfying the narrow band approximation assumption. Then

$$\|u - u_h\|_{H^1(\Gamma)} \leq ch.$$

Another narrow band method. We use $\phi = d$, d being the signed distance function to Γ, as the level set function to create a polyhedral approximation D_h of Γ. We interpolate d on $\mathcal{T}_h$ and define

$$D_h = \{x \in \Omega \mid |I_h d(x)| < h\}.$$

The finite element problem is to find $u_h \in S_h$ such that

$$\int_{D_h} \nabla u_h \cdot \nabla \eta_h + u_h \eta_h \, \mathrm{d}x = \int_{D_h} f^e \eta_h \, \mathrm{d}x, \quad \text{for all } \eta_h \in S_h.$$

This is the same method as that of Deckelnick *et al.* (2010), except that it uses full instead of projected gradients. This allows simpler assembly of mass and stiffness matrices. See Deckelnick *et al.* (2013) for an error analysis and computational results.

Observe that in these unfitted finite element narrow band methods the computational domain D_h has a polygonal boundary and that $\mathcal{T}_h|_{D_h}$, the restriction of $\mathcal{T}_h$ to D_h, is not shape-regular and can have arbitrary small elements: see Figure 9.1. Note that the region of integration involves parts of elements. Dealing with this is the price to pay for avoiding the triangulation of the surface. To implement the method one can divide the irregular regions into simplices and use appropriate quadrature rules. The term *cut cell* is used in this context (Engwer and Heimann 2012).

9.2. Sharp interface method using a bulk finite element space

Another method for the model equation (9.1) is to use a bulk unfitted mesh independent of the surface and seek an approximate solution in a finite element space on that bulk mesh using a variational equation formulated on the zero level set of an approximate level set function. This is the method of Section 9.1 with $\gamma = 0$. This has been proposed and analysed in Olshanskii *et al.* (2009) and Olshanskii and Reusken (2010). An extension of this method has been proposed by Deckelnick *et al.* (2013).

The method calculates a sharp interface approximation to (9.2) using integrals over an approximation to Γ. We define $\widetilde{\mathcal{T}}_h$ to be an unfitted regular triangulation, consisting of closed simplices, of an arbitrary region containing Γ. We restrict this triangulation to $\mathcal{T}_h$ given by

$$\mathcal{T}_h = \{T \in \widetilde{\mathcal{T}}_h \mid \mathcal{H}^n(T \cap \Gamma) > 0\},$$

where $\mathcal{H}^n(\omega)$ denotes the n-dimensional measure of ω. We impose further that $\Gamma \subseteq \bigcup_{T\in\mathcal{T}_h} T$ and

$$\mathcal{H}^n(\Gamma) = \sum_{T\in\mathcal{T}_h} \mathcal{H}^n(T \cap \Gamma).$$

This restriction means that in the case where Γ is the face between two elements we must make an arbitrary, but fixed, choice of one of the two adjoining simplices. We define h to be the diameter of the largest element in $\mathcal{T}_h$. We let U_h denote the interior of the region covered by these elements:

$$\overline{U}_h = \bigcup_{T\in\mathcal{T}_h} T.$$

We assume h is sufficiently small that $U_h \subseteq U_\delta$.

We now define a finite element space S_h on the triangulation $\mathcal{T}_h$ given by

$$S_h = \{\eta_h \in C^0(U_h) \mid \eta_h|_T \text{ is linear affine, for each } T \in \mathcal{T}_h\}.$$

We use the distance function d to construct a polyhedral approximation of Γ_h of Γ using the following procedure. We interpolate d on $\mathcal{T}_h$ using the bulk piecewise linear finite element space S_h, and define

$$\Gamma_h = \{x \in U_h \mid I_h d(x) = 0\}.$$

This defines a polyhedral surface Γ_h consisting of lines if $n = 1$ and triangles and quadrilaterals if $n = 2$. Notice that in general, $\mathcal{T}_h|_{\Gamma_h}$, the restriction of U_h to Γ_h, is not shape-regular and can have arbitrary small elements.

This method could be implemented using a level set function instead of a distance function.

Both full gradients and tangential gradients may be used in the discrete weak formulation yielding two methods. The method of Deckelnick *et al.* (2013) is to find $u_h \in S_h$ such that

$$\int_{\Gamma_h} \nabla u_h \cdot \nabla \eta_h + u_h \eta_h \,\mathrm{d}A_h = \int_{\Gamma_h} f^e \eta_h \,\mathrm{d}A_h, \quad \text{for all } \eta_h \in S_h.$$

The method of Olshanskii *et al.* (2009) uses the tangential gradient to Γ_h in the discrete variational form and a slightly different finite element space S_{Γ_h} consisting of functions on Γ_h which are the trace of functions in S_h. Both methods yields errors of order h and h^2 in the H^1 and L^2 norms, respectively.

As an illustration of the motivation for unfitted finite element methods we consider the complicated two-dimensional surface $\Gamma = \{x \in \Omega \mid \phi(x) = 0\}$ given by

$$\begin{aligned}\phi(x) = (x_1^2 + x_2^2 - 4)^2 + (x_3^2 - 1)^2 + (x_2^2 + x_3^2 - 4)^2 + (x_1^2 - 1)^2 \\ + (x_3^2 + x_1^2 - 4)^2 + (x_2^2 - 1)^2 - 3.\end{aligned}$$

The surface is shown in Figure 9.2 and is shaded to reflect the solution of the equation with right-hand side $f(x) = 100 \sum_{j=1}^{4} \mathrm{e}^{-|x-x_{(j)}|^2}$, with

$$x_{(1)} = (-1.0, 1.0, 2.04), \quad x_{(2)} = (1.0, 2.04, 1.0),$$
$$x_{(3)} = (2.04, 0.0, 1.0), \quad x_{(4)} = (-0.5, -1.0, -2.04).$$

The points $x_{(j)}$ are close to the surface Γ. In Figure 9.3 we see the intersection of the surface with a uniform bulk grid of tetrahedra. In this case the numerical method to obtain this solution is the one described in Section 9.2.

9.3. Diffuse interface method

Here we describe a diffuse interface method for the parabolic equation (5.13) modelling conservation on a moving hypersurface $\Gamma(t)$:

$$\partial^\bullet u + u \nabla_\Gamma \cdot v - \nabla_\Gamma \cdot (\mathcal{A} \nabla_\Gamma u) = 0. \tag{9.6}$$

For simplicity we take $\mathcal{A} = \mathcal{D}_c I$, $\mathcal{D}_c > 0$ and the velocity to be purely in the normal direction $v = V\nu$.

The approach is based on approximating $\Gamma(t)$ by an evolving thin interfacial layer $\Gamma(\varepsilon, t)$ of thickness ε on which a bulk advection–diffusion equation is solved. The motivation for this diffuse interface approach arises from modelling evolving surface problems using phase field methods. A thin diffuse interfacial layer of width $O(\varepsilon)$ across which a phase field variable φ_ε has a steep transition from the bulk values $\approx \pm 1$ on either side of the interface approximates the sharp interface: see Caginalp (1989) and Blowey and Elliott (1993). Diffuse interfaces with compact support occur naturally when the double-obstacle phase field models are employed (Blowey and Elliott 1991, 1993). In these models the bulk values are identically ± 1 and the diffuse layer layer is sharp. This leads to the sharp diffuse interface front-tracking method formulated in Elliott and Styles (2003) and Deckelnick *et al.* (2005).

The bulk advection–diffusion equation is constructed using a family of non-negative, differentiable functions ρ with compact spatial support which defines the evolving diffuse interface $\Gamma(\varepsilon, t)$. To be definite we set $\rho = 1 - \varphi_\varepsilon^2$, where φ_ε is continuously differentiable double-obstacle phase field function such that the diffuse interface

$$\Gamma(\varepsilon, t) = \{x \mid |\varphi_\varepsilon(x,t)| < 1\}$$

is of finite small width approximately $\pi\varepsilon/2$. Phase field asymptotics indicate that the double-obstacle phase field approach yields the following approximation to the phase field variable (Blowey and Elliott 1993):

$$\varphi_\varepsilon(x,t) \approx \sin\left(\frac{d(x,t)}{\varepsilon}\right) \quad \text{for } |d(x,t)| \lessapprox \frac{\pi}{2}\varepsilon,$$

$$\varphi_\varepsilon(x,t) = \pm 1 \quad \text{for } |d(x,t)| \gtrapprox \frac{\pi}{2}\varepsilon,$$

Figure 9.2. Example of the solution of an elliptic equation on a complicated surface.

Figure 9.3. Example of the intersection of a bulk mesh with a complicated surface.

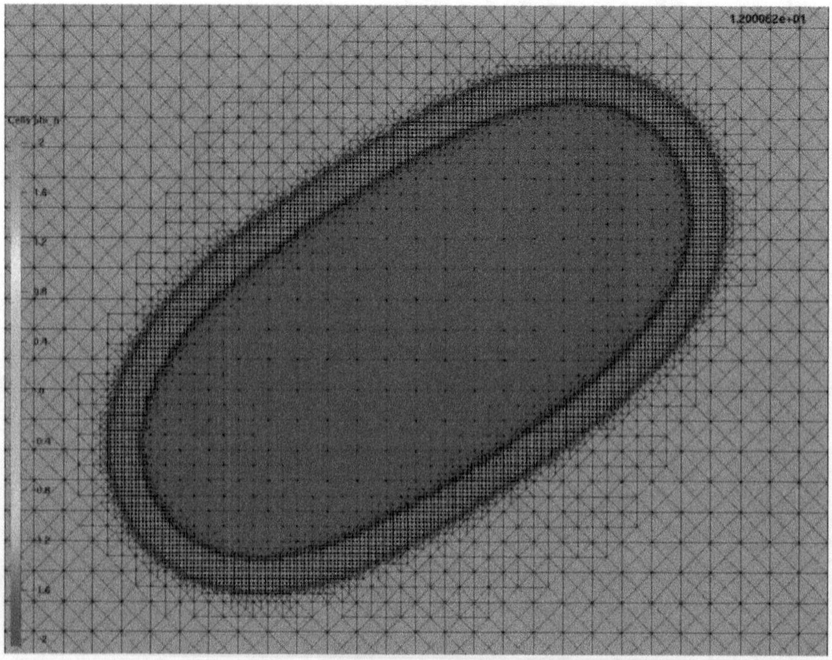

Figure 9.4. Example of a mesh arising in a diffuse interface computation.

where $d(\cdot, t)$ is the signed distance function to $\Gamma(t)$. If $\Gamma(t)$ is known through a level set function or distance function then the above formula for φ_ε can be applied directly to define $\Gamma(\varepsilon, t)$. The normal velocity may be extended to all of $\Gamma(\varepsilon, t)$ by

$$v = d_t \nabla d \quad \text{or} \quad v = \frac{\varphi_{\epsilon t} \nabla \varphi_\epsilon}{|\nabla \varphi_\epsilon|^2}$$

In practical applications the surface $\Gamma(t)$ may be unknown and an approximation is determined by some phase field model. One calculates a phase field function φ_ε, and in that case the second of the above equation is appropriate to use.

Figure 9.4 displays a mesh used in a simulation of advection and diffusion of a surfactant on an interface separating a drop of one fluid in another (Elliott *et al.* 2011). The fluid equations were modelled using a double-obstacle Cahn–Hilliard variational inequality (Blowey and Elliott 1992, 1993), coupling with the Navier–Stokes equations. The double-obstacle approach generates a sharp diffuse interface which is used to solve the parabolic equation on the moving diffuse layer as described above. In the figure one can observe the diffuse interface transition layer, which has a finer triangulation than the two bulk triangulated regions it separates. In this simulation the diffuse layer evolves in time and mesh adaptivity is used.

Description of diffuse interface bulk equation. Our goal is now to solve the parabolic equation

$$(\rho u)_t + \nabla \cdot (\rho u v) - \nabla \cdot \big(\mathcal{D}_c \rho \nabla u\big) = 0 \quad \text{on } \Gamma(\varepsilon, t), \tag{9.7}$$

which involves degenerating coefficients since ρ vanishes on $\partial\Gamma$. The conserved bulk quantity ρu is transported with an appropriate extension of the velocity field v away from the moving surface Γ. As analysed in Elliott and Stinner (2009) for curves, the equation (9.7) indeed approximates the surface equation (9.6) as $\varepsilon \to 0$. We remark that a degenerate equation of the form (9.7) appeared in a phase field model of diffusion-induced grain boundary motion (Fife, Cahn and Elliott 2001, Deckelnick, Elliott and Styles 2001).

Another diffuse interface approach in which the diffuse interface does not have compact support was proposed in Rätz and Voigt (2006), where the underlying phase field method is based on the classical double-well potential.

In the following description we assume that the function ρ is given. The equation is to be solved in the time interval $I = [0, t_f)$ with $t_f > 0$ in a spatial domain $\Omega \subset \mathbb{R}^n$, $n = 2, 3$, which is an appropriate domain into which the evolving closed hypersurface $\Gamma(t)$ is embedded at all times and $\Gamma(\varepsilon, t) \subset \Omega$. The initial values for the sharp equation (9.6), denoted by u_0 and the velocity v, need to be extended to all of $\Gamma(\varepsilon, t)$. A choice is to extend u_0 constantly in the normal direction away from Γ to obtain initial values for (9.7). On $\Omega\backslash\Gamma(\varepsilon, t)$ we set $u_0 = 0$.

Weak solution. A function $u : \Omega \times [0, t_f) \to \mathbb{R}$ with

$$u(x, t) = 0 \quad \text{if } x \notin \Gamma(\varepsilon, t)$$

is a weak solution to (9.7) if it satisfies

$$\int_\Omega ((\rho u)_t \chi - \rho u\, v \cdot \nabla \chi + \mathcal{D}_c \rho \nabla u \cdot \nabla \chi)\, \mathrm{d}x = 0 \tag{9.8}$$

for all test functions $\chi : \Omega \to \mathbb{R}$, and if

$$u(\cdot, 0) = u_0(\cdot) \quad \text{on } \Omega.$$

Time and space discretization. For simplicity let $\tau = \frac{t_f}{N_f}$, for an integer $N_f \in \mathbb{N}$, be a uniform time step, and define $t_n = n\tau$, $n = 0, \ldots, N_f$. Function evaluations or approximations of functions at time t_n will be denoted by an upper index n, and we define a discrete time derivative by

$$\partial_\tau f^n = \frac{f^{n+1} - f^n}{\tau}.$$

Again for simplicity we suppose that the bulk triangulation is fixed. Let $\mathcal{T}_h$ be a triangulation of the domain $\Omega \in \mathbb{R}^n, n = 2, 3$ consisting of simplices with maximum mesh size $h = \max_{e \in \mathcal{T}_h} \operatorname{diam}(e)$. Let J be the number and $\mathcal{N}$ be the set of vertex indices. The vertex coordinates are denoted by $\{X_1, \dots, X_J\}$. For an index $i \in \mathcal{N}$ let ω_i denote the neighbouring vertices connected to vertex i via an edge and let $\mathcal{T}_i = \{e \in \mathcal{T}_h \mid X_i \in e\}$ be the set of elements which have i as a vertex. Further, let $\mathcal{N}_e = \{j \in \mathcal{N} \mid X_j \in e\}$ be the set of vertices belonging to an element $e \in \mathcal{T}_h$.

The numerical solution is to be found in the discrete finite element space defined by

$$\mathcal{S}^h = \{\eta \in C^0(\bar{\Omega}) \mid \eta \text{ is a linear affine on each } e \in \mathcal{T}_h\}.$$

We define an interpolation operator $\Pi^h : C^0(\bar{\Omega}) \to \mathcal{S}^h$ by

$$\Pi^h(\eta) = \sum_{i=1}^{J} \eta(X_i)\chi_i. \tag{9.9}$$

Here $\chi_1, \dots, \chi_J$ denote the standard basis functions of $\mathcal{S}^h$, that is, $\chi_i \in C^0(\bar{\Omega})$ and $\chi_i|_e \in P_1(e)$ for all $e \in \mathcal{T}_h$ satisfying $\chi_i(X_j) = \delta_{ij}$ for all $i, j = 1, \dots, J$.

Evolving discrete diffuse interface. The computational domain is defined using the interpolant of the given support function ρ. We set $\rho^n(X_k) := \rho(X_k, t^n)$. The *discrete interface* at time t_n is defined by

$$\Gamma_h^n = \{e \in \mathcal{T}_h \mid \mathcal{N}_e \subset \mathcal{N}_h^n\}, \tag{9.10}$$

where

$$\mathcal{N}_h^n = \{i \in \mathcal{N} \mid \text{ there is } j \in \omega_i \text{ such that } \rho^n(X_j) > 0\}. \tag{9.11}$$

Note that this is a small subset of elements because of the underlying assumption of the small compact support of ρ.

The index set $\mathcal{N}_h^n$ is split up as follows:

$$\begin{aligned}
\mathcal{N}_h^n &= \mathcal{N}_{I,h}^n \cup \mathcal{N}_{B,h}^n, \\
\mathcal{N}_{I,h}^n &= \{i \in \mathcal{N}_h^n \mid \rho^n(X_i) > 0\}, \\
\mathcal{N}_{B,h}^n &= \{i \in \mathcal{N}_h^n \mid \rho^n(X_i) = 0\}.
\end{aligned} \tag{9.12}$$

In order for the numerical scheme to be well defined the following assumption is required.

Discrete interface assumption. It holds for all $n = 0, \dots, N_f - 1$ that if an index $i \in \mathcal{N}_h^n$ does not belong to $\mathcal{N}_h^{n+1}$, then $\rho^n(X_i) = 0$ (*i.e.*, $i \in \mathcal{N}_{B,h}^n$). This is a restriction on the time step depending on the magnitude of the interface velocity, which may be guaranteed by applying an adaptive time-stepping strategy or a condition of the form $\tau \le Ch/(\|v\|_\infty)$.

Numerical scheme. In order to formulate a numerical method for (9.8) we introduce the following forms for functions $\xi, \eta \in \mathcal{S}^h$:

$$\mathcal{M}(\xi,\eta)_h^n = \int_\Omega \Pi^h(\rho^n \xi\eta)\,\mathrm{d}x, \tag{9.13}$$

$$\mathcal{C}(\xi,\eta)_h^n = \int_\Omega \Pi^h(\rho^n \xi)\Pi^h(v^n)\cdot\nabla\eta\,\mathrm{d}x, \tag{9.14}$$

$$\mathcal{A}(\xi,\eta)_h^n = \int_\Omega \mathcal{D}_c \Pi^h(\rho^n)\,\nabla\xi\cdot\nabla\eta\,\mathrm{d}x. \tag{9.15}$$

We denote by U_h^n the finite element approximation to $u(\cdot,t_n)$. Let u_0^e be an extension to Ω of the initial data u_0 on $\Gamma(0)$. We define U_h^0 by

$$U_h^0(X_k) = u_0^e(X_k) \quad \text{if } k \in \mathcal{N}_h^0, \tag{9.16}$$

$$U_h^0(X_k) = 0 \qquad\qquad \text{if } k \notin \mathcal{N}_h^0. \tag{9.17}$$

For example, we may extend constantly in the normal direction and cut off to zero outside the narrow band corresponding to the initial diffuse layer defined by the compact support of $\rho(\cdot,0)$.

Algorithm 9.2. For each $n = 0,\dots,N_f-1$ we seek a function $U_h^{n+1} \in \mathcal{S}^h$ such that

$$U_h^{n+1}(X_k) = 0, \quad \text{if } k \notin \mathcal{N}_h^{n+1}, \tag{9.18}$$

and satisfying

$$\partial_\tau \mathcal{M}(U_h^n,\eta)_h^n - \mathcal{C}(U_h^{n+1},\eta)_h^{n+1} + \mathcal{A}(U_h^{n+1},\eta)_h^{n+1} = 0, \quad \text{for all } \eta \in \mathcal{S}^h. \tag{9.19}$$

The discrete interface assumption together with the *boundary condition* (9.18) implies that the variational scheme (9.19) may be localized so that the nodal variables at the new time level $U_h^{n+1}(X_i)$ for $i \in \mathcal{N}_h^{n+1}$ are determined from

$$\begin{aligned}&\frac{1}{\tau}\left(\int_\Omega \{(\Pi^h(\rho^{n+1}U_h^{n+1}\chi_j) - \Pi^h(\rho^n U_h^n \chi_j)\}\,\mathrm{d}x\right) \\ &\quad + \int_\Omega \{(-\Pi^h(\rho^{n+1}U_h^{n+1})\Pi^h(v^{n+1}) + \mathcal{D}_c\Pi^h(\rho^{n+1})\,\nabla U_h^{n+1})\cdot\nabla\chi_j\}\,\mathrm{d}x = 0,\end{aligned} \tag{9.20}$$

for all $j \in \mathcal{N}_h^{n+1}$.

Unique solvability and mass conservation. The following estimate is useful for showing that the above system (9.20) is solvable.

Lemma 9.3. For $n = 1,\dots,N_f$, $\delta > 0$, and $\xi,\eta \in S_h$ we have that

$$|\mathcal{C}(\xi,\eta)_h^n| \le \delta\mathcal{A}(\eta,\eta)_h^n + \frac{\|v^n\|_{\infty,\Omega_I}^2}{4\delta\mathcal{D}_c}\mathcal{M}(\xi,\xi)_h^n. \tag{9.21}$$

Proof. The desired estimate follows from

$$
\begin{aligned}
|\mathcal{C}(\xi,\eta)_h^n| &= \left|\int_\Omega \Pi^h(\rho^n\xi)\Pi^h(v^n)\cdot\nabla\eta \,\mathrm{d}x\right| \\
&\le \int_\Omega \|v^n\|_{\infty,\Omega}\left|\sum_{i=1}^N \rho_i^n \xi_i\chi_i\right||\nabla\eta|\,\mathrm{d}x \\
&\le \sum_{i=1}^N \int_\Omega (\rho_i^n\chi_i)^{1/2}|\nabla\eta|\,\|v^n\|_\infty(\rho_i^n\chi_i)^{1/2}|\xi_i|\,\mathrm{d}x \\
&\le \sum_{i=1}^N \int_\Omega \left(\mathcal{D}_c\delta\rho_i^n\chi_i|\nabla\eta|^2 + \frac{\|v^n\|_{\infty,\Omega}^2}{4\delta\mathcal{D}_c}\rho_i^n\chi_i\xi_i^2\right)\mathrm{d}x \\
&= \delta\mathcal{D}(\eta,\eta)_h^n + \frac{\|v^n\|_{\infty,\Omega}^2}{4\delta\mathcal{D}_c}\mathcal{M}(\xi,\xi)_h^n,
\end{aligned}
$$

where we use the abbreviation $\rho_i^n = \rho(X_i,t_n)$. □

Proposition 9.4. If $\tau < 4\mathcal{D}_c/\|v\|_{\infty,\Omega_I}^2$ then the scheme (9.19) has a unique solution.

It holds that for every n

$$\mathcal{M}(U_h^n,1)_h^n = \mathcal{M}(U_h^0,1)_h^0.$$

Proof. Proceeding by induction, given the assumption on the initial datum U_h^0, we may suppose that $U_h^n(X_k) = 0$ if $k \notin \mathcal{N}_h^n$. Taking W_h^{n+1} to be the difference of two possible solutions, it is sufficient to show that $W_h^{n+1} = 0$ is the only solution of

$$\mathcal{M}(W_h^{n+1},\eta)_h^{n+1} - \tau\mathcal{C}(W_h^{n+1},\eta)_h^{n+1} + \tau\mathcal{A}(W_h^{n+1},\eta)_h^{n+1} = 0, \quad \text{for all } \eta\in\mathcal{S}^h.$$

Taking $\eta = W_h^{n+1}$ and using (9.21) gives

$$\left(1-\tau\frac{\|v^{n+1}\|_{\infty,\Omega}^2}{4\delta\mathcal{D}_c}\right)\mathcal{M}(W_h^{n+1},W_h^{n+1})_h^{n+1} + (1-\delta)\mathcal{D}(W_h^{n+1},W_h^{n+1})_h^{n+1} \le 0,$$

which, upon taking δ arbitrarily close to 1, yields for $\tau < 4\mathcal{D}_c/\|v^{n+1}\|_{\infty,\Omega}^2$,

$$\mathcal{M}(W_h^{n+1},W_h^{n+1})_h^{n+1} = 0, \tag{9.22}$$

$$\mathcal{D}(W_h^{n+1},W_h^{n+1})_h^{n+1} = 0. \tag{9.23}$$

It follows from (9.22) that $W_h^{n+1}(X_i) = 0$ for all $i\in\mathcal{N}_{I,h}^{n+1}$. By (9.23) we have that $\nabla W_h^{n+1} = 0$ in every finite element $e \in \Gamma_h^{n+1}$, whence we conclude that also $W_h^{n+1}(X_i) = 0$ for $i \in \mathcal{N}_{B,h}^{n+1}$. The mass conservation is immediate. □

Discrete equations. Let J_n be the number of nodes in $\mathcal{N}_h^n$. We can decompose U_h^n as

$$U_h^n = \sum_{j \in \mathcal{N}_h^n} \alpha_j^n \chi_j = \sum_{j \in \mathcal{N}_h^{n+1}} \widetilde{\alpha}_j^n \chi_j + \sum_{j \in \mathcal{N}_h^n \setminus \mathcal{N}_h^{n+1}} \alpha_j^n \chi_j,$$

where we note that

$$\widetilde{\alpha}_j^n = \alpha_j^n \quad \text{for } j \in \mathcal{N}_h^{n+1} \cap \mathcal{N}_h^n \quad \text{and} \quad \widetilde{\alpha}_j^n = 0 \quad \text{for } j \in \mathcal{N}_h^{n+1} \setminus \mathcal{N}_h^n.$$

We now define the matrices $M^{n+1}, S^{n+1}, C^{n+1}, \widetilde{M}^{n+1} \in \mathbb{R}^{J_{n+1} \times J_{n+1}}$ with the entries

$$\begin{aligned}
M_{i,j}^{n+1} &= \int_\Omega \Pi^h(\rho_h^{n+1} \chi_i \chi_j) \, \mathrm{d}x, \\
S_{i,j}^{n+1} &= \int_\Omega \mathcal{D}_c \Pi^h(\rho_h^{n+1}) \nabla \chi_i \cdot \nabla \chi_j \, \mathrm{d}x, \\
C_{i,j}^{n+1} &= \int_\Omega \Pi^h(\rho_h^{n+1} \chi_i) \Pi^h(v^{n+1}) \cdot \nabla \chi_j \, \mathrm{d}x, \\
\widetilde{M}_{i,j}^n &= \int_\Omega \Pi^h(\rho_h^n \chi_i \chi_j) \, \mathrm{d}x,
\end{aligned}$$

where the indices i, j belong to the set $\mathcal{N}_h^{n+1}$. To evaluate these forms it is useful to employ quadrature.

At each time step $n+1$ the solution U_h^{n+1} is obtained by solving the system

$$\left(\tfrac{1}{\tau} M^{n+1} + S^{n+1} - C^{n+1}\right) \underline{\alpha}^{n+1} = \tfrac{1}{\tau} \widetilde{M}^n \underline{\widetilde{\alpha}}^n$$

of J_{n+1} linear equations which have a unique solution (Proposition 9.4). Because ρ^{n+1} is zero at boundary points of $\mathcal{N}_h^{n+1}$ there may be some equations with small diagonal elements, which leads to ill-conditioning. This may be remedied by preconditioning using the inverse diagonal of $\frac{1}{\tau} M^{n+1} + S^{n+1} - C^{n+1}$.

Numerical tests in Elliott *et al.* (2011) indicate that reducing the width of the interface together with the mesh size h in a constant ratio can lead to errors in the L^2 norm on the surface of order h^2. The concentration profile becomes flat as the approximation parameters tend to zero. This is in accordance with the rigorous ε asymptotic analysis of Elliott and Stinner (2009).

9.4. Other methods

Thick interface method. Let $\Gamma_\varepsilon = \{x \in U \mid |d(x)| < \varepsilon\}$. Consider the following problem: find $u^\varepsilon\colon \Gamma_\varepsilon \to \mathbb{R}$ such that

$$\begin{aligned} -\Delta u^\varepsilon + u^\varepsilon &= f^e \quad \text{in } \Gamma_\varepsilon, \\ \frac{\partial u^\varepsilon}{\partial \nu} &= 0 \quad \text{on } \partial\Gamma_\varepsilon. \end{aligned}$$

It can be shown that the solution of this bulk equation converges to the solution of the surface equation as $\varepsilon \to 0$. This is the basis of a method proposed in Schwartz *et al.* (2005). It is a narrow band approximation analogous to the phase field method in Section 9.3, but only for stationary interfaces, based on choosing ρ to be the characteristic function of Γ in the equation (9.7).

Closest point approach. The closest point method for partial differential equations on stationary surfaces (Ruuth and Merriman 2008, Macdonald and Ruuth 2008, 2009, Macdonald, Brandman and Ruuth 2011, Marz and Macdonald 2013) is based on considering $u(a(x))$, where $a(x) \in \Gamma$ is the point closest to x. The surface partial differential equation is then embedded and discretized in a neighbourhood of Γ using $u(a(x))$. Implementation requires the knowledge or calculation of the closest point $a(x)$. In the cited references this approach has been used to solve a wide variety of equations on stationary surfaces.

10. Applications

10.1. Coupling geometric evolution to diffusion on a surface

A generic example of geometric coupling to surface physics is diffusion on a surface coupled to a geometric equation for the evolution of the surface. A general system might take the following form: find a scalar field u and an evolving hypersurface $\Gamma(t) \subset \mathbb{R}^3$ such that u solves the conservation law

$$\partial^\bullet u - \nabla_\Gamma \cdot q + u\nabla_\Gamma \cdot v = f(V, u) \quad \text{on } \Gamma(t) \tag{10.1a}$$

and $\Gamma(t)$ evolves with the velocity law

$$V = g(\nu, H, u), \tag{10.1b}$$

with q being a flux for u, $v = V\nu + v_\tau$, where V is the normal velocity of $\Gamma(t)$, v_τ is a material tangential velocity, $\partial^\bullet$ is the associated material derivative and H and ν are the mean curvature and normal to $\Gamma(t)$. This system (10.1) is strongly coupled as the evolution of $\Gamma(t)$ is determined by the solution of the PDE that holds on $\Gamma(t)$. We look at a couple of examples in Sections 10.4 and 10.5.

10.2. Surfactants

Surface partial differential equations often arise as subproblems in complex systems of partial differential equations in which surface processes couple geometry and physics. A strongly motivating example is that of the advection and diffusion of a surface active agent on a fluid interface. A particular situation concerns two immiscible viscous fluids with a drop of one fluid inside the other separated by an energetic interface. We suppose that there is a surfactant which is insoluble in the fluids and is confined to the fluid interface, and that the surface energy depends on the concentration of the surfactant and thus leads to a concentration-dependent surface tension and the Marangoni effect. The mathematical model then comprises a moving interface problem for the Navier–Stokes equations coupled to an advection–diffusion equation on the interface. The problem is to find two fluid domains $\Omega^+(t)$, $\Omega^-(t)$ separated by a surface $\Gamma(t)$, a fluid velocity v and a surfactant concentration u. Within the bulk fluid domains we have the Navier–Stokes system

$$\nabla \cdot v = 0, \quad \partial_t v + (v \cdot \nabla) v = -\nabla p + \frac{1}{Re} \Delta v.$$

On the unknown moving hypersurface $\Gamma(t)$ we require the mass and momentum balances

$$[v]_-^+ = 0, \quad v \cdot \nu = V,$$

$$\left[-pI + \frac{2}{Re} D(v) \right]_-^+ \nu = -\frac{1}{Re\, Ca} \big(\sigma(u) H + \nabla_\Gamma \sigma(u) \big)$$

to be satisfied, where ν is the unit normal pointing into Ω^+, p is the pressure within the fluids, H is the mean curvature of $\Gamma(t)$, $\sigma(u)$ is the concentration-dependent surface tension, Re and Ca are dimensionless numbers derived from physical parameters, V is the normal velocity of Γ in this direction, and

$$D(v) = \tfrac{1}{2}(\nabla v + (\nabla v)^T).$$

The concentration U satisfies the advection–diffusion equation

$$\partial^\bullet u - \nabla_\Gamma \cdot q + u \nabla_\Gamma \cdot v = 0 \quad \text{on } \Gamma(t),$$

where v is the material fluid velocity.

We refer to James and Lowengrub (2004), Xu, Li, Lowengrub and Zhao (2006), Lowengrub, Xu and Voigt (2007), Lai, Tseng and Huang (2008), Ganesan and Tobiska (2009), Elliott *et al.* (2011), Ganesan, Hahn, Held and Tobiska (2012) and Xu, Yang and Lowengrub (2012) for a variety of numerical methods for solving this problem. These include level set, phase field and surface finite element approaches.

10.3. Phase separation on a surface

As an example of an advection–diffusion equation with a fourth-order elliptic operator we may consider a Cahn–Hilliard equation on a moving surface. The equation for conservation of mass is

$$\partial^\circ u + HVu + \nabla_\Gamma \cdot q = 0,$$

which is of the form (10.1a) with $f = 0$.

Taking a constitutive law for the flux to be of Cahn–Hilliard type which allows up-hill diffusion and leads to phase separation yields

$$\begin{aligned} q &= -b(u)\nabla_\Gamma w, \\ w &= -\gamma\Delta_\Gamma u + \psi'(u), \end{aligned}$$

where w is the chemical potential, $b(\cdot)$ is a concentration-dependent mobility, and $\psi(\cdot)$ is a double-well free energy. Associated with this model is the surface Cahn–Hilliard surface free energy functional

$$E(c) = \int_{\Gamma(t)} \left(\frac{\gamma}{2}|\nabla_\Gamma u|^2 + \psi(u)\right),$$

where γ is a gradient energy coefficient. The classical quartic double-well potential is

$$\psi(u) = \frac{1}{4}(u^2-1)^2.$$

Placing ourselves in the context of the evolving surface finite element method described in Section 5, we formulate a semidiscretization of the Cahn–Hilliard equation in the case $b(u) = 1$. Given $U_{h0} \in S_h(0)$, find

$$U_h \in S_h^T = \{\phi_h \text{ and } \partial_h^\bullet \phi_h \in C^0(\mathcal{G}_T^h) \mid \phi_h(\cdot,t) \in S_h(t)\ t \in [0,T]\}$$

and

$$W_h \in \widetilde{S}_h^T = \{\phi_h \in C^0(\mathcal{G}_T^h) \mid \phi_h(\cdot,t) \in S_h(t)\ t \in [0,T]\}$$

such that, for all $\phi_h \in S_h^T$ and $t \in (0,T]$,

$$\frac{\mathrm{d}}{\mathrm{d}t} m_h(U_h,\phi_h) + a_h(W_h,\phi_h) = m_h(U_h,\partial_h^\bullet\phi_h), \quad U_h(\cdot,0) = U_{h0},$$

and

$$m_h(W_h,\phi_h) = \gamma a_h(U_h,\phi_h) + m_h(\Pi_h\psi'(U_h),\phi_h),$$

which is the natural analogue of the splitting into second-order equations of Elliott *et al.* (1989). For a stationary surface an analysis of the error is given in Du, Ju and Tian (2011). See also Schonborn and Desai (1997) and Marenduzzo and Orlandini (2013) for models and computations on stationary surfaces. Mercker *et al.* (2012) consider a more complicated version of the Cahn–Hilliard equation, which involves an energy depending on mean

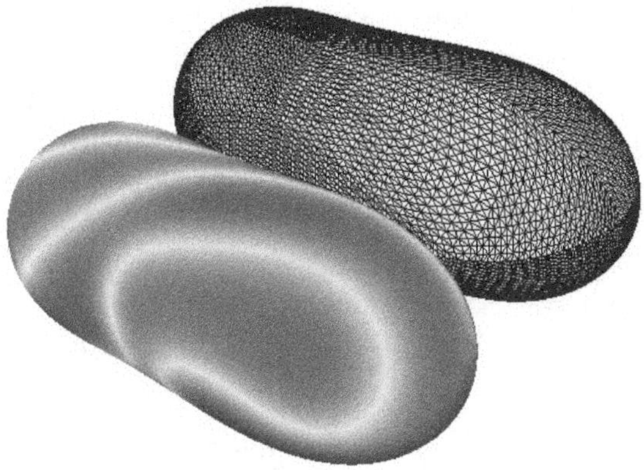

Figure 10.1. Phase separation on a growing surface.

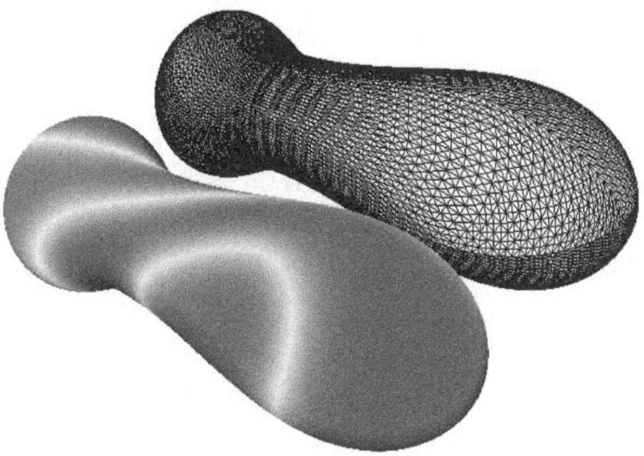

Figure 10.2. Phase separation at a later time than Figure 10.4.

curvature of the surface, in the context of modelling curvature modulated sorting in biomembranes.

Examples of a numerical experiment for an implementation of this scheme is shown in Figures 10.1 and 10.2. In each figure the evolving surface is shaded with respect to the values of U_h and the associated triangulation is shown by the side.

10.4. Surface dissolution

An example of coupling a surface evolution with a surface process which leads to a highly nonlinear system with complex surface morphology is a model for the etching of silver in an Ag–Au alloy whose surface is immersed in an electrolyte. The dissolution of silver atoms occurs at the surface, leaving a monolayer of gold adatoms on the surface which are mobile within the electrolyte. The gold adatoms agglomerate in clusters exposing the next layer of silver atoms for dissolution. This may lead to the growth of porosity into the bulk and formation of nanoporosity: see Erlebacher *et al.* (2001) and Erlebacher and McCue (2012). The model developed in Erlebacher *et al.* (2001) and also Eilks and Elliott (2008) comprises an evolution equation for the time-dependent dissolving surface $\Gamma(t)$ and an equation for conservation of mass. It leads to a system of the form (10.1).

The geometric surface evolution equation is of forced mean curvature type,

$$V = -J_{\text{diss}} = v_0(c)(1 - \delta H) \tag{10.2}$$

where $V\nu$ is the normal velocity of the surface and the rate of dissolution is denoted by J_{diss} which depends on a surface-concentration-dependent etching rate $v_0(c) \geq 0$ and the mean curvature of the surface, H. Physically, in portions where the surface is completely covered by gold, etching cannot take place because the electrolyte is no longer in contact with silver atoms and the etching rate satisfies $v_0(1) = 0$. We use c to denote the surface concentration of gold atoms in the binary mixture of gold and electrolyte adatoms in the surface monolayer, C_0 to be the bulk gold concentration and q to denote the diffusive surface flux of adatoms.

Then the equation for conservation of mass is

$$\partial^\circ c + HVc = VC_0 - \nabla_\Gamma \cdot q, \tag{10.3}$$

which is of the form (10.1a) with $f = VC_0$. This latter term is due to the picking up on the surface of gold adatoms from the bulk where diffusion is neglected.

The constitutive law for the flux is of Cahn–Hilliard type, which allows up-hill diffusion and leads to phase separation so that

$$\begin{aligned} q &= -b(c)\nabla_\Gamma w, \\ w &= -\gamma\Delta_\Gamma c + \psi'(c), \end{aligned}$$

where μ is the chemical potential, $b(\cdot)$ is a concentration-dependent mobility, and $\psi(\cdot)$ is the double-well free energy occurring in Cahn–Hilliard theory. Assuming a regular solution model for homogeneous free energy of the binary mixture of gold and electrolyte adatoms in the monolayer leads to the logarithmic function

$$\psi(c) = \frac{\theta_{cr}}{2}c(1-c) + \frac{\theta}{4}(c\log(c) + (1-c)\log(1-c)).$$

This precludes values of c outside the interval $[-1, 1]$ and a typical form for the mobility is the degenerate mobility

$$b(c) = B(1 - c^2).$$

Thus we are led to the study of the highly nonlinear coupled system of forced mean curvature flow for an evolving surface with the forcing depending on the solution of a degenerate Cahn–Hilliard equation. On the surface we expect surface phase separation to take place, yielding gold-rich and gold-depleted regions on the surface monolayer $\Gamma(t)$. Numerical simulations are shown in Figure 10.3. See Eilks and Elliott (2008) for a description of the computational methods and further simulations. These computations are based on the evolving surface finite element method for the solution of the Cahn–Hilliard equations and a variant of the scheme of Dziuk (1991) for forced mean curvature flow. The surfaces shown in Figure 10.3 are triangulated surfaces. Adaptivity is used to refine and coarsen depending on the solution of the Cahn–Hilliard equation, and from time to time a new triangulation is required in order to avoid degenerate shapes for the triangles.

10.5. Pattern formation on biological surfaces

Following the classical paper of Turing (1952), pattern formation on a growing biological surface may be modelled by reaction–diffusion equations which exhibit diffusion-driven instability of spatially uniform structures and thus lead to spatial patterns. In biological applications it is natural to consider this mechanism for pattern formation to hold on evolving curved surfaces which form the boundaries of growing three-dimensional domains. The numerical solution of a number of models using the evolving surface finite element method was proposed in Barreira *et al.* (2011). A Lagrangian particle method using level set ideas for reaction–diffusion equations on moving surfaces was presented in Bergdorf, Sbalzarini and Koumoutsakos (2010).

An example of an application is presented by Chaplain, Ganesh and Graham (2001), who propose a model for the growth of solid tumours. In the model the evolution of the solid bulk tumour is determined by a concentration of a growth-promoting factor causing growth on the surface. The resulting mathematical problem is that of determining $\Gamma(t) \subset \mathbb{R}^3$ modelling the tumour surface and scalar functions $u(x,t)$ and $w(x,t)$ that diffuse and react on the surface, which satisfy

$$\begin{aligned} \partial^\bullet u + u\nabla_\Gamma \cdot V\nu &= \Delta_\Gamma u + f_1(u,w) \quad \text{on } \Gamma(t), \\ \partial^\bullet w + w\nabla_\Gamma \cdot V\nu &= D_w\Delta_\Gamma w + f_2(u,w) \quad \text{on } \Gamma(t), \end{aligned}$$

with normal velocity $V\nu$ and

$$V = -\varepsilon H + \delta u. \tag{10.4}$$

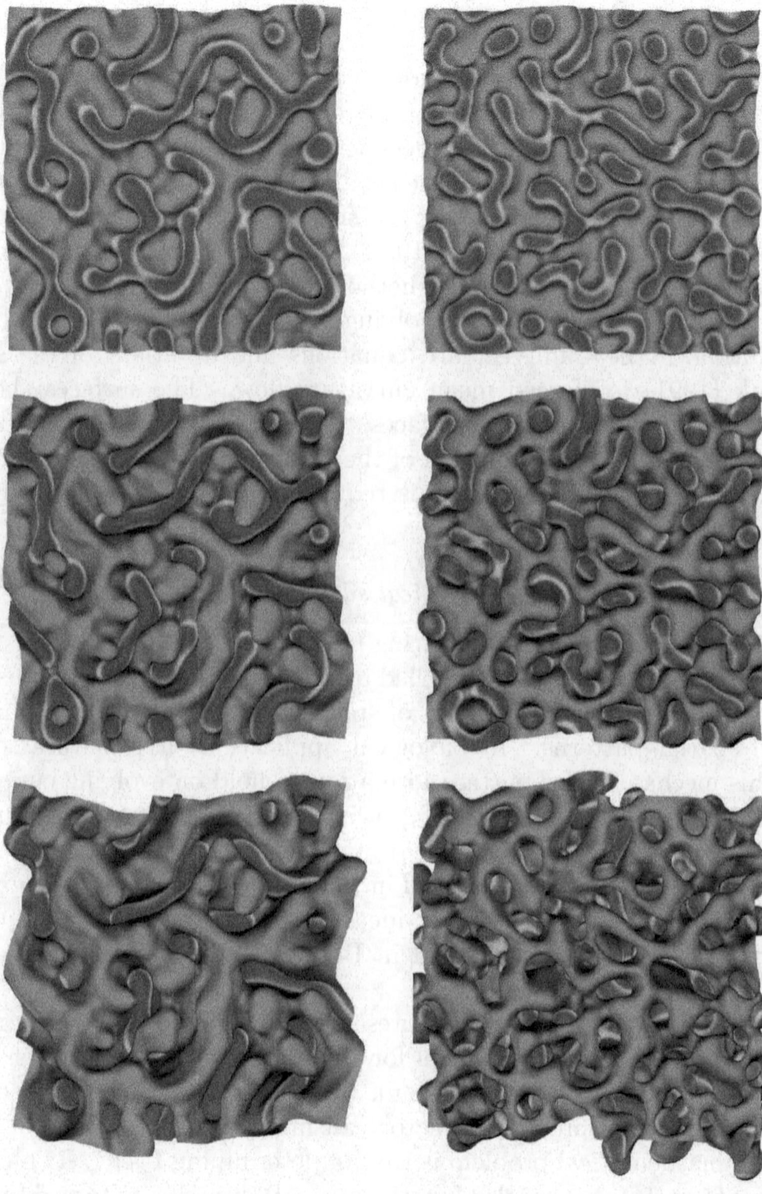

(a) (b)

Figure 10.3. The early stages of spinodal decomposition on a dissolving surface. (a) The surface with constant mobility and (b) a surface with degenerate mobility, $t = 0.1$, $t = 0.2$ and $t = 0.3$.

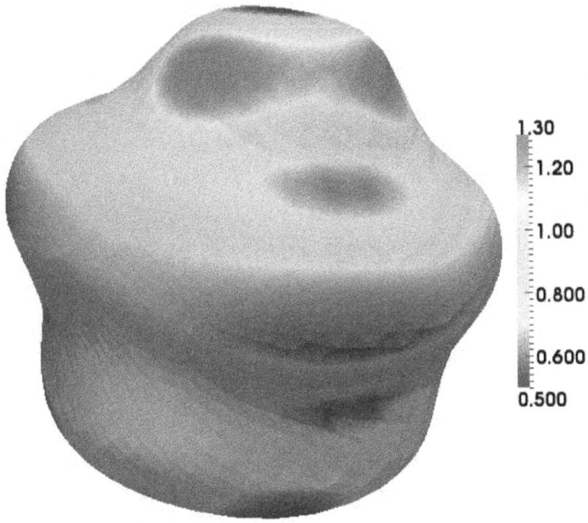

Figure 10.4. Pattern formation on growing surface.

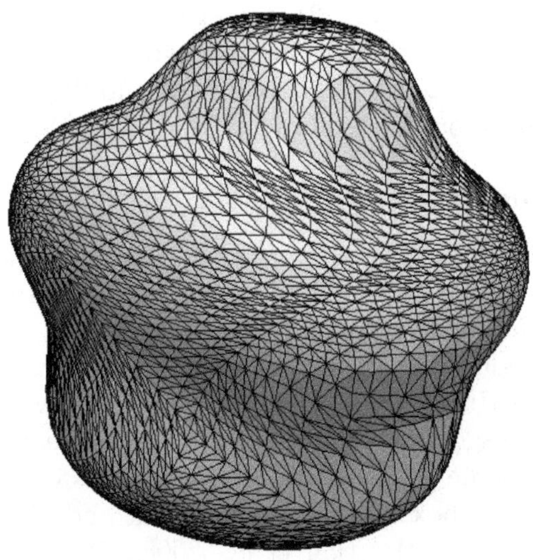

Figure 10.5. Associated triangulation for Figure 10.4.

In the velocity law (10.4) the second term on the right comes from the suggestion in Crampin, Gaffney and Maini (1999) that growth should be faster in regions of higher concentration of the growth-promoting factor u, and the first term on the right is a surface tension term which has the effect of yielding smooth surfaces in the short term. Finally D_w, δ and ε are positive parameters. The reaction functions f_1 and f_2 model the interactions between the two surface concentrations. An example is the activator-depleted substrate model (Schnakenberg 1979), known as the Brusselator model, in which

$$f_1(u,w) = \gamma(a - u + u^2 w) \quad \text{and} \quad f_2(u,w) = \gamma(b - u^2 w),$$

where γ, a and b are positive constants.

An example of a simulation together with the associated triangulation is given in Figures 10.4 and 10.5. The shading corresponds to the computed values of u. The simulation was carried out with the ALE evolving-surface finite element method (Elliott and Styles 2012), coupled with the method of Barrett, Garcke and Nürnberg (2008*a*).

For examples of studies of reaction–diffusion pattern formation on surfaces we refer to Aragón, Barrio and Varea (1999), Barrio *et al.* (2004), Barreira (2009) and Rätz and Röger (2012). Other examples in computational biology which involve the coupling of surface evolution to diffusion and phase separation on the surface are models for cell motility (Elliott *et al.* 2012) and phase separation on biomembranes (Elliott and Stinner 2010).

Acknowledgements

The authors wish to thank Carsten Eilks, Thomas Ranner, Björn Stinner and Vanessa Styles for providing computational results. This article was completed whilst one of the authors was visiting the Freie Universität Berlin in association with the award of a Humboldt Forschungspreis by the Alexander von Humboldt Foundation. The work was supported by the Deutsche Forschungsgemeinschaft via SFB/TR 71 'Geometric partial differential equations'. Some of the computational results were obtained by using the finite element toolbox ALBERTA (Schmidt and Siebert 2005) and the DUNE software package (Bastian *et al.* 2008*a*, 2008*b*) and visualized using the visualization applications ParaView (Henderson 2007) and GRAPE (Rumpf, Schmidt, *et al.* 1990).

REFERENCES[1]

D. Adalsteinsson and J. A. Sethian (2003), 'Transport and diffusion of material quantities on propagating interfaces via level set methods', *J. Comput. Phys.* **185**, 271–288.

J. Aragón, R. A. Barrio and C. Varea (1999), 'Turing patterns on a sphere', *Phys. Rev. E* **60**, 4588–4592.

R. Barreira (2009), Numerical solution of nonlinear partial differential equations on triangulated surfaces. DPhil thesis, University of Sussex.

R. Barreira, C. M. Elliott and A. Madzvamuse (2011), 'The surface finite element method for pattern formation on evolving biological surfaces', *J. Math. Biology* **63**, 1095–1119.

J. W. Barrett and C. M. Elliott (1982), A finite element method on a fixed mesh for the Stefan problem with convection in a saturated porous medium. In *Numerical Methods for Fluid Dynamics, Conference proceedings: University of Reading, 29-31 March, 1982* (K. W. Morton and M. J. Baines, eds), Conference Series (Institute of Mathematics and its Applications), Academic Press, pp. 389–409.

J. W. Barrett and C. M. Elliott (1984), 'A finite-element method for solving elliptic equations with Neumann data on a curved boundary using unfitted meshes', *IMA J. Numer. Anal.* **4**, 309–325.

J. W. Barrett and C. M. Elliott (1985), 'Fixed mesh finite element approximations to a free boundary problem for an elliptic equation with an oblique derivative boundary condition', *Comput. Math. Appl.* **11**, 335–345.

J. W. Barrett and C. M. Elliott (1987*a*), 'Fitted and unfitted finite-element methods for elliptic equations with smooth interfaces', *IMA J. Numer. Anal.* **7**, 283–300.

J. W. Barrett and C. M. Elliott (1987*b*), 'A practical finite element approximation of a semi-definite Neumann problem on a curved domain', *Numer. Math.* **51**, 23–36.

J. W. Barrett and C. M. Elliott (1988), 'Finite-element approximation of elliptic equations with a Neumann or Robin condition on a curved boundary', *IMA J. Numer. Anal.* **8**, 321–342.

J. W. Barrett, H. Garcke and R. Nürnberg (2007), 'A parametric finite element method for fourth order geometric evolution equations', *J. Comput. Phys.* **222**, 441–467.

J. W. Barrett, H. Garcke and R. Nürnberg (2008*a*), 'On the parametric finite element approximation of evolving hypersurfaces in R^3', *J. Comput. Phys.* **227**, 4281–4307.

J. W. Barrett, H. Garcke and R. Nürnberg (2008*b*), 'Parametric approximation of Willmore flow and related geometric evolution equations', *SIAM J. Sci. Comput.* **31**, 225–253.

[1] The URLs cited in this work were correct at the time of going to press, but the publisher and the authors make no undertaking that the citations remain live or are accurate or appropriate.

R. A. Barrio, P. K. Maini, P. Padilla, R. G. Plaza and Sánchez-Garduno (2004), 'The effect of growth and curvature on pattern formation', *J. Dynam. Diff. Equations* **16**, 1093–1121.

P. Bastian and C. Engwer (2009), 'An unfitted finite element method using discontinuous Galerkin', *Internat. J. Numer. Methods Engng* **79**, 1557–1576.

P. Bastian, M. Blatt, A. Dedner, C. Engwer, R. Klofkorn, M. Oldberger and O. Sander (2008*a*), 'A generic grid interface for parallel and adaptive scientific computing I: Abstract framework', *Computing* **82**, 103–119.

P. Bastian, M. Blatt, A. Dedner, C. Engwer, R. Klofkorn, M. Oldberger and O. Sander (2008*b*), 'A generic grid interface for parallel and adaptive scientific computing II: Implementation and tests in DUNE', *Computing* **82**, 121–138.

M. Bergdorf, I. Sbalzarini and P. Koumoutsakos (2010), 'A Lagrangian particle method for reaction–diffusion systems on deforming surfaces', *J. Math. Biology* **61**, 649–663.

M. J. Berger, D. A. Calhoun, C. Helzel and R. J. Leveque (2009), 'Logically rectangular finite volume methods with adaptive refinement on the sphere', *Phil. Trans. Royal Soc. London A* **367**, 4483–4496.

M. Bertalmío, L.-T. Cheng, S. J. Osher and G. Sapiro (2001), 'Variational problems and partial differential equations on implicit surfaces', *J. Comput. Phys.* **174**, 759–780.

J. F. Blowey and C. M. Elliott (1991), 'The Cahn–Hilliard gradient theory for phase separation with non-smooth free energy I: Mathematical analysis', *European J. Appl. Math.* **2**, 233–280.

J. F. Blowey and C. M. Elliott (1992), 'The Cahn–Hilliard gradient theory for phase separation with non-smooth free energy II: Numerical analysis', *European J. Appl. Math.* **3**, 147–179.

J. F. Blowey and C. M. Elliott (1993), Curvature dependent phase boundary motion and parabolic obstacle problems. In *Degenerate Diffusion* (W.-M. Ni, L. A. Peletier and J. L. Vasquez, eds), Vol. 47 of *IMA Volumes in Mathematics and its Applications*, Springer, pp. 19–60.

M. Burger (2009), 'Finite element approximation of elliptic partial differential equations on implicit surfaces', *Comput. Vis. Sci.* **12**, 87–100.

G. Caginalp (1989), 'Stefan and Hele-Shaw type models as asymptotic limits of the phase field equations', *Phys. Rev. A* **39**, 5887–5896.

D. A. Calhoun and C. Helzel (2009), 'A finite volume method for solving parabolic equations on logically Cartesian curved surface meshes', *SIAM J. Sci. Comput.* **31**, 4066–4099.

P. Cermelli, E. Fried and M. E. Gurtin (2005), 'Transport relations for surface integrals arising in the formulation of balance laws for evolving fluid interfaces', *J. Fluid Mech.* **544**, 339–351.

M. A. J. Chaplain, M. Ganesh and I. G. Graham (2001), 'Spatio-temporal pattern formation on spherical surfaces: Numerical simulation and application to solid tumour growth', *J. Math. Biology* **42**, 387–423.

P. G. Ciarlet (1978), *The Finite Element Method for Elliptic Problems*, North-Holland.

M. I. M. Copetti and C. M. Elliott (1992), 'Numerical analysis of the Cahn–Hilliard equation with a logarithmic free energy', *Numer. Math.* **63**, 39–65.

E. J. Crampin, E. A. Gaffney and P. K. Maini (1999), 'Reaction and diffusion on growing domains: Scenarios for robust pattern formation', *Bull. Math. Biology* **61**, 1093–1120.

K. Deckelnick, G. Dziuk and C. M. Elliott (2005), Computation of geometric partial differential equations and mean curvature flow. In *Acta Numerica*, Vol. 14, Cambridge University Press, pp. 139–232.

K. Deckelnick, G. Dziuk, C. M. Elliott and C.-J. Heine (2010), 'An h-narrow band finite element method for implicit surfaces', *IMA J. Numer. Anal.* **30**, 351–376.

K. Deckelnick, C. M. Elliott and T. Ranner (2013), Unfitted finite element methods using bulk meshes for surface partial differential equations. In preparation.

K. Deckelnick, C. M. Elliott and V. Styles (2001), 'Numerical diffusion-induced grain boundary motion', *Interfaces Free Bound.* **6**, 329–349.

A. Dedner, P. Madhavan and B. Stinner (2013), 'Analysis of the discontinuous Galerkin method for elliptic problems on surfaces', *IMA J. Numer. Anal.*, to appear.

A. Demlow (2009), 'Higher-order finite element methods and pointwise error estimates for elliptic problems on surfaces', *SIAM J. Numer. Anal.* **47**, 805–827.

A. Demlow and G. Dziuk (2007), 'An adaptive finite element method for the Laplace–Beltrami operator on surfaces', *SIAM J. Numer. Anal.* **45**, 421–442.

A. Demlow and M. Olshanskii (2012), 'An adaptive surface finite element method based on volume meshes', *SIAM J. Numer. Anal.* **50**, 1624–1647.

Q. Du, L. Ju and L. Tian (2011), 'Finite element approximation of the Cahn–Hilliard equation on surfaces', *Comput. Methods Appl. Mech. Engng* **200**, 2458–2470.

G. Dziuk (1988), Finite elements for the Beltrami operator on arbitrary surfaces. In *Partial Differential Equations and Calculus of Variations* (S. Hildebrandt and R. Leis, eds), Vol. 1357 of *Lecture Notes in Mathematics*, Springer, pp. 142–155.

G. Dziuk (1991), 'An algorithm for evolutionary surfaces', *Numer. Math.* **58**, 603–611.

G. Dziuk and C. M. Elliott (2007*a*), 'Finite elements on evolving surfaces', *IMA J. Numer. Anal.* **25**, 385–407.

G. Dziuk and C. M. Elliott (2007*b*), 'Surface finite elements for parabolic equations', *J. Comput. Math.* **25**, 385–407.

G. Dziuk and C. M. Elliott (2008), 'Eulerian finite element method for parabolic PDEs on implicit surfaces', *Interfaces Free Bound.* **10**, 119–138.

G. Dziuk and C. M. Elliott (2010), 'An Eulerian approach to transport and diffusion on evolving implicit surfaces', *Comput. Vis. Sci.* **13**, 17–28.

G. Dziuk and C. M. Elliott (2012), 'Fully discrete evolving surface finite element method', *SIAM J. Numer. Anal.* **50**, 2677–2694.

G. Dziuk and C. M. Elliott (2013), 'L^2 estimates for the evolving surface finite element method', *Math. Comp.* **82**, 1–24.

G. Dziuk, D. Kröner and T. Müller (2012), Scalar conservation laws on moving hypersurfaces. Technical report, Freiburg.

G. Dziuk, C. Lubich and D. Mansour (2012), 'Runge–Kutta time discretization of parabolic differential equations on evolving surfaces', *IMA J. Numer. Anal.* **32**, 394–416.

C. Eilks and C. M. Elliott (2008), 'Numerical simulation of dealloying by surface dissolution via the evolving surface finite element method', *J. Comput. Phys.* **227**, 9727–9741.

C. M. Elliott and T. Ranner (2013), 'Finite element analysis for a coupled bulk–surface partial differential equation', *IMA J. Numer. Anal.*, to appear.

C. M. Elliott and B. Stinner (2009), 'Analysis of a diffuse interface approach to an advection diffusion equation on a moving surface', *Math. Mod. Methods Appl. Sci.* **5**, 787–802.

C. M. Elliott and B. Stinner (2010), 'Modeling and computation of two phase geometric biomembranes using surface finite elements', *J. Comput. Phys.* **229**, 6585–6612.

C. M. Elliott and B. Stinner (2012), 'Computation of two-phase biomembranes with phase dependent material parameters using surface finite elements', *Commun. Comput. Phys.* **13**, 325–360.

C. M. Elliott and V. Styles (2003), 'Computations of bidirectional grain boundary dynamics in thin metallic films', *J. Comput. Phys.* **187**, 524–543.

C. M. Elliott and V. M. Styles (2012), 'An ALE ESFEM method for solving PDEs on evolving surfaces', *Milan J. Math.* **80**, 469–501.

C. M. Elliott, D. A. French and F. A. Milner (1989), 'A second order splitting method for the Cahn–Hilliard equation', *Numer. Math.* **54**, 575–590.

C. M. Elliott, B. Stinner and C. Venkataraman (2012), 'Modelling cell motility and chemotaxis with evolving surface finite elements', *J. Royal Society Interface* **9**, 3027–3044.

C. M. Elliott, B. Stinner, V. Styles and R. Welford (2011), 'Numerical computation of advection and diffusion on evolving diffuse interfaces', *IMA J. Numer. Anal.* **31**, 786–812.

C. Engwer and F. Heimann (2012), DUNE-UDG: A cut-cell framework for unfitted discontinuous Galerkin methods. In *Advances in DUNE*, Springer, pp. 89–100.

J. Erlebacher and I. McCue (2012), 'Geometric characterization of nanoporous metals', *Acta Materialia* **60**, 6164–6174.

J. Erlebacher, M. Aziz, A. Karma, N. Dimitrov and K. Sieradzki (2001), 'Evolution of nanoporosity in dealloying', *Nature* **410**, 450–453.

L. C. Evans (1998), *Partial Differential Equations*, first edition, Graduate Studies in Mathematics, AMS.

P. C. Fife, J. W. Cahn and C. M. Elliott (2001), 'A free-boundary model for diffusion-induced grain boundary motion', *Interfaces Free Bound.* **3**, 291–336.

S. Ganesan and L. Tobiska (2009), 'A coupled arbitrary Lagrangian–Eulerian and Lagrangian method for computation of free-surface flows with insoluble surfactants', *J. Comput. Phys.* **228**, 2859–2873.

S. Ganesan, A. Hahn, K. Held and L. Tobiska (2012), An accurate numerical method for computation of two-phase flows with surfactants. In *European Congress on Computational Methods in Applied Sciences and Engineering* (J. Eberhardsteiner *et al.*, eds), Vol. CD-ROM, TU Vienna. office@eccomas2012.tuwien.ac.at

D. Gilbarg and N. S. Trudinger (1998), *Elliptic Partial Differential Equations of Second Order*, Grundlehren der Mathematischen Wissenschaften, Springer.

J. B. Greer (2006), 'An improvement of a recent Eulerian method for solving PDEs on general geometries', *J. Sci. Comput.* **29**, 321–352.

J. B. Greer, A. L. Bertozzi and G. Sapiro (2006), 'Fourth order partial differential equations on general geometries', *J. Comput. Phys.* **216**, 216–246.

A. Hansbo and P. Hansbo (2002), 'An unfitted finite element method, based on Nitsche's method, for elliptic interface problems', *Comput. Methods Appl. Mech. Engng* **191**, 5537–5552.

A. Henderson (2007), *ParaView Guide: A Parallel Visualization Application*, Kitware.

A. J. James and J. Lowengrub (2004), 'A surfactant-conserving volume-of-fluid method for interfacial flows with insoluble surfactant', *J. Comput. Phys.* **201**, 685–722.

L. Ju and Q. Du (2009), 'A finite volume method on general surfaces and its error estimates', *J. Math. Anal. Appl.* **352**, 645–668.

L. Ju, L. Tian and D. Wang (2009), '*A posteriori* error estimates for finite volume approximations of elliptic equations on general surfaces', *Comput. Methods Appl. Mech. Engng* **198**, 716–726.

M.-C. Lai, Y.-H. Tseng and H. Huang (2008), 'An immersed boundary method for interfacial flows with insoluble surfactant', *J. Comput. Phys.* **227**, 7279–7293.

J. Lowengrub, J.-J. Xu and A. Voigt (2007), 'Surface phase separation and flow in a simple model of multicomponent drops and vesicles', *Fluid Dynamics and Material Processing* **3**, 1–20.

C. Lubich and D. Mansour (2012), Variational discretisation of linear wave equation on evolving surfaces. Technical report, Tübingen.

C. B. Macdonald and S. J. Ruuth (2008), 'Level set equations on surfaces via the closest point method', *J. Sci. Comput.* **35**, 219–240.

C. B. Macdonald and S. J. Ruuth (2009), 'The implicit closest point method for the numerical solution of partial differential equations on surfaces.', *SIAM J. Sci. Comput.* **31**, 4330–4350.

C. B. Macdonald, J. Brandman and S. J. Ruuth (2011), 'Solving eigenvalue problems on curved surfaces using the closest point method.', *J. Comput. Phys.* **230**, 7944–7956.

D. Marenduzzo and E. Orlandini (2013), 'Phase separation dynamics on curved surfaces', *Soft Matter* **9**, 1178–1187.

T. Marz and C. B. Macdonald (2013), 'Calculus on surfaces with general closest point functions', *SIAM J. Numer. Anal.* **50**, 3303–3328.

M. Mercker, M. Ptashnyk, J. Kühnle, D. Hartmann, M. Weiss and W. Jäger (2012), 'A multiscale approach to curvature modulated sorting in biological membranes', *J. Theoret. Biology* **301**, 67–82.

J. Nedelec (1976), 'Curved finite element methods for the solution of singular integral equations on surfaces in R^3', *Comput. Methods Appl. Mech. Engng* **8**, 61–80.

I. L. Novak, F. Gao, Y.-S. Choi, D. Resasco, J. C. Schaff and B. M. Slepchenko (2007), 'Diffusion on a curved surface coupled to diffusion in the volume: Application to cell biology', *J. Comput. Phys.* **226**, 1271–1290.

M. A. Olshanskii and A. Reusken (2010), 'A finite element method for surface PDEs: Matrix properties', *Numer. Math.* **114**, 491–520.

M. A. Olshanskii, A. Reusken and J. Grande (2009), 'A finite element method for elliptic equations on surfaces', *SIAM J. Numer. Anal.* **47**, 3339–3358.

S. Osher and R. Fedkiw (2003), *Level Set Methods and Dynamic Implicit Surfaces*, Vol. 153 of *Applied Mathematical Sciences*, Springer.

A. Rätz and M. Röger (2012), 'Turing instabilities in a mathematical model for signaling networks', *J. Math. Biology* **65**, 1215–1244.

A. Rätz and A. Voigt (2006), 'PDEs on surfaces: A diffuse interface approach', *Commun. Math. Sci.* **4**, 575–590.

M. Rumpf, A. Schmidt *et al.* (1990), GRAPE: Graphics programming environment. SFB 256 Report 8, Universität Bonn.

S. J. Ruuth and B. Merriman (2008), 'A simple embedding method for solving partial differential equations on surfaces', *J. Comput. Phys.* **227**, 1943–1961.

A. Schmidt and K. G. Siebert (2005), *Design of Adaptive Finite Element Software: The Finite Element Toolbox ALBERTA*, Vol. 42 of *Lecture Notes in Computational Science and Engineering*, Springer.

J. Schnakenberg (1979), 'Simple chemical reaction systems with limit cycle behaviour', *J. Theoret. Biology* **81**, 389–400.

O. Schonborn and R. C. Desai (1997), 'Phase-ordering kinetics on curved surfaces', *Physica A* **239**, 412–419.

P. Schwartz, D. Adalsteinsson, P. Colella, A. P. Arkin and M. Onsum (2005), 'Numerical computation of diffusion on a surface', *Proc. Nat. Acad. Sci.* **102**, 11151–11156.

J. A. Sethian (1999), *Level Set Methods and Fast Marching Methods*, Cambridge Monographs on Applied and Computational Mathematics, Cambridge University Press.

H. A. Stone (1990), 'A simple derivation of the time dependent convective diffusion equation for surfactant transport along a deforming interface', *Phys. Fluids A* **2**, 111–112.

A. M. Turing (1952), 'The chemical basis of morphogenesis', *Phil. Trans. Royal Soc. London B* **237**, 37–72.

J.-J. Xu and H.-K. Zhao (2003), 'An Eulerian formulation for solving partial differential equations along a moving interface', *J. Sci. Comput.* **19**, 573–594.

J.-J. Xu, Z. Li, J. Lowengrub and H. Zhao (2006), 'A level-set method for interfacial flows with surfactant', *J. Comput. Phys.* **212**, 590–616.

J.-J. Xu, Y. Yang and J. Lowengrub (2012), 'A level-set continuum method for two-phase flows with insoluble surfactant', *J. Comput. Phys.* **231**, 5897–5909.

Acta Numerica (2013), pp. 397–508
doi:10.1017/S0962492913000068

Atomistic-to-continuum coupling

Mitchell Luskin
School of Mathematics,
University of Minnesota,
MN 55455, USA
E-mail: luskin@umn.edu

Christoph Ortner
Mathematics Institute,
University of Warwick,
Coventry CV4 7AL, UK
E-mail: c.ortner@warwick.ac.uk

Atomistic-to-continuum (a/c) coupling methods are a class of computational multiscale schemes that combine the accuracy of atomistic models with the efficiency of continuum elasticity. They are increasingly being utilized in materials science to study the fundamental mechanisms of material failure such as crack propagation and plasticity, which are governed by the interaction between crystal defects and long-range elastic fields.

In the construction of a/c coupling methods, various approximation errors are committed. A rigorous numerical analysis approach that classifies and quantifies these errors can give confidence in the simulation results, as well as enable optimization of the numerical methods for accuracy and computational cost. In this article, we present such a numerical analysis framework, which is inspired by recent research activity.

CONTENTS

1. Introduction

Crystal defects such as grain boundaries, cracks, or dislocations play a central role in determining material behaviour, and their study represents a substantial component of materials science research. Molecular simulation provides a unique way to study complex material behaviour at the nanoscale, in particular how defects affect macroscopic processes such as elasticity, plasticity, and fracture.

The key difficulty in atomistic simulation is that crystal defects affect elastic fields far beyond their immediate atomic neighbourhood; that is, they give rise to *strongly coupled multiscale problems.* Computational materials scientists therefore face a compromise between inaccurate atomistic models and inaccurate representations of the crystal environment.

Since defects occupy only a small proportion of bulk crystals, one may attempt to model the elastic fields using more efficient models of continuum elasticity. This idea naturally leads to the construction of concurrent couplings between atomistic descriptions of defect cores and continuum elasticity descriptions of the elastic fields. We henceforth refer to techniques of this kind as atomistic-to-continuum coupling methods, or simply a/c methods. By employing coarse discretizations (*e.g.*, finite elements) of the continuum elasticity model they achieve a considerable reduction in computational cost compared to full atomistic descriptions, and potentially circumvent the compromise between model accuracy and system size.

The present review article develops a numerical analysis framework within which to assess and control the approximation errors committed when replacing the fully atomistic model of a crystal (the 'exact' model) with an a/c coupling scheme, and to evaluate the relative accuracy and efficiency of different types of a/c coupling. Our aim is to provide a reference that will motivate further work by applied mathematicians in general and numerical analysts in particular, and a transfer of modern numerical analysis methodologies.

During the past century, physicists and engineers have developed the subject of micromechanics to give a fundamental understanding to the mechanism of material failure. The building blocks of micromechanics are the nucleation and movement of point, line, and surface defects and their long-range elastic interactions. These mechanisms were first verified against macroscopic experiments and then more recently against experiments at the nanoscale. Computational micromechanics has begun to extend the predictive scope of theoretical micromechanics, but mathematical theory able to assess the accuracy and efficiency of multiscale methods is needed for computational micromechanics to reach its full potential. We hope that this article will help to nucleate a much wider research effort to establish rigorous mathematical underpinnings for (computational) micromechanics.

1.1. Brief history

Remarkably, the history of atomistic-to-continuum multiscale methods begins at a time when the existence of atoms had not even been confirmed: Cauchy (1882), in a search for simplified stress–strain relations for continuum elasticity, postulated a pair interaction law between atoms arranged in a crystalline structure and derived symmetry relations for the elastic constants known as *Cauchy relations*. Various similar connections between atomistic and continuum descriptions of matter were used in solid state physics throughout the twentieth century: see Born and Huang (1954) for a standard reference.

As the critical role of crystal defects became more widely understood, the first ideas of a/c coupling emerged. Numerous authors in the 1950s and 1960s employed continuum linear elasticity to obtain boundary conditions for an atomistic simulation of a defect core; see Kanzaki (1957) for an early example. Eventually, it was recognized that, conversely, the defect core also provides boundary conditions for the elastic field and hence the two descriptions were coupled to interact concurrently. The first instance of such a method that we are aware of was developed by Sinclair (1971), and we may consider this as the first a/c coupling scheme.

The first occurrences of a/c couplings employing finite element methodology to discretize the continuum model can be found in the works of Baskes, Melius and Wilson (1981), Mullins and Dokainish (1982), Fischmeister *et al.* (1989) and Kohlhoff, Gumbsch and Fischmeister (1991). Another critical step was the introduction of the variational framework and the introduction of the Cauchy–Born model by Tadmor, Ortiz and Phillips (1996). The latter reference coined the popular term 'quasicontinuum method'.

By this point, it had became widely understood that the interface (or handshake region) treatment was critical in the construction of accurate and reliable a/c couplings. Various new ideas were put forward, most prominently, the iterative ghost force correction method (Shenoy *et al.* 1999), force-based a/c coupling (Curtin and Miller 2003), blending schemes (Xiao and Belytschko 2004), and the quasi-nonlocal coupling (Shimokawa, Mortensen, Schiøtz and Jacobsen 2004).

The first numerical analysis contributions to the field of a/c coupling are the works of Lin (2003) and Blanc, Le Bris and Legoll (2005). The first analyses that focused on the effect of different a/c interface treatments on the global error can be found in Dobson and Luskin (2009*a*, 2009*b*) and Ming and Yang (2009). The nonlinear analysis framework that we use in this article was introduced in Ortner and Süli (2008) and Ortner (2011). From here on, the numerical analysis of a/c coupling has turned into a rapidly developing field. We will introduce some of the key ideas and survey the various contributions throughout the remainder of this paper.

1.2. Outline and reading guide

This article is essentially divided into two parts, which can be read independently. The first part, which comprises only Section 2, gives a rapid formal introduction to the main ideas in the simplest possible non-trivial setting of second neighbour Lennard-Jones interaction in one dimension. Already in this simple setting, many interesting aspect of a/c coupling can be discussed. Section 2 is intended to provide a first glimpse of the subject, or as the basis of a short series of lectures, or simply for readers who prefer explicit computations in a relatively simple setting over general mathematical structure and rigorous proofs. We summarize the main conclusions of Section 2 in Table 2.1.

In the remainder of the article we develop a complete theory of a/c coupling for static defect computations in 1D. Although a theory of a/c coupling in 2D/3D is now beginning to emerge, there are too many gaps and open questions to present a unified picture. Instead, we have chosen to present the 1D theory in a way that allows us to discuss existing generalizations to 2D/3D, as well as point out gaps. Moreover, with the exception of the reflection method introduced in Section 4.5, we only analyse a/c couplings whose formulations translate verbatim to 2D/3D.

We begin by presenting a rigorous formulation of an atomistic model problem for an infinite chain. This infinite-dimensional problem allows us to discuss the *rates of convergence* of various approximation schemes.

Atomistic-to-continuum couplings can simply be considered as an approximation scheme for this atomistic model. To analyse the errors committed, we employ the two fundamental concepts of numerical analysis: *consistency and stability* (see Section 4.7 for a formal outline of the approximation error analysis). The two central sections in this article are Section 6, where we establish the *consistency* of a/c couplings, and Section 7, where we establish their *stability*. These results are then combined and further refined in Section 8 into *a priori error estimates in terms of computational cost.*

The consistency error estimates and the *a priori* error estimates are summarized, respectively, in Tables 6.1 and 8.1.

Throughout the article, we discuss relevant background literature, existing extensions of our presentation to 2D/3D, and open problems. In addition, in Section 9 we give brief summaries of various extensions of the presentation in this article and a discussion of pressing open problems.

1.3. Omissions

No review article on a field of research as rich as a/c coupling can be complete. Our choice of topics reflects our personal view on the most important generic ideas in a/c coupling: the Cauchy–Born approximation, ghost forces at local/nonlocal model interfaces, ghost force reduction via blending, ghost force removal by conservative interface corrections, and force-based (non-conservative) a/c coupling.

We exclude many variants of how to combine these ingredients into practical a/c coupling schemes, as well as other classes of a/c multiscale methods. The analytical tools developed in this article for a/c coupling can potentially provide a framework for analysing these omitted methods, and we would consider the writing of this article a success if it motivates new research in this direction. The reader should refer to Section 9 for further discussions.

1.4. Notation

Notation is introduced throughout the manuscript where it is natural to do so. We have included a table of symbols in Appendix C.

Here, we merely mention that our notation for ℓ^p and L^p norms is standard. For a function v with discrete domain $\mathcal{X}$ and $\mathcal{A} \subset \mathcal{X}$, we define

$$\|v\|_{\ell^p(\mathcal{A})} := \begin{cases} \left(\sum_{\xi\in\mathcal{A}} |v(\xi)|^p\right)^{1/p}, & 1 \le p < \infty, \\ \sup_{\xi\in\mathcal{A}} |v(\xi)|, & p = \infty. \end{cases}$$

If we omit the domain in $\|v\|_{\ell^p}$, then the norm is computed on the entire domain of definition $\mathcal{X}$.

Analogously, for a Lebesgue-measurable function $v : X \to \mathbb{R}$, where $X \subset \mathbb{R}$ is measurable, and for $A \subset X$, we define

$$\|v\|_{L^p(A)} := \begin{cases} (\int_A |v(x)|^p \, dx)^{1/p}, & 1 \le p < \infty, \\ \operatorname{ess\,sup}_{x \in A} |v(x)|, & p = \infty. \end{cases}$$

If we omit the domain in $\|v\|_{L^p}$, then the norm is computed on the entire domain of definition X.

When we write $f \lesssim g$, we mean that there exists a constant $C > 0$, independent of the solution and approximation parameters, such that $f \le Cg$.

2. Atomistic-to-continuum Lennard-Jones next-nearest neighbour models

In this section, we offer a brief introduction in the simplest non-trivial setting, the Lennard-Jones next-nearest neighbour model. For the sake of simplicity of presentation, we keep this section entirely formal. Except where stated otherwise, all statements can be made rigorous. In fact, we present rigorous proofs for most of the statements in greater generality in subsequent sections.

We emphasize from the outset that the 1D Lennard-Jones model hides many fundamental issues one has to face when dealing with general many-body and 2D/3D situations, which are not merely of a technical nature. We will occasionally comment on such discrepancies.

2.1. The atomistic Lennard-Jones next-nearest neighbour model

We consider an infinite atomistic chain, indexed by $\mathbb{Z}$. A displacement of the chain is a map $u : \mathbb{Z} \to \mathbb{R}$, where we will assume that $u(\xi) \to 0$ as $|\xi| \to \infty$. The reference lattice is $\mathsf{A}\mathbb{Z}$, where $\mathsf{A} > 0$ is a macroscopic stretch, so the corresponding deformation of the atomistic chain is given by the mapping $\xi \in \mathbb{Z}$ to $y(\xi) := \mathsf{A}\xi + u(\xi)$. We normally assume that y is strictly increasing, that is, $y(\xi) - y(\xi - 1) > 0$, and hence $u(\xi) - u(\xi - 1) > -1$ for all ξ.

We assume in this introductory section that first and second neighbours interact via the Lennard-Jones potential $\phi(r) = r^{-12} - 2r^{-6}$ (or a similar pair potential): see Figure 2.1. The energy of a displacement u can then be written as

$$\begin{aligned} \mathcal{E}^{\mathrm{a}}(u) &:= \sum_{\xi \in \mathbb{Z}} \{\phi(y(\xi) - y(\xi - 1)) + \phi(y(\xi + 1) - y(\xi - 1))\} \\ &= \sum_{\xi \in \mathbb{Z}} \{\phi(\mathsf{A} + u'_\xi) + \phi(2\mathsf{A} + u'_\xi + u'_{\xi+1})\} \\ &= \sum_{\xi \in \mathbb{Z}} \{\phi_1(u'_\xi) + \phi_2(u'_\xi + u'_{\xi+1})\}, \end{aligned} \tag{2.1}$$

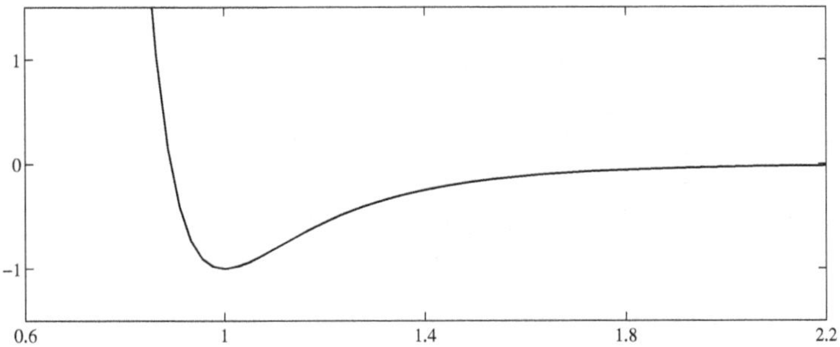

Figure 2.1. The Lennard-Jones potential.

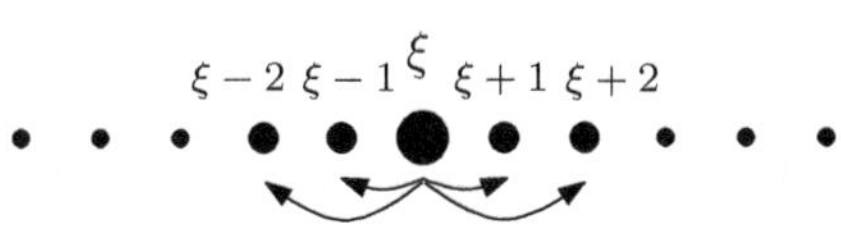

Figure 2.2. Interactions in a 1D second-neighbour model.

where

$$\phi_i(s) := \phi(i\mathsf{A} + s) \quad \text{and} \quad u'_\xi := u(\xi) - u(\xi - 1)$$

(and hence $u'_\xi + u'_{\xi+1} = u(\xi+1) - u(\xi-1)$).

More realistic molecular interactions are usually modelled via many-body potentials (see Section 3.2). Hence, to appreciate certain design decisions in coarse-graining schemes, it is important that we sometimes write $\mathcal{E}^{\mathrm{a}}$ in a form that can be generalized:

$$\mathcal{E}^{\mathrm{a}}(u) = \sum_{\xi \in \mathbb{Z}} \Phi^{\mathrm{a}}_\xi(u), \quad \text{where the } \textit{atomistic site energy} \text{ is} \tag{2.2}$$

$$\Phi^{\mathrm{a}}_\xi(u) := \tfrac{1}{2}\big\{\phi_1(u'_\xi) + \phi_1(u'_{\xi+1}) + \phi_2(u'_{\xi-1} + u'_\xi) + \phi_2(u'_{\xi+1} + u'_{\xi+2})\big\}.$$

See Figure 2.2 for an illustration.

We will see in Section 3.2 that if we assume without loss of generality that $\phi_1(0) + \phi_2(0) = 0$, then $\mathcal{E}^{\mathrm{a}}$ is well-defined on a suitable function space of displacements. We will assume in this section that all displacement trial and test functions belong to this space, which we will later denote by $\mathcal{U}$.

For simplicity, we consider only dead load external forces throughout this article. Let $f : \mathbb{Z} \to \mathbb{R}$ with $f(\xi) \to 0$ as $|\xi| \to \infty$. Then we seek a solution of

$$u^{\mathrm{a}} \in \arg\min\big\{\mathcal{E}^{\mathrm{a}}(u) - \langle f, u\rangle_{\mathbb{Z}}\big\}, \tag{2.3}$$

where $\langle f, u\rangle_{\mathbb{Z}} := \sum_{\xi\in\mathbb{Z}} f(\xi)u(\xi)$. Solutions of (2.3) solve the Euler–Lagrange equation

$$\langle \delta\mathcal{E}^{\mathrm{a}}(u^{\mathrm{a}}), v\rangle = \langle f, v\rangle_{\mathbb{Z}} \quad \text{for all } v, \quad \text{where} \tag{2.4}$$

$$\begin{aligned}\langle \delta\mathcal{E}^{\mathrm{a}}(u), v\rangle &= \sum_{\xi\in\mathbb{Z}} \{\phi_1'(u_\xi')v_\xi' + \phi_2'(u_\xi' + u_{\xi+1}')(v_\xi' + v_{\xi+1}')\} \\ &= \sum_{\xi\in\mathbb{Z}} \{\phi_1'(u_\xi') + \phi_2'(u_{\xi-1}' + u_\xi') + \phi_2'(u_\xi' + u_{\xi+1}')\}v_\xi' \qquad (2.5) \\ &= \sum_{\xi\in\mathbb{Z}} \mathsf{S}^{\mathrm{a}}_\xi(u)v_\xi'\end{aligned}$$

for an *atomistic stress* $\mathsf{S}^{\mathrm{a}}_\xi(u)$ defined by

$$\mathsf{S}^{\mathrm{a}}_\xi(u) := \phi_1'(u_\xi') + \phi_2'(u_{\xi-1}' + u_\xi') + \phi_2'(u_\xi' + u_{\xi+1}'). \tag{2.6}$$

Remark 2.1. We think of (2.3) as the 'thermodynamic limit' (at zero temperature) of a chain of N atoms as $N \to \infty$. The advantage over the 'scaling limit', in which one would let the atomic spacing tend to zero, is that the thermodynamic limit keeps the atomistic detail we are interested in, but removes the dependence on boundary conditions. Intuitively, we can simply think of (2.3) as an infinite chain approximation of a large but finite chain.

2.2. The Cauchy–Born approximation

To approximate the atomistic description, we wish to model the infinite chain $\mathbb{Z}$ using a continuum elasticity model with an energy functional of the form

$$\mathcal{E}^{\mathrm{c}}(u) = \int_{\mathbb{R}} W(\nabla u)\,\mathrm{d}x, \tag{2.7}$$

where $W : (-1, \infty) \to \mathbb{R}$ is a suitable strain energy function and u is now defined for $x \in \mathbb{R}$. Interpreting ∇u as a homogeneous strain applied to the infinite crystal $\mathbb{Z}$, and hence $W(\nabla u)$ as the resulting energy per unit volume corresponding to the atomistic model (2.1), we obtain the *Cauchy–Born strain energy density function*

$$W(\mathsf{F}) := \phi_1(\mathsf{F}) + \phi_2(2\mathsf{F}). \tag{2.8}$$

We note that $W(0) = 0$ since $\phi_1(0)+\phi_2(0) = 0$, and (2.7) is thus well-defined on a suitable function space of displacements that decay at infinity.

In the remainder of Section 2, we consider a finite element Cauchy–Born model, taking the atomistic chain as the set of nodes; that is, for discrete displacements $u : \mathbb{Z} \to \mathbb{R}$, we (re-)define

$$\mathcal{E}^{\mathrm{c}}(u) = \sum_{\xi\in\mathbb{Z}} W(u_\xi'),$$

which is in fact equivalent to (2.7) when $u(x) : \mathbb{R} \to \mathbb{R}$ is the continuous piecewise linear interpolant of $u : \mathbb{Z} \to \mathbb{R}$.

For this discretized Cauchy–Born model, we seek

$$u^{\mathrm{c}} \in \arg\min \{\mathcal{E}^{\mathrm{c}}(u) - \langle f, u\rangle_{\mathbb{Z}}\}. \tag{2.9}$$

Solutions of (2.9) solve the Euler–Lagrange equation

$$\langle \delta\mathcal{E}^{\mathrm{c}}(u^{\mathrm{c}}), v\rangle = \langle f, v\rangle_{\mathbb{Z}} \quad \text{for all } v, \quad \text{where} \tag{2.10}$$

$$\langle \delta\mathcal{E}^{\mathrm{c}}(u), v\rangle = \sum_{\xi\in\mathbb{Z}} \partial_{\mathsf{F}} W(u'_\xi) v'_\xi, \tag{2.11}$$

where $\partial_{\mathsf{F}} W(u'_\xi)$ is the 1D variant of the first Piola–Kirchhoff stress tensor.

Suppose that $\mathcal{E}^{\mathrm{c}}$ is uniformly *stable* on $\{su^{\mathrm{a}} + (1-s)u^{\mathrm{c}} \,|\, 0 \le s \le 1\}$ in the sense that there exists $\gamma > 0$ such that

$$\gamma\|v'\|_{\ell^2}^2 \le \langle \delta^2\mathcal{E}^{\mathrm{c}}(su^{\mathrm{a}} + (1-s)u^{\mathrm{c}})v,\, v\rangle \quad \text{for all } v,\ 0 \le s \le 1 \tag{2.12}$$

(see (2.16) for an analysis of stability of the reference state and Section 7 for a rigorous general analysis). We then obtain

$$\begin{aligned}
\gamma\|(u^{\mathrm{a}} - u^{\mathrm{c}})'\|_{\ell^2}^2 &\le \biggl\langle \int_0^1 \delta^2\mathcal{E}^{\mathrm{c}}(su^{\mathrm{a}} + (1-s)u^{\mathrm{c}})\,\mathrm{d}s\,(u^{\mathrm{a}} - u^{\mathrm{c}}), u^{\mathrm{a}} - u^{\mathrm{c}} \biggr\rangle \\
&= \biggl\langle \int_0^1 \frac{\mathrm{d}}{\mathrm{d}s}\delta\mathcal{E}^{\mathrm{c}}(su^{\mathrm{a}} + (1-s)u^{\mathrm{c}})\,\mathrm{d}s\,, u^{\mathrm{a}} - u^{\mathrm{c}} \biggr\rangle \\
&= \bigl\langle \delta\mathcal{E}^{\mathrm{c}}(u^{\mathrm{a}}) - \delta\mathcal{E}^{\mathrm{c}}(u^{\mathrm{c}}), u^{\mathrm{a}} - u^{\mathrm{c}} \bigr\rangle \\
&= \bigl\langle \delta\mathcal{E}^{\mathrm{c}}(u^{\mathrm{a}}) - \delta\mathcal{E}^{\mathrm{a}}(u^{\mathrm{a}}), u^{\mathrm{a}} - u^{\mathrm{c}} \bigr\rangle,
\end{aligned}$$

where, in the last equality, we have employed (2.4) and (2.10) (Galerkin orthogonality). Dividing through by $\gamma\|(u^{\mathrm{a}} - u^{\mathrm{c}})'\|_{\ell^2}$, we arrive at

$$\|(u^{\mathrm{a}} - u^{\mathrm{c}})'\|_{\ell^2} \le \gamma^{-1} \sup_{\|v'\|_{\ell^2}=1} \bigl\langle \delta\mathcal{E}^{\mathrm{c}}(u^{\mathrm{a}}) - \delta\mathcal{E}^{\mathrm{a}}(u^{\mathrm{a}}), v \bigr\rangle, \tag{2.13}$$

which is reminiscent of the 'variational crimes' (or, simply, *consistency error*) studied in the finite element literature. We call the right-hand side of (2.13) the *modelling error* of the Cauchy–Born model.

Applying (2.5) and (2.11), we obtain

$$\begin{aligned}
\bigl\langle \delta\mathcal{E}^{\mathrm{c}}(u) - \delta\mathcal{E}^{\mathrm{a}}(u), v \bigr\rangle &= \sum_{\xi\in\mathbb{Z}} \bigl\{\partial_{\mathsf{F}} W(u'_\xi) - \mathsf{S}^{\mathrm{a}}_\xi(u)\bigr\} v'_\xi \\
&= \sum_{\xi\in\mathbb{Z}} \bigl\{2\phi'_2(2u'_\xi) - \phi'_2(u'_\xi + u'_{\xi+1}) - \phi'_2(u'_{\xi-1} + u'_\xi)\bigr\} v'_\xi.
\end{aligned}$$

A Taylor expansion of the stress error $\partial_{\mathsf{F}} W(u'_\xi) - \mathsf{S}^{\mathrm{a}}_\xi(u)$ to second order and an application of the Cauchy–Schwarz inequality yield the modelling error estimate

$$\langle \delta\mathcal{E}^{\mathrm{c}}(u) - \delta\mathcal{E}^{\mathrm{a}}(u), v\rangle \lesssim \big(\|u'''\|_{\ell^2} + \|u''\|_{\ell^4}^2\big)\|v'\|_{\ell^2}, \tag{2.14}$$

where $u''_\xi := u'_{\xi+1} - u'_\xi$ and $u'''_\xi := u''_\xi - u''_{\xi-1}$. Inserting this result into (2.13), we obtain the second-order error estimate

$$\|(u^{\mathrm{a}} - u^{\mathrm{c}})'\|_{\ell^2} \lesssim \|(u^{\mathrm{a}})'''\|_{\ell^2} + \|(u^{\mathrm{a}})''\|_{\ell^4}^2. \tag{2.15}$$

This error estimate states that, if $(u^{\mathrm{a}})'_\xi$ varies slowly relative to the atomic spacing, then the Cauchy–Born solution is a good approximation to the atomistic solution. A fully analogous result, valid in 2D/3D and for general many-body interactions, is given by Ortner and Theil (2013); see also E and Ming (2007) and Makridakis and Süli (2013) for related results in this direction.

Remark 2.2 (scaling). All of the formulations and results in this paper can be given from the point of view of the *scaling limit* rather than the *thermodynamic limit* as described in Remark 2.1. For example, rescaling space through $\xi \rightsquigarrow \epsilon\xi$ and $u(\xi) \rightsquigarrow \epsilon u(\xi)$, where ϵ is the atomic spacing, gives the second-order estimate

$$\|(u^{\mathrm{a}} - u^{\mathrm{c}})'\|_{\ell^2_\epsilon} \lesssim \epsilon^2\|(u^{\mathrm{a}})'''\|_{\ell^2_\epsilon} + \epsilon^2\|(u^{\mathrm{a}})''\|_{\ell^4_\epsilon}^2,$$

where $\|v\|_{\ell^p_\epsilon} := \big(\sum_{\xi\in\epsilon\mathbb{Z}} \epsilon|v_\xi|^p\big)^{1/p}$.

We prefer to use atomic units since they enable us to focus on the atomistic details, as in applications the geometry of the defect core is a quantity of interest.

Stability of atomistic and Cauchy–Born models

In the formal error analysis above, we have seen how the stability, that is, positive-definiteness of the Hessian of the approximate model, comes into play. To indicate why we would expect (2.12) to hold, we briefly analyse the atomistic and Cauchy–Born Hessians at the reference state.

The Cauchy–Born and atomistic Hessians are, respectively, given by

$$\langle \delta^2\mathcal{E}^{\mathrm{c}}(0)v, v\rangle = W''(0)\sum_{\xi\in\mathbb{Z}} |v'_\xi|^2, \tag{2.16}$$

$$\langle \delta^2\mathcal{E}^{\mathrm{a}}(0)v, v\rangle = \phi_1''(0)\sum_{\xi\in\mathbb{Z}} |v'_\xi|^2 + \phi_2''(0)\sum_{\xi\in\mathbb{Z}} \big|v'_\xi + v'_{\xi+1}\big|^2.$$

Noting that $W''(0) = \phi_1''(0) + 4\phi_2''(0)$ and applying the parallelogram identity

$$\big|v'_\xi + v'_{\xi+1}\big|^2 = 2|v'_\xi|^2 + 2|v'_{\xi+1}|^2 - |v''_\xi|^2 \tag{2.17}$$

(recall that $v''_\xi = v'_{\xi+1} - v'_\xi$), we observe that

$$\langle \delta^2\mathcal{E}^{\mathrm{a}}(0)v, v\rangle = \langle \delta^2\mathcal{E}^{\mathrm{c}}(0)v, v\rangle - \phi_2''(0)\sum_{\xi\in\mathbb{Z}} |v''_\xi|^2.$$

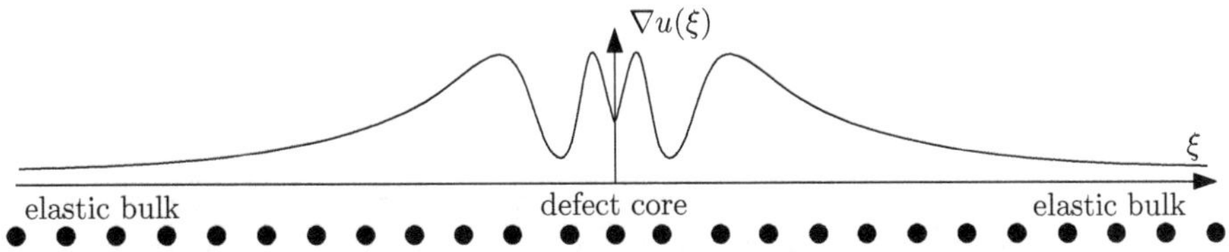

Figure 2.3. 1D analogy of a crystal defect: the deformation field varies rapidly in the defect core but is 'smooth' in the elastic bulk.

For Lennard-Jones type potentials we expect that $\phi_2''(0) < 0$, and therefore $\langle \delta^2\mathcal{E}^{\mathrm{a}}(0)v, v\rangle \geq \langle \delta^2\mathcal{E}^{\mathrm{c}}(0)v, v\rangle$. *Vice versa*, one can give general arguments that

$$\inf_{\|v'\|_{\ell^2}=1} \langle \delta^2\mathcal{E}^{\mathrm{c}}(0)v, v\rangle \geq \inf_{\|v'\|_{\ell^2}=1} \langle \delta^2\mathcal{E}^{\mathrm{a}}(0)v, v\rangle \tag{2.18}$$

(see Section 7.1; in the present case this can also be checked by a direct calculation (Dobson, Luskin and Ortner 2010*a*)). We can summarize these results (for $\phi_2''(0) < 0$) as

$$\inf_{\|v'\|_{\ell^2}=1} \langle \delta^2\mathcal{E}^{\mathrm{a}}(0)v, v\rangle = \inf_{\|v'\|_{\ell^2}=1} \langle \delta^2\mathcal{E}^{\mathrm{c}}(0)v, v\rangle. \tag{2.19}$$

Thus, we have proved that *the reference state $u = 0$ is stable in the atomistic model if and only if it is stable in the Cauchy–Born model.* We stress that, while (2.18) is a generic result, the equivalence (2.19) is specific to 1D Lennard-Jones-type interactions (Hudson and Ortner 2012, Li and Luskin 2013).

2.3. The need for atomistic-to-continuum coupling

In 2D/3D settings, crystalline solids exhibit many types of defects including, for example, impurities (an atom is replaced with an atom of a different species), vacancies (an atom is missing from a lattice site), dislocations (topological defects with high mobility that are the mechanism for crystal plasticity), or cracks. Defects themselves cannot normally be described with a continuum model. In addition, they generate elastic fields, which can be thought of as being singular at the defect core.

Due to the discrete setting, there are of course no singularities. However, in the notation of our 1D model, $(u^{\mathrm{a}})''_\xi$ will be of order $O(1)$ in a neighbourhood of the defect (see Figure 2.3), and the Cauchy–Born model therefore commits an $O(1)$ error. To obtain an accurate solution for problems with defects, a/c coupling methods decompose the reference lattice $\mathbb{Z}$ into an atomistic region $\mathcal{A}$ that contains the neighbourhood of defects where the exact atomistic model is used and a continuum region $\mathcal{C}$ where $(u^{\mathrm{a}})'_\xi$ varies slowly. See Figure 2.4 for an illustration of this decomposition.

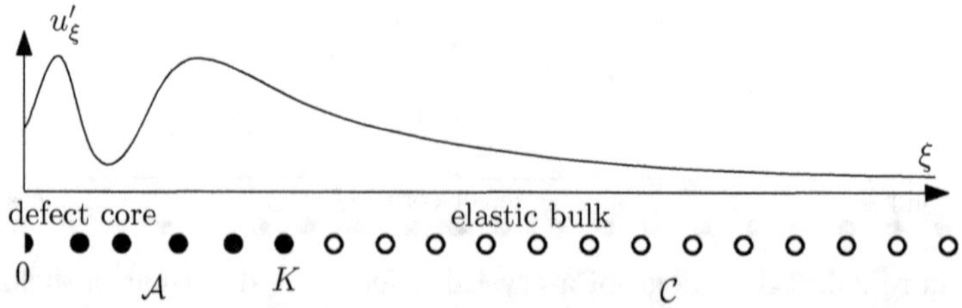

Figure 2.4. Decomposition of the atomistic chain into an atomistic region $\mathcal{A}$ and a continuum region $\mathcal{C}$, as employed in the QCE and QCF methods.

The solution of an a/c method can be expected to satisfy the error estimate

$$\|(u^{\mathrm{a}} - u^{\mathrm{ac}})'\|_{\ell^2} \lesssim \|(u^{\mathrm{a}})'''\|_{\ell^2(\mathcal{C})} + \|(u^{\mathrm{a}})''\|^2_{\ell^4(\mathcal{C})} + \text{coupling error}, \tag{2.20}$$

where u^{ac} is a minimizer of some a/c coupling functional $\mathcal{E}^{\mathrm{ac}} - \langle f, \cdot \rangle_{\mathbb{Z}}$. In this estimate, the error depends only on the 'smoothness' of u^{a} in the continuum region $\mathcal{C}$. If there is a defect in the atomistic region, that is, $(u^{\mathrm{a}}_\xi)'' = O(1)$ for some $\xi < K$, then this does not affect the error in an a/c coupling.

2.4. QCE coupling

The hybrid energy used in the energy-based quasicontinuum (QCE) coupling (Tadmor *et al.* 1996) is given by

$$\mathcal{E}^{\mathrm{qce}}(u) = \sum_{\xi \in \mathcal{A}} \Phi^{\mathrm{a}}_\xi(u) + \sum_{\xi \in \mathcal{C}} \Phi^{\mathrm{c}}_\xi(u), \tag{2.21}$$

where $\Phi^{\mathrm{a}}_\xi(u)$ is the *atomistic site energy* given in (2.2) and

$$\Phi^{\mathrm{c}}_\xi(u) := \tfrac{1}{2} W(u'_\xi) + \tfrac{1}{2} W(u'_{\xi+1})$$

is the *Cauchy–Born site energy*: see Figure 2.4. In the QCE method, we seek

$$u^{\mathrm{qce}} \in \arg\min \big\{ \mathcal{E}^{\mathrm{qce}}(u) - \langle f, u \rangle_{\mathbb{Z}} \big\}. \tag{2.22}$$

For simplicity and clarity of analysis, we will consider problems with antisymmetric forces and corresponding antisymmetric displacements

$$f_\xi = -f_{-\xi} \quad \text{and} \quad u_\xi = -u_{-\xi},$$

and a single symmetric atomistic region $\mathcal{A} := \{-K \le \xi \le K\}$. We will then define atomistic and QCE energies for $u : \mathbb{Z}_+ := \{0, \ldots, \infty\} \to \mathbb{R}$ to be half of their value when defined on $\mathbb{Z}$:

$$\begin{aligned}\mathcal{E}^{\mathrm{a}}_{+}(u) &:= \frac{1}{2}\Phi^{\mathrm{a}}_0(u) + \sum_{\xi=1}^{\infty}\Phi^{\mathrm{a}}_{\xi}(u),\\ \mathcal{E}^{\mathrm{qce}}_{+}(u) &:= \frac{1}{2}\Phi^{\mathrm{a}}_0(u) + \sum_{\xi=1}^{K}\Phi^{\mathrm{a}}_{\xi}(u) + \sum_{\xi=K+1}^{\infty}\Phi^{\mathrm{c}}_{\xi}(u)\end{aligned} \tag{2.23}$$

(see Figure 2.4). We then have the identity $\mathcal{E}^{\mathrm{a}}(u) = 2\mathcal{E}^{\mathrm{a}}_{+}(u)$ and $\mathcal{E}^{\mathrm{qce}}(u) = 2\mathcal{E}^{\mathrm{qce}}_{+}(u)$ for antisymmetric $u : \mathbb{Z} \to \mathbb{R}$. We note that we evaluate $\Phi^{\mathrm{a}}_0(u)$, $\Phi^{\mathrm{a}}_1(u)$, and $\Phi^{\mathrm{c}}_0(u)$ in our definition of $\mathcal{E}^{\mathrm{a}}_{+}(u)$ and $\mathcal{E}^{\mathrm{qce}}_{+}(u)$ by extending $u : \mathbb{Z}_{+} \to \mathbb{R}$ to antisymmetric $u : \mathbb{Z} \to \mathbb{R}$, so $u'_0 = -u_{-1} = u_1 = u'_1$, and so forth. Since, from now on, we will analyse only the antisymmetric problem, we will drop the subscript from $\mathcal{E}^{\mathrm{a}}_{+}(u)$ and $\mathcal{E}^{\mathrm{qce}}_{+}(u)$.

The rationale behind QCE is that Φ^{c}_{ξ} is exact under homogeneous deformation, that is, $\Phi^{\mathrm{a}}_{\xi}(u^{\mathsf{F}}) = \Phi^{\mathrm{c}}_{\xi}(u^{\mathsf{F}})$ for all $\mathsf{F} \in \mathbb{R}$ where $u^{\mathsf{F}}_{\xi} := \mathsf{F}\xi$. Hence, one may expect that $\mathcal{E}^{\mathrm{qce}}(u) \approx \mathcal{E}^{\mathrm{a}}(u)$ if the displacement u is 'smooth' in the continuum region $\mathcal{C}$. Indeed, if we compute the energy error, we obtain

$$\begin{aligned}\mathcal{E}^{\mathrm{qce}}(u) - \mathcal{E}^{\mathrm{a}}(u) &= \sum_{\xi=K+1}^{\infty}\{\Phi^{\mathrm{c}}_{\xi}(u) - \Phi^{\mathrm{a}}_{\xi}(u)\}\\ &= \sum_{\xi=K+1}^{\infty}\frac{1}{2}\{\phi_2(2u'_{\xi}) + \phi_2(2u'_{\xi+1}) - \phi_2(u'_{\xi-1}+u'_{\xi}) - \phi_2(u'_{\xi+1}+u'_{\xi+2})\},\end{aligned}$$

from which it is easy to obtain, by Taylor expansion of ϕ_2, that the energy error can be bounded by

$$|\mathcal{E}^{\mathrm{a}}(u) - \mathcal{E}^{\mathrm{qce}}(u)| \lesssim \|u'''\|_{\ell^1(\tilde{\mathcal{C}})} + \|u''\|^2_{\ell^2(\tilde{\mathcal{C}})},$$

where $\tilde{\mathcal{C}} := \{K \le \xi < \infty\}$.

On the other hand, following the arguments in Section 2.2 we may estimate the error between the atomistic and QCE solutions in terms of the error in the first variation. Computing the error in the first variation of QCE gives

$$\begin{aligned}&\langle\delta\mathcal{E}^{\mathrm{qce}}(u) - \delta\mathcal{E}^{\mathrm{a}}(u), v\rangle\\ &\quad= -\tfrac{1}{2}\phi'_2(u'_K + u'_{K+1})v'_K + \tfrac{1}{2}\phi'_2(u'_{K+1} + u'_{K+2})v'_{K+2}\\ &\qquad+ \tfrac{1}{2}\{2\phi'_2(2u'_{K+1}) - \phi'_2(u'_K + u'_{K+1}) - \phi'_2(u'_{K+1} + u'_{K+2})\}v'_{K+1}\\ &\qquad+ \sum_{\xi=K+2}^{\infty}\{2\phi'_2(2u'_{\xi}) - \phi'_2(u'_{\xi} + u'_{\xi+1}) - \phi'_2(u'_{\xi-1} + u'_{\xi})\}v'_{\xi},\end{aligned} \tag{2.24}$$

from which we can obtain by Taylor's theorem and the Cauchy–Schwarz inequality that

$$\langle \delta\mathcal{E}^{\rm qce}(u) - \delta\mathcal{E}^{\rm a}(u), v \rangle \lesssim \big(c_g + \|u'''\|_{\ell^2(\tilde{\mathcal{C}})} + \|u''\|^2_{\ell^4(\tilde{\mathcal{C}})}\big)\|v'\|_{\ell^2}, \tag{2.25}$$

where we (crudely) estimated the coupling error by

$$-\tfrac{1}{2}\phi_2'(u_K' + u_{K+1}')v_K' + \tfrac{1}{2}\phi_2'(u_{K+1}' + u_{K+2}')v_{K+2}' \lesssim c_g\|v'\|_{\ell^2}.$$

Here, c_g is some constant that is an estimate for the force acting on second-neighbour bonds.

To see that this upper bound is in fact attained, we consider for simplicity the case $f \equiv 0$. Then it is easy to check that $u^{\rm a} = u^{\rm c} = 0$ are solutions of, respectively, (2.4) and (2.10) (in this case the Cauchy–Born model is exact). However, it follows from (2.24) and $\delta\mathcal{E}^{\rm a}(0) = 0$ that

$$\langle \delta\mathcal{E}^{\rm qce}(0), v \rangle = \frac{\phi_2'(0)}{2}(v_{K+2}' - v_K') = \sum_{\xi=1}^{\infty} G_\xi^K v_\xi', \tag{2.26}$$

where

$$G_\xi^K := \frac{\phi_2'(0)}{2}(\delta_{K+2,\xi} - \delta_{K,\xi}) \tag{2.27}$$

is called the 'ghost force' ($\delta_{i,j}$ is the Kronecker delta). Ghost forces are spurious forces observed in most energy-based a/c methods, which are entirely due to the coupling mechanism rather than a mismatch between the atomistic and continuum descriptions. They are one of the most widely discussed issues both in the engineering and mathematical a/c literature. Often, the notion of the ghost force as conjugate to displacement rather than conjugate to strain, $\mathcal{F}_\xi^{\text{ghost force}} := -(G_\xi^{K+1} - G_\xi^K)$, is used.

Upon testing with a compactly supported virtual displacement $\hat{v}$ satisfying $\hat{v}_0 = 0$ and $\|\hat{v}'\|_{\ell^2} = 1$ defined by

$$\hat{v}_\xi' := \frac{\operatorname{sign}(\phi_2'(0))}{\sqrt{2}}(\delta_{K+2,\xi} - \delta_{K,\xi}),$$

we obtain for $\phi_2'(0) \neq 0$ that

$$\sup_{\|v'\|_{\ell^2}=1} \langle \delta\mathcal{E}^{\rm qce}(0) - \delta\mathcal{E}^{\rm a}(0), v \rangle \geq \langle \delta\mathcal{E}^{\rm qce}(0) - \delta\mathcal{E}^{\rm a}(0), \hat{v} \rangle = \frac{|\phi_2'(0)|}{\sqrt{2}} \gtrsim c_g.$$

We now return to the case of nonlinear deformation, $f \neq 0$. If we assume, as in fact it is easy to prove, that $\delta\mathcal{E}^{\rm qce}$ is Lipschitz-continuous, then we obtain that

$$\begin{aligned} c_g &\lesssim \langle \delta\mathcal{E}^{\rm qce}(u^{\rm a}) - \delta\mathcal{E}^{\rm a}(u^{\rm a}), \hat{v} \rangle = \langle \delta\mathcal{E}^{\rm qce}(u^{\rm a}) - \delta\mathcal{E}^{\rm qce}(u^{\rm qce}), \hat{v} \rangle \\ &\lesssim \|(u^{\rm a} - u^{\rm qce})'\|_{\ell^2}\|\hat{v}'\|_{\ell^2} = \|(u^{\rm a} - u^{\rm qce})'\|_{\ell^2}. \end{aligned}$$

Thus, the interfacial 'ghost forces' are responsible for the fact that the QCE method creates an $O(1)$ error in the strains. A more detailed analysis of the decay of the displacement and strain error away from the QCE interface is given by Dobson and Luskin (2009*a*) and by Ming and Yang (2009).

Stability of QCE

As in the case of the Cauchy–Born method, we briefly discuss the stability of the QCE method by focusing on homogeneous deformation. We can obtain

$$\begin{aligned}\langle \delta^2\mathcal{E}^{\mathrm{qce}}(0)v, v\rangle &= \phi_1''(0)\sum_{\xi=1}^{\infty}|v_\xi'|^2 + \phi_2''(0)\sum_{\xi=K+1}^{\infty}\tfrac12\big(|2v_\xi'|^2 + |2v_{\xi+1}'|^2\big)\\ &\quad+\phi_2''(0)\sum_{\xi=1}^{K}\tfrac12\big(|v_{\xi-1}'+v_\xi'|^2 + |v_{\xi+1}'+v_{\xi+2}'|^2\big) + \phi_2''(0)\tfrac12|v_1'+v_2'|^2\\ &= \langle \delta^2\mathcal{E}^{\mathrm{c}}(0)v, v\rangle\\ &\quad+\phi_2''(0)\sum_{\xi=1}^{K}\tfrac12\big(|v_{\xi-1}'+v_\xi'|^2 + |v_{\xi+1}'+v_{\xi+2}'|^2 - |2v_\xi'|^2 - |2v_{\xi+1}'|^2\big)\\ &\quad+\phi_2''(0)\tfrac12\big(|v_1'+v_2'|^2 - |2v_1'|^2\big).\end{aligned}$$

Applying the parallelogram identity (2.17), we deduce that

$$\begin{aligned}&\tfrac12\big(|v_{\xi-1}'+v_\xi'|^2 + |v_{\xi+1}'+v_{\xi+2}'|^2 - |2v_\xi'|^2 - |2v_{\xi+1}'|^2\big)\\ &\quad= |v_{\xi-1}'|^2 - |v_\xi'|^2 - |v_{\xi+1}'|^2 + |v_{\xi+2}'|^2 - \tfrac12|v_{\xi-1}''|^2 - \tfrac12|v_{\xi+1}''|^2,\end{aligned} \tag{2.28}$$

and inserting this formula above, we arrive at

$$\begin{aligned}\langle \delta^2\mathcal{E}^{\mathrm{qce}}(0)v, v\rangle &= \langle \delta^2\mathcal{E}^{\mathrm{c}}(0)v, v\rangle - \phi_2''(0)\big(|v_K'|^2 - |v_{K+2}'|^2\big)\\ &\qquad - \phi_2''(0)\left(\sum_{\xi=1}^{K-1}|v_\xi''|^2 + \tfrac12|v_K''|^2 + \tfrac12|v_{K+1}''|^2\right).\end{aligned}$$

Thus, assuming that $\phi_2''(0) < 0$, we obtain the lower bound

$$\langle \delta^2\mathcal{E}^{\mathrm{qce}}(0)v, v\rangle \geq \langle \delta^2\mathcal{E}^{\mathrm{c}}(0)v, v\rangle + \phi_2''(0)\|v'\|_{\ell^2}^2.$$

To obtain an upper bound for $\langle \delta^2\mathcal{E}^{\mathrm{qce}}(0)v, v\rangle/\|v'\|_{\ell^2}^2$, we test with the virtual displacement $\hat v$ satisfying $\hat v_0 = 0$ and $\hat v_\xi' := (\phi_2''(0))\delta_{K+2,\xi}$, and compute that

$$\langle \delta^2\mathcal{E}^{\mathrm{qce}}(0)\hat v, \hat v\rangle \leq \langle \delta^2\mathcal{E}^{\mathrm{c}}(0)\hat v, \hat v\rangle - \tfrac12|\phi_2''(0)|\,\|\hat v\|_{\ell^2}^2. \tag{2.29}$$

Noting from (2.19) that $\delta^2\mathcal{E}^{\mathrm{c}}(0)$ exactly reproduces the stability of $\delta^2\mathcal{E}^{\mathrm{a}}(0)$, we obtain (for $\phi_2''(0) < 0$) the result

$$\inf_{\|v'\|_{\ell^2}=1}\langle \delta^2\mathcal{E}^{\mathrm{qce}}(0)v, v\rangle \leq \inf_{\|v'\|_{\ell^2}=1}\langle \delta^2\mathcal{E}^{\mathrm{a}}(0)v, v\rangle - \tfrac12|\phi_2''(0)|, \tag{2.30}$$

that is, the coercivity constant of the QCE Hessian is strictly smaller than that of the atomistic model. In particular, it may happen that the QCE method is unstable even when both the atomistic and Cauchy–Born models are stable. We expand on this observation in the following remark.

Remark 2.3 ((in-)accuracy of the critical strain). Recalling from (2.1) that $\phi_i(s) := \phi(i\mathsf{A} + s)$, we see that $\phi_i''(0) = \phi''(i\mathsf{A})$. Reverting to this notation, and also recalling that $\phi''(2\mathsf{A})$ is always assumed to be negative, we have shown in (2.19) and (2.30) that

$$\begin{aligned}\inf_{\|v'\|_{\ell^2}=1} \langle \delta^2\mathcal{E}^{\mathrm{a}}(0)v, v\rangle &= \phi''(\mathsf{A}) + 4\phi''(2\mathsf{A}) \quad \text{and} \\ \inf_{\|v'\|_{\ell^2}=1} \langle \delta^2\mathcal{E}^{\mathrm{c}}(0)v, v\rangle &= \phi''(\mathsf{A}) + 4\phi''(2\mathsf{A}), \quad \text{but} \\ \inf_{\|v'\|_{\ell^2}=1} \langle \delta^2\mathcal{E}^{\mathrm{qce}}(0)v, v\rangle &\le \phi''(\mathsf{A}) + 4.5\phi''(2\mathsf{A}).\end{aligned} \tag{2.31}$$

Consider now a quasi-static loading scenario where we increase the macroscopic strain A until the chain becomes unstable, that is, when the Hessian of the lattice energy functional becomes indefinite and the system reaches a bifurcation point. For the atomistic and Cauchy–Born models, it follows from (2.31) that this *critical strain* $\mathsf{A}_{\mathrm{crit}}$ is given by the condition $\phi''(\mathsf{A}_{\mathrm{crit}}) + 4\phi''(2\mathsf{A}_{\mathrm{crit}}) = 0$. This instability can be considered to be a simple model for fracture.

We see from (2.31) that the loss of positive-definiteness of $\delta^2\mathcal{E}^{\mathrm{qce}}(0)$ occurs at a critical strain $\mathsf{A}^{\mathrm{qce}}_{\mathrm{crit}} \le \mathsf{B}$, where $\phi''(\mathsf{B}) + 4.5\phi''(2\mathsf{B}) = 0$, and it therefore follows that $\mathsf{A}^{\mathrm{qce}}_{\mathrm{crit}} < \mathsf{A}_{\mathrm{crit}}$. That is, the QCE method incorrectly predicts the load at which fracture occurs.

This effect is further exacerbated by the fact that the reference state is not an equilibrium of the QCE energy, and hence its stability is not characterized simply by the positive-definiteness of $\delta^2\mathcal{E}^{\mathrm{qce}}(0)$, but that of $\delta^2\mathcal{E}^{\mathrm{qce}}(u^{\mathrm{qce}})$. An analysis that also takes this effect into account is given in Dobson *et al.* (2010*a*). This analysis provides a theoretical basis to explain the numerical experiments reported by Miller and Tadmor (2009) and Van Koten, Li, Luskin and Ortner (2012), who observed that the QCE method incorrectly predicts lattice instability at a significantly reduced applied load.

2.5. Energy blending (B-QCE)

To overcome the significant interfacial error committed in the QCE method, Xiao and Belytschko (2004) proposed replacing the sharp transition from the atomistic to the continuum model with a smooth blending. We should not expect to remove the ghost forces entirely in this way, but we can hope that 'spreading' them will reduce the error.

Here, we present a variant of the blending approach that fits more closely into our general framework (Van Koten and Luskin 2011). We choose a

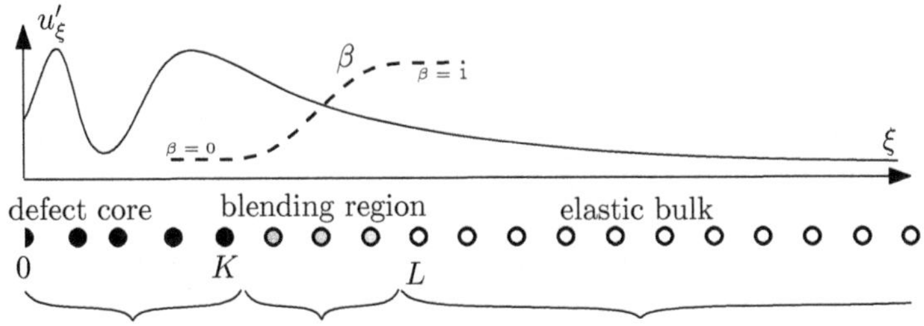

B-QCE	Φ^{a}	$(1-\beta)\Phi^{\mathrm{a}} + \beta\Phi^{\mathrm{c}}$	Φ^{c}
B-QCF	$\mathcal{F}^{\mathrm{a}}$	$(1-\beta)\mathcal{F}^{\mathrm{a}} + \beta\mathcal{F}^{\mathrm{c}}$	$\mathcal{F}^{\mathrm{c}}$

Figure 2.5. Illustration of the B-QCE and B-QCF methods.

blending function $\beta : \mathbb{Z}_+ \to [0,1]$ and define

$$\mathcal{E}^{\mathrm{bqce}}(u) := \sum_{\xi=0}^{\infty} \big\{(1-\beta(\xi))\Phi_\xi^{\mathrm{a}}(u) + \beta(\xi)\Phi_\xi^{\mathrm{c}}(u)\big\}. \tag{2.32}$$

For the purpose of implementation, we choose $K < L$ and assume that $\beta(\xi) = 0$ in $\{0, \dots, K\}$ and $\beta(\xi) = 1$ in $\{L, L+1, \dots\}$: that is, the smooth transition takes place between $\xi = K$ and L. We call $L - K$ the *blending width*. See Figure 2.5 for an illustration of this construction.

It is easy to rewrite (2.32) in the form

$$\mathcal{E}^{\mathrm{bqce}}(u) = \sum_{\xi=0}^{L}(1-\beta(\xi))\Phi_\xi^{\mathrm{a}}(u) + \int_K^\infty I\beta \cdot W(\nabla u)\,\mathrm{d}x, \tag{2.33}$$

where $I\beta(x)$ is the continuous piecewise affine interpolant of $\beta(\xi)$. The QCE energy is a special case of the B-QCE energy with β being given by the indicator function $\beta(\xi) = 0$ in $\{0, \dots, K\}$ and $\beta(\xi) = 1$ in $\{K+1, \dots\}$. In the B-QCE method, we seek

$$u^{\mathrm{bqce}} \in \arg\min \big\{\mathcal{E}^{\mathrm{bqce}}(u) - \langle f, u\rangle_{\mathbb{Z}_+}\big\}. \tag{2.34}$$

Reduction of ghost forces

We wish to explore, briefly, whether B-QCE indeed reduces the ghost forces, that is, the coupling error. To that end, it is convenient to use the fact that the B-QCE energy is an average of QCE energies with different interface positions (Van Koten and Luskin 2011). To see this, let $\mathcal{E}_\eta^{\mathrm{qce}}(u)$ denote the QCE energy with a/c interface at $K = \eta$. Then one readily checks that

$$\mathcal{E}^{\mathrm{bqce}}(u) = \sum_{\eta=K}^{L-1}(\beta(\eta+1) - \beta(\eta))\mathcal{E}_\eta^{\mathrm{qce}}(u).$$

We can then calculate that

$$\begin{aligned}\langle \delta\mathcal{E}^{\mathrm{bqce}}(0), v\rangle &= \sum_{\eta=K}^{L-1} \langle (\beta(\eta+1)-\beta(\eta))\delta\mathcal{E}_\eta^{\mathrm{qce}}(0), v\rangle \\ &= \sum_{\eta=K}^{L-1}\sum_{\xi=1}^{\infty}(\beta(\eta+1)-\beta(\eta))G_\xi^\eta v'_\xi \\ &=: \sum_{\xi=1}^{\infty} G_\xi^{\mathrm{bqce}} v'_\xi.\end{aligned} \tag{2.35}$$

It follows from (2.27) that the B-QCE ghost force is given by

$$G_\xi^{\mathrm{bqce}} = \sum_{\eta=K}^{L-1}(\beta(\eta+1)-\beta(\eta))G_\xi^\eta = -\tfrac{1}{2}\phi_2'(0)\big(\beta''_{\xi-1}+\beta''_\xi\big).$$

We have thus shown that

$$\langle \delta\mathcal{E}^{\mathrm{bqce}}(0), v\rangle = -\tfrac{1}{2}\phi_2'(0)\sum_{\xi=1}^{\infty}\big(\beta''_{\xi-1}+\beta''_\xi\big)v'_\xi, \tag{2.36}$$

from which we deduce that the B-QCE modelling error at $u^{\mathrm{a}}=0$ is bounded by

$$\sup_{\|v'\|_{\ell^2}=1} \big\langle \delta\mathcal{E}^{\mathrm{bqce}}(0) - \delta\mathcal{E}^{\mathrm{a}}(0), v\big\rangle \le |\phi_2'(0)|\|\beta''\|_{\ell^2}. \tag{2.37}$$

If we choose a piecewise linear blending function,

$$\beta(\xi) := \begin{cases} 0, & \xi \le K, \\ \frac{\xi-K}{L-K}, & K<\xi<L, \\ 1, & \xi \ge L, \end{cases}$$

then we obtain an $(L-K)^{-1}$ rate of modelling error reduction,

$$\sup_{\|v'\|_{\ell^2}=1} \big\langle \delta\mathcal{E}^{\mathrm{bqce}}(0) - \delta\mathcal{E}^{\mathrm{a}}(0), v\big\rangle \lesssim (L-K)^{-1} \sup_{\|v'\|_{\ell^2}=1} \big\langle \delta\mathcal{E}^{\mathrm{qce}}(0), v\big\rangle.$$

(Note that $\beta''_\xi = 0$, except at $\xi \in \{K, L\}$ where $\beta''_\xi = (L-K)^{-1}$.) That is, the blending approach does not remove but reduces the consistency error due to the ghost forces.

To improve the rate of error reduction, it follows from (2.37) that we should choose β to minimize $\|\beta''\|_{\ell^2}$ subject to the boundary conditions $\beta(\xi)=0$ for $\xi \le K$ and $\beta(\xi)=1$ for $\xi \ge L$, which is achieved for a cubic blending function with zero derivative at $\xi = K$ and $\xi = L$ (Van Koten and Luskin 2011, Luskin, Ortner and Van Koten 2013). The improved optimal rate of error reduction can be directly calculated by scaling to be $(L-K)^{-3/2}$.

Finally, we remark that a complete consistency and stability error analysis (see Van Koten and Luskin (2011), and Sections 6.2 and 7.2) reveals that

$$\|(u^{\mathrm{a}}-u^{\mathrm{bqce}})'\|_{\ell^2} \lesssim \|\beta''\|_{\ell^2}+\|\beta'\cdot(u^{\mathrm{a}})''\|_{\ell^2}+\|\beta\cdot(u^{\mathrm{a}})'''\|_{\ell^2}+\|\beta\cdot(u^{\mathrm{a}})''\|_{\ell^4}^2. \tag{2.38}$$

Stability of B-QCE

To understand why we should expect the B-QCE scheme to be stable, we present a brief analysis of the B-QCE Hessian at the reference state. More details are given in Van Koten and Luskin (2011). By a careful calculation, we can write it in the form

$$\begin{aligned}
\langle \delta^2\mathcal{E}^{\mathrm{bqce}}(0)v,v\rangle &= \phi_1''(0)\sum_{\xi=1}^{\infty}|v_\xi'|^2 + \phi_2''(0)\sum_{\xi=1}^{\infty}\beta(\xi)\tfrac{1}{2}\big(|2v_\xi'|^2+|2v_{\xi+1}'|^2\big)\\
&\quad + \phi_2''(0)\sum_{\xi=1}^{\infty}(1-\beta(\xi))\tfrac{1}{2}\big(|v_{\xi-1}'+v_\xi'|^2+|v_{\xi+1}'+v_{\xi+2}'|^2\big)\\
&\quad + \phi_2''(0)\tfrac{1}{2}|v_1'+v_2'|^2\\
&= \langle \delta^2\mathcal{E}^{\mathrm{c}}(0)v,v\rangle + \phi_2''(0)\bigg\{\tfrac{1}{2}|v_1'+v_2'|^2+\sum_{\xi=1}^{\infty}(1-\beta(\xi))\\
&\qquad\times\tfrac{1}{2}\big(|v_{\xi-1}'+v_\xi'|^2+|v_{\xi+1}'+v_{\xi+2}'|^2-|2v_\xi'|^2-|2v_{\xi+1}'|^2\big).
\end{aligned}$$

Applying the identity (2.28), we arrive at

$$\begin{aligned}
\langle \delta^2\mathcal{E}^{\mathrm{bqce}}(0)v,v\rangle &= \langle \delta^2\mathcal{E}^{\mathrm{c}}(0)v,v\rangle - \phi_2''(0)\sum_{\xi=1}^{\infty}(\beta_{\xi-1}''+\beta_\xi'')|v_\xi'|^2\\
&\quad - \phi_2''(0)\sum_{\xi=1}^{\infty}\big(1-\tfrac{1}{2}\beta(\xi-1)-\tfrac{1}{2}\beta(\xi+1)\big)|v_\xi''|^2.
\end{aligned}$$

Thus, assuming again that $\phi_2''(0)<0$ and $\beta\in[0,1]$, we obtain that

$$\langle \delta^2\mathcal{E}^{\mathrm{bqce}}(0)v,v\rangle \geq \langle \delta^2\mathcal{E}^{\mathrm{c}}(0)v,v\rangle - 2|\phi_2''(0)|\|\beta''\|_{\ell^\infty}\|v'\|_{\ell^2}^2. \tag{2.39}$$

Using the equivalence of the stability of $\delta^2\mathcal{E}^{\mathrm{a}}(0)$ and $\delta^2\mathcal{E}^{\mathrm{c}}(0)$ from (2.19), we can summarize these results as

$$\inf_{\|v'\|_{\ell^2}=1}\langle \delta^2\mathcal{E}^{\mathrm{bqce}}(0)v,v\rangle \geq \inf_{\|v'\|_{\ell^2}=1}\langle \delta^2\mathcal{E}^{\mathrm{a}}(0)v,v\rangle - 2|\phi_2''(0)|\|\beta''\|_{\ell^\infty}. \tag{2.40}$$

By choosing β quasi-optimally as discussed above, we obtain that $\|\beta''\|_{\ell^\infty}\lesssim (L-K)^{-2}$, and hence we conclude that up to an error controllable by the blending width, we can guarantee stability of $\delta^2\mathcal{E}^{\mathrm{bqce}}(0)$.

2.6. Quasi-nonlocal (QNL) coupling

To conclude our discussion of ghost forces we show how, in the second-neighbour chain model, they can be removed altogether. The first method

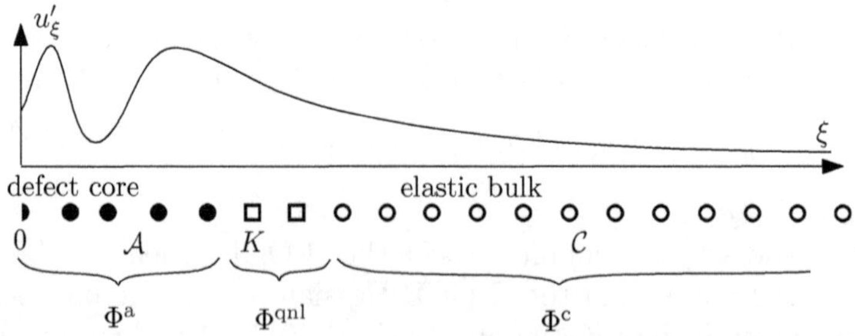

Figure 2.6. Illustration of the construction of the QNL method.

of this kind was formulated by Shimokawa *et al.* (2004). Here, we present a construction valid only for 1D pair interactions (Ortner 2011), employing an approximation of bonds rather than site energies:

$$\phi_2(u'_\xi + u'_{\xi+1}) \approx \tfrac{1}{2}\phi_2(2u'_\xi) + \tfrac{1}{2}\phi_2(2u'_{\xi+1}).$$

Taking the first variation on both sides above, at $u = 0$, we see that they are in fact equal:

$$\phi'_2(0)(v'_\xi + v'_{\xi+1}) = \tfrac{1}{2}\big\{\phi'_2(0)2v'_\xi + \phi'_2(0)2v'_{\xi+1}\big\}.$$

This immediately implies that the following construction exhibits no ghost forces: $\delta\mathcal{E}^{\rm qnl}(0) = 0$, where

$$\mathcal{E}^{\rm qnl}(u) := \mathcal{E}^{\rm a}(u) + \sum_{\xi=K+1}^{\infty} \big[\tfrac{1}{2}\{\phi_2(2u'_\xi) + \phi_2(2u'_{\xi+1})\} - \phi_2(u'_\xi + u'_{\xi+1})\big]. \tag{2.41}$$

Upon rearranging the sums, one may rewrite the QNL energy in the form

$$\mathcal{E}^{\rm qnl}(u) = \sum_{\xi=0}^{K-1} \Phi^{\rm a}_\xi(u) + \sum_{\xi=K}^{K+1} \Phi^{\rm qnl}_\xi(u) + \sum_{\xi=K+2}^{\infty} \Phi^{\rm c}_\xi(u), \tag{2.42}$$

where

$$\begin{aligned}\Phi^{\rm qnl}_K(u) &:= \tfrac{1}{2}\big\{\phi_1(u'_K) + \phi_2(u'_{K-1} + u'_K) + W(u'_{K+1})\big\}, \quad \text{and}\\ \Phi^{\rm qnl}_{K+1}(u) &:= \tfrac{1}{2}\big\{\phi_1(u'_{K+1}) + \phi_2(u'_K + u'_{K+1}) + W(u'_{K+2})\big\}\end{aligned}$$

(see Figures 2.6 and 2.7). Equation (2.42) is the form of the QNL scheme as originally proposed by Shimokawa *et al.* (2004).

Thus, the QNL method has a highly localized interface correction, and in particular we observe that the QNL energy indeed reduces to the atomistic model in the atomistic region and to the Cauchy–Born model in the continuum region. The *interface atoms* $\{K, K+1\}$ are labelled 'quasi-nonlocal' (Shimokawa *et al.* 2004) since their interaction with the atomistic region is nonlocal but their interaction with the continuum region is local. We will

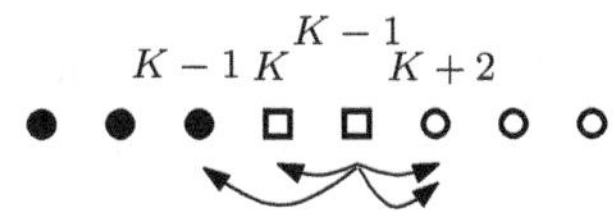

Figure 2.7. Modified interaction stencil used to construct the QNL method. Compare with the atomistic stencil of Figure 2.2.

consider couplings of the form (2.42) for more general interatomic potentials in Section 4.5.

To obtain an error estimate for the QNL method, we again compute the error in the first variation,

$$\begin{aligned}\langle \delta\mathcal{E}^{\mathrm{qnl}}(u) - \delta\mathcal{E}^{\mathrm{a}}(u), v\rangle &\\ = \sum_{\xi=K+2}^{\infty} \big\{2\phi_2'(2u_\xi') - \phi_2'(u_\xi' + u_{\xi+1}') - \phi_2'(u_{\xi-1}' + u_\xi')\big\} v_\xi' &\\ + \big\{\phi_2'(2u_{K+1}') - \phi_2'(u_{K+1}' + u_{K+2}')\big\} v_{K+1}', &\end{aligned}$$

and following the arguments leading to (2.15), we now obtain the first-order error estimate

$$\|(u^{\mathrm{a}} - u^{\mathrm{qnl}})'\|_{\ell^2} \lesssim \|(u^{\mathrm{a}})'''\|_{\ell^2(\mathcal{C})} + \|(u^{\mathrm{a}})''\|_{\ell^4(\mathcal{C})}^2 + |(u^{\mathrm{a}})_{K+1}''|, \tag{2.43}$$

where now $\mathcal{C} = \{K+1, \dots\}$. This is a clear potential improvement over the error in the B-QCE method, where there is an additional $\|\beta''\|_{\ell^2}$ term in the error estimate. We will more carefully distinguish the relative accuracy of B-QCE and QNL in Section 8.

Stability of QNL

As in the case of the Cauchy–Born and B-QCE models, we need to guarantee stability of the QNL method in order to make the error estimates discussed above rigorous. For the purpose of illustration, we again investigate stability of the QNL Hessian in the reference state,

$$\begin{aligned}\langle \delta^2\mathcal{E}^{\mathrm{qnl}}(0)v, v, \rangle = \phi_1''(0)\sum_{\xi=1}^{\infty} |v_\xi'|^2 + \phi_2''(0)|2v_1'|^2 + \phi_2''(0)\sum_{\xi=1}^{K} |v_\xi' + v_{\xi+1}'|^2 &\\ + \phi_2''(0) \sum_{\xi=K+1}^{\infty} \big(\tfrac{1}{2}|v_\xi'|^2 + \tfrac{1}{2}|v_{\xi+1}'|^2\big). &\end{aligned}$$

Applying the parallelogram identity (2.17) yields

$$\langle \delta^2\mathcal{E}^{\mathrm{qnl}}(0)v, v\rangle = \langle \delta^2\mathcal{E}^{\mathrm{c}}(0)v, v\rangle - \phi_2''(0)\sum_{\xi=1}^{K} |v_\xi''|^2.$$

For Lennard-Jones type interactions, we assume that $\phi_2''(0) < 0$, and hence it is easy to show (as in the Cauchy–Born case) that

$$\inf_{\|v'\|_{\ell^2}=1} \langle \delta^2 \mathcal{E}^{\rm qnl}(0)v, v\rangle = \inf_{\|v'\|_{\ell^2}=1} \langle \delta^2 \mathcal{E}^{\rm c}(0)v, v\rangle = \inf_{\|v'\|_{\ell^2}=1} \langle \delta^2 \mathcal{E}^{\rm a}(0)v, v\rangle.$$

That is, the QNL Hessian $\delta^2\mathcal{E}^{\rm qnl}(0)$ is stable if and only if the atomistic and Cauchy–Born Hessians are stable.

2.7. Force-based coupling (QCF)

An a/c hybrid energy without ghost forces for general 2D/3D interface geometries has yet to be constructed and implemented. A popular alternative has been to construct a/c approximations based on coupling forces. In the most basic variant of this approach, one computes the (negative of the) force of each atom from either the atomistic or the continuum model:

$$\mathcal{F}^{\rm qcf}_\xi(u) := \begin{cases} \dfrac{\partial \mathcal{E}^{\rm a}(u)}{\partial u_\xi} = -\mathsf{S}^{\rm a}_{\xi+1}(u) + \mathsf{S}^{\rm a}_{\xi}(u), & \text{for } \xi = 1, \dots, K, \\ \dfrac{\partial \mathcal{E}^{\rm c}(u)}{\partial u_\xi} = -\partial_{\mathsf{F}} W(u'_{\xi+1}) + \partial_{\mathsf{F}} W(u'_\xi), & \text{for } \xi \geq K+1. \end{cases} \tag{2.44}$$

One then solves the nonlinear system

$$\mathcal{F}^{\rm qcf}_\xi(u) = f_\xi \quad \text{for all } \xi = 1, \dots, \infty. \tag{2.45}$$

It can be checked that $\mathcal{F}^{\rm qcf}_\xi(u)$ is not a conservative force (Dobson and Luskin 2008*a*), so it is not given by the Euler–Lagrange equation for an energy. Nonetheless, we can formulate (2.45) in 'variational form' as

$$\langle \mathcal{F}^{\rm qcf}(u), v\rangle_{\mathbb{Z}_+} = \langle f, v\rangle_{\mathbb{Z}_+} \quad \text{for all } v,\ v(0) = 0.$$

We estimate the modelling error following Makridakis, Ortner and Süli (2011). We obtain from summation by parts that

$$\begin{aligned} &\langle \mathcal{F}^{\rm qcf}(u), v\rangle_{\mathbb{Z}_+} \\ &\quad = -\sum_{\xi=1}^{K} \{\mathsf{S}^{\rm a}_{\xi+1}(u) - \mathsf{S}^{\rm a}_\xi(u)\} v_\xi - \sum_{\xi=K+1}^{\infty} \{\partial_{\mathsf{F}} W(u'_{\xi+1}) - \partial_{\mathsf{F}} W(u'_\xi)\} v_\xi \\ &\quad = \sum_{\xi=1}^{K+1} \mathsf{S}^{\rm a}_\xi(u) v'_\xi + \sum_{\xi=K+2}^{\infty} \partial_{\mathsf{F}} W(u'_\xi) v'_\xi + \big(\partial_{\mathsf{F}} W(u'_{K+1}) - \mathsf{S}^{\rm a}_{K+1}(u)\big) v_{K+1}. \end{aligned} \tag{2.46}$$

We thus obtain the second-order modelling error bound

$$\begin{aligned} &\langle \mathcal{F}^{\rm qcf}(u), v\rangle_{\mathbb{Z}_+} - \langle \delta\mathcal{E}^{\rm a}(u), v\rangle \\ &\quad = \sum_{\xi=K+2}^{\infty} \{\partial_{\mathsf{F}} W(u'_\xi) - \mathsf{S}^{\rm a}_\xi(u)\} v'_\xi + \big(\partial_{\mathsf{F}} W(u'_{K+1}) - \mathsf{S}^{\rm a}_{K+1}(u)\big) v_{K+1} \end{aligned} \tag{2.47}$$

$$
\begin{aligned}
&= \sum_{\xi=K+2}^{\infty} \{2\phi_2'(2u_\xi') - \phi_2'(u_{\xi-1}' + u_\xi') - \phi_2'(u_\xi' + u_{\xi+1}')\} v_\xi' \\
&\qquad + \{2\phi_2'(2u_{K+1}') - \phi_2'(u_K' + u_{K+1}') - \phi_2'(u_{K+1}' + u_{K+2}')\} v_{K+1} \\
&\lesssim \{\|u'''\|_{\ell^\infty(\tilde{\mathcal{C}})} + \|u''\|_{\ell^\infty(\tilde{\mathcal{C}})}^2\} \|v'\|_{\ell^1},
\end{aligned}
$$

where $\tilde{\mathcal{C}} = \{K, \dots\}$, since we can use $v_0 = 0$ to bound

$$
|v_{K+1}| \leq \sum_{\xi=1}^{K+1} |v_\xi'| \leq \|v'\|_{\ell^1}.
$$

We note that the above estimate does not give a second-order modelling bound uniform in K in a discrete $W^{-1,2}$-seminorm since $|v_{K+1}|$ cannot be bounded by $\|v'\|_{\ell^2}$.

Even if we could prove a second-order modelling error bound in a discrete $W^{-1,2}$-seminorm that is uniform in K, this would not lead to the analogous error estimate. Indeed, it was shown in Dobson, Luskin and Ortner (2010*b*) that $\delta\mathcal{F}^{\mathrm{qcf}}(u)$ is not stable in the discrete $W^{1,p}$-seminorm for any $p < \infty$. It is somewhat technical to prove this, so instead we establish the slightly weaker result that $\delta\mathcal{F}^{\mathrm{qcf}}(u)$ is nearly always indefinite.

To see this, we take again $u = 0$, to obtain after some careful algebra that

$$
\langle \delta\mathcal{F}^{\mathrm{qcf}}(0)v, v\rangle_{\mathbb{Z}_+} = \langle \delta^2\mathcal{E}^{\mathrm{c}}(0)v, v\rangle - 4\phi_2''(0) \sum_{\xi=1}^{K} |v''(\xi)|^2 \tag{2.48}
$$

$$
+ \phi_2''(0)(v_{K+2}' - 2v_{K+1}' + v_K')v_{K+1}. \tag{2.49}
$$

Under the usual assumption that $\phi_2''(0) < 0$, the right-hand side of (2.48) is positive definite if and only if $\phi_1''(0) + 4\phi_2''(0) > 0$, but this is not sufficient to prove that $\langle \delta\mathcal{F}^{\mathrm{qcf}}(0)v, v\rangle$ is positive since (2.49) cannot be bounded by $\|v'\|_{\ell^2}^2$. Precisely, using the fact that

$$
\sup_{\|v'\|_{\ell^2}, v_0=0} |v_{K+1}| \approx K^{1/2},
$$

one can show that there exist positive constants C_1, C_2 such that

$$
\inf_{\|v'\|_{\ell^2}=1, v_0=0} \langle \delta\mathcal{F}^{\mathrm{qcf}}(0)v, v\rangle_{\mathbb{Z}_+} \leq C_1 - C_2 K^{1/2}.
$$

That is, for K sufficiently large (*i.e.*, for a sufficiently large atomistic region), $\delta\mathcal{F}^{\mathrm{qcf}}(0)$ cannot be positive definite.

However, using more general stability arguments, it can be shown that $\delta\mathcal{F}^{\mathrm{qcf}}(u)$ is in fact stable in the discrete $W^{1,\infty}$-seminorm (Dobson *et al.* 2010*b*) and hence it follows from (2.47) that

$$
\|(u^{\mathrm{a}} - u^{\mathrm{qcf}})'\|_{\ell^\infty} \lesssim \|(u^{\mathrm{a}})'''\|_{\ell^\infty(\mathcal{C})} + \|(u^{\mathrm{a}})''\|_{\ell^\infty(\mathcal{C})}^2. \tag{2.50}
$$

Remark 2.4. The non-symmetric and indefinite structure of the QCF method also presented a challenge to the efficient and reliable iterative solution of the QCF equations. Preconditioned linear stationary and GMRES methods have been studied in Luskin and Ortner (2012), Dobson, Luskin and Ortner (2011) and Dobson, Ortner and Shapeev (2012).

Ghost force correction (GFC)

At the time this article was written, the most popular publicly available a/c coupling software is the *Quasicontinuum Code*, which can be obtained from http://qcmethod.org/. At the core of this code lies the *ghost force correction* (GFC) *iteration*, which we briefly put into our context.

The GFC scheme is normally applied for computing quasi-static evolutions. In our context, we are given a family of external forces $\{f(t)\}_{t\in[0,1]}$, say, and aim to compute a continuous path

$$u^{\mathrm{a}}(t) \in \arg\min \left\{ \mathcal{E}^{\mathrm{a}}(v) - \langle f(t), v\rangle \right\} \quad \text{for } t \in [0,1].$$

To compute this, Shenoy *et al.* (1999) propose the following iteration. Given an initial condition $u^{\mathrm{gfc}}(0)$ and a time step k, let

$$u_k^{\mathrm{gfc}}(jk) \in \arg\min \left\{ \mathcal{E}^{\mathrm{qce}}(v) - \langle f(jk), v\rangle - \langle g_{j-1}, v\rangle \right\} \quad \text{for } j = 1, \ldots, 1/k.$$

where g_{j-1} is a dead-load correction that is intended to remove the ghost forces. Since the QCF operator has no ghost forces, a natural choice is

$$g(t_{j-1}) := \delta\mathcal{E}^{\mathrm{qce}}(u^{\mathrm{gfc}}(t_{j-1})) - \mathcal{F}^{\mathrm{qcf}}(u^{\mathrm{gfc}}(t_{j-1})),$$

which in particular ensures that g_{j-1} is concentrated in a small neighbourhood of the a/c interface.

Formally, it is straightforward to see that, if we interpolate u_k^{gfc} in time and if $u_k^{\mathrm{gfc}}(t) \to u(t)$ as $k \to 0$, then this limit solves the quasi-static QCF evolution,

$$\langle \mathcal{F}^{\mathrm{qcf}}(u(t)), v\rangle = \langle f(t), v\rangle \quad \text{for } t \in [0,1].$$

We therefore consider the GFC method as an iterative scheme to solve the QCF equation, and our analyses in the remainder of the paper will only focus on the latter.

This observation was first made by Dobson and Luskin (2008*b*) and Dobson *et al.* (2011), where further rigorous results on the GFC iteration and further discussions of the interesting challenges associated with solving the QCF equations can be found.

2.8. Force-based blending (B-QCF)

We can construct a blended version of QCF (B-QCF) that is positive definite with respect to the discrete $W^{1,2}$-seminorm (Li, Luskin and Ortner 2012) (see Lu and Ming (2013) for a stability result in other topologies) by

$$\mathcal{F}_\xi^{\mathrm{bqcf}}(u) := \beta(\xi)\mathcal{F}_\xi^{\mathrm{a}}(u) + (1-\beta(\xi))\mathcal{F}_\xi^{\mathrm{c}}(u).$$

The consistency of the B-QCF method is intuitively clear since there is no interface coupling error. However we shall see in Section 6.4 that there are non-trivial technical issues to overcome, hence we postpone this discussion.

Here, we focus instead on the stability of the B-QCF method. With some work, following in part the discussions of stability of the Cauchy–Born, B-QCE and QCF methods, we can derive the identity (the details of this computation are also given in Section 7.3 and Li *et al.* (2012))

$$\langle \delta\mathcal{F}^{\mathrm{bqcf}}(0)u, u\rangle = \langle \delta^2\mathcal{E}^{\mathrm{bqce}}(0)u, u\rangle + \phi_2''(0)\big(\mathbf{R}+\mathbf{S}\big), \tag{2.51}$$

where $|\mathbf{R}| \lesssim \|\beta''\|_{\ell^\infty}\|u'\|_{\ell^2}^2$ and the (crucial) error term $\mathbf{S}$ is given by

$$\mathbf{S} := \sum_{\xi=1}^{\infty}\big(\beta'''_{\xi+1}\big)u_\xi u'_{\xi+1}, \tag{2.52}$$

which can be bounded by

$$|\mathbf{S}| \le \|\beta'''\|_{\ell^\infty}\|u\|_{\ell^2[K,L]}\|u'\|_{\ell^2}. \tag{2.53}$$

We can construct blending functions by the scaling $\beta(\xi) = \hat\beta\big(\frac{\xi-K}{L-K}\big)$, where $\hat\beta \in C^3$, that satisfy

$$\|\beta''\|_{\ell^\infty} \lesssim (L-K)^{-2} \quad\text{and}\quad \|\beta'''\|_{\ell^\infty} \lesssim (L-K)^{-3}. \tag{2.54}$$

At this stage, we require a Poincaré-type inequality for u to bound $\|u\|_{\ell^2[K,L]}$ in (2.53). Recalling that $u_0 = 0$ from antisymmetry, we obtain

$$\|u\|_{\ell^2[K,L]} \le (L-K)^{1/2}\|u-u_0\|_{\ell^\infty[K,L]} \lesssim (L-K)^{1/2}L^{1/2}\|u'\|_{\ell^2}.$$

We thus obtain that, for some non-negative constants C_R and C_S,

$$\begin{aligned}\langle \delta\mathcal{F}^{\mathrm{bqcf}}(0)u, u\rangle &\ge \langle \delta^2\mathcal{E}^{\mathrm{bqce}}(0)u, u\rangle \\ &\quad - |\phi_2''(0)|\big(C_R(L-K)^{-2} + C_S(L-K)^{-5/2}L^{1/2}\big)\|u'\|_{\ell^2}^2.\end{aligned} \tag{2.55}$$

Utilizing the estimate (2.40) for the stability of $\delta^2\mathcal{E}^{\mathrm{bqce}}(0)$, we finally obtain

$$\begin{aligned}\inf_{\|v'\|_{\ell^2}=1}\langle \delta\mathcal{F}^{\mathrm{bqcf}}(0)u, u\rangle &\ge \inf_{\|v'\|_{\ell^2}=1}\langle \delta^2\mathcal{E}^{\mathrm{a}}(0)v, v\rangle \\ &\quad - C\big\{(L-K)^{-2} + (L-K)^{-5/2}L^{1/2}\big\}.\end{aligned} \tag{2.56}$$

We can conclude that, by choosing a sufficiently large blending region and imposing the mild requirement $(L-K) \gg L^{1/5}$, then, up to a controllable error, $\delta\mathcal{F}^{\mathrm{bqcf}}(0)$ is positive-definite if and only if $\delta^2\mathcal{E}^{\mathrm{a}}(0)$ is positive-definite.

Detailed numerical tests for 1D and 2D problems exploring the dependence of stability of the B-QCF method on the blending width are reported

Table 2.1. Summary of stability results and error estimates formally derived in Section 2. Here, $e := u^{\mathrm{a}} - u^{\mathrm{ac}}$, and $\eta_p^{\mathrm{cb}} := \|(u^{\mathrm{a}})'''\|_{\ell^p(\tilde{\mathcal{C}})} + \|(u^{\mathrm{a}})''\|^2_{\ell^{2p}(\tilde{\mathcal{C}})}$.

Method	Stability	Error estimate
QCE (Section 2.4)	reduced stability in energy norm	$c_g \lesssim \|e'\|_{\ell^2} \lesssim c_g + \eta_2^{\mathrm{cb}}$ (zeroth-order at interface)
B-QCE (Section 2.5)	controllable stability error in energy norm	$\|e'\|_{\ell^2} \lesssim \|\beta''\|_{\ell^2} + \eta_2^{\mathrm{cb}}$ (blending dictates error)
QNL (Section 2.6)	sharp stability in energy norm	$\|e'\|_{\ell^2} \lesssim \|(u^{\mathrm{a}})''_K\| + \eta_2^{\mathrm{cb}}$ (first-order at interface)
QCF (Section 2.7)	unstable in discrete $W^{1,p}$-norms, $p < \infty$	$\|e'\|_{\ell^\infty} \lesssim \eta_\infty^{\mathrm{cb}}$ (no interface error)
B-QCF (Section 2.8)	controllable stability error in energy norm	$\|e'\|_{\ell^2} \lesssim \eta_2^{\mathrm{cb}}$ (no interface error; this is not proved in Section 2)

in Li, Luskin, Ortner and Shapeev (2013). Sharp stability results for the B-QCF method in a discrete H^2-type norm are established by Lu and Ming (2013).

2.9. Summary of results

We briefly summarize the main results that we derived in this section in Table 2.1. To simplify the presentation, we introduce notation for the error $e := u^{\mathrm{a}} - u^{\mathrm{ac}}$, and for the Cauchy–Born consistency error, $\eta_p^{\mathrm{cb}} := \|(u^{\mathrm{a}})'''\|_{\ell^p(\tilde{\mathcal{C}})} + \|(u^{\mathrm{a}})''\|^2_{\ell^{2p}(\tilde{\mathcal{C}})}$, for a suitably defined (method-dependent) continuum region $\tilde{\mathcal{C}}$.

3. Atomistic simulation

In this section, we introduce an atomistic model for an infinite crystal. Some care must be taken with the formulation of the appropriate function space setting, but it is well suited to discussion of *rates of convergence* for defect simulations. We will then approximate the full atomistic problem by a reduced atomistic problem in a finite domain, which can be solved numerically, and estimate the resulting error in terms of the computational cost.

As in Section 2, we shall again consider only antisymmetric displacements. In the set-up of the model, this will require some additional notation. However, this initial investment will pay off in our subsequent discussion and analysis of a/c couplings.

The proofs of the results presented in this chapter are not of immediate interest to a/c coupling. Hence we have postponed some of the proofs to the Appendix.

3.1. Lattice functions

We identify a lattice function $v : \mathbb{Z} \to \mathbb{R}$ with its continuous piecewise affine interpolant, with weak derivative ∇v, which is also the pointwise derivative in each interval $(\xi - 1, \xi)$.

In order to measure smoothness of discrete functions, we introduce a second interpolant $\tilde{v} \in C^{2,1}(\mathbb{R})$ as follows: $\tilde{v}$ is the unique continuous function such that, for all $\xi \in \mathbb{Z}$,

$$\tilde{v}|_{(\xi-1,\xi)} \text{ is a polynomial of degree 5,} \tag{3.1}$$

$$\tilde{v}(\xi) = v(\xi), \tag{3.2}$$

$$\nabla\tilde{v}(\xi) = \tfrac{1}{2}\big(v(\xi+1) - v(\xi-1)\big), \quad \text{and} \tag{3.3}$$

$$\nabla^2\tilde{v}(\xi) = v(\xi+1) - 2v(\xi) + v(\xi-1). \tag{3.4}$$

There exists a unique interpolant satisfying these conditions. It is easy to see that $\tilde{v}$ is quasi-optimal among all possible $C^{2,1}$-interpolants of v, in the following sense.

Proposition 3.1. Let $v : \mathbb{Z} \to \mathbb{R}$, and let $\hat{v} \in C^{2,1}$ satisfy $\hat{v}(\xi) = v(\xi)$ for all $\xi \in \mathbb{Z}$. Then

$$\|\nabla^j \tilde{v}\|_{L^p(\xi-1,\xi)} \lesssim \|\nabla^j \hat{v}\|_{L^p(\xi-2,\xi+1)} \quad \text{for all } p \in [1,\infty],\ \xi \in \mathbb{Z},\ j = 1,2,3.$$

Moreover,

$$\|\nabla v\|_{L^p(\xi-1,\xi)} \le \|\nabla\tilde{v}\|_{L^p(\xi-1,\xi)} \lesssim \|\nabla v\|_{L^p(\xi-2,\xi+1)} \quad \text{for all } p \in [1,\infty],\ \xi \in \mathbb{Z}.$$

Proof. See Appendix A. □

Remark 3.2. Our definition of $\tilde{v}$ is natural, but also arbitrary in that many alternative interpolants could be employed that yield qualitatively the same results. Our choice was made only for its simplicity.

Antisymmetric lattice displacements are functions $u : \mathbb{Z}_+ \to \mathbb{R}$, where $\mathbb{Z}_+ := \{0, 1, 2, \dots\}$. Throughout, we identify them with lattice functions on $\mathbb{Z}$ through the convention that $u(-\xi) = -u(\xi)$ for $\xi \in \mathbb{Z}_+$. We define the *spaces* of antisymmetric displacements by

$$\mathcal{U}_0 := \big\{u : \mathbb{Z}_+ \to \mathbb{R} \,\big|\, \mathrm{supp}(u) \text{ is bounded and } u(0) = 0\big\}, \quad \text{and}$$

$$\mathcal{U} := \big\{u : \mathbb{Z}_+ \to \mathbb{R} \,\big|\, \nabla u \in L^2 \text{ and } u(0) = 0\big\}.$$

We equip the space $\mathcal{U}$ with the H^1-seminorm $|u|_{H^1} = \|\nabla u\|_{L^2}$, which makes it a Hilbert space. After defining the energy functional in the next section, we will motivate why an H^1-type function space is natural.

Proposition 3.3. *$\mathcal{U}_0$ is dense in $\mathcal{U}$.*

Proof. See Appendix A. □

We conclude this section on lattice functions with definitions of the finite difference operators. We fix an *interaction range* $\mathcal{R} = \{\pm 1, \dots, \pm r_{\rm cut}\} \subset \mathbb{Z}$, where $r_{\rm cut} \in \mathbb{N}$ is fixed throughout. (We typically think of $2 \le r_{\rm cut} \le 4$.) For each $u \in \mathcal{U}$ and $\xi \in \mathbb{Z}_+$, we define the interaction stencil

$$Du(\xi) := \big(D_\rho u(\xi)\big)_{\rho\in\mathcal{R}}, \quad \text{where} \quad D_\rho u(\xi) := u(\xi+\rho) - u(\xi),$$

and where we employ the antisymmetric convention $u(-\xi) := -u(\xi)$; for example, $D_{-\rho}u(0) = -D_\rho u(0)$ and $D_{-2}u(1) = -2u(1)$.

Remark 3.4. We can relate the finite difference operators u'_ξ used in Section 2 for nearest neighbour and next-nearest neighbour interactions ($r_{\rm cut} = 2$) to the finite difference operators $D_\rho u(\xi) := u(\xi+\rho) - u(\xi)$ for more general interaction ranges $\rho \in \mathcal{R}$ by

$$\begin{aligned} u'_\xi &:= u(\xi) - u(\xi-1) = D_1 u(\xi-1) = -D_{-1}u(\xi), \\ u'_\xi + u'_{\xi-1} &= D_2 u(\xi-2) = -D_{-2}u(\xi). \end{aligned} \tag{3.5}$$

3.2. Atomistic energy

Let $V \in C^k(\mathbb{R}^{\mathcal{R}})$, $k \ge 2$, be the interatomic many-body site potential. For compact displacements $u \in \mathcal{U}_0$, we can then define

$$\mathcal{E}^{\rm a}(u) := \sum_{\xi=0}^{\infty} \Phi^{\rm a}_\xi(u), \quad \text{where} \quad \Phi^{\rm a}_\xi(u) := \begin{cases} V(Du(\xi)) - V(\mathbf{0}), & \xi \ge 1, \\ \tfrac12\big(V(Du(\xi)) - V(\mathbf{0})\big), & \xi = 0. \end{cases}$$

Remark 3.5. For the next-nearest neighbour Lennard-Jones model studied in Section 2, we have from (2.2) and (3.5) that

$$\begin{aligned} V(g_{-2}, g_{-1}, g_1, g_2) &= \frac12 \sum_{\rho=1}^{2} [\phi(\rho\mathsf{A} + g_\rho) + \phi(\rho\mathsf{A} - g_{-\rho})] \\ &= \tfrac12\big\{\phi_1(-g_{-1}) + \phi_1(g_1) + \phi_2(-g_{-2}) + \phi_2(g_2)\big\}, \end{aligned} \tag{3.6}$$

where $\phi_i(s) := \phi(i\mathsf{A} + s)$ and $\mathsf{A} > 0$ is the macroscopic stretch described in Section 2.1.

For the embedded atom model (Daw and Baskes 1984), we have

$$
\begin{aligned}
V(g_{-r_{\mathrm{cut}}}, \dots, g_{r_{\mathrm{cut}}}) = \frac{1}{2} \sum_{\rho=1}^{r_{\mathrm{cut}}} \big[\phi(\rho\mathsf{A} + g_\rho) + \phi(\rho\mathsf{A} - g_{-\rho})\big] \\
+ \psi\Bigg(\sum_{\rho=1}^{r_{\mathrm{cut}}} \big[\eta(\rho\mathsf{A} + g_\rho) + \eta(\rho\mathsf{A} - g_{-\rho})\big]\Bigg),
\end{aligned} \tag{3.7}
$$

where ψ is the embedding energy function and η is the electron density function.

Remark 3.6. The definition of $\mathcal{E}^{\mathrm{a}}$ employs the convention that u is extended to an antisymmetric displacement. Hence, we should confirm that this coincides with the natural definition

$$
\mathcal{E}^{\mathrm{a}}(u) = \frac{1}{2} \sum_{\xi \in \mathbb{Z}} \big\{V(Du(\xi)) - V(\mathbf{0})\big\}. \tag{3.8}
$$

One can readily check that

$$
D_\rho u(\xi) = -D_{-\rho} u(-\xi). \tag{3.9}
$$

Hence, to obtain identity (3.8), we require that $V(\{g_\rho\}) = V(\{-g_{-\rho}\})$. Indeed, this is a natural condition, originating from lattice symmetries, that we will state as a standing assumption in (3.15).

Since Φ^{a}_ξ is an energy difference from the reference state, the lattice sum is finite for compact displacements, and hence $\mathcal{E}^{\mathrm{a}} : \mathcal{U}_0 \to \mathbb{R}$ is well-defined. $\mathcal{E}^{\mathrm{a}}(u)$ should be understood as the energy difference between the deformed state $y = y^{\mathsf{A}} + u$ and the reference state $y^{\mathsf{A}} = \mathsf{A}\mathbb{Z}$ (or $y^{\mathsf{A}}(\xi) = \mathsf{A}\xi$; similarly we will use $u^{\mathsf{F}}(\xi) = \mathsf{F}\xi$). The macroscopic stretch A is encoded in the definition of the potential V (see (2.1)).

We impose the far-field boundary condition $y(\xi) \sim y^{\mathsf{A}}(\xi)$ as $\xi \to \infty$, which, in terms of u, corresponds to $u(\xi) \sim 0$ as $\xi \to \infty$. Hence, $\mathcal{U}_0$ should be contained in any admissible displacement space. We now briefly explain why it is natural to take its closure with respect to the H^1-seminorm. From the proofs that we present in Appendix A, one can readily deduce that $|\mathcal{E}^{\mathrm{a}}(u)| \lesssim \|\nabla u\|_{L^2}^2$ for all $u \in \mathcal{U}_0$. Moreover, for the choice $V(Du(\xi)) = |D_1 u(\xi)|^2 + |D_2 u(\xi)|^2$ this estimate is sharp (in general, it is sharp for all 'small' displacements). Consequently, $\mathcal{U}$ is the largest space to which $\mathcal{E}^{\mathrm{a}}$ can be extended continuously.

Proposition 3.7. $\mathcal{E}^{\mathrm{a}} : \mathcal{U}_0 \to \mathbb{R}$ is continuous. In particular, there exists a unique continuous extension of $\mathcal{E}^{\mathrm{a}}$ to $\mathcal{U}$, which we still denote by $\mathcal{E}^{\mathrm{a}}$.

Proof. See Appendix A. □

From now on, we understand $\mathcal{E}^{\mathrm{a}}$ to be defined on $\mathcal{U}$. Next, we establish the differentiability of $\mathcal{E}^{\mathrm{a}}$. To that end, we introduce some notation for partial derivatives of V and Φ^{a}_{ξ}. For $\boldsymbol{\rho} = (\rho_1, \dots, \rho_j) \in \mathcal{R}^j$, $1 \le j \le k$, for $\boldsymbol{g} \in \mathbb{R}^{\mathcal{R}}$ and for $\xi \in \mathbb{Z}$, we define

$$V_{\boldsymbol{\rho}}(\boldsymbol{g}) := \frac{\partial^j V(\boldsymbol{g})}{\partial g_{\rho_1} \dots \partial g_{\rho_j}}, \quad \text{and analogously}$$

$$\Phi^{\mathrm{a}}_{\xi,\boldsymbol{\rho}}(u) := \begin{cases} V_{\boldsymbol{\rho}}(Du(\xi)), & \xi \in \mathbb{Z}_+ \setminus \{0\}, \\ \frac{1}{2} V_{\boldsymbol{\rho}}(Du(0)), & \xi = 0. \end{cases}$$

Note that for $\xi \ge r_{\mathrm{cut}}$, $\Phi^{\mathrm{a}}_{\xi,\rho}(u) = \partial \Phi^{\mathrm{a}}_{\xi}(u)/\partial u(\xi + \rho)$, and so forth, but for $\xi < r_{\mathrm{cut}}$ this is false.

We define bounds on partial derivatives

$$m(\boldsymbol{\rho}) := \prod_{i=1}^{j} |\rho_i| \sup_{\boldsymbol{g} \in \mathbb{R}^{\mathcal{R}}} \|V_{\boldsymbol{\rho}}(\boldsymbol{g})\| \quad \text{for } \boldsymbol{\rho} \in \mathcal{R}^j, \quad \text{and}$$

$$M^{(j,s)} := \sum_{\boldsymbol{\rho} \in \mathcal{R}^j} m(\boldsymbol{\rho}) |\boldsymbol{\rho}|_\infty^s,$$

where $\|\cdot\|$ denotes the ℓ^2-operator norm of a multilinear form and $|\boldsymbol{\rho}|_\infty := \max_{i=1,\dots,j} |\rho_i|$.

We shall assume throughout that the constants $M^{(j,s)}$ for $j = 1, \dots, k$ and $s \ge 0$ are finite. This assumption is in stark contrast to realistic interatomic potentials, which become singular as the distance between any two nuclei tends to zero. We make this assumption purely for the sake of convenience; removing it is not analytically challenging (see, *e.g.*, Ortner (2011) and Ortner and Theil (2013)) but complicates both the formulations and proofs of our results without introducing interesting conceptual ideas. To weaken the assumption, one may simply take the supremum of $\boldsymbol{g}$ over a *regime of interest*.

Remark 3.8. For most of our error analysis we will show how the constants in the estimates depend on the interaction potential by explicitly including prefactors of the form $M^{(j,s)}$. These prefactors in the estimates can be obtained by carefully tracing the dependence on the interaction potential in the proofs. Because this is tedious, we will in fact never carry out these details. However, it is important to be aware of this dependence, since the constants in the error estimates can be large (or even infinite) when the interaction decays slowly.

Lemma 3.9. $\mathcal{E}^{\mathrm{a}}$ is k times continuously Fréchet-differentiable, with variations

$$\langle \delta^j \mathcal{E}^{\mathrm{a}}(u)\boldsymbol{v}\rangle = \sum_{\xi\in\mathbb{Z}_+}\sum_{\boldsymbol{\rho}\in\mathcal{R}^j} \Phi^{\mathrm{a}}_{\xi,\boldsymbol{\rho}}(u)\prod_{i=1}^{j} D_{\rho_i}v_i(\xi) \quad \text{for all } \boldsymbol{v}\in\mathcal{U}_0^j, \tag{3.10}$$

for $j = 1,\dots,k$. Moreover, if $\sum_{i=1}^{j}\frac{1}{p_i} = 1$, then $\delta^j\mathcal{E}^{\mathrm{a}}$ satisfies the bound

$$\langle \delta^j \mathcal{E}^{\mathrm{a}}(u)\boldsymbol{v}\rangle \le M^{(j,0)}\prod_{i=1}^{j}\|\nabla v_i\|_{L^{p_i}} \quad \text{for all } \boldsymbol{v}\in\mathcal{U}_0^j. \tag{3.11}$$

Proof. See Appendix A. □

Remark 3.10. For test functions $v\in\mathcal{U}$, second and higher variations are well-defined via (3.10). First variations can be written as

$$\langle \delta\mathcal{E}^{\mathrm{a}}(u), v\rangle = \sum_{\xi\in\mathbb{Z}_+}\sum_{\rho\in\mathcal{R}}\big(\Phi^{\mathrm{a}}_{\xi,\rho}(u) - \Phi^{\mathrm{a}}_{\xi,\rho}(0)\big)D_\rho v(\xi). \tag{3.12}$$

Note that $\sum_{\xi\in\mathbb{Z}} D_\rho u(\xi) = 0$ for $u\in\mathcal{U}_0$, hence the constant $\Phi^{\mathrm{a}}_{\xi,\rho}(0)$ need not be subtracted. For $u\in\mathcal{U}$, (3.10) with $j=1$ may be ill-defined.

Remark 3.11 (far-field boundary condition). For $u\in\mathcal{U}$, the Cauchy–Schwarz inequality immediately implies that

$$|u(\xi)| \le \xi^{1/2}\|\nabla u\|_{L^2}. \tag{3.13}$$

Moreover, by checking that the displacement $u(\xi) := |\xi|^{1/2-\epsilon}$ belongs to $\mathcal{U}$ for $0<\epsilon<1/2$, we see that this bound is essentially sharp. This raises the question in which sense the far-field boundary condition $u\sim 0$ is still observed in the space $\mathcal{U}$. We interpret this as follows.

Let $y(\xi) := \mathsf{A}\xi + u(\xi)$ be the atomistic *deformation*, where $\mathsf{A}>0$ is a prescribed macroscopic stretch. Then (3.13) implies that

$$y(\xi) = \mathsf{A}\xi + u(\xi) \sim \mathsf{A}\xi \quad \text{as } |\xi|\to\infty, \tag{3.14}$$

which formally translates to $u(\xi)\sim 0$.

We remark that, in 2D, elements in the closure of $\mathcal{U}_0$ may still exhibit logarithmic growth, while in 3D they decay to zero uniformly due to the Gagliardo–Nirenberg–Sobolev embedding.

Symmetries of V

Since we are considering a single species of atoms it follows that permuting the atom positions should not change the total energy. Moreover, inversion of the atom positions about any point should also leave the energy invariant. It is therefore natural to assume that the site potential satisfies

$$V\big(\{-g_{-\rho}\}_{\rho\in\mathcal{R}}\big) = V\big(\{g_\rho\}_{\rho\in\mathcal{R}}\big) \quad \text{for all } \boldsymbol{g}\in\mathbb{R}^{\mathcal{R}}. \tag{3.15}$$

See, for example, the Lennard-Jones (3.6) and EAM (3.7) models. As a matter of fact, if only the global energy satisfies such a symmetry, but not the site potential, then one can construct a modified site potential that does satisfy (3.15): see Van Koten and Ortner (2012).

From (3.15), and the fact that $-\mathsf{F}(-\rho) = \mathsf{F}\rho$, it is easy to deduce that

$$V_{-\boldsymbol{\rho}}(\{\mathsf{F}\rho\}_{\rho\in\mathcal{R}}) = (-1)^j V_{\boldsymbol{\rho}}(\{\mathsf{F}\rho\}_{\rho\in\mathcal{R}}) \quad \text{for all } \boldsymbol{\rho} \in \mathcal{R}^j,\ \mathsf{F} \in \mathbb{R}. \tag{3.16}$$

This will be a useful property to exploit later on.

3.3. *The variational problem*

For the sake of simplicity, we consider only (antisymmetric) dead load external forces. This allows us to focus on the properties of the internal energies. We say that a lattice function $f : \mathbb{Z}_+ \to \mathbb{R}$ belongs to $\mathcal{U}^*$ if there exists a constant $\|f\|_{\mathcal{U}^*}$ such that

$$\langle f, v\rangle_{\mathbb{Z}_+} \le \|f\|_{\mathcal{U}^*} \|\nabla v\|_{L^2} \quad \text{for all } v \in \mathcal{U}_0.$$

In this case, the linear form $\langle f, \cdot\rangle_{\mathbb{Z}_+}$ can be extended to a bounded linear functional on $\mathcal{U}$. It is easy to see that $f \in \mathcal{U}^*$ if and only if there exists $g \in \ell^2(\mathbb{Z}_+)$ such that $D_{-1}g(\xi) = f(\xi)$ for $\xi \ge 1$.

Given $f \in \mathcal{U}^*$, we seek

$$u^{\mathrm{a}} \in \arg\min \big\{\mathcal{E}^{\mathrm{a}}(u) - \langle f, u\rangle_{\mathbb{Z}_+} \,\big|\, u \in \mathcal{U}\big\}. \tag{3.17}$$

We understand (3.17) as a *local minimization problem* that may have several or no solutions. If u^{a} solves (3.17), then it satisfies the Euler–Lagrange equation

$$\langle \delta\mathcal{E}^{\mathrm{a}}(u), v\rangle = \langle f, v\rangle_{\mathbb{Z}_+} \quad \text{for all } v \in \mathcal{U}. \tag{3.18}$$

Vice versa, if $u^{\mathrm{a}} \in \mathcal{U}$ satisfies (3.18) and in addition the second-order sufficient optimality condition

$$\langle \delta^2\mathcal{E}^{\mathrm{a}}(u^{\mathrm{a}})v, v\rangle \ge c_0\|\nabla v\|_{L^2}^2 \quad \text{for all } v \in \mathcal{U}, \tag{3.19}$$

for some $c_0 > 0$, then u^{a} is a solution to (3.17). We call such a solution a *strong local minimizer.* (Note that (3.19) is not a strong requirement: if u^{a} solves (3.17) then it automatically satisfies (3.19) with $c_0 = 0$.)

We shall not be concerned with a general existence theory of solutions to (3.17). However, we state the following trivial result.

Proposition 3.12. Suppose that $k \ge 3$ and that the reference state is strongly stable; that is, there exists $c_0 > 0$ such that

$$\langle \delta^2\mathcal{E}^{\mathrm{a}}(0)v, v\rangle \ge c_0\|\nabla v\|_{L^2}^2 \quad \text{for all } v \in \mathcal{U}.$$

Then there exists $\epsilon > 0$ such that, for all $f \in \mathcal{U}^*$ with $\|f\|_{\mathcal{U}^*} < \epsilon$, (3.17) has a strong local minimizer.

Proof. This result is an immediate consequence of the inverse function theorem. □

In order to analyse approximations to (3.17), we will need to exploit regularity properties of the solution $u^{\rm a}$. To that end, we shall make the following *assumption*.

(DH) *Decay hypothesis.* There exists a strong local minimizer $u^{\rm a} \in \mathcal{U}$ of (3.17) and $\alpha > 1/2$, such that

$$\left|\nabla^j \tilde{u}^{\rm a}(x)\right| \le C_{\mathbf{DH}} x^{-\alpha+1-j}, \quad \text{for } x > r_0,\ j = 0,1,2,3. \tag{3.20}$$

Moreover, we assume that $|\nabla^j \tilde{f}(x)| \le C_{\mathbf{DH}} x^{-\alpha-1-j}$ for $j = 0, 1, 2$.

Remark 3.13. **(1)** The requirement that $\alpha > 1/2$ is due to the fact that, if $|\nabla u(x)| \sim x^{-\alpha}$ as $x \to \infty$, for $\alpha \le 1/2$, then $u \notin \mathcal{U}$. Thus, **(DH)** does not represent a significant restriction of generality.

(2) Formally, we expect that (3.20) holds provided that $|\nabla^j \tilde{f}(x)| \lesssim x^{-\alpha-1-j}$ for $j = 0, 1$. A rigorous proof of this result is tedious, but not difficult. *Vice versa*, it is easy to see that (3.20) implies the stated decay for f, $j = 0, 1$. The natural additional assumption that $|\nabla^2 \tilde{f}(x)| \lesssim x^{-\alpha-3}$ will be required in the external consistency estimate derived in Section 5.3.

3.4. Approximation from a finite domain

Since $\mathcal{U}$ is infinite-dimensional, (3.17) cannot be solved directly, hence we reduce the problem to a finite domain and estimate the resulting error. We fix $N \in \mathbb{N}$ and define the finite-dimensional displacement space

$$\mathcal{U}_N := \big\{ u \in \mathcal{U}_0 \,|\, u(\xi) = 0 \text{ for } \xi \ge N \big\}.$$

We approximate the exact variational problem (2.3) with

$$u^N \in \arg\min \big\{ \mathcal{E}^{\rm a}(u) - \langle f, u \rangle_{\mathbb{Z}_+} \,\big|\, u \in \mathcal{U}_N \big\}. \tag{3.21}$$

This formulation effectively reduces the infinite domain $\mathbb{Z}$ to the finite domain $\{0, \dots, N\}$.

Since (3.21) is a pure Galerkin approximation, we expect that the error is equivalent to the best approximation error. We define an approximation operator $\Pi_N : \mathcal{U} \to \mathcal{U}_N$ as follows:

$$\Pi_N u(\xi) := \begin{cases} u(\xi) - \frac{\xi}{N} u(N), & \xi = 0, \dots, N, \\ 0, & \xi > N. \end{cases}$$

It is straightforward to see that, if u satisfies **(DH)**, then

$$\|\nabla u - \nabla \Pi_N u\|_{L^2} \le C N^{1/2-\alpha}, \tag{3.22}$$

where C depends on α and on $C_{\mathbf{DH}}$. This motivates the following result.

Theorem 3.14. Let u^{a} be a strong local minimizer of (3.17) satisfying **(DH)** and (3.19). Then there exists $N_0 \in \mathbb{N}$ such that, for all $N \geq N_0$, there exists a strong local minimizer u^N of (3.21) satisfying

$$\|\nabla u^{\mathrm{a}} - \nabla u^N\|_{L^2} \leq \frac{C}{c_0} N^{1/2-\alpha},$$

where C depends on $M^{(2,0)}, \alpha$, and $C_{\mathbf{DH}}$; and c_0 is the coercivity constant from (3.19).

Proof. The result is an easy consequence of the quantitative inverse function theorem stated in Appendix B.

The Euler–Lagrange equation for (3.21) is

$$\langle \delta\mathcal{E}^{\mathrm{a}}(u^N), v\rangle = (f, v)_{\mathbb{Z}_+} \quad \text{for all } v \in \mathcal{U}_N,$$

which can be written equivalently as

$$\langle \mathcal{G}_N(e), v\rangle := \langle \delta\mathcal{E}^{\mathrm{a}}(\Pi_N u^{\mathrm{a}} + e) - \delta\mathcal{E}^{\mathrm{a}}(u^{\mathrm{a}}), v\rangle = 0 \quad \text{for all } v \in \mathcal{U}_N, \tag{3.23}$$

where $e := u^N - \Pi_N u^{\mathrm{a}}$. We aim to solve (3.23) in $\mathcal{U}_N$. Since $\delta\mathcal{G}_N = \delta^2\mathcal{E}^{\mathrm{a}}(\Pi_N u^{\mathrm{a}} + \cdot)$ is Lipschitz (according to Lemma 3.9) we only need to show that $\|\mathcal{G}_N(0)\|_{\mathcal{U}_N^*}$ can be made arbitrarily small and that $\delta\mathcal{G}_N(0)$ is positive definite.

Using (3.11) and (3.22), we bound the residual by

$$\begin{aligned}\|\mathcal{G}_N(0)\|_{\mathcal{U}_N^*} &= \|\delta\mathcal{E}^{\mathrm{a}}(\Pi_N u^{\mathrm{a}}) - \delta\mathcal{E}^{\mathrm{a}}(u^{\mathrm{a}})\|_{\mathcal{U}_N^*} \\ &\leq M^{(2,0)}\|\nabla\Pi_N u^{\mathrm{a}} - \nabla u^{\mathrm{a}}\|_{L^2} \leq CM^{(2,0)}N^{1/2-\alpha}.\end{aligned}$$

In particular, $\|\mathcal{G}_N(0)\|_{\mathcal{U}_N^*} \to 0$ as $N \to \infty$.

Using again (3.11) and (3.22), and also the fact that $\|\nabla w\|_{L^\infty} \leq \|\nabla w\|_{L^2}$ for all $w \in \mathcal{U}$, we can also bound the coercivity constant of $\delta\mathcal{G}_N(0)$:

$$\begin{aligned}\langle \delta\mathcal{G}_N(0)v, v\rangle &= \langle \delta^2\mathcal{E}^{\mathrm{a}}(\Pi_N u^{\mathrm{a}})v, v\rangle \\ &\geq \langle \delta^2\mathcal{E}^{\mathrm{a}}(u^{\mathrm{a}})v, v\rangle - M^{(3,0)}\|\nabla\Pi_N u^{\mathrm{a}} - \nabla u^{\mathrm{a}}\|_{L^\infty}\|\nabla v\|_{L^2}^2 \\ &\geq \big(c_0 - CM^{(3,0)}N^{1/2-\alpha}\big)\|\nabla v\|_{L^2}^2.\end{aligned}$$

In particular, if N is sufficiently large, then

$$\langle \delta\mathcal{G}_N(0)v, v\rangle \geq \frac{c_0}{2}\|\nabla v\|_{L^2}^2 \quad \text{for all } v \in \mathcal{U}_N.$$

Combining these estimates and applying the inverse function theorem, Lemma B.1, we obtain the stated result. □

Remark 3.15. **(1)** Theorem 3.14 is an error estimate in terms of the *computational cost* required to solve (3.21). This motivates us to analyse a/c methods in terms of their computational cost as well.

(2) Theorem 3.14 can also be understood to establish that (3.17) is the *thermodynamic limit* of (3.21) as $N \to \infty$ and generalizations are therefore interesting. (i) If u^{a} does not satisfy the decay hypothesis **(DH)**, then Theorem 3.14 is still true, but without an explicit convergence rate. (ii) If we were to apply periodic boundary conditions to (3.21), then the approximation operator Π_N would still be admissible and hence, with only minor modifications, the result can be extended to this case. (iii) Moreover, the result can be readily extended to 2D/3D.

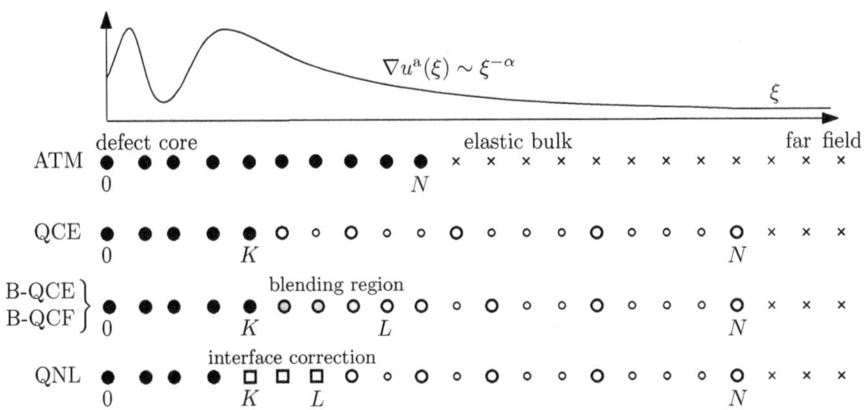

Figure 3.1. Illustration of the a/c coupling methods introduced in Section 4, and of the decay hypothesis **(DH)**. Black discs (•) indicate atoms described by the atomistic model; white discs (○) indicate atoms modelled by the Cauchy–Born model; grey discs indicate the blending region used in the B-QCE and B-QCF methods; squares □ indicate atoms receiving special interface corrections as in QNL-type constructions. Finally, small white discs are atoms that are 'slaved' to the rep-atoms (large white discs) via P1 interpolation, and small crosses × are atoms removed via the far-field boundary condition.

4. Many-body finite-range atomistic-to-continuum coupling

In this section, we introduce the a/c couplings that we will analyse in the context of the atomistic model introduced in the previous section. We will point out the main approximations being made in the various a/c couplings and give an outline of our analytical approach, to motivate the subsequent analysis.

An illustration and summary of all a/c coupling methods, including the purely atomistic model (ATM), is given in Figure 3.1.

4.1. The quasicontinuum idea: Galerkin projection

In Section 2 we introduced a/c methods via the idea of coupling an atomistic to a continuum model. Here, we take an alternative point of view: we first project $\mathcal{E}^{\mathrm{a}}$ onto a coarse subspace, and then apply further approximations to render the coarse optimization problem computable. Ultimately, the two philosophies lead to the same a/c coupling formulations.

We restrict displacements again to a computational domain $[0, N]$, $N \in \mathbb{N}$: see Section 3.4. Let $\mathcal{T}_h = \{T\}$ be a regular partition of $[0, N]$ into closed intervals T, with vertices $\mathcal{N}_h \subset \mathbb{Z}_+$ (or, *rep-atoms* in the language of the quasicontinuum method). We define the coarse displacement space by

$$\mathcal{U}_h := \{u_h \in \mathcal{U}_N \mid u_h \text{ is piecewise affine with respect to } \mathcal{T}_h\}. \tag{4.1}$$

Elements of $\mathcal{U}_h$ are defined pointwise, but give rise to lattice functions through point evaluation. Since finite element nodes lie on lattice sites, this is compatible with our interpolation of lattice functions.

The philosophy of the *quasicontinuum* (QC) *method* is to retain the atomistic description, but restrict the admissible set to the *coarse space* $\mathcal{U}_h$:

$$u_h^{\mathrm{qc}} \in \arg\min \{\mathcal{E}^{\mathrm{a}}(u_h) - \langle f, u_h\rangle_{\mathbb{Z}_+} \mid u_h \in \mathcal{U}_h\}. \tag{4.2}$$

Typically, one would let all atoms in the neighbourhood of a defect (in our case the origin) be finite element vertices (atomistic region), and rapidly coarsen the mesh in the surrounding bulk (continuum region). Thus, one would fully resolve the defect, but transition seamlessly to a coarse 'continuum-like' description of the far field.

The method (4.2) is formulated in the spirit of Galerkin approximations for nonlinear variational problems. Applying the standard arguments for approximations of this type, one can see that, if a solution u^{a} of (3.17) exists, and its best approximation from $\mathcal{U}_h$ lies 'sufficiently close', then a solution u_h^{qc} of (4.2) exists, satisfying Céa's lemma.

In 1D, the implementation of (4.2) can be achieved with $O(\#\mathcal{N}_h)$ complexity since the element interfaces are particularly simple. However, in 2D/3D the evaluation of $\mathcal{E}^{\mathrm{a}}(u_h)$ and $\delta\mathcal{E}^{\mathrm{a}}(u_h)$ for $u_h \in \mathcal{U}_h$ still requires summation over a number of lattice points proportional to the area of the element faces, which is much larger than $\#\mathcal{N}_h$. To make this precise: while displacement gradients ∇u_h are homogeneous inside elements $T \in \mathcal{T}_h$, the interaction stencils $Du_h(\xi)$ are only homogeneous ($Du_h(\xi) = \nabla u_h|_T \cdot \mathcal{R}$) in subsets of the elements T. Near element interfaces, one must resort to exact summation over all lattice sites. Thus, the cost of evaluating $\mathcal{E}^{\mathrm{a}}(u_h)$ and $\delta\mathcal{E}^{\mathrm{a}}(u_h)$ scales with the surface area of the mesh, which in 1D is $O(\#\mathcal{N}_h)$, but is significantly higher in 2D/3D.

In conclusion, while (4.2) is attractive from a theoretical perspective, it is not a practical scheme.

4.2. Finite element notation

Before we introduce the various many-body a/c couplings in the following subsections, we summarize a unified notation for finite element meshes.

We fix three mesh parameters K, L, $N \in \mathbb{N}$, $K \le L \le N/2$, where the computational domain is $[0, N]$. We choose a set of finite element nodes $\mathcal{N}_h = \{\nu_0, \dots, \nu_{N_{\mathcal{T}_h}}\}$, for some $N_{\mathcal{T}_h} \in \mathbb{N}$, such that $\{0, \dots, L, N\} \subset \mathcal{N}_h \subset \{0, \dots, N\}$. The set of interior nodes is $\mathcal{N}_h^\circ := \mathcal{N}_h \setminus \{0, N\}$. The finite elements are given by $\mathcal{T}_h = \{[\nu_{j-1}, \nu_j] \,|\, j = 1, \dots, J\}$. For each $T \in \mathcal{T}_h$, let $h_T := \operatorname{diam} T$. For each $x \in [0, N]$, $x \in \operatorname{int} T$, let $h(x) := h_T$. For $x > N$, let $h(x) := 1$.

The space of all continuous piecewise affine functions on $[0, N]$ is given by $\mathrm{P1}(\mathcal{T}_h)$, and the space of piecewise constant functions by $\mathrm{P0}(\mathcal{T}_h)$. The admissible finite element space $\mathcal{U}_h$ is defined by (4.1), and imposes the boundary condition $u_h(0) = u_h(N) = 0$.

Finally, for any function $v : \mathcal{N}_h \to \mathbb{R}$, let $I_h v : [0, N] \to \mathbb{R}, I_h v \in \mathrm{P1}(\mathcal{T}_h)$, denote its continuous piecewise affine interpolant,

$$I_h v(\zeta) := v(\zeta) \quad \text{for all } \zeta \in \mathcal{N}_h.$$

Coarse-graining the external forces
For $f, g : \mathcal{N}_h \to \mathbb{R}^J$, we define

$$\langle f, g \rangle_h := \int_0^N I_h(f \cdot g)\,\mathrm{d}x = \sum_{j=1}^{J} \tfrac{1}{2} h_T \{ f(\nu_{j-1}) \cdot g(\nu_{j-1}) + f(\nu_j) \cdot g(\nu_j) \}.$$

In all our a/c couplings, we shall approximate the potential of the external forces, $\langle f, u_h \rangle_{\mathbb{Z}_+}$, by its (coarse-grained) trapezoidal rule approximation, $\langle f, u_h \rangle_h$.

4.3. Cauchy–Born approximation and QCE method

In Section 4.1 we constructed a coarse-graining scheme that reduces the number of degrees of freedom, but still has a high computational cost for evaluating the energy and forces (at least its 2D/3D variants do). An analogy in continuum mechanics is that integrals occurring in Galerkin projections of nonlinear equations cannot be evaluated explicitly. In continuum mechanics, the solution to this issue is to employ quadrature rules to approximate the variational forms.

In the present case, since we are using P1 finite elements, the midpoint rule is a natural choice. For each 'large' element $T \in \mathcal{T}_h$ we choose an atomic site ξ_T near the centre of T and formally approximate

$$\sum_{\xi \in T} \Phi_\xi^{\mathrm{a}}(u_h) \approx \#(T \cap \mathbb{Z}) \Phi_{\xi_T}^{\mathrm{a}}(u_h) \approx |T| \Phi_{\xi_T}^{\mathrm{a}}(u_h),$$

where the issue of multiple counting of atoms at element boundaries is resolved at the second approximation step. If T is sufficiently large that

$\xi_T + \mathcal{R} \subset T$, then we obtain

$$\sum_{\xi \in T} \Phi^{\mathrm{a}}_\xi(u_h) \approx |T| W(\nabla u_{h|_T}), \quad \text{where} \quad W(\mathsf{F}) := V(Du^{\mathsf{F}}(0)) - V(\mathbf{0}), \tag{4.3}$$

(recall that $u^{\mathsf{F}}(\xi) := \mathsf{F}\xi$), which is called the *Cauchy–Born strain energy density function*, generalizing the definition (2.8) for the NNN Lennard-Jones model. This approximation is also suitable for smaller elements provided that u_h is approximately affine in a neighbourhood of each element T (*i.e.*, u_h is 'smooth'). This motivates the following construction.

In the energy-based quasicontinuum method (QCE method) of Tadmor *et al.* (1996), previously introduced in Section 2.4, we fix $r_{\mathrm{cut}} \leq K = L < N$. For each site $\xi = K+1, K+2, \dots$, we define the Cauchy–Born site energy

$$\Phi^{\mathrm{c}}_\xi(u_h) := \int_{\xi-1/2}^{\xi+1/2} W(\nabla u_h)\,\mathrm{d}x.$$

Then the energy functional of the QCE method is defined by $\mathcal{E}^{\mathrm{qce}} : \mathcal{U}_h \to \mathbb{R}$, where

$$\mathcal{E}^{\mathrm{qce}}(u_h) := \sum_{\xi=0}^{K} \Phi^{\mathrm{a}}_\xi(u_h) + \sum_{\xi=K+1}^{N} \Phi^{\mathrm{c}}_\xi(u_h). \tag{4.4}$$

Since the QCE energy is given as a finite sum of C^k potentials, it follows that $\mathcal{E}^{\mathrm{qce}} \in C^k(\mathcal{U}_h)$.

The energy (4.4) can be rewritten in a format more natural for its implementation:

$$\mathcal{E}^{\mathrm{qce}}(u_h) = \sum_{\xi=0}^{K} \Phi^{\mathrm{a}}_\xi(u_h) + \int_{K+1/2}^{N} W(\nabla u_h)\,\mathrm{d}x. \tag{4.5}$$

In the QCE method, we approximate the atomistic problem (3.17) by

$$u_h^{\mathrm{qce}} \in \arg\min \big\{ \mathcal{E}^{\mathrm{qce}}(u_h) - \langle f, u_h \rangle_h \,\big|\, u_h \in \mathcal{U}_h \big\}. \tag{4.6}$$

The patch test and the ghost force problem

A key motivation for the QCE method is that

$$\Phi^{\mathrm{c}}_\xi(u^{\mathsf{F}}) = \Phi^{\mathrm{a}}_\xi(u^{\mathsf{F}}) \quad \text{for all } \mathsf{F} \in \mathbb{R},\ \xi \geq K+1. \tag{4.7}$$

That is, the QCE site energies are exact under (locally) homogeneous deformations. We call (4.7) *local energy consistency*. As demonstrated in Section 2.4, it implies that $\mathcal{E}^{\mathrm{qce}}(u_h) \approx \mathcal{E}^{\mathrm{a}}(u)$ if u is 'smooth' in the continuum region and u_h is a 'good approximation' to u.

However, in order to ensure that minimizers of (4.6) approximate minimizers of (3.17), we require more stringent conditions (consistency and stability, which we touched upon in Section 2 and will discuss in more detail in Section 4.7). Indeed, as shown in Section 2.4, the QCE method generally fails the *force patch test*, $\delta\mathcal{E}^{\mathrm{qce}}(0) \neq 0$, that is, $u_h = 0$ is not a critical point of (4.6). In other words, even if u^{a} is 'smooth', a solution u_h^{qce} to (4.6) develops an oscillation at the a/c interface that lowers the energy further. This effect *always* creates an $O(1)$ error in the QCE methods (Dobson and Luskin 2009*a*, Ming and Yang 2009).

Further remarks

(1) Whereas earlier literature on a/c coupling exists, the introduction of the QCE method by Tadmor *et al.* (1996) spawned a surge of activity in the field. In particular, the Cauchy–Born approximation appears to have been used for the first time in this context. An iterative correction of the interface inconsistency ('ghost forces') was suggested by the same group (Shenoy *et al.* 1999), which leads naturally into force-based coupling schemes, discussed in Sections 2.7, 2.8 and 4.6.

(2) We may also consider the 'pure Cauchy–Born' approximation,

$$u^{\mathrm{c}} \in \arg\min \big\{\mathcal{E}^{\mathrm{c}}(u) - \langle f, u\rangle_{\mathbb{R}_+} \,\big|\, u \in \dot{H}^1\big\}, \tag{4.8}$$

where

$$\mathcal{E}^{\mathrm{c}}(u) = \int_0^\infty W(\nabla u)\,\mathrm{d}x,$$

$\langle f, u\rangle_{\mathbb{R}_+} = \int_0^\infty f u \,\mathrm{d}x$ provided that $f \cdot u$ is integrable, and $\dot{H}^1$ is a homogeneous Sobolev space ($u \in \dot{H}^1$ if and only if $u \in H^1_{\mathrm{loc}}$ and $\nabla u \in L^2$). This model goes back to the work of Cauchy (1882), who used it to derive symmetries of the tensor of linearized elasticity (the Cauchy relations). In the modern calculus of variations and PDE literature, the Cauchy–Born model has been analysed and discussed, for instance by Blanc, Le Bris and Lions (2002), Friesecke and Theil (2002), E and Ming (2007), and Makridakis and Süli (2013). Recently, Ortner and Theil (2013) have proved for both static and dynamic problems and general many-body interactions that the Cauchy–Born approximation is second-order accurate in a natural scaling limit.

(3) In this subsection we have only discussed the consequences of choosing a midpoint-type summation rule. However, we have approximated this further by the Cauchy–Born stored energy density. Lin (2007) shows that taking a cluster of atoms near the centre of each element leads to a first-order accurate method. Similarly, Gunzburger and Zhang (2010) consider discrete variants of higher-order quadrature rules. In both cases the transition to a fully refined mesh, which introduces substantial new algorithmic and analytical challenges, is not taken into account.

Knap and Ortiz (2001) as well as Eidel and Stukowski (2009) consider an entirely different class of summation rules, which are node-based rather than element-based. These rules are, in principle, suited to transition to full atomistic resolution. However, Luskin and Ortner (2009) have shown that these node-based summation rules result in prohibitively large and uncontrollable consistency errors.

4.4. Energy blending (B-QCE)

We now generalize the energy blending approach described in Section 2.5. If we can 'spread' the QCE ghost force error over an interface region,

$$G(\xi) := \begin{cases} (L-K)^{-1}, & \xi = K, \dots, L-1, \\ 0, & \text{otherwise}, \end{cases} \quad \text{then } \|G\|_{\ell^2} = (L-K)^{-1/2}.$$

(However, $\|G\|_{\ell^1} = 1$ for all K, L.) In our consistency analysis in Section 6.2, we will in fact see that exploiting symmetries of the interaction potential reduces the coupling error by a factor of $(L-K)^{-3/2}$, provided that the blending function is chosen in an optimal way.

We now choose a blending function $\beta : [0, N] \to [0, 1]$ and define for $u_h \in \mathcal{U}_h$

$$\mathcal{E}^{\mathrm{bqce}}(u_h) := \sum_{\xi=0}^{N} \{(1 - \beta(\xi))\Phi_\xi^{\mathrm{a}}(u_h) + \beta(\xi)\Phi_\xi^{\mathrm{c}}(u_h)\}. \tag{4.9}$$

We choose $r_{\mathrm{cut}} < K < L < N$ and assume that $\beta(\xi) = 0$ in $[0, K]$ and $\beta(\xi) = 1$ in $[L, N]$; that is, the transition takes place between $\xi = K$ and L. It is then easy to rewrite (4.9) in the form

$$\mathcal{E}^{\mathrm{bqce}}(u_h) := \sum_{\xi=0}^{L} (1 - \beta(\xi))\Phi_\xi^{\mathrm{a}}(u_h) + \int_K^N I_h\beta \cdot W(\nabla u_h)\,\mathrm{d}x. \tag{4.10}$$

Again, we observe that $\mathcal{E}^{\mathrm{bqce}} \in C^k(\mathcal{U}_h)$.

In the B-QCE method, we seek

$$u_h^{\mathrm{bqce}} \in \arg\min \{\mathcal{E}^{\mathrm{bqce}}(u_h) - \langle f, u_h\rangle_h \,|\, u_h \in \mathcal{U}_h\}. \tag{4.11}$$

Further remarks

B-QCE shares the idea of a blending region with the bridging domain method (Belytschko and Xiao 2003), the AtC coupling method (Badia *et al.* 2008), the Arlequin method (Bauman *et al.* 2008, Prudhomme *et al.* 2008), and overlapping domain methods (Parks, Bochev and Lehoucq 2008, Seleson and Gunzburger 2010).

4.5. QNL-type methods

Concurrently with the introduction of blending methods, Shimokawa *et al.* (2004) introduced an approach that promised not just to reduce the ghost forces but to remove them altogether. Their idea was to create a narrow transition region of 'quasi-nonlocal atoms' with modified site energies, so that their interaction with the atomistic region would have atomistic qualities (nonlocal) and their interaction with the continuum region would have continuum qualities (local). They proposed the *ansatz*

$$\begin{aligned}\mathcal{E}^{\mathrm{qnl}}(u_h) &:= \sum_{\xi=0}^{K} \Phi^{\mathrm{a}}_\xi(u_h) + \sum_{\xi=K+1}^{L-1} \Phi^{\mathrm{i}}_\xi(u_h) + \sum_{\xi=L}^{N} \Phi^{\mathrm{c}}_\xi(u_h) \\ &= \sum_{\xi=0}^{K} \Phi^{\mathrm{a}}_\xi(u_h) + \sum_{\xi=K+1}^{L-1} \Phi^{\mathrm{i}}_\xi(u_h) + \int_{L-1/2}^{N} W(\nabla u_h)\,\mathrm{d}x\end{aligned}$$

where Φ^{i}_ξ is a modified site potential to be defined for $\xi = K+1, \dots, L-1$.

In the remainder of this article, we will allow the following general form of $\mathcal{E}^{\mathrm{qnl}}$, which also requires the definition of a site potential Φ^{i}_L:

$$\mathcal{E}^{\mathrm{qnl}}(u_h) = \sum_{\xi=0}^{K} \Phi^{\mathrm{a}}_\xi(u_h) + \sum_{\xi=K+1}^{L} \Phi^{\mathrm{i}}_\xi(u_h) + \int_{L}^{N} W(\nabla u_h)\,\mathrm{d}x. \tag{4.12}$$

(The previous definition of $\mathcal{E}^{\mathrm{qnl}}$ fits into this general form by defining the rightmost site potential to be $\Phi^{\mathrm{i}}_L(u_h) := \int_{L-1/2}^{L} W(\nabla u_h)\,\mathrm{d}x$.)

We have already seen an explicit example in Section 2.6, which suggests that this formulation can potentially lead to substantially improved accuracy. The key advantage over the B-QCE method is that we may be able to choose the interface width $L - K$ to be much smaller. However, this comes at the cost of a far more complex construction. Indeed, at the time of writing this article it is still not known whether methods of the type discussed in this section can be constructed for general many-body potentials and general interface geometries in 2D/3D.

The key new ingredient in (4.12) is the interface site energy Φ^{i}_ξ. We will later prove that it is sufficient to construct Φ^{i}_ξ in such a way that the resulting energy satisfies the following force consistency condition.

(F) *Force consistency.* $\mathcal{F}^{\mathrm{qnl}}_\zeta(u^{\mathsf{F}}) = 0$ for all $\zeta \in \mathcal{N}_h^\circ$, $\mathsf{F} \in \mathbb{R}$,

where, here and throughout, we define

$$\mathcal{F}^{\mathrm{qnl}}_\zeta(u_h) := \frac{\partial \mathcal{E}^{\mathrm{qnl}}(u_h)}{\partial u_h(\zeta)},$$

and remark that due to the finite interaction range, $\mathcal{F}^{\mathrm{qnl}}_\xi(u^{\mathsf{F}})$ is well-defined despite the fact that $u^{\mathsf{F}} \notin \mathcal{U}_h$. (It is more common to define a force as $\mathcal{F}_\zeta = -\frac{\partial \mathcal{E}}{\partial u_h(\zeta)}$, but for notational convenience we have not followed this convention.)

The geometry reconstruction idea

The second key idea of Shimokawa *et al.* (2004) was to define the interface site energy Φ^{i} in terms of the original atomistic site energy Φ^{a}, but with 'reconstructed' atom positions. Rephrased in our notation, they proposed setting

$$\Phi^{\mathrm{i}}_\xi(u_h) := V\big(\tilde{D}u_h(\xi)\big) - V(\mathbf{0}) \quad \text{for } \xi = K+1, \dots, L-1, \tag{4.13}$$

where $\tilde{D}u_h(\xi) = \{\tilde{D}_\rho u_h(\xi)\}_{\rho\in\mathcal{R}}$ is a modified finite difference stencil. (Note the abuse of notation: $\tilde{D}_\rho$ may in fact be ξ-dependent.) The rationale is to allow interface atoms to interact via atomistic rules (nonlocally) with their neighbours in the atomistic region, and by continuum rules (locally) with their neighbours in the continuum region.

For example, suppose that $\mathcal{R} = \{\pm1, \pm2\}$. Then this intuition leads us to define $L = K+3$, and

$$\tilde{D}u(\xi) = \big\{D_{-2}u(\xi), D_{-1}u(\xi), D_1u(\xi), 2D_1u(\xi)\big\} \quad \text{for } \xi = K+1, K+2. \tag{4.14}$$

It is straightforward to check that the resulting a/c coupling satisfies the force consistency condition **(F)**.

Note, however, that this idea does not extend to third neighbour interactions: if $\mathcal{R} = \{\pm1, \pm2, \pm3\}$, $L = K+4$, and

$$\tilde{D}u(\xi) = \big\{D_{-3}u(\xi), D_{-2}u(\xi), D_{-1}u(\xi), D_1u(\xi), 2D_1u(\xi), 3D_1u(\xi)\big\},$$

then there is generally a ghost force acting on the atom $\xi = K+1$, and possibly others.

E, Lu and Yang (2006) proposed a more general form of the stencil $\tilde{D}u(\xi)$ with free parameters that are then to be fitted in order to satisfy **(F)**. This idea was further generalized by Ortner and Zhang (2012) to

$$\tilde{D}u(\xi) = \left\{\sum_{\varsigma\in\mathcal{R}} C_{\xi,\rho,\varsigma} D_\varsigma u(\xi)\right\}_{\rho\in\mathcal{R}}.$$

It turns out that, with this *ansatz*, **(F)** reduces to a linear system of equations that is independent of F and V. It is unknown at present whether this system has a solution for general a/c interfaces in 2D/3D; see E *et al.* (2006) and Ortner and Zhang (2012) for more details.

The reflection method

We now present a construction proposed by Ortner, Shapeev and Zhang (2013) (in a slightly different context this was previously proposed by Langwallner, Ortner and Süli (2013)) which is valid for general many-body interactions in 1D. Unfortunately there appears to be no straightforward generalization to 2D/3D.

The idea is to apply a reflection boundary condition to the atomistic region, thus decoupling it from the continuum region. We fix $K > 0$, set $L := K + r_{\rm cut}$, and for each $v \in \mathcal{U}$ define its antisymmetric reflection about the interface L,

$$\begin{cases} v^*(\xi) := v(\xi), & 0 \le \xi \le L, \\ v^*(L+\rho) := 2v(L) - v(L-\rho), & 1 \le \rho \le r_{\rm cut}. \end{cases} \tag{4.15}$$

Then we define

$$\mathcal{E}^{\rm rfl}(u) := \mathcal{E}^*(u) + \int_L^\infty W(\nabla u)\,{\rm d}x, \quad \text{where} \tag{4.16}$$

$$\mathcal{E}^*(u) := \sum_{\xi=0}^{L-1} \Phi^{\rm a}_\xi(u^*) + \tfrac{1}{2}\Phi^{\rm a}_L(u^*).$$

It is easy to see that $\mathcal{E}^{\rm rfl}$ can be rewritten in the form (4.12).

We will now rewrite $\mathcal{E}^*$ in a more convenient form, in terms of a periodic cell energy, from which it will become immediately obvious that $\mathcal{E}^{\rm rfl}$ satisfies the force-consistency condition **(F)**. This construction follows Langwallner *et al.* (2013, Section 3.3).

First, we redefine the reflection operator by

$$u^*(\xi + 2L\eta) := 2\eta u(L) + u(\xi), \quad \text{for all } \xi \in \{-L, \dots, L\} \text{ and } \eta \in \mathbb{Z}. \tag{4.17}$$

One can readily check that (4.17) is an extension of the earlier definition:

$$u^*(L+\rho) = u^*\big([-L+\rho] + 2L\big) = 2u(L) + u(-L+\rho) = 2u(L) - u(L-\rho).$$

To understand (4.17), we rewrite $u(\xi) = \mathsf{G}\xi + w(\xi)$ where $\mathsf{G} = u(L)/L$ and $w(\pm L) = 0$ (recall the antisymmetric extension of u across the origin). Then w^* is simply a $2L$-periodic extension of $w|_{\{-L+1,\dots,L\}}$ and

$$u^*(\xi + 2L\eta) = \mathsf{G}(\xi + 2L\eta) + w(\xi).$$

Thus, we see that $Du^*(\xi)$ is $2L$-periodic.

Lemma 4.1. Let u^* be defined by (4.17), then

$$\mathcal{E}^*(u) = \tfrac{1}{2} \sum_{\xi=-L+1}^{L} V(Du^*(\xi)) + C^*, \tag{4.18}$$

where $C^* = -2LV(0)$.

Proof. We have already seen in (3.9) that $D_\rho u(\xi) = -D_{-\rho}u(-\xi)$ for all $\xi \in \mathbb{Z}$ due to the antisymmetry of the displacement. One readily checks that the extension u^* preserves antisymmetry. It now follows immediately from the symmetry of V (3.15) that

$$\mathcal{E}^*(u) = \tfrac{1}{4}V(Du^*(-L)) + \tfrac{1}{2}\sum_{\xi=-L+1}^{L-1} V(Du^*(\xi)) + \tfrac{1}{4}V(Du^*(L)) + C^*.$$

Since Du^* is $2L$-periodic, it follows that $V(Du^*(-L)) = V(Du^*(L))$ and hence the result follows. □

We now use the representation (4.18) to prove **(F)**. In Section 7 we will also see that this representation leads to a straightforward proof of stability of $\mathcal{E}^{\rm rfl}$.

Corollary 4.2. $\mathcal{E}^{\rm rfl}$ satisfies **(F)**.

Proof. Fix $v \in \mathcal{U}_0$. Using the form of $\mathcal{E}^*$ derived in Lemma 4.1, we obtain

$$\langle \delta\mathcal{E}^*(u^{\mathsf{F}}), v\rangle = \tfrac{1}{2}\sum_{\xi=-L+1}^{L}\sum_{\rho\in\mathcal{R}} V_\rho(\mathsf{F}\mathcal{R})D_\rho v^*(\xi). \tag{4.19}$$

Now let $v(\xi) = \mathsf{G}\xi + w(\xi)$, where $\mathsf{G} = v(L)/L$. Then w^* is periodic and hence (4.19) implies that $\langle \delta\mathcal{E}^*(u^{\mathsf{F}}), w\rangle = 0$. Hence, we obtain

$$\begin{aligned}\langle \delta\mathcal{E}^*(u^{\mathsf{F}}), v\rangle &= \langle \delta\mathcal{E}^*(u^{\mathsf{F}}), u^{\mathsf{G}}\rangle + \langle \delta\mathcal{E}^*(u^{\mathsf{F}}), w\rangle \\ &= \tfrac{1}{2}(2L)\sum_{\rho\in\mathcal{R}} V_\rho \mathsf{G}\rho = L\partial_{\mathsf{F}}W(\mathsf{F})\mathsf{G} = \partial_{\mathsf{F}}W(\mathsf{F})v(L).\end{aligned}$$

Inserting this result into the definition of $\delta\mathcal{E}^{\rm rfl}$, we obtain

$$\langle \delta\mathcal{E}^{\rm rfl}(u^{\mathsf{F}}), v\rangle = \partial_{\mathsf{F}}W(\mathsf{F})v(L) + \int_L^\infty \nabla v\,\mathrm{d}x = 0.$$

This clearly implies **(F)**. □

Further remarks

(1) One often also demands that the QNL interface correction satisfies a local energy consistency condition (see E *et al.* (2006) and Ortner and Zhang (2012)):

(E) *Local energy consistency.* $\Phi^{\rm i}_\xi(u^{\mathsf{F}}) = \Phi^{\rm a}_\xi(u^{\mathsf{F}})$ for all $\mathsf{F} \in \mathbb{R}$, $\xi = K+1, \ldots, L-1$.

However, it is straightforward to show (Ortner 2012, § 6.4.3) that **(F)** implies a weaker global version of **(E)**. It is therefore unclear whether the condition **(E)** is required at all, except of course to approximate absolute energies. Indeed, we shall not use it in our subsequent 1D analysis.

(2) As remarked above, it is unknown whether the general geometric consistency equations have a solution. It is clear from comparing the number of parameters to the number of equations that the corresponding linear operator must have a non-trivial kernel. It would therefore be interesting to augment the consistency equations with additional constraints that induce further desirable properties of the resulting a/c coupling.

(3) For 2D pair interactions, Shapeev (2011) has provided a remarkably simple explicit construction of an a/c coupling satisfying **(F)**, which does not use the ideas of geometric reconstruction. Unfortunately the construction cannot be applied to general many-body interactions, but it has proved useful for the *analysis* of a/c couplings of many-body interactions (Ortner 2012). We will also frequently apply 1D variants of these ideas throughout our consistency analysis. A recent idea applicable to 3D pair interactions is presented in Makridakis, Mitsoudis and Rosakis (2012).

4.6. Force-based a/c coupling

We have so far described two a/c couplings that reduce or remove the ghost force. While the construction of the B-QCE method is straightforward for general a/c interfaces in 2D/3D, the construction of energy-based a/c couplings without ghost forces in 2D/3D remains a challenging open problem.

An alternative is to construct a/c couplings for the forces, and to accept a non-conservative force field. The simplest method of this type is the force-based quasicontinuum (QCF) method. Let $K = L$, and define

$$\mathcal{F}^{\mathrm{qcf}}_{\zeta}(u_h) := \begin{cases} \dfrac{\partial \mathcal{E}^{\mathrm{a}}(u_h)}{\partial u_h(\zeta)}, & \zeta \in \{1, \dots, K\}, \\[2mm] \dfrac{\partial \mathcal{E}^{\mathrm{c}}(u_h)}{\partial u_h(\zeta)}, & \zeta \in \mathcal{N}_h^{\circ} \setminus \{1, \dots, K\}. \end{cases} \tag{4.20}$$

Then we aim to solve the nonlinear system

$$\big\langle \mathcal{F}^{\mathrm{qcf}}(u_h), v_h \big\rangle = \langle f, v_h \rangle_h \quad \text{for all } v_h \in \mathcal{U}_h, \tag{4.21}$$

where

$$\langle \mathcal{F}^{\mathrm{qcf}}(u_h), v_h \rangle = \sum_{\zeta \in \mathcal{N}_h^{\circ}} \mathcal{F}^{\mathrm{qcf}}_{\zeta}(u_h) v_h(\zeta).$$

This scheme is naturally free of ghost forces: $\mathcal{F}^{\mathrm{qcf}}_{\zeta}(u^{\mathsf{F}}) = 0$ for all finite element nodes $\zeta \in \mathcal{N}_h^{\circ}$.

The first analysis of the QCF method (Dobson *et al.* 2010*b*) revealed that while the operator is naturally consistent, its stability is a subtle issue. For example, it was shown that $\delta\mathcal{F}^{\mathrm{qcf}}(0)$ is nearly *always* indefinite and never uniformly stable as an operator from $\mathcal{U}_h$ to $\mathcal{U}_h^*$. This not only makes the stability analysis in 2D/3D particularly challenging, but in fact it is currently unknown whether the QCF method is stable in any suitable function space setting.

To overcome this difficulty, Lu and Ming (2013) proposed applying the blending idea behind the B-QCE method to the QCF method. For earlier variants of force-based blending see Fischmeister *et al.* (1989), Kohlhoff *et al.* (1991) and Badia *et al.* (2007). To introduce this scheme, let $K < L$ and $\beta : \mathbb{R} \to [0,1]$ with $\beta = 0$ in $[0, K]$ and $\beta = 1$ in $[L, N]$, and define

$$\mathcal{F}^{\rm bqcf}_\zeta(u_h) := (1-\beta(\zeta))\frac{\partial \mathcal{E}^{\rm a}(u_h)}{\partial u_h(\zeta)} + \beta(\zeta)\frac{\partial \mathcal{E}^{\rm c}(u_h)}{\partial u_h(\zeta)}. \tag{4.22}$$

In the blended QCF (B-QCF) method, we aim to solve the nonlinear system

$$\langle \mathcal{F}^{\rm bqcf}_\zeta(u_h), v_h\rangle = \langle f, v_h\rangle_h \quad \text{for all } v_h \in \mathcal{U}_h.$$

Like the QCE and B-QCE methods, the QCF and the B-QCF methods are straightforward to implement, using purely atomistic and purely continuum finite element assembly techniques.

4.7. Formal outline of the error analysis

The fundamental theorem of numerical analysis states, loosely speaking, that *consistency* and *stability* imply convergence. A key motivation of this article is to flesh out this principle in the non-standard setting of a/c coupling methods. In the present section, we outline a general framework, in order to motivate the analysis of Sections 5–8.

Let $\mathcal{F}^{\rm ac} \in C^{k-1}(\mathcal{U}_h; \mathcal{U}_h^*)$ be the force operator associated with one of the a/c coupling methods that we constructed in the previous sections; that is, solutions of the a/c coupling under consideration satisfy

$$\langle \mathcal{F}^{\rm ac}(u_h^{\rm ac}), v_h\rangle = \langle f, v_h\rangle_h \quad \text{for all } v_h \in \mathcal{U}_h. \tag{4.23}$$

We now turn (4.23) into an *error equation.* We first assume the existence of an atomistic solution $u^{\rm a}$ to (3.17). Next, we define a quasi-best approximation $\Pi_h u^{\rm a} \in \mathcal{U}_h$ (Section 5). Accepting the decay hypothesis (3.20), assuming that K is sufficiently large and the finite element mesh sufficiently fine, we will obtain that $\Pi_h u^{\rm a}$ is 'close' to $u^{\rm a}$. Let $e_h := u_h^{\rm ac} - \Pi_h u^{\rm a}$; then (4.23) is equivalent to

$$\begin{aligned}\big\langle \mathcal{F}^{\rm ac}(\Pi_h u^{\rm a} + e_h) - \mathcal{F}^{\rm ac}(\Pi_h u^{\rm a}), v_h\big\rangle &= \big\langle \delta\mathcal{E}^{\rm a}(u^{\rm a}) - \mathcal{F}^{\rm ac}(\Pi_h u^{\rm a}), v_h\big\rangle \\ &\quad - \big\{\langle f, v_h\rangle_{\mathbb{Z}_+} - \langle f, v_h\rangle_h\big\} \\ &=: \langle \eta^{\rm ac}_{\rm int}, v_h\rangle - \langle \eta_{\rm ext}, v_h\rangle,\end{aligned} \tag{4.24}$$

where $\eta^{\rm ac}_{\rm int}$ is the consistency error arising from the approximation of the internal energy and $\eta_{\rm ext}$ the consistency error arising from the approximation of the external forces.

If $\eta^{\rm ac}_{\rm int} = \eta_{\rm ext} = 0$, then a solution to (4.24) is $e_h = 0$. Thus, if $\eta^{\rm ac}_{\rm int}$ and $\eta_{\rm ext}$ are sufficiently 'small', and if the linearization of the right-hand side of

Table 4.1.

Error contribution	Origin	Approximation parameter(s)
far-field error	reduction to finite domain	N
coarsening error	finite element discretization	h
Cauchy–Born modelling error	replacing the atomistic model with continuum model in $[L, N]$	L
coupling error	interface treatment	K, L, β

(4.24) is an isomorphism (we will prove in Section 7 that $\langle \delta\mathcal{F}^{\mathrm{ac}}(\Pi_h u^{\mathrm{a}})v, v\rangle \geq (C^{\mathrm{stab}})^{-1}\|\nabla v\|_{L^2}^2$), then we expect from the inverse function theorem that a locally unique solution e_h to (4.24) exists, and that

$$\|\nabla e_h\|_{L^2} \leq C^{\mathrm{stab}}\big(\|\eta_{\mathrm{int}}^{\mathrm{ac}}\|_{\mathcal{U}_h^*} + \|\eta_{\mathrm{ext}}\|_{\mathcal{U}_h^*}\big), \tag{4.25}$$

where

$$\|\eta\|_{\mathcal{U}_h^*} := \sup_{v_h \in \mathcal{U}_h \setminus \{0\}} \frac{\langle \eta, v_h\rangle}{\|\nabla v_h\|_{L^2}}. \tag{4.26}$$

We will analyse the consistency errors in Section 6. We will split the error contributions as shown in Table 4.1. While the coarsening error is a standard component of classical finite element error analysis, the combination with the continuum modelling error and in particular the interfacial error requires new ideas. We will obtain consistency error estimates, similar to those in Section 2, depending on the smoothness of u^{a}, the mesh coarsening, the computational domain size, and possibly other approximation parameters.

Next, we will analyse the stability of a/c schemes in Section 7. We shall prove for the B-QCE, B-QCF, and QNL methods that the approximation parameters can be chosen in such a way that positivity of $\delta^2\mathcal{E}^{\mathrm{a}}(u^{\mathrm{a}})$ (see (3.19)) implies positivity of $\delta\mathcal{F}^{\mathrm{ac}}(\Pi_h u^{\mathrm{a}})$.

Next, we will optimize all approximation parameters to obtain consistency error estimates, purely in terms of the computational cost,

$$\|\eta_{\mathrm{int}}^{\mathrm{ac}}\|_{\mathcal{U}_h^*} + \|\eta_{\mathrm{ext}}\|_{\mathcal{U}_h^*} \lesssim N_{\mathcal{T}_h}^{-\beta},$$

for some exponent $\beta > 0$. We will also translate the stability results from Section 7 to this setting, showing that if $N_{\mathcal{T}_h}$ is sufficiently large then $\delta\mathcal{F}^{\mathrm{ac}}(\Pi_h u^{\mathrm{a}})$ is stable.

Finally, in Section 8 we will make the formal argument presented in this subsection fully rigorous. We will obtain results of the form: Let u^{a} be a strongly stable solution of (3.17) satisfying **(DH)**. Then for $N_{\mathcal{T}_h}$ sufficiently large there exists a solution u^{ac} of (4.23) satisfying $\|\nabla u^{\mathrm{ac}} - \nabla u^{\mathrm{a}}\|_{L^2} \lesssim N_{\mathcal{T}_h}^{-\beta}$.

5. Coarsening error

Our first step in the error analysis is an analysis of the finite element coarsening error. The results in this section are fairly standard, and hence we only give brief summaries of certain auxiliary results that we require later on. Proofs are given, for the sake of completeness, in the Appendix.

5.1. *Best approximation operator*

Recall the definition of the nodal interpolation operator I_h from Section 4.2. Since I_h does not map $\mathcal{U}$ to $\mathcal{U}_h$, and to avoid any error contributions from the atomistic region (and possibly a neighbourhood of the interface region), we define

$$\Pi_h u(x) := \begin{cases} I_h u(x), & x \in [0, L + r_{\rm cut}], \\ I_h u(x) - \frac{x - L - r_{\rm cut}}{N - L - r_{\rm cut}} u(N), & x \in [L + r_{\rm cut}, N]. \end{cases}$$

With this definition, $\Pi_h u \in \mathcal{U}_h$ for all $u : \mathcal{N}_h \to \mathbb{R}$.

In our coarsening error analysis below, it will be useful to split the interpolant into $\Pi_h u = I_h u + (\Pi_h u - I_h u)$. Hence, we separately estimate the interpolation errors as follows.

Lemma 5.1. Let $T \in \mathcal{T}_h$, $T \subset [L, N]$ and $u \in \mathcal{U}$. Then

$$\|\nabla \tilde{u} - \nabla I_h u\|_{L^2(T)} \lesssim h_T \|\nabla^2 \tilde{u}\|_{L^2(T)}. \tag{5.1}$$

If, in addition, u satisfies **(DH)**, $L \leq N/2$ and $N \geq r_0$, then

$$\big|\nabla I_h u(x) - \nabla \Pi_h u(x)\big| \lesssim \begin{cases} 0, & x \in [0, L + r_{\rm cut}], \\ N^{-\alpha}, & x \in [L + r_{\rm cut}, N]. \end{cases} \tag{5.2}$$

Proof. The first result follows from Poincaré's inequality.

To prove the second estimate, we first note that $\nabla I_h u = \nabla \Pi_h u = \nabla u$ in $[0, L + r_{\rm cut}]$. In $[L + r_{\rm cut}, N]$ we have

$$\big|\nabla I_h u - \nabla \Pi_h u\big| = (N - L)^{-1} |u(N)| \lesssim (N - L)^{-1} N^{1-\alpha} \lesssim N^{-\alpha}. \quad \square$$

5.2. *Coarsening error of the internal forces*

The first variation of the continuum energy contribution $\int_L^N W(\nabla u)\,\mathrm{d}x$ is given by

$$v \mapsto \int_L^N \partial_{\mathsf{F}} W(\nabla u) \nabla v \,\mathrm{d}x.$$

The following lemma estimates the error contribution from this operator induced by finite element coarsening and reduction to a finite domain.

Lemma 5.2. Let $u \in \mathcal{U}$ satisfy **(DH)**, $N > r_0$, and $0 < L \le N/2$. Then

$$\left| \int_L^N \big(\partial_{\mathsf{F}} W(\nabla \Pi_h u) - \partial_{\mathsf{F}} W(\nabla \tilde{u})\big) \nabla v_h \, \mathrm{d}x \right|$$
$$\lesssim M^{(2,0)} \big(\|h \nabla^2 \tilde{u}\|_{L^4}^2 + N^{-\alpha+1/2} \big) \|\nabla v_h\|_{L^2} \quad \text{for all } v_h \in \mathrm{P1}(\mathcal{T}_h).$$

Proof. See Appendix A.2. □

Remark 5.3. **(1)** We observe that we require at least $\alpha > 1/2$ in order to control the far-field error, which is precisely the assumption we made in (3.20). Recall though that this is no severe restriction since a displacement gradient $\nabla u(x)$ with $\nabla u(x) \sim x^{-\alpha}$ as $x \to \infty$ belongs to L^2 if and only if $\alpha > 1/2$.

(2) The nodal interpolation error enters the coarsening error only as a quadratic term $\|h\nabla^2 \tilde{u}\|_{L^4}^2$. This effect is related to the well-known 1D superconvergence property that the nodal interpolant is at the same time the Ritz projection.

5.3. *Coarsening error of external forces*

We now turn towards the consistency error due to the approximation of the external potential $\langle f, v_h \rangle_{\mathbb{Z}_+}$ by the trapezoidal rule $\langle f, v_h \rangle_h$. The main challenge in this analysis is to avoid the use of the Poincaré inequality $\|v_h\|_{L^2} \lesssim N \|\nabla v_h\|_{L^2}$ at all costs (we will point out the relevant step in the proof of Proposition 5.4 in Appendix A.2). A common approach in unbounded domains is to employ weighted Poincaré inequalities instead. This yields the following result.

Proposition 5.4. Let $L > 1$ and suppose that $h(x) \le \kappa x$ for almost every $x \in [L, N]$. Then there exists a constant C_κ such that

$$\|\eta_{\mathrm{ext}}\|_{\mathcal{U}_h^*} = \big\| \langle f, \cdot \rangle_{\mathbb{Z}_+} - \langle f, \cdot \rangle_h \big\|_{\mathcal{U}_h^*}$$
$$\lesssim \big\| h^2 \nabla \tilde{f} \big\|_{L^2(L,\infty)} + \frac{C_\kappa}{\log L} \big\| h^2 \omega \nabla^2 \tilde{f} \big\|_{L^2(L,\infty)},$$

where $\omega(x) := x \log x$.

Proof. See Appendix A.2. □

Remark 5.5. The proof of Proposition 5.4 reveals that for practical meshes one should expect $C_\kappa \approx \kappa$, hence it is important to control the bound $h(x) \le \kappa x$. In Section 8, we shall optimize the finite element grid to balance the far-field error and interpolation error contributions with the best approximation error. For the quasi-optimal mesh scaling that we will derive, we will be able to make this bound explicit.

6. Consistency

Recall from Section 4.7 that the *internal consistency error*, associated with a displacement $u \in \mathcal{U}$, is given by a linear form

$$\langle \eta^{\rm ac}_{\rm int}(u), v_h \rangle := \langle \delta\mathcal{E}^{\rm a}(u) - \mathcal{F}^{\rm ac}(\Pi_h u), v_h \rangle,$$

where $\mathcal{F}^{\rm ac}$ is the force operator associated with one of the a/c coupling methods constructed in Section 4.

In the present section, we establish consistency error bounds associated with the model approximations (reminiscent of variational crimes in finite element methods) in the various a/c coupling schemes. These will then be combined with the coarsening error bounds derived in Section 5.2 to yield qualitatively sharp bounds on $\eta^{\rm ac}_{\rm int}$ in the $\mathcal{U}_h^*$-norm defined in (4.26) (a discrete H^{-1}-type norm).

6.1. *Weak form of the atomistic model and Cauchy–Born modelling error*

For a sharp consistency error analysis, it is useful to develop convenient 'weak forms' of the first variations of the atomistic, continuum, and a/c models. The canonical representation (3.12) of $\delta\mathcal{E}^{\rm a}$ is not immediately useful since the occurrence of the test function is nonlocal. In the following, we derive an alternative weak formulation, employing an atomistic notion of stress.

The idea is to localize the finite differences $D_\rho v(\xi)$ by rewriting them as *bond integrals*. A conceptually straightforward mechanism to achieve this is to simply take the derivatives of $\mathcal{E}^{\rm a}$ with respect to the discrete strains u'_ξ. However, this technique is restricted to 1D. Instead, we present a simplified 1D variant of a construction that was developed for the construction and analysis of a/c coupling methods in 2D/3D (see Section 6.6 for more details on this discussion).

We define the weighted characteristic function of a bond $(\xi, \xi+\rho)$ by

$$\chi^{\mathbb{R}}_{\xi,\rho}(x) := \begin{cases} |\rho|^{-1}, & x \in \operatorname{int}(\operatorname{conv}\{\xi, \xi+\rho\}), \\ \tfrac{1}{2}|\rho|^{-1}, & x \in \{\xi, \xi+\rho\}, \\ 0, & \text{otherwise.} \end{cases}$$

We then obtain for $D_\rho v(\xi) := v(\xi+\rho) - v(\xi)$ that

$$D_\rho v(\xi) = \int_\xi^{\xi+\rho} \frac{\rho}{|\rho|} \nabla v \,\mathrm{d}x = \int_{\mathbb{R}} \rho \chi^{\mathbb{R}}_{\xi,\rho} \nabla v \,\mathrm{d}x. \tag{6.1}$$

To express $D_\rho v(\xi)$ as an integral over $\mathbb{R}_+$, we use that $\nabla v(-x) = \nabla v(x)$ for all $v \in \mathcal{U}$ to obtain

$$D_\rho v(\xi) = \int_{\mathbb{R}} \rho \chi^{\mathbb{R}}_{\xi,\rho} \nabla v \,\mathrm{d}x = \int_{\mathbb{R}_+} \rho \chi_{\xi,\rho} \nabla v \,\mathrm{d}x, \tag{6.2}$$

where

$$\chi_{\xi,\rho}(x) := \chi^{\mathbb{R}}_{\xi,\rho}(x) + \chi^{\mathbb{R}}_{\xi,\rho}(-x).$$

Inserting this representation into (3.12) yields

$$\langle \delta\mathcal{E}^{\mathrm{a}}(u), v\rangle = \int_{\mathbb{R}_+} \bigg\{ \sum_{\rho\in\mathcal{R}} \rho \sum_{\xi\in\mathbb{Z}_+} \Phi^{\mathrm{a}}_{\xi,\rho}(u)\chi_{\xi,\rho}(x) \bigg\} \nabla v \,\mathrm{d}x.$$

Thus we obtain the following result.

Proposition 6.1. Let $u \in \mathcal{U}$, and $v \in \mathcal{U}_0$. Then

$$\langle \delta\mathcal{E}^{\mathrm{a}}(u), v\rangle = \int_{\mathbb{R}} \mathsf{S}^{\mathrm{a}}(u;x)\nabla v(x)\,\mathrm{d}x,$$

where the *atomistic stress function* S^{a} is defined by

$$\mathsf{S}^{\mathrm{a}}(u;x) := \sum_{\rho\in\mathcal{R}}\sum_{\xi\in\mathbb{Z}_+} \rho\Phi^{\mathrm{a}}_{\xi,\rho}(u)\chi_{\xi,\rho}(x). \tag{6.3}$$

Since the sum defining $\mathsf{S}^{\mathrm{a}}(u;x)$ is finite (the sum over ξ only needs to be taken over $\mathbb{Z}_+ \cap [x - r_{\mathrm{cut}}, x + r_{\mathrm{cut}}]$), S^{a} is well-defined and $k-1$ times continuously differentiable with respect to u. It is also clear from the definition that $x \mapsto \mathsf{S}^{\mathrm{a}}(u;x)$ is constant on each interval $(\eta, \eta+1)$, $\eta \in \mathbb{Z}_+$.

Moreover, if u is locally homogeneous, that is, $u(\xi) = \mathsf{F}\xi$ in $[x - 2r_{\mathrm{cut}}, x + 2r_{\mathrm{cut}}]$, then

$$\begin{aligned}\mathsf{S}^{\mathrm{a}}(u;x) = \mathsf{S}^{\mathrm{a}}(u^{\mathsf{F}};x) &= \sum_{\rho\in\mathcal{R}}\sum_{\xi\in\mathbb{Z}_+} \rho V_\rho(\mathsf{F}\mathcal{R})\chi_{\xi,\rho}(x) \\ &= \sum_{\rho\in\mathcal{R}} \rho V_\rho(\mathsf{F}\mathcal{R}) = \partial_{\mathsf{F}} W(\mathsf{F}),\end{aligned} \tag{6.4}$$

where we used the fact that $\sum_{\xi\in\mathbb{Z}_+}\chi_{\xi,\xi+\rho}(x) = 1$ (see Lemma 6.3 below). Since $\mathsf{S}^{\mathrm{a}} = \partial_{\mathsf{F}} W$ under locally homogeneous deformations, it follows easily that the Cauchy–Born stress is first-order consistent with the atomistic stress. By exploiting symmetries of the interaction potential, we can even prove that it is second-order consistent at bond midpoints.

Theorem 6.2. Let $u \in W^{3,\infty}_{\mathrm{loc}}$ and $x \in \frac{1}{2} + \mathbb{Z}_+$, $x > 2r_{\mathrm{cut}}$. Then

$$\big|\mathsf{S}^{\mathrm{a}}(u;x) - \partial_{\mathsf{F}} W(\nabla u(x))\big| \lesssim M^{(2,2)}\|\nabla^3 u\|_{L^\infty(\nu_x)} + M^{(3,2)}\|\nabla^2 u\|^2_{L^\infty(\nu_x)},$$

where $\nu_x := [x + \frac{1}{2} - 2r_{\mathrm{cut}}, x - \frac{1}{2} + 2r_{\mathrm{cut}}]$.

Proof. We note that $\mathsf{S}^{\mathrm{a}}(u;x)$ is defined for $x \in \mathbb{R}_+$ from the restriction of $u \in W^{3,\infty}_{\mathrm{loc}}$ to the nodes $\xi \in \mathbb{Z}_+$. The result is proved by Taylor expansion of the nonlocal terms in the definition of S^{a}, and exploiting the symmetries of the interaction potential.

The sum over ξ in (6.3) is only taken over $\xi \in N_x := \{x+\frac{1}{2}-r_{\rm cut}, \dots, x-\frac{1}{2}+r_{\rm cut}\}$, but the finite differences $D_\rho u(\xi)$, for $\xi \in N_x$ depend on the neighbourhoods ν_x defined in the statement of the theorem. In order to keep the notation concise, we define

$$\epsilon_2 := \|\nabla^2 u\|_{L^\infty(\nu_x)} \quad \text{and} \quad \epsilon_3 := \|\nabla^3 u\|_{L^\infty(\nu_x)}.$$

We will suppress all arguments where possible. For example, we define $\Phi^{\rm a}_{\xi,\rho} := \Phi^{\rm a}_{\xi,\rho}(u)$ and $V_\rho := V_\rho(\nabla u(x)\cdot\mathcal{R})$. Moreover, we often suppress dependence of constants on the potential, which can be readily reconstructed.

We begin by expanding $D_\varsigma u(\xi)$, for $\xi \in N_x, \varsigma \in \mathcal{R}$,

$$D_\varsigma u(\xi) = \nabla_\varsigma u(x) + \nabla^2 u(x)\varsigma(\xi - x + \tfrac{1}{2}\varsigma) + O(\epsilon_3),$$

which, together with the fact that $D_\varsigma u(\xi) - \nabla_\varsigma u(x) = O(\epsilon_2)$, allows us to expand $\Phi^{\rm a}_{\xi,\rho}$:

$$\begin{aligned}\Phi^{\rm a}_{\xi,\rho} &= V_\rho + \sum_{\varsigma\in\mathcal{R}} V_{\rho\varsigma}\cdot(D_\varsigma u(\xi) - \nabla_\varsigma u(x)) + O(\epsilon_2^2) \\ &= V_\rho + \nabla^2 u \sum_{\varsigma\in\mathcal{R}} \varsigma V_{\rho\varsigma}\cdot(\xi - x + \tfrac{1}{2}\varsigma) + O(\epsilon_2^2+\epsilon_3). \end{aligned} \tag{6.5}$$

Inserting (6.5) into (6.3), recalling (6.4) and applying (6.6) and (6.7) with $\eta := x - 1/2$, yields

$$\begin{aligned}\mathsf{S}^{\rm a}(u;x) - \partial_{\mathsf{F}} W(\nabla u(x)) &= \Big\{\sum_{\rho\in\mathcal{R}} \rho V_\rho \sum_{\xi\in\mathbb{Z}} \chi_{\xi,\rho} - \partial_{\mathsf{F}} W(\nabla u)\Big\} \\ &\qquad + \nabla^2 u \sum_{\rho,\varsigma\in\mathcal{R}} \rho\varsigma V_{\rho\varsigma} \sum_{\xi\in\mathbb{Z}} \chi_{\xi,\rho}\Big(\xi - x + \frac{\varsigma}{2}\Big) + O(\epsilon_2^2+\epsilon_3) \\ &= \Big\{\sum_{\rho\in\mathcal{R}} \rho V_\rho - \partial_{\mathsf{F}} W(\nabla u)\Big\} \\ &\qquad + \nabla^2 u \sum_{\rho,\varsigma\in\mathcal{R}} \rho\varsigma V_{\rho\varsigma}\Big(x - \frac{\rho}{2} - x + \frac{\varsigma}{2}\Big) + O(\epsilon_2^2+\epsilon_3) \\ &= \nabla^2 u \frac{1}{2} \sum_{\rho\varsigma\in\mathcal{R}} \rho\varsigma V_{\rho\varsigma}(\varsigma - \rho) + O(\epsilon_2^2+\epsilon_3). \end{aligned}$$

Applying the symmetry $V_{\rho,\varsigma} = V_{-\rho,-\varsigma}$ (see (3.16)) we obtain that

$$\sum_{\rho\varsigma\in\mathcal{R}} \rho\varsigma V_{\rho\varsigma}(\varsigma - \rho) = \sum_{\rho\in\mathcal{R}_+}\sum_{\varsigma\in\mathcal{R}} \rho\varsigma V_{\rho\varsigma}\big[(\varsigma-\rho) + (-\varsigma+\rho)\big] = 0.$$

This concludes the proof of the result. □

Lemma 6.3. Let $x \in (\eta, \eta+1)$ for some $\eta \in \mathbb{Z}_+$ and $\rho \in \mathbb{Z} \setminus \{0\}$. Then

$$\sum_{\xi \in \mathbb{Z}_+} \chi_{\xi,\rho}(x) = 1. \tag{6.6}$$

If, moreover, $x > r_{\mathrm{cut}}$, then

$$\sum_{\xi \in \mathbb{Z}_+} \chi_{\xi,\rho}(x) \cdot \xi = \left(\eta + \frac{1}{2} - \frac{\rho}{2}\right). \tag{6.7}$$

Proof. Suppose first that $x > r_{\mathrm{cut}}$. If $\rho > 0$, then

$$\sum_{\xi \in \mathbb{Z}_+} \chi_{\xi,\rho}(x) = \frac{1}{\rho}\sum_{j=1}^{\rho} 1 = 1 \quad \text{and}$$

$$\sum_{\xi \in \mathbb{Z}_+} \chi_{\xi,\rho}(x)\xi = \frac{1}{\rho}\sum_{j=1}^{\rho}(\eta - j) = \frac{1}{\rho}\left(\rho\eta - \frac{\rho(\rho+1)}{2}\right) = \eta + \frac{1}{2} - \frac{\rho}{2}.$$

For $\rho < 0$ an analogous argument applies.

If $0 < x < r_{\mathrm{cut}}$ and $\rho < 0$, then the same argument gives

$$\sum_{\xi \in \mathbb{Z}} \chi^{\mathbb{R}}_{\xi,\rho}(x) = 1.$$

Moreover, one readily checks that

$$\sum_{\xi \in \mathbb{Z}} \chi^{\mathbb{R}}_{\xi,\rho}(x) = \sum_{\xi \in \mathbb{Z}_+} \chi_{\xi,\rho}(x). \qquad \square$$

Corollary 6.4 (Cauchy–Born modelling error). If $u \in \mathcal{U}$ and $L \geq 2r_{\mathrm{cut}}$, then

$$\int_L^N \big(\mathsf{S}^{\mathrm{a}}(u) - \partial_{\mathsf{F}} W(\nabla \tilde{u})\big)\nabla v_h \,\mathrm{d}x$$
$$\lesssim \big(M^{(3,3)}\|\nabla^3 \tilde{u}\|_{L^2(\nu)} - M^{(2,3)}\|\nabla^2 \tilde{u}\|^2_{L^4(\nu)}\big)\|\nabla v_h\|_{L^2(L,N)} \quad \text{for all } v_h \in \mathcal{U}_h,$$

where $\nu := (L - 2r_{\mathrm{cut}} + 1, N + 2r_{\mathrm{cut}})$.

Proof. Let $\bar{x} := \eta - 1/2$, $\eta \in \mathbb{Z}_+$. Then, recalling that $\mathsf{S}^{\mathrm{a}}(u)$ is piecewise constant and that $\tilde{u}|_{\mathbb{Z}_+} = u$,

$$\int_{\eta-1}^{\eta} \big(\mathsf{S}^{\mathrm{a}}(u) - \partial_{\mathsf{F}} W(\nabla \tilde{u})\big)\nabla v_h \,\mathrm{d}x = \bigg\{\big[\mathsf{S}^{\mathrm{a}}(\tilde{u};\bar{x}) - \partial_{\mathsf{F}} W(\nabla\tilde{u}(\bar{x}))\big]$$
$$+ \int_{\eta-1}^{\eta} \big[\partial_{\mathsf{F}} W(\nabla\tilde{u}(\bar{x})) - \partial_{\mathsf{F}} W(\nabla\tilde{u}(x))\big]\,\mathrm{d}x\bigg\}\nabla v_h. \tag{6.8}$$

Applying Theorem 6.2, the local norm-equivalence for polynomials of degree 5,

$$\|\nabla^j \tilde{u}\|_{L^\infty(\eta-1,\eta)} \lesssim \|\nabla^j \tilde{u}\|_{L^2(\eta-1,\eta)},$$

and the embedding $\ell^2 \subset \ell^\infty$, we obtain

$$\begin{aligned}\big|\mathsf{S}^{\mathrm{a}}(\tilde{u};\bar{x}) - \partial_{\mathsf{F}} W(\nabla\tilde{u}(\bar{x}))\big| &\lesssim \|\nabla^3\tilde{u}\|_{L^\infty(\nu_{\bar{x}})} + \|\nabla^2\tilde{u}\|_{L^\infty(\nu_{\bar{x}})}^2 \\ &\lesssim \|\nabla^3\tilde{u}\|_{L^2(\nu_{\bar{x}})} + \|\nabla^2\tilde{u}\|_{L^4(\nu_{\bar{x}})}^2.\end{aligned}$$

Moreover, using the fact that the midpoint rule is second-order accurate,

$$\begin{aligned}\left|\int_{\eta-1}^{\eta} \big(\partial_{\mathsf{F}} W(\nabla\tilde{u}(\bar{x})) - \partial_{\mathsf{F}} W(\nabla\tilde{u}(x))\big)\,\mathrm{d}x\right| &\lesssim \|\nabla^2\partial_{\mathsf{F}} W(\nabla\tilde{u}(x))\|_{L^1(\eta-1,\eta)} \\ &\lesssim \|\nabla^3\tilde{u}\|_{L^1(\eta-1,\eta)} + \|\nabla^2\tilde{u}\|_{L^2(\eta-1,\eta)}^2 \\ &\lesssim \|\nabla^3\tilde{u}\|_{L^2(\eta-1,\eta)} + \|\nabla^2\tilde{u}\|_{L^4(\eta-1,\eta)}^2.\end{aligned}$$

Inserting these estimates into (6.8) yields

$$\begin{aligned}&\left|\int_{\eta-1}^{\eta} \big(\mathsf{S}^{\mathrm{a}}(u) - \partial_{\mathsf{F}} W(\nabla\tilde{u})\big)\nabla v_h\,\mathrm{d}x\right| \\ &\qquad\lesssim \big(\|\nabla^3\tilde{u}\|_{L^2(\nu_{\bar{x}})} + \|\nabla^2\tilde{u}\|_{L^4(\nu_{\bar{x}})}^2\big)\|\nabla v_h\|_{L^2(\eta-1,\eta)},\end{aligned}$$

where we recall that $\nu_{\bar{x}} = (\eta - 2r_{\mathrm{cut}}, \eta - 1 + 2r_{\mathrm{cut}})$.

Summing over all $\eta = L+1, \dots, N$, applying Hölder's inequality, and noting that any interval $(\xi - 1, \xi)$ intersects with at most $4r_{\mathrm{cut}}$ neighbourhoods $\nu_{\bar{x}}$, we obtain

$$\int_L^N \big(\mathsf{S}^{\mathrm{a}}(u) - \partial_{\mathsf{F}} W(\nabla\tilde{u})\big)\nabla v_h\,\mathrm{d}x \lesssim \big(\|\nabla^3\tilde{u}\|_{L^2(\nu)} + \|\nabla^2\tilde{u}\|_{L^4(\nu)}^2\big)\|\nabla v_h\|_{L^2(L,N)},$$

where $\nu := \bigcup_{\bar{x}\in[L,N]} \nu_{\bar{x}}$. This definition of ν is equivalent to the one given in the statement of the result. □

Further remarks

(1) The consistency proof in this section mimics an analogous proof valid in general dimensions by Ortner and Theil (2013), which builds on ideas of Shapeev (2011) and Ortner (2012). The atomistic stress function S^{a} and its 2D/3D generalizations are closely connected to a notion of stress defined by Hardy (1982). A recent general account of stress in molecular mechanics simulations is given by Admal and Tadmor (2010).

(2) The prefactors $M^{(j,3)}$ cannot be immediately derived from Theorem 6.2, but can only be obtained if the dependence of the interaction neighbourhoods of tuples of pairs and triples of bonds occurring in the proof of Theorem 6.2 are carefully traced. We refer to Ortner and Theil (2013) for further details.

6.2. Consistency of the B-QCE method

Recall from the definition of the B-QCE method in Section 4.4 that we assumed $0 < K < L \le N/2$ and that $\nabla\beta = 0$ in $[0, K] \cup [L, N]$. In addition, we will assume that $K \ge r_0 + 2r_{\rm cut} + 1$.

Motivated by the derivation of the atomistic stress in Section 6.1, we derive a similar notion of stress for the BQCE method. The natural form of the first variation of $\mathcal{E}^{\rm bqce}$ is

$$\langle \delta\mathcal{E}^{\rm bqce}(u_h), v_h\rangle = \sum_{\xi=0}^{L}(1-\beta(\xi))\sum_{\rho\in\mathcal{R}} \Phi^{\rm a}_{\xi,\rho}(u_h)D_\rho v_h(\xi) \\ + \int_K^N I_h\beta\cdot\partial_{\sf F}W(\nabla u_h)\cdot\nabla v_h\,{\rm d}x \quad \text{for all } u_h, v_h\in\mathcal{U}_h.$$

Following closely the calculations in Section 6.1 yields the following result. Note that, since we assumed that $\{0,\dots,L\}\subset\mathcal{N}_h$, the apparent h-dependence in the definition of $\mathsf{S}^{\rm bqce}$ through $I_h\beta$ is not an actual dependence.

Proposition 6.5. Let $u_h, v_h \in \mathcal{U}_h$. Then

$$\langle \delta\mathcal{E}^{\rm bqce}(u_h), v_h\rangle = \int_0^N \mathsf{S}^{\rm bqce}(u_h)\nabla v_h\,{\rm d}x, \quad \text{where} \tag{6.9}$$

$$\mathsf{S}^{\rm bqce}(u_h;x) := \sum_{\rho\in\mathcal{R}}\sum_{\xi\in\mathbb{Z}_+}(1-\beta(\xi))\chi_{\xi,\rho}(x)\Phi^{\rm a}_{\xi,\rho}(u_h)\rho \\ + I_h\beta(x)\cdot\partial_{\sf F}W(\nabla u_h(x)).$$

We immediately obtain a result stating that the consistency error vanishes in a core atomistic region.

Lemma 6.6. Let $u \in \mathcal{U}$. Then

$$\mathsf{S}^{\rm a}(u;x) = \mathsf{S}^{\rm bqce}(\Pi_h u;x) \quad \text{for } x\in(0, K - r_{\rm cut}+1).$$

Proof. This result follows immediately from (6.9) and the fact that $u = \Pi_h u$ in $[0, L]$. □

We begin by analysing the error in a simplified version of the B-QCE stress, where $I_h\beta$ is replaced with β.

Lemma 6.7 (stress consistency). For $\tilde u \in W^{1,\infty}_{\rm loc}$ and $x\in\mathbb{R}$, let

$$\tilde{\mathsf{S}}^{\rm bqce}(\tilde u;x) := \sum_{\rho\in\mathcal{R}}\sum_{\xi\in\mathbb{Z}_+}(1-\beta(\xi))\chi_{\xi,\rho}(x)\Phi^{\rm a}_{\xi,\rho}(\tilde u)\rho + \beta(x)\partial_{\sf F}W(\nabla\tilde u(x)).$$

If $\tilde u \in W^{3,\infty}_{\mathrm{loc}}$ and $x \in \frac12 + \mathbb{Z}$, $x \geq K - r_{\mathrm{cut}} + 1$, then

$$\big|\mathsf{S}^{\mathrm{a}}(\tilde u; x) - \tilde{\mathsf{S}}^{\mathrm{bqce}}(\tilde u; x)\big| \lesssim M^{(1,2)}\|\nabla^2\beta\|_{L^\infty(\nu_x)} + M^{(2,2)}|\nabla\beta(x)\nabla^2\tilde u(x)| \\ + M^{(3,2)}\|\nabla^2\tilde u\|^2_{L^\infty(\nu_x)} + M^{(2,2)}\|\nabla^3\tilde u\|_{L^\infty(\nu_x)},$$

where we recall that $\nu_x = [x + \frac12 - 2r_{\mathrm{cut}}, x - \frac12 + 2r_{\mathrm{cut}}]$.

Proof. We begin by writing out

$$\mathsf{S}^{\mathrm{a}} - \tilde{\mathsf{S}}^{\mathrm{bqce}} = \sum_{\rho\in\mathcal{R}} \rho \sum_{\xi\in\mathbb{Z}} \beta(\xi)\Phi^{\mathrm{a}}_{\xi,\rho}\chi_{\xi,\rho} - \beta\partial_{\mathsf{F}}W, \tag{6.10}$$

and observe the similarity to the Cauchy–Born stress error $\mathsf{S}^{\mathrm{a}} - \partial_{\mathsf{F}}W$. In the present case, we need to expand both $\Phi^{\mathrm{a}}_{\xi,\rho}$, using (6.5), and

$$\beta(\xi) = \beta + \nabla\beta\cdot(\xi - x) + O(\delta_2),$$

where $\delta_2 := \|\nabla^2\beta\|_{L^\infty(\nu_x)}$, which yields

$$\begin{aligned}\mathsf{S}^{\mathrm{a}} - \tilde{\mathsf{S}}^{\mathrm{bqce}} &= \sum_{\rho\in\mathcal{R}} \rho \sum_{\xi\in\mathbb{Z}_+} [\beta + \nabla\beta\cdot(\xi - x)]\,\Phi^{\mathrm{a}}_{\xi,\rho}\chi_{\xi,\rho} - \beta\partial_{\mathsf{F}}W + O(\delta_2) \\ &= \beta[\mathsf{S}^{\mathrm{a}} - \partial_{\mathsf{F}}W] + \nabla\beta\sum_{\rho\in\mathcal{R}}\rho\sum_{\xi\in\mathbb{Z}_+}(\xi - x)\Phi^{\mathrm{a}}_{\xi,\rho}\chi_{\xi,\rho} + O(\delta_2).\end{aligned}$$

Inserting the expansion (6.5) for $\Phi^{\mathrm{a}}_{\xi,\rho}$, recalling that $V_\rho = V_\rho(D\tilde u(\xi))$, and applying Theorem 6.2 to estimate $\mathsf{S}^{\mathrm{a}} - \partial_{\mathsf{F}}W$ gives

$$\begin{aligned}\mathsf{S}^{\mathrm{a}} - \tilde{\mathsf{S}}^{\mathrm{bqce}} = \nabla\beta &\sum_{\rho\in\mathcal{R}} \rho V_\rho \sum_{\xi\in\mathbb{Z}_+} \chi_{\xi,\rho}\cdot(\xi - x) \\ &+ \nabla\beta\,\nabla^2 y \sum_{\rho,\varsigma\in\mathcal{R}} \rho\varsigma V_{\rho\varsigma} \sum_{\xi\in\mathbb{Z}_+} \chi_{\xi,\rho}\cdot(\xi - x)(\xi - x + \tfrac{\varsigma}{2}) \\ &+ O(\delta_2 + \epsilon_2^2 + \epsilon_3).\end{aligned}$$

Performing similar computations as in the proof of Theorem 6.2, in particular taking into account the symmetry $V_{-\rho} = -V_\rho$, we observe that the first group on the right-hand side vanishes. Thus, we obtain

$$\big|\mathsf{S}^{\mathrm{a}} - \tilde{\mathsf{S}}^{\mathrm{bqce}}\big| \lesssim \delta_2 + |\nabla\beta\nabla^2 y| + \epsilon_2^2 + \epsilon_3,$$

which is the required result. □

We can now easily derive the global consistency error estimate for the B-QCE method.

Theorem 6.8. Let $u \in \mathcal{U}$ satisfy **(DH)**. Then

$$\begin{aligned}\|\eta_{\rm int}^{\rm bqce}(u)\|_{\mathcal{U}_h^*} \lesssim{}& M^{(1,3)}\|\nabla^2\beta\|_{L^2} + M^{(2,2)}\|\nabla\beta\nabla^2\tilde{u}\|_{L^2} \\ &+ M^{(2,3)}\|\nabla^3\tilde{u}\|_{L^2(\Omega^{\rm c})} + M^{(3,3)}\|h\nabla^2\tilde{u}\|_{L^4(\Omega^{\rm c})}^2 + M^{(2,0)}N^{1/2-\alpha},\end{aligned}$$

where $\Omega^{\rm c} := (K - 2r_{\rm cut}, \infty)$. (Recall that $h(x) = 1$ for $x > N$.)

Proof. Let $K' := K - r_{\rm cut} + 1$ and assume that $\|\nabla v_h\|_{L^2} = 1$. Using Proposition 6.5 and Lemma 6.6, we split the consistency error into

$$\begin{aligned}\langle \delta\mathcal{E}^{\rm a}(u) - \delta\mathcal{E}^{\rm bqce}(\Pi_h u), v_h\rangle &= \int_{K'}^{N} \big(\mathsf{S}^{\rm a}(u) - \mathsf{S}^{\rm bqce}(\Pi_h u)\big)\nabla v_h\,{\rm d}x \\ &= \int_{K'}^{N} \big(\tilde{\mathsf{S}}^{\rm bqce}(\tilde{u}) - \mathsf{S}^{\rm bqce}(\Pi_h u)\big)\nabla v_h\,{\rm d}x \\ &\quad + \int_{K'}^{N} \big(\mathsf{S}^{\rm a}(u) - \tilde{\mathsf{S}}^{\rm bqce}(\tilde{u})\big)\nabla v_h\,{\rm d}x \\ &=: \eta_1 + \eta_2. \end{aligned} \tag{6.11}$$

Estimating η_1. Since the nodal values of β and $I_h\beta$ are the same, the discrete components of $\mathsf{S}^{\rm bqce}$ and $\tilde{\mathsf{S}}^{\rm bqce}$ are identical, and only the continuous components remain. Hence, and also using $\Pi_h u = u = I_h\tilde{u}$ in $[0, L + 2r_{\rm cut}]$, we can split the group η_1 further into

$$\begin{aligned}\eta_1 &= \int_{K}^{N} \big(\beta\,\partial_{\mathsf{F}}W(\nabla\tilde{u}) - I_h\beta\,\partial_{\mathsf{F}}W(\nabla\Pi_h u)\big)\nabla v_h\,{\rm d}x \\ &= \int_{K}^{L} (\beta - I_h\beta)\partial_{\mathsf{F}}W(\nabla u)\nabla v_h\,{\rm d}x \\ &\quad + \int_{K}^{L} \beta\big(\partial_{\mathsf{F}}W(\nabla\tilde{u}) - \partial_{\mathsf{F}}W(\nabla I_h\tilde{u})\big)\nabla v_h\,{\rm d}x \\ &\quad + \int_{L}^{N} \big(\partial_{\mathsf{F}}W(\nabla\tilde{u}) - \partial_{\mathsf{F}}W(\nabla\Pi_h u)\big)\nabla v_h\,{\rm d}x \\ &=: \eta_{1,1} + \eta_{1,2} + \eta_{1,3}.\end{aligned}$$

Using an interpolation error estimate for β, the term $\eta_{1,1}$ can be bounded by

$$\eta_{1,1} \lesssim \|\nabla^2\beta\|_{L^2}.$$

The term $\eta_{1,3}$ is analysed in Lemma 5.2. The term $\eta_{1,2}$ can be estimated similarly as the coarsening error in Lemma 5.2, but now the first term in the expansion must be taken into account, resulting in

$$\eta_{1,2} \lesssim \|\nabla\beta\,\nabla^2\tilde{u}\|_{L^2(K,L)} + \|\nabla^2\tilde{u}\|_{L^4(K,L)}^2.$$

Combining these estimates yields

$$\eta_1 \lesssim \|\nabla^2\beta\|_{L^2} + \|\nabla\beta\,\nabla^2\tilde{u}\|_{L^2} + \|h\nabla^2\tilde{u}\|_{L^4(K,N)}^2.$$

Estimating η_2. Applying Lemma 6.7 and then arguing analogously as in the proof of Theorem 6.4, we obtain the bound

$$\eta_2 \lesssim \|\nabla^2\beta\|_{L^2} + \|\nabla\beta\nabla^2\tilde{u}\|_{L^2} + \|\nabla^3\tilde{u}\|_{L^2(\Omega^c)} + \|\nabla^2\tilde{u}\|^2_{L^4(\Omega^c)}.$$

We obtain the stated result by combining the estimates for η_1 and for η_2. □

Optimizing the blending function

Under the decay hypothesis **(DH)**, one can readily check that the dominant consistency error contribution in the B-QCE method is the interface coupling error term $\|\nabla^2\beta\|_{L^2}$. This term originates from the smeared-out ghost forces. A natural question at this point is how one should choose β to minimize this consistency error contribution.

Recall that we have assumed that $\beta = 0$ in $[0, K]$ and $\beta = 1$ in $[L, N]$. Thus, we must solve the variational problem

$$\min \|\nabla^2\beta\|_{L^2(K,L)} \quad \text{subject to} \quad \beta(K) = \beta'(K) = \beta'(L) = 0, \text{ and } \beta(L) = 1.$$

The solution to this problem is the cubic polynomial

$$\beta^*(x) = \begin{cases} 0, & x \in [0, K], \\ \hat{\beta}\big(\frac{x-K}{L-K}\big), & x \in [K, L], \\ 1, & x > L, \end{cases} \quad \text{where} \quad \hat{\beta}(s) = 3s^2 - 2s^3. \tag{6.12}$$

One may now readily check that

$$\|\nabla^2\beta^*\|_{L^2} = (L-K)^{-3/2}\|\nabla^2\hat{\beta}\|_{L^2(0,1)} = \sqrt{12}(L-K)^{-3/2}. \tag{6.13}$$

In particular, we see again that the consistency error can be controlled by increasing the blending width.

We summarize further properties of β that will be useful throughout the remainder of the paper.

Lemma 6.9. Let β^* be given by (6.12) and $p \in [1, \infty]$. Then

$$\|\nabla\beta^*\|_{L^p} \lesssim (L-K)^{-1/p'}, \qquad \|\nabla^2\beta^*\|_{L^p} \lesssim (L-K)^{-1-1/p'}, \tag{6.14}$$

$$\|\nabla\sqrt{\beta^*}\|_{L^\infty} \lesssim (L-K)^{-1}, \qquad \|\nabla\sqrt{1-\beta^*}\|_{L^\infty} \lesssim (L-K)^{-1}. \tag{6.15}$$

Proof. The two estimates in (6.14) can be shown with the same scaling argument as (6.13).

To establish (6.15), we first note that, for $s \in [0, 1]$,

$$\sqrt{\hat{\beta}(s)} = s\sqrt{3-2s} \quad \text{and} \quad \sqrt{1-\hat{\beta}(s)} = (1-s)\sqrt{1+2s},$$

that is, $\sqrt{\hat{\beta}}$ and $\sqrt{1-\hat{\beta}}$ are both Lipschitz. Hence, (6.15) follows from an analogous scaling argument. □

Setting $\beta = \beta^*$ in Theorem 6.8, and inserting (6.14) into the consistency estimate (dropping the constants for simplicity), we obtain

$$\begin{aligned}\|\eta_{\text{int}}^{\text{bqce}}\|_{\mathcal{U}_h^*} &\lesssim (L-K)^{-3/2} + (L-K)^{-1}\|\nabla^2\tilde{u}\|_{L^2(K,L)} \\ &\quad + \|\nabla^3\tilde{u}\|_{L^2(\Omega^c)} + \|h\nabla^2\tilde{u}\|_{L^4(\Omega^c)}^2 + N^{1/2-\alpha}.\end{aligned} \tag{6.16}$$

Further remarks

(1) The first analysis of the B-QCE method (for 1D pair interactions) was given by Van Koten and Luskin (2011). In that paper, the consistency estimates follow the lines of the analysis in Ortner (2011) and Section 2. Our approach in the present section is a simplified 1D variant of the 2D/3D analysis in Li *et al.* (2013).

(2) Since we only attempted to obtain qualitative estimates, our Theorem 6.8 hides the fact that nearest-neighbour interactions do not generate a ghost force in 1D and hence do not contribute to the modelling error. Thus, the prefactor for the dominant error term, $\|\nabla^2\beta\|_{L^2}$, can be considered small (Van Koten and Luskin 2011). However, in 2D/3D, this observation is no longer valid (Ortner and Zhang 2012).

(3) To obtain (6.14) it is not necessary to use the 'optimal' blending function (6.12). Through a straightforward scaling argument one can see that $\hat{\beta}$ may be replaced by any function that is twice continuously differentiable and satisfies the same boundary conditions. However, the bounds (6.15) crucially enter the stability analysis.

6.3. Consistency of QNL-type methods

In Section 4.5 we introduced a general setting for the construction of force-consistent methods and have presented some explicit constructions as well, in particular the second-neighbour QNL method (4.14) and the reflection method (4.16). Rather than focusing on any particular variant, we shall show that *all* methods of the form (4.12), satisfying force-consistency **(F)** and the following additional technical assumptions, are first-order consistent at the interface.

We shall require two additional restrictions on the interface site energies, which we call the locality and scaling conditions.

(L) *Locality.* $\Phi^{\rm i}_\xi(u_h)$ is a k times continuously differentiable function of the interaction stencil $Du_h(\xi) = (D_\rho u_h(\xi))_{\rho\in\mathcal{R}}$. Moreover, $\Phi^{\rm i}_{\xi,\rho} \equiv 0$ for $\xi + \rho > L$.

(S) *Scaling.* $|\Phi^{\rm i}_{\xi,\boldsymbol{\rho}}(u)| \lesssim c(\boldsymbol{\rho})$, for $\boldsymbol{\rho} \in \mathcal{R}^j$, $j = 2, 3$, where

$$\sum_{\boldsymbol{\rho}\in\mathcal{R}^j} |\boldsymbol{\rho}|^s \prod_{i=1}^{j} |\rho_i| c(\boldsymbol{\rho}) \lesssim M^{(j,s)} \quad \text{for } 0 \le s \le 3.$$

Both the locality and scaling conditions ensure that the qualitative analytical properties of the interface potentials are similar to those of the atomistic potential. Both conditions can be weakened, but they are, at least in some form, necessary to obtain the following results (Ortner 2012). Both conditions are satisfied for the reflection method (4.16) and for the geometric reconstruction scheme discussed in Section 4.5, provided that the reconstruction coefficients $C_{\xi,\rho\varsigma}$ satisfy a uniform bound.

We begin, once again, by deriving a weak form of the first variation.

Proposition 6.10. For $u_h \in \mathcal{U}_h$, we have that

$$\langle \delta\mathcal{E}^{\mathrm{qnl}}(u_h), v_h\rangle = \int_0^N \mathsf{S}^{\mathrm{qnl}}(u_h; x)\nabla v_h \,\mathrm{d}x \quad \text{for all } v_h \in \mathcal{U}_h, \tag{6.17}$$

where, for $u \in W^{1,\infty}_{\mathrm{loc}}$, we define

$$\mathsf{S}^{\mathrm{qnl}}(u;x) := \begin{cases} \sum_{\rho\in\mathcal{R}}\sum_{\xi=0}^{K}\chi_{\xi,\rho}(x)\rho\Phi^{\mathrm{a}}_{\xi,\rho}(u) \\ +\sum_{\rho\in\mathcal{R}}\sum_{\xi=K+1}^{L}\chi_{\xi,\rho}(x)\rho\Phi^{\mathrm{i}}_{\xi,\rho}(u), & \text{for a.e. } x < L, \\ \partial_{\mathsf{F}}W(\nabla u(x)), & \text{for a.e. } x > L. \end{cases} \tag{6.18}$$

Proof. The proof of this result is analogous to the derivation of the atomistic stress, using the form (4.12) of the QNL energy, and the locality assumption **(L)**. □

The next lemma establishes the key properties of $\mathsf{S}^{\mathrm{qnl}}$ required to derive the modelling error estimate.

Lemma 6.11. Let $u \in W^{1,\infty}_{\mathrm{loc}}$. Then

$$\mathsf{S}^{\mathrm{qnl}}(u;x) = \mathsf{S}^{\mathrm{a}}(u;x), \qquad \text{for a.e. } x \in [0, K - r_{\mathrm{cut}}], \tag{6.19}$$

$$\mathsf{S}^{\mathrm{qnl}}(u;x) = \partial_{\mathsf{F}}W(\nabla u), \quad \text{for a.e. } x \in [L, N], \tag{6.20}$$

$$\mathsf{S}^{\mathrm{qnl}}(u^{\mathsf{F}};x) = \partial_{\mathsf{F}}W(\mathsf{F}), \qquad \text{for a.e. } x \in \mathbb{R}_+. \tag{6.21}$$

Proof. Throughout, we assume that $x \notin \mathbb{Z}$. To prove (6.19), we note that, for $x \le K - r_{\mathrm{cut}}$, $\chi_{\xi,\rho}(x) \neq 0$ only when $\xi \le K$. Hence, (6.3) and (6.18) imply (6.19).

The property (6.20) follows immediately from the definition (6.18) of $\mathsf{S}^{\mathrm{qnl}}$.

To prove property (6.21), we observe that the force consistency condition **(F)** on page 437 can be restated as

$$\langle \delta\mathcal{E}^{\mathrm{qnl}}(u^{\mathsf{F}}), v_h\rangle = \int_0^N \mathsf{S}^{\mathrm{qnl}}(u^{\mathsf{F}})\nabla v_h \,\mathrm{d}x = 0 \quad \text{for all } v_h \in \mathcal{U}_h,\ \mathsf{F} \in \mathbb{R}.$$

Since $\mathsf{S}^{\mathrm{qnl}}$ is a piecewise constant function, this is only possible if $\mathsf{S}^{\mathrm{qnl}}(u^{\mathsf{F}};x)$ is constant. Since $\mathsf{S}^{\mathrm{qnl}}(u^{\mathsf{F}};x) = \partial_{\mathsf{F}}W(\mathsf{F})$ in the continuum region, (6.21) follows. □

Next, we show that (6.21) implies a first-order error estimate on the a/c stress function. Note, however, that the stress error is second-order in the continuum region and vanishes in the atomistic region.

Lemma 6.12. Let $u \in W^{2,\infty}_{\rm loc}$. Then for almost every $x \in [K - r_{\rm cut}, L]$,

$$|\mathsf{S}^{\rm qnl}(u;x) - \mathsf{S}^{\rm a}(u;x)| \lesssim M^{(2,1)} \|\nabla^2 u\|_{L^\infty(\nu_x \setminus [0, K-r_{\rm cut}])}. \tag{6.22}$$

Proof. The result is a consequence of the Lipschitz estimate

$$\big|\mathsf{S}^{\rm qnl}(u;x) - \mathsf{S}^{\rm qnl}(v;x)\big| \lesssim \|\nabla u - \nabla v\|_{L^\infty(\nu_x)}.$$

However, to remove the dependence on the atomistic region, we proceed more carefully. Since the bounds on derivatives of $\Phi^{\rm i}$ are now part of the approximation, we have to trace this dependence carefully. We assume, without loss of generality, that $|\rho||\varsigma|c(\rho,\varsigma) \gtrsim m(\rho,\varsigma)$ (see condition **(S)**).

Let $x \in [K - r_{\rm cut}, L] \setminus \mathbb{Z}$. Then

$$\begin{aligned}\mathsf{R}(u;x) &:= \mathsf{S}^{\rm qnl}(u;x) - \mathsf{S}^{\rm a}(u;x) \\ &= \sum_{\rho\in\mathcal{R}} \rho \sum_{\xi=K+1}^{L} \chi_{\xi,\rho}(x)\big(\Phi^{\rm i}_{\xi,\rho}(u) - \Phi^{\rm a}_{\xi,\rho}(u)\big).\end{aligned}$$

From (6.21), we know that $\mathsf{R}(u^{\mathsf{F}};x) = 0$ for all $\mathsf{F} \in \mathbb{R}$, and hence we can instead estimate

$$\begin{aligned}&\big|\mathsf{R}(u;x)\big| = \big|\mathsf{R}(u;x) - \mathsf{R}(u^{\mathsf{F}};x)\big| \\ &\quad\le \sum_{\rho\in\mathcal{R}} |\rho| \sum_{\xi=K+1}^{L} \chi_{\xi,\rho}\big\{\big|\Phi^{\rm i}_{\xi,\rho}(u) - \Phi^{\rm i}_{\xi,\rho}(u^{\mathsf{F}})\big| + \big|\Phi^{\rm a}_{\xi,\rho}(u) - \Phi^{\rm a}_{\xi,\rho}(u^{\mathsf{F}})\big|\big\}.\end{aligned}$$

With the choice $\mathsf{F} := \nabla u(x)$, and for $\xi \in \nu_x \setminus [0, K]$, we can further bound

$$\begin{aligned}\big|\Phi^{\rm i}_{\xi,\rho}(u) - \Phi^{\rm i}_{\xi,\rho}(u^{\mathsf{F}})\big| &\le \sum_{\varsigma\in\mathcal{R}} c(\rho,\varsigma)\big|D_\varsigma u(\xi) - \nabla u(x)\varsigma\big| \\ &\lesssim |\xi - x| \sum_{\varsigma\in\mathcal{R}} |\varsigma| c(\rho,\varsigma) \|\nabla^2 u\|_{L^\infty(\nu_x\setminus[0,K-r_{\rm cut}])}.\end{aligned}$$

An analogous bound is satisfied by $|\Phi^{\rm a}_{\xi,\rho}(u) - \Phi^{\rm a}_{\xi,\rho}(u^{\mathsf{F}})|$.

Combining these estimates, and applying **(S)**, we obtain

$$\begin{aligned}\big|\mathsf{R}(u;x)\big| &\lesssim \|\nabla^2 u\|_{L^\infty(\nu_x\setminus[0,K-r_{\rm cut}])} \sum_{\rho,\varsigma\in\mathcal{R}} |\rho||\varsigma|c(\rho,\varsigma) \sum_{\xi=K+1}^{\infty} \chi_{\xi,\rho}|\xi - x| \\ &\lesssim M^{(2,1)} \|\nabla^2 u\|_{L^\infty(\nu_x\setminus[0,K-r_{\rm cut}])}.\end{aligned}$$

This concludes the proof of the result. □

Theorem 6.13. Let $u \in \mathcal{U}$ satisfy **(DH)** and let $K > r_{\rm cut}, N > r_0$. Then

$$\|\eta^{\rm qnl}_{\rm int}\|_{\mathcal{U}_h^*} \lesssim M^{(2,2)}\|\nabla^2\tilde{u}\|_{L^2(K-r_{\rm cut},L)} + M^{(2,3)}\|\nabla^3\tilde{u}\|_{L^2(L-2r_{\rm cut},\infty)} + M^{(3,3)}\|h\nabla^2\tilde{u}\|^2_{L^4(L-2r_{\rm cut},\infty)} + M^{(2,0)}N^{1/2-\alpha}.$$

Proof. The result can be proved along similar lines to Theorem 6.8. We simply remark that the term $M^{(2,2)}\|\nabla^2\tilde{u}\|_{L^2([K-r_{\rm cut},L])}$ is due to the first-order consistency in the a/c interface established in Lemma 6.12, while the remaining terms are due to the Cauchy–Born modelling and coarsening errors. □

Further remarks

The key result in this section is (6.21). Although it is straightforward to prove in 1D, it is in fact non-trivial. In particular, this property of the a/c stress is in general false in 2D/3D, due to the fact that $\mathsf{S}^{\rm qnl}$ may contain divergence-free components (Ortner and Zhang 2012). A divergence-free 'corrector' must therefore be added to the the a/c stress $\mathsf{S}^{\rm qnl}$, which leads to a modified stress function $\hat{\mathsf{S}}^{\rm qnl}(u)$. In 2D, one can exploit the characterization of discrete divergence-free tensor fields due to Arnold and Falk (1989) to characterize the difference $\mathsf{S}^{\rm qnl}(u^{\mathsf{F}}) - \mathsf{S}^{\rm a}(u^{\mathsf{F}})$ and construct such a corrector. This was carried out explicitly for a model problem by Ortner and Zhang (2012). A general result was established by Ortner (2012), which established first-order consistency for a general class of a/c couplings satisfying the force consistency **(F)**, locality **(L)**, and scaling conditions **(S)**.

It is, at present, an open question whether a similar result holds in 3D too. Moreover, even the 2D result (Ortner 2012) is not completely general. For example, it requires that the atomistic region is connected. (Although our framework includes this assumption as well, it is in fact not required in 1D.)

6.4. Consistency of the B-QCF method

We now turn to the consistency analysis of the B-QCF method, the only non-conservative scheme that we consider. We can again employ many of the ideas we developed in the previous sections. However, some interesting new effects appear. These can already be seen in the weak form of the B-QCF force operator in the following lemma.

We remark that we can replace I_h with I_1 (the nodal interpolant with respect to the atomistic grid $\mathbb{Z}_+$) in the following lemma.

Lemma 6.14. Let $u_h, v_h \in \mathcal{U}_h$. Then

$$\begin{aligned}\langle \mathcal{F}^{\mathrm{bqcf}}(u_h), v_h\rangle &= \langle \delta\mathcal{E}^{\mathrm{a}}(u_h), I_h((1-\beta)v_h)\rangle + \langle \delta\mathcal{E}^{\mathrm{c}}(u_h), I_h(\beta v_h)\rangle \qquad (6.23)\\ &= \int_0^N \big\{\mathsf{S}^{\mathrm{a}}(u_h)\nabla I_h\big((1-\beta)v_h\big) + \partial_{\mathsf{F}}W(\nabla u_h)\nabla I_h(\beta v_h)\big\}\,\mathrm{d}x.\end{aligned}$$

Proof. From (4.22) and the definition of the weak form below (4.21), we obtain

$$\begin{aligned}\langle \mathcal{F}^{\mathrm{bqcf}}(u_h), v_h\rangle &= \sum_{\varsigma\in\mathcal{N}_h}\bigg\{(1-\beta(\varsigma))\frac{\partial\mathcal{E}^{\mathrm{a}}(u_h)}{\partial u_h(\varsigma)} + \beta(\varsigma)\frac{\partial\mathcal{E}^{\mathrm{c}}(u_h)}{\partial u_h(\varsigma)}\bigg\}\cdot v_h(\varsigma)\\ &= \sum_{\varsigma\in\mathcal{N}_h}\frac{\partial\mathcal{E}^{\mathrm{a}}(u_h)}{\partial u_h(\varsigma)}\cdot I_h\big((1-\beta)v_h\big)(\varsigma) + \sum_{\varsigma\in\mathcal{N}_h}\frac{\partial\mathcal{E}^{\mathrm{c}}(u_h)}{\partial u_h(\varsigma)}\cdot I_h(\beta v_h)(\varsigma)\\ &= \langle \delta\mathcal{E}^{\mathrm{a}}(u_h), I_h((1-\beta)v_h)\rangle + \langle \delta\mathcal{E}^{\mathrm{c}}(u_h), I_h(\beta v_h)\rangle.\end{aligned}$$

The representation of $\delta\mathcal{E}^{\mathrm{a}}$ given in terms of the stress S^{a} in Proposition 6.1 and the standard weak form of $\delta\mathcal{E}^{\mathrm{c}}$ yield the stated result. □

As a consequence, we obtain the following form of the internal B-QCF consistency error.

Lemma 6.15. Let $u \in \mathcal{U}$ and suppose that $\{0, \ldots, L + r_{\mathrm{cut}}\} \subset \mathcal{N}_h$. Then

$$\langle \eta^{\mathrm{bqcf}}_{\mathrm{int}}(u), v_h\rangle = \int_K^N \big(\mathsf{S}^{\mathrm{a}}(u) - \partial_{\mathsf{F}}W(\nabla\Pi_h u)\big)\nabla I_h(\beta v_h)\,\mathrm{d}x \quad \text{for all } v_h \in \mathcal{U}_h.$$

Proof. Under the assumption that $\{0, \ldots, L + r_{\mathrm{cut}}\} \subset \mathcal{N}_h$, we have that $\Pi_h u = u$ in $[0, L + r_{\mathrm{cut}}]$, which implies that $\mathsf{S}^{\mathrm{a}}(\Pi_h u; x) = \mathsf{S}^{\mathrm{a}}(u; x)$ for $x \in [0, L]$. Hence, Lemma 6.14 yields

$$\begin{aligned}\langle \eta^{\mathrm{bqcf}}_{\mathrm{int}}(u), v_h\rangle &= \langle \mathcal{F}^{\mathrm{bqcf}}(\Pi_h u) - \delta\mathcal{E}^{\mathrm{a}}(u), v_h\rangle\\ &= \int_0^N \big\{\mathsf{S}^{\mathrm{a}}(u)\nabla I_h\big((1-\beta)v_h\big) + \partial_{\mathsf{F}}W(\nabla\Pi_h u)\nabla I_h(\beta v_h) - \mathsf{S}^{\mathrm{a}}(u)\nabla v_h\big\}\,\mathrm{d}x\\ &= \int_0^N \big(\partial_{\mathsf{F}}W(\nabla\Pi_h u) - \mathsf{S}^{\mathrm{a}}(u)\big)\nabla I_h(\beta v_h)\,\mathrm{d}x.\end{aligned}$$

Since $\beta = 0$ in $[0, K]$, the result follows. □

We see from Lemma 6.15 that we can use our techniques from previous sections to bound the residual in terms of $\|\nabla(\beta v_h)\|_{L^2}$. However, we need a bound in terms of $\|\nabla v_h\|_{L^2}$. This is provided by the next lemma.

Lemma 6.16. Suppose that β is defined globally with $\beta \in W^{1,\infty}$, and suppose that $|\nabla\beta| \le 1$. Then

$$\|\nabla I_h(\beta v_h)\|_{L^2} \lesssim \big\{1 + \|x^{1/2}\nabla\beta\|_{L^2}\big\}\|\nabla v_h\|_{L^2} \quad \text{for all } v_h \in \mathcal{U}_h.$$

Proof. In order to evaluate the gradient of the product, we add and subtract $\nabla(v_h I_h \beta)$:

$$\|\nabla I_h(\beta v_h)\|_{L^2} \le \|\nabla(v_h I_h\beta)\|_{L^2} + \|\nabla(v_h I_h\beta) - \nabla I_h(v_h\beta)\|_{L^2}. \qquad (6.24)$$

The error term can be controlled in terms of $\nabla^2(v_h I_h \beta)$ (understood in the piecewise sense),

$$\begin{aligned}\|\nabla(v_h I_h\beta) - \nabla I_h(v_h\beta)\|_{L^2} &\lesssim \|\nabla^2(v_h I_h\beta)\|_{L^2} = 2\|\nabla v_h \nabla I_h\beta\|_{L^2} \\ &\lesssim \|\nabla I_h\beta\|_{L^\infty}\|\nabla v_h\|_{L^2} \le \|\nabla\beta\|_{L^\infty}\|\nabla v_h\|_{L^2} \\ &\le \|\nabla v_h\|_{L^2},\end{aligned}$$

where in the last line we used the assumption that $|\nabla\beta| \le 1$.

The first term on the right-hand side of (6.24) can be bounded by

$$\|\nabla(v_h I_h\beta)\|_{L^2} \le \|\nabla\beta\cdot v_h\|_{L^2(K,L)} + \|\nabla v_h\|_{L^2}.$$

Using the fact that $v_h(0) = 0$, Hölder's inequality implies

$$|v_h(x)| \le x^{1/2}\|\nabla v_h\|_{L^2(0,x)}.$$

Hence we estimate

$$\|\nabla\beta\cdot v_h\|_{L^2(K,L)} \le \|x^{1/2}\nabla\beta\|_{L^2(K,L)}\|\nabla v_h\|_{L^2}.$$

Combining the computations above, we obtain the stated result. □

We are now ready to prove the consistency result for the B-QCF method.

Theorem 6.17. Let $u \in \mathcal{U}$ satisfy the decay hypothesis **(DH)**, let $|\nabla\beta| \le 1$ and let $L > K > 2r_{\rm cut}$. Then

$$\begin{aligned}\big\|\eta_{\rm int}^{\rm bqcf}(u)\big\|_{\mathcal{U}_h^*} \le \big(1 + \|x^{1/2}\nabla\beta\|_{L^2}\big)\big\{M^{(2,3)}\|\nabla^3\tilde u\|_{L^2(K-2r_{\rm cut},\infty)} \\ + M^{(3,3)}\|h\nabla^2\tilde u\|^2_{L^4(K-2r_{\rm cut},\infty)} + M^{(2,0)}N^{1/2-\alpha}\big\}.\end{aligned}$$

Proof. Applying Corollary 6.4 and Lemma 5.2 to the representation of $\eta_{\rm int}^{\rm bqcf}$ derived in Lemma 6.15, we obtain

$$\begin{aligned}\langle\eta_{\rm int}^{\rm bqcf}(u), v_h\rangle \lesssim \big\{\|\nabla^3\tilde u\|_{L^2(K-2r_{\rm cut},N+2r_{\rm cut})} + \|h\nabla^2\tilde u\|^2_{L^4(K-2r_{\rm cut},N+2r_{\rm cut})} \\ + N^{1/2-\alpha}\big\}\|\nabla(\beta v_h)\|_{L^2}.\end{aligned}$$

Employing Lemma 6.16 yields the stated consistency error estimate. □

Suppose now that we use the blending function $\beta = \beta^*$ defined in (6.12). Then we have

$$\|\nabla\beta\|_{L^2} \lesssim (L-K)^{-1/2}$$

and hence

$$\|x^{1/2}\nabla\beta\|_{L^2} \lesssim L^{1/2}(L-K)^{-1/2},$$

and this cannot be improved upon. Thus, using the optimal blending function, the estimate (6.14) becomes

$$\begin{aligned}\|\eta_{\rm int}^{\rm bqcf}\|_{\mathcal{U}_h^*} \lesssim\ & (1 + L^{1/2}(L-K)^{-1/2})\big\{\|\nabla^3 \tilde{u}\|_{L^p(K-2r_{\rm cut},\infty)} \\ & + \|h\nabla^2 \tilde{u}\|_{L^{2p}(K-2r_{\rm cut},\infty)}^2 + N^{1/2-\alpha}\big\}.\end{aligned} \tag{6.25}$$

In order to remove the prefactor dependence on the blending width, we must choose $L \geq cK$, $c > 1$. We will return to this point when we optimize all the approximation parameters in Section 8.

Further remarks

(1) The consistency analysis of the B-QCF method follows the more general multi-dimensional analysis in Li *et al.* (2013). Previous results in this direction can be found in Dobson *et al.* (2010*b*), Makridakis *et al.* (2011) and Lu and Ming (2013).

(2) Since H^1-stability for the QCF method (without blending) is false (Dobson *et al.* 2010*b*), the first error estimates for the QCF method were given in the $W^{1,\infty}$-seminorm and hence the consistency error was estimated in $W^{-1,\infty}$. Similarly, the consistency analysis of Lu and Ming (2013) is performed in the L^2-norm, which (in our scaling) is even weaker than the $W^{-1,\infty}$-norm.

The main difficulty in these analytical settings is to extend the analyses to realistic 2D/3D situations with defects, where $W^{1,\infty}$-regularity or H^2-regularity is likely to be false. Conversely, in the convenient (and natural) energy norm, the original QCF method is unstable. However, Li *et al.* (2012) showed that B-QCF is stable in the energy norm, which motivates our choice of analysing B-QCF in the framework of the present paper.

(3) If we had not imposed antisymmetry of the displacements, then we would not have the boundary condition $v_h(0) = 0$ available, but only $v_h(N) = 0$, and hence Lemma 6.16 would become

$$\|\nabla I_h(\beta v_h)\|_{L^2} \lesssim (1 + N^{1/2}(L-K)^{-1/2})\|\nabla v_h\|_{L^2}.$$

This is clearly a problem, since it would require an $O(N)$ blending region to control this prefactor. However, it is purely a 1D artefact. Indeed, analogous estimates in 2D and 3D (Li *et al.* 2013) show that the prefactor can be efficiently controlled.

6.5. Summary of consistency estimates

For the reader's convenience, we summarize all consistency estimates in Table 6.1. For each entry, we write the coupling error in the first line, the coarsening and Cauchy–Born modelling errors in the second line, and the

Table 6.1. Summary of consistency error estimates. For each entry, we write the coupling error in the first line, the coarsening and Cauchy–Born modelling errors in the second line, and the far-field error in the third line.

a/c method	$\|\eta^{\mathrm{ac}}_{\mathrm{int}}(u)\|_{\mathcal{U}_h^*} \lesssim$	Reference
B-QCE (Section 4.4)	$\|\nabla^2\beta\|_{L^2} + \|\nabla\beta\nabla^2\tilde{u}\|_{L^2}$ $+\|\nabla^3\tilde{u}\|_{L^2(\Omega^{\mathrm{c}})} + \|h\nabla^2\tilde{u}\|^2_{L^2(\Omega^{\mathrm{c}})}$ $+N^{1/2-\alpha}$	Theorem 6.8
QNL (Section 4.5)	$\|\nabla^2\tilde{u}\|_{L^2(K-r_{\mathrm{cut}},L)}$ $+\|\nabla^3\tilde{u}\|_{L^2(\Omega^{\mathrm{c}})} + \|h\nabla^2\tilde{u}\|^2_{L^2(\Omega^{\mathrm{c}})}$ $+N^{1/2-\alpha}$	Theorem 6.13
B-QCF (Section 4.6)	0 $+\|\nabla^3\tilde{u}\|_{L^2(\Omega^{\mathrm{c}})} + \|h\nabla^2\tilde{u}\|^2_{L^2(\Omega^{\mathrm{c}})}$ $+N^{1/2-\alpha}$	Theorem 6.17 and (6.25)

far-field error in the third line. For the B-QCF method, we assume that a blending function and choice of K, L was used for which the prefactor becomes $O(1)$.

6.6. *Remarks on extensions to 2D/3D*

The key concept in our consistency analyses is the representations of first variations of the atomistic and coupling energies in terms of stress functions ('weak forms'). To derive these, we employed a splitting of long-range finite differences in terms of local gradients (see (6.1)),

$$D_\rho v(\xi) = \int_{\mathbb{R}_+} \rho\,\chi_{\xi,\rho} \nabla v(x)\,\mathrm{d}x. \tag{6.26}$$

As a matter of fact, this technicality is not entirely necessary (though still convenient) in 1D since the stress in 1D can be simply defined as the derivative with respect to the gradient, for example,

$$\mathsf{S}^{\mathrm{a}}\big(u;\eta+\tfrac{1}{2}\big) = \frac{\partial\mathcal{E}^{\mathrm{a}}(u)}{\partial D_1 u(\eta)},$$

due to the fact that there is a one-to-one mapping between u and $D_1 u$. This was employed, for example, by Makridakis *et al.* (2011) for the analysis of the QCF method and the construction of a non-conservative scheme in divergence form. Our main reason for employing (6.26) in the present work is to mimic the extension of our results to 2D/3D.

The main difficulty in extending (6.26) to 2D/3D is that line integrals and volume integrals are no longer identical. A striking result of Shapeev (2011) showed that this need not be a problem: if T is a triangle with vertices belonging to $\mathbb{Z}^2$, $\rho \in \mathbb{Z}^2$, and $\chi_T(x) = 1$ if $x \in \mathrm{int}T$, $\chi_T(x) = 1/2$ if $x \in \partial T$, then

$$\sum_{\xi \in \mathbb{Z}^2} \int_{t=0}^{1} \chi_T(\xi + t\rho)\,\mathrm{d}t = |T|.$$

This *bond-density formula* enables the generalization of various 1D results to 2D, including the construction and rigorous analysis of the bond-splitting scheme (2.41); see Shapeev (2011) and Ortner and Shapeev (2013).

Unfortunately, the result is false in 3D, and moreover, there appears to be no analogue of (6.7) even in 2D, which makes it more difficult to exploit the symmetries of the interaction potential. An attempt to generalize it through the use of mollification operators (Ortner and Shapeev 2012) proved highly useful for the analysis of the Cauchy–Born approximation (Ortner and Theil 2013) as well as the B-QCE and B-QCF methods (Li *et al.* 2013), but has not led to a practical a/c coupling even for pair potentials. A promising approach based on yet another variant of writing finite differences as volume integrals and employing a family of interpolants of discrete functions (rather than a single interpolant) was proposed by Makridakis *et al* (2012). The techniques introduced therein may also provide an alternative avenue for the analysis of a/c coupling in 2D/3D.

7. Stability

The previous section was devoted to consistency estimates for the three main a/c methods. This section is devoted to the second fundamental ingredient for the error analysis: the stability of the linearized a/c operators $\delta\mathcal{F}^{\mathrm{ac}} \in \{\delta^2\mathcal{E}^{\mathrm{bqce}}, \delta^2\mathcal{E}^{\mathrm{qnl}}, \delta\mathcal{F}^{\mathrm{bqcf}}\}$. Assuming that $u \in \mathcal{U}$ is a strongly stable state satisfying (3.19), that is,

$$\langle \delta^2\mathcal{E}^{\mathrm{a}}(u^{\mathrm{a}})v, v\rangle \geq c_0\|\nabla v\|_{L^2}^2 \quad \text{for all } v \in \mathcal{U},$$

we will show that

$$\langle \delta\mathcal{F}^{\mathrm{ac}}(u)v, v\rangle \geq \big(c_0 - \epsilon\big)\|\nabla v\|_{L^2}^2,$$

where ϵ is a stability error that is controllable in terms of the approximation parameters.

7.1. Stability of the continuum model

In this preparatory section, we review the classical observation that stability of the atomistic model implies stability of the Cauchy–Born model. To state the result, we extend the definition of Hessians to general homogeneous displacements $u = u^{\mathsf{F}}$ and $v \in \mathcal{U}$, by the same formula (3.10). This is well-defined since we have assumed that the second partial derivatives of V are bounded. We then define

$$\gamma^{\mathrm{a}}(\mathsf{F}) := \inf_{\substack{v\in\mathcal{U}_0 \\ \|\nabla v\|_{L^2}=1}} \langle \delta^2\mathcal{E}^{\mathrm{a}}(u^{\mathsf{F}})v, v\rangle.$$

Lemma 7.1. $\gamma^{\mathrm{a}}(\mathsf{F}) \le \partial_{\mathsf{F}}^2 W(\mathsf{F})$ for all $\mathsf{F} \in \mathbb{R}$.

Proof. Recall from (4.8) that $\mathcal{E}^{\mathrm{c}}(u) = \int_0^\infty W(\nabla u)\,\mathrm{d}x$. Then

$$\langle \delta^2\mathcal{E}^{\mathrm{c}}(u^{\mathsf{F}})v, v\rangle = \int_0^\infty \partial_{\mathsf{F}}^2 W(\mathsf{F})|\nabla v|^2\,\mathrm{d}x \quad \text{for } v \in W^{1,\infty}_{\mathrm{loc}},\ \nabla v \in L^2.$$

The proof uses the fact that $\mathcal{E}^{\mathrm{c}}$ is the scaling limit of $\mathcal{E}^{\mathrm{a}}$ and the scale invariance of the continuum model.

We fix an antisymmetric function $v \in C_0^\infty(\mathbb{R}), v \neq 0$, compactly supported, and define

$$v_R(x) := Rv(x/R).$$

Then $\nabla v_R(x) = (\nabla v)(x/R)$ and $\nabla^2 v_R(x) = R^{-1}(\nabla^2 v)(x/R)$. Using these properties, it is easy to show (see Hudson and Ortner (2012) for details) that

$$\left| \frac{\langle \delta^2\mathcal{E}^{\mathrm{a}}(u^{\mathsf{F}})v_R, v_R\rangle}{\|\nabla v_R\|_{L^2}^2} - \frac{\langle \delta^2\mathcal{E}^{\mathrm{c}}(u^{\mathsf{F}})v_R, v_R\rangle}{\|\nabla v_R\|_{L^2}^2} \right| \to 0 \quad \text{as } R \to \infty,$$

$$\text{and} \quad \frac{\langle \delta^2\mathcal{E}^{\mathrm{c}}(u^{\mathsf{F}})v_R, v_R\rangle}{\|\nabla v_R\|_{L^2}^2} = \frac{\langle \delta^2\mathcal{E}^{\mathrm{c}}(u^{\mathsf{F}})v, v\rangle}{\|\nabla v\|_{L^2}^2} = \partial_{\mathsf{F}}^2 W(\mathsf{F}).$$

Hence, we obtain

$$\gamma^{\mathrm{a}}(\mathsf{F}) \le \lim_{R\to\infty} \frac{\langle \delta^2\mathcal{E}^{\mathrm{a}}(u^{\mathsf{F}})v_R, v_R\rangle}{\|\nabla v_R\|_{L^2}^2} = \partial_{\mathsf{F}}^2 W(\mathsf{F}). \qquad \square$$

Corollary 7.2. Suppose there exists *any* $u \in \mathcal{U}$ satisfying the strong stability condition (3.19). Then $\partial_{\mathsf{F}}^2 W(0) \ge c_0$.

Proof. Let $v \in \mathcal{U}_0$ with $v(\xi) = 0$ for $0 \le \xi \le r_{\mathrm{cut}}$, and let

$$v_R(\xi) := \begin{cases} 0, & 0 \le \xi \le R, \\ v(\xi - R), & \xi \ge R. \end{cases}$$

Then it is easy to see that

$$\langle \delta^2\mathcal{E}^{\mathrm{a}}(u)v_R, v_R\rangle \to \langle \delta^2\mathcal{E}^{\mathrm{a}}(0)v, v\rangle \quad \text{as } R \to \infty.$$

Hence, we deduce that for any v such that $v(\xi) = 0$ for $0 \le \xi \le r_{\rm cut}$,

$$\langle \delta^2 \mathcal{E}^{\rm a}(0) v, v\rangle \ge c_0 \|\nabla v\|_{L^2}^2.$$

Slightly modifying the proof of Lemma 7.1, we obtain that $\partial_{\mathsf{F}}^2 W(0) \ge c_0$. $\square$

Further remarks

As mentioned above, the result $\gamma^{\rm a}(\mathsf{F}) \le \partial_{\mathsf{F}}^2 W(\mathsf{F})$ is classical and can be found in standard textbooks, for example Wallace (1998). The proof reducing the argument to only two ingredients (the fact that the Cauchy–Born model is the scaling limit of the atomistic model, and the scale-invariance of the continuum model) was given by Hudson and Ortner (2012). An interesting question that seems to have gone largely unnoticed is under which conditions the atomistic and continuum stability regions coincide. That is, when does $\partial_{\mathsf{F}}^2 W(\mathsf{F}) > 0$ imply $\gamma^{\rm a}(\mathsf{F}) > 0$? For materials where this condition holds, the Cauchy–Born model can be expected to correctly predict the onset of defect nucleation or other types of singularities.

7.2. Stability of B-QCE

The second variation of $\mathcal{E}^{\rm bqce}$ is given by

$$\langle \delta^2 \mathcal{E}^{\rm bqce}(u) v, v\rangle = \sum_{\xi \in \mathbb{Z}_+} (1-\beta(\xi)) \sum_{\rho,\varsigma \in \mathcal{R}} \Phi_{\xi,\rho\varsigma}(u) D_\rho v(\xi) D_\varsigma v(\xi) + \int_0^\infty I_h\beta \cdot \partial_{\mathsf{F}}^2 W(\nabla u) |\nabla v|^2 \,{\rm d}x. \tag{7.1}$$

If we assume that $\delta^2 \mathcal{E}^{\rm a}(u) > 0$, then this suggests that the atomistic component of $\delta^2 \mathcal{E}^{\rm bqce}(u)$ might be positive as well. Moreover, Lemma 7.1 suggests that $\partial_{\mathsf{F}}^2 W(\nabla u) > 0$, in which case the continuum component of $\delta^2 \mathcal{E}^{\rm bqce}(u)$ would also be positive. To make this intuition precise, we will 'split' the test function v into an atomistic and continuum component, using the following lemma.

Lemma 7.3. Let $\beta \in C^{1,1}(0,\infty)$, with $0 \le \beta \le 1$. For each $v \in \mathcal{U}$, there exists $v^{\rm a}, v^{\rm c} \in \mathcal{U}$ such that

$$\big|\sqrt{1-\beta(\xi)} D_\rho v(\xi) - D_\rho v^{\rm a}(\xi)\big| \le C_1 \big\|\nabla \sqrt{1-\beta}\big\|_{L^\infty} \|\nabla v\|_{L^2(\nu_\xi)}, \tag{7.2}$$

$$\big|\sqrt{I_h\beta(x)} \nabla v(x) - \nabla v^{\rm c}(x)\big| \le C_2 \big\|\nabla \sqrt{\beta}\big\|_{L^\infty} |\nabla v(x)|, \quad \text{and} \tag{7.3}$$

$$|\nabla v^{\rm a}|^2 + |\nabla v^{\rm c}|^2 = |\nabla v|^2, \tag{7.4}$$

where C_1 may depend on $r_{\rm cut}$, but C_2 is a generic constant. In particular, $\nabla v^{\rm a}(x) = 0$ for $x \ge L$ and $\nabla v^{\rm c}(x) = 0$ for $x \le K$.

Proof. Let $\psi(x) := \sqrt{1-\beta(x)}$ and assume, without loss of generality, that $\rho > 0$. Then,

$$\sqrt{1-\beta(\xi)}D_\rho v(\xi) = \psi(\xi)\sum_{\eta=\xi}^{\xi+\rho-1} D_1 v(\eta)$$
$$= \sum_{\eta=\xi}^{\xi+\rho-1} \psi(\eta)D_1 v(\eta) + \sum_{\eta=\xi}^{\xi+\rho-1} \big(\psi(\xi)-\psi(\eta)\big)D_1 v(\eta).$$

If we define v^{a} by $v^{\mathrm{a}}(0) = 0$ and $D_1 v^{\mathrm{a}}(\eta) = \psi(\eta)D_1 v(\eta)$, then we obtain

$$\sum_{\eta=\xi}^{\xi+\rho-1} \psi(\eta)D_1 v(\eta) = D_\rho v^{\mathrm{a}}(\xi),$$

and hence

$$\big|\sqrt{1-\beta}D_\rho v(\xi) - D_\rho v^{\mathrm{a}}(\xi)\big| \le |\rho|^{3/2}\|\nabla\psi\|_{L^\infty}\|\nabla v\|_{L^2(\xi,\xi+\rho)}.$$

This establishes (7.2). The proof of (7.3) is analogous, with v^{c} defined by $D_1 v^{\mathrm{c}}(\xi) = \sqrt{\beta(\xi)}D_1 v(\xi)$.

With these definitions (7.4) is an immediate consequence. □

It is of course possible that $\|\nabla\sqrt{1-\beta}\|_{L^\infty}$ or $\|\nabla\sqrt{\beta}\|_{L^\infty}$ are unbounded. However, if we choose $\beta = \beta^*$ defined in (6.12), then, employing (6.15), the estimates (7.2) and (7.3) become

$$\big|\sqrt{1-\beta(\xi)}D_\rho v(\xi) - D_\rho v^{\mathrm{a}}(\xi)\big| \lesssim (L-K)^{-1}\|\nabla v\|_{L^2(\nu_\xi)}, \quad \text{and}$$
$$\big|\sqrt{I_h\beta(x)}\nabla v(x) - \nabla v^{\mathrm{c}}(x)\big| \lesssim (L-K)^{-1}|\nabla v(x)|.$$

Thus, the following result shows that the coercivity constant of the B-QCE Hessian can be controlled in terms of the atomistic region size and the blending width.

Theorem 7.4. Let $u \in \mathcal{U}$ satisfy **(DH)** as well as the stability condition (3.19) and let $K \ge r_0$. Then

$$\big\langle \delta^2\mathcal{E}^{\mathrm{bqce}}(u)v, v\big\rangle \ge \big[c_0 - C\epsilon^{\mathrm{bqce}}\big]\|\nabla v\|_{L^2}^2 \quad \text{for all } v \in \mathcal{U}_0,$$

where

$$\epsilon^{\mathrm{bqce}} := \max\big(\|\nabla\sqrt{1-\beta}\|_{L^\infty}, \|\nabla\sqrt{\beta}\|_{L^\infty}, K^{-\alpha}\big),$$

and $C \lesssim M^{(2,2)} + C_{\mathbf{DH}}M^{(3,0)}$ is independent of all approximation parameters.

If $\beta = \beta^*$, then $\epsilon^{\mathrm{bqce}} \lesssim \max((L-K)^{-1}, K^{-\alpha})$.

Proof. Let $\epsilon_1 := \max(\|\nabla\sqrt{1-\beta}\|_{L^\infty}, \|\nabla\sqrt{\beta}\|_{L^\infty})$. Applying the estimates (7.2) and (7.3) to (7.1), we obtain

$$\langle \delta^2\mathcal{E}^{\mathrm{bqce}}(u)v, v\rangle \geq \sum_{\xi\in\mathbb{Z}_+}\sum_{\rho,\varsigma\in\mathcal{R}} \big\{\Phi_{\xi,\rho\varsigma}(u)D_\rho v^{\mathrm{a}}(\xi)D_\varsigma v^{\mathrm{a}}(\xi) - C'_{1,\rho\varsigma}\epsilon_1\|\nabla v\|^2_{L^2(\nu_\xi)}\big\}$$
$$+ \int_0^\infty \big\{\partial^2_{\mathsf{F}}W(\nabla u) - C'_2\epsilon_1\big\}|\nabla v^{\mathrm{c}}|^2\,\mathrm{d}x, \tag{7.5}$$

where $C'_{1,\rho\varsigma}, C'_2$ are independent of any approximation parameters. Since $\nabla v^{\mathrm{c}} = 0$ in $(0, K)$, and using also Lemma 7.1,

$$\partial^2_{\mathsf{F}}W(\nabla u) \geq \partial^2_{\mathsf{F}}W(0) - C'_3K^{-\alpha} \geq c_0 - C'_3K^{-\alpha},$$

where C'_3 depends only on $M^{(3,0)}$ and on $C_{\mathbf{DH}}$.

Estimating the overlaps from the terms $\|\nabla v^{\mathrm{a}}\|_{L^2(\nu_\xi)}$, and using the fact that $|\nabla v^{\mathrm{c}}| \leq |\nabla v|$, gives

$$\langle \delta^2\mathcal{E}^{\mathrm{bqce}}(u)v, v\rangle \geq \langle \delta^2\mathcal{E}^{\mathrm{a}}(u)v^{\mathrm{a}}, v^{\mathrm{a}}\rangle + c_0\|\nabla v^{\mathrm{c}}\|^2_{L^2} - C\epsilon\|\nabla v\|^2_{L^2},$$

where $\epsilon := \max(\epsilon_1, K^{-\alpha})$. Now we use the stability of the atomistic model and property (7.4), to conclude that

$$\begin{aligned}\langle \delta^2\mathcal{E}^{\mathrm{bqce}}(u)v, v\rangle &\geq c_0\|\nabla v^{\mathrm{a}}\|^2_{L^2} + c_0\|\nabla v^{\mathrm{c}}\|^2_{L^2} - C\epsilon\|\nabla v\|^2_{L^2}\\ &= \big[c_0 - C\epsilon\big]\|\nabla v\|^2_{L^2}.\end{aligned}$$

□

7.3. Stability of B-QCF

The stability analysis of the B-QCF method uses similar ideas to that of the B-QCE method. The key difference is where the blending function occurs in the Jacobian. From (6.23) we immediately obtain, for $u_h, v_h \in \mathcal{U}_h$,

$$\begin{aligned}\big\langle \delta\mathcal{F}^{\mathrm{bqcf}}(u_h)v_h, v_h\big\rangle &= \big\langle \delta^2\mathcal{E}^{\mathrm{a}}(u_h)v_h, (1-\beta)v_h\big\rangle + \big\langle \delta^2\mathcal{E}^{\mathrm{c}}(u_h)v_h, I_h(\beta v_h)\big\rangle\\ &= \sum_{\xi\in\mathbb{Z}_+}\sum_{\rho,\varsigma\in\mathcal{R}} \Phi_{\xi,\rho\varsigma}(u)D_\rho v_h(\xi)D_\varsigma\big[(1-\beta)v_h\big](\xi)\\ &\quad + \int_{\mathbb{R}+} \partial^2_{\mathsf{F}}W(\nabla u)\nabla v_h\nabla I_h[\beta v_h]\,\mathrm{d}x.\end{aligned}$$

Next, we note that $I_h[\beta v_h] = I_1[\beta v_h]$ for all $v_h \in \mathcal{U}_h$, where I_1 denotes the piecewise affine interpolant with respect to the atomistic mesh $\mathbb{Z}_+$. Therefore, we can extend the definition of $\delta\mathcal{F}^{\mathrm{bqcf}}$ naturally to $u, v \in \mathcal{U}$, by

$$\begin{aligned}\big\langle \delta\mathcal{F}^{\mathrm{bqcf}}(u)v, v\big\rangle &= \big\langle \delta^2\mathcal{E}^{\mathrm{a}}(u)v, (1-\beta)v\big\rangle + \big\langle \delta^2\mathcal{E}^{\mathrm{c}}(u)v, I_1(\beta v)\big\rangle \qquad (7.6)\\ &= \sum_{\xi\in\mathbb{Z}_+}\sum_{\rho,\varsigma\in\mathcal{R}} \Phi_{\xi,\rho\varsigma}(u)D_\rho v(\xi)D_\varsigma\big[(1-\beta)v\big](\xi)\\ &\quad + \int_{\mathbb{R}+} \partial^2_{\mathsf{F}}W(\nabla u)\nabla v\nabla I_1[\beta v]\,\mathrm{d}x.\end{aligned}$$

This will simplify the subsequent stability analysis.

We will now rewrite $\delta\mathcal{F}^{\rm bqce}$ in terms of $\delta^2\mathcal{E}^{\rm bqce}$ and a controllable error. The intuition behind the first intermediate step is the simple observation that $D_\varsigma[(1-\beta)v(\xi)] \approx -\varsigma\nabla\beta(\xi)v(\xi) + (1-\beta(\xi))D_\varsigma v(\xi)$.

Lemma 7.5. Let $u \in \mathcal{U}$ satisfy **(DH)** with $K - r_{\rm cut} \geq r_0$, and let $\|\nabla\beta\|_{L^\infty} \leq 1$. Then, for $v \in \mathcal{U}_h$,

$$\langle \delta\mathcal{F}^{\rm bqcf}(u)v, v\rangle \geq \langle \delta^2\mathcal{E}^{\rm bqce}(u)v, v\rangle + \langle Sv, v\rangle - C_1\epsilon_1\|\nabla v\|_{L^2}^2,$$

where $\epsilon_1 := \max(\|\nabla\beta\|_{L^\infty}K^{-\alpha}, \|\nabla^2\beta\|_{L^\infty})L^{1/2}(L-K)^{1/2}$, $C_1 \lesssim M^{(2,2)} + M^{(3,1)}$, and

$$\langle Sv, v\rangle := \sum_{\xi\in\mathbb{Z}_+} \nabla\beta(\xi)v(\xi) \sum_{\rho,\varsigma\in\mathcal{R}} \varsigma V_{\rho\varsigma}(0)\{\rho D_1 v(\xi) - D_\rho v(\xi)\}. \tag{7.7}$$

Proof. Let $\delta_j := \|\nabla^j\beta\|_{L^\infty}$. Using the Taylor expansion

$$\beta(\xi+\varsigma) = \beta(\xi) + \varsigma\nabla\beta(\xi) + O(\delta_2),$$

we can compute

$$\begin{aligned} D_\varsigma[(1-\beta)v](\xi) = (1-\beta(\xi))D_\varsigma v(\xi) - \varsigma\nabla\beta(\xi)v(\xi) \\ - \varsigma\nabla\beta(\xi)D_\varsigma v(\xi) - O(\delta_2)\cdot\big(v(\xi) + D_\varsigma v(\xi)\big). \end{aligned}$$

Inserting this estimate into (7.6), it is easy to get

$$\begin{aligned} \langle \delta^2\mathcal{E}^{\rm a}(u)v, (1-\beta)v\rangle = \sum_{\xi\in\mathbb{Z}_+}(1-\beta(\xi)) \sum_{\rho,\varsigma\in\mathcal{R}} \Phi_{\xi,\rho\varsigma}(u)D_\rho v(\xi)D_\varsigma v(\xi) \\ - \sum_{\xi\in\mathbb{Z}_+} \nabla\beta(\xi)v(\xi) \sum_{\rho,\varsigma\in\mathcal{R}} \varsigma\Phi_{\xi,\rho\varsigma}(u)D_\rho v(\xi) \\ - C_1'(\delta_1+\delta_2)\|\nabla v\|_{L^2}^2 - C_1'\delta_2\|v\|_{L^2(K,L)}\|\nabla v\|_{L^2}, \end{aligned} \tag{7.8}$$

where $C_1' \lesssim M^{(2,2)}$.

Applying **(DH)** to estimate the coefficients $\Phi_{\xi,\rho\varsigma}(u)$, and the fact that

$$\|v\|_{L^2(K,L)} \leq L^{1/2}(L-K)^{1/2}\|\nabla v\|_{L^2} \quad \text{for all } v \in \mathcal{U}, \tag{7.9}$$

we obtain

$$\begin{aligned} - \sum_{\xi\in\mathbb{Z}_+} \nabla\beta(\xi)v(\xi) \sum_{\rho,\varsigma\in\mathcal{R}} \varsigma\Phi_{\xi,\rho\varsigma}(u)D_\rho v(\xi) \\ \geq - \sum_{\xi\in\mathbb{Z}_+} \nabla\beta(\xi)v(\xi) \sum_{\rho,\varsigma\in\mathcal{R}} \varsigma V_{\rho\varsigma}(0)D_\rho v(\xi) \\ - C_2'K^{-\alpha}\|\nabla\beta\|_{L^\infty}\|v\|_{L^2(K,L)}\|\nabla v\|_{L^2} \end{aligned}$$

$$\geq - \sum_{\xi\in\mathbb{Z}_+} \nabla\beta(\xi)v(\xi) \sum_{\rho,\varsigma\in\mathcal{R}} \varsigma V_{\rho\varsigma}(0)D_\rho v(\xi)$$
$$- C_2'\delta_1 K^{-\alpha}L^{1/2}(L-K)^{1/2}\|\nabla v\|_{L^2}^2,$$

where $C_2' \lesssim M^{(3,1)}$. Combining the last estimate with (7.8), and again applying (7.9) to the term $\|v\|_{L^2(K,L)}$ in (7.8), we obtain

$$\langle\delta^2\mathcal{E}^{\mathrm{a}}(u)v,(1-\beta)v\rangle \geq \sum_{\xi\in\mathbb{Z}_+}(1-\beta(\xi))\sum_{\rho,\varsigma\in\mathcal{R}}\Phi_{\xi,\rho\varsigma}(u)D_\rho v(\xi)D_\varsigma v(\xi) \quad (7.10)$$
$$- \sum_{\xi\in\mathbb{Z}_+} \nabla\beta(\xi)v(\xi) \sum_{\rho,\varsigma\in\mathcal{R}} \varsigma V_{\rho\varsigma}(0)D_\rho v(\xi)$$
$$- C_3'\epsilon_1'\|\nabla v\|_{L^2}^2,$$

where $C_3' \lesssim C_1' + C_2'$ and

$$\epsilon_1' := \max(\delta_1 K^{-\alpha}L^{1/2}(L-K)^{1/2}, \delta_1+\delta_2, \delta_2 L^{1/2}(L-K)^{1/2})$$
$$\lesssim \max(\delta_1 K^{-\alpha},\delta_2)L^{1/2}(L-K)^{1/2} = \epsilon_1.$$

Similarly, we observe that, if $x\in(\xi,\xi+1)$, then $\nabla I_1[\beta v](x) = D_1[\beta v](\xi)$, and hence

$$\nabla I_1[\beta v](x) = \beta\nabla v(x) + \nabla\beta(\xi)v(\xi) + \nabla\beta(\xi)D_1 v(\xi)$$
$$+ O(\delta_2)\cdot(v(\xi)+D_1 v(\xi)).$$

We continue to argue similarly as in the proof of (7.10), and also use the fact that $\partial_{\mathsf{F}}^2 W(\mathsf{F}) = \sum_{\rho,\varsigma}\rho\varsigma V_{\rho\varsigma}(\mathsf{F}\cdot\mathcal{R})$, to obtain

$$\langle\delta^2\mathcal{E}^{\mathrm{c}}(u)v, I_1(\beta v)\rangle \geq \int_{\mathbb{R}_+} I_1\beta\cdot\partial_{\mathsf{F}}^2 W(\nabla u)|\nabla v|^2\,\mathrm{d}x \quad (7.11)$$
$$+ \sum_{\xi\in\mathbb{Z}_+}\nabla\beta(\xi)v(\xi)\sum_{\rho,\varsigma}\varsigma\rho V_{\rho\varsigma}(0)D_1 v(\xi)$$
$$- C_3'\epsilon_1'\|\nabla v\|_{L^2}^2.$$

Combining (7.10) and (7.11) yields the stated result. □

The stability of $\delta^2\mathcal{E}^{\mathrm{bqce}}(u)$, in terms of the approximation parameters, is established in Theorem 7.4. We will see in Section 8 that the second error term, ϵ_1, can be made arbitrarily small by choosing β appropriately and increasing K and $L-K$. It remains to investigate the first error term, $\langle Sv,v\rangle$, occurring in Lemma 7.5.

To motivate the next result it is instructive to think of $\langle Sv,v\rangle$ as being formally of the same structure as

$$\langle\hat{S}v,v\rangle = \int \nabla\beta\cdot v\cdot\nabla^2 v\,\mathrm{d}x.$$

Integrating by parts gives

$$\begin{aligned}\langle \hat{S}v, v\rangle &= -\int \nabla^2\beta \cdot v \cdot \nabla v \,\mathrm{d}x - \int \nabla\beta \cdot |\nabla v|^2 \,\mathrm{d}x \\ &\gtrsim -\|\nabla^2\beta\|_{L^\infty} L^{1/2}(L-K)^{1/2}\|\nabla v\|_{L^2}^2 - \|\nabla\beta\|_{L^\infty}\|\nabla v\|_{L^2}^2,\end{aligned}$$

which is of a similar form to ϵ_1 and is also controllable. We now extend this argument to $\langle Sv, v\rangle$.

Lemma 7.6. Let S be defined by (7.7). Then, for all $v \in \mathcal{U}$,

$$\big|\langle Sv, v\rangle\big| \lesssim M^{(2,0)}\big(\|\nabla^2\beta\|_{L^\infty} L^{1/2}(L-K)^{1/2} + \|\nabla\beta\|_{L^\infty}\big)\|\nabla v\|_{L^2}^2.$$

Proof. We write $S = \sum_{\rho,\varsigma\in\mathcal{R}} \sigma V_{\rho\varsigma}(0) S_\rho$, where

$$\langle S_\rho v, v\rangle = \sum_{\xi\geq K} \nabla\beta(\xi) v(\xi)\big[\rho D_1 v(\xi) - D_\rho v(\xi)\big].$$

Now suppose, without loss of generality, that $\rho > 0$. Then

$$\big[\rho D_1 v(\xi) - D_\rho v(\xi)\big] = \sum_{j=1}^{\rho-1}\big[D_1 v(\xi) - D_1 v(\xi + j)\big],$$

and hence $S_\rho = \sum_{j=1}^{\rho-1} S_{\rho,j}$, where

$$\langle D_{\rho,j} v, v\rangle = \sum_{\xi\geq K} \nabla\beta(\xi) v(\xi)\big[D_1 v(\xi) - D_1 v(\xi + j)\big].$$

Summation by parts gives

$$\begin{aligned}\langle S_{\rho,j} v, v\rangle &= \sum_{\xi\geq K}\big[\nabla\beta(\xi) v(\xi) - \nabla\beta(\xi - j) v(\xi - j)\big] D_1 v(\xi) \\ &= \sum_{\xi\geq K}\big[\nabla\beta(\xi) - \nabla\beta(\xi - j)\big] v(\xi) D_1 v(\xi) \\ &\qquad + \sum_{\xi\geq K} \nabla\beta(\xi - j)\big[v(\xi) - v(\xi - j)\big] D_1 v(\xi).\end{aligned}$$

Using the same arguments as in the proof of Lemma 7.5, it is now easy to show that

$$\big|\langle S_{\rho,j} v, v\rangle\big| \lesssim |\rho|\big(\|\nabla^2\beta\|_{L^\infty} L^{1/2}(L-K)^{1/2} + \|\nabla\beta\|_{L^\infty}\big)\|\nabla v\|_{L^2}^2.$$

Inserting this result into the definition of S_ρ and then S concludes the proof. □

Combining Lemma 7.5 and Lemma 7.6 yields the following stability result.

Theorem 7.7. Let $u \in \mathcal{U}$ satisfy **(DH)** and the stability condition (3.19), $K - r_{\mathrm{cut}} \geq r_0$. Then

$$\langle \delta \mathcal{F}^{\mathrm{bqcf}}(u)v, v\rangle \geq \big(c_0 - C\epsilon^{\mathrm{bqcf}}\big)\|\nabla v\|_{L^2}^2,$$

where

$$\epsilon^{\mathrm{bqcf}} = \max(\epsilon_1, \|\nabla\beta\|_{L^\infty}),$$

ϵ_1 is defined in Lemma 7.5, and $C \lesssim \max(M^{(2,2)}, M^{(3,1)})$. If $\beta = \beta^*$, then

$$\epsilon^{\mathrm{bqcf}} \lesssim \max\big\{(L-K)^{-1}, (L-K)^{-1/2}L^{1/2}K^{-\alpha}, (L-K)^{-3/2}L^{1/2}\big\}.$$

Further remarks

(1) By exploiting symmetries one can improve the estimate of Lemma 7.6, but not the prefactor in the consistency estimate. For stability, and with a smoother blending function, we only need $L - K \gg K^{1/5}$: see Li *et al.* (2012) for more details.

(2) The first proof of stability of the original QCF method, restricted to 1D, was in a discrete $W^{1,\infty}$-type norm (Dobson *et al.* 2010*b*). However, the proof suggested a smaller stability region than the numerical experiments. In the same work, the authors proved that QCF is never uniformly stable in $W^{1,p}$ for any $p < \infty$. Dobson *et al.* (2012) proved a sharp stability result for the QCF method in a discrete H^2-norm, still restricted to 1D, exploiting an interesting spectral equivalence with a QNL-type operator. To date, it is unknown whether the original QCF method is stable in 2D/3D; however, we note that some recent progress for a planar interface has been made by Lu and Ming (2012).

(3) A first proof of stability of the B-QCF method, valid in up to three dimensions, was given in an H^2-type norm by Lu and Ming (2013). This proof uses the machinery of pseudo-difference operators and requires that β is smooth at a macroscopic scale. In our framework, this would require a blending width of order $O(N)$ and thus no improvement in the order of the computational cost as compared to a purely atomistic computation. Li *et al.* (2012) showed that, in fact, the B-QCF method is positive definite in 2D provided that the blending width is much larger than $(\log N)^{1/5}K^{1/5}$ and an associated B-QCE operator is stable. (We expect that in the corresponding 3D result the log-factor is dropped.)

Finally, we remark that it still has to be demonstrated that an error analysis of the B-QCF (or any other a/c coupling) method in an H^2-like setting can incorporate crystal defects. Since the associated elastic fields have singularities in the continuum limit, which appears to prevent H^2-regularity, we see no path towards achieving this. (The QCF analysis of Lu and Ming (2012) and the BQCF analysis of Lu and Ming (2013) only admits smooth elastic solutions.)

7.4. Stability of QNL-type methods

Let $\mathcal{E}^{\rm qnl}$ be of the form (4.12) and assume that it satisfies the three key technical conditions **(F)**, **(L)**, and **(S)**.

Our strategy in the proof of stability of the B-QCE method was to split the test function into two components, and hence the Hessian into a purely atomistic and a purely continuum contribution. Although QNL-type methods do not have a 'built-in' blending, we can nevertheless apply the splitting technique to separate out the interface and continuum from the stability of the atomistic region (*i.e.*, the defect core). We call the following result the *conditional QNL stability theorem.*

Theorem 7.8. Let $u \in \mathcal{U}$ satisfy the strong stability condition (3.19) and suppose that there exists $\gamma^{\rm qnl} \in \mathbb{R}$ such that

$$\langle \delta^2 \mathcal{E}^{\rm qnl}(0)v, v\rangle \geq \gamma^{\rm qnl} \|\nabla v\|_{L^2}^2 \quad \text{for all } v \in \mathcal{U}_0. \tag{7.12}$$

Then

$$\langle \delta^2 \mathcal{E}^{\rm qnl}(u)v, v\rangle \geq \big(\min(c_0, \gamma^{\rm qnl}) - CK^{-\min(1,\alpha)}\big)\|\nabla v\|_{L^2}^2 \quad \text{for all } v \in \mathcal{U}_0,$$

where $C \lesssim M^{(2,0)} + C_{\mathbf{DH}} M^{(3,0)}$.

Proof. The idea behind the result is to artificially insert the blending function β^* into the QNL method in the fully atomistic region, with blending width $O(K)$. Then the result follows by closely mimicking the proof of Theorem 7.4. The details of the argument are given in Appendix A.3. □

As a matter of fact, Theorem 7.8 applies to any a/c coupling, including the B-QCE and B-QCF method. However, the natural occurrence of the blending function in those methods allowed for a more natural decomposition of the atomistic and continuum contributions.

Theorem 7.8 reduces the question of stability of a QNL-type method to stability at homogeneous deformations. We will consider two explicit examples.

The reflection method

The reflection method was proposed by Ortner *et al.* (2013) as a 'universally stable' method. Indeed, for this method we are able to provide a simple and general stability result.

Proposition 7.9. Let $\mathcal{E}^{\rm rfl}$ be the energy defined in (4.16). Then

$$\langle \delta^2 \mathcal{E}^{\rm rfl}(0)v, v\rangle \geq \gamma^{\rm a}(0)\|\nabla v\|_{L^2}^2 \quad \text{for all } v \in \mathcal{U}_0.$$

Proof. From the definition of $\mathcal{E}^{\rm rfl}$, we obtain

$$\langle \delta^2 \mathcal{E}^{\rm rfl}(0)v, v\rangle = \langle \delta^2 \mathcal{E}^*(0)v, v\rangle + \partial_{\sf F}^2 W(0)\|\nabla v\|_{L^2(L,\infty)}^2.$$

Since $\mathcal{E}^*$ depends only on $v(\xi)$, $\xi \le L$, and since we know from Lemma 4.1 that $\mathcal{E}^*$ can be understood as a periodic version of $\mathcal{E}^{\mathrm{a}}$, it is reasonable to conjecture that

$$\langle \delta^2 \mathcal{E}^*(0) v, v \rangle \ge \gamma^{\mathrm{a}}(0) \|\nabla v\|^2_{L^2(0,L)}. \tag{7.13}$$

If we can prove this, then we obtain the stated result.

To prove (7.13), we first write out $\langle \delta^2 \mathcal{E}^*(0) v, v \rangle$, using Lemma 4.1,

$$\langle \delta^2 \mathcal{E}^*(0) v, v \rangle = \frac{1}{2} \sum_{\rho,\varsigma \in \mathcal{R}} V_{\rho\varsigma} \sum_{\xi=-L+1}^{L} D_\rho v^*(\xi) D_\varsigma v^*(\xi).$$

Using $2L$-periodicity of v^*, we can equivalently write this as

$$\langle \delta^2 \mathcal{E}^*(0) v, v \rangle = \frac{1}{2m} \sum_{\rho,\varsigma \in \mathcal{R}} V_{\rho\varsigma} \sum_{\xi=-mL+1}^{mL} D_\rho v^*(\xi) D_\varsigma v^*(\xi) =: \langle H_m v^*, v^* \rangle,$$

for any $m \in \mathbb{N}$.

Extending the definition 'symbolically' to all displacements $w \in \mathcal{U}_0$ with $w(\xi) = 0$ for $|\xi| \ge mL - r_{\mathrm{cut}}$, we obtain

$$\langle H_m w, w \rangle \ge \frac{1}{m} \gamma^{\mathrm{a}}(0) \|\nabla w\|^2_{L^2}.$$

Now define v_m by

$$v_m(\xi) := \begin{cases} v^*(\xi), & |\xi| < mL - r_{\mathrm{cut}}, \\ 0, & |\xi| \ge mL - r_{\mathrm{cut}}. \end{cases}$$

Then

$$\begin{aligned} \langle H_m v^*, v^* \rangle &= \langle H_m v_m, v_m \rangle + O(m^{-1}) \\ &\ge \frac{1}{m} \gamma^{\mathrm{a}}(0) \|\nabla v_m\|^2_{L^2} + O(m^{-1}) \\ &= \frac{1}{m} \gamma^{\mathrm{a}}(0) \|\nabla v^*\|^2_{L^2(-mL,mL)} + O(m^{-1}). \end{aligned}$$

Since $\|\nabla v^*\|^2_{L^2(-mL,mL)} = 2m \|\nabla v\|^2_{L^2(0,L)}$, we obtain

$$\langle H_m v^*, v^* \rangle \ge \gamma^{\mathrm{a}}(0) \|\nabla v\|^2_{L^2(0,L)} + O(m^{-1}),$$

and letting $m \to \infty$ yields (7.13). □

Stability of the 'classical' QNL method

In Section 7.9 we have seen that the reflection method is 'universally stable'. That is, up to controllable errors, $\delta^2 \mathcal{E}^{\mathrm{rfl}}$ is positive if and only if $\delta^2 \mathcal{E}^{\mathrm{a}}$ is positive. For other QNL-type methods, this is not the case but depends crucially on the set-up of the problem.

First, we consider the original second-neighbour QNL method defined by (4.14), which reduces to (2.41) in the case of second-neighbour pair interactions.

Proposition 7.10. In the case of second-neighbour two-body interactions, we have that

$$\langle \delta^2 \mathcal{E}^{\mathrm{qnl}}(0)v, v\rangle \geq \gamma^{\mathrm{a}}(0)\|\nabla v\|_{L^2}^2 \quad \text{for all } v \in \mathcal{U}_0, \tag{7.14}$$

where $\mathcal{E}^{\mathrm{qnl}}$ is the energy functional defined in (2.41).

Proof. To discuss this case, we again adopt the notation introduced in Section 2 and also define $\phi_j'' \equiv \phi_j''(0)$ and $\partial_{\mathsf{F}}^2 W \equiv \partial_{\mathsf{F}}^2 W(0)$. In Section 2.6 we have already shown that if $\phi_2'' \leq 0$, then

$$\langle \delta^2 \mathcal{E}^{\mathrm{qnl}}(0)v, v\rangle \geq \partial_{\mathsf{F}}^2 W \|v'\|_{\ell^2}^2 \geq \gamma^{\mathrm{a}}(0)\|v'\|_{\ell^2}^2.$$

Using a completely analogous calculation, we can show that if $\phi_2'' \geq 0$, then the QNL Hessian can be written in the form

$$\begin{aligned}\langle \delta^2 \mathcal{E}^{\mathrm{qnl}}(0)v, v\rangle &= \langle \delta^2 \mathcal{E}^{\mathrm{a}}(0)v, v\rangle + \phi_2'' \sum_{\xi=K+1}^{\infty} |v_\xi''|^2 \\ &\geq \langle \delta^2 \mathcal{E}^{\mathrm{a}}(0)v, v\rangle \geq \gamma^{\mathrm{a}}(0)\|v'\|_{\ell^2}^2,\end{aligned}$$

and we again obtain the stated result. □

Stabilizing QNL-type methods

We have seen in Proposition 7.10 that, for second-neighbour pair interactions, the original QNL method is stable provided that the atomistic model is stable. Unfortunately, for many-body interactions, this cannot be guaranteed. An explicit example can be constructed of an interaction potential for which $\gamma^{\mathrm{a}}(0) > 0$, but $\delta^2\mathcal{E}^{\mathrm{qnl}}(0)$ is indefinite (Ortner *et al.*, 2013). (Since the construction of this example is not very instructive, we do not give the details here.)

It turns out, however, that, at least in 1D, all QNL-type coupling schemes can be *stabilized*. We only give a brief outline of the argument and refer to Ortner *et al.* (2013) for further details. It is relatively straightforward to see that, for *any* QNL-type a/c coupling, the Hessian can be written in the form

$$\begin{aligned}\langle \delta^2 \mathcal{E}^{\mathrm{qnl}}(0)v, v\rangle &= \sum_{\xi=0}^{\infty} c_0(\xi)|D_1 v(\xi)|^2 \\ &\quad + \sum_{j=1}^{2r_{\mathrm{cut}}-1} \sum_{\xi=0}^{\infty} c_j(\xi)\big|D_1 v(\xi+j) - D_1 v(\xi)\big|^2.\end{aligned}$$

For the reflection scheme $\mathcal{E}^{\mathrm{qnl}} = \mathcal{E}^{\mathrm{rfl}}$, we denote the coefficients by $c_j \equiv c_j^{\mathrm{rfl}}$, whereas for a general QNL scheme we just denote them by c_j, as above.

Next, one observes that, if force-consistency **(F)** is satisfied, then $c_0(\xi) \equiv \partial_{\mathsf{F}}^2 W(0)$ for all ξ. Thus, we obtain that

$$\begin{aligned}\langle \delta^2 \mathcal{E}^{\rm qnl}(0)v, v\rangle &- \langle \delta^2 \mathcal{E}^{\rm rfl}(0)v, v\rangle \\ &= \sum_{j=1}^{2r_{\rm cut}-1} \sum_{\xi=1}^{\infty} \big(c_j(\xi) - c_j^{\rm rfl}(\xi)\big)\big|D_1 v(\xi+j) - D_1 v(\xi)\big|^2.\end{aligned}$$

Since both the QNL energy and the reflection energy reduce to the atomistic model in the atomistic region and to the Cauchy–Born model in the continuum region, it follows that $c_j(\xi) = c_j^{\rm rfl}(\xi)$ except possibly in a small region surrounding the interface $\xi \in \{K, \ldots, L\}$. To be precise, it is fairly straightforward to deduce that there exists a constant $C_1 \geq 0$ that depends only on the coefficients $c_j(\xi) - c_j^{\rm rfl}(\xi)$, such that

$$\langle \delta^2 \mathcal{E}^{\rm qnl}(0)v, v\rangle - \langle \delta^2 \mathcal{E}^{\rm rfl}(0)v, v\rangle \geq -C_1 \sum_{\xi=K-r_{\rm cut}+1}^{L-1} |v''_\xi|^2,$$

where we recall that $v''_\xi = D_1 v(\xi) - D_1 v(\xi-1)$.

Let us now define

$$\mathcal{E}^{\rm stab}(u) := \mathcal{E}^{\rm qnl}(u) + C_1 \sum_{\xi=K-r_{\rm cut}+1}^{L-1} |v''_\xi|^2.$$

Then we obtain that $\mathcal{E}^{\rm stab}$ obviously satisfies the force-consistency **(F)** and locality **(L)** conditions. The scaling condition **(S)** follows from the fact that C_1 is essentially a sum over the coefficients $c_j(\xi) - c_j^{\rm rfl}(\xi)$. In addition to these conditions, we now also obtain that

$$\langle \delta^2 \mathcal{E}^{\rm stab}(0)v, v\rangle \geq \langle \delta^2 \mathcal{E}^{\rm rfl}(0)v, v\rangle \quad \text{for all } v \in \mathcal{U}_0.$$

Since $\mathcal{E}^{\rm rfl}$ is 'universally stable', it follows that $\mathcal{E}^{\rm stab}$ is stable as well. That is, $\mathcal{E}^{\rm stab}$ satisfies all the conditions of the conditional stability Theorem 7.8.

A similarly general construction in 2D/3D is open, but it is already clear from the initial results presented in Ortner *et al.* (2013) that stabilization cannot in general be as straightforward as in the 1D case.

8. *A priori* error estimates and computational cost

We are now ready to put into practice the strategy outlined in Section 4.7. Based on the stability analysis in Section 7 we now have strong reasons to expect that

$$\|\nabla u^{\rm a} - \nabla u_h^{\rm ac}\|_{L^2} \lesssim \tfrac{1}{c_0}\big(\|\eta_{\rm int}^{\rm ac}\|_{\mathcal{U}_h^*} + \|\eta_{\rm ext}\|_{\mathcal{U}_h^*}\big),$$

with suitable upper bounds on the two consistency components, derived in Sections 5 and 6. We shall now *optimize* the approximation parameters in such a way that these upper bounds tend to zero as we increase the number of degrees of freedom. For sufficiently small consistency errors, we can then apply an inverse function theorem to make the error estimate fully rigorous.

8.1. Optimizing the finite element grid

The finite mesh size h is the first approximation parameter that we will optimize. In this section, we use a classical technique to optimize the mesh grading.

The two terms occurring in the coarsening and consistency analysis that depend on $\mathcal{T}_h$ are the best approximation and the coarsening error terms

$$\|h\nabla^2\tilde{u}\|_{L^2(L,N)} + \|h\nabla^2\tilde{u}\|^2_{L^4(L,N)}.$$

It is easy to see that, for L sufficiently large, the best approximation error $\|h\nabla^2\tilde{u}\|_{L^2(L,N)}$ is the dominant contribution. Thus, we optimize this term only. A discussion of optimizing a/c methods for the superconvergent error component, $\|h\nabla^2\tilde{u}\|^2_{L^4(L,N)}$, is given in Section 8.6.

Suppose that $u \in \mathcal{U}$ satisfies **(DH)** and $L > r_0$. Then

$$\|h\nabla^2\tilde{u}\|_{L^2(L,N)} \lesssim \|hx^{-\alpha-1}\|_{L^2(L,N)}.$$

We wish to choose h to minimize this quantity, subject to fixing the number of degrees of freedom, $N_{\mathcal{T}_h} - 1$, which is given by

$$N_{\mathcal{T}_h} = \sum_{j=1}^{N_{\mathcal{T}_h}} 1 = \sum_{j=1}^{N_{\mathcal{T}_h}} h_j \frac{1}{h_j} = \int_0^1 \frac{1}{h}\,\mathrm{d}x.$$

We ignore the discreteness of the mesh size function and solve

$$\min \|hx^{-\alpha-1}\|^2_{L^2(L,N)} \quad \text{subject to} \int_L^N \frac{1}{h}\,\mathrm{d}x = \text{const}.$$

The solution to this variational problem satisfies

$$h(x) = \lambda x^{\frac{2}{3}(\alpha+1)} \quad \text{for } x \in (L,N),$$

for some constant $\lambda > 0$. This gives us an optimal scaling of the mesh size function.

We now impose the condition $h(L) \approx 1$, which yields

$$h(x) \approx \left(\frac{x}{L}\right)^{\frac{2}{3}(\alpha+1)} =: \tilde{h}(x) \quad \text{for } x \in (L,N). \tag{8.1}$$

If $\alpha' := \frac{2}{3}(\alpha+1)$, then $\alpha' > 1$, and hence this choice of h yields

$$\int_L^N \frac{1}{h}\,\mathrm{d}x \approx \frac{L^{\alpha'}(N^{1-\alpha'} - L^{1-\alpha'})}{1-\alpha'} \approx \frac{L}{\alpha'-1}. \tag{8.2}$$

Thus, the choice (8.1) gives a comparable number of degrees of freedom in the continuum region to that in the atomistic and interface regions. The resulting best approximation error bound can be estimated by

$$\|hx^{-\alpha-1}\|_{L^2(L,N)} \approx \frac{L^{\frac{1}{2}-(\alpha+1)}}{(\alpha'-1)^{1/2}} \approx L^{-\frac{1}{2}-\alpha}. \tag{8.3}$$

We can now also determine an optimal balance between the choices for L, N and $\mathcal{T}_h$. To balance the interpolation error with the far-field error, we should choose N such that

$$L^{-1/2-\alpha} \approx N^{1/2-\alpha}, \quad \text{that is,} \quad N \approx L^{\frac{\alpha+1/2}{\alpha-1/2}}.$$

We now turn this formal motivation into an explicit construction of a finite element mesh.

Algorithm T.

(1) Set $N := \lceil L^{\frac{\alpha+1/2}{\alpha-1/2}} \rceil$.
(2) Set $\mathcal{N}_h := \{0, \ldots, L\}$.
(3) While $n := \max \mathcal{N}_h < N$
  Set $\mathcal{N}_h := \mathcal{N}_h \cup \{\min(N, n + \lfloor \tilde{h}(n) \rfloor)\}$.

Meshes constructed via Algorithm T satisfy qualitatively the same properties as predicted by the formal computations (8.2) and (8.3).

Lemma 8.1. Let $u \in \mathcal{U}$ satisfy **(DH)** and let N and $\mathcal{T}_h$ be constructed by Algorithm T. Then, for L sufficiently large, $N_{\mathcal{T}_h} \leq C_1 L$,

$$\|h\nabla^2\tilde{u}\|_{L^2(L,N)} + N^{1/2-\alpha} \leq C_2 N_{\mathcal{T}_h}^{-1/2-\alpha}, \quad \text{and} \tag{8.4}$$

$$\|h\nabla^2\tilde{u}\|_{L^4(L,N)}^2 \leq C_3 N_{\mathcal{T}_h}^{-1/2-2\alpha}, \tag{8.5}$$

where C_1 depends on α and C_2, C_3 depend on α and on $C_{\mathbf{DH}}$.

Proof. The proof is given in Appendix A.4. □

Cost estimate for the external potential

We now turn the external consistency error estimate into an estimate in terms of $N_{\mathcal{T}_h}$ as well. Let f satisfy **(DH)** and suppose that $\mathcal{T}_h$ and N are constructed using Algorithm T. Since $h(x) \leq x/2$ (see Lemma A.3), a straightforward computation yields

$$\begin{aligned}\|\eta_{\rm ext}\|_{\mathcal{U}_h^*} &\lesssim \|h^2\nabla f\|_{L^2(L,\infty)} + \frac{C_1}{\log L}\|h^2\omega\nabla^2 f\|_{L^2(L,\infty)} \\ &\lesssim L^{-\alpha-3/2} + \frac{L^{-\alpha-3/2}\log^2 N}{\log L}.\end{aligned}$$

We insert $N \lesssim L^{\frac{2}{3}(\alpha+1)}$ to obtain the following result. In particular, we can conclude that the external consistency error is dominated by the best approximation error.

Proposition 8.2. Let f satisfy **(DH)** and let $\mathcal{T}_h, N$ be chosen by Algorithm T. Then

$$\|\eta_{\rm ext}\|_{\mathcal{U}_h^*} \le CL^{-\alpha-3/2}\log L,$$

where C depends on α and on $C_{\mathbf{DH}}$.

Cost estimates for the Cauchy–Born error

Although not part of the coarsening error estimates, we still briefly derive corresponding estimates for the Cauchy–Born modelling error, since this is another generic component used in estimating the consistency error of all a/c couplings.

The only approximation parameter available to control the Cauchy–Born modelling error is the size of the atomistic region (recall that in Algorithm T we fixed N in terms of L). A straightforward calculation yields the following result, which we will later use to conveniently estimate the corresponding terms in the consistency error estimates.

Proposition 8.3. Let $u \in \mathcal{U}$ satisfy **(DH)** and let $L' \ge r_0$. Then

$$\|\nabla^2\tilde{u}\|^2_{L^4(L',\infty)} + \|\nabla^3\tilde{u}\|_{L^2(L',\infty)} \le C(L')^{-\alpha-3/2}, \tag{8.6}$$

where C depends on α and on $C_{\mathbf{DH}}$.

Proof. The result follows from a straightforward computation. □

We notice, in particular, that if $L' \gtrsim L \gtrsim N_{\mathcal{T}_h}$ (in practice we will need to choose $K \gtrsim L$), then the Cauchy–Born modelling error is dominated by the best approximation error.

8.2. Error estimate for QNL-type methods

Fix a strongly stable atomistic solution $u^{\rm a} \in \mathcal{U}$ satisfying the decay hypothesis **(DH)**. We wish to approximate this equilibrium using a QNL-type approximation

$$u_h^{\rm qnl} \in \arg\min\big\{\mathcal{E}^{\rm qnl}(u_h) - \langle f, u_h\rangle_h \,\big|\, u_h \in \mathcal{U}_h\big\}, \tag{8.7}$$

where $\mathcal{E}^{\rm qnl}$ is of the form (4.12). We now derive a quasi-optimal balance of the approximation parameters K, N and h. In the case of the QNL method, this will follow directly from Section 8.1.

Given an atomistic core size K, we define L as determined by the particular QNL method that we employ. (In practice, $L - K \le Cr_{\rm cut} \lesssim 1$; we state this in Theorem 8.4 as an assumption.) Next, we construct N and $\mathcal{T}_h$ using Algorithm T. Thus, we obtain

$$N_{\mathcal{T}_h} \lesssim L \lesssim K.$$

We can now apply Lemma 8.1, Lemma 8.3, and the decay hypothesis **(DH)**, to deduce that the QNL consistency error derived in Theorem 6.13 can be bounded above by

$$\begin{aligned}
\|\eta^{\rm qnl}_{\rm int}\|_{\mathcal{U}_h^*} &\lesssim \|\nabla^2 \tilde{u}^{\rm a}\|_{L^2([K-r_{\rm cut}, L])} + \|h\nabla^2 \tilde{u}^{\rm a}\|^2_{L^{2p}(L,\infty)} \\
&\quad + \|\nabla^3 \tilde{u}^{\rm a}\|_{L^p(L,\infty)} + N^{1/2-\alpha} \\
&\lesssim K^{-\alpha-1} + N_{\mathcal{T}_h}^{-1/2-2\alpha} + L^{-\alpha-3/2} + L^{-1/2-\alpha} \\
&\lesssim N_{\mathcal{T}_h}^{-1/2-\alpha} \lesssim K^{-1/2-\alpha}. \qquad (8.8)
\end{aligned}$$

This means that, by choosing K sufficiently large, we can make the consistency error arbitrarily small. This motivates the following theorem.

Theorem 8.4. Let $u^{\rm a} \in \mathcal{U}$ be a strongly stable atomistic solution satisfying **(DH)**. Consider the QNL problem (8.7) with quasi-optimal choice of $N, \mathcal{T}_h$ constructed by Algorithm T. Suppose, moreover, that $\mathcal{E}^{\rm qnl}$ is stable in the reference state (7.12), and that $L - K$ remains bounded as $K \to \infty$. (For example, this is always satisfied for $\mathcal{E}^{\rm qnl} = \mathcal{E}^{\rm rfl}$.)

Then there exists K_0 such that, for all $K \geq K_0$, (8.7) has a locally unique, strongly stable solution $u_h^{\rm qnl}$ which satisfies

$$\|\nabla u^{\rm a} - \nabla u_h^{\rm qnl}\|_{L^2} \lesssim N_{\mathcal{T}_h}^{-1/2-\alpha}.$$

Proof. We will use the quantitative inverse function theorem, Lemma B.1, with

$$\mathcal{G}_h(u_h) := \delta\mathcal{E}^{\rm qnl}(u_h) - \langle f, \cdot\rangle_h,$$

with $\bar{u}_h = \Pi_h u^{\rm a}$.

We first remark that the scaling condition **(S)** implies a Lipschitz bound for $\delta^2\mathcal{E}^{\rm qnl}$,

$$\|\delta^2\mathcal{E}^{\rm qnl}(u) - \delta^2\mathcal{E}^{\rm qnl}(v)\|_{\mathcal{L}(\mathcal{U},\mathcal{U}^*)} \leq M\|\nabla u - \nabla v\|_{L^\infty} \quad \text{for all } u, v \in \mathcal{U}, \quad (8.9)$$

where $M \lesssim M^{(3,0)}$. The proof of (8.9) is analogous to the proof of (3.11) and is therefore omitted. Since $\|\cdot\|_{L^\infty} \lesssim \|\cdot\|_{L^2}$, we can also replace the L^∞-norm on the right-hand side with the L^2-norm. This establishes condition (B.1) in Lemma B.1.

The residual estimate (8.8) and the external residual estimate of Proposition 8.2 give

$$\|\mathcal{G}_h(\Pi_h u^{\rm a})\|_{\mathcal{U}_h^*} \lesssim K^{-1/2-\alpha},$$

which establishes the second condition (B.2) in Lemma B.1.

From Theorem 7.8 we obtain that

$$\langle \delta^2 \mathcal{E}^{\mathrm{qnl}}(u^{\mathrm{a}})v_h, v_h\rangle \geq \big(\min(c_0, \gamma^{\mathrm{qnl}}) - CK^{-\min(1,\alpha)}\big)\|\nabla v_h\|_{L^2}^2.$$

Let $\gamma := \frac{1}{2}\min(c_0, \gamma^{\mathrm{qnl}})$. Applying the Lipschitz bound (8.9) and the best approximation error estimate (8.4), we obtain

$$\langle \delta^2 \mathcal{E}^{\mathrm{qnl}}(\Pi_h u^{\mathrm{a}})v_h, v_h\rangle \geq \big(2\gamma - CK^{-\min(1,\alpha)} - CK^{-1/2-\alpha}\big)\|\nabla v_h\|_{L^2}^2.$$

Hence, for K sufficiently large, we obtain that

$$\langle \delta \mathcal{G}_h(\Pi_h u^{\mathrm{a}})v_h, v_h\rangle \geq \gamma\|\nabla v_h\|_{L^2}^2 \quad \text{for all } v_h \in \mathcal{U}_h,$$

which establishes the third condition (B.3) in Lemma B.1.

To check the final condition (B.4), we observe that

$$\frac{2M\eta}{\gamma^2} \lesssim \frac{2MK^{-1/2-\alpha}}{\gamma^2},$$

and hence, for sufficiently large K, (B.4) is satisfied as well.

Thus, we deduce the existence of u_h^{qnl} satisfying $\mathcal{G}_h(u_h^{\mathrm{qnl}}) = 0$. The stability estimate (B.6) guarantees that u_h^{qnl} is in fact a strongly stable solution to (8.7). The error estimate (B.5) implies

$$\|\nabla u_h^{\mathrm{qnl}} - \nabla\Pi_h u^{\mathrm{a}}\|_{L^2} \lesssim K^{-1/2-\alpha} \approx N_{\mathcal{T}_h}^{-1/2-\alpha}.$$

Applying the best approximation error estimate

$$\|\nabla\Pi_h u^{\mathrm{a}} - \nabla u^{\mathrm{a}}\|_{L^2} \lesssim N_{\mathcal{T}_h}^{-1/2-\alpha},$$

which follows from (8.4), we finally obtain the stated error bound. □

Remark 8.5. **(1)** Theorem 8.4 provides a rigorous convergence result for the reflection method (4.15) and for stabilized QNL-type methods (see Section 7.4). For more general QNL-type methods it depends on the details of the potential and the macroscopic strain, whether the stability assumption (7.12) holds. We have provided some examples and counterexamples in Section 7.8.

(2) The convergence rate $N_{\mathcal{T}_h}^{-1/2-\alpha}$ is, in some sense, optimal since it is the rate of the best approximation error. Indeed, it can be checked that the remaining consistency error terms decay at a higher order than the best approximation error. We comment further on the issue of optimal convergence rates in Section 9.1.

8.3. Error estimate for B-QCE

We follow the same strategy as in Section 8.2. Let $u^{\mathrm{a}} \in \mathcal{U}$ be a strongly stable atomistic solution, which we aim to approximate using the B-QCE method,

$$u_h^{\mathrm{bqce}} \in \arg\min\big\{\mathcal{E}^{\mathrm{bqce}}(u_h) - \langle f, u_h\rangle_h \,\big|\, u_h \in \mathcal{U}_h\big\}. \tag{8.10}$$

Before we state the error estimate, we again turn the consistency estimate of Theorem 6.8 into an estimate in terms of $N_{\mathcal{T}_h}$. Let $K < L$ be given and $N, \mathcal{T}_h$ constructed by Algorithm T, and let $\beta = \beta^*$ defined in (6.12). Then, applying **(DH)**, (8.5) and (8.6) to (6.16), we obtain

$$\begin{aligned} \|\eta_{\rm int}^{\rm bqce}\|_{\mathcal{U}_h^*} &\lesssim (L-K)^{-3/2} + (L-K)^{-1}\|\nabla^2\tilde{u}^{\rm a}\|_{L^2(K,L)} \\ &\quad + \|\nabla^3\tilde{u}\|_{L^2(K,\infty)} + \|h\nabla^2\tilde{u}\|_{L^4(K,\infty)}^2 + N^{1/2-\alpha} \\ &\lesssim (L-K)^{-3/2} + (L-K)^{-1/2}K^{-\alpha-1} + K^{-\alpha-3/2} + L^{-1/2-\alpha}. \end{aligned} \tag{8.11}$$

It remains to determine the (quasi-)optimal ratio between L and K. We think of L as fixed (and determining the number of degrees of freedom), and aim to optimize the choice of K. We consider three regimes.

(1) If $c_1K \le L - K \le c_2K$ for some $c_1, c_2 > 0$, then we also obtain $L - K \approx K \approx L$, and hence
$$\|\eta_{\rm int}^{\rm bqce}\|_{\mathcal{U}_h^*} \lesssim L^{-3/2} + L^{-3/2-\alpha} + L^{-\alpha-1/2} \approx L^{-\min(3/2,\alpha+1/2)}.$$

(2) If $(L-K)/K \to \infty$ as $K \to \infty$, then the upper bound in (8.11) is bounded below by $L^{-\min(3/2,\alpha+1/2)}$. That is, we cannot obtain an improved convergence rate over the earlier choice, but require more degrees of freedom.

(3) If $(L-K)/K \to 0$ as $K \to \infty$, then we clearly obtain a reduced convergence rate, but the number of degrees of freedom only improves by a constant factor.

In practice, in the B-QCE method we first choose the size of the atomistic region K and then define $L = 2K$. Thus, we obtain $N_{\mathcal{T}_h} \approx L \approx K \approx (L-K)$ and
$$\|\eta_{\rm int}^{\rm bqce}\|_{\mathcal{U}_h^*} \lesssim N_{\mathcal{T}_h}^{-\min(3/2,\alpha+1/2)} \approx K^{-\min(3/2,\alpha+1/2)}. \tag{8.12}$$
Again, we observe that by choosing K sufficiently large we can make the consistency error arbitrarily small.

According to the discussion above, this choice is asymptotically quasi-optimal, but it may be possible to improve on it quantitatively, as well as in the pre-asymptotic regime.

Theorem 8.6. Let $u^{\rm a} \in \mathcal{U}$ be a strongly stable atomistic solution satisfying **(DH)**. Consider the B-QCE method (8.10) with quasi-optimal choice of approximation parameters $L, N, \mathcal{T}_h, \beta$ for given K, as described above.

Then there exists K_0 such that, for all $K \ge K_0$, (8.10) has a locally unique, strongly stable solution $u_h^{\rm bqce}$, which satisfies
$$\|\nabla u^{\rm a} - \nabla u_h^{\rm bqce}\|_{L^2} \lesssim N_{\mathcal{T}_h}^{-\min(3/2,1/2+\alpha)}.$$

Proof. The proof is analogous to the proof of Theorem 8.4. The Lipschitz bound for $\delta^2\mathcal{E}^{\mathrm{bqce}}$ is again easy to establish. The consistency estimate in terms of K is given in (8.12). Finally, combining Theorem 7.4 with the Lipschitz bound for $\delta^2\mathcal{E}^{\mathrm{bqce}}$ ensures stability,

$$\langle \delta^2\mathcal{E}^{\mathrm{bqce}}(\Pi_h u^{\mathrm{a}})v_h, v_h\rangle \geq \tfrac{1}{2}c_0\|\nabla v_h\|_{L^2}^2,$$

provided that K is sufficiently large. (With our choice of K, L, $\beta = \beta^*$, the stability error is bounded by $\epsilon^{\mathrm{bqce}} \lesssim K^{-\min(1,\alpha)}$.) □

Remark 8.7. The B-QCE method has a reduced convergence rate $N_{\mathcal{T}_h}^{-3/2}$ for $\alpha > 1$ (fast decaying elastic field), but it has the optimal convergence rate $N_{\mathcal{T}_h}^{-1/2-\alpha}$ for $\alpha \in (1/2, 1]$ (slowly decaying elastic field). In the latter case it would be sufficient to choose $L - K \approx K^{(1+2\alpha)/3} \ll K$ to obtain the optimal rate.

8.4. Error estimate for B-QCF

In the B-QCF method, we approximate a strongly stable atomistic equilibrium u^{a} by the solution to

$$\langle \mathcal{F}^{\mathrm{bqcf}}(u_h^{\mathrm{bqcf}}), v_h\rangle = \langle f, v_h\rangle_h \quad \text{for all } v_h \in \mathcal{U}_h, \tag{8.13}$$

where $\mathcal{F}^{\mathrm{bqcf}}$ is given by (4.22). We will again use the blending function $\beta = \beta^*$ defined in (6.12) and, given K and L, define N and $\mathcal{T}_h$ via Algorithm T.

In the consistency error estimate given in Theorem 6.16 the blending function occurs again, but this time as a prefactor. We slightly rewrite (6.25) as

$$\begin{aligned}\|\eta_{\mathrm{int}}^{\mathrm{bqcf}}\|_{\mathcal{U}_h^*} \lesssim \big(1 + L^{1/p}(L-K)^{-1/p}\big)\\ \cdot \big(\|\nabla^3\tilde{u}^{\mathrm{a}}\|_{L^2(K-r_{\mathrm{cut}},\infty)} + \|\nabla^2\tilde{u}^{\mathrm{a}}\|_{L^4(K-2r_{\mathrm{cut}},\infty)}^2\\ + \|h\nabla^2\tilde{u}^{\mathrm{a}}\|_{L^4(L,N)}^2 + N^{1/2-\alpha}\big).\end{aligned}$$

Applying Lemma 8.1 and Lemma 8.3 to (6.25), and assuming that $K - 2r_{\mathrm{cut}} \gtrsim K$, we obtain

$$\|\eta_{\mathrm{int}}^{\mathrm{bqcf}}\|_{\mathcal{U}_h^*} \lesssim \big(1 + L^{1/p}(L-K)^{-1/p}\big)\cdot\big(K^{-\alpha-3/2} + N_{\mathcal{T}_h}^{-1/2-\alpha}\big). \tag{8.14}$$

We remark that choosing $\beta = \beta^*$ was not necessary, but that $\hat{\beta}$ could have been replaced by any Lipschitz function.

Before we state the B-QCF convergence result, we recall from Section 2.8 that we must optimize β not only for consistency but also for stability:

$$\langle \delta\mathcal{F}^{\mathrm{bqcf}}(u^{\mathrm{a}})v, v\rangle \geq (c_0 - C\epsilon^{\mathrm{bqcf}})\|\nabla v\|_{L^2}^2,$$

where

$$\epsilon^{\mathrm{bqcf}} \lesssim \max\big\{(L-K)^{-1}, (L-K)^{-1/2}L^{1/2}K^{-\alpha}, (L-K)^{-3/2}L^{1/2}\big\}. \tag{8.15}$$

Table 8.1. Summary of *a priori* error estimates in terms of computational cost.

a/c method	$\|\nabla u^{\mathrm{a}} - \nabla u_h^{\mathrm{ac}}\|_{L^2} \lesssim$	Reference
ATM (Section 3.4)	$N^{-\alpha+1/2}$	Theorem 3.14
B-QCE (Section 4.4)	$N_{\mathcal{T}_h}^{-\min(3/2,\alpha+1/2)}$	Theorem 8.6
QNL (Section 4.5)	$N_{\mathcal{T}_h}^{-\alpha-1/2}$	Theorem 8.4
B-QCF (Section 4.6)	$N_{\mathcal{T}_h}^{-\alpha-1/2}$	Theorem 8.8

We now observe that by choosing $L - K \approx K \approx L \approx N_{\mathcal{T}_h}$, (8.14) and (8.15) become

$$\|\eta_{\mathrm{int}}^{\mathrm{bqcf}}\|_{\mathcal{U}_h^*} \lesssim K^{-\alpha-3/2} + N_{\mathcal{T}_h}^{-1/2-\alpha} \approx N_{\mathcal{T}_h}^{-1/2-\alpha} \approx K^{-1/2-\alpha}, \quad \text{and} \tag{8.16}$$

$$\epsilon^{\mathrm{bqcf}} \lesssim K^{-\min(1,\alpha)}. \tag{8.17}$$

In particular, the consistency error is of the same order of magnitude as the best approximation error and the stability error tends to zero as $K \to \infty$.

Thus, as in the B-QCE method, we first choose K, then define $L = 2K$ and then construct $N, \mathcal{T}_h$ using Algorithm T, to obtain a quasi-optimal implementation of the B-QCF method. This yields the following result.

Theorem 8.8. Let $u^{\mathrm{a}} \in \mathcal{U}$ be a strongly stable atomistic solution satisfying **(DH)**. Consider the B-QCF method (8.13) with quasi-optimal choice of approximation parameters $L, N, \mathcal{T}_h, \beta$ for given K, as described above.

Then there exists K_0 such that, for all $K \geq K_0$, (8.13) has a locally unique, strongly stable solution u_h^{bqcf}, which satisfies

$$\|\nabla u^{\mathrm{a}} - \nabla u_h^{\mathrm{bqcf}}\|_{L^2} \lesssim N_{\mathcal{T}_h}^{-1/2-\alpha}.$$

Proof. The proof is analogous to the proof of Theorems 8.4 and 8.6. The Lipschitz bound for $\delta\mathcal{F}^{\mathrm{bqcf}}$ is again easy to establish, employing Lemma 6.16. The consistency estimate in terms of K is given in (8.16) and the stability error in terms of K is given in (8.17). □

8.5. Summary of a priori error estimates

For the reader's convenience we summarize all *a priori* error estimates, in terms of computational cost in Table 8.1. We see in particular from this table that the coarsening error dominates both in the QNL and B-QCF methods. In Section 9.1 (optimal a/c couplings) we discuss how to improve upon these rates. Recall also from **(DH)** that $\alpha > 1/2$.

8.6. *Remark on superconvergence*

In the analyses of the QCF and B-QCF methods by Dobson *et al.* (2010*b*), Makridakis *et al.* (2011) and Lu and Ming (2013), it is shown that the error is of a higher order than for energy-based a/c couplings. The reason we have not observed this here is because the consistency errors are dominated by the best approximation error.

Alternatively we could consider the superconvergent error $\nabla\Pi_h u^{\mathrm{a}} - \nabla u_h^{\mathrm{bqcf}}$, for which we might expect a better rate of convergence. Carrying out the analysis for this error leads to the same rates of convergence as the 'optimal rates' discussed in Section 9.1.

8.7. *Numerical tests*

We perform a selection of numerical tests to confirm the convergence rates we predicted in the preceding sections.

The model

We choose a one-dimensional variant (3.7) of the *embedded atom method* (EAM) (Daw and Baskes 1984), which is a popular atomistic model for solids. In this model ((3.7) for macroscopic stretch $\mathsf{A} = 1$), the site potential is given by

$$V(g) = \frac{1}{2}\sum_{\rho\in\mathcal{R}} \phi(|\rho + g_\rho|) + \psi\biggl(\sum_{\rho\in\mathcal{R}} \eta(|\rho + g_\rho|)\biggr),$$

where ϕ is a pair potential, η an electron density function and ψ an embedding energy (typically convex). Mimicking popular variants of EAM, we define

$$\phi(r) = 2\mathrm{e}^{-2a(|r|-1)} - 4\mathrm{e}^{-a(|r|-1)},$$
$$\eta(r) = \mathrm{e}^{-b|r|},$$
$$\psi(t) = c\bigl((t-t_0)^2 + (t-t_0)^4\bigr),$$

where a, b, c, t_0 are parameters of the model. Throughout our computational experiments we choose

$$r_{\mathrm{cut}} = 2, \quad a = 3, \quad b = 3, \quad c = 5 \quad \text{and} \quad t_0 = \sum_{\rho\in\mathcal{R}} \mathrm{e}^{-0.95b|\rho|}.$$

Nonlinear solver

Let $\mathcal{F}^{\mathrm{ac}} : \mathcal{U}_h \to \mathcal{U}_h^*$ denote the force-operator associated with an a/c coupling. To compute the solution of the variational problem

$$\langle \mathcal{F}^{\mathrm{ac}}(u_h^{\mathrm{ac}}), v_h\rangle = \langle f, v_h\rangle_h \quad \text{for all } v_h \in \mathcal{U}_h,$$

we use a preconditioned steepest descent method with fixed step size s. Given an iterate $u_h^{[\ell]}$ we compute

$$\int_0^N \big(\nabla u_h^{[\ell+1]} - \nabla u_h^{[\ell]}\big) \cdot \nabla v_h \,\mathrm{d}x = -s\big\{\langle \mathcal{F}^{\mathrm{ac}}(u_h^{[\ell]}), v_h\rangle - \langle f, v_h\rangle_h\big\}$$

for all $v_h \in \mathcal{U}_h$. The step size s is chosen by trial end error to ensure stability of the iteration. As a starting guess, we choose $\Pi_h u^{\mathrm{a}}$, where u^{a} is a prescribed exact solution (see next paragraph). The iteration is terminated when the ℓ^∞-norm of the residual falls below the tolerance 10^{-8}.

Finally, we remark that in the case of the B-QCF method the iteration is not a steepest descent method (since there is no associated energy), but that it is nevertheless convergent due to the fact that $\delta\mathcal{F}^{\mathrm{bqcf}}$ is positive definite in a neighbourhood of the B-QCF solution.

More sophisticated and more efficient nonlinear solvers should of course be employed for 2D/3D problems.

Convergence rates

We fix an exact solution

$$u^{\mathrm{a}}(\xi) := \tfrac{1}{10}(1+\xi^2)^{-\alpha/2}\xi,$$

and compute the external forces $f(\xi)$ to be equal to the internal forces under the displacement u^{a}. The parameter α is a prescribed decay exponent. One may readily check that this solution and the associated external forces satisfy the decay hypothesis **(DH)**.

We then solve the atomistic method (ATM) with far-field boundary condition (3.21), the QCE method (4.6), the B-QCE method (8.10), the B-QCF method (8.13) and the reflection method (QNL(REFL)) (8.7) with $\mathcal{E}^{\mathrm{qnl}} = \mathcal{E}^{\mathrm{rfl}}$, and compute the error in the energy norm.

The results for $\alpha = 1.5$ (rapid decay) are shown in Figure 8.1 and the results for $\alpha = 0.8$ (slow decay) are shown in Figure 8.2. We observe precisely the rates predicted in the preceding sections.

Moreover, we note that, in the case $\alpha = 1.5$, the B-QCE scheme is indeed less efficient than the B-QCF and QNL schemes. However, since the convergence of *all* schemes (including the naive ATM scheme) is fairly fast, this may not be of great significance.

By contrast, in the case $\alpha = 0.8$ where a/c couplings yield much more significant computational savings, the best approximation error dominates the coupling error in the B-QCE scheme and hence the computational cost of B-QCE, B-QCF and QNL are essentially equivalent. For a similar reason, the error in the QCE method is initially close to the best approximation error and only stagnates once the elastic field is well resolved.

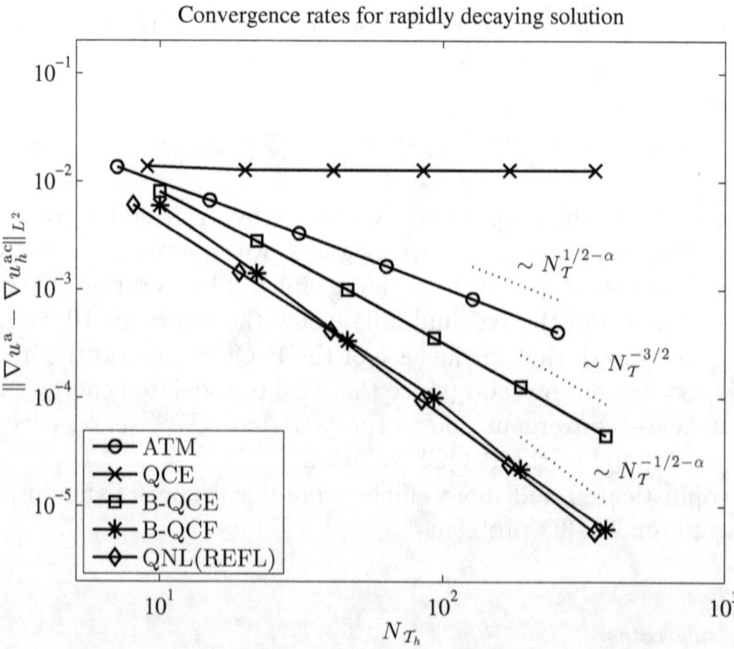

Figure 8.1. Convergence rates of various a/c couplings for a rapidly decaying atomistic solution ($\alpha = 1.5$).

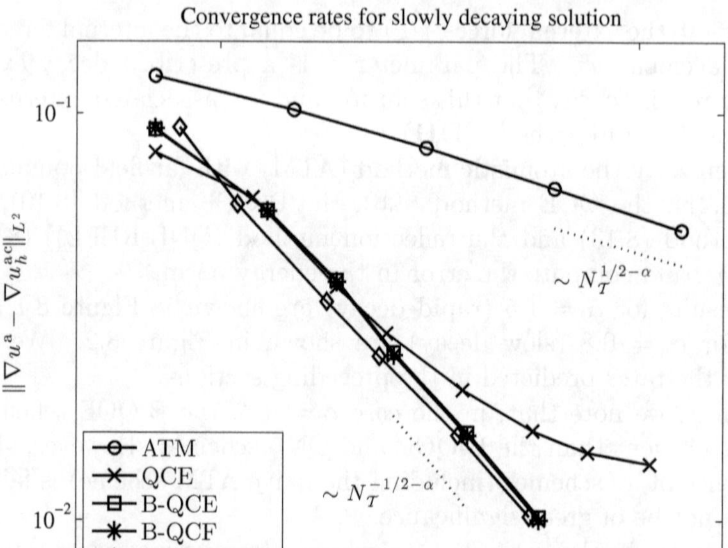

Figure 8.2. Convergence rates of various a/c couplings for a slowly decaying atomistic solution ($\alpha = 0.8$).

9. Extensions and open problems

In this article we have presented a numerical analysis of the fundamental a/c coupling methods. To conclude, we will briefly comment on several further important topics that we skipped or only touched upon, as well as open problems that motivate further research.

9.1. 'Optimal' a/c coupling

Suppose, for a moment, that we need not discretize the continuum region, but can solve the Cauchy–Born model at negligible computational cost. In this case, we may ignore both the far-field and the coarsening errors, and would obtain, using the notation and concepts from Section 8, the following error estimates for the B-QCE, QNL and B-QCF methods (we drop the subscripts h since we have not discretized the continuum model):

$$\begin{aligned} \|\nabla u^{\mathrm{a}} - \nabla u^{\mathrm{bqce}}\|_{L^2} &\lesssim \|\nabla^2 \beta\|_{L^2} + \|\nabla \beta \nabla^2 \tilde{u}^a\|_{L^2} + \eta^{\mathrm{cb}}, \\ \|\nabla u^{\mathrm{a}} - \nabla u^{\mathrm{qnl}}\|_{L^2} &\lesssim \|\nabla^2 \tilde{u}^{\mathrm{a}}\|_{L^2(K - r_{\mathrm{cut}}, L)} + \eta^{\mathrm{cb}}, \\ \|\nabla u^{\mathrm{a}} - \nabla u^{\mathrm{bqcf}}\|_{L^2} &\lesssim \eta^{\mathrm{cb}}, \end{aligned}$$

where $\eta^{\mathrm{cb}} := \|\nabla^3 \tilde{u}^a\|_{L^2(K - r_{\mathrm{cut}}, \infty)} + \|\nabla^2 \tilde{u}^a\|^2_{L^4(K - r_{\mathrm{cut}}, \infty)}$, and we assumed that in the B-QCF method β, K, L are chosen such that the prefactor in the estimate is of order $O(1)$.

In these estimates, we again clearly observe the improvements in interface contribution to the error that the QNL and B-QCF methods represent. Applying the decay hypothesis **(DH)** to converting the estimates into estimates in terms of the number of degrees of freedom in the atomistic and interface region (which we have just assumed is now proportional to computational cost), we obtain for $\alpha > 1/2$ that

$$\begin{aligned} \|\nabla u^{\mathrm{a}} - \nabla u^{\mathrm{bqce}}\|_{L^2} &\lesssim K^{-3/2}, \\ \|\nabla u^{\mathrm{a}} - \nabla u^{\mathrm{qnl}}\|_{L^2} &\lesssim K^{-\alpha - 1}, \\ \|\nabla u^{\mathrm{a}} - \nabla u^{\mathrm{bqcf}}\|_{L^2} &\lesssim K^{-\alpha - 3/2}, \end{aligned} \tag{9.1}$$

where we assumed that $\beta = \beta^*$ and $L - K \approx K$ in the B-QCE and B-QCF methods, and that $L - K = O(1)$ in the QNL method.

These estimates are natural limits that cannot be improved upon by changing the type of continuum model discretization. Moreover, we note that a recent result of Dobson (2011) implies that the estimate for the QNL method is optimal, that is, there exists no QNL-type scheme, coupling an atomistic to the Cauchy–Born model, that has an accuracy higher than first-order in the interface region.

We note two interesting challenges. First, it would be highly desirable to construct practical a/c schemes that achieve the optimal rates stated above. Second, by admitting more general (higher-order) continuum models, it may be possible to construct a/c schemes that can in fact improve upon the rates in (9.1).

9.2. Stability

As is the case with all approximation schemes, the two central pillars of the *a priori* error analysis of a/c couplings are *consistency* and *stability*. The stability of a/c couplings in 2D/3D is an entirely open question. We are aware of only three results in this direction, all of which are unsatisfactory in our view. Ortner and Shapeev (2013) obtain stability results in a 2D pair interaction setting by assuming that nearest-neighbour interactions dominate, which would break down for large deformations. Lu and Ming (2013) prove stability for the B-QCF Jacobian operator as a map from H^2 to L^2, but achieve this only for an impractical $O(N)$ blending width. Moreover, it is not clear how to incorporate crystal defects into their analysis. Finally, Li *et al.* (2012) prove positivity of the B-QCF Jacobian operator in 2D, but only under the assumption that the B-QCE Hessian is positive definite.

The approach to split test functions, which we employed in Section 7, is unlikely to generalize to 2D/3D. It was only possible to employ this in 1D because we could split the displacement *gradients* instead of the displacements themselves.

The result of Ortner and Shapeev (2013) would be generalizable to any of the a/c couplings we considered as well as to 3D, but this would not be satisfactory as it cannot be readily generalized to many-body interactions, and there is even less hope of extending it in such a way that stability of a/c couplings can be proved up to bifurcation points.

An entirely new approach may be required to obtain a satisfactory understanding of the stability of a/c couplings in 2D/3D.

9.3. Mesh and model adaptivity

We have presented an *a priori* error analysis approach to designing a/c couplings with a balance of approximation parameters that is quasi-optimal in the asymptotic regime. However, for many real defects, there is a significant pre-asymptotic regime; see, for example, Luskin *et al.* (2013). For practical computations, it is therefore desirable to develop quantitative *a posteriori* error estimates and corresponding mesh and model adaptation schemes.

Adaptive methods to determine the region needing atomistic accuracy and to select an efficient mesh in the continuum region have been considered (Miller and Tadmor 2003) and goal-oriented approaches have been studied by Prudhomme, Bauman and Oden (2006) and Arndt and Luskin

(2007, 2008*a*, 2008*b*). An adaptive algorithm that is conceptually similar to the framework of the present article has been developed by Ortner and Wang (2013).

Another key open question in this context is a rigorous criterion for the prediction of defect nucleation in the continuum region. For example, in a nano-indentation experiment (Miller and Rodney 2008), an indenter is pushed into a crystal surface in a quasi-static loading scenario. At some critical load, a dislocation dipole forms under the indenter, but the precise location of the nucleation is unknown, and in fact, the atomistic solution up to that point is sufficiently regular to employ the Cauchy–Born approximation. Predicting the correct nucleation load and nucleation region requires augmentation of the Cauchy–Born model with a suitable atomistic stability criterion.

9.4. Electronic structure

Classical interatomic potentials of the form employed in the present article are normally obtained by postulating a functional form, which is usually guided by physical intuition, with free parameters that are subsequently fitted to experimental data or simulations obtained from a more accurate model, such as density functional theory (DFT). Consequently, these models cannot account for all possible configurations and indeed are likely to be inaccurate in defect cores (unless, of course, the model has been fitted to the specific defect under consideration).

It is therefore of considerable interest to materials science research to extend the construction and analysis of a/c couplings (or possibly other kinds of multiscale schemes) to DFT models, or other more accurate quantum mechanics based models. Some recent work in this direction can be found in Gavini, Bhattacharya and Ortiz (2007) and Garcia-Cervera, Lu and E (2007). A closely related alternative approach is the DFT/linear elasticity coupling developed, for example, by Woodward and Rao (2002) and by Trinkle (2008) (see also Section 9.6 for other examples of a/c couplings employing linear elasticity as the continuum model).

As far as we are aware, there exists no systematic error analysis of these types of a/c couplings, with the exception of a highly simplified model by Langwallner *et al.* (2013). However, see Blanc *et al.* (2002) or E and Lu (2013) for analyses of the Cauchy–Born approximation of some DFT models, which might provide a starting point for the analysis of a/c couplings.

9.5. Finite temperature equilibrium and dynamic quasicontinuum methods

The computation of finite temperature equilibrium properties is usually done within the framework of statistical physics. The goal is then to compute the canonical average of observables (such as thermal expansion) that only depend on the rep-atoms. To this end, it is enough to know the free energy associated with the rep-atoms, which is obtained by integrating the canonical Gibbs measure over all of the constrained atoms (Tadmor *et al.* 2013, Blanc, Le Bris, Legoll and Patz 2010). However, this free energy, termed the coarse-grained free energy in what follows, is a high-dimensional integral (since the number of constrained atoms is large), and it must therefore be approximated for computation.

To approximate the coarse-grained free energy, two asymptotic regimes have been considered in the literature. Dupuy, Tadmor, Miller and Phillips (2005) used a local harmonic approximation, which corresponds to a low temperature regime, and the Cauchy–Born rule to derive an approximate coarse-grained free energy of the form

$$\mathcal{E}^{c}(u_h, \theta) = \int_0^N W(\nabla u_h, \theta)\,\mathrm{d}x, \tag{9.2}$$

where $W(\mathsf{F}, \theta)$ is a Cauchy–Born elastic free energy density. The hot-QC energy of Tadmor *et al.* (2013) then uses a QCE coupling of the atomistic model with this coarse-grained free energy,

$$\mathcal{E}^{\mathrm{qce}}(u_h, \theta) = \sum_{\xi=0}^{K} \Phi_\xi^{\mathrm{a}}(u_h) + \int_{K+1/2}^{N} W(\nabla u_h, \theta)\,\mathrm{d}x, \tag{9.3}$$

to compute the canonical average of observables that depend only on the atomistic region.

Another possibility is to consider the thermodynamic limit when the number of constrained atoms goes to infinity, while the temperature remains fixed and is not necessarily small. This regime has been investigated in the one-dimensional case and in some two-dimensional cases in Blanc *et al.* (2010) and Blanc and Legoll (2013). These studies provide insight into the accuracy of the harmonic approximation previously mentioned.

We conclude this section by noting that many interesting phenomena actually involve dynamical effects. This goes beyond the equilibrium setting described above. For instance, quasi-static fracture and plastic deformation generally evolve by a sequence of rare events in which the system oscillates in an energy well for a long time scale before moving to a new energy well.

To handle such problems, one approach is to approximate the exact atomistic dynamics by a non-equilibrium dynamics using the free energy (9.3) in order to reduce the number of degrees of freedom. One possibility is to use a Langevin equation where the potential energy is given by (9.3). However, such direct molecular dynamics applied to a/c energies such as (9.3) still cannot reach the time scales necessary for useful information, which requires the system to visit many states. Thus, acceleration of the time dynamics is needed concurrently with coarse-graining in space. An example of such a scheme is the hyper-QC method Kim *et al.* (2013), which has been developed to utilize the hyperdynamics method of Voter (1997) to achieve long-time dynamics with hot-QC.

9.6. *Other types of a/c coupling schemes*

The recent article of Miller and Tadmor (2009) reviews and benchmarks a list of at least fourteen different a/c couplings. Most of these schemes have at their core some aspects of the three fundamental a/c coupling concepts considered in the present article: energy-based blending, force-based coupling (and blending) and ghost force removal (QNL-type ideas). However, they are presented in a primarily algorithmic format and it would be interesting to distinguish the various approximate models from nonlinear solver and implementation issues to provide a classification more suitable to a numerical analysis of these methods.

It is likely that there are other fundamentally different and competitive approaches to a/c coupling that we have not considered in the present paper. As potential competitors we mention the field-based a/c coupling developed by Iyer and Gavini (2011), the approach of Anitescu *et al.* (2009), which is based on an optimization framework, the stress-based a/c coupling of Makridakis *et al.* (2011) or the method of Olson, Bochev, Luskin and Shapeev (2013), which minimizes the difference between the atomistic and continuum solutions in an overlap region. The latter has been shown to be stable (and positive definite), although no practical extension to 2D/3D has yet been formulated.

We briefly describe below a small selection of alternative a/c coupling methods. Some of these methods are formulated as finite temperature equilibrium or dynamics methods; the static methods can potentially be extended to finite temperature equilibrium and dynamics by the ideas described in Section 9.5.

The coupled atomistic and discrete dislocation method (CADD) couples an atomistic region with a region modelled by dislocations that interact through their linear elastic fields (Shilkrot, Miller and Curtin 2004), thereby allowing the simulation of atomistic systems with a large number of dislocations. Recent extensions have further coupled the discrete dislocation region with an even larger region modelled by continuum crystal plasticity (Wallin, Curtin, Ristinmaa and Needleman 2008).

The bridging scale method (BSM) decomposes the atomistic displacement into a coarse-grid displacement and a fine-grid displacement by projection (Liu *et al.* 2006*b*, Tang, Hou and Liu 2006). In the 'continuum' region, the fine-grid displacement is modelled rather than computed, thus giving a closed coarse-grained model, and in the 'atomistic region' the fine-grid displacement is computed, giving fully atomistic resolution.

The coarse-grained molecular dynamics (CGMD) method uses a coarse-grained Hamiltonian (Rudd and Broughton 2005, Rudd and Broughton 2000) defined for continuous, piecewise linear displacements with respect to a coarse mesh. The coarse-grained Hamiltonian is given by an approximation of the canonical average of the molecular Hamiltonian constrained to the space of atomistic displacements with mean square projection onto a given continuous, piecewise linear displacement.

If linear elasticity is used as the continuum model, then boundary integral methods can be employed instead of finite element methods to compute the elastic fields. The use of fast solution methods for the boundary integral and approximations of the boundary integral can give efficient methods (Haq, Movchan and Rodin 2007, Li 2009). These methods are in spirit close to the predecessors of a/c coupling such as the method of Sinclair (1971).

Peridynamics seeks to model atomistic systems with defects by a nonlocal continuum theory (Du, Gunzburger, Lehoucq and Zhou 2013, Silling and Lehoucq 2010). As a continuum theory it must be discretized for numerical implementation, thus leading to a nonlocal particle model in regions where the mesh is refined to length scales less than the 'horizon', which plays the role of a cut-off radius in a nonlocal atomistic model. To reach larger length scales (Seleson, Beneddine and Prudhomme 2013), the mesh must be coarsened beyond the length scale of the horizon, which leads to local approximations when quadrature points are added to the interior of elements (this can be seen by adapting the derivation in Section 4.3 for a nonlocal atomistic model to a nonlocal peridynamic continuum model). The mathematical framework and results presented in this article for a/c coupling can thus potentially be applied to the analysis of numerical approximations of peridynamics models.

An evaluation within the framework given in this article of the cost of each of the above methods for the computation of defects would enable their relative merits to be assessed. Further discussion of alternative a/c coupling methods can be found in the books by Tadmor and Miller (2011), Liu, Karpov and Park (2006*a*), and E (2011), and in the review article by Curtin and Miller (2003).

9.7. Benchmarks

The accuracy of atomistic-to-continuum coupling methods has generally been studied by comparing to the solution of fully atomistic models (Miller and Tadmor 2009, Van Koten *et al.* 2012). Since these approaches do not always distinguish the many potential sources of error (atomistic core size, blending function, coupling error, continuum modelling error, continuum mesh, far-field error, iterative solution error), they do not provide a definite test of the relative accuracy of the various atomistic-to-continuum coupling methods. It would be useful to develop a more comprehensive, theory-based approach to benchmarking a/c multiscale methods that can also provide further insight into the optimization of the atomistic core size, blending functions, continuum meshes, and other approximation parameters. Some initial steps in this direction have been taken in Luskin *et al.* (2013) and Li *et al.* (2013), but a more systematic benchmark protocol on a much wider range of model problems is required.

A. Proofs

A.1. Proofs from Section 3

Proof of Proposition 3.1. The proof uses standard arguments for interpolation error estimates.

Let $\hat{w}$ be a quadratic polynomial on $[\xi-2, \xi+1]$ with nodal values $w(\xi)$. Then it is straightforward to see that $\tilde{w}(x) = \hat{w}(x)$ on $[\xi-1, \xi]$.

Using this fact and the stability of the interpolation $v \mapsto \tilde{v}$, we estimate, for any quadratic $\hat{w}$,

$$\begin{aligned}\|\nabla^j(\tilde{v}-\hat{v})\|_{L^p(\xi-1,\xi)} &\lesssim \|\nabla^j(\tilde{v}-\tilde{w})\|_{L^p(\xi-1,\xi)} + \|\nabla^j(\hat{v}-\hat{w})\|_{L^p(\xi-1,\xi)} \\ &\lesssim \|\hat{v}-\hat{w}\|_{W^{1,p}(\xi-2,\xi+1)} + \|\nabla^j(\hat{v}-\hat{w})\|_{L^p(\xi-1,\xi)} \\ &\lesssim \|\hat{v}-\hat{w}\|_{W^{j,p}(\xi-2,\xi+1)}.\end{aligned}$$

One can now readily show that, for suitable choice of $\hat{w}$,

$$\|\hat{v}-\hat{w}\|_{W^{j,p}(\xi-2,\xi+1)} \lesssim \|\nabla^j\hat{v}\|_{L^p(\xi-2,\xi+1)}.$$

This establishes the first result.

The second inequality of the second result can be deduced by an analogous argument, employing a linear $\hat{w}$.

The first inequality of the second result follows from Hölder's inequality and the observation that $\int_{\xi-1}^{\xi} \nabla\tilde{v}(x)\,\mathrm{d}x = \nabla v|_{(\xi-1,\xi)}$. □

Proof of Proposition 3.3. Let $\eta \in \mathbb{Z}_+$ and let $u \in \mathcal{U}$ be defined by $u'_\xi = 1$ for $\xi = \eta$ and $u'_\xi = 0$ otherwise. Since linear combinations of such functions are dense in $\mathcal{U}$, it is sufficient to show that they belong to the closure of $\mathcal{U}_0$. To that end, let $u_n \in \mathcal{U}_0$ be defined by

$$u'_{n,\xi} := \begin{cases} 1, & \xi = \eta, \\ -1/n, & \xi = \eta+1, \ldots, \eta+n, \\ 0, & \text{otherwise.} \end{cases}$$

Then it is straightforward to show that $\|\nabla u_n - \nabla u\|_{L^2} = n^{-1/2} \to 0$. □

Lemma A.1. For any $\rho \in \mathbb{Z} \setminus \{0\}$, we have

$$\|D_\rho u\|_{\ell^2} \le \|\nabla_\rho u\|_{L^2} = |\rho| \|\nabla u\|_{L^2} \quad \text{for all } u \in \mathcal{U}. \tag{A.1}$$

Proof. For each $u \in \mathcal{U}$, we have

$$\begin{aligned} \|D_\rho u\|_{\ell^2}^2 &= \sum_{\xi \in \mathbb{Z}} \big| u(\xi + \rho) - u(\xi) \big|^2 = \sum_{\xi \in \mathbb{Z}} \bigg| \int_{t=0}^1 \nabla_\rho u(\xi + t\rho)\, \mathrm{d}t \bigg|^2 \\ &\le \sum_{\xi \in \mathbb{Z}} \int_{t=0}^1 |\nabla_\rho u(\xi + t\rho)|^2 \, \mathrm{d}t = \|\nabla_\rho u\|_{L^2}^2. \end{aligned}$$

The last equality is due to a simple counting argument. □

Proof of Proposition 3.7 and Lemma 3.9. In this proof, it is convenient to revert to displacements of $\mathbb{Z}$, rather than $\mathbb{Z}_+$. To account for this, we obtain a factor $\frac{1}{2}$ in front of the energy functional. The ℓ^p norms are now defined on all of $\mathbb{Z}$ and the L^p norms on all of $\mathbb{R}$.

We prove the continuity of $\mathcal{E}^{\mathrm{a}}$ together with the Fréchet differentiability. We begin by proving that (3.12) and (3.10) are well-defined expressions, and at the same time prove the bounds (3.11). For second and higher variations, we apply the generalized Cauchy–Schwarz inequality and (A.1) to obtain

$$\begin{aligned} \big|\langle \delta^j \mathcal{E}^{\mathrm{a}}(u) \boldsymbol{v} \rangle\big| &\le \frac{1}{2} \sum_{\boldsymbol{\rho} \in \mathcal{R}^j} \frac{m(\boldsymbol{\rho})}{\prod_{i=1}^j |\rho_i|} \sum_{\xi \in \mathbb{Z}} \prod_{i=1}^j |D_{\rho_i} v_i(\xi)| \\ &\le \frac{1}{2} \sum_{\boldsymbol{\rho} \in \mathcal{R}^j} \frac{m(\boldsymbol{\rho})}{\prod_{i=1}^j |\rho_i|} \prod_{i=1}^j \|D_{\rho_i} v_i\|_{\ell^{p_i}} \\ &\le \frac{1}{2} \sum_{\boldsymbol{\rho} \in \mathcal{R}^j} \frac{m(\boldsymbol{\rho})}{\prod_{i=1}^j |\rho_i|} \prod_{i=1}^j \|\nabla_{\rho_i} v_i\|_{L^{p_i}} \le \frac{1}{2} M^{(j)} \prod_{i=1}^j \|\nabla v_i\|_{L^{p_i}}. \end{aligned}$$

Thus we have shown (3.11). Moreover, if we choose $p_i \ge 2$, then we have $\|\nabla v_i\|_{L^{p_i}} \le \|\nabla v_i\|_{L^2}$, and hence it follows that (3.10) is indeed well-defined for any $\boldsymbol{v} \in \mathcal{U}^j$.

To show that (3.12) is also well-defined, we simply note that

$$\Phi^{\mathrm{a}}_{\xi,\rho}(u) - \Phi^{\mathrm{a}}_{\xi,\rho}(0) = \langle \delta \Phi^{\mathrm{a}}_{\xi,\rho}(\theta u), u \rangle = \sum_{\varsigma \in \mathcal{R}} \Phi^{\mathrm{a}}_{\xi,\rho\varsigma}(\theta u) D_\varsigma u(\xi),$$

where $\theta \in (0,1)$, and apply the above argument (for the case $j = 2$) again.

To prove continuity of $\mathcal{E}^{\mathrm{a}}$, we define the modified potential

$$\Psi_\xi(u) := \Phi^{\mathrm{a}}_\xi(u) - \langle \delta \Phi^{\mathrm{a}}_\xi(0), u \rangle = \Phi^{\mathrm{a}}_\xi(u) - \sum_{\rho \in \mathcal{R}} \Phi^{\mathrm{a}}_{\xi,\rho}(0) D_\rho u(\xi).$$

Since $\sum_{\xi\in\mathbb{Z}} D_\rho u(\xi) = 0$ for all $u \in \mathcal{U}_0$, we have

$$\mathcal{E}^{\mathrm{a}}(u) = \frac{1}{2}\sum_{\xi\in\mathbb{Z}^d} \Psi_\xi(u) \quad \text{for all } u \in \mathcal{U}_0. \tag{A.2}$$

We will show that this definition of the energy is in fact well-formed for all $u \in \mathcal{U}$, as well as continuous and k times differentiable with variations given by (3.12) and (3.10). Since the new form agrees with the formal definition on $\mathcal{U}_0$, the result then follows.

To prove that (A.2) is well-defined, we expand

$$\Psi_\xi(u) = \Psi_\xi(0) + \sum_{\rho\in\mathcal{R}} \Psi_{\xi,\rho}(0) D_\rho u(\xi) + \sum_{\rho,\varsigma\in\mathcal{R}} \Psi_{\xi,\rho\varsigma}(\theta u) D_\rho u(\xi) D_\varsigma u(\xi),$$

where $\theta \in (0,1)$. Since $\Psi_\xi(0) = \Psi_{\xi,\rho}(0) = 0$, it is now easy to show, using the fact that $Du(\xi) \in \ell^2$, that $(\Psi_\xi(u))_{\xi\in\mathbb{Z}} \in \ell^1$ for all $u \in \mathcal{U}$.

To prove differentiability, we first note that, since $V \in C^k$,

$$\begin{aligned}\Psi_\xi(u+h) = \Psi_\xi(u) + \langle \delta\Psi_\xi(u), h\rangle + \cdots + \tfrac{1}{k!}\langle \delta^k \Psi_\xi(u)(h,\ldots,h)\rangle \\ + o(\|\nabla h\|^k_{L^2(\xi - r_{\mathrm{cut}}, \xi + r_{\mathrm{cut}})}).\end{aligned}$$

The bounds (3.11) show that the sum over $\xi \in \mathbb{Z}$ is well-defined, and one can readily deduce that $\mathcal{E}^{\mathrm{a}}$ is k times Fréchet-differentiable. □

A.2. Proofs from Section 5 (coarsening)

Proof of Lemma 5.2. For each $T \in \mathcal{T}_h$, since ∇v_h is constant in T, we have

$$\begin{aligned}\int_T \big(\partial_{\mathsf{F}} W(\nabla \Pi_h u) - \partial_{\mathsf{F}} W(\nabla \tilde u)\big)\nabla v_h \,\mathrm{d}x \\ = \int_T \big(\partial_{\mathsf{F}} W(\nabla I_h u) - \partial_{\mathsf{F}} W(\nabla \tilde u)\big)\,\mathrm{d}x \cdot \nabla v_h|_T \\ + \int_T \big(\partial_{\mathsf{F}} W(\nabla \Pi_h u) - \partial_{\mathsf{F}} W(\nabla I_h u)\big)\,\mathrm{d}x \cdot \nabla v_h|_T.\end{aligned} \tag{A.3}$$

Using (5.2) we obtain

$$|\partial_{\mathsf{F}} W(\nabla \Pi_h u) - \partial_{\mathsf{F}} W(\nabla I_h u)| \le M^{(2,0)} |\nabla \Pi_h u - \nabla I_h u| \lesssim N^{-\alpha}.$$

Moreover, using the fact that $\int_T \nabla I_h u \,\mathrm{d}x = \int_T \nabla \tilde u \,\mathrm{d}x$, and the interpolation error estimate for $I_h u$ from Lemma 5.1, we have

$$\begin{aligned}\left|\int_T \big(\partial_{\mathsf{F}} W(\nabla I_h u) - \partial_{\mathsf{F}} W(\nabla \tilde u)\big)\,\mathrm{d}x\right| \le \left|\int_T \partial_{\mathsf{F}} W'(\nabla I_h u)(\nabla I_h u - \nabla \tilde u)\,\mathrm{d}x\right| \\ + M^{(2,0)} \int_T |\nabla I_h u - \nabla \tilde u|^2 \,\mathrm{d}x \\ \lesssim \|\nabla I_h u - \nabla \tilde u\|^2_{L^2(T)} \lesssim \|h \nabla^2 \tilde u\|^2_{L^2(T)}.\end{aligned}$$

Combining these estimates with (A.3) we obtain

$$\left|\int_T \big(\partial_{\mathsf{F}} W(\nabla \Pi_h u) - \partial_{\mathsf{F}} W(\nabla \tilde{u})\big)\nabla v_h \,\mathrm{d}x\right| \\ \lesssim \big(\|h\nabla^2\tilde{u}\|_{L^4(T)}^2 + h_T^{1/2}N^{-\alpha}\big)\|\nabla v_h\|_{L^2(T)}.$$

Summing over $T \in \mathcal{T}_h$, $T \subset [L, N]$ and again applying Hölder's inequality yields the stated result. □

Lemma A.2. Let $\omega(x) := x\log(x)$ and $L > 1$. Then

$$\|\omega^{-1}v\|_{L^2(L,\infty)} \le \frac{1}{\log L}\|\nabla v\|_{L^2} \quad \text{for all } v \in \mathcal{U}.$$

Proof. For each $x > K$ we have

$$v(x) = \int_0^x \nabla v(s)\,\mathrm{d}s \le x^{1/2}\|\nabla v\|_{L^2},$$

hence

$$\|\omega^{-1}v\|_{L^2(L,\infty)} \le \|\omega^{-1}x^{1/2}\|_{L^2(L,\infty)}\|\nabla v\|_{L^2}.$$

Integrating the constant gives

$$\int_K^\infty \omega^{-2}x\,\mathrm{d}x = \int_L^\infty \frac{1}{x\log^2 x}\,\mathrm{d}x = \frac{1}{\log L}. \qquad \square$$

Proof of Proposition 5.4. We begin by adding and subtracting $\langle I_h f, v_h\rangle_{\mathbb{Z}_+}$:

$$\big|\langle f, v_h\rangle_{\mathbb{Z}_+} - \langle f, v_h\rangle_h\big| \le \big|\langle f - I_h f, v_h\rangle_{\mathbb{Z}_+}\big| + \big|\langle I_h f, v_h\rangle_{\mathbb{Z}_+} - \langle f, v_h\rangle_h\big|. \tag{A.4}$$

Using a similar argument as in the proof of Lemma 6.16, we can estimate the second group on the right-hand side of (A.4) by

$$\big|\langle I_h f, v_h\rangle_{\mathbb{Z}_+} - \langle f, v_h\rangle_h\big| \lesssim \|h^2\nabla\tilde{f}\|_{L^2}\|\nabla v_h\|_{L^2}. \tag{A.5}$$

The estimate of the first term on the right-hand side of (A.4) requires some care. First, we reduce it to a more conventional continuum object: if $g := \tilde{f} - I_h\tilde{f}$ and $\mathbb{Z}_{L,N} := \{L, \ldots, N\}$, then

$$\begin{aligned}\big|\langle g, v_h\rangle_{Z_+}\big| &= \big|\langle g, v_h\rangle_{\mathbb{Z}_{L,N}}\big| \\ &\le \left|\langle g, v_h\rangle_{\mathbb{Z}_{L,N}} - \int_L^N g v_h\,\mathrm{d}x\right| + \left|\int_L^N g v_h\,\mathrm{d}x\right|. \end{aligned}\tag{A.6}$$

To estimate the first group on the right-hand side of (A.6) we note that $\langle g, v_h\rangle_{\mathbb{Z}_{L,N}}$ is a trapezoidal rule approximation to $\int_L^N g v_h\,\mathrm{d}x$ with mesh size $h = 1$, which yields

$$\left|\langle g, v_h\rangle_{\mathbb{Z}_{L,N}} - \int_L^N g v_h\,\mathrm{d}x\right| \lesssim \|\nabla g\nabla v_h\|_{L^1(L,N)} + \|\nabla^2 g v_h\|_{L^1(L,N)}.$$

We write $\nabla^2 g v_h = (\omega\nabla^2 g)(\omega^{-1} v_h)$, and apply the Cauchy–Schwarz and weighted Poincaré inequalities to obtain

$$\begin{aligned}\Big|\langle g, v_h\rangle_{\mathbb{Z}_{L,N}} - \int_L^N g v_h \,\mathrm{d}x\Big| \\ &\lesssim \|\nabla g\|_{L^2(L,N)}\|\nabla v_h\|_{L^2} + \|\omega\nabla^2 g\|_{L^2(L,N)}\|\omega^{-1}v_h\|_{L^2(L,N)} \\ &\lesssim \big(\|\nabla g\|_{L^2(L,N)} + \|\omega^{-1}\nabla^2 g\|_{L^2(L,N)}\big)\|\nabla v_h\|_{L^2}.\end{aligned}$$

Finally, using the fact that $\nabla^2 g = \nabla^2 \tilde{f}$ and $\|\nabla g\|_{L^2} \le \|\nabla \tilde{f}\|_{L^2}$ (since $\nabla I_h \tilde{f}$ is the L^2-projection of $\nabla \tilde{f}$), we obtain

$$\begin{aligned}\Big|\langle g, v_h\rangle_{\mathbb{Z}_{L,N}} - \int_L^N g v_h \,\mathrm{d}x\Big| \\ &\lesssim \Big\{\|\nabla \tilde{f}\|_{L^2(L,N)} + \frac{1}{\log L}\|\omega^{-1}\nabla^2 \tilde{f}\|_{L^2(L,N)}\Big\}\|\nabla v_h\|_{L^2}.\end{aligned} \tag{A.7}$$

We are left to estimate the second group on the right-hand side of (A.6). Fix an element $T \subset [L, N]$. Then

$$\begin{aligned}\Big|\int_T (\tilde{f} - I_h\tilde{f}) v_h \,\mathrm{d}x\Big| &\le \|h^2\nabla^2 \tilde{f}\|_{L^2(T)}\|v_h\|_{L^2(T)} \\ &\le \max_T \omega^{-1} \max_T \omega \,\|\omega h^2 \nabla^2 \tilde{f}\|_{L^2(T)}\|\omega^{-1} v_h\|_{L^2(T)}.\end{aligned}$$

One readily checks that, due to the assumption that $h(x) \lesssim \kappa x$, we have

$$\max_T \omega^{-1} \max_T \omega \lesssim \frac{(1+\kappa)\log^2(1+\kappa)}{\log^2(\min_T x)} + (1+\kappa) =: C_\kappa.$$

(Note that for practical meshes we expect that $C_\kappa \approx \kappa$.) Summing over $T \in \mathcal{T}_h, T \subset [L, N]$ and applying a Poincaré inequality followed by the weighted Poincaré inequality (Lemma A.2), we arrive at

$$\Big|\int_T (\tilde{f} - I_h\tilde{f}) v_h \,\mathrm{d}x\Big| \lesssim \frac{C_\kappa}{\log L}\|\omega h^2\nabla^2 \tilde{f}\|_{L^2(L,N)}\|\nabla v_h\|_{L^2}. \tag{A.8}$$

Inserting (A.8) and (A.7) into (A.6) and the resulting estimate as well as (A.5) into (A.4) yields the stated result. □

A.3. Proof of conditional QNL stability

Proof of Theorem 7.8. Let $K' := \lfloor K/2 \rfloor < K$, and let

$$\beta(x) := \begin{cases} 0, & 0 \le x \le K', \\ \hat{\beta}\big(\frac{x-K'}{K-K'}\big), & K' \le x \le K, \\ 1, & K \le x, \end{cases}$$

where $\hat{\beta}$ is the quasi-optimal blending function defined in (6.12). Recall from Section 8.3 that

$$\|\nabla\sqrt{\beta}\|_{L^\infty} + \|\nabla\sqrt{1-\beta}\|_{L^\infty} \lesssim K^{-1}.$$

We can now write

$$\begin{aligned}\langle\delta^2\mathcal{E}^{\rm qnl}(u)v,v\rangle &= \sum_{\xi\in\mathbb{Z}_+}\sum_{\rho,\varsigma\in\mathcal{R}} \Phi^{\rm qnl}_{\xi,\rho\varsigma}(u)D_\rho v(\xi)D_\varsigma v(\xi) \\ &= \sum_{\xi\in\mathbb{Z}_+}\sum_{\rho,\varsigma\in\mathcal{R}} \Phi^{\rm a}_{\xi,\rho\varsigma}(u)(1-\beta(\xi))D_\rho v(\xi)D_\varsigma v(\xi) \\ &\quad + \sum_{\xi\in\mathbb{Z}_+}\sum_{\rho,\varsigma\in\mathcal{R}} \Phi^{\rm qnl}_{\xi,\rho\varsigma}(u)\beta(\xi)D_\rho v(\xi)D_\varsigma v(\xi),\end{aligned}$$

where we also used the fact that, according to our definition of β, the first sum ranges only over those sites where $\Phi^{\rm qnl}_\xi = \Phi^{\rm a}_\xi$.

We now define $v^{\rm a}, v^{\rm c}$ according to Lemma 7.3 to obtain

$$\langle\delta^2\mathcal{E}^{\rm qnl}(u)v,v\rangle \geq \langle\delta^2\mathcal{E}^{\rm a}(u)v^{\rm a},v^{\rm a}\rangle + \langle\delta^2\mathcal{E}^{\rm qnl}(u)v^{\rm c},v^{\rm c}\rangle - CK^{-1}\|\nabla v\|^2_{L^2}. \quad (A.9)$$

If K is sufficiently large, then $\delta^2\mathcal{E}^{\rm qnl}(u)$ can be estimated using the decay hypothesis **(DH)** (see the proof of Theorem 7.4 for a similar argument) by

$$\langle\delta^2\mathcal{E}^{\rm qnl}(u)v^{\rm c},v^{\rm c}\rangle \geq \langle\delta^2\mathcal{E}^{\rm qnl}(0)v^{\rm c},v^{\rm c}\rangle - CK^{-\alpha}\|\nabla v^{\rm c}\|^2_{L^2}. \quad (A.10)$$

Combining (A.9) and (A.10) and the assumptions of the theorem gives the stated result. □

A.4. Proofs from Section 8 (mesh optimization)

To prove Lemma 8.1 we first establish a crucial auxiliary result.

Lemma A.3. Let $N, \mathcal{T}_h$ be constructed using Algorithm T, then, for L sufficiently large, $h(x) \leq x/2$ for all $x \in [L, N]$.

Proof. For $x \in (\nu_j, \nu_{j+1})$, $j \geq L$, we have

$$\frac{h(x)}{x} \leq \frac{\tilde{h}(\nu_j)}{\nu_j},$$

where $\tilde{h}(x) = (x/L)^{\frac{2}{3}(\alpha+1)}$ was defined in (8.1). It is easy to see that $\tilde{h}(x)/x$ is monotonically increasing, hence it is sufficient to check that $\tilde{h}(N)/N \leq 1/2$.

In Algorithm T we defined $N := \lceil L^{(\alpha+1/2)/(\alpha-1/2)}\rceil \approx L^{(\alpha+1/2)/(\alpha-1/2)}$. An easy calculation shows that, in this case,

$$\frac{\tilde{h}(N)}{N} \approx L^{\left[\frac{\alpha+1/2}{\alpha-1/2}-1\right]\frac{2}{3}(\alpha+1)} L^{-\frac{\alpha+1/2}{\alpha-1/2}} = L^{-1/3}.$$

Thus, choosing L sufficiently large, we obtain $h(x) \le x/2$ for all $x \in [L, N]$. □

Proof of Lemma 8.1. Algorithm T ensures that $h(x) \le \tilde{h}(x)$ for all $x \in [L, N] \setminus \mathcal{N}_h$. From this, it is easy to prove the estimates (8.4) and (8.5) provided we replace $N_{\mathcal{T}_h}$ with L on the right-hand sides.

If we can prove that

$$h(x) \ge c\tilde{h}(x), \tag{A.11}$$

then the formal estimate (8.2) becomes rigorous (*i.e.*, we obtain $N_{\mathcal{T}_h} \le C_1 L$) and we also obtain (8.4) and (8.5) as stated.

Let $\alpha' = \frac{2}{3}(\alpha + 1)$ and let $x \in (L, N)$ belong to an element (ν_j, ν_{j+1}). Then

$$\begin{aligned} h(x) = \tilde{h}(x)\frac{h(x)}{\tilde{h}(x)} &= \tilde{h}(x)\frac{\tilde{h}(\nu_j)}{\tilde{h}(x)} \ge \tilde{h}(x)\frac{\tilde{h}(\nu_j)}{\tilde{h}(\nu_{j+1})} \\ &= \tilde{h}(x)\left(\frac{\nu_j}{\nu_{j+1}}\right)^{\alpha'} = \tilde{h}(x)\left(\frac{\nu_{j+1} - \tilde{h}(\nu_j)}{\nu_{j+1}}\right)^{\alpha'} = \tilde{h}(x)\left(1 - \frac{\tilde{h}(\nu_j)}{\nu_{j+1}}\right)^{\alpha'}. \end{aligned}$$

From Lemma A.3 we know that, for L sufficiently large, $\tilde{h}(\nu_{j+1}) \le \frac{1}{2}\nu_{j+1}$, and hence

$$1 - \frac{\tilde{h}(\nu_j)}{\nu_{j+1}} \ge 1 - \frac{\tilde{h}(\nu_{j+1})}{\nu_{j+1}} \ge 1 - \frac{1}{2} = \frac{1}{2}.$$

Thus, we obtain

$$h(x) \ge \tilde{h}(x)\left(\frac{1}{2}\right)^{\alpha'}.$$

This concludes the proof of (A.11) and hence of the lemma. □

B. Inverse function theorem

All rigorous error estimates that we derive in the main text use the following quantitative version of the inverse function theorem.

Lemma B.1. Let $\mathcal{U}_h$ be a subspace of $\mathcal{U}$, equipped with $\|\nabla \cdot\|_{L^2}$, and let $\mathcal{G}_h \in C^1(\mathcal{U}_h, \mathcal{U}_h^*)$ with Lipschitz-continuous derivative $\delta\mathcal{G}_h$:

$$\|\delta\mathcal{G}_h(u_h) - \delta\mathcal{G}_h(v_h)\|_{\mathcal{L}} \le M\|\nabla u_h - \nabla v_h\|_{L^2} \quad \text{for all } u_h, v_h \in \mathcal{U}_h, \tag{B.1}$$

where $\|\cdot\|_{\mathcal{L}}$ denotes the $\mathcal{L}(\mathcal{U}_h, \mathcal{U}_h^*)$-operator norm.

Let $\bar{u}_h \in \mathcal{U}_h$ satisfy

$$\|\mathcal{G}_h(\bar{u}_h)\|_{\mathcal{U}_h^*} \le \eta, \tag{B.2}$$

$$\langle \delta\mathcal{G}_h(\bar{u}_h)v_h, v_h\rangle \ge \gamma\|\nabla v_h\|_{L^2}^2 \quad \text{for all } v_h \in \mathcal{U}_h, \tag{B.3}$$

such that L, η, γ satisfy the relation

$$\frac{2M\eta}{\gamma^2} < 1. \tag{B.4}$$

Then there exists a (locally unique) $u_h \in \mathcal{U}_h$ such that $\mathcal{G}_h(u_h) = 0$,

$$\|\nabla u_h - \nabla \bar{u}_h\|_{L^2} \le 2\frac{\eta}{\gamma}, \quad \text{and} \tag{B.5}$$

$$\langle \delta\mathcal{G}_h(u_h)v_h, v_h\rangle \ge \left(1 - \frac{2M\eta}{\gamma^2}\right)\gamma\|\nabla v_h\|_{L^2}^2 \quad \text{for all } v_h \in \mathcal{U}_h. \tag{B.6}$$

Proof. The result is a simplified and specialized version of Lemma 2.2 of Ortner (2011), but similar statements can be obtained from most proofs of the inverse function theorem. □

C. List of symbols

Π_h	Quasi-best approximation operator for $\mathcal{U}_h$, page 444
Π_N	Best approximation operator for $\mathcal{U}_N$, page 429
I_h	Nodal interpolation operator, page 433
$\mathcal{N}_h, \mathcal{N}_h^\circ$	Finite element nodes, page 433
$N_{\mathcal{T}_h}$	Number of degrees of freedom, page 433
ν_j	Finite element node, page 433
$\mathrm{P1}(\mathcal{T}_h)$	Space of (continuous) piecewise affine functions, page 433
$\mathrm{P0}(\mathcal{T}_h)$	Space of piecewise constant functions, page 433
r_{cut}	Interaction cut-off in the reference configuration, page 424
$\mathcal{R}$	Interaction range $\mathcal{R} = \{\pm 1, \dots, \pm r_{\mathrm{cut}}\}$, page 424
S^{a}	Atomistic stress function, page 447
$\mathsf{S}^{\mathrm{bqce}}$	B-QCE stress function, page 451
$\mathsf{S}^{\mathrm{qnl}}$	QNL stress function, page 456
$\mathcal{T}_h$	Finite element mesh (partition), page 433
$h_T, h(x)$	Mesh size function, page 433
$\tilde{h}(x)$	Quasi-optimal mesh size function, page 476
$\tilde{\mathsf{S}}^{\mathrm{bqce}}$	Modified B-QCE stress function, page 451
$\mathcal{U}$	Space of antisymmetric displacements with finite energy norm (discrete H^1), page 423
$\mathcal{U}_h$	Finite element displacement space, page 432
$\mathcal{U}_N$	Displacement space for the reduced atomistic scheme, page 429
$\mathcal{U}_0$	Space of compact antisymmetric displacements, page 423
$\mathbb{Z}_+$	$\{0, 1, 2, \dots\}$, page 423
$D_\rho u(\xi)$	Directional finite difference, page 424
$Du(\xi)$	Interaction stencil, page 424
K, L	Atomistic-to-continuum interface parameters, page 433
$m(\boldsymbol{\rho})$	Bounds on partial derivatives of V, page 426
$M^{(j,s)}$	Weighted bounds on partial derivatives of V, page 426
N	Radius of computational domain, page 429
V	Interaction potential, page 424
$V_{\boldsymbol{\rho}}, \Phi^{\mathrm{a}}_{\xi,\boldsymbol{\rho}}$	Partial derivatives of site potential and site energy, page 426
W	Cauchy–Born strain energy density, page 434
$y^{\mathsf{A}}, u^{\mathsf{F}}$	Affine maps $y^{\mathsf{A}}(\xi) = \mathsf{A}\xi, u^{\mathsf{F}}(\xi) = \mathsf{F}\xi$, page 425
(DH)	Decay hypothesis, page 429
(F)	Force-consistency (patch test) condition, page 437
(L)	Locality condition for QNL methods, page 455
(S)	Scaling condition for QNL methods, page 455

Acknowledgements

The work of ML was supported in part by the PIRE grant OISE-0967140, DOE Award DE-SC0002085, and AFOSR Award FA9550-12-1-0187. ML is also grateful for the opportunity to present this work at the University of Heidelberg as the Werner Romberg Guest Professor.

CO was supported by EPSRC grant EP/H003096 'Analysis of Atomistic-to-Continuum Coupling Methods'.

ML and CO would also like to acknowledge the support of the Institute for Pure & Applied Mathematics and the suggestions and advice of Andrew Binder, Matthew Dobson, Frederic Legoll, Xingjie Helen Li, Derek Olson, which helped us to eradicate several mistakes and unclear passages.

REFERENCES[1]

N. C. Admal and E. B. Tadmor (2010), 'A unified interpretation of stress in molecular systems', *J. Elasticity* **100**, 63–143.

M. Anitescu, D. Negrut, P. Zapol and A. El-Azab (2009), 'A note on the regularity of reduced models obtained by nonlocal quasi-continuum-like approaches', *Math. Program. Ser. A* **118**, 207–236.

M. Arndt and M. Luskin (2007), 'Goal-oriented atomistic-continuum adaptivity for the quasicontinuum approximation', *Internat. J. Multiscale Comput. Engng* **5**, 407–415.

M. Arndt and M. Luskin (2008*a*), 'Error estimation and atomistic-continuum adaptivity for the quasicontinuum approximation of a Frenkel–Kontorova model', *SIAM J. Multiscale Model. Simul.* **7**, 147–170.

M. Arndt and M. Luskin (2008*b*), 'Goal-oriented adaptive mesh refinement for the quasicontinuum approximation of a Frenkel–Kontorova model', *Comput. Methods Appl. Mech. Engng* **197**, 4298–4306.

D. N. Arnold and R. S. Falk (1989), 'A uniformly accurate finite element method for the Reissner–Mindlin plate', *SIAM J. Numer. Anal.* **26**, 1276–1290.

S. Badia, P. Bochev, R. Lehoucq, M. L. Parks, J. Fish, M. Nuggehally and M. Gunzburger (2007), 'A force-based blending model for atomistic-to-continuum coupling', *Internat. J. Multiscale Comput. Engng* **5**, 387–406.

S. Badia, M. Parks, P. Bochev, M. Gunzburger and R. Lehoucq (2008), 'On atomistic-to-continuum coupling by blending', *Multiscale Model. Simul.* **7**, 381–406.

M. I. Baskes, C. F. Melius and W. D. Wilson (1981), Solubility and diffusivity of hydrogen and helium at dislocations and in the stress-field near a crack tip. In *Hydrogen Effects in Metals* (I. M. Bernstein and A. W. Thompson, eds), TMS-AIME, pp. 67–75.

[1] The URLs cited in this work were correct at the time of going to press, but the publisher and the authors make no undertaking that the citations remain live or are accurate or appropriate.

P. T. Bauman, H. Ben Dhia, N. Elkhodja, J. T. Oden and S. Prudhomme (2008), 'On the application of the Arlequin method to the coupling of particle and continuum models', *Comput. Mech.* **42**, 511–530.

T. Belytschko and S. P. Xiao (2003), 'Coupling methods for continuum model with molecular model', *Internat. J. Multiscale Comput. Engng* **1**, 115–126.

X. Blanc and F. Legoll (2013), 'A numerical strategy for coarse-graining two-dimensional atomistic models at finite temperature: The membrane case', *Comput. Mater. Sci.* **66**, 84–95.

X. Blanc, C. Le Bris and F. Legoll (2005), 'Analysis of a prototypical multiscale method coupling atomistic and continuum mechanics', *M2AN: Math. Model. Numer. Anal.* **39**, 797–826.

X. Blanc, C. Le Bris and P.-L. Lions (2002), 'From molecular models to continuum mechanics', *Arch. Ration. Mech. Anal.* **164**, 341–381.

X. Blanc, C. Le Bris, F. Legoll and C. Patz (2010), 'Finite-temperature coarse-graining of one-dimensional models: Mathematical analysis and computational approaches', *J. Nonlinear Sci.* **20**, 241–275.

M. Born and K. Huang (1954), *Dynamical Theory of Crystal Lattices*, Oxford University Press.

A.-L. Cauchy (1882), De la pression ou tension dans un système de points matériels. In *Oeuvres complétes d'Augustin Cauchy*, Vol. 2ᵉ Sèrie, tome VIII, Gauthier-Villars, pp. 253–277. First published in 1828.

W. A. Curtin and R. M. Miller (2003), 'Atomistic/continuum coupling in computational materials science', *Model. Simul. Mater. Sci.* **11**, R33–R68.

M. S. Daw and M. I. Baskes (1984), 'Embedded-atom method: Derivation and application to impurities, surfaces, and other defects in metals', *Phys. Rev. B* **20**, 6443–6453.

M. Dobson (2011), There is no pointwise consistent quasicontinuum energy. arXiv:1109.1897

M. Dobson and M. Luskin (2008*a*), 'Analysis of a force-based quasicontinuum approximation', *Math. Model. Numer. Anal.* **42**, 113–139.

M. Dobson and M. Luskin (2008*b*), 'Iterative solution of the quasicontinuum equilibrium equations with continuation', *J. Sci. Comput.* **37**, 19–41.

M. Dobson and M. Luskin (2009*a*), 'An analysis of the effect of ghost force oscillation on quasicontinuum error', *Math. Model. Numer. Anal.* **43**, 591–604.

M. Dobson and M. Luskin (2009*b*), 'An optimal order error analysis of the one-dimensional quasicontinuum approximation', *SIAM. J. Numer. Anal.* **47**, 2455–2475.

M. Dobson, M. Luskin and C. Ortner (2010*a*), 'Accuracy of quasicontinuum approximations near instabilities', *J. Mech. Phys. Solids* **58**, 1741–1757.

M. Dobson, M. Luskin and C. Ortner (2010*b*), 'Stability, instability, and error of the force-based quasicontinuum approximation', *Arch. Ration. Mech. Anal.* **197**, 179–202.

M. Dobson, M. Luskin and C. Ortner (2011), 'Iterative methods for the force-based quasicontinuum approximation: Analysis of a 1D model problem', *Comput. Methods Appl. Mech. Engng* **200**, 2697–2709.

M. Dobson, C. Ortner and A. V. Shapeev (2012), 'The spectrum of the force-based quasicontinuum operator for a homogeneous periodic chain', *Multiscale Model. Simul.* **10**, 744–765.

Q. Du, M. Gunzburger, R. Lehoucq and K. Zhou (2013), 'Analysis of the volume-constrained peridynamic Navier equation of linear elasticity', *J. Elasticity*, to appear.

L. M. Dupuy, E. B. Tadmor, R. E. Miller and R. Phillips (2005), 'Finite temperature quasicontinuum: Molecular dynamics without all the atoms', *Phys. Rev. Lett.* **95**, 060202.

W. E (2011), *Principles of Multiscale Modeling*, Cambridge University Press.

W. E and J. Lu (2013), 'The Kohn–Sham equation for deformed crystals', *Mem. Amer. Math. Soc.* **221**, to appear.

W. E and P. Ming (2007), 'Cauchy–Born rule and the stability of crystalline solids: Static problems', *Arch. Ration. Mech. Anal.* **183**, 241–297.

W. E, J. Lu and J. Z. Yang (2006), 'Uniform accuracy of the quasicontinuum method', *Phys. Rev. B* **74**, 214115.

B. Eidel and A. Stukowski (2009), 'A variational formulation of the quasicontinuum method based on energy sampling in clusters', *J. Mech. Phys. Solids* **57**, 87–108.

H. Fischmeister, H. Exner, M.-H. Poech, S. Kohlhoff, P. Gumbsch, S. Schmauder, L. S. Sigi and R. Spiegler (1989), 'Modelling fracture processes in metals and composite materials', *Z. Metallkde.* **80**, 839–846.

G. Friesecke and F. Theil (2002), 'Validity and failure of the the Cauchy–Born hypothesis in a two-dimensional mass-spring lattice', *J. Nonlin. Sci.* **12**, 445–478.

C. J. Garcia-Cervera, J. Lu and W. E (2007), 'A sub-linear scaling algorithm for computing the electronic structure of materials', *Commun. Math. Sci.* **5**, 999–1026.

V. Gavini, K. Bhattacharya and M. Ortiz (2007), 'Quasi-continuum orbital-free density-functional theory: A route to multi-million atom non-periodic DFT calculation', *J. Mech. Phys. Solids* **55**, 697–718.

M. Gunzburger and Y. Z. Zhang (2010), 'A quadrature-rule type approximation to the quasi-continuum method', *Multiscale Model. Simul.* **8**, 571–590.

S. Haq, A. B. Movchan and G. J. Rodin (2007), 'Lattice Green's functions in nonlinear analysis of defects', *J. Appl. Mech.* **74**, 686–690.

R. Hardy (1982), 'Formulas for determining local properties in molecular dynamics simulations: Shock waves', *J. Chem. Phys.* **76**, 622–628.

T. Hudson and C. Ortner (2012), 'On the stability of Bravais lattices and their Cauchy–Born approximations', *M2AN: Math. Model. Numer. Anal.* **46**, 81–110.

M. Iyer and V. Gavini (2011), 'A field theoretical approach to the quasi-continuum method', *J. Mech. Phys. Solids* **59**, 1506–1535.

H. Kanzaki (1957), 'Point defects in face-centred cubic lattice I: Distortion around defects', *J. Phys. Chem. Solids* **2**, 24–36.

W. K. Kim, M. Luskin, D. Perez, E. B. Tadmor and A. F. Voter (2012), Hyper-QC: An accelerated finite-temperature quasicontinuum method using hyperdynamics. Submitted.

J. Knap and M. Ortiz (2001), 'An analysis of the quasicontinuum method', *J. Mech. Phys. Solids* **49**, 1899–1923.

S. Kohlhoff, P. Gumbsch and H. F. Fischmeister (1991), 'Crack propagation in b.c.c. crystals studied with a combined finite-element and atomistic model', *Phil. Mag. A* **64**, 851–878.

B. Langwallner, C. Ortner and E. Süli (2013), 'Atomistic-to-continuum coupling approximation of a one-dimensional toy model for density functional theory', *Multiscale Model. Simul.* **11**, 59–91.

X. Li (2009), 'Efficient boundary condition for molecular statics models of solids', *Phys. Rev. B* **80**, 104112.

X. H. Li and M. Luskin (2013), 'Lattice stability for atomistic chains modeled by local approximations of the embedded atom method', *Comput. Mater. Sci.* **66**, 96–103.

X. H. Li, M. Luskin and C. Ortner (2012), 'Positive-definiteness of the blended force-based quasicontinuum method', *SIAM J. Multiscale Model. Simul.* **10**, 1023–1045.

X. H. Li, M. Luskin, C. Ortner and A. Shapeev (2013), Theory-based benchmarking of the blended force-based quasicontinuum method. Manuscript.

X. H. Li, M. Luskin, C. Ortner, A. Shapeev and B. Vankoten (2013). Manuscript.

P. Lin (2003), 'Theoretical and numerical analysis for the quasi-continuum approximation of a material particle model', *Math. Comp.* **72**, 657–675.

P. Lin (2007), 'Convergence analysis of a quasi-continuum approximation for a two-dimensional material without defects', *SIAM J. Numer. Anal.* **45**, 313–332.

W. K. Liu, E. G. Karpov and H. S. Park (2006*a*), *Nano Mechanics and Materials: Theory, Multiscale Methods and Applications*, Wiley.

W. K. Liu, H. Park, D. Qian, E. G. Karpov, H. Kadowaki and G. J. Wagner (2006*b*), 'Bridging scale methods for nanomechanics and materials', *Comput. Methods Appl. Mech. Engng* **195**, 1407–1421.

J. Lu and P. Ming (2012), Stability of a force-based hybrid method in three dimensions with sharp interface. arXiv:1212.3643

J. Lu and P. Ming (2013), 'Convergence of a force-based hybrid method in three dimensions', *Commun. Pure Appl. Math.* **66**, 83–108.

M. Luskin and C. Ortner (2009), 'An analysis of node-based cluster summation rules in the quasicontinuum method', *SIAM J. Numer. Anal.* **47**, 3070–3086.

M. Luskin and C. Ortner (2012), Linear stationary iterative methods for the force-based quasicontinuum approximation. In *Numerical Analysis and Multiscale Computations* (B. Engquist, O. Runborg and R. Tsai, eds), Vol. 82 of *Lecture Notes in Computer Science and Engineering*, Springer, pp. 331–368.

M. Luskin, C. Ortner and B. Van Koten (2013), 'Formulation and optimization of the energy-based blended quasicontinuum method', *Comput. Methods Appl. Mech. Engng* **253**, 160–168.

C. Makridakis and E. Süli (2013), 'Finite element analysis of Cauchy–Born approximations to atomistic models', *Arch. Ration. Mech. Anal.* **207**, 813–843.

C. Makridakis, D. Mitsoudis and P. Rosakis (2012), On atomistic-to-continuum couplings without ghost forces in three dimensions. arXiv:1211.7158

C. Makridakis, C. Ortner and E. Süli (2011), '*A priori* error analysis of two force-based atomistic/continuum models of a periodic chain', *Numer. Math.* **119**, 83–121.

R. Miller and D. Rodney (2008), 'On the nonlocal nature of dislocation nucleation during nanoindentation', *J. Mech. Phys. Solids* **56**, 1203–1223.

R. Miller and E. Tadmor (2003), 'The quasicontinuum method: Overview, applications and current directions', *J. Computer-Aided Mater. Design* **9**, 203–239.

R. Miller and E. Tadmor (2009), 'A unified framework and performance benchmark of fourteen multiscale atomistic/continuum coupling methods', *Model. Simul. Mater. Sci. Engng* **17**, 053001.

P. Ming and J. Z. Yang (2009), 'Analysis of a one-dimensional nonlocal quasicontinuum method', *Multiscale Model. Simul.* **7**, 1838–1875.

M. Mullins and M. Dokainish (1982), 'Simulation of the (0 0 1) plane crack in α-iron employing a new boundary scheme', *Phil. Mag. A* **46**, 771–787.

D. Olson, P. Bochev, M. Luskin and A. Shapeev (2013), An optimization-based atomistic-to-continuum coupling method. Manuscript.

C. Ortner (2011), '*A priori* and *a posteriori* analysis of the quasinonlocal quasicontinuum method in 1D', *Math. Comp.* **80**, 1265–1285.

C. Ortner (2012), 'The role of the patch test in 2D atomistic-to-continuum coupling methods', *ESAIM: Math. Model. Numer. Anal.* **46**, 1275–1319.

C. Ortner and A. Shapeev (2012), Interpolants of lattice functions for the analysis of atomistic/continuum multiscale methods. arXiv:1204.3705

C. Ortner and A. V. Shapeev (2013), 'Analysis of an energy-based atomistic/continuum coupling approximation of a vacancy in the 2D triangular lattice', *Math. Comp.*, to appear.

C. Ortner and E. Süli (2008), 'Analysis of a quasicontinuum method in one dimension', *M2AN: Math. Model. Numer. Anal.* **42**, 57–91.

C. Ortner and F. Theil (2013), 'Justification of the Cauchy–Born approximation of elastodynamics', *Arch. Ration. Mech. Anal.* **207**, 1025–1073.

C. Ortner and H. Wang (2013), '*A posteriori* error control for a quasicontinuum approximation of a periodic chain', *Math. Model. Methods Appl. Sci.*, to appear.

C. Ortner and L. Zhang (2012), Construction and sharp consistency estimates for atomistic/continuum coupling methods with general interfaces: A 2D model problem, *SIAM J. Numer. Anal.* **50**, 2940–2965.

C. Ortner, A. Shapeev and L. Zhang (2013), Manuscript.

M. L. Parks, P. B. Bochev and R. B. Lehoucq (2008), 'Connecting atomistic-to-continuum coupling and domain decomposition', *Multiscale Model. Simul.* **7**, 362–380.

S. Prudhomme, P. T. Bauman and J. T. Oden (2006), 'Error control for molecular statics problems', *Internat. J. Multiscale Comput. Engng* **4**, 647–662.

S. Prudhomme, H. Ben Dhia, P. T. Bauman, N. Elkhodja and J. T. Oden (2008), 'Computational analysis of modeling error for the coupling of particle and continuum models by the Arlequin method', *Comput. Methods Appl. Mech. Engng* **197**, 3399–3409.

R. Rudd and J. Broughton (2000), 'Concurrent coupling of length scales in solid state systems', *Physica Status Solidi B* **217**, 251–291.

R. E. Rudd and J. Q. Broughton (2005), 'Coarse-grained molecular dynamics: Nonlinear finite elements and finite temperature', *Phys. Rev. B* **72**, 144104.

P. Seleson, S. Beneddine and S. Prudhomme (2013), 'A force-based coupling scheme for peridynamics and classical elasticity', *Comput. Mater. Sci.* **66**, 34–49.

P. Seleson and M. Gunzburger (2010), 'Bridging methods for atomistic-to-continuum coupling and their implementation', *Commun. Comput. Phys.* **7**, 831–876.

A. V. Shapeev (2011), 'Consistent energy-based atomistic/continuum coupling for two-body potentials in one and two dimensions', *Multiscale Model. Simul.* **9**, 905–932.

V. B. Shenoy, R. Miller, E. B. Tadmor, D. Rodney, R. Phillips and M. Ortiz (1999), 'An adaptive finite element approach to atomic-scale mechanics: The quasicontinuum method', *J. Mech. Phys. Solids* **47**, 611–642.

L. E. Shilkrot, R. E. Miller and W. A. Curtin (2004), 'Multiscale plasticity modeling: Coupled atomistics and discrete dislocation mechanics', *J. Mech. Phys. Solids* **52**, 755–787.

T. Shimokawa, J. J. Mortensen, J. Schiøtz and K. W. Jacobsen (2004), 'Matching conditions in the quasicontinuum method: Removal of the error introduced at the interface between the coarse-grained and fully atomistic region', *Phys. Rev. B* **69**, 214104.

S. Silling and R. Lehoucq (2010), Peridynamic theory of solid mechanics. In *Advances in Applied Mechanics* (H. Aref and E. van der Giessen, eds), Vol. 44 of *Advances in Applied Mechanics*, Elsevier, pp. 73–168.

J. E. Sinclair (1971), 'Improved atomistic model of a bcc dislocation core', *J. Appl. Phys.* **42**, 5231.

E. B. Tadmor and R. E. Miller (2011), *Modeling Materials: Continuum, Atomistic and Multiscale Techniques*, Cambridge University Press.

E. B. Tadmor, F. Legoll, W. K. Kim, L. M. Dupuy and R. E. Miller (2013), 'Finite-temperature quasicontinuum', *Appl. Mech. Rev.*, to appear.

E. B. Tadmor, M. Ortiz and R. Phillips (1996), 'Quasicontinuum analysis of defects in solids', *Philosophical Magazine A* **73**, 1529–1563.

S. Tang, T. Y. Hou and W. K. Liu (2006), 'A mathematical framework of the bridging scale method', *Internat. J. Numer. Methods Engng* **65**, 1688–1713.

D. R. Trinkle (2008), 'Lattice Green function for extended defect calculations: Computation and error estimation with long-range forces', *Phys. Rev. B* **78**, 014110.

B. Van Koten and M. Luskin (2011), 'Analysis of energy-based blended quasicontinuum approximations', *SIAM J. Numer. Anal.* **49**, 2182–2209.

B. Van Koten and C. Ortner (2012), Symmetries of 2-lattices and second order accuracy of the Cauchy–Born model. arXiv:1203.5854

B. Van Koten, X. H. Li, M. Luskin and C. Ortner (2012), A computational and theoretical investigation of the accuracy of quasicontinuum methods. In *Numerical Analysis of Multiscale Problems* (I. Graham, T. Hou, O. Lakkis and R. Scheichl, eds), Vol. 83 of *Lecture Notes in Computer Science and Engineering*, Springer, pp. 67–96.

A. F. Voter (1997), 'Hyperdynamics: Accelerated molecular dynamics of infrequent events', *Phys. Rev. Lett.* **78**, 3908–3911.

D. Wallace (1998), *Thermodynamics of Crystals*, Dover.

M. Wallin, W. Curtin, M. Ristinmaa and A. Needleman (2008), 'Multi-scale plasticity modeling: Coupled discrete dislocation and continuum crystal plasticity', *J. Mech. Phys. Solids* **56**, 3167–3180.

C. Woodward and S. Rao (2002), 'Flexible *ab initio* boundary conditions: Simulating isolated dislocations in bcc Mo and Ta', *Phys. Rev. Lett.* **88**, 216402.

S. P. Xiao and T. Belytschko (2004), 'A bridging domain method for coupling continua with molecular dynamics', *Comput. Methods Appl. Mech. Engng* **193**, 1645–1669.

Acta Numerica (2013), pp. 509–575
doi:10.1017/S096249291300007X

On first-order algorithms for ℓ_1/nuclear norm minimization

Yurii Nesterov*
Center for Operations Research and Econometrics,
34 voie du Roman Pays, 1348, Louvain-la-Neuve, Belgium
E-mail: yurii.nesterov@uclivain.be

Arkadi Nemirovski†
School of Industrial and Systems Engineering,
Georgia Institute of Technology,
Atlanta, Georgia 30332, USA
E-mail: arkadi.nemirovski@isye.gatech.edu

In the past decade, problems related to l_1/nuclear norm minimization have attracted much attention in the signal processing, machine learning and optimization communities. In this paper, devoted to ℓ_1/nuclear norm minimization as 'optimization beasts', we give a detailed description of two attractive first-order optimization techniques for solving problems of this type. The first one, aimed primarily at lasso-type problems, comprises fast gradient methods applied to composite minimization formulations. The second approach, aimed at Dantzig-selector-type problems, utilizes saddle-point first-order algorithms and reformulation of the problem of interest as a *generalized bilinear saddle-point problem.* For both approaches, we give complete and detailed complexity analyses and discuss the application domains.

CONTENTS

* Partly supported by Direction de la Recherche Scientifique, Communauté Française de Belgique, via grant Action de Recherche Concerté ARC 04/09-315, and by the Laboratory of Structural Methods of Data Analysis in Predictive Modelling via the RF government grant 11.G34.31.0073c.

† Supported in part by NSF via grants DMS-0914785 and CMMI 1232623, by BSF via grant 2008302, and by ONR via grant N000140811104.

1. Introduction

1.1. Motivation

The goal of this paper is to review recent progress in applications of first-order optimization methods to problems of ℓ_1/nuclear norm minimization. For us, the problems of primary interest are the *lasso-type problems*,

$$\mathrm{Opt} = \min_{x \in E} \{ \lambda \|x\| + \|Ax - b\|_2^2 \}, \tag{1.1a}$$

and the *Dantzig-selector-type problems*,

$$\mathrm{Opt} = \min_{x \in E} \{ \|x\| : \|Ax - b\|_p \leq \delta \}. \tag{1.1b}$$

Here E is a Euclidean space, $\|\cdot\|$ is a given norm on E (often different from the Euclidean norm), $x \mapsto Ax$ is a linear mapping from E to some $\mathbb{R}^m$, and $b \in \mathbb{R}^m$ is a given vector. In (1.1a), $\lambda > 0$ is a given *penalty parameter*. In (1.1b), $\delta \geq 0$, and p is either 2 ('ℓ_2-fit'), or ∞ ('uniform fit').[1] We will be mainly (but not exclusively) interested in the following cases:

- *ℓ_1 minimization*, where $E = \mathbb{R}^n$ and $\|\cdot\| = \|\cdot\|_1$,
- *nuclear norm minimization*, where $E = \mathbb{R}^{\mu \times \nu}$ is the space of $\mu \times \nu$ matrices and $\|\cdot\| = \|\cdot\|_{\mathrm{nuc}}$ is the *nuclear norm*, the sum of the singular values of a matrix.

The main source of ℓ_1 minimization problems of the form (1.1a,b) is *sparsity-oriented signal processing*, where the goal is to recover an unknown 'signal' $x \in \mathbb{R}^n$ from its noisy observations $y = Bx + \xi$, where ξ is the observation error. The signal is known to be *sparse*, *i.e.*, it has at most $s \ll n$ non-zero entries. The most interesting case here is *compressive sensing*, where the observations are 'deficient' ($m = \dim y < n = \dim x$). Hence the observations themselves, even noiseless ones, do not allow recovery of x unless some additional information on x is utilized, *e.g.*, sparsity.

The conceptually simplest way to recover a sparse x from noiseless observations is offered by ℓ_0 minimization, where we minimize the ℓ_0 'norm' of $x \in \mathbb{R}^n$ (*i.e.*, the number of non-zero entries in x) under the restriction $Bx = y$. It is immediately seen that if $y = Bx_*$ for some x_* with at most s non-zero entries, and if every $2s$ columns of B are linearly independent, then the ℓ_0 minimization does indeed recover x_*. The independence condition is true, for example, for any $m \times n$ matrix B in 'general position' provided that $m \geq 2s$.

[1] The restriction $p \in \{2, \infty\}$ is met in all applications of Dantzig selector models known to us. However, the following results can be straightforwardly extended to the case of $p \in [2, \infty]$ (but *not* to the case of $1 \leq p < 2$).

This procedure, however, is heavily computationally intractable. To make it tractable, we can replace the ℓ_0 norm by its 'closest' convex approximation $\|\cdot\|_1$. Thus we come to the recovery problem $\widehat{x} \in \arg\min_x\{\|x\|_1 : Bx = y\}$, *i.e.*, problem (1.1b) with $A = B$, $b = y$ and $\delta = 0$.

A natural way to account for the observation noise ξ having $\|\cdot\|_2$ magnitude not exceeding a given δ is by passing from minimization of $\|\cdot\|_1$ on the subspace $Bx = y$ to minimization of this norm over the set $\{x : \|Bx - y\|_2 \le \delta\}$ of all signals compatible with our observations. Thus, we come to problem (1.1b) with $p = 2$. Note that the lasso problem (1.1a) with $\|\cdot\| = \|\cdot\|_1$ can be viewed as a re-parametrization of (1.1b) with $p = 2$. Indeed, geometrically, the curve of solutions to (1.1b) parametrized by δ is the same as the curve of solutions to (1.1a) parametrized by λ. In some situations (*e.g.*, when the magnitude of the observation error is not known), the latter parametrization is better than the former one.

For random observation error ξ (*e.g.*, white noise $\xi \sim \mathcal{N}(0, \sigma^2 I_m)$), it would be too conservative to measure the noise in terms of its $\|\cdot\|_2$ magnitude, since this approach results in growth of δ with m as $\sigma\sqrt{m}$, and essentially ignores the random nature of the noise. From the statistical viewpoint, in this case it is much better to pass to 'artificial' observations $b = H^T y$, where H is a properly selected 'contrast matrix' with columns $h^1, \ldots, h^K \in \mathbb{R}^m$. Then $b = (H^T B)x + \tilde{\xi}$ with the *uniform* norm of the new observation error $\tilde{\xi} = H^T\xi$ of order $\sigma \max_j \|h^j\|_2$. We can now apply the above recovery routine to the artificial observations, to obtain the estimator

$$\widehat{x} \in \arg\min_x \left\{ \|x\|_1 : \|\underbrace{H^T B}_{A} x - b\|_\infty \le \delta = \delta(\epsilon) := \sigma \max_i \|h_i\|_2 \operatorname{Erf\,Inv}\left(\frac{\epsilon}{K}\right) \right\},$$

where $\operatorname{Erf\,Inv}(\cdot)$ is the inverse error function, and $1 - \epsilon$ is the desired confidence. Note that our $\delta(\epsilon)$, with probability at least $1 - \epsilon$, is an upper bound on the uniform norm $\|H^T\xi\|_\infty$ of the new observation error. This bound is 'nearly independent' of ϵ and the dimensions of the problem. The resulting recovery routine is exactly (1.1b) with $p = \infty$.[2]

Nuclear norm minimization can be viewed as the 'matrix analogue' of sparsity-oriented signal processing, with the $\mu \times \nu$ matrix x playing the role of the n-dimensional signal vector, and the rank of the matrix (*i.e.*, the ℓ_0 norm of the vector of its singular values) replacing the number of non-zero entries in a vector. As in the case for the usual sparsity, the nuclear norm approach can be viewed as the 'best convex approximation' of the rank, which gives rise to problems (1.1a,b) with $\|\cdot\| = \|\cdot\|_{\text{nuc}}$.

[2] The actual Dantzig selector as defined by Candès and Tao (2007) corresponds to the situation when all columns in B are normalized to have unit $\|\cdot\|_2$ norm, and $H = B$. For other choices of the contrast matrix, see Juditsky and Nemirovski (2011*c*) and Juditsky, Kılınç Karzan, Nemirovski and Polyak (2013*b*).

1.2. *Justification of ℓ_1/nuclear norm recovery*

We have already mentioned that ℓ_1/nuclear norm minimizations are just 'convex approximations' of ℓ_0/rank minimization. This gives rise to the crucial question of whether and when these approximations can be 'justified', that is, whether and when they admit meaningful error bounds in the case of noisy observations and nearly sparse/low-rank signals. Here, 'meaningful' means that the recovery error should tend to zero as the magnitude of noise and the deviation of the signal from sparse/low rank tend to zero. While the 'justification issues' go beyond the scope of this paper, which is devoted to (1.1a,b) as 'optimization beasts', we believe it makes sense to comment on them very briefly. The numerous related results in the literature (see Candès 2006, Candès and Tao 2006 and Candès, Romberg and Tao 2006, and references therein) show that by imposing appropriate conditions on the sensing matrix, the outlined procedures can indeed be justified. A typical result is as follows.

Theorem (Juditsky *et al.* 2013*b*). Let s be a positive integer, and let a sensing matrix $B \in \mathbb{R}^{m\times n}$ satisfy the *restricted isometry property* with parameters κ and k, meaning that for every k-sparse signal x we have

$$(1-\kappa)\|x\|_2^2 \le x^T B^T Bx \le (1+\kappa)\|x\|_2^2.$$

Let $x_* \in \mathbb{R}^n$ be the 'true signal', and let $y = Bx_* + \xi$ be the observation. Assume that $k \ge 2s$ and $\kappa \le 0.1$.

(i) Let the observation error ξ satisfy $\|\xi\|_2 \le \delta$ with a given $\delta \ge 0$. Then the optimal solution $\widehat{x}$ to the Dantzig selector problem

$$\min_x \{ \|x\|_1 : \|Bx - y\|_2 \le \delta \}$$

satisfies

$$\|\widehat{x} - x_*\|_r \le O(1)\big[s^{\frac{1}{r}-\frac{1}{2}}\delta + s^{\frac{1}{r}-1}\|x_* - x_*^s\|_1\big], \quad 1 \le r \le 2,$$

where x_*^s is the best s-sparse approximation of x_* obtained by replacing all entries of x_* with zeros except the s largest in magnitude.[3]

(ii) Let the observation error ξ be Gaussian, $\xi \sim \mathcal{N}(0, \sigma^2 I)$, and let $\epsilon \in (0, 1/2)$. Then the optimal solution $\widehat{x}$ to the Dantzig selector problem

$$\min_x \{ \|x\|_1 : \|B^T(Bx - y)\|_\infty \le \delta := 2\sigma\sqrt{2\ln(m/\epsilon)} \}$$

satisfies

$$\|\widehat{x} - x_*\|_r \le O(1)\big[s^{\frac{1}{r}}\delta + s^{\frac{1}{r}-1}\|x_* - x_*^s\|_1\big], \quad 1 \le r \le 2,$$

with probability at least $1 - \epsilon$.

[3] Here and in what follows, $O(1)$ stands for positive absolute constants.

It is also known (see, *e.g.*, Candès 2006, Candès and Tao 2006 and Candès, Romberg and Tao 2006) that for a 'good' probability distribution $\mathbb{P}$ on $\mathbb{R}^{m\times n}$ (*e.g.*, the entries are independent $\mathcal{N}(0, 1/m)$ random variables, or they take the values $\pm 1/\sqrt{m}$ independently with probability 1/2), if m and n grow, then the matrix drawn from $\mathbb{P}$ satisfies, with probability approaching one, the restricted isometry property with parameters $\kappa = 0.1$, and k as large as $O(1)m/\ln(2n/m)$.

Similar results hold for nuclear norm recovery, but with rank in the role of the sparsity parameter s, the Frobenius norm replacing $\|\cdot\|_2$, and $\|x_* - x_*^s\|_{\text{nuc}}$ replacing $\|x_* - x_*^s\|_1$.

In problems (1.1a,b), the most frequently used norms are of course ℓ_1 and the nuclear norm. However, there are other important norms (the list below is not exhaustive).

- *Block ℓ_1/ℓ_2 norm* $\|x\| = \sum_{j=1}^n \|x^j\|_2$, where $x^j \in \mathbb{R}^{k_j}$ are non-overlapping blocks in the $(k_1 + \cdots + k_n)$-dimensional signal x. This norm is used when there are reasons to believe that the true signal is *block-sparse*, *i.e.*, the number of non-zero blocks in an *a priori* known partition of the signal is relatively small.
- *Total variation* (the so-called TV norm) is defined on the space $\mathbb{R}^{\mu\times\nu}$ of $\mu \times \nu$ images x, normalized to have $x_{1,1} = 0$. Then

$$\mathrm{TV}(x) = \sum_{i=1}^{\mu-1}\sum_{j=1}^{\nu} |x_{i+1,j} - x_{i,j}| + \sum_{i=1}^{\mu}\sum_{j=1}^{\nu-1} |x_{i,j+1} - x_{i,j}|.$$

  This is a discrete analogue of the L_1 norm of the gradient field of a function. It is very popular for de-noising and de-blurring images.

The literature on design, justification, and applications of problems (1.1a) and (1.1b) with the outlined norms is really huge and rapidly growing. To gain some impression of it, we refer the reader to the papers by Bach, Mairal and Ponce (2008), Baraniuk, Cevher, Duarte and Hegde (2010), Bickel, Ritov and Tsybakov (2008), Cai, Candès and Shen (2008), Candès (2006), Candès and Recht (2008), Candès and Tao (2006, 2007, 2009), Candès, Romberg and Tao (2006), Chandrasekaran, Recht, Parrilo and Willsky (2012), Chandrasekaran, Sanghavi, Parrilo and Willsky (2011), Chesneau and Hebiri (2008), Cohen, Dahmen and DeVore (2009), Donoho and Tanner (2005), Donoho, Elad and Temlyakov (2006), Eldar, Kuppinger and Bölcskei (2010), Elhamifar and Vidal (2012), Huang and Zhang (2010), Juditsky and Nemirovski (2011*a*, 2011*c*), Juditsky, Kılınç Karzan and Nemirovski (2011*a*), Juditsky *et al.* (2013*b*), Juditsky, Kılınç Karzan, Nemirovski and Polyak (2011*b*), Liu and Zhang (2009), Liu, Shang, Jiang and Liu (2010), Lounici, Pontil, van de Geer and Tsybakov (2011), Meier, van de Geer and Bühlmann (2008), Meinshausen and Yu (2009), Nardi and Rinaldo (2008),

Recht, Fazel and Parrilo (2010), Recht, Xuy and Hassibi (2011*b*), Rudin, Osher and Fatemi (1992), Tibshirani (1996), Tropp (2006) and Yuan and Lin (2006) for design and justification of recovery routines, and Herman and Strohmer (2007), Lustig, Donoho and Pauly (2007), Men *et al.* (2011), Dai, Sheikh, Milenkovic and Baraniuk (2009), Parvaresh, Vikalo, Misra and Hassibi (2008), Santosa and Symes (1986), Studer *et al.* (2012), Taylor, Banks and McCoy (1979), Vasanawala *et al.* (2010) and Wagner, Schmieder, Stern and Hoch (1993), and references therein, for applications in imaging, including biological and hyperspectral imaging, radar control, magnetic resonance, bioinformatics, design of photonic crystals, and inverse problems in geophysics, to name just a few.[4]

1.3. Algorithms

Problems (1.1a) and (1.1b) are well-structured convex optimization problems. Hence, they can be solved in a theoretically efficient fashion. For example, all these problems are within the 'theoretical grasp' of polynomial-time interior-point methods (IPMs) for linear programming (ℓ_1/total variation minimization with uniform fit), conic quadratic programming (ℓ_1/total variation minimization with ℓ_2 fit, and ℓ_1/ℓ_2 minimization) and semidefinite programming (nuclear norm minimization). The 'interior-point option' should definitely be kept in mind when processing (1.1a,b) numerically. The difficulty, however, is that numerous applications lead to extremely large-scale problems, with sizes of A in the range of tens and hundreds of thousands, or even millions. In addition, it is necessary to operate with highly dense analytically given sensing matrices A. For example, those matrices typically arising in compressive sensing are randomly selected submatrices of discrete Fourier/Hadamard transform matrices, or matrices with independent entries drawn from Gaussian or binomial distributions. As a result, in most large-scale applications our problems of interest go beyond the 'practical grasp' of IPMs with their time-consuming iterations (in the dense case their complexity has cubic dependence on design dimension).

There seems to exist a consensus that the only methods in the modern 'convex optimization toolbox' that are applicable to the large-scale problems of ℓ_1/nuclear norm minimization are the *first-order methods* (FOMs). As far as problems (1.1a,b) are concerned, one can define FOMs to be algorithms based only on computing matrix–vector products involving matrices A and A^T, and avoiding the matrix–matrix multiplications/Cholesky decompositions with 'large-scale' operands needed by IPMs. One can think of an FOM as an algorithm which, instead of direct access to the data matrix A, is allowed to call a 'multiplication oracle' which, given on input a

[4] In addition to our (by far incomplete) list of references, we refer the reader to the depository *Compressive Sensing Resources* at http://dsp.rice.edu/cs.

vector of appropriate dimension, returns its products with A or A^T. In the large-scale case, iterations of typical FOMs are vastly cheaper than those of IPMs.

This is the main argument in favour of FOMs in large-scale applications. In the context of ℓ_1/nuclear norm minimization problems, an additional argument is that the geometry of these problems is especially well suited to first-order minimization. As applied to such a problem, properly designed FOMs exhibit dimension-independent rates of convergence (or nearly so). Unfortunately, this rate of convergence is only sublinear. In contrast to the linear convergence of IPMs, the inaccuracy tends to zero with the number of iterations t as $O(1/t^2)$ at best, or even $O(1/t)$ (see below). However, in the majority of applications of ℓ_1/nuclear norm minimization, we are not interested in high-accuracy solutions, only in medium-accuracy ones. Thus the relatively slow convergence of FOMs is more than compensated for by their insensitivity to problem size. FOMs are especially attractive in the case of ℓ_1 minimization, where the computational effort per iteration reduces essentially to $O(1)$ matrix–vector multiplication involving A and A^T. As a result, any acceleration of matrix–vector multiplications allowed by sparsity or the special structure of A directly translates into overall acceleration of the solution process.[5] In order to gain an impression of what can be achieved in this way, look at the data in Table 1.1.

1.4. Scope and organization of the paper

In the literature we find an extremely wide spectrum of FOMs aimed at our problems of interest. They range from adaptations of general-purpose convex optimization techniques (*e.g.*, a multitude of proximal-point algorithms, numerous methods based on augmented Lagrangians and alternating directions, active set methods, *etc.*) capable of solving (1.1a,b) independently of any assumptions on sparsity of their optimal solutions, to highly specialized algorithms based on thresholding and matching pursuit, and aimed at recovery of *sparse/low-rank* near-optimal solutions. Some impression of the related literature can be obtained from Beck and Teboulle (2009*a*, 2009*b*), Becker, Candès and Grant (2010), Chambolle (2004), Goldfarb and Ma (2011), Goldfarb, Scheinberg and Xi (2011), Goldfarb and Yin (2009), Huang, Ma and Goldfarb (2013), Jaggi and Sulovský (2010), Lee *et al.* (2010), Liu, Sun and Toh (2012), Ma, Goldfarb and Chen (2011), Nesterov (2007*b*), Recht, Ré, Wright and Niu (2011*a*), Osher, Mao, Dong and Yin (2010), Qin and Goldfarb (2012), Recht and Ré (2011), Scheinberg, Ma

[5] Recall that in many applications A is a submatrix of the DFT/Hadamard transform, or represents a discrete convolution, and as such allows for fast matrix–vector multiplications.

Table 1.1. IPM versus FOM on ℓ_1 minimization problem (1.1b) with ℓ_2 fit and $\delta = 0.006$. A is the $m \times n$ matrix comprising real and imaginary parts of randomly selected $m/2$ rows in the unitary $n \times n$ DFT matrix; x_{res} is the solution found by the method. Platform: Intel Core i7 CPU M620 @ 2.66 GHz 2.66 GHz 8 GB RAM under Windows 7 64-bit.

	IPM†		FOM‡	
$m \times n$	$\lVert x_{\text{res}} \rVert_1$	CPU (sec)	$\lVert x_{\text{res}} \rVert_1$	CPU (sec)
256×512	0.971	1.2	0.977	2.0
512×1024	0.963	9.1	0.967	2.3
1024×2048	0.951	71.3	0.954	10.7
2048×4096	0.946	644.9	0.957	20.1
4096×8192	0.933	5768.5	0.937	53.0
8192×16384	out of memory		0.931	53.5

† Commercial IPM solver mosekopt.
‡ FOM (Mirror Prox: see Section 3) using FFT.

and Goldfarb (2010), Shi, Yin, Osher and Sajda (2010), Wen, Yin, Goldfarb and Zhang (2012), Wen, Yin, Zhang and Goldfarb (2012), Yang and Yuan (2013) and Yin, Osher, Goldfarb and Darbon (2008), and references therein. Although our sample is far from complete, we hope it is sufficiently representative. A more detailed overview of the extremely wide spectrum of FOMs proposed for ℓ_1/nuclear norm minimization seems to be an impossible task, going far beyond the scope of this paper. Here we intend to consider just two approaches in depth:

- *composite minimization* for lasso problems (1.1a),
- *saddle-point $O(1/t)$-converging first-order algorithms* for Dantzig selector problems (1.1b).

The rationale behind this selection is twofold. One consideration is that this selection is the optimal fit to the professional expertise and research experience of the authors, presented in Lemarechal, Nemirovski and Nesterov (1995), Nemirovski (2004), Nesterov (2005*a*, 2007*a*, 2007*b*, 2009), Juditsky and Nemirovski (2011*b*) and Juditsky, Kılınç Karzan and Nemirovski (2013*a*). Another consideration is that the selected methods are highly instructive examples of first-order algorithms with the *currently best known theoretical complexity guarantees* in our context (large-scale ℓ_1/nuclear norm minimization). Specifically, assuming that the norm $\|\cdot\|$ in (1.1a,b) is one of those mentioned in Section 1.1, the outlined algorithms exhibit $O(1/t^2)$ (lasso) or $O(1/t)$ (Dantzig selector) rate of convergence with 'natural' and nearly dimension-independent hidden factors in $O(\cdot)$. For example, we shall see the following.

- Applied to the ℓ_1/nuclear norm lasso problem, composite minimization ensures that inaccuracy in terms of the objective after $t = 1, 2, \ldots$ iterations does not exceed $O(1)\ln(\dim x)L^2\|x^*\|^2/t^2$, where x^* is an optimal solution to the problem and $L = \max\{\|Ax\|_2 : \|x\| \le 1\}$.
- Applied to the ℓ_1/nuclear norm Dantzig selector problem, saddle-point methods allow us to find a solution of accuracy ϵ in at most
$$O(1)\sqrt{\ln(\dim x)}[\ln(\dim Ax)]^{1/2-1/p}(L \cdot \mathrm{Opt}/\epsilon)\ln(L \cdot \mathrm{Opt}/\epsilon)$$
iterations, where Opt is the optimal value of the problem and
$$L = \max\{\|Ax\|_p : \|x\| \le 1\}.$$

To put these results into proper perspective, note that there are strong reasons to believe that these convergence rates are the best achievable by FOMs in the large-scale case. Specifically, it is known (Nemirovski 1992) that when solving the least-squares problem
$$\min_{x\in\mathbb{R}^n}\{\|Ax - b\|_2 : \|x\|_2 \le R\}$$
by a multiplication-oracle-based algorithm, *with zero optimal value* and spectral norm of A not exceeding L, the number of oracle calls needed to achieve accuracy ϵ in terms of the objective in the large-scale case $n \ge LR/\epsilon$ cannot be smaller than $O(1)LR/\epsilon$. It immediately follows that the outlined $O(1/t^2)$ and $O(1/t)$ rates of convergence are indeed unimprovable when solving large-scale lasso/Dantzig selector problems with $\|\cdot\|_2$ in the role of $\|\cdot\|$. Rigorously speaking, this observation does not imply similar consequences for ℓ_1/nuclear norm minimization. However, common sense, in full accordance with what is known on the subject so far, strongly suggests that the latter problems can hardly be much easier than 'ℓ_2 minimization'. Thus, the algorithms we intend to focus on are indeed most probably optimal in the large-scale case.

The rest of the paper is organized as follows. The constructions and results related to composite minimization in general, and its application to lasso problems in particular, are presented in Section 2. This section also contains a description of *proximal set-ups* underlying all algorithms considered in the paper. Section 3 is devoted to saddle-point algorithms and their applications to Dantzig selector problems. To make the paper self-contained, we present all necessary proofs, independent of whether their reference substitutions are available. All proofs are relegated to the Appendix.

Henceforth we shall operate with elements from direct products of Euclidean spaces $E = E_1 \times \cdots \times E_K$. An element x of E will usually be denoted by $x = [x^1; \dots; x^K]$, where $x^j \in E_j$ is the jth block of x. The inner product of two vectors x, y from a Euclidean space E is denoted by $\langle x, y \rangle$. It is always clear from the context which Euclidean space is meant. The inner product on $E = E_1 \times \cdots \times E_K$ is always inherited from the inner products on the E_j, that is,

$$\langle [x^1; \dots; x^K], [y^1; \dots; y^K] \rangle = \sum_j \langle x^j, y^j \rangle.$$

For any norm $\|\cdot\|_E$ on a Euclidean space E, its conjugate norm is denoted by $\|\cdot\|_{E,*}$:

$$\|x\|_{E,*} = \max_y \{ \langle x, y \rangle : \|y\|_E \le 1 \}. \tag{1.2}$$

For a linear map $x \mapsto Ax : E \to F$, where E and F are Euclidean spaces, the conjugate map $y \mapsto A^*y : F \to E$ is given by the identity

$$\langle y, Ax \rangle = \langle A^*y, x \rangle, \quad x \in E, \, y \in F.$$

Given norms $\|\cdot\|_E$, $\|\cdot\|_F$ on Euclidean spaces E and F, the induced norm of a linear map $x \mapsto Ax : E \to F$ is defined by

$$\|A\|_{\|\cdot\|_E, \|\cdot\|_F} = \max_x \{ \|Ax\|_F : \|x\|_E \le 1 \}. \tag{1.3}$$

Note the identity

$$\|A\|_{\|\cdot\|_E, \|\cdot\|_F} = \|A^*\|_{\|\cdot\|_{F,*}, \|\cdot\|_{E,*}}.$$

2. Composite minimization and lasso problems

2.1. Preliminaries: proximal set-up

The algorithms we are about to consider are of proximal point type. It makes sense to start by describing the *proximal set-up* underlying the methods of this family. Let Z be a closed convex domain in Euclidean space E. A *proximal set-up* for Z is given by a norm $\|\cdot\|_E$ on E (not necessarily the Euclidean one), and a *distance-generating function* (DGF) $\omega(z) : Z \to R$, with the following properties.

- $\omega(\cdot)$ is a continuous convex function on Z.
- $\omega(\cdot)$ admits a selection of subgradients which is continuous on $Z^o = \{z \in Z : \partial\omega(z) \neq \emptyset\}$: there exists a continuous vector field $\omega'(z) : Z^o \to E$ such that $\omega'(z) \in \partial\omega(z)$ for all $z \in Z^o$.
- $\omega(\cdot)$ is strongly convex with modulus 1, with respect to $\|\cdot\|_E$, that is,
$$\langle \omega'(z) - \omega'(z'), z - z' \rangle \ge \|z - z'\|_E^2, \quad \text{for all } z, z' \in Z^o.$$

The proximal set-up gives rise to several concepts heavily exploited in the associated proximal-point algorithms, as follows.

- The *ω-centre* $z_\omega = \arg\min_{z\in Z}\omega(z) \in Z^o$.
- The *Bregman distance* $V_z(w)$, defined for $z \in Z^o$, $w \in Z$ by the relation

$$V_z(w) = \omega(w) - \omega(z) - \langle \omega'(z), w - z\rangle \geq \frac{1}{2}\|w - z\|_E^2, \qquad (2.1)$$

  where the final inequality is due to the strong convexity of $\omega(\cdot)$.
- The *prox-mapping* $\text{Prox}_z(\xi) : E \to Z^o$, parametrized by the *prox-centre* $z \in Z^o$ and defined by

$$\text{Prox}_z(\xi) = \arg\min_{w\in Z}\big[\langle \xi, w\rangle + V_z(w)\big] = \arg\min_{w\in Z}\big[\langle \xi - \omega'(z), w\rangle + \omega(w)\big].$$

  Perhaps the most important property of this mapping is the inequality

$$\langle \xi, z_+ - w\rangle \leq V_z(w) - V_{z_+}(w) - V_z(z_+), \qquad (2.2)$$

  for all $w \in Z$, where $z \in Z^o$, $\xi \in E$ and $z_+ = \text{Prox}_z(\xi)$.
- The *ω-radius*. Let Z' be a *bounded* subset of Z containing z_ω. The ω-radius of Z' is the quantity

$$\Omega[Z'] = \sqrt{2\Big[\max_{z\in Z'}\omega(z) - \min_{z\in Z}\omega(z)\Big]}. \qquad (2.3)$$

  The name stems from the immediate observation that

$$\|z - z_\omega\|_E \leq \sqrt{2V_{z_\omega}(z)} \leq \Omega[Z'], \qquad (2.4)$$

  for all $z \in Z'$. Indeed, we clearly have $\text{Prox}_{z_\omega}(\omega'(z_\omega)) = z_\omega$, whence by (2.2), $\langle \omega'(z_\omega), z - z_\omega\rangle \geq 0$ for all $z \in Z$, and thus

$$\frac{1}{2}\|z - z_\omega\|_E^2 \leq V_{z_\omega}(z) \leq \omega(z) - \omega(z_\omega),$$

  for all $z \in Z$.

2.1.1. Some useful proximal set-ups

Our selection of proximal set-ups is primarily motivated by the desire to obtain the currently best known complexity estimates in the applications we are interested in (for more details on this subject, see Section A.6 in the Appendix). Note that this goal can be achieved not only with the set-ups we intend to present but with other set-ups too. For the sake of brevity we do not discuss these alternatives, referring the interested reader to Juditsky and Nemirovski (2011*b*) (where one should ignore a completely incorrect version of the ℓ_1/ℓ_2 set-up). The structure of our presentation is as follows. We consider a Euclidean space E and a norm $\|\cdot\|_E$ on E. For every such normed space we suggest a 'good' DGF for two 'canonical' sets Z, namely $Z_E(1) = \{z \in E : \|z\|_E \leq 1\}$, and $Z = E$ (we will see that these are exactly the cases we are primarily interested in).

It turns out that all our needs in this paper can be covered just by three proximal set-ups.

The ℓ_1/ℓ_2 norm. Let $E = \mathbb{R}^{k_1 \times \cdots \times k_n}$, so that $z \in E$ is a block vector: $z = [z^1; \ldots; z^n]$ with blocks $z^j \in \mathbb{R}^{k_j}$. Let us equip E with the ℓ_1/ℓ_2 norm $\|z = [z^1; \ldots; z^n]\|_E = \sum_{j=1}^n \|z^j\|_2$. Note that this situation covers, as extreme cases,

- the plain ℓ_2 norm on $\mathbb{R}^k$, which occurs when $n = 1$,
- the plain ℓ_1 norm on $\mathbb{R}^n$, which occurs when $k_1 = \cdots = k_n = 1$.

Theorem 2.1. Given E and $\|\cdot\|_E$ as specified, let

$$Z = Z_E(1) := \{z \in E : \|z\|_E \le 1\}.$$

Then the function $\omega : E \to \mathbb{R}$ defined by

$$\omega(z) = \frac{1}{p\gamma} \sum_{j=1}^n \|z^j\|_2^p, \tag{2.5}$$

$$\text{where } p = \begin{cases} 2, & n \le 2, \\ 1 + \frac{1}{\ln n}, & n \ge 3, \end{cases} \quad \text{and } \gamma = \begin{cases} 1, & n = 1, \\ 1/2, & n = 2, \\ \frac{1}{\mathrm{e}\ln(n)}, & n > 2, \end{cases}$$

is a DGF for Z compatible with $\|\cdot\|_E$, and

$$\Omega[Z] \le \sqrt{2\mathrm{e}\ln(n+1)}. \tag{2.6}$$

Corollary 2.2. Let E and $\|\cdot\|_E$ be as in Theorem 2.1. Then the function $\widehat{\omega} : E \to \mathbb{R}$, given by

$$\widehat{\omega}(z) = \frac{n^{(p-1)(2-p)/p}}{2\gamma} \left[\sum_{j=1}^n \|z^j\|_2^p \right]^{\frac{2}{p}}, \tag{2.7}$$

with p, γ given by (2.5), is a DGF for $Z = E$ compatible with $\|\cdot\|_E$, and

$$\Omega[\{z \in E : \|z\|_E \le R\}] \le O(1)\sqrt{\ln(n+1)}R, \quad \text{for all } R > 0. \tag{2.8}$$

The Euclidean set-up. In the case of $n = 1$ the above construction yields $\|\cdot\|_E = \|\cdot\|_2$, $\omega(z) = \frac{1}{2}z^T z$ for both $Z = Z_E(1)$ and $Z = E$. In fact, for every Euclidean space E, the Euclidean norm $\|z\|_E = \|z\|_2 := \sqrt{\langle z, z\rangle}$ and DGF $\omega(z) = \frac{1}{2}\langle z, z\rangle$ yield a proximal set-up for E. Restricting $\omega(\cdot)$ to a closed convex domain $Z \subset E$, we get a proximal Euclidean set-up for Z.

The nuclear norm. Let E be the Euclidean space $\mathbb{R}^{\mu \times \nu}$ of $\mu \times \nu$ matrices equipped with the Frobenius inner product. We assume without loss of generality that $\mu \le \nu$. Let $\sigma(z) = [\sigma_1(z); \ldots; \sigma_\mu(z)]$ be the vector of singular values of $z \in E$, taken in non-ascending order. Let $\|\cdot\|_E$ be the nuclear norm $\|\cdot\|_{\text{nuc}}(z) = \|\sigma(z)\|_1$ on E.

Theorem 2.3. With E and $\|\cdot\|_E$ as specified, let

$$Z = Z_E(1) = \{z \in E : \|z\|_E \leq 1\}.$$

Then the function $\omega : Z \to \mathbb{R}$ defined by

$$\omega(z) = \frac{4\sqrt{\mathrm{e}}\ln(2\mu)}{2^q(1+q)}\sum_{i=1}^{\mu}\sigma_i^{1+q}(z), \quad \text{where } q = \frac{1}{2\ln(2\mu)}, \tag{2.9}$$

is a DGF for Z compatible with $\|\cdot\|_E$, and

$$\Omega[Z] \leq 2\sqrt{2\sqrt{\mathrm{e}}\ln(2\mu)} \leq 4\sqrt{\ln(2\mu)}. \tag{2.10}$$

Corollary 2.4. Let E, μ, ν, $\|\cdot\|_E$ be as in Theorem 2.3. Then the function $\hat{\omega} : E \to \mathbb{R}$, given by

$$\widehat{\omega}(z) = 2\mathrm{e}\ln(2\mu)\left[\sum_{j=1}^{\mu}\sigma_j^{1+q}(z)\right]^{\frac{2}{1+q}}, \quad \text{where } q = \frac{1}{2\ln(2\mu)}, \tag{2.11}$$

is a DGF for $Z = E$ compatible with $\|\cdot\|_E$, and

$$\Omega[\{z \in E : \|z\|_E \leq R\}] \leq O(1)\sqrt{\ln(2\mu)}R, \quad \text{for all } R > 0. \tag{2.12}$$

Computational issues. For all the following algorithms, an iteration requires $O(1)$ computations of the prox-mappings. Therefore, except for the complexity bounds of the associated algorithms (*i.e.*, bounds expressing the decay of the error of approximate solutions as a function of the number of iterations), a particular proximal set-up affects the complexity of an iteration. Let us now look at the computational cost of a single computation of this type for the outlined set-ups and our 'canonical' Z. Observe that computing the value of prox-mapping at a given input essentially reduces to finding, for a given $\xi \in E$, the point z_+ defined by

$$z_+ = \arg\min_{z \in Z}\{\omega(z) + \langle \xi, z\rangle\}. \tag{2.13}$$

(A) When $\|[z^1; \ldots; z^n]\|_E = \sum_j \|z^j\|_2$ is the ℓ_1/ℓ_2 norm on $E = \mathbb{R}^{k_1} \times \cdots \times \mathbb{R}^{k_n}$ and the DGF is either (2.5), or (2.7), then solving (2.13) is easy. Indeed, it is immediately seen that the blocks z_+^j in z_+ are multiples, with non-positive coefficients, of the blocks ξ^j in ξ. This reduces finding z_+ to solving either a problem of the form

$$\min_{t \in \mathbb{R}^n}\left\{\frac{1}{p}\sum_{j=1}^{n} t_j^p - \alpha_j t_j : t \geq 0, \sum_{j=1}^{n} t_j \leq 1\right\} \tag{2.14}$$

($Z = Z_E(1)$ and DGF (2.5)), or a problem of the form

$$\min_{t\in\mathbb{R}^n}\left\{\frac{1}{2}\left(\sum_{j=1}^{n} t_j^p\right)^{2/p} - \sum_{j=1}^{n}\alpha_j t_j : t \geq 0\right\} \tag{2.15}$$

($Z = E$ and DGF (2.7)). In both cases, $p \in (1, 2]$ and $\alpha_j \geq 0$. Now, (2.14) is a simple convex problem with a separable objective function and one separable linear constraint. As such, it can be solved to machine accuracy in just $O(n)$ arithmetic operations.[6] Problem (2.15) is even easier. It admits the closed-form solution $t = \|\alpha\|_q^{(p-2)/(p-1)}\tau$, where

$$\tau = \left[\alpha_1^{1/(p-1)};\ldots;\alpha_n^{1/(p-1)}\right] \quad \text{and} \quad q = \frac{p}{p-1}.$$

As a result, in the cases under consideration the complexity of computing the prox-mapping is linear in $\dim E$.

(B) When $\|\cdot\|_E$ is the nuclear norm on $E = \mathbb{R}^{\mu\times\nu}$ and the DGF $\omega(\cdot)$ is given by (2.9) or by (2.11), solving (2.13) essentially reduces to computing the singular value decomposition (SVD) $\xi = UDV^T$ of the matrix ξ. The main observation is that, with our DGFs, z_+ is of the form $z_+ = UGV^T$, where G is a diagonal matrix with non-positive diagonal entries $-t_j$, $1 \leq j \leq \mu$.[7] Computing the t_j clearly reduces to solving a problem of the form (2.14) when $Z = Z_E(1)$, or (2.15) when $Z = E$, with μ in the role of n and $\sigma_j(\xi)$ in the role of α_j. Thus, in the case of the nuclear norm, our set-up remains practical in the range of sizes μ, ν allowing for reasonably fast SVD.[8]

With the Euclidean proximal set-up, prox-mapping reduces to the metric projection onto Z. In situations (A) and (B), the computational cost of the Euclidean prox-mapping is the same as for the set-ups we have just considered.

[6] Indeed, for $\sum_j \alpha_j^{1/(p-1)}$, the optimal solution to (2.14) is given by $t_j = \alpha_j^{1/(p-1)}$. Otherwise it is $t_j = t_j(\lambda^*)$, where $t_j(\lambda) = \max[\alpha_j - \lambda, 0]^{1/(p-1)}$ and λ_* is the unique positive root of the equation $\sum_j t_j(\lambda) = 1$. This root can be rapidly approximated to high accuracy, *e.g.*, by bisection.

[7] Indeed, due to rotational invariance of $\omega(\cdot)$ ($\omega(PzQ^T) = \omega(z)$ for orthogonal P, Q), it suffices to verify that if ξ is diagonal then so is z_+. Assuming ξ to be diagonal, both the feasible set and the objective of (2.13) are invariant with respect to the transformations $z \mapsto \mathrm{Diag}\{\epsilon_1,\ldots,\epsilon_\mu\}z\mathrm{Diag}\{\epsilon_1,\ldots,\epsilon_\nu\}$ with $\epsilon_i = \pm 1$. Since problem (2.13) is convex with unique optimal solution z_+ (recall that $\omega(\cdot)$ is strongly convex), this solution is invariant with respect to the outlined transformations, implying that z_+ is indeed diagonal.

[8] With the existing hardware and software and $\mu \approx \nu$, μ, ν can be in the range of a few thousand.

Remark 2.5. It can be seen immediately that when a DGF $\omega(\cdot)$ for $Z \subset E$ is compatible with a given norm $\|\cdot\|_E$ on E, and $Z' \subset Z$ is a closed convex set intersecting the relative interior of Z, then the restriction of $\omega(\cdot)$ onto Z' is a DGF for the latter set, still compatible with $\|\cdot\|_E$. With this in mind, the particular DGFs we have presented give rise to DGFs compatible with $\|\cdot\|_E$, for *all* closed convex and non-empty sets $Z' \subset Z$, not only for our 'canonical' unit $\|\cdot\|_E$-ball and all of E. An immediate question here is: When does this conversion not increase the computational complexity of the prox-mapping too much? Here are some good examples of this type associated with our 'basic' proximal set-ups.

(1) Let $E = \mathbb{R}^{k_1} \times \cdots \times \mathbb{R}^{k_n}$, $\|z\|_E = \sum_j \|z^j\|_2$. If $Z = Z_E(1)$ and the DGF for Z is (2.5), the computational complexity of prox-mapping remains basically intact when Z' is obtained from Z by imposing upper bounds $\|z^j\|_2 \leq r_j$ on some of the blocks z^j of $z \in Z'$, and, further, imposing restrictions on signs of some of the entries in z. This also holds when $Z = E$ and the DGF for Z is (2.7).

(2) Let $E = \mathbb{R}^{\mu\times\nu}$, $\|\cdot\|_E = \|\cdot\|_{\rm nuc}$, and let $\widehat{E}$ be the subspace of E comprising all block-diagonal matrices $z \in E$ with a prescribed number n of diagonal blocks $z^j \in \mathbb{R}^{\mu_j\times\nu_j}$, $1 \leq j \leq n$, with some prescribed μ_j, ν_j. Restricting the DGF (2.11) to $\widehat{E}$ and specifying $\|\cdot\|_{\widehat{E}}$ to be the nuclear norm on $\widehat{E}$, we obtain a proximal set-up for $Z' = \widehat{E}$, and do not increase the complexity of the prox-mapping,[9] since both the SVD of $z \in \widehat{E}$ and computing this SVD are decomposable along the blocks z^j. We can further restrict the blocks z^j of $z \in \widehat{E}$ by (a) imposing upper bounds on the spectral norms of some of the blocks, or (b) requiring some of the blocks z^j to be symmetric (and perhaps also positive semidefinite).[10] All of this can be done without greatly complicating the prox-mapping. We can proceed in the same fashion with the DGF (2.9), getting the proximal set-ups for the nuclear norm ball in the space of block-diagonal matrices and for the sets obtained from this ball by operations (a) and (b). Again, this does not greatly complicate the prox-mapping. In particular, we can thereby obtain reasonably good proximal set-ups for the space of symmetric matrices equipped with the trace norm, and for the positive semidefinite (PSD) cone in this space, and for spectrahedra, *i.e.*, intersections of the PSD-cone with trace norm balls centred at the origin.

Remark 2.5 is fully applicable to Euclidean set-ups as well.

[9] In fact it decreases, and quite significantly so when $\mu_j \ll \mu$ or $\nu_j \ll \nu$.
[10] Of course, (b) relates only to square blocks z^j.

Remark 2.6. A discussion of proximal set-ups cannot be complete without mentioning the *entropy* DGF:

$$\eta(z) = \sum_{i=1}^{n} z_i \ln z_i, \quad z_i \geq 0,\ i = 1,\ldots,n.$$

Its absence from our discussion is explained by the unconstrained nature of the main problems (1.1a,b). Indeed, the natural feasible set for this function is the simplex

$$\Delta_n = \left\{ z \in R^n_+ : \sum_{i=1}^{n} z_i = 1 \right\}.$$

On this set, $\eta(\cdot)$ is a DGF for the $\|\cdot\|_1$ norm with $\Omega[\Delta_n] = \sqrt{2\ln n}$. All the results of this paper can be easily adapted to the entropy DGF provided that we have additional simplex constraints in problems (1.1a,b).

2.2. Composite minimization via gradient methods

2.2.1. Problem formulation

The general problem of composite minimization is as follows. We want to solve the convex program

$$\min_{x\in Z} \left[\phi(x) \stackrel{\text{def}}{=} f(x) + \Psi(x)\right], \tag{2.16}$$

where Z is a closed convex set in a Euclidean space E equipped with a norm $\|\cdot\|_E$ (not necessarily the Euclidean one), $\Psi(z) : Z \to \mathbb{R} \cup \{+\infty\}$ is a lower-semicontinuous convex function which is finite on the relative interior of Z, and $f : Z \to \mathbb{R}$ is a convex function with Lipschitz-continuous gradient, that is,

$$\|\nabla f(x) - \nabla f(y)\|_{E,*} \leq L_f \|x - y\|_E, \quad x, y \in Z. \tag{2.17}$$

Recall that $\|\cdot\|_{E,*}$ is the norm conjugate to $\|\cdot\|_E$ (see (1.2)).

Condition (2.17) ensures that

$$f(y) - f(x) - \langle \nabla f(x), y - x\rangle \leq \frac{L_f}{2}\|x - y\|_E^2, \quad \text{for all } x, y \in Z,$$

whence by (2.1)

$$f(y) - f(x) - \langle \nabla f(x), y - x\rangle \leq L_f V_x(y), \quad \text{for all } x \in Z^o,\ y \in Z, \tag{2.18a}$$

$$f(y) - f(x) - \langle \nabla f(x), y - x\rangle \leq L_f V_y(x), \quad \text{for all } y \in Z^o,\ x \in Z. \tag{2.18b}$$

To avoid unnecessary complications, we assume the constant $L_f > 0$ is known. At the end of Section 2 (see Algorithm 2.15) we discuss an efficient strategy for using its updated approximations.

From now on, we assume that Z is equipped with a proximal set-up comprising our norm $\|\cdot\|_E$ on E and a DGF $\omega(z)$ for Z compatible with this norm. We associate with this function the *composite* prox-mapping

$$\mathrm{Prox}_{\Psi,z}(\xi) = \arg\min_{w\in Z}[\langle \xi, w\rangle + L_f V_z(w) + \Psi(w)],$$

where $z \in Z^o = \{z \in Z : \partial\omega(z) \neq \emptyset\}$ and $\xi \in E$. Under our assumptions this mapping is clearly well defined, since $Z \neq \emptyset$ is closed and convex, $\omega(\cdot)$ is continuous and strongly convex, with modulus 1, with respect to $\|\cdot\|_E$, on Z, while Ψ is convex lower-semicontinuous and finite on int Z. One can easily check (see Lemma A.1) that it takes its values in Z^o. To be useful in practical methods, the composite prox-mapping should be easy to compute (either by an explicit formula, or by a cheap computational procedure); we assume this from now on. Finally, we assume that (2.16) is solvable.

2.2.2. Gradient algorithms for composite minimization

In this subsection we present several methods for minimizing composite functions. For the Euclidean DGF these methods were suggested in Nesterov (2007*b*). The fast gradient method with Bregman distances was proposed in Nesterov (2007*b*). The most general form of these methods with stochastic oracle was justified by Devolder (2011). In the Appendix we give simplified versions of the corresponding proofs.

Preliminaries. Analysis of the subsequent composite minimization methods is based on the following simple observation.

Lemma 2.7. Let $f : Z \to \mathbb{R} \cup \{+\infty\}$ be a convex function such that the difference $f - L\omega$ for some $L > 0$ is convex lower-semicontinuous on Z and finite on the relative interior of Z. Let $\bar{w} = \arg\min_{w\in Z} f(w)$. Then $\bar{w} \in Z^o$, and for all $w \in Z$ we have

$$f(w) \geq f(\bar{w}) + LV_{\bar{w}}(w). \tag{2.19}$$

The primal gradient method. Inequality (2.19) justifies the following algorithm.

Algorithm 2.8 (primal gradient method).

Choose $x_0 \in Z^o$

For $t \geq 0$ **do**

$x_{t+1} = \mathrm{Prox}_{\Psi,x_t}(\nabla f(x_t))$ ◇

Let x^* denote an optimal solution to problem (2.16).

Theorem 2.9. Let the sequence $\{x_t\}_{t\geq 0}$ be constructed by Algorithm 2.8. Then $\phi(x_{t+1}) \leq \phi(x_t)$, and for every $T \geq 1$ we have

$$\sum_{t=1}^{T}(\phi(x_t) - \phi(x^*)) \leq L_f V_{x_0}(x^*). \tag{2.20}$$

As a result, for every $t \geq 1$, setting $x^t = \frac{1}{t}\sum_{\tau=1}^{t} x_\tau$, we have

$$\max\bigl[\phi(x_t), \phi(x^t)\bigr] - \phi(x^*) \leq \frac{L_f V_{x_0}(x^*)}{t} \tag{2.21}$$

and $x_t, x^t \in Z$.

The dual gradient method. In contrast to the primal gradient method (Algorithm 2.8), which ensures a sequence of monotonically decreasing function values $\{\phi(x_t)\}_{t\geq 0}$, our next method is not monotone.

Algorithm 2.10 (dual gradient method).

Set $s_0 = 0 \in E$

For $t \geq 0$ **do**

$x_t = \mathrm{Prox}_{t\Psi, z_\omega}(s_t)$

$y_t = \mathrm{Prox}_{\Psi, x_t}(\nabla f(x_t))$

$s_{t+1} = s_t + \nabla f(x_t)$ ◇

Theorem 2.11. Let $\{x_t\}_{t\geq 0}$ be constructed by Algorithm 2.10. Then $x_t \in Z^o$, so that the auxiliary points y_t, $t \geq 0$, are well defined and belong to Z^o. For every $T \geq 1$ we have

$$\sum_{t=0}^{T-1}(\phi(y_t) - \phi(x^*)) \leq L_f V_{z_\omega}(x^*). \tag{2.22}$$

As a result, letting $\bar{y}^t$ denote the best of the points $y_0, \ldots, y_{t-1}$, in terms of the values of ϕ, and letting $y^t = \frac{1}{t}\sum_{s=0}^{t-1} y_s$ be the average of these points, we have for all $t \geq 1$

$$\max\bigl[\phi(\bar{y}^t), \phi(y^t)\bigr] - \phi(x^*) \leq \frac{L_f V_{z_\omega}(x^*)}{t} \tag{2.23}$$

and $\bar{y}^t, y^t \in Z$.

Remark 2.12. It can be easily verified that, when redefining the points $\{y_t\}_{t\geq 0}$ in Algorithm 2.10 according to the usual composite gradient mapping, specifically,

$$y_t = \arg\min_{w\in Z}\{f(x_t) + \langle\nabla f(x_t), w - x_t\rangle + \tfrac{1}{2}L_f\|w - x_t\|_E^2 + \Psi(w)\}, \tag{2.24}$$

one preserves efficiency estimates (2.22) and (2.23).

It is interesting that for justification of both Algorithms 2.8 and 2.10 we use only inequality (2.18), which is weaker than the condition (2.17).

Fast composite gradient minimization. The methods we have described so far solve the composite minimization problem (2.16) with the $O(1/t)$ efficiency estimate. We are about to describe a 'fast' method for the same problem, with the $O(1/t^2)$ efficiency estimate.

The algorithm is as follows.

Algorithm 2.13 (fast gradient method).

Initialization $\psi_0(w) = L_f V_{z_\omega}(w),\ y_0 = z_\omega$

For $t \geq 0$ **do**

(a) Compute $z_t = \arg\min_{w\in Z} \psi_t(w)$. Set $\tau_t = \dfrac{2(t+2)}{(t+1)(t+4)}$.

(b) Choose $x_{t+1} = \tau_t z_t + (1-\tau_t) y_t$.

(c) Compute $\hat{x}_{t+1} = \arg\min_{w\in Z} \left[\langle \nabla f(x_{t+1}), w\rangle + \Psi(w) + \dfrac{2L_f}{t+2} V_{z_t}(w) \right]$.

(d) Update $y_{t+1} = \tau_t \hat{x}_{t+1} + (1-\tau_t) y_t$,

$$\psi_{t+1}(w) = \psi_t(w) + \frac{t+2}{2}\left[f(x_{t+1}) + \langle \nabla f(x_{t+1}), w - x_{t+1}\rangle + \Psi(w)\right].$$

◇

Note that $\psi_t(w) = \langle \xi_t, w\rangle + \gamma_t \Psi(w) + L_f V_{z_\omega}(w)$ with $\gamma_t > 0$, that is, operations (a) and (c) reduce to computing the values of composite prox-mappings with positive multiples of Ψ in the role of Ψ.

Theorem 2.14. Let the sequence $\{y_t\}_{t\geq 0}$ be generated by Algorithm 2.13. Then, for all $t \geq 1$ we have $y_t \in Z$ and

$$\phi(y_t) - \phi(x^*) \leq \frac{4L_f V_{z_\omega}(x^*)}{t(t+3)}. \tag{2.25}$$

2.3. Composite minimization and lasso-type problems

The algorithms presented in Section 2.2.2 are well suited to solving lasso-type problems (1.1a) associated with the norms $\|\cdot\|$ of our primary interest. Indeed, (1.1a) is simply problem (2.16) with

$$\Psi(\cdot) \equiv \lambda\|\cdot\|, \quad f(x) \equiv \|Ax - b\|_2^2, \tag{2.26}$$

and $Z = E$. It is assumed from now on that the starting point x_0 in the primal gradient method is chosen as the ω-centre z_ω of $Z = E$, which makes the efficiency estimates (2.21) and (2.23) of the primal and the dual gradient methods identical to each other.

Lasso with the ℓ_1/ℓ_2 norm. The first case we are interested in is that of $E = \mathbb{R}^{k_1} \times \cdots \times \mathbb{R}^{k_n}$ and the ℓ_1/ℓ_2 norm

$$\|[z^1;\ldots;z^n]\| = \sum_{j=1}^{n} \|z^j\|_2.$$

Note that this covers the cases of plain ℓ_1 and plain ℓ_2 norms (see what happens when $k_j = 1$ for all j, respectively, when $n = 1$).

(A) When using $\|\cdot\|_E = \|\cdot\|$ and the DGF $\omega(\cdot)$ given by (2.7), the three gradient methods from Section 2.2.2 ensure the efficiency estimates[11]

$$\begin{aligned} &[\lambda\|x^t\| + \|Ax^t - b\|_2^2] - \min_{x\in Z}[\lambda\|x\| + \|Ax - b\|_2^2] \qquad (2.27)\\ &\qquad \le O(1)\frac{\ln(n+1)\|x^*\|^2\|A\|^2_{\|\cdot\|,\|\cdot\|_2}}{t^\kappa},\\ &\text{where } \kappa = \begin{cases} 1, & \text{primal and dual gradient methods,}\\ 2, & \text{fast gradient method,}\end{cases}\\ &\text{and } \|A\|_{\|\cdot\|,\|\cdot\|_2} = \max_{x:\|x\|\le 1}\|Ax\|_2,\end{aligned}$$

where x^t is an approximate solution built after the first t steps, and x^* is an optimal solution to problem (2.16).

(B) An alternative is to use the Euclidean proximal set-up,

$$\|x\|_E = \|x\|_2 := \sqrt{\langle x, x\rangle}, \quad \omega(x) = \frac{1}{2}\langle x, x\rangle,$$

which results in

$$\begin{aligned} &[\lambda\|x^t\| + \|Ax^t - b\|_2^2] - \min_{x\in Z}[\lambda\|x\| + \|Ax - b\|_2^2] \le O(1)\frac{\|x^*\|_2^2\|A\|^2_{\|\cdot\|_2,\|\cdot\|_2}}{t^\kappa},\\ &\qquad \text{where } \|A\|_{\|\cdot\|_2,\|\cdot\|_2} = \max_{x:\|x\|_2\le 1}\|Ax\|_2, \qquad (2.28)\end{aligned}$$

with the same κ as in (2.27).

Except for the efficiency estimates, the question of primary importance is the computational effort per iteration. From the description of the algorithms, in all three of them this effort is dominated by the need to compute $O(1)$ values of $\nabla f(\cdot)$, which requires $O(1)$ matrix–vector multiplications $x \mapsto Ax \mapsto A^*(Ax - b)$, and computation of the values of $O(1)$ composite prox-mappings. For both set-ups we have considered so far, the latter task is

[11] See (2.21), (2.23) and (2.25) and take into account that we have $L_f = 2\|A\|^2_{\|\cdot\|,\|\cdot\|_2}$.

easy. Let us show that it requires $O(\dim E)$ arithmetic operations. Indeed, in our present situation, and for both proximal set-ups in question, computing composite prox-mapping reduces to solving the following analogue of problem (2.13):

$$z_+ = \arg\min_{z\in Z}\left\{\omega(z) + \beta\sum_j \|z^j\|_2 + \sum_j \langle z^j, \xi^j\rangle\right\}, \quad \text{for } \beta \geq 0. \tag{2.29}$$

Now the two DGFs we are considering are of the form $\gamma(\sum_j \|z^j\|_2^p)^{2/p}$, where $\gamma > 0$ and $p \in (1, 2]$. With this in mind, it is clear that, at the minimum, z^j should be of the form $z^j = -t_j\xi^j/\|\xi^j\|_2$, $t_j \geq 0$. Thus (2.29) is reduced to the following problem:

$$\min_t\left\{\gamma\left(\sum_j t_j^p\right)^{2/p} + \sum_j [\beta - \|\xi^j\|_2]t_j : t \geq 0\right\}. \tag{2.30}$$

It can be seen immediately that, at the minimum, $t_j = 0$ for every j such that $\beta \geq \|\xi^j\|_2$. Eliminating these j, we end up with a problem of type (2.15) which, as we have seen in Section 2.1.1, admits an easily computable closed-form solution.

Lasso with the TV norm. The total variation on the space of $\mu \times \nu$ images x (normalized to have $x_{1,1} = 0$) can be viewed as follows. The space of discrete gradients of $\mu \times \nu$ images is a subspace E of the space

$$\mathbb{R}^N = \mathbb{R}^{(\mu-1)\times\nu} \times \mathbb{R}^{\mu\times(\nu-1)}$$

of first-order finite differences of an image $[x_{ij}]$ taken along the index $i \leq \mu$ ($\mathbb{R}^{(\mu-1)\times\nu}$) and along the index $j \leq \nu$ ($\mathbb{R}^{\mu\times(\nu-1)}$). The space E is cut from $\mathbb{R}^N$ by a system of homogeneous linear equations expressing the potentiality of the discrete gradient field of an image. Thus, the TV norm on E is a restriction of the standard ℓ_1 norm on $\mathbb{R}^N$ onto $E \subset \mathbb{R}^N$. This allows us to equip $Z = E$ with a DGF $\omega(\cdot)$ (namely, the restriction to E of the DGF (2.7) associated with the $\|\cdot\|_1$ on $\mathbb{R}^N$), which leads to a reasonably good iteration complexity bound. As an alternative, we could use the Euclidean proximal set-up for E.

The difficulty, however, is that we now need to take into account the linear constraints cutting E from $\mathbb{R}^N$. Hence both the plain and the composite prox-mappings associated with the outlined proximal set-ups are much more computationally demanding than in the case of lasso with the ℓ_1/ℓ_2 norm. As a result, the 'straightforward' TVversions of the methods we have presented are hardly advisable.

Lasso with the nuclear norm. In the second important case, we have $E = \mathbb{R}^{\mu\times\nu}$, $\|\cdot\| = \|\cdot\|_{\text{nuc}}$, and $Z = E$.

(A) Using our three basic methods and the proximal set-up with $\|\cdot\|_E = \|\cdot\|$ and with $\omega(\cdot)$ given by (2.11), we come to the efficiency estimates

$$[\lambda\|x^t\|_{\rm nuc} + \|Ax^t - b\|_2^2] - \min_{x\in Z}[\lambda\|x\|_{\rm nuc} + \|Ax - b\|_2^2] \le O(1)\frac{\ln(\mu+1)\|x^*\|_{\rm nuc}^2\|A\|^2_{\|\cdot\|_{\rm nuc},\|\cdot\|_2}}{t^\kappa}, \tag{2.31}$$

$$\text{where } \kappa = \begin{cases} 1, & \text{primal and dual gradient methods,} \\ 2, & \text{fast gradient method,} \end{cases}$$

$$\text{and } \|A\|_{\|\cdot\|_{\rm nuc},\|\cdot\|_2} = \max_{x:\|x\|_{\rm nuc}\le 1} \|Ax\|_2,$$

where x^t is the approximate solution obtained after the first t iterations, and x^* is the optimal solution.

(B) An alternative is to use the Euclidean set-up,

$$\|x\|_E = \|x\|_{\rm Fro} := \sqrt{{\rm Tr}(x^Tx)}, \quad \omega(x) = \frac{1}{2}\langle x, x\rangle = \frac{1}{2}{\rm Tr}(x^Tx).$$

The corresponding efficiency estimate is

$$[\lambda\|x^t\|_{\rm nuc} + \|Ax^t - b\|_2^2] - \min_{x\in Z}[\lambda\|x\|_{\rm nuc} + \|Ax - b\|_2^2] \le O(1)\frac{\|x^*\|_{\rm Fro}^2\|A\|^2_{\|\cdot\|_{\rm Fro},\|\cdot\|_2}}{t^\kappa}, \tag{2.32}$$

$$\text{where } \|A\|_{\|\cdot\|_{\rm Fro},\|\cdot\|_2} = \max_{x:\|x\|_{\rm Fro}\le 1} \|Ax\|_2.$$

The computational effort per iteration is now dominated by the need to carry out $O(1)$ matrix–vector multiplications $x \mapsto Ax \mapsto A^*(Ax - b)$ along with $O(1)$ SVDs of $\mu \times \nu$ matrices. Arguments virtually identical to those used in (B) at the end of Section 2.1.1 show that computing composite prox-mapping, modulo a single SVD, for both our proximal set-ups reduces to solving a problem of the form (2.30). Thus it is easy.

It should be pointed out that the efficiency estimates (2.27), (2.28) and (2.31) are nearly independent of the sizes of corresponding lasso problems.

Finally, observe that Remark 2.5 is fully applicable to composite prox-mappings associated with the DGFs and corresponding norms $\|\cdot\|_E$. Thus the methods we have developed can be applied to lasso problems on more general domains than $Z \equiv E$, with the efficiency estimates and complexity of iterations being preserved.

Updating the Lipschitz constant for the gradient. In this subsection we assume that the Lipschitz constant L_f is known. However, it is easy to introduce into the methods a simple updating procedure, which allows us to adjust an estimate to L_f in accordance with the local curvature of the objective function. We present below the corresponding modification of the primal gradient method (Algorithm 2.8).

Algorithm 2.15 (modified primal gradient method).

Choose $x_0 \in Z^o$, $L_0 \leq L_f$.

For $t \geq 0$ **do**

(a) Choose the smallest $i_t \geq 0$ such that, for $L = 2^{i_t} L_t$ and

$$w_t = \arg\min_{w \in Z} \left[\langle \nabla f(x_t), w\rangle + LV_{x_t}(w) + \Psi(w)\right],$$

we have

$$\phi(w_t) \leq f(x_t) + \langle \nabla f(x_t), w_t - x_t\rangle + LV_{x_t}(w_t) + \Psi(w_t).$$

(b) Set $x_{t+1} = w_t$, $L_{t+1} = 0.5 \cdot 2^{i_t} L_t$. ◇

It is easy to show that such a strategy ensures the rate of convergence (*cf.* (2.21))

$$\phi(x_t) - \phi(x^*) \leq \frac{2L_f V_{x_0}(x^*)}{t}, \quad t \geq 1.$$

At the same time, it allows flexible dynamics for estimates L_t. It can be shown that the average number of 'trials' $i_t + 1$ per iteration in such a *backtracking strategy with recourse* is at most two (Nesterov and Polyak 2006). The dual gradient method (Algorithm 2.10) and the fast gradient method (Algorithm 2.13) can be modified in a similar way.

3. Saddle-point algorithms for Dantzig selector problems

The algorithmic scheme we intend to use for Dantzig-selector-type problems (1.1b) originates from Juditsky *et al.* (2013*a*). To allow for non-trivial extensions, we now remove the restriction that the 'model discrepancy' $Ax - b$ be measured in the $\|\cdot\|_p$ norm with $p = 2$ or $p = \infty$, and focus on a slightly more general problem,

$$\text{Opt} = \min_{x \in E} \left\{ \|x\| : \pi(Ax - b) \leq \delta \right\}, \tag{3.1}$$

where $E = E_u$ is a Euclidean space,[12] $\|\cdot\|$ is a given norm on E, $x \mapsto Ax - b$ is a given affine mapping from E_x into another Euclidean space E_v, and $\pi(\cdot)$ is a given norm on E_v. Aside from the structural discrepancy between (3.1) and the associated lasso-type problem

$$\min_x \left\{ \lambda\|x\| + \pi^2(Ax - b) \right\} \tag{3.2}$$

[12] The reason for the notation $E = E_u$, instead of the seemingly more natural $E = E_x$, will become clear in a while.

(this discrepancy disappears when δ and λ are treated as parameters rather than given constants), the major difference (from the optimization viewpoint) between what we intend to do now and what was done in Sections 2.2 and 2.3 stems from the fact that we no longer impose any smoothness restrictions on the norm $\pi(\cdot)$, which makes it impossible to process (3.2) by the composite minimization technique from Section 2. On the other hand, allowing for non-smooth $\pi(\cdot)$ is instrumental when solving Dantzig-selector-type problems with uniform fit.

From now on we make the following assumption.

Standing Assumption. Optimization problem (3.1) is feasible (and therefore solvable) and $\mathrm{Opt} > 0$ (equivalently $\pi(b) > \delta$).

3.1. Strategy

Our strategy when solving (3.1) is motivated by the desire to minimize, to the best of our abilities, the theoretical complexity of the resulting algorithm, and is as follows. Substituting $x = \rho^{-1}u$ with $\rho > 0$ and $u \in E_u$, $\|u\| \leq 1$ and taking into account that $\mathrm{Opt} > 0$, it is immediately seen that (3.1) can be equivalently rewritten as

$$\frac{1}{\mathrm{Opt}} = \rho_* := \max\left\{\rho \geq 0 : \Phi(\rho) := \min_{u\in U} \overbrace{\max\nolimits_{v\in V}\left[\langle v, Au - \rho b\rangle - \rho\delta\right]}^{=\pi(Au-\rho b)-\rho\delta} \leq 0\right\},$$
$$\text{for} \quad U = \{u \in E_u : \|u\| \leq 1\}, \quad V = \{v \in E_v : \pi_*(v) \leq 1\}, \tag{3.3}$$

where $\pi_*(\cdot)$ is the norm on E_v conjugate to $\pi(\cdot)$.

Problem (3.3) (a specific *generalized bilinear saddle-point problem* in the terminology of Juditsky *et al.* 2013*a*) is a univariate problem. As is easily seen, under our Standing Assumption $\Phi(\rho)$ is a Lipschitz-continuous (with constant $2\pi(b)$) *convex* function of $\rho \geq 0$ with exactly two roots $\rho = 0$ and $\rho = \rho_* > 0$. Further, $\Phi(\rho) \to \infty$ as $\rho \to \infty$, and $\lim_{\rho\to\infty} \Phi'(\rho) = \pi(b) - \delta$.

Our goal is to find an ϵ-*solution* to (3.3), defined as a pair ρ_ϵ, u_ϵ satisfying

$$\rho_\epsilon \geq \rho_*, \quad u_\epsilon \in U \quad \text{and} \quad \pi(Au_\epsilon - \rho_\epsilon b) - \rho_\epsilon\delta \leq \rho_\epsilon\epsilon, \tag{3.4}$$

where $\epsilon > 0$ is a given tolerance. As is seen immediately, such a solution gives rise to the solution

$$x_\epsilon = \rho_\epsilon^{-1} u_\epsilon$$

to the problem (3.1) such that

$$\|x_\epsilon\| \leq \mathrm{Opt} \;\&\; \pi(Ax_\epsilon - b) \leq \delta + \epsilon. \tag{3.5}$$

In other words, x_ϵ is a 'super-optimal ϵ-feasible' solution to the problem of interest; from now on, we refer to a solution x_ϵ satisfying (3.5) as an ϵ-solution to (3.1).

Now, (3.3) is just a problem of finding the largest root of a univariate function $\Phi(\cdot)$; the only difficulty is that this function is given implicitly, as the saddle-point value of a parametric bilinear saddle-point problem. Our strategy for solving (3.3) originates from Lemarechal *et al.* (1995) and reduces to applying to (3.3) a Newton-type root-finding procedure, the (approximate) first-order information for the procedure being given by a first-order saddle-point algorithm applied to the auxiliary bilinear saddle-point problems

$$\Phi(\rho_s) = \min_{u\in U}\max_{v\in V}\big[\phi^{\rho_s}(u,v) := \langle v, Au - \rho_s b\rangle - \rho_s\delta\big], \tag{3.6}$$

the corresponding monotone operators being $F^s : Z := U \times V \to E := E_u \times E_v$, given by

$$F^s(u,v) = \tag{3.7}$$
$$[F^s_u(v) := \nabla_u\phi^{\rho_s}(u,v) = A^*v; F^s_v(u) := -\nabla_v\phi^{\rho_s}(u,v) = \rho^s b - Au].$$

Specifically, we work in *stages* $s = 1, 2, \ldots$. At stage s, we have at our disposal a current approximation $\rho_s \geq \rho_*$ to ρ_*, and apply to (3.6) a saddle-point algorithm $\mathcal{B}$, based only on the values of F^s, and thus only on a multiplication oracle computing matrix–vector products $(u,v) \mapsto (A^*v, Au)$. For the time being, the only requirement on the algorithm is that in $t = 1, 2, \ldots$ iterations, with $O(1)$ calls to the multiplication oracle per iteration, it produces approximate solutions $z^t = [u^t; v^t] \in Z$ to (3.6)[13] along with the values $F^s(z^t)$, such that the *duality gap* converges to zero as $t \to \infty$, where

$$\begin{aligned}\mathrm{DualityGap}^s(z^t) &:= \max_{v\in V}\phi^{\rho_s}(u^t,v) - \min_{u\in U}\phi^{\rho_s}(u,v^t) \qquad (3.8)\\ &= \underbrace{[\pi(F^s_v(u^t)) - \rho_s\delta]}_{\Phi_t^+} - \underbrace{[-\|F^s_u(v^t)\|_* - \rho_s\langle v^t, b\rangle - \rho_s\delta]}_{\Phi_t^-(\rho)\big|_{\rho=\rho_s}}\\ &= [\pi(Au^t - \rho_s b) - \rho_s\delta] - [-\|A^*v^t\|_* - \rho_s\langle v^t, b\rangle - \rho_s\delta],\end{aligned}$$

and $\|\cdot\|_*$ is the dual norm to $\|\cdot\|$.

Observe the following.

(I) Given z^t and $F^s(z^t)$, it is easy to compute the quantity Φ_t^+ and the affine function $\Phi_t^-(\cdot)$ (specifically, these computations reduce to single evaluation of $\pi(\cdot)$ and single evaluation of $\|\cdot\|_*$).

(II) The quantity $\Phi_t^+ = \pi(Au^t - \rho_s b) - \rho_s\delta$ is an upper bound on $\Phi(\rho_s)$, while the affine function

$$\Phi_t^-(\rho) = [-\|A^*v^t\|_* - \rho\langle v^t, b\rangle - \rho\delta] = \min_{u\in U}\phi^{\rho}(u, v^t)$$

[13] Of course, z^t depends on s and not only on t. To simplify the notation we suppress the explicit dependence on s, which does no harm when speaking about a particular stage.

underestimates $\Phi(\rho)$ for all $\rho \geq 0$. As a result, the quantities

$$\Phi^{t,+} = \min_{1 \leq \tau \leq t} \Phi_\tau^+, \quad \Phi^{t,-} = \max_{1 \leq \tau \leq t} \Phi_\tau^-(\rho_s)$$

form upper bounds on $\Phi(\rho_s)$ which are non-increasing functions of t (resp. lower bounds non-decreasing with t). Further, because $\rho_s \geq \rho_*$, the affine function $\Phi_t^-(\cdot)$ underestimates $\Phi(\cdot)$, and $\Phi(\rho_*) = 0$, for every t such that $\Phi_t^-(\rho_s) > 0$, the slope of the affine function $\Phi_t^-(\cdot)$ is positive, and the only root r_t of this function belongs to $[\rho_*, \rho_s)$. Finally, we clearly have $\Phi^{t,+} - \Phi^{t,-} \leq \Phi_t^+ - \Phi_t^-(\rho_s) = \mathrm{DualityGap}^s(z^t)$, so the gap $\Phi^{t,+} - \Phi^{t,-}$ goes to zero as $t \to \infty$.

From (II) it follows that eventually one of the following two possibilities will be met.

(A) We arrive at $\Phi^{t,+} \leq \epsilon\rho_s$. Recalling that $\Phi^{t,+} = \pi(Au^\tau - \rho_s b) - \rho_s\delta$ with some $\tau \leq t$ and known $u^\tau \in U$, in the case in question we terminate the solution process and output $\rho_\epsilon = \rho_s$ and $u_\epsilon = u^\tau$. The resulting pair ρ_ϵ, u_ϵ clearly satisfies (3.4).

(B) We arrive at $\Phi^{t,-} > 0$ and $\Phi^{t,+}/\Phi^{t,-} \leq \theta$, where $\theta \in (1, 2)$ is a parameter of our construction. Since $\Phi^{t,-} > 0$, we have at our disposal a non-empty collection R_t of reals $r_\tau \in [\rho_*, \rho_s)$ coming from the iterations $\tau \leq t$ with $\Phi_\tau^-(\rho_s) > 0$. We define ρ_{s+1} as the minimal element of R_t, thus ensuring that $\rho_* \leq \rho_{s+1} < \rho_s$, terminate stage s and pass to stage $s + 1$.

Validity and complexity. We use the following result of Lemarechal *et al.* (1995).

Lemma 3.1. Let the root-finding routine described above, with parameter $\theta \in (1, 2)$, be initialized by some upper bound ρ_1 on ρ_*. Then the routine terminates in at most

$$S = \frac{1}{\ln(2/\theta)} \ln\left(1 + \frac{3\pi(b)\rho_1}{\epsilon\rho_*}\right) + 1 \tag{3.9}$$

stages. Further, suppose that the saddle-point method $\mathcal{B}$ used to solve problems (3.6) ensures that at every stage, for every $t = 1, 2, \ldots$, we have

$$\mathrm{DualityGap}^s(z^t) \leq \varepsilon_t \tag{3.10}$$

for some sequence ε_t (the same sequence for all stages) satisfying $\varepsilon_t \to 0$ as $t \to \infty$. Then the number of iterations at every stage does not exceed

$$t(\epsilon) = \min\left\{t : \varepsilon_t \leq \frac{\theta - 1}{\theta}\epsilon\rho_*\right\}. \tag{3.11}$$

Initialization. Our root finding should start with some upper bound ρ_1 on ρ_*. To get such a bound, note that every feasible solution x to (3.1) clearly satisfies $\pi(Ax) \geq \pi(b) - \delta > 0$, implying that

$$\rho_* = \frac{1}{\text{Opt}} \leq \rho_1 := \frac{\|A\|_{\|\cdot\|,\pi(\cdot)}}{\pi(b) - \delta}, \quad \text{where } \|A\|_{\|\cdot\|,\pi(\cdot)} = \max_x\{\pi(Ax) : \|x\| \leq 1\}. \tag{3.12}$$

3.2. Workhorse: Mirror Prox algorithm

To implement the outlined strategy, we need an efficient first-order algorithm for solving bilinear saddle-point problems. We intend to use to this end the $O(1/t)$-converging *Mirror Prox* (MP) algorithm from Nemirovski (2004). This algorithm is applicable to a convex–concave saddle-point problem

$$\text{SadVal} = \min_{u \in U} \max_{v \in V} \phi(u, v), \tag{3.13}$$

where $U \subset E_u$, $V \subset E_v$ are closed convex subsets of Euclidean spaces, and $\phi : Z := U \times V \to \mathbb{R}$ is a function convex in $u \in U$ and concave in $v \in V$ with Lipschitz-continuous gradient. We associate with (3.13) the vector field $F : Z := X \times Y \to E = E_u \times E_v$, given by

$$F(z = (u, v)) = [F_u(z) = \nabla_u \phi(u, v); F_v(z) = -\nabla_v \phi(u, v)]. \tag{3.14}$$

Recall that (3.13) gives rise to two convex optimization problems:

$$(P) \quad \text{Opt}(P) = \min_{u \in U} \Big[\overline{\phi}(u) := \sup_{v \in V} \phi(u, v)\Big], \tag{3.15a}$$

$$(D) \quad \text{Opt}(D) = \max_{v \in V} \Big[\underline{\phi}(v) := \inf_{u \in U} \phi(u, v)\Big], \tag{3.15b}$$

with $\text{Opt}(P) \geq \text{Opt}(D)$. Further, ϕ possesses a saddle point on $U \times V$ if and only if both (3.15a) and (3.15b) are solvable with equal optimal values, for which we have $\text{Opt}(P) = \text{Opt}(D) = \text{SadVal}$. Moreover, saddle points of ϕ, when they exist, are exactly the pairs $[u_*; v_*]$, where u_*, v_* are optimal solutions to the respective problems (3.15a) and (3.15b). Finally, when U and V are bounded and convex–concave ϕ is continuous (as is the case with the problems (3.6)), ϕ definitely has a saddle point. Assuming that the latter is the case, a natural inaccuracy measure of a candidate solution $[u; v] \in Z$ to (3.13) is given by (*cf.* (3.8))

$$\text{DualityGap}(u, v) = \overline{\phi}(u) - \underline{\phi}(v) = [\overline{\phi}(u) - \text{Opt}(P)] + [\text{Opt}(D) - \underline{\phi}(v)].$$

The Mirror Prox algorithm. The Mirror Prox (MP) algorithm is associated with a proximal set-up $(\|\cdot\|_E, \omega(\cdot))$ for the domain $Z = U \times V \subset E = E_u \times E_v$ (see Section 2.1), and is given by the following recurrence, where $\gamma_t > 0$ are the step sizes.

Algorithm 3.2 (MP algorithm).

Choose $z_1 = z_\omega = \arg\min_Z \omega(\cdot)$

For $0 \le \tau \le t-1$ **do**

$w_\tau = [u_\tau; v_\tau] := \mathrm{Prox}_{z_\tau}(\gamma_\tau F(z_\tau))$

$z_{\tau+1} = \mathrm{Prox}_{z_\tau}(\gamma_\tau F(w_\tau))$

$z^t = [u^t; v^t] := \left[\sum_{\tau=1}^t \gamma_\tau\right]^{-1} \sum_{\tau=1}^t \gamma_\tau z_\tau$ ◇

The efficiency estimate. The efficiency estimate of an MP is given by the following theorem.

Theorem 3.3. Let the Mirror Prox algorithm (Algorithm 3.2) be applied to a convex–concave saddle-point problem (3.13) associated with the Lipschitz-continuous vector field F, *i.e.*,

$$\|F(z) - F(z')\|_{E,*} \le L\|z - z'\|_E, \quad \text{for all } z, z' \in Z, \tag{3.16}$$

where $\|\cdot\|_{E,*}$ is the norm conjugate to the norm $\|\cdot\|_E$. Further, let the step size policy ensure that

$$\gamma_\tau \ge L^{-1} \ \&\ \delta_\tau := \gamma_\tau \langle F(w_\tau), w_\tau - z_{\tau+1}\rangle - V_{z_\tau}(z_{\tau+1}) \le 0 \tag{3.17}$$

for all τ; this is definitely so when $\gamma_\tau = L^{-1}$ for all τ. Finally, let $Z' = U' \times V'$, where the sets $U' \subset U$ and $V' \subset V$ are closed and bounded and such that $z_\omega \in Z'$. Then, for every t we have $z^t = [u^t; v^t] \in Z$ and

$$\varepsilon_{Z'}(z^t) := \max_{v \in V'} \phi(u^t, v) - \min_{u \in U'} \phi(u, v^t) \le \frac{\Omega^2[Z']}{2\sum_{\tau=1}^t \gamma_\tau} \le \frac{\Omega^2[Z']L}{2t}, \tag{3.18}$$

with $\Omega[\cdot]$ defined by (2.3).

Discussion. Let the premises of Theorem 3.3 hold.

(A) Assume that U and V are bounded, as is the case with problems (3.6). Then we can set $Z' = Z$, which results in $\varepsilon_{Z'} \equiv \mathrm{DualityGap}$ and provides us with the bound

$$\mathrm{DualityGap}(u^t, v^t) \le \frac{\Omega^2[Z]L}{2t}, \quad \text{for } t = 1, 2, \ldots, \tag{3.19}$$

with L given by (3.16).

(B) Assume that V is bounded, and ϕ admits a saddle point $[u_*; v_*]$. Setting $V' = V$, $U' = [u_\omega, u_*]$, where u_ω is the u-component of z_ω, we extract from (3.18) that

$$\varepsilon_{Z'}(z^t) = \overline{\phi}(u^t) - \min_{u \in U'} \phi(u, v^t) \ge \overline{\phi}(u^t) - \phi(u_*, v^t) \ge \overline{\phi}(u^t) - \phi(u_*, v_*),$$

where the final inequality follows from the fact that $\phi(u_*, v) \leq \phi(u_*, v_*)$ by the definition of a saddle point. As explained above, the existence of a saddle point $[u_*; v_*]$ implies that $\phi(u_*, v_*) = \mathrm{Opt}(P) = \mathrm{Opt}(D)$, and we obtain

$$\overline{\phi}(u^t) - \mathrm{Opt}(P) \leq \varepsilon_{Z_*}(z^t) \leq \frac{\Omega^2[Z_*]L}{2t}, \quad \text{where } Z_* = [u_\omega, u_*] \times V. \tag{3.20}$$

3.3. *Assembling partial proximal set-ups*

Our strategy for solving the Dantzig selector problem (3.1) reduces this task to using the Mirror Prox algorithm (Algorithm 3.2) to solve a 'small series' of bilinear saddle-point problems (3.6) on a common domain $Z = U \times V$ with *bounded* U, V, specifically the unit balls of the norms $\|\cdot\|$ and $\pi_*(\cdot)$, respectively. To implement this strategy, we need a good proximal set-up for Z, and we intend to 'assemble' this set-up from the basic set-ups listed in Section 2.1.1. The assembling to be implemented originates from Nemirovski (2004) and is as follows. Consider a convex–concave saddle-point problem (3.13) with smooth convex–concave cost function ϕ and *bounded* U, V, and let U and V be given as direct products of other closed and bounded convex domains:

$$U = Z^1 \times \cdots \times Z^p, \quad V = Z^{p+1} \times \cdots \times Z^K,$$

where $Z^k \subset E_k$ are convex, closed and bounded subsets of Euclidean spaces. We clearly have

$$Z := U \times V = Z^1 \times \cdots \times Z^K \subset E = E_1 \times \cdots \times E_K.$$

Now, let every pair Z^k, E_k be equipped with proximal set-up $\|\cdot\|_{E_k}$, $\omega_k(z^k)$. Consider an assembling of those set-ups into a proximal set-up for Z, E. Specifically, given positive α_k, $1 \leq k \leq K$ (the parameters of the construction), we set

$$\|z = [z^1; \ldots; z^K]\|_E = \sqrt{\sum_{k=1}^K \alpha_k^{-1} \|z^k\|_{E_k}^2}, \tag{3.21a}$$

$$\omega(z = [z^1; \ldots; z^K]) = \sum_{k=1}^K \alpha_k^{-1} \omega_k(z^k). \tag{3.21b}$$

It is immediately seen that this is indeed a proximal set-up for Z, E, and that computation of the associate proximal mapping reduces to computing proximal mappings associated with Z^k and our 'partial' DGFs $\omega_k(\cdot)$, $1 \leq k \leq K$. It is natural to choose the 'assembling parameters' in such a way as to obtain the best efficiency estimate of the MP algorithm allowed by our scheme as applied to (3.13). To this end, note that in view of the direct product structure of E, the vector field (3.14) associated with (3.13) can be written as $F(z) = [F_1(z); \ldots; F_K(z)]$ with $F_k(z) \in E_k$. Since this field is Lipschitz-continuous, we can find 'partial Lipschitz constants' $L_{k\ell}$, $1 \le k, \ell \le K$ such that, for all $z = [z^1; \ldots; z^K] \in Z$, $w = [w^1; \ldots; w^K] \in Z$, we have

$$\|F_k(z) - F_k(w)\|_{E_k,*} \le \sum_{\ell=1}^{K} L_{k\ell}\|z^\ell - w^\ell\|_{E_\ell}, \quad \text{for all } k \le K,$$

where $\|\cdot\|_{E_k,*}$ is the norm on E_k conjugate to $\|\cdot\|_{E_k}$. Note that we can always enforce $L_{k\ell} = L_{\ell k}$, which we assume from now on.

Example. When ϕ is bilinear, $\phi(u,v) = \langle a, u\rangle + \langle b, v\rangle + \langle v, Au\rangle + c$, the mapping F is affine,

$$F([z^1; \ldots; z^K]) = [a; -b] + \mathcal{A} \cdot [z^1; \ldots; z^K],$$

with linear mapping $\mathcal{A}$ given by

$$[\mathcal{A} \cdot [z^1; \ldots; z^K]]^k = \sum_{\ell=1}^{K} \mathcal{A}^{k\ell} z^\ell, \quad 1 \le k \le K,$$

where the $\mathcal{A}^{k\ell}$ are linear mappings from E_ℓ to E_k. It is evident that a characteristic property of $\mathcal{A}$ is that it is *skew-symmetric*: the mapping conjugate to $\mathcal{A}^{k\ell}$ is $-\mathcal{A}^{\ell k}$. In addition, $\mathcal{A}^{k\ell}$ should be zero when both Z^k and Z^ℓ are either factors in U or factors in V. In this case, a natural choice of $L_{k\ell}$ is

$$L_{k\ell} = \|\mathcal{A}^{k\ell}\|_{\|\cdot\|_{E_\ell}, \|\cdot\|_{E_k,*}} := \max_{z^\ell \in E_\ell} \{\|\mathcal{A}^{k\ell} z^\ell\|_{E_k,*} : \|z^\ell\|_{E_\ell} \le 1\},$$

and we indeed have $L_{k\ell} = L_{\ell k}$.

Now, it is immediately seen that with the assembling (3.21), the ω-radius $\Omega[Z]$ of $Z = Z^1 \times \cdots \times Z^K$ is given by

$$\Omega^2[Z] = \sum_{k=1}^{K} \alpha_k^{-1} \Omega^2[Z^k],$$

where $\Omega[Z^k] = \sqrt{2[\max_{Z^k} \omega_k(\cdot) - \min_{Z_k} \omega_k(\cdot)]}$ is the ω_k-radius of Z^k. Further, the (natural upper bound on the) $(\|\cdot\|_E, \|\cdot\|_{E,*})$ Lipschitz constant of the vector field F is

$$\mathcal{L} = \sigma_{\max}\big([L_{k\ell}\sqrt{\alpha_k \alpha_\ell}]_{1 \le k,\ell \le K}\big),$$

where $\sigma_{\max}(\cdot)$ is the maximal singular value of a matrix. The efficiency estimate of Mirror Prox associated with the resulting proximal set-up is $(\Omega^2[Z]\mathcal{L})/2t$ (see (3.19)), and the optimal choice of α_k should minimize $\Omega^2[Z]\mathcal{L}$ with the $\Omega[Z]$, $\mathcal{L}$ just defined. An optimal solution to the latter problem is given by

$$\alpha_k = \frac{1}{2}\Omega^2[Z^k]\frac{\sum_{\kappa,\ell} L_{\kappa\ell}\Omega[Z^\kappa]\Omega[Z^\ell]}{\sum_\ell L_{k\ell}\Omega[Z^k]\Omega[Z^\ell]}, \quad \text{for } 1 \le k \le K, \tag{3.22}$$

$$\text{which results in} \quad \Omega^2[Z] = 2, \quad \mathcal{L} = \frac{1}{2}\sum_{k,\ell} L_{k\ell}\Omega[Z^k]\Omega[Z^\ell],$$

and the resulting Mirror Prox efficiency estimate is

$$\text{DualityGap}(u^t, v^t) \le \frac{\sum_{k,\ell} L_{k\ell}\Omega[Z^k]\Omega[Z^\ell]}{2t}. \tag{3.23}$$

3.4. Implementing the strategy

In summary of our developments we obtain the following algorithm for solving (3.1) to a given accuracy ϵ.

Algorithm 3.4 (solving (3.1) to given accuracy ϵ).

(1) The set-up for our algorithm is given by a proximal set-up $\|\cdot\|_{E_u}, \omega_U(\cdot)$ for $U = \{u \in E_u : \|u\| \le 1\}, E_u$ and by a proximal set-up $\|\cdot\|_{E_v}, \omega_V(\cdot)$ for $V = \{v \in E_v : \pi_*(v) \le 1\}, E_v$.

(2) Problem (3.1) is solved by the root-finding routine presented in Section 3.1 with ρ_1 chosen according to (3.12), and the auxiliary bilinear saddle-point problems (3.6) solved by the Mirror Prox algorithm (Algorithm 3.2).

(3) The proximal set-up for MP is obtained from those indicated in (1) by the construction presented in Section 3.3. In the notation from this subsection, we have $K = 2$, $Z^1 = U$, $Z^2 = V$. Further, the affine vector fields $F = F^s$ associated with problems (3.6), $s = 1, 2, \ldots$ differ from each other by constant terms (see (3.7)), implying that the linear mappings $\mathcal{A}$ associated with different stages are identical to each other:

$$\mathcal{A} = \begin{bmatrix} & A^* \\ -A & \end{bmatrix}.$$

As a result, the assembled proximal set-up is the same for all stages and is given by (*cf.* (3.22))

$$\|[u; v]\|_E = \sqrt{\frac{\|u\|_{E_u}^2}{\Omega^2[U]} + \frac{\|v\|_{E_v}^2}{\Omega^2[V]}}, \quad \omega(u, v) = \frac{\omega_U(u)}{\Omega^2[U]} + \frac{\omega_V(v)}{\Omega^2[V]}, \tag{3.24}$$

where $\Omega[U]$, $\Omega[V]$ are the ω_U-radius and ω_V-radius of U, V, respectively. For every $t = 1, 2, \ldots$, let z_s^t denote the approximate solution obtained in t iterations by the MP, Algorithm 3.2, as applied to (3.6). The complexity bound (3.23) then implies that, for every s,

$$\text{DualityGap}^s(z_s^t) \leq \varepsilon_t := \frac{\Omega[U]\Omega[V]\|A\|_{\|\cdot\|_{E_u},\|\cdot\|_{E_v,*}}}{t}, \tag{3.25}$$

with $\|A\|_{\|\cdot\|_{E_u},\|\cdot\|_{E_v,*}}$ given by (1.3). ◇

The computational effort per iteration in the 'stand-alone' MP algorithm (Algorithm 3.2) is dominated by the need to compute two values of the vector field F to which the algorithm is applied, together with two values of the prox-mapping. With the strategy from Section 3.1, this should be augmented by computing $F^s(z^t)$ and the bounds Φ_t^+ and $\Phi_t^-(\cdot)$. These additional computations keep the complexity of an iteration basically intact. Since F^s is affine, $F^s(z^t)$ is readily given by the vectors $F^s(w_\tau)$, $1 \leq \tau \leq t$, which are computed in any case. Given $F^s(z^t)$, computing the above bounds reduces to maximizing two given linear forms, one depending on u and the other on v, over U, V, respectively, which is certainly no more complicated than computing a single prox-mapping. Combining these observations, relation (3.25) and Lemma 3.1, we arrive at the following result.

Theorem 3.5. Under the Standing Assumption, given $\epsilon > 0$ and solving (3.1) by the outlined algorithm, an ϵ-solution to (3.1) is found in no more than

$$\frac{1}{\ln(2/\theta)} \ln\left(1 + \frac{3\pi(b)}{\pi(b) - \delta} \cdot \frac{1}{\nu}\right), \quad \nu = \frac{\epsilon}{\text{Opt} \cdot \|A\|_{\|\cdot\|,\pi(\cdot)}} \tag{3.26}$$

stages, with at most

$$N = \text{Ceil}\left(\frac{\varkappa\Omega[U]\Omega[V]}{\widetilde{\nu}}\right), \quad \varkappa = \frac{\theta}{\theta - 1}, \quad \widetilde{\nu} = \frac{\epsilon}{\text{Opt} \cdot \|A\|_{\|\cdot\|_{E_u},\|\cdot\|_{E_v,*}}} \tag{3.27}$$

Mirror Prox iterations per stage. Here $\theta \in (1, 2)$ is the parameter of the root-finding procedure (see Section 3.1), and $\Omega[U], \Omega[V]$ are the $\omega_U(\cdot)$ radius of U and $\omega_V(\cdot)$ radius of V, respectively.

The computational effort per Mirror Prox iteration is dominated by the need to carry out two matrix–vector multiplications $(u, v) \mapsto (A^*v, Au)$, and to compute two values of the prox-mapping associated with the proximal set-up from step (3) of Algorithm 3.4.

3.5. Discussion

Let us look at what we get in the case of the Dantzig selector problems (3.1) of our primary interest. We shall assume that a suitable $\pi(\cdot)$ in (3.1) is the $\|\cdot\|_\infty/\|\cdot\|_2$ norm, that is, the image space E_v of the linear mapping A is $\mathbb{R}^{\ell_1} \times \cdots \times \mathbb{R}^{\ell_m}$, and for $v = [v^1; \ldots; v^m] \in E_v$ we have $\pi(v) = \max_{1 \leq i \leq m} \|v^i\|_2$. This choice allows us to treat in a unified fashion the cases of our primary interest, as specified in the Introduction, specifically those of the uniform fit (where $\ell_1 = \cdots = \ell_m = 1$) and the ℓ_2 fit (where $m = 1$). Moreover, the ℓ_∞/ℓ_2 fit is important in its own right: it arises in some Dantzig-selector-type recoveries dealing with block-sparse signals (see Juditsky *et al.* 2013*b*). In what follows, we treat the parameter $\theta \in (1,2)$ of the algorithm as fixed and thus as an absolute constant.

(A) We start with the case when the proximal set-ups underlying the algorithm use

$$\|\cdot\|_{E_u} = \|\cdot\|, \quad \|\cdot\|_{E_v} = \pi_*(\cdot). \tag{3.28}$$

Note that in this case

$$\|A\|_{\|\cdot\|_{E_u},\|\cdot\|_{E_v,*}} = \|A\|_{\|\cdot\|,\pi(\cdot)} = \max_u\{\pi(Au) : \|u\| \le 1\}, \quad \widetilde{\nu} = \nu. \tag{3.29}$$

With our $\pi(\cdot)$, the norm $\|\cdot\|_{E_v} \equiv \pi_*(\cdot)$ is the ℓ_1/ℓ_2 norm, and V is the unit ball of this norm. The latter allows us to specify $\omega_V(\cdot)$ as in Theorem 2.1, so that

$$\|[v^1;\ldots;v^m]\|_{E_v} = \sum_{i=1}^m \|v^i\|_2, \tag{3.30}$$

$$\omega_V(v^1,\ldots,v^m) = \frac{1}{p\gamma}\sum_{i=1}^m \|v^i\|_2^p, \quad \Omega[V] \le O(1)\sqrt{\ln(m+1)},$$

with γ and p given by (2.5) (where one should replace n with m).

The case of the ℓ_1/ℓ_2 norm $\|\cdot\|$. Here $E_u = \mathbb{R}^{k_1} \times \cdots \times \mathbb{R}^n$, and

$$\|[u^1;\ldots;u^n]\| \equiv \|u := [u^1;\ldots;u^n]\|_{E_u} := \sum_{j=1}^n \|u^j\|_2,$$

with $u^j \in \mathbb{R}^{k_j}$. Further, U is the unit ball of $\|\cdot\|_{E_u}$, and we can take as $\omega_U(\cdot)$ the DGF given by Theorem 2.1, thus arriving at

$$\|[u^1;\ldots;u^n]\|_{E_u} = \sum_{j=1}^n \|u^j\|_2, \tag{3.31}$$

$$\omega_U(u^1,\ldots,u^n) = \frac{1}{p\gamma}\sum_{j=1}^n \|u^j\|_2^p, \quad \Omega[U] \le O(1)\sqrt{\ln(n+1)},$$

with γ and p given by (2.5). With these choices, the quantity

$$\|A\|_{\|\cdot\|_{E_u}\|\cdot\|_{E_v,*}} = \|A\|_{\|\cdot\|,\pi(\cdot)}$$

is as follows. A is a linear mapping from $E_u = \mathbb{R}^{k_1} \times \cdots \times \mathbb{R}^{k_n}$ to $E_v = \mathbb{R}^{\ell_1} \times \cdots \times \mathbb{R}^{\ell_m}$, and as such can be naturally identified with a block matrix with $m \times n$ blocks $A^{ij} \in \mathbb{R}^{\ell_i \times k_j}$. It is immediately seen that

$$\|A\|_{\|\cdot\|,\pi(\cdot)} = \max_{i \le m, j \le n} \sigma_{\max}(A^{ij}). \tag{3.32}$$

Applying Theorem 3.5, finding an ϵ-solution to (3.1) requires at most S stages, where

$$S = O(1) \ln\left(1 + \frac{3 \max_i \|b^i\|_2}{\max_i \|b^i\|_2 - \delta} \cdot \frac{1}{\nu}\right), \tag{3.33}$$

$$\nu := \frac{\epsilon}{\texttt{Opt} \cdot \|A\|_{\|\cdot\|,\pi(\cdot)}} = \frac{\epsilon}{\texttt{Opt} \cdot \max_{i,j} \sigma_{\max}(A^{ij})},$$

with at most

$$O(1)\sqrt{\ln(m+1)\ln(n+1)}\text{Ceil}(1/\nu) \tag{3.34}$$

iterations per stage. Note that ν can be naturally interpreted as relative accuracy.

The computational effort per iteration is dominated by the need to compute $O(1)$ matrix–vector products $(u, v) \mapsto (A^* v, Au)$, plus a 'computational overhead' of $O(1)(\dim u + \dim v)$ arithmetic operations.

In attempting to utilize the outlined approach for the Dantzig selector with the TV norm in the role of ℓ_1/ℓ_2, one runs into the same difficulties as in the case of lasso problems: see Section 2.3.

The case of $\|\cdot\| = \|\cdot\|_{\text{nuc}}$. Now consider the case when $E_u = \mathbb{R}^{\mu \times \nu}$ and $\|\cdot\|$ is the nuclear norm on E_u. Without loss of generality we can assume $2 \le \mu \le \nu$. Taking into account that U is the unit ball of $\|\cdot\|_{E_u} \equiv \|\cdot\|$, we can choose $\omega_U(\cdot)$ according to Theorem 2.3, thus arriving at

$$\|u\|_{E_u} = \|u\|_{\text{nuc}} = \sum_{i=1}^{\mu} \sigma_i(u), \tag{3.35}$$

$$\omega_U(u) = \alpha \sum_{i=1}^{\mu} \sigma_i^{1+q}(u), \quad \Omega[U] \le O(1)\sqrt{\ln(\mu+1)},$$

with α and q readily given by (2.9). The quantity

$$\|A\|_{\|\cdot\|_{E_u}, \|\cdot\|_{E_v,*}} = \|A\|_{\|\cdot\|_{\text{nuc}}, \pi(\cdot)}$$

is as follows: the linear mapping $A : \mathbb{R}^{\mu \times \nu} \to E_v$ can be split into 'blocks' $A^i : \mathbb{R}^{\mu \times \nu} \to \mathbb{R}^{\ell_i}$, according to the direct product structure of E_v, and $\|A\|_{\|\cdot\|_{\text{nuc}}, \pi(\cdot)} = \max_i \|A^i\|_{\|\cdot\|_{\text{nuc}}, \|\cdot\|_2}$. For example, in the plain matrix completion problem, where the mapping $u \mapsto Au$ is the restriction of u to a given m-element pattern of cells and $\pi(\cdot)$ is the ℓ_∞/ℓ_2 norm, one can safely bound $\|A\|_{\|\cdot\|_{\text{nuc}}, \pi(\cdot)}$ from above by 1.

The algorithm we obtain finds an ϵ-solution to problem (3.1) in at most S stages, where

$$S = O(1)\ln\left(1 + \frac{3\max_i \|b^i\|_2}{\max_i \|b^i\|_2 - \delta}\cdot\frac{1}{\nu}\right), \tag{3.36}$$

$$\nu := \frac{\epsilon}{\text{Opt}\cdot\|A\|_{\|\cdot\|_{\text{nuc}},\pi}} = \frac{\epsilon}{\text{Opt}\cdot\max_i\|A^i\|_{\|\cdot\|_{\text{nuc}},\|\cdot\|_2}},$$

with at most

$$O(1)\sqrt{\ln(m+1)\ln(\mu+1)}\text{Ceil}(1/\nu) \tag{3.37}$$

iterations per stage.

The computational effort per iteration is dominated by the need to compute $O(1)$ matrix–'vector' products $(u,v)\mapsto(A^*v, Au)$ and $O(1)$ SVDs of matrices from $\mathbb{R}^{\mu\times\nu}$, plus a computational overhead of $O(1)(\mu\nu+\dim E_v)$ arithmetic operations.

Observe that Remark 2.5 is fully applicable to our present situation, implying the possibility of extending the scope of the MP-based techniques beyond the case of the $\|\cdot\|$-unit ball in the role of U, as discussed in Section 2.3.

(B) Coming back to 'plain' Dantzig-selector-type problems (3.1) with ℓ_1/ℓ_2 or the nuclear norm in the role of $\|\cdot\|$ and the ℓ_∞/ℓ_2 norm in the role of $\pi(\cdot)$, note that so far we have been considering the case of (3.28) and proximal set-ups for U, E_u and V, E_v developed in Section 2.1.1. An alternative, just as in the composite minimization case, is to use the plain Euclidean set-ups for U, V, that is, to use $\|u\|_{E_u} = \sqrt{\langle u,u\rangle}$, $\omega_U(u) = \frac{1}{2}\langle u,u\rangle$, and similarly for V, E_v. With our U and V, this results in $\Omega[U] = \Omega[V] = 1$. The assembled set-up for $Z = U\times V, E = E_u\times E_v$ is in this case again the plain Euclidean set-up. When passing from our original set-ups to the Euclidean one, the complexity per iteration remains basically intact. Inequality (3.25) becomes

$$\text{DualityGap}^s(z_s^t) \le \varepsilon_t := \frac{\|A\|_{2,2}}{t}, \quad \|A\|_{2,2} = \max_u\{\|Au\|_2 : \|u\|_2 \le 1\} \tag{3.38}$$

(for $\|\cdot\| = \|\cdot\|_{\text{nuc}}$, $\|u\|_2$ is the Frobenius norm of u), the upper bounds (3.33) and (3.36) on the number of stages remain intact, and the upper bounds (3.34) and (3.37) on the number of iterations per stage become

$$O(1)\text{Ceil}(1/\widetilde{\nu}), \quad \widetilde{\nu} = \frac{\epsilon}{\text{Opt}\cdot\|A\|_{2,2}}. \tag{3.39}$$

Note that $\|A\|_{2,2}$ is never less than the norms $\|A\|_{\|\cdot\|,\pi}$ appearing in (3.33) and (3.36), and can be significantly larger. For example, when $\|\cdot\|$ is the plain $\|\cdot\|_1$ norm on $\mathbb{R}^n$, the ratio $\|A\|_{2,2}/\|A\|_{\|\cdot\|,\pi}$ can be as large as $\sqrt{n}$ when $\pi(\cdot)$ is the plain ℓ_2 norm, and as large as $\sqrt{mn}$ when π is the plain ℓ_∞ norm on $\mathbb{R}^m$. When $\|\cdot\|$ is the nuclear norm on $\mathbb{R}^{\mu\times\nu}$, the ratios in question can be as large as $O(1)\sqrt{\mu}$ (π is the ℓ_2 norm) and $O(1)\sqrt{\mu m}$ (π is the ℓ_∞ norm). With this in mind, looking at (3.34), (3.37) and (3.39) we conclude that in our context, from the worst-case perspective, the Euclidean set-up is significantly outperformed by our original set-ups. Moreover, the original set-ups result in the best efficiency estimates achievable in the framework of our algorithmic scheme, up to logarithmic dependence on the problem size: see Section A.6. This being said, note that the worst-case-oriented analysis is not exactly the same as computational practice, where the Euclidean set-up behaves surprisingly well in many cases.

3.6. Comments

Some comments on the results of this section are in order.

3.6.1. On-line adjustable step sizes

(See the concluding remarks in Section 2.) Theorem 3.3 states that whenever the step sizes γ_τ ensure (3.17), the accuracy estimate (3.18) holds true; looking at this estimate, we see that the larger the step sizes the better. Theorem 3.3 also states that (3.17) is ensured by choosing $\gamma_\tau = L^{-1}$. The question is whether we could use step sizes which are larger than the 'safe' L^{-1}. The simplest way to achieve this goal is as follows. At the beginning of iteration τ of the MP algorithm, we have at our disposal a certain 'guessed step size' $\widehat{\gamma}_\tau \geq L^{-1}$, with $\widehat{\gamma}_1 = L^{-1}$. We start iteration τ with an attempt to use $\widehat{\gamma}_\tau$ as γ_τ. If this step size results in $\delta_\tau \leq 0$ (see (3.17)), we set $\widehat{\gamma}_{\tau+1} = \vartheta\widehat{\gamma}_\tau$ and pass to step $\tau+1$; here $\vartheta > 1$ is the method's parameter. If $\gamma_\tau = \widehat{\gamma}_\tau$ results in $\delta_\tau > 0$, we reset γ_τ according to $\gamma_\tau := \min[\mu\gamma_\tau, 1/L]$, where $\mu \in (0,1)$ is another method's parameter, and repeat the step with this reduced step size. If we get $\delta_\tau > 0$ with the reduced value of γ_τ also, we iterate, reducing the 'trial step size' once more, and perform a new trial step, and so on until $\delta_\tau \leq 0$ is obtained (this must eventually happen, since $\gamma_\tau = L^{-1}$ ensures $\delta_\tau \leq 0$). When $\delta_\tau \leq 0$ is obtained, we pass to step $\tau+1$, using the latest value of γ_τ as $\widehat{\gamma}_{\tau+1}$. Extensive numerical experimentation (where we use $\vartheta = 1.2$ and $\mu = 0.8$) demonstrates that, with this on-line step size policy, the total number of values of F (or equivalently the total number of prox-mappings) we need to compute to arrive at a desired value of the duality gap is never significantly larger than the similar quantity for the off-line worst-case-oriented step sizes $\gamma_\tau \equiv L^{-1}$, and is typically orders of magnitude less. Note that this policy, with obvious modifications, can be utilized in the case when we know that F is Lipschitz-continuous but do not know the corresponding Lipschitz constant.

3.6.2. Nearly dimension-independent rate of convergence

It should be stressed that in the situations we have considered, the resulting efficiency estimates, *i.e.*, the total iteration count as a function of the desired relative accuracy ν given by (3.33) or (3.36), up to a *logarithmic* factor in $1/\nu$ and the problem size, is just $1/\nu$. Logarithmic dependence of the efficiency estimate on the problem size reflects the 'nice geometry' of the norms $\|\cdot\|$, $\pi_*(\cdot)$ in question. Along with these and several other 'nice geometry' norms, there exist 'bad geometry' ones, *e.g.*, the norm $\|\cdot\|_\infty$ on $\mathbb{R}^n$ and its matrix analogue: the spectral norm $\sigma_{\max}(\cdot)$. For the two latter norms in the roles of $\|\cdot\|$ or $\pi_*(\cdot)$, the approach we have developed does *not* yield 'nearly dimension-independent' efficiency estimates, but this is equally true for any other approach known to us.

3.6.3. More general norms $\|\cdot\|$

The scope of the approach we have developed is not restricted to the norms $\|\cdot\|$, $\pi(\cdot)$ considered so far; all we need are 'good proximal set-ups' for the unit balls U, V of the norms $\|\cdot\|$, $\pi_*(\cdot)$, *i.e.*, moderate $\Omega[U]$ and $\Omega[V]$ and computationally tractable prox-mappings. As an instructive example, consider recovery of a matrix ('image') x from noisy observations of Px. In some cases it is natural to assume that the true image is the sum of sparse and low-rank images.[14] The Dantzig selector form of the problem is

$$\min_{y^1,y^2\in\mathbb{R}^{\mu\times\nu}} \{\mu_1\|y^1\|_1+\mu_2\|y^2\|_{\rm nuc} : \pi(P(y^1+y^2)-b)\le\delta\},$$

where $\mu_1>0$, $\mu_2>0$ are the penalty parameters and $\|y^1\|_1=\sum_{i,j}|y^1_{ij}|$. Equivalently, the recovery problem is

$$\min_{x=(x^1,x^2)\in E_u:=\mathbb{R}^{\mu\times\nu}\times\mathbb{R}^{\mu\times\nu}} \{\|x\| : \pi(Ax-b:=\mu_1^{-1}Px^1+\mu_2^{-1}Px^2-b)\le\delta\},$$

$$\|x:=(x^1,x^2)\|\equiv\|x^1\|_1+\|x^2\|_{\rm nuc}.$$

All we need to process the resulting problem is a good proximal set-up for the unit ball U of the norm $\|\cdot\|$, and such a set-up is readily given by the results of Section 2.1.1:

$$\|\cdot\|_{E_u}=\|\cdot\|,\quad \omega_U(u^1,u^2)=2\omega_1(\mathrm{Vec}(u^1))+2\omega_2(u^2),$$

where $\mathrm{Vec}(u^1)$ is the 'vectorization' of u^1, and the functions $\omega_1(\cdot)$, $\omega_2(\cdot)$ are given by (2.5) (where $n=\mu\nu$) and (2.9) respectively; note that the ω_U-radius of U does not exceed $O(1)\sqrt{\ln(\mu\nu)}$. When $\pi(\cdot)=\|\cdot\|_2$, we are also interested in processing the lasso version of the problem,

$$\min_{x=(x^1,x^2)} \{\|(x^1,x^2)\|+\|Ax-b\|_2^2\}.$$

[14] As a toy example, imagine the façade of a building, a black rectangle ('wall') with a regular grid of white rectangular holes ('windows'); this image has rank 2. Adding a few irregularities ('architectural elements') occupying a relatively small part of the façade, we get an image which is the sum of low-rank and sparse images.

As above, the required proximal set-up is found to be

$$\|\cdot\|_E = \|\cdot\|, \quad \omega_E(u^1, u^2) = 2\widehat{\omega}_1(u^1) + 2\widehat{\omega}_2(u^2),$$

with $\widehat{\omega}_1(\cdot), \widehat{\omega}_2(\cdot)$ given by (2.7) (where $n = \mu\nu$) and (2.11), respectively. This set-up ensures that the quantity $V_{z_\omega}(x^*)$ appearing in the bounds (2.21), (2.23) and (2.25) does not exceed $O(1)\ln(\mu\nu)\|x^*\|^2$. Finally, the computational cost of the prox-mapping associated with $\|\cdot\|$ is basically the same as the cost of computing the SVD of a matrix from $\mathbb{R}^{\mu\times\nu}$.

3.6.4. Plain saddle-point forms of ℓ_1/nuclear norm minimization

Along with lasso and Dantzig selector recovery routines, sparsity/low-rank oriented signal processing considers other recovery routines, *e.g.*, the *penalized recovery* (Juditsky and Nemirovski 2011*c*, Juditsky *et al.* 2013*b*, Juditsky *et al.* 2011*b*)

$$\widehat{x} \in \arg\min_{x\in E_x}\big\{\|x\| + \lambda\pi(Ax - b)\big\}, \quad \text{for } \lambda > 0, \tag{3.40}$$

possessing attractive statistical properties. Assuming for the sake of definiteness that $\|\cdot\|$ is either the ℓ_1/ℓ_2 norm on $E_x = \mathbb{R}^{k_1+\cdots+k_n}$ or the nuclear norm on $E_x = \mathbb{R}^{\mu\times\nu}$, and $\pi(\cdot)$ is the ℓ_∞/ℓ_2 norm on $E_v = \mathbb{R}^{\ell_1+\cdots+\ell_m}$ (this is indeed the case in typical applications), and rewriting the problem equivalently as

$$\mathrm{SadVal} = \min_{u=[x;\xi]\in U}\max_{v\in V}\big[\phi(u := [x;\xi], v) = \xi + \lambda\langle v, Ax - b\rangle\big], \tag{3.41}$$

$$U = \{[x;\xi] : \xi \ge \|x\|\} \subset E_u := E_x \times \mathbb{R}, \quad V = \{v \in E_v : \pi_*(v) \le 1\},$$

we see that in this case we need to solve just one bilinear saddle-point problem rather than a 'small series' of them, so that a single application of MP is enough. Equipping V, E_v with the same ℓ_1/ℓ_2 proximal set-up $\|\cdot\|_{E_v} = \pi_*(v)$, $\omega_V(\cdot)$ as above, and U, E_u with the proximal set-up

$$\|u := [x;\xi]\|_{E_u} = \|x\| + |\xi|, \quad \omega_U(u := [x;\xi]) = 2\omega(u) + \xi^2,$$

where $\omega(u)$ is given by (2.7) or (2.11), depending on whether $\|\cdot\|$ is the ℓ_1/ℓ_2 or the nuclear norm, we can assemble these 'partial' proximal set-ups for U, E_u and V, E_v into a proximal set-up $\|\cdot\|_E$, $\omega(\cdot)$ for $Z = U \times V$, $E = E_u \times E_v$ according to

$$\|[[x;\xi], v]\|_E = \sqrt{\alpha^{-2}(\|x\| + |\xi|)^2 + \pi_*^2(v)},$$
$$\omega([x;\xi], v) = \alpha^{-2}[2\omega(x) + \xi^2] + \omega_V(v),$$

where $\alpha > 0$ is a positive parameter. It is immediately seen that this assembling indeed results in a proximal set-up. Applying (2.6), (2.8), (2.12) and (3.20), we obtain the efficiency estimate

$$
\begin{aligned}
&\overline{\phi}(x^t) - \min_x \overline{\phi}(x) \le O(1)\frac{\alpha^{-1}(\alpha\kappa_1 R_* + \kappa_2)^2\lambda\|A\|_{\|\cdot\|,\pi(\cdot)}}{t}, \quad t = 1,2,\ldots,\\
&\text{where}\quad \overline{\phi}(x) = \|x\| + \lambda\pi(Ax-b), \quad R_* = \min_{x_*\in\arg\min_{x\in E_x}\overline{\phi}(x)} \|x\|, \qquad (3.42)\\
&\kappa_1 = \begin{cases} \sqrt{\ln(n+1)}, & E_x = \mathbb{R}^{k_1+\cdots+k_n}, \|x := [x^1;\ldots;x^n]\| = \sum_j \|x^j\|_2,\\ \sqrt{\ln(\mu+1)}, & E_x = \mathbb{R}^{\mu\times\nu}(\mu\le\nu), \|x\| = \|x\|_{\text{nuc}}, \end{cases}\\
&\kappa_2 = \sqrt{\ln(m+1)}.
\end{aligned}
$$

Here x^t is the x-component of the approximate solution obtained by t MP iterations. Assuming $R_* > 0$ and optimizing the efficiency estimate over the 'assembling parameter' α, we obtain $\alpha = \alpha_* = \kappa_2/(\kappa_1 R_*)$ and the efficiency estimate

$$\overline{\phi}(x^t) - \min_x \overline{\phi}(x) \le O(1)\frac{\kappa_1\kappa_2\lambda\|A\|_{\|\cdot\|,\pi(\cdot)}R_*}{t}.$$

The outlined choice of α is usually unrealistic; a realistic choice is $\alpha = \kappa_2/(\kappa_1 R)$, where R is our guess for R_*. The efficiency estimate associated with this α is within the factor $O(1)\max[R/R_*, R_*/R]$ of the 'optimal' efficiency. Finally, note that the computational price of the prox mapping associated with our set-up is basically the same as in the case when U is the $\|\cdot\|$-unit ball of E_u: see Section 2.1.1.

3.6.5. *Accelerating the MP algorithm*

Under favourable structural assumptions on the cost function ϕ of (3.13), the MP algorithm can be accelerated (Juditsky and Nemirovski 2011*b*, Section 6.4); in particular, in the case of the ℓ_1/ℓ_2 or nuclear norm $\|\cdot\|$, a properly accelerated MP algorithm allows us to solve lasso-type problems (1.1a) with the same $O(1/t^2)$ efficiency estimate as the fast gradient method from Section 2.

3.6.6. *Randomization*

When solving ℓ_1/nuclear norm minimization problems (1.1a,b) to medium relative accuracy, the 'practical grasp' of first-order algorithms is limited by our abilities to carry out, in reasonable time, several hundred (or a few thousand) matrix–vector multiplications involving A and A^* and, in the case of nuclear norm minimization and $x \in \mathbb{R}^{\mu\times\nu}$, $\mu \le \nu$, the ability to compute the same number of SVDs of $\mu\times\nu$ matrices. While these limitations are essentially less restrictive than those for IPMs, where one needs to assemble and solve a few tens of linear systems with $\dim x$ unknowns,[15] they cannot be neglected. Let us discuss the related difficulties and options.

In ℓ_1 or ℓ_1/ℓ_2 minimization, with their inexpensive prox-mappings, difficulties arise when the required matrix–vector multiplications become prohibitively time-consuming. There is, however, a way to overcome these difficulties to some extent, by *randomizing* matrix–vector multiplications.

The simplest way to randomize the matrix–vector multiplication $u \mapsto Bu : \mathbb{R}^N \to \mathbb{R}^M$ is as follows. Let us associate with $u \in \mathbb{R}^N$ a stochastic vector with the entries $|u_j|/\|u\|_1$. Treat this vector as a probability distribution on the set $\{B_1, \dots, B_N\}$ of columns of B, draw from this distribution a random column B_j, and return the vector $\xi = \|u\|_1 \mathrm{sign}(u_j) B_j$. It is immediately seen that the resulting random estimate ξ of Bu is unbiased: $\mathbb{E}\{\xi\} = Bu$. Generating this estimate costs $O(M+N)$ arithmetic operations,[16] which compares favourably with the cost of $O(1)MN$ arithmetic operations of the deterministic computation of the precise value of Bu.[17] Of course, the unbiased estimate ξ of Bu is noisy, with the magnitude of the noise, measured in some norm $\|\cdot\|$, bounded by $\|u\|_1 \|B\|_{\|\cdot\|_1, \|\cdot\|} = \|u\|_1 \max_j \|B_j\|$.

It turns out that the strategy we have developed for (3.1), and its MP-based implementation, 'survive' replacing precise matrix–vector products $(u, v) \mapsto (A^* v, Au)$ by their unbiased random estimates, albeit at the price of a certain growth in the iteration count and sensitivity of the approach to the geometry of the domains U, V underlying (3.6).[18] The potential of the outlined randomization in the context of ℓ_1 minimization problems (1.1b) with the uniform or the ℓ_2 fit was investigated by Juditsky *et al.* (2013*a*) (see also Juditsky and Nemirovski 2011*b*), whose results can be summarized as follows.

- For ℓ_1 minimization with the uniform and ℓ_2 fit, and with a dense 'general' $m \times n$ matrix A, building an ϵ-solution to (3.1) with the *deterministic* algorithm we have developed takes ν^{-1} iterations,[19] with

[15] Indeed, assuming that the image and the argument dimensions of the mapping $x \to Ax$ are of the order of $n := \dim x$, a single matrix–vector multiplication costs $O(1)n^2$ arithmetic operations (or less when A is 'nicely structured', *e.g.*, sparse, or discrete-convolution-like or discrete Fourier/wavelet transform), and a single SVD costs $O(1)\mu^2\nu \le O(1)n^{3/2}$, while the cost of a single IPM iteration is of the order of n^3 in at least some of the applications we are interested in.

[16] Assuming, as is usually the case, an $O(1)M$-arithmetic-operation cost of extracting a column from B given its index.

[17] In this comparison, we assume that B has no 'nice structure' allowing for fast matrix–vector multiplications

[18] This sensitivity primarily stems from the fact that the noisiness of the unbiased estimate of Bu is in proportion to $\|u\|_1$ and can thus be large when U or V contains vectors of large $\|\cdot\|_1$ norm.

[19] Here and in the discussion to follow, we skip log-factors.

mn arithmetic operations per iteration, where the A_j are the columns of A, $\nu = \epsilon(\mathrm{Opt} \cdot \max_{1 \le j \le n} \|A_j\|_p)^{-1}$ is the relative accuracy, and $p = \infty$ (uniform fit) or $p = 2$ (ℓ_2 fit).[20] The overall arithmetic complexity is therefore $mn\nu^{-1}$, which, as far as we know, is the best possible complexity bound achievable with existing *deterministic* algorithms for the large-scale problems under consideration.

- For ℓ_1 minimization with uniform fit, randomization allows us to reduce the cost of an iteration to $m + n$ at the price of increasing the iteration count to ν^{-2}. Thus the overall complexity becomes $(m+n)\nu^{-2}$. We see that, with ν fixed and $n = O(m)$ growing, the randomized algorithm eventually progressively outperforms the deterministic ones. Numerical experiments demonstrate that randomization results in fairly significant acceleration (by a factor of around 5.4 in terms of CPU time) for 'not too large' problems (specifically, those with $8\,000 \times 16\,000$ randomly generated matrices). It should be added that when $\ln(n) = O(\ln(m))$ and m, n are large, the randomized algorithm generates an ϵ-solution with high confidence by inspecting a negligible fraction (tending to zero as $m, n \to \infty$) of the data, thus exhibiting *sublinear time* behaviour.[21]
- For ℓ_1 minimization with ℓ_2 fit, the situation is the same as that of the uniform fit, modulo two points: first, δ in (1.1b) should be 'small enough' (specifically, $\le O(1)\|b\|_2/\sqrt{m}$);[22] second, we need $|A_{ij}| \le O(1)\max_j \|A_j\|_2\sqrt{\ln(m)/m}$ (roughly speaking, the entries in A should be of nearly the same magnitude). Juditsky *et al.* (2013*a*) showed that the latter assumption can be ensured by an appropriate randomized preprocessing which costs $mn\ln(m)$ arithmetic operations; note that when $\nu \ll 1$ is fixed and m, n are large, the cost of this preprocessing is negligibly small compared to the cost of the deterministic algorithm. Finally, for the ℓ_2 fit, the experimentally observed acceleration due to randomization in Juditsky *et al.* (2013*a*), although smaller than for the uniform fit, is still quite significant (by a factor of 2.4 on $8\,000 \times 16\,000$ problems).

[20] See (3.34) and (3.26), and take into account that we are in a situation where either for every j the blocks A^{ij} are just entries in A (uniform fit), or for every j there exists only one block A^{ij}, specifically the entire column A_j (ℓ_2 fit).

[21] For ℓ_1 minimization with uniform fit, the saddle-point problems (3.6) can be reduced to matrix games. The possibility of solving matrix games in sublinear time was discovered in Grigoriadis and Khachiyan (1995). The *ad hoc* randomized matrix game algorithm proposed in this reference is in hindsight pretty similar – though not identical – to the 'uniform fit' version of the algorithm from Juditsky *et al.* (2013*a*). In particular, both algorithms share the same complexity bound.

[22] This restriction can be lifted: see Juditsky and Nemirovski (2011*b*).

Table 3.1. Randomization making a difference: $32\,000 \times 64\,000$ matrix A, $\delta = 0.005$, termination on either achieving accuracy $\epsilon = 0.0025$ or reaching the CPU limit of 7 200 sec. The computational platform is the same as in Table 1.1.

	Method	Steps	CPU (sec)	$\|Ax_\epsilon - b\|_p$
$p = \infty$	DMP†	31	7 363	0.1598
	SMP‡	7 501	5 352	0.0074
$p = 2$	DMP	30	7 536	0.0248
	SMP	2 602	2 350	0.0072

†Deterministic MP with on-line adjustable step size policy.
‡Stochastic, or randomized, version of MP (Juditsky *et al.* 2013*a*).

For an impression of the potential of randomization, see Table 3.1. The data are taken from Juditsky *et al.* (2013*a*).

Now, in the case of nuclear norm minimization, the difficulties can come from the impossibility of carrying out in a reasonable time the required number of either (a) SVDs, or (b) matrix–'vector' multiplications $(u, v) \mapsto (A^*v, Au)$. While the difficulties arising in case (a) can sometimes be circumvented (Arora, Hazan and Kale 2005, Arora and Kale 2007, d'Aspremont 2011, Baes, Bürgisser and Nemirovski 2013), we would say that typically case (a) rules out proximal-point algorithms as we know them today. In contrast to this, in case (b) one can try to save the day by randomization. For the sake of definiteness, assume that we are speaking about nuclear norm minimization with uniform fit, so that $u \in \mathbb{R}^{\mu \times \nu}$ and $Au \in \mathbb{R}^m$. Assume also that for every i, the ith entry in Au is given by

$$(Au)_i = \sum_{\ell=1}^{k_i} p_{i\ell}^T u q_{i\ell},$$

where $p_{i\ell} \in \mathbb{R}^\mu$ and $q_{i\ell} \in \mathbb{R}^\nu$. Letting $\#(r)$ denote the number of non-zero entries in the vector r, the straightforward computation of

$$\left(A^*v = \sum_{i=1}^{m} v_i \sum_{\ell=1}^{k_i} p_{i\ell} q_{i\ell}^T, Au \right)$$

for given u, v costs

$$\mathcal{C}_d = O(1) \sum_{i=1}^{m} \sum_{\ell=1}^{k_i} \#(p_{i\ell}) \#(q_{i\ell}) \text{ arithmetic operations.}$$

Similarly, an unbiased random estimate η of A^*v can be constructed using $\|v\|_1 \text{sign}(v_\imath) \sum_{\ell=1}^{k_\imath} p_{\imath\ell} q_{\imath\ell}^T$, where $\imath \in \{1, \ldots, m\}$ is drawn at random according to $\text{Prob}\{\imath = i\} = |v_i|/\|v\|_1$. The cost of generating a realization of η is

$$\mathcal{C}_\eta = O(1) \left[m + \max_i \sum_{\ell=1}^{k_i} \#(p_{i\ell}) \#(q_{i\ell}) \right] \text{ arithmetic operations.}$$

To randomize computing Au, one can use a 'matrix analogue' of the randomization used earlier. Specifically, given the SVD $u = \sum_{j=1}^{\mu} \sigma_j(u) e_j f_j^T$ of u,[23] we draw $\jmath \in \{1, \ldots, \mu\}$ at random according to

$$\text{Prob}\{\jmath = j\} = \sigma_j(u)/\|u\|_{\text{nuc}}, \quad 1 \leq j \leq \mu,$$

and return $\xi \in \mathbb{R}^m$ defined by its components

$$\xi_i = \|u\|_{\text{nuc}} \sum_{\ell=1}^{k_i} (p_{i\ell}^T e_\jmath)(q_{i\ell}^T f_\jmath), \quad 1 \leq i \leq m.$$

It is immediately seen that ξ is an unbiased random estimate of Au, and generating a realization of ξ given the SVD of u costs

$$\mathcal{C}_\xi = O(1) \left[\mu + \sum_{i=1}^{m} \sum_{\ell=1}^{k_i} [\#(p_{i\ell}) + \#(q_{i\ell})] \right] \text{ arithmetic operations.}$$

To gain an impression of what can be saved here, assume that $k_i = k$, $\#(p_{i\ell}) = \#(q_{i\ell}) = p$ for all i, ℓ. Then, skipping $O(1)$ factors, we obtain

$$\Theta := \frac{\mathcal{C}_\eta + \mathcal{C}_\xi}{\mathcal{C}_d} = \frac{1}{kp^2} + \frac{1}{m} + \frac{\mu}{kmp^2} + \frac{1}{p}.$$

In a meaningful range of the sizes μ, m, k and p, Θ tends to zero as the sizes increase,[24] meaning that randomization can indeed accelerate the solution process.

Finally, we remark that the noise of our random estimates of matrix–vector products can be reduced by taking the average of several realizations of the estimate. For more details on this subject, see Juditsky *et al.* (2013*a*) and Baes *et al.* (2013).

[23] Looking at the description of the MP algorithm, we see that all the u for which Au should be computed are the values (already computed when Au is to be computed) of the prox-mapping associated either with the DGF (2.9) or with the DGF (2.11). As explained in Section 2.1.1, when computing such a value u we in fact get the SVD of u. Thus, in our context, when computing Au, the SVD of u is indeed readily available.

[24] Let, for example, $\mu = \nu$, $m = O(\mu^2)$, $p = O(\mu)$ (square u, the number of observations is of order of $\dim u$, $p_{i\ell}$, $q_{i\ell}$ are dense). In this case, $\mathcal{C}_d = O(1)k\mu^4$, $\Theta \leq O(1)\mu^{-1}$ tends to zero as μ grows. In addition, in the case under consideration, the cost per iteration of computing prox-mappings is $O(1)[\mu^3 + m] \leq O(1)\mu^3 \leq O(1)\mu^{-1}\mathcal{C}_d$, implying that the randomization reduces arithmetic complexity of an iteration by a factor of $O(\mu)$.

Appendix: Proofs

A.1. Relation (2.2)

We need the following simple lemma.

Lemma A.1. Let Z be a closed convex domain in Euclidean space E, let $\|\cdot\|_E, \omega(\cdot)$ be a proximal set-up for Z, E, let $\Psi : Z \to \mathbb{R} \cup \{+\infty\}$ be a lower-semicontinuous convex function which is finite on the relative interior of Z, and let $\phi(\cdot) : Z \to \mathbb{R}$ be a convex continuously differentiable function. Assume that z_+ is a minimizer of the function $\omega(w) + \phi(w) + \Psi(w)$ on Z. Then $z_+ \in Z^o$, and there exists $\Psi' \in \partial\Psi(z_+)$ such that $\langle \omega'(z_+) + \nabla\phi(z_+) + \Psi', w - z_+\rangle \geq 0$ for all $w \in Z$. If $\Psi(\cdot)$ is differentiable at z_*, one can take $\Psi' = \nabla\Psi(z_+)$.

Proof. Let us set $\widehat{\omega}(w) = \omega(w) + \phi(w)$. Since ϕ is continuously differentiable on Z, $\widehat{\omega}$ possesses the same smoothness properties as ω: specifically, $\widehat{\omega}$ is convex and continuous on Z,

$$\{w \in Z : \partial\widehat{\omega}(w) \neq \emptyset\} = Z^o := \{w : \partial\omega(w) \neq \emptyset\},$$

and $\widehat{\omega}'(w) := \omega'(w) + \nabla\phi(w) : Z^o \to E$ is a continuous selection of subgradients of $\widehat{\omega}$.

Without loss of generality we can assume that $\operatorname{int} Z \neq \emptyset$ and that $z_+ = 0$. Note that z_+ is a minimizer of $\widehat{\omega}(w) + \Psi(w)$ over $w \in Z$. Let $\bar{w} \in \operatorname{int} Z$, and let $w_t = t\bar{w}$, so that $w_t \in \operatorname{int} Z$ for $0 < t \leq 1$.

Let us prove first that $z_+ = 0 \in Z^o$. To this end it suffices to verify that the vectors $\widehat{\omega}'(w_t)$ remain bounded as $t \to +0$. Indeed, in this case the set of limiting points of $\widehat{\omega}'(w_t)$ as $t \to +0$ is non-empty. Since $\widehat{\omega}(\cdot)$ is continuous on Z, every one of these limiting points belongs to $\partial\widehat{\omega}(0)$, and thus $\partial\widehat{\omega}(0) \neq \emptyset$, meaning that $0 \in Z^o$. To prove that $\widehat{\omega}'(w_t)$ is bounded as $t \to +0$, let $r > 0$ be such that with $B = \{h \in E : \|h\|_E \leq r\}$ we have $B^+ = \bar{w} + B \subset \operatorname{int} Z$. All we need to prove is that the linear forms $\langle \widehat{\omega}'(w_t), h\rangle$ of $h \in E$ are upper-bounded on B uniformly in $t \in (0, 1/2)$. The convexity of $\widehat{\omega}$ says that these forms are uniformly upper-bounded on B^+, since with $\|h\|_E \leq r$ we have

$$\begin{aligned}\langle \widehat{\omega}'(w_t), \bar{w} + h\rangle &= (1-t)^{-1}\langle \widehat{\omega}'(w_t), [\bar{w} + (1-t)h] - w_t\rangle \\ &\leq (1-t)^{-1}\Big[\max_{w \in B^+} \widehat{\omega}(w) - \min_{0 \leq t \leq 1} \widehat{\omega}(w_t)\Big] \\ &< \infty.\end{aligned}$$

All we need to derive upper bounds on the linear forms $f_t(h) = \langle \widehat{\omega}'(w_t), h\rangle$ on B^+, uniform in $0 < t \leq 1/2$, is to verify lower bounds on $\langle \widehat{\omega}'(w_t), \bar{w}\rangle$ which are uniform in $0 < t \leq 1/2$. Assume, to the contrary, that for some sequence $t_i > 0$ converging to 0 as $i \to \infty$, we have $p_i := \langle \widehat{\omega}'(w_{t_i}), \bar{w}\rangle \to -\infty$, and let $q = \langle \Psi'(\bar{w}), \bar{w}\rangle$ with $\Psi'(\bar{w}) \in \partial\Psi(\bar{w})$. Then, by convexity of Ψ and due to $w_t = t\bar{w}$, we have $\Psi(0) \geq \Psi(w_t) - qt$, and by convexity of $\widehat{\omega}(\cdot)$ we have $\widehat{\omega}(0) \geq \widehat{\omega}(w_t) - t\langle\widehat{\omega}'(w_t), \bar{w}\rangle$, so that $0 \geq \widehat{\omega}(0) + \Psi(0) - [\widehat{\omega}(w_{t_i}) + \Psi(w_{t_i})] \geq -t_i(p_i + q)$ (recall that $z_+ = 0$ is the minimizer of $\widehat{\omega} + \Psi$ on Z) for all i, which contradicts $p_i \to -\infty$, $i \to \infty$.

Now, since $0 = z_+ \in \operatorname{Dom}\Psi$ and $\operatorname{Dom}\Psi$ contains the interior of Z, from lower-semicontinuity and convexity of Ψ on the segment $[0, \bar{w}]$, it follows that Ψ is continuous on this segment. Setting $\psi(t) = \Psi(w_t)$, $\phi(t) = \widehat{\omega}(w_t)$, observe that from the smoothness properties of $\widehat{\omega}(\cdot)$ it follows that ϕ is continuously differentiable on $[0, 1]$ with the derivative $\phi'(t) = \langle\widehat{\omega}'(w_t), \bar{w}\rangle$. Since $\psi(t)+\phi(t)$ attains its minimum on $[0, 1]$ at $t = 0$, it immediately follows that the right derivative $\psi'(0)$ of ψ at $t = 0$ is $\geq -\phi'(0) = -\langle\widehat{\omega}'(0), \bar{w}\rangle$, whence $\Psi(\bar{w}) - \Psi(0) = \psi(1) - \psi(0) \geq \langle-\widehat{\omega}'(0), \bar{w}\rangle$. The relation $\Psi(\bar{w}) - \Psi(0) \geq -\langle\widehat{\omega}'(0), \bar{w}\rangle$ holds true for all $\bar{w} \in \operatorname{int} Z$ and therefore, by convexity of Ψ, for all $w \in Z$, meaning that $\Psi' := -\widehat{\omega}'(0) \in \partial\Psi(0)$, and of course $\langle\Psi'+\widehat{\omega}'(0), w\rangle \geq 0$ for all w. Finally, when $\Psi(\cdot)$ is differentiable at $z_+ = 0$, we have $\psi'(0) = \langle\nabla\Psi(0), \bar{w}\rangle$. Since we have already seen that $\psi'(0)+\phi'(0) \geq 0$, we get $\langle\nabla\Psi(0) + \widehat{\omega}'(0), \bar{w}\rangle \geq 0$. This inequality holds true for all $\bar{w} \in \operatorname{int} Z$ and thus for all $\bar{w} \in Z$. Recalling that $\widehat{\omega}'(0) = \omega'(0) + \nabla\phi(0)$, the proof is complete. □

Proof of relation (2.2). Applying Lemma A.1 with $\Psi(\cdot) \equiv 0$ and $\phi(w) = \langle\xi - \omega'(z), w\rangle$, so that $z_+ := \operatorname{Prox}_z(\xi)$ is the minimizer of $\omega(w)+\phi(w)+\Psi(w)$ over $w \in Z$, we get $z_+ \in Z^o$ and

$$\langle\omega'(z_+) + \xi - \omega'(z), w - z_+\rangle \geq 0, \quad \text{for all } w \in Z. \tag{A.1}$$

Plugging into (2.2) the definitions of Bregman distances, we immediately see that this is exactly (A.1). □

A.2. Proximal set-ups

Proof of Theorem 2.1. For $z \in Z' = \{z \in Z : z^j \neq 0\,\forall j\}$ we have

$$\gamma D\omega(z)[h] = \sum_{j=1}^{n} \|z^j\|_2^{p-2}\langle z^j, h^j\rangle$$

and

$$\begin{aligned}
\gamma D^2\omega(z)[h, h] &= -(2-p)\sum_{j=1}^{n} \|z^j\|_2^{p-4}[\langle z^j, h^j\rangle]^2 + \sum_{j=1}^{n} \|z^j\|_2^{p-2}\|h^j\|_2^2 \\
&\geq \sum_{j=1}^{n} \|z^j\|_2^{p-2}\|h^j\|_2^2 - (2-p)\sum_{j=1}^{n} \|z^j\|_2^{p-4}\|z^j\|_2^2\|h^j\|_2^2 \\
&\geq (p-1)\sum_{j=1}^{n} \|z^j\|_2^{p-2}\|h^j\|_2^2.
\end{aligned}$$

At the same time,

$$\begin{aligned}\left[\sum_j \|h^j\|_2\right]^2 &= \left[\sum_{j=1}^n [\|h^j\|_2\|z^j\|_2^{\frac{p-2}{2}}]\|z^j\|_2^{\frac{2-p}{2}}\right]^2 \\ &\le \left[\sum_{j=1}^n \|h^j\|_2^2\|z^j\|_2^{p-2}\right]\left[\sum_{j=1}^n \|z^j\|_2^{2-p}\right],\end{aligned}$$

and hence

$$\left[\sum_j \|h^j\|_2\right]^2 \le \left[\sum_{j=1}^n \|z^j\|_2^{2-p}\right]\frac{\gamma}{p-1}D^2\omega(z)[h,h].$$

Setting $t_j = \|z^j\|_2 \ge 0$, we have $\sum_j t_j \le 1$, whence due to $0 \le 2-p \le 1$ we have $\sum_j t_j^{2-p} \le n^{p-1}$. Thus

$$\left[\sum_j \|h^j\|_2\right]^2 \le n^{p-1}\frac{\gamma}{p-1}D^2\omega(z)[h,h], \tag{A.2}$$

while

$$\max_{z\in Z}\omega(z) - \min_{z\in Z}\omega(z) \le \frac{1}{\gamma p}. \tag{A.3}$$

With p, γ as stated in Theorem 2.1, when $n \ge 3$ we get $\frac{\gamma}{p-1}n^{p-1} = 1$, and similarly for $n = 1, 2$. Consequently,

$$\left[\sum_{j=1}^n \|h^j\|_2\right]^2 \le D^2\omega(z)[h,h], \quad \text{for all } z \in Z', h. \tag{A.4}$$

Since $\omega(\cdot)$ is continuously differentiable and the complement of Z' in Z is the union of finitely many proper linear subspaces of E, (A.4) implies that ω is strongly convex on Z, with modulus 1, with respect to the ℓ_1/ℓ_2 norm $\|\cdot\|_E$. Moreover, we have

$$\frac{1}{\gamma p}\left\{\begin{array}{ll} = \frac{1}{2}, & n = 1 \\ = 1, & n = 2 \\ \le \mathrm{e}\ln(n), & n \ge 3 \end{array}\right\} \le O(1)\ln(n+1),$$

which combines with (A.3) to imply (2.6). □

Proof of Corollary 2.2. Let

$$w(z) = \frac{1}{\gamma p}\sum_{j=1}^n \|z^j\|_2^p : E := \mathbb{R}^{k_1} \times \cdots \times \mathbb{R}^{k_n} \to \mathbb{R}, \tag{A.5}$$

with γ, p given by (2.5); note that $p \in (1,2]$. As we know from Theorem 2.1, the restriction of $w(\cdot)$ onto the unit ball of the ℓ_1/ℓ_2 norm $\|\cdot\|$ is continuously differentiable and strongly convex, with modulus 1, with respect to $\|\cdot\|$. Further, $w(\cdot)$ is clearly non-negative and positively homogeneous of degree p, *i.e.*, $w(\lambda z) \equiv |\lambda|^p w(z)$. Note that the outlined properties clearly imply that $w(\cdot)$ is convex and continuously differentiable on all of E. We need the following simple lemma.

Lemma A.2. Let H be a Euclidean space, let $\|\cdot\|_H$ be a norm (not necessarily the Euclidean one) on H, and let $Y_H = \{y \in H : \|y\|_H \leq 1\}$. Further, let $\zeta : H \to \mathbb{R}$ be a non-negative continuously differentiable function, absolutely homogeneous of a degree $p \in (1,2]$, and strongly convex, with modulus 1, with respect to $\|\cdot\|_H$, on Y_H. Then the function

$$\widehat{\zeta}(y) = \zeta^{2/p}(y) : H \to \mathbb{R}$$

is continuously differentiable, homogeneous of degree 2 and strongly convex, with modulus

$$\beta = \frac{2}{p}\left[\min_{y:\|y\|_H=1} \zeta^{2/p-1}(y)\right],$$

with respect to $\|\cdot\|_H$, on all of H.

Proof. It is evident that $\widehat{\zeta}(\cdot)$ is continuously differentiable, convex, and homogeneous of degree 2 on H. All we need is to verify that the function is strongly convex, with modulus β, with respect to $\|\cdot\|_H$:

$$\langle \widehat{\zeta}'(y) - \widehat{\zeta}'(y'), y - y' \rangle \geq \beta \|y - y'\|_H^2, \quad \text{for all } y, y'. \tag{A.6}$$

It is immediately seen that for a continuously differentiable function $\widehat{\zeta}$, the target property is local: to establish (A.6) it suffices to verify that every $\bar{y} \in H \backslash \{0\}$ admits a neighbourhood U such that the inequality in (A.6) holds for all $y, y' \in U$. Because of homogeneity, establishing the latter property for every $\bar{y} \in H \backslash \{0\}$ is the same as proving it when $\|\bar{y}\|_H = 1 - \epsilon$, for a fixed $\epsilon \in (0,1)$. Thus, let us fix an $\epsilon \in (0, 2/3)$ and $\bar{y} \in H$ with $\|\bar{y}\|_H = 1 - \epsilon$, and let $U = \{y' : \|y' - \bar{y}\|_H \leq \epsilon/2\}$, so that U is a compact set contained in the interior of Y_H. Now let $\phi(\cdot)$ be a non-negative C^∞ function on H which vanishes away from Y_H and has unit integral with respect to the Lebesgue measure on H. For $0 < r \leq 1$, let us set

$$\phi_r(y) = r^{-\dim H} \phi(y/r), \quad \zeta_r(\cdot) = (\zeta * \phi_r)(\cdot), \; \widehat{\zeta}_r(\cdot) = \zeta_r^{2/p}(\cdot),$$

where $*$ stands for convolution. As $r \to +0$, the non-negative C^∞ function $\zeta_r(\cdot)$ converges with the first-order derivative, uniformly on U, to $\zeta(\cdot)$, and $\widehat{\zeta}_r(\cdot)$ converges with the first-order derivative, uniformly on U, to $\widehat{\zeta}(\cdot)$. Further, when $r < \epsilon/2$, the function $\zeta_r(\cdot)$ is, along with $\zeta(\cdot)$, strongly convex, with modulus 1, with respect to $\|\cdot\|_H$, on U (since $\zeta_r'(\cdot)$ is the convolution of $\zeta'(\cdot)$ and $\phi_r(\cdot)$). Since ζ_r is C^∞, we conclude that when $r < \epsilon/2$ we have

$$D^2\zeta_r(y)[h,h] \geq \|h\|_H^2, \quad \text{for all } y \in U,\ h \in H,$$

whence also

$$\begin{aligned} D^2\widehat{\zeta}_r(y)[h,h] &= \frac{2}{p}\left(\frac{2}{p}-1\right)\zeta_r^{2/p-2}(y)[D\zeta_r(y)[h]]^2 + \frac{2}{p}\zeta_r^{2/p-1}(y)D^2\zeta_r[h,h] \\ &\geq \frac{2}{p}\zeta_r^{2/p-1}(y)\|h\|_H^2, \quad \text{for all } h \in H,\ y \in U. \end{aligned}$$

It follows that when $r < \epsilon/2$ we have

$$\langle \widehat{\zeta}_r'(y) - \widehat{\zeta}_r'(y'), y - y'\rangle \geq \beta_r[U]\|y - y'\|_H^2, \quad \text{for all } y, y' \in U,$$
$$\text{where } \beta_r[U] = \frac{2}{p}\min_{u\in U}\zeta_r^{2/p-1}(u).$$

As $r \to +0$, $\widehat{\zeta}_r'(\cdot)$ converges to $\widehat{\zeta}'(\cdot)$ uniformly on U, and $\beta_r[U]$ converges to $\beta[U] = \frac{2}{p}\min_{u\in U}\zeta^{2/p-1}(u) \geq (1-3\epsilon/2)^{2-p}\beta$ (recall the origin of β and U and note that ζ is absolutely homogeneous of degree p), we arrive at the relation

$$\langle \widehat{\zeta}'(y) - \widehat{\zeta}'(y'), y - y'\rangle \geq (1-3\epsilon/2)^{2-p}\beta\|y - y'\|_H^2, \quad \text{for all } y, y' \in U.$$

As we have already explained, the recently established validity of the latter relation for every $U = \{y : \|y - \bar{y}\|_H \leq \epsilon/2\}$ and every $\bar{y}$, $\|\bar{y}\|_H = 1 - \epsilon$ implies that $\widehat{\zeta}$ is strongly convex, with modulus $(1-3\epsilon/2)^{2-p}\beta$, on all of H. Since $\epsilon \in (0, 2/3)$ is arbitrary, Lemma A.2 is proved. □

Applying Lemma A.2 to E in the role of H, the ℓ_1/ℓ_2 norm $\|\cdot\|$ in the role of $\|\cdot\|_H$ and the function $w(\cdot)$ given by (A.5) in the role of ζ, we conclude that the function $\widehat{w}(\cdot) \equiv w^{2/p}(\cdot)$ is strongly convex, with modulus

$$\beta = \frac{2}{p}\left[\min_y\{w(y) : y \in E, \|y\| = 1\}\right]^{2/p-1},$$

with respect to $\|\cdot\|$, on the entire space E, whence the function $\beta^{-1}\widehat{w}(z)$ is continuously differentiable and strongly convex, with modulus 1, with respect to $\|\cdot\|$, on all of E. This observation, in view of the evident relation $\beta = \frac{2}{p}(n^{1-p}\gamma^{-1}p^{-1})^{2/p-1}$ and the origin of γ, p, immediately implies all the claims in Corollary 2.2. □

Proof of Theorem 2.3. Let $\mathbb{S}^n$ be the space of $n \times n$ symmetric matrices equipped with the Frobenius inner product. For $y \in \mathbb{S}^n$, let $\lambda(y)$ be the vector of eigenvalues of y (taken with their multiplicities in non-ascending order), and let $|y|_1 = \|\lambda(y)\|_1$ be the trace norm.

Lemma A.3. Let $N \geq M \geq 2$, and let F be a linear subspace in $\mathbb{S}^N$ such that every matrix $y \in F$ has at most M non-zero eigenvalues. Let $q = 1/(2\ln(M))$, so that $0 < q < 1$, and let $\widehat{\omega} : \mathbb{S}^N \to \mathbb{R}$ be given by

$$\widehat{\omega}(y) = \frac{4\sqrt{\mathrm{e}}\ln(M)}{1+q}\sum_{j=1}^{N}|\lambda_j(y)|^{1+q}.$$

The function $\widehat{\omega}(\cdot)$ is continuously differentiable, convex, and its restriction on the set $Y_F = \{y \in F : |y|_1 \leq 1\}$ is strongly convex, with modulus 1, with respect to $|\cdot|_1$.

Proof. **(1)** Let $0 < q < 1$. Consider the following function of $y \in \mathbb{S}^N$:

$$\chi(y) = \frac{1}{1+q}\sum_{j=1}^{N}|\lambda_j(y)|^{1+q} = \mathrm{Tr}(f(y)), \quad f(s) = \frac{1}{1+q}|s|^{1+q}. \tag{A.7}$$

(2) The function $f(s)$ is continuously differentiable on the axis and twice continuously differentiable away from the origin. Consequently, we can find a sequence of polynomials $f_k(s)$ converging to f, as $k \to \infty$, along with their first derivatives, uniformly on every compact subset of $\mathbb{R}$ and, further, converging to f uniformly, along with the first and second derivatives, on every compact subset of $\mathbb{R}\backslash\{0\}$. Now let $y, h \in \mathbb{S}^N$, let $y = u\,\mathrm{Diag}\{\lambda\}u^T$ be the eigenvalue decomposition of y, and let $h = u\widehat{h}u^T$. For a polynomial $p(s) = \sum_{k=0}^{K} p_k s^k$, setting $P(w) = \mathrm{Tr}(\sum_{k=0}^{K} p_k w^k)$, for $\omega \in \mathbb{S}^N$, and letting γ denote a closed contour in $\mathbb{C}$ enclosing the spectrum of y, we have

$$P(y) = \mathrm{Tr}(p(y)) = \sum_{j=1}^{N} p(\lambda_j(y)), \tag{A.8a}$$

$$DP(y)[h] = \sum_{k=1}^{K} kp_k \mathrm{Tr}(y^{k-1}h) = \mathrm{Tr}(p'(y)h) = \sum_{j=1}^{N} p'(\lambda_j(y))\widehat{h}_{jj}, \tag{A.8b}$$

$$\begin{aligned} D^2P(y)[h,h] &= \frac{\mathrm{d}}{\mathrm{d}t}\Big|_{t=0} DP(y+th)[h] = \frac{\mathrm{d}}{\mathrm{d}t}\Big|_{t=0}\mathrm{Tr}(p'(y+th)h) \\ &= \frac{\mathrm{d}}{\mathrm{d}t}\Big|_{t=0}\frac{1}{2\pi\mathrm{i}}\oint_\gamma \mathrm{Tr}(h(zI-(y+th))^{-1})p'(z)\,\mathrm{d}z \\ &= \frac{1}{2\pi\mathrm{i}}\oint_\gamma \mathrm{Tr}(h(zI-y)^{-1}h(zI-y)^{-1})p'(z)\,\mathrm{d}z \\ &= \frac{1}{2\pi\mathrm{i}}\oint_\gamma \sum_{i,j=1}^{N}\widehat{h}_{ij}^2\frac{p'(z)}{(z-\lambda_i(y))(z-\lambda_j(y))}\,\mathrm{d}z = \sum_{i,j=1}^{n}\widehat{h}_{ij}^2\Gamma_{ij}, \end{aligned} \tag{A.8c}$$

$$\text{where}\quad \Gamma_{ij} = \begin{cases} \frac{p'(\lambda_i(y))-p'(\lambda_j(y))}{\lambda_i(y)-\lambda_j(y)}, & \lambda_i(y) \neq \lambda_j(y), \\ p''(\lambda_i(y)), & \lambda_i(y) = \lambda_j(y). \end{cases}$$

We conclude from (A.8a) and (A.8b) that as $k \to \infty$, the real-valued polynomials $F_k(\cdot) = \mathrm{Tr}(f_k(\cdot))$ on $\mathbb{S}^N$ converge, along with their first-order derivatives, uniformly on every bounded subset of $\mathbb{S}^N$, and the limit of the sequence, by (A.8a), is exactly $\chi(\cdot)$. Thus, $\chi(\cdot)$ is continuously differentiable, and (A.8b) states that

$$D\chi(y)[h] = \sum_{j=1}^{N} f'(\lambda_j(y))\widehat{h}_{jj}. \tag{A.9}$$

Further, (A.8) states that if U is a closed convex set in $\mathbb{S}^N$ which does not contain singular matrices, then $F_k(\cdot)$ converges, as $k \to \infty$, uniformly on every compact subset of U, together with its first and second derivatives. Thus $\chi(\cdot)$ is twice continuously differentiable on U, and at every point $y \in U$ we have

$$D^2\chi(y)[h,h] = \sum_{i,j=1}^{N} \widehat{h}_{ij}^2 \Gamma_{ij}, \tag{A.10}$$

$$\text{where} \quad \Gamma_{ij} = \begin{cases} \frac{f'(\lambda_i(y)) - f'(\lambda_j(y))}{\lambda_i(y) - \lambda_j(y)}, & \lambda_i(y) \neq \lambda_j(y), \\ f''(\lambda_i(y)), & \lambda_i(y) = \lambda_j(y), \end{cases}$$

and in particular $\chi(\cdot)$ is convex on U.

(3) We intend to prove that (i) $\chi(\cdot)$ is convex, and (ii) its restriction to the set Y_F is strongly convex, with a certain modulus $\alpha > 0$, with respect to the trace norm $|\cdot|_1$. Since χ is continuously differentiable, all we need to prove (i) is to verify that

$$\langle \chi'(y') - \chi'(y''), y' - y'' \rangle \geq 0 \tag{A.11}$$

for a dense subset (y', y'') in $\mathbb{S}^N \times \mathbb{S}^N$, *e.g.*, those with non-singular $y' - y''$. If $y' - y''$ is non-singular, then the polynomial $q(t) = \mathrm{Det}(y' + t(y'' - y'))$, $t \in \mathbb{R}$, is not identically zero, and thus has finitely many roots in $[0,1]$. In other words, we can find finitely many points $t_0 = 0 < t_1 < \cdots < t_n = 1$ such that all 'matrix intervals' $\Delta_i = (y_i, y_{i+1})$, $y_k = y' + t_k(y'' - y')$, $1 \leq i \leq n-1$, contain only non-singular matrices. Therefore χ is convex on every closed segment contained in one of the Δ_i, and since χ is continuously differentiable, (A.11) follows.

(4) Now let us prove that with properly defined $\alpha > 0$ we have

$$\langle \chi'(y') - \chi'(y''), y' - y'' \rangle \geq \alpha |y' - y''|_1^2, \quad \text{for all } y', y'' \in Y_F. \tag{A.12}$$

Let $\epsilon > 0$, and let Y^ϵ be a convex and open in $Y = \{y : |y|_1 \leq 1\}$ subset of Y containing Y_F and such that, for all $y \in Y^\epsilon$, at most M eigenvalues of y are of magnitude $> \epsilon$. We intend to prove that for some $\alpha_\epsilon > 0$ we have

$$\langle \chi'(y') - \chi'(y''), y' - y'' \rangle \geq \alpha_\epsilon |y' - y''|_1^2, \quad \text{for all } y', y'' \in Y^\epsilon. \tag{A.13}$$

As above, it suffices to verify this relation for a dense set of pairs $y', y'' \in Y^\epsilon$ in $Y^\epsilon \times Y^\epsilon$, *e.g.*, those pairs $y', y'' \in Y^\epsilon$ for which $y' - y''$ is non-singular. Defining matrix intervals Δ_i as above and taking into account continuous differentiability of χ, it suffices to verify that, if $y \in \Delta_i$ and $h = y' - y''$, then $D^2\chi(y)[h,h] \geq \alpha_\epsilon |h|_1^2$. To this end, observe that by (A.10) all we have to prove is that

$$D^2\chi(y)[h,h] = \sum_{i,j=1}^{N} \widehat{h}_{ij}^2 \Gamma_{ij} \geq \alpha_\epsilon |h|_1^2. \tag{A.14}$$

(5) Setting $\lambda_j = \lambda_j(y)$, observe that $\lambda_i \neq 0$ for all i due to the origin of y. We claim that if $|\lambda_i| \geq |\lambda_j|$, then $\Gamma_{ij} \geq q|\lambda_i|^{q-1}$. Indeed, the latter relation definitely holds true when $\lambda_i = \lambda_j$. Now, if λ_i and λ_j are of the same sign, then

$$\Gamma_{ij} = \frac{|\lambda_i|^q - |\lambda|_j^q}{|\lambda_i| - |\lambda_j|} \geq q|\lambda_i|^{q-1},$$

since the derivative of the concave function t^q of $t > 0$ is positive and non-increasing (recall that $0 < q \leq 1$). If λ_i and λ_j are of different signs, then

$$\Gamma_{ij} = \frac{|\lambda_i|^q + |\lambda_j|^q}{|\lambda_i| + |\lambda_j|} \geq |\lambda_i|^{q-1}$$

due to $|\lambda_j|^q \geq |\lambda_j||\lambda_i|^{q-1}$, and therefore $\Gamma_{ij} \geq q|\lambda_i|^{q-1}$. Thus, our claim is justified.

Without loss of generality we can assume that the positive reals $\mu_i = |\lambda_i|$, $i = 1, \ldots, N$, form a non-decreasing sequence, so that, by the above, $\Gamma_{ij} \geq q\mu_j^{q-1}$ when $i \leq j$. Further, at most M of μ_j are $\geq \epsilon$, since $y', y'' \in Y^\epsilon$ and therefore $y \in Y^\epsilon$ by convexity of Y^ϵ. Hence

$$D^2\chi(y)[h,h] \geq 2q \sum_{i<j\leq N} \widehat{h}_{ij}^2 \mu_j^{q-1} + q \sum_{j=1}^{N} \widehat{h}_{jj}^2 \mu_j^{q-1},$$

or equivalently, by symmetry of $\widehat{h}$, if

$$h^j = \begin{bmatrix} & & & \widehat{h}_{1j} \\ & & & \widehat{h}_{2j} \\ & & & \vdots \\ \widehat{h}_{j1} & \widehat{h}_{j2} & \cdots & \widehat{h}_{jj} \end{bmatrix}$$

and $H_j = \|h^j\|_{\text{Fro}}$ is the Frobenius norm of h^j, then

$$D^2\chi(y)[h,h] \geq q \sum_{j=1}^{N} H_j^2 \mu_j^{q-1}. \tag{A.15}$$

Observe that the h^j are of rank ≤ 2, whence $|h^j|_1 \leq \sqrt{2}\|h^j\|_{\rm Fro} = \sqrt{2}H_j$, so that

$$
\begin{aligned}
|h|_1^2 = |\widehat{h}|_1^2 &\leq \left(\sum_{j=1}^N |h^j|_1\right)^2 \leq 2\left(\sum_{j=1}^N H_j\right)^2 = 2\left(\sum_j [H_j\mu_j^{(q-1)/2}]\mu_j^{(1-q)/2}\right)^2 \\
&\leq 2\left(\sum_{j=1}^N H_j^2\mu_j^{q-1}\right)\left(\sum_{j=1}^N \mu_j^{1-q}\right) \\
&\leq (2/q)D^2\chi(y)[h,h]\left(\sum_{j=1}^N \mu_j^{1-q}\right) \quad \text{(by (A.15))} \\
&\leq (2/q)D^2\chi(y)[h,h]\left((N-M)\epsilon^{1-q} + [M^{-1}\sum_{j=N-M+1}^N \mu_j]^{1-q}M\right) \\
&\qquad \text{(due to } 0<q<1 \text{ and } \mu_j \leq \epsilon,\ j \leq N-M) \\
&\leq (2/q)D^2\chi(y)[h,h]((N-M)\epsilon^{1-q} + M^q) \\
&\qquad \text{(since } \textstyle\sum_j \mu_j \leq 1 \text{ due to } y \in Y^\epsilon \subset \{y : |y|_1 \leq 1\}),
\end{aligned}
$$

and we see that (A.14) holds with $\alpha_\epsilon = q/(2[(N-M)\epsilon + M^q])$. As a result, with the α_ϵ just defined, relation (A.13) holds true, whence (A.12) is satisfied with $\alpha = \lim_{\epsilon\to+0}\alpha_\epsilon = \frac{q}{2}M^{-q}$, that is, χ is a continuously differentiable convex function which is strongly convex, with modulus α, with respect to $|\cdot|_1$, on Y_F. Recalling the definition of q, we see that $\widehat{\omega}(\cdot) = \alpha^{-1}\chi(\cdot)$, so that $\widehat{\omega}$ satisfies the conclusion of Lemma A.3. □

We are now ready to prove Theorem 2.3. Under the premises and in the notation of the theorem, let

$$
N = \mu + \nu, \quad M = 2\mu, \quad \mathcal{A}z = \frac{1}{2}\begin{bmatrix} & z \\ z^T & \end{bmatrix} \in \mathbb{S}^N, \quad \text{for } z \in \mathbb{R}^{\mu\times\nu}. \tag{A.16}
$$

Observe that the image space F of $\mathcal{A}$ is a linear subspace of $\mathbb{S}^N$, and that the eigenvalues of $y = \mathcal{A}z$ are the 2μ reals $\pm\sigma_i(z)/2$, $1 \leq i \leq \mu$, and $N-M$ zeros, so that $\|z\|_{\rm nuc} \equiv |\mathcal{A}z|_1$ and M, F satisfy the premises of Lemma A.3. Setting

$$
\omega(z) = \widehat{\omega}(\mathcal{A}z) = \frac{4\sqrt{\mathrm{e}}\ln(2\mu)}{2^q(1+q)}\sum_{i=1}^{\mu}\sigma_i^{1+q}(z), \quad q = \frac{1}{2\ln(2\mu)},
$$

and invoking Lemma A.3, we see that ω is a convex continuously differentiable function on $\mathbb{R}^{\mu\times\nu}$ which, due to the identity $\|z\|_{\rm nuc} \equiv |\mathcal{A}z|_1$, is strongly convex, with modulus 1, with respect to $\|\cdot\|_{\rm nuc}$, on the $\|\cdot\|_{\rm nuc}$-unit ball Z. This function clearly satisfies (2.10). □

Proof of Corollary 2.4. This corollary is obtained from Theorem 2.3 in exactly the same way as Corollary 2.2 was derived from Theorem 2.1. □

A.3. Proofs for Section 2.2.2

Proof of Lemma 2.7. Let $F(w) = f(w) - L\omega(w)$. Lemma A.1 as applied to $\phi \equiv 0$, $\Psi \equiv L^{-1}F$, implies that $\bar{w} \in Z^o$ and that there exists $F' \in \partial F(\bar{w})$ such that $\langle F' + L\omega'(\bar{w}), w - \bar{w}\rangle \geq 0$ for all $w \in Z$. Therefore,

$$\begin{aligned} f(w) &= F(w) + L\omega(w) \geq F(\bar{w}) + \langle F', w - \bar{w}\rangle + L\omega(w) \\ &\geq F(\bar{w}) - L\langle \omega'(\bar{w}), w - \bar{w}\rangle + L\omega(w) = F(\bar{w}) + L\omega(\bar{w}) + LV_{\bar{w}}(w) \\ &= f(\bar{w}) + LV_{\bar{w}}(w). \end{aligned}$$

□

Proof of Theorem 2.9. Let

$$\ell_t(x) \stackrel{\text{def}}{=} f(x_t) + \langle \nabla f(x_t), x - x_t\rangle + L_f V_{x_t}(x) + \Psi(x) \stackrel{(2.18)}{\geq} \phi(x).$$

Thus, $\ell_t(\cdot)$ satisfies the premises of Lemma 2.7 (note the definition of $V_{x_t}(x)$). Since x_{t+1} is the minimizer of this expression over $x \in Z$, Lemma 2.7 states that $x_{t+1} \in Z^o$ (so that the algorithm is well defined), and for every $w \in Z$ we have

$$\ell_t(x_{t+1}) + L_f V_{x_{t+1}}(w) \leq \ell_t(w).$$

Substituting $w = x_t$, we obtain

$$\phi(x_t) = \ell_t(x_t) \geq \ell_t(x_{t+1}) + L_f V_{x_{t+1}}(x_t) \geq \phi(x_{t+1}).$$

Thus, $\phi(x_{t+1}) \leq \phi(x_t)$, as claimed. Substituting $w = x^*$, we obtain

$$\ell_t(x_{t+1}) + L_f V_{x_{t+1}}(x^*) \leq \ell_t(x^*) \leq \phi(x^*) + L_f V_{x_t}(x^*),$$

whence $\phi(x_{t+1}) + L_f V_{x_{t+1}}(x^*) \leq \phi(x^*) + L_f V_{x_t}(x^*)$. Summing up these inequalities over $t = 0, \ldots, T-1$, we find

$$\sum_{t=1}^{T}(\phi(x_t) - \phi(x^*)) \leq L_f V_{x_0}(x^*) - L_f V_{x_T}(x^*) \leq L_f V_{x_0}(x^*),$$

which proves (2.20). The remaining claims of Theorem 2.9 are immediate corollaries of (2.20) due to the monotonicity of the algorithm, the inclusions $x_t \in Z$ and the convexity of ϕ. □

Proof of Theorem 2.11. Define the functions

$$\psi_t(x) = \sum_{i=0}^{t}[f(x_i) + \langle \nabla f(x_i), x - x_i\rangle] + (t+1)\Psi(x) + L_f V_{z_\omega}(x).$$

Let $\psi_t^* = \min_{w \in Z}\psi_t(w)$. Note that $\psi_t^* = \psi_t(x_{t+1})$ by construction of the algorithm, and that $x_t, y_t \in Z^o$ for all t by Lemma A.1.

Taking into account that $\psi_t(\cdot) - L_f\omega(\cdot)$ is convex, and that $x_{t+1} \in \arg\min_Z \psi_t$, by Lemma 2.7 we have

$$\psi_t^* + L_f V_{x_{t+1}}(x) \le \psi_t(x) \le (t+1)\phi(x) + L_f V_{z_\omega}(x), \quad \text{for } x \in Z. \tag{A.17}$$

Let us now prove that

$$\psi_t^* \ge \sum_{i=0}^{t} \phi(y_i). \tag{A.18}$$

For $t = 0$, we have $x_0 = z_\omega$. Therefore,

$$\begin{aligned} \psi_0^* &= \min_{w\in Z}[f(x_0) + \langle \nabla f(x_0), w - x_0\rangle + \Psi(x) + L_f V_{x_0}(w)] \\ &= f(x_0) + \langle \nabla f(x_0), y_0 - x_0\rangle + \Psi(y_0) + L_f V_{x_0}(y_0) \\ &\overset{(2.18)}{\ge} \phi(y_0). \end{aligned}$$

Assume now that (A.18) is true for some $t \ge 0$. Then

$$\begin{aligned} \psi_{t+1}^* &= \min_{w\in Z}\{\psi_t(w) + f(x_{t+1}) + \langle \nabla f(x_{t+1}), w - x_{t+1}\rangle + \Psi(w)\} \\ &\overset{(A.17)}{\ge} \min_{w\in Z}\{\psi_t^* + L_f V_{x_{t+1}}(w) + f(x_{t+1}) + \langle \nabla f(x_{t+1}), w - x_{t+1}\rangle + \Psi(w)\} \\ &\overset{(A.18)}{\ge} \sum_{i=0}^{t}\phi(y_i) + \min_{w\in Z}\{L_f V_{x_{t+1}}(w) + f(x_{t+1}) \\ &\qquad\qquad + \langle \nabla f(x_{t+1}), w - x_{t+1}\rangle + \Psi(w)\} \\ &= \sum_{i=0}^{t}\phi(y_i) + f(x_{t+1}) + \langle \nabla f(x_{t+1}), y_{t+1} - x_{t+1}\rangle \\ &\qquad\qquad + L_f V_{x_{t+1}}(y_{t+1}) + \Psi(y_{t+1}) \\ &\overset{(2.18)}{\ge} \sum_{i=0}^{t+1}\phi(y_i). \end{aligned}$$

Thus, we prove (A.18) for all $t \ge 0$.

Combining (A.18) and (A.17) and setting $x = x^*$, we arrive at (2.22). The remaining claims of Theorem 2.11 are immediate corollaries of (2.22) due to the inclusions $y_t \in Z$ and the convexity of ϕ. □

Proof of Theorem 2.14. Observe that by construction

$$\psi_t(\cdot) = L_f\omega(\cdot) + \ell_t(\cdot) + \alpha_t\Psi(\cdot),$$

with non-negative α_t and affine $\ell_t(\cdot)$, whence the algorithm is well defined and maintains the inclusions $z_t \in Z^o$, $x_t, y_t \in Z$. Further, let us introduce a convenient notation for the coefficients of Algorithm 2.13. Namely, let

$$a_{t+1} = \frac{t+2}{2}, \quad A_t = \sum_{i=1}^{t} a_i = \frac{t(t+3)}{4}, \quad t \geq 0.$$

Then $A_0 = 0$ and $\tau_t = \frac{a_{t+1}}{A_{t+1}}$. It is important that

$$\tau_t^2 A_{t+1} \leq 1, \quad t \geq 0. \tag{A.19}$$

Let $\psi_t^* = \psi_t(z_t)$. Let us prove by induction that

$$\psi_t^* \geq A_t \phi(y_t), \quad t \geq 0. \tag{A.20}$$

For $t = 0$ this inequality is valid. Suppose that it is true for some $t \geq 0$. Let us prove that it is valid for index $t+1$.

Taking into account the structure of $\psi_t(\cdot)$ just outlined, and applying Lemma A.1, for all $w \in Z$ we have

$$\begin{aligned}\psi_t(w) &\overset{(2.19)}{\geq} \psi_t^* + L_f V_{z_t}(w) \overset{(A.20)}{\geq} A_t \phi(y_t) + L_f V_{z_t}(w) \\ &\geq A_t\big[f(x_{t+1}) + \langle \nabla f(x_{t+1}), y_t - x_{t+1}\rangle + \Psi(y_t)\big] + L_f V_{z_t}(w).\end{aligned}$$

Therefore, for all $w \in Z$,

$$\begin{aligned}\psi_{t+1}(w) \geq L_f V_{z_t}(w) &+ A_t\big[f(x_{t+1}) + \langle \nabla f(x_{t+1}), y_t - x_{t+1}\rangle + \Psi(y_t)\big] \\ &+ a_{t+1}\big[f(x_{t+1}) + \langle \nabla f(x_{t+1}), w - x_{t+1}\rangle + \Psi(w)\big].\end{aligned}$$

Since $A_t(y_t - x_{t+1}) - a_{t+1}x_{t+1} \overset{\text{Alg. 2.13(b)}}{=} -a_{t+1}z_t$, we conclude that

$$\begin{aligned}\psi_{t+1}(w) &\geq L_f V_{z_t}(w) + A_{t+1} f(x_{t+1}) \\ &\quad + A_t \Psi(y_t) + a_{t+1}[\langle \nabla f(x_{t+1}), w - z_t\rangle + \Psi(w)]\end{aligned}$$

for all $w \in Z$. Thus,

$$\begin{aligned}\psi_{t+1}^* &\geq \min_{w\in Z}\big\{L_f V_{z_t}(w) + A_{t+1} f(x_{t+1}) + A_t \Psi(y_t) \\ &\qquad\qquad + a_{t+1}[\langle \nabla f(x_{t+1}), w - z_t\rangle + \Psi(w)]\big\} \\ &\overset{\text{Alg. 2.13(c)}}{=} L_f V_{z_t}(\hat{x}_{t+1}) + A_{t+1} f(x_{t+1}) + A_t \Psi(y_t) \\ &\qquad\qquad + a_{t+1}[\langle \nabla f(x_{t+1}), \hat{x}_{t+1} - z_t\rangle + \Psi(\hat{x}_{t+1})] \\ &\overset{\text{Alg. 2.13(d)}}{\geq} \frac{L_f}{2}\|\hat{x}_{t+1} - z_t\|_E^2 + A_{t+1} f(x_{t+1}) \\ &\qquad\qquad + a_{t+1}\langle \nabla f(x_{t+1}), \hat{x}_{t+1} - z_t\rangle + A_{t+1}\Psi(y_{t+1}).\end{aligned}$$

It remains to note that $\hat{x}_{t+1} - z_t \overset{\text{Alg. 2.13(b,d)}}{=} (y_{t+1} - x_{t+1})/\tau_t$. Therefore,

$$\begin{aligned}\psi_{t+1}^* &\geq \frac{L_f}{2\tau_t^2}\|y_{t+1} - x_{t+1}\|_E^2 + A_{t+1}f(x_{t+1}) \\ &\qquad + A_{t+1}\langle \nabla f(x_{t+1}), y_{t+1} - x_{t+1}\rangle + A_{t+1}\Psi(y_{t+1}) \\ &\overset{(A.19)}{\geq} A_{t+1}\left[\frac{L_f}{2}\|y_{t+1} - x_{t+1}\|_E^2 + f(x_{t+1})\right. \\ &\qquad \left. + \langle \nabla f(x_{t+1}), y_{t+1} - x_{t+1}\rangle + \Psi(y_{t+1})\right] \\ &\overset{(2.17)}{\geq} A_{t+1}\phi(y_{t+1}).\end{aligned}$$

Thus, we have proved inequality (A.20) for all $t \geq 0$. Since $\psi_t^* \leq A_t\phi(x^*) + L_f V_{z_\omega}(x^*)$, we obtain the estimate (2.25). □

A.4. Proof of Lemma 3.1

Verifying (3.9). Let $\bar{s} \geq 1$ be such that termination does not occur at stage $\bar{s}$. For $1 \leq s \leq \bar{s}$, let t_s be the last iteration of stage s, and let $\Phi_s = \Phi^{t_s,+}$, $\phi_s = \Phi^{t_s,-}$, so that $\phi_s \geq \Phi_s/\theta > 0$. By construction of $\Phi^{t,-}$, we have $\phi_s = \Phi_\tau^-(\rho_s)$ for some $\tau = \tau(s) \leq t_s$. Taking into account that $\Phi_\tau^-(\cdot)$ underestimates $\Phi(\cdot)$, $\Phi(\rho_*) = 0$, $\rho_s > \rho_*$ and $\Phi_\tau^-(\rho_s) > 0$ and setting $p_s = \frac{\mathrm{d}}{\mathrm{d}\rho}\Phi_\tau^-(\cdot)$, we get $p_s > 0$, and the root $r_\tau = \rho_s - \Phi_\tau^-(\rho_s)/p_s = \rho_s - \phi_s/p_s$ of $\Phi_\tau^-(\cdot)$ is a point of R_{t_s}, whence $\rho_{s+1} \leq r_\tau$. Further, assuming $s > 1$, we have $\phi_s + p_s(\rho_{s-1} - \rho_s) = \Phi_\tau^-(\rho_{s-1}) \leq \Phi(\rho_{s-1}) \leq \Phi_{s-1}$. The bottom line is that, for $1 \leq s \leq \bar{s}$, we have

$$\begin{aligned} &0 < \phi_s \leq \Phi(\rho_s) \leq \Phi_s, && (a_s) \\ &\Phi_s/\phi_s \leq \theta, && (b_s) \\ &\rho_s - \rho_{s+1} \geq \phi_s/p_s, && (c_s) \\ &\Phi_{s-1} \geq \phi_s + p_s(\rho_{s-1} - \rho_s) \quad \text{when } s > 1. && (d_s) \end{aligned} \tag{A.21}$$

As a result, for $1 \leq s < \bar{s}$ and using the notation of (A.21) henceforth,

$$\Phi_s - \phi_{s+1} \underset{(d_{s+1})}{\geq} p_{s+1}(\rho_s - \rho_{s+1}) \underset{(c_s)}{\geq} p_{s+1}\phi_s/p_s \underset{(b_s)}{\geq} \theta^{-1}p_{s+1}\Phi_s/p_s,$$

which taken together with (a_{s+1}) implies

$$\Phi_s - \theta^{-1}\Phi_{s+1} \geq \theta^{-1}p_{s+1}\Phi_s/p_s,$$

and hence

$$\Phi_s \geq \theta^{-1}\frac{p_{s+1}\Phi_s + p_s\Phi_{s+1}}{p_s} \geq 2\theta^{-1}\sqrt{p_{s+1}\Phi_s p_s\Phi_{s+1}}/p_s,$$

and finally

$$\Phi_{s+1}p_{s+1} \leq [\theta/2]^2\Phi_s p_s.$$

Thus,

$$\Phi_{\bar{s}} p_{\bar{s}} \le [\theta/2]^{2(\bar{s}-1)} \Phi_1 p_1. \tag{A.22}$$

Now $\Phi(\cdot)$ is Lipschitz-continuous with constant $2\pi(b)$, $\Phi_1 \le \theta\phi_1 \le \theta\Phi(\rho_1) \le 2\theta\pi(b)(\rho_1 - \rho_*)$ (since $\Phi(\rho_*) = 0$), whence $\Phi_1 \le 2\theta\pi(b)\rho_1$. Further, by construction $p_1 > 0$ is the slope of an affine function which underestimates $\Phi(\cdot)$ on the ray $\rho \ge 0$, whence $p_1 \le \lim_{\rho\to\infty} \Phi'(\rho) \le \pi(b) - \delta \le \pi(b)$, and thus $\Phi_1 p_1 \le 2\theta\pi^2(b)\rho_1$. Further, invoking ($c_{\bar{s}}$) and ($b_{\bar{s}}$) we have $p_{\bar{s}}(\rho_{\bar{s}} - \rho_{\bar{s}+1}) \ge \phi_{\bar{s}} \ge \theta^{-1}\Phi_{\bar{s}}$, so that $p_{\bar{s}} \ge \theta^{-1}\Phi_{\bar{s}}/(\rho_{\bar{s}} - \rho_{\bar{s}+1}) \ge \theta^{-1}\Phi_{\bar{s}}/\rho_1$, and thus $\Phi_{\bar{s}} p_{\bar{s}} \ge \theta^{-1}\Phi_{\bar{s}}^2/\rho_1$. We see that (A.22) implies the bound

$$\Phi_{\bar{s}}^2 \le 2\theta^2 [\theta/2]^{2(\bar{s}-1)} \pi^2(b)\rho_1^2.$$

On the other hand, our root finding does not terminate at stage $\bar{s}$, implying, due to the termination rule, that $\Phi_{\bar{s}} > \epsilon\rho_{\bar{s}} \ge \epsilon\rho_*$. We arrive at the bound

$$[2/\theta]^{\bar{s}} \le 2\sqrt{2}\pi(b)\rho_1/(\epsilon\rho_*).$$

Taking into account that $2/\theta > 1$ and $\bar{s}$ is the number of an arbitrary non-terminal stage, the total number of stages admits the bound (3.9).

Verifying (3.11). If t is the number of a non-terminal iteration of stage s, then by construction $\Phi^{t,+} > \epsilon\rho_s$ and $\Phi^{t,-} < \theta^{-1}\Phi^{t,+}$, implying that

$$\Phi^{t,+} - \Phi^{t,-} > \frac{\theta - 1}{\theta}\epsilon\rho_s \ge \frac{\theta - 1}{\theta}\epsilon\rho_*.$$

As we have seen when describing our root-finding routine, the left-hand side of this inequality does not exceed $\mathrm{DualityGap}^s(z^t)$, and we see that for our t we have

$$[\varepsilon_t \ge]\mathrm{DualityGap}^s(z^t) > \frac{\theta - 1}{\theta}\epsilon\rho_*,$$

which implies the bound (3.11). □

A.5. Proof of Theorem 3.3

We shall apply (2.2), first setting (a) $z = z_\tau$, $\xi = \gamma_\tau F(w_\tau)$ (so that $z_+ = z_{\tau+1}$), and then (b) $z = z_\tau, \xi = \gamma_\tau F(z_\tau)$ (so that $z_+ = w_\tau$). For (a), we obtain, for all $w \in Z$,

$$\gamma_\tau \langle F(w_\tau), z_{\tau+1} - w\rangle \le V_{z_\tau}(w) - V_{z_{\tau+1}}(w) - V_{z_\tau}(z_{\tau+1}),$$

and hence

$$\begin{aligned}&\gamma_\tau \langle F(w_\tau), w_\tau - w\rangle \\ &\qquad \le V_{z_\tau}(w) - V_{z_{\tau+1}}(w) + \underbrace{[\gamma_\tau \langle F(w_\tau), w_\tau - z_{\tau+1}\rangle - V_{z_\tau}(z_{\tau+1})]}_{=\delta_\tau},\end{aligned}$$

For (b), we find

$$\gamma_\tau\langle F(z_\tau), w_\tau - z_{\tau+1}\rangle \le V_{z_\tau}(z_{\tau+1}) - V_{w_\tau}(z_{\tau+1}) - V_{z_\tau}(w_\tau) \qquad \text{(A.23)}$$

and so

$$\begin{aligned}
\delta_\tau &= \gamma_\tau\langle F(w_\tau) - F(z_\tau), w_\tau - z_{\tau+1}\rangle - V_{z_\tau}(z_{\tau+1}) + \gamma_\tau\langle F(z_\tau), w_\tau - z_{\tau+1}\rangle \\
&\le \gamma_\tau\langle F(w_\tau) - F(z_\tau), w_\tau - z_{\tau+1}\rangle - V_{w_\tau}(z_{\tau+1}) - V_{z_\tau}(w_\tau) \quad \text{(by (A.23))} \\
&\le \gamma_\tau\|F(w_\tau) - F(z_\tau)\|_{E,*}\|w_\tau - z_{\tau+1}\|_E - \frac{1}{2}\|w_\tau - z_{\tau+1}\|_E^2 - \frac{1}{2}\|w_\tau - z_\tau\|_E^2 \\
&\le \frac{\gamma_\tau^2}{2}\|F(w_\tau) - F(z_\tau)\|_{E,*}^2 - \frac{1}{2}\|w_\tau - z_\tau\|_E^2 \\
&\le \frac{1}{2}[\gamma_\tau^2 L^2 - 1]\|w_\tau - z_\tau\|_E^2 \quad \text{(by (3.16))}.
\end{aligned}$$

We see that, first, $\gamma_\tau \le 1/L$ indeed ensures that $\delta_\tau \le 0$, and, second, whenever the latter relation holds for all τ we have

$$\gamma_\tau\langle F(w_\tau), w_\tau - w\rangle \le V_{z_\tau}(w) - V_{z_{\tau+1}}(w), \quad \text{for all } \tau, w \in Z.$$

Summing up the latter inequalities over $\tau = 1, \ldots, t$ and taking into account that $V_{z_1}(w) = V_{z_\omega}(w) \le \frac{1}{2}\Omega^2[Z']$ for $w \in Z'$, while $V_{z_{t+1}}(w) \ge 0$, we obtain

$$\max_{w\in Z'}\sum_{\tau=1}^{t}\lambda_\tau^t\langle F(w_\tau), w_\tau - w\rangle \le \frac{\frac{1}{2}\Omega^2[Z']}{\sum_{\tau=1}^t \gamma_\tau}, \quad \lambda_\tau^t = \frac{\gamma_\tau}{\sum_{s=1}^t \gamma_s}, \qquad \text{(A.24)}$$

for all t. Denoting $w_\tau = [u_\tau; v_\tau]$, $z^t = [u^t; v^t]$, we have

$$\begin{aligned}
&\max_{w=[u;v]\in Z'}\sum_{\tau=1}^{t}\lambda_\tau^t\langle F(w_\tau), w_\tau - w\rangle \\
&\quad = \max_{[u;v]\in Z'}\sum_{\tau=1}^{t}\lambda_\tau^t\left[\langle\nabla_u\phi(u_\tau, v_\tau), u_\tau - u\rangle + \langle\nabla_v\phi(u_\tau, v_\tau), v - v_\tau\rangle\right] \\
&\quad \ge \max_{[u;v]\in Z'}\sum_{\tau=1}^{t}\lambda_\tau^t\left[[\phi(u_\tau, v_\tau) - \phi(u, v_\tau)] + [\phi(u_\tau, v) - \phi(u_\tau, v_\tau)]\right] \\
&\qquad\qquad \text{(since } \phi \text{ is convex–concave)} \\
&\quad = \max_{[u;v]\in Z'}\sum_{\tau=1}^{t}\lambda_\tau^t\left[\phi(u_\tau, v) - \phi(u, v_\tau)\right] \\
&\quad \ge \max_{[u;v]\in Z'}\left[\phi(u^t, v) - \phi(u, v^t)\right] \\
&\qquad\qquad \text{(since } \phi \text{ is convex–concave and by definition of } z^t\text{)}.
\end{aligned}$$

The resulting inequality combines with (A.24) to imply (3.18). □

A.6. Good proximal set-ups

Our goal here is to understand to what extent the set-ups introduced in Section 2.1.1 are the 'best' ones for the applications considered in this paper. Let Z be a closed and bounded convex subset of a Euclidean space E which affinely spans E, so that the set $Z^+ = \frac{1}{2}(Z - Z)$ is the unit ball of some norm $\|\cdot\|_Z$ on E uniquely defined by Z. Also, let $\|\cdot\|_{Z,*}$ be the norm conjugate to $\|\cdot\|_Z$. We need the following simple observation.

Lemma A.4. With Z and E as above, let $\|\cdot\|_E$, $\omega(\cdot)$ be a proximal set-up for Z, E, let $F(x) = Ax+b : E \to E$ be an affine vector field, and let $L_E(F)$, $L_Z(F)$ be the Lipschitz constants of $F\big|_Z$ with respect to $(\|\cdot\|_E, \|\cdot\|_{E,*})$ and with respect to $(\|\cdot\|_Z, \|\cdot\|_{Z,*})$, respectively:

$$L_E(F) = \sup_{x,y\in Z, x\neq y} \frac{\|F(x)-F(y)\|_{E,*}}{\|x-y\|_E},$$
$$L_Z(F) = \sup_{x,y\in Z, x\neq y} \frac{\|F(x)-F(y)\|_{Z,*}}{\|x-y\|_Z}.$$

Then

$$L_E(F)\Omega^2[Z,\omega(\cdot)] \geq L_Z(F),$$

where $\Omega[Z,\omega(\cdot)]$ is the $\omega(\cdot)$-radius of Z.

Proof. Since Z affinely spans E and $F(x) = Ax + b$ is affine, we clearly have for all $x, y, u, v \in Z$:

$$\langle A(x-y), u-v\rangle \leq L_E(F)\|x-y\|_E\|u-v\|_E \leq 4L_E(F)\Omega^2[Z,\omega(\cdot)], \quad \text{(A.25)}$$

where the concluding inequality is given by (2.4) with $Z' = Z$. On the other hand, as x, y, u, v run through Z, $(x - y, u - v)$ can be made equal to any desired pair (g, h) with $g, h \in 2Z^+$. Consequently, the maximum of the left-hand side of (A.25) over $x, y, u, v \in Z$ is $4\max_{g\in E:\|g\|_Z\leq 1}\|Ag\|_{Z,*} \geq 4L_Z(F)$, and we obtain $4L_Z(F) \leq 4L_E(F)\Omega^2[Z,\omega(\cdot)]$, as claimed. □

Now assume that we are given a proximal set-up $\|\cdot\|_E, \omega(\cdot)$ for Z, E such that $\omega'(z_w) = 0$,[25] and

- either we are using the set-up in question to solve a problem (2.16) with the domain Z and a convex quadratic $f(x)$ by one of the methods from Section 2.2.2 (with the primal gradient method initialized by $z_0 = z_\omega$)
- or we are using the set-up to solve a *bilinear* saddle-point problem (3.13) with the domain $Z = U \times V$ by the MP algorithm from Section 3.2.

[25] This is without loss of generality, since we lose nothing (in fact we even gain by reducing the ω-radius of Z) when passing from the original DGF $\omega(\cdot)$ to the DGF $\omega(z) - \langle\omega'(z_\omega), z - z_\omega\rangle$.

In the first case, the guaranteed efficiency estimate of the algorithm is

$$\epsilon_t \le O(1)\Omega^2[Z,\omega(\cdot)]L_E(F)t^{-\kappa}, \tag{A.26}$$

where ϵ_t is the error after t steps, in terms of the objective of (2.16). Here $F(z) \equiv \nabla f(z)$, $\kappa = 1$ for the primal and dual gradient methods, and $\kappa = 2$ for the fast gradient method.[26] In the second case, invoking (3.19), the efficiency estimate is again (A.26), now with $\epsilon_t = \mathrm{DualityGap}(z^t)$, $\kappa = 1$ and the vector field (3.14) in the role of F. Note that this field is affine, due to the assumed bilinearity of ϕ.

Now assume that we have at our disposal an alternative proximal set-up $\|\cdot\|, \omega_Z(\cdot)$ for Z, E, with $\|\cdot\|$ θ-compatible with $\|\cdot\|_Z$:

$$\theta^{-1}\|z\| \le \|z\|_Z \le \theta\|z\|, \quad \text{for all } z.$$

Let us say that this alternative set-up is *good* if both θ and the quantity $\Omega_Z = \Omega[Z,\omega_Z(\cdot)]$ are 'moderate', *i.e.*, absolute constants, as is the case for the Euclidean set-up and the unit Euclidean ball Z ($\theta = \Omega_Z = 1$), or are universal functions of $\dim Z$ which grow with $\dim Z$ *logarithmically*, as is the case for the set-ups from Section 2.1.1, where $\theta = 1$ and $\Omega_Z \le O(1)\sqrt{\ln(\dim Z)}$. What happens when we pass from the proximal set-up $\|\cdot\|_E$, $\omega(\cdot)$ to this alternative set-up? The answer is clear: since $\|\cdot\|$ is within a factor θ of $\|\cdot\|_Z$, the Lipschitz constant L of F with respect to $\|\cdot\|, \|\cdot\|_*$ does not exceed $\theta^2 L_Z(F)$, and the right-hand side $O(1)\Omega_Z L t^{-\kappa}$ in the efficiency estimate associated with our alternative set-up will be $\le O(1)(\theta\Omega_Z)^2 L_Z(F)t^{-\kappa}$. That is, it will be at worst a factor $O(1)\theta^2\Omega_Z^2$ larger than the right-hand side of (A.26): see Lemma A.4. Since $\theta\Omega_Z$ is moderate, passing from the original set-up to the alternative one cannot 'spoil' the efficiency estimate by more than a factor of 'nearly $O(1)$'. In other words, as far as our algorithms are concerned, a good proximal set-up for the domain of interest is nearly optimal in terms of the efficiency estimate, with the given formal interpretation of the word 'nearly'.

The analysis above justifies the proximal set-ups used to handle the ℓ_1/ℓ_2 and nuclear norm Dantzig selector problems (1.1b). Indeed, using the approach of Section 3 we obtain bilinear saddle-point problems on bounded domains Z; the set-ups we use are good for these domains (and in fact result in the smallest values of $\theta\Omega_Z$ known to us, up to a factor of $O(1)$), and are thus nearly optimal within the framework of our algorithmic scheme.

The situation with the ℓ_1/ℓ_2 and nuclear norm lasso problems (1.1a), solved via composite minimization, is less clear. When posed in the form (2.16), these problems 'have trivial domain' $Z = E$, which makes the above

[26] To see that (A.26) is indeed the case, look at (2.21), (2.23) and (2.25), and note that x^* can be an arbitrary point of Z, while $\max_{x^*\in Z} V_{z_\omega}(x^*) = \frac{1}{2}\Omega^2[Z,\omega(\cdot)]$ due to $\omega'(z_w) = 0$.

analysis inapplicable and indeed makes the choice of proximal set-up a difficult task. Basically all gradient methods for lasso-type problems proposed in the literature, *e.g.*, the very popular algorithm FISTA of Beck and Teboulle (2009*a*), operate with the standard Euclidean set-up $\|z\| = \|z\|_2 := \sqrt{\langle z, z\rangle}$, $\omega(z) = \frac{1}{2}\langle z, z\rangle$. We intend to show, however, that the 'non-Euclidean' set-ups used in Section 2, while not being 'provably nearly optimal', in some cases have significant potential advantages compared to the Euclidean set-up. For the sake of definiteness, let us focus on the ℓ_1 lasso problem (1.1a) with $m \times n$ matrix A (the situations with the ℓ_1/ℓ_2 and nuclear norms are similar). Assume that the true signal x_* underlying observations b is sparse, *i.e.*, it has at most $s \ll n = \dim x_*$ non-zero entries, which is the standard motivation when using the ℓ_1 lasso.[27] For the s-sparse signal x_* we have $\|x_*\|_1 \le \sqrt{s}\|x_*\|_2$, and we can expect something similar from the optimal solution x^* to the lasso problem – at least in the low-noise regime, where x^* should be close to x_*, for otherwise why bother using lasso in the first place? Thus, it is natural to assume that $\|x^*\|_1 \le O(1)\sqrt{s}\|x^*\|_2$ with some $s \ll n$. This assumption allows us to compare the efficiency estimates

$$\epsilon_t \le O(1)\big(\sqrt{\ln(n)}\|x^*\|_1\|A\|_{\|\cdot\|_1,\|\cdot\|_2}\big)^2 t^{-\kappa}$$

(see (2.27)) of the gradient methods associated with the 'ℓ_1 proximal set-up' (the one given by Corollary 2.2 in the case of one-dimensional blocks z^j) with the estimates

$$\epsilon_t \le O(1)\big(\|x^*\|_2\|A\|_{\|\cdot\|_2,\|\cdot\|_2}\big)^2 t^{-\kappa}$$

(see (2.28)) associated with the Euclidean set-up. The ratio of the right-hand sides of these efficiency estimates is

$$\mathcal{R} = O(1)\Bigg(\underbrace{\frac{\sqrt{\ln(n)}\|x^*\|_1}{\|x^*\|_2}}_{P}\ \underbrace{\frac{\max_j \|\mathrm{Col}_j[A]\|_2}{\sigma_1(A)}}_{Q}\Bigg)^2.$$

The factor P is always ≥ 1 and thus is 'against' the ℓ_1 set-up. However, we are in the situation when P is not too large: $P \le O(1)\sqrt{\ln(n)s}$. In contrast, the factor Q is 'in favour of' the ℓ_1 set-up: it is always ≤ 1 and can be as small as $1/\sqrt{n}$ (look what happens when A is the $m \times n$ all-ones matrix). Thus, from the worst-case perspective, in the range $\sqrt{\ln(n)s} \ll \sqrt{n}$ the ℓ_1 set-up is significantly preferable to the Euclidean one. This being said, note that in some applications, *e.g.*, in compressive sensing, Q is typically significantly larger than its worst-case sharp lower bound $1/\sqrt{n}$. For example, matrices popular in compressive sensing are filled by independent

[27] An alternative motivation is to get a 'nearly sparse' approximation to x_* by penalizing the approximant with a 'sparsity-promoting' ℓ_1 norm; this motivation leads to the same conclusions as the one where x_* itself is sparse.

realizations of $\mathcal{N}(0, 1/m)$ random variables, or random variables taking values $\pm 1/\sqrt{m}$ with probability 1/2. For an $m \times n$ matrix A of this type, a typical value of $\max_j \|\mathrm{Col}_j[A]\|_2$ is $O(1)$, while a typical value of $\sigma_1(A)$ is $O(1)\sqrt{n/m}$, which yields $Q = O(1)\sqrt{m/n}$. When $m/n = O(1)$, which is often the case, $Q = O(1)\sqrt{m/n}$ certainly resolves the 'competition' between the ℓ_1 and the Euclidean set-ups in favour of the latter.

REFERENCES[28]

A. Argyriou, T. Evgeniou and M. Pontil (2008), 'Convex multi-task machine learning', *Mach. Learning* **73**, 243–272.

S. Arora and S. Kale (2007), A combinatorial, primal–dual approach to semidefinite programs. In *Proc. 39th Annual ACM Symposium on Theory of Computing*, ACM, pp. 227–236.

S. Arora, E. Hazan and S. Kale (2005), Fast algorithms for approximate semidefinite programming using the multiplicative weights update method. In *Proc. 46th Annual IEEE Symposium on Foundations of Computer Science*, IEEE Computer Society, pp. 339–348.

A. W. d'Aspremont (2011), 'Subsampling algorithms for semidefinite programming', *Stochastic Systems* **1**, 274–305.

F. R. Bach, J. Mairal and J. Ponce (2008), Convex sparse matrix factorizations. Preprint, Laboratoire d'Informatique de l'Ecole Normale Supérieure. arXiv:0812.1869v1

M. Baes, M. Bürgisser and A. Nemirovski (2013), Randomized Mirror-Prox method for solving structured large-scale matrix saddle-point problems. *SIAM J. Optim.*, to appear.

R. Baraniuk, V. Cevher, M. F. Duarte and C. Hegde (2010), 'Model-based compressive sensing', *IEEE Trans. Inform. Theory* **56**, 1982–2001.

A. Beck and M. Teboulle (2003), 'Mirror descent and nonlinear projected subgradient methods for convex optimization', *Oper. Res. Lett.* **31**, 167–175.

A. Beck and M. Teboulle (2009*a*), 'A fast iterative shrinkage-thresholding algorithm for linear inverse problems', *SIAM J. Imaging Sci.* **2**, 183–202.

A. Beck and M. Teboulle (2009*b*), 'Fast gradient-based algorithms for constrained total variation image denoising and deblurring problems', *IEEE Trans. Image Processing* **18**, 2419–2434.

S. Becker, E. Candès and M. Grant (2010), 'Templates for convex cone problems with applications to sparse signal recovery', *Math. Prog. Comput.* **3**, 165–218.

P. Bickel, Y. Ritov and A. Tsybakov (2008), 'Simultaneous analysis of Lasso and Dantzig selector', *Ann. Statist.* **37**, 1705–1732.

J.-F. Cai, E. Candès and Z. Shen (2008), 'A singular value thresholding algorithm for matrix completion', *SIAM J. Optim.* **20**, 1956–1982.

[28] The URLs cited in this work were correct at the time of going to press, but the publisher and the authors make no undertaking that the citations remain live or are accurate or appropriate.

E. Candès (2006), Compressive sampling. In *Proc. International Congress of Mathematicians 2006*, Vol. 3, *Invited Lectures* (M. Sanz-Solé, J. Soria, J. L. Varona and J. Verdera, eds), European Mathematical Society, pp. 1433–1452.

E. Candès and B. Recht (2008), 'Exact matrix completion via convex optimization', *Found. Comput. Math.* **9**, 717–772.

E. Candès and T. Tao (2006), 'Decoding by linear programming', *IEEE Trans. Inform. Theory* **51**, 4203–4215.

E. Candès and T. Tao (2007), 'The Dantzig selector: Statistical estimation when p is much larger than n', *Ann. Statist.* **35**, 2313–23516.

E. Candès and T. Tao (2009), 'The power of convex relaxation: Near-optimal matrix completion', *IEEE Trans. Inform. Theory* **56**, 2053–2080.

E. Candès, J. Romberg and T. Tao (2006), 'Stable signal recovery from incomplete and inaccurate measurements', *Comm. Pure Appl. Math.* **59**, 1207–1223.

A. Chambolle (2004), 'An algorithm for total variation minimization and applications', *J. Math. Imaging Vision* **20**, 89–107.

V. Chandrasekaran, B. Recht, P. Parrilo and A. Willsky (2012), 'The convex geometry of linear inverse problems', *Found. Comput. Math.* **12**, 805–849.

V. Chandrasekaran, D. Sanghavi, P. Parrilo and A. Willsky (2011), 'Rank-sparsity incoherence for matrix decomposition', *SIAM J. Optim.* **21**, 572–596.

C. Chesneau and M. Hebiri (2008), 'Some theoretical results on the grouped variables lasso', *Math. Methods Statist.* **27**, 317–326.

A. Cohen, W. Dahmen and R. DeVore (2009), 'Compressed sensing and best k-term approximation', *J. Amer. Math. Soc.* **22**, 211–231.

W. Dai, M. Sheikh, O. Milenkovic, and R. Baraniuk (2009), 'Compressive sensing DNA microarrays', *EURASIP J. Bioinformatics and Systems Biology* **2009**.

O. Devolder (2011), Stochastic first order methods in smooth convex optimization. CORE Discussion paper 2011/70.

D. Donoho and J. Tanner (2005), 'Sparse nonnegative solutions of underdetermined linear equations by linear programming', *Proc. Natl Acad. Sci. USA* **102**, 9446–9451.

D. Donoho, M. Elad and V. Temlyakov (2006), 'Stable recovery of sparse overcomplete representations in the presence of noise', *IEEE Trans. Inform. Theory* **53**, 6–18.

Y. Eldar, P. Kuppinger and H. Bölcskei (2010), 'Block-sparse signals: uncertainty relations and efficient recovery', *IEEE Trans. Signal Processing* **58**, 3042–3054.

E. Elhamifar and R. Vidal (2012), 'Block-sparse recovery via convex optimization', *IEEE Trans. Signal Processing* **60** 4094–4107.

D. Goldfarb and S. Ma (2011), 'Convergence of fixed-point continuation algorithms for matrix rank minimization', *Found. Comput. Math.* **11**, 183–210.

D. Goldfarb and W. Yin (2009), 'Parametric maximum flow algorithms for fast total variation minimization', *SIAM J. Sci. Comput.* **31**, 3712–3743.

D. Goldfarb, K. Scheinberg and B. Xi (2011), Fast first order methods for composite convex optimization with backtracking. http://www.optimization-online.org/DB_HTML/2011/04/3004.html

M. D. Grigoriadis and L. G. Khachiyan (1995), 'A sublinear-time randomized approximation algorithm for matrix games', *Oper. Res. Lett.* **18**, 53–58.

M. Herman and T. Strohmer (2007), 'High-resolution radar via compressed sensing', *IEEE Trans. Signal Process.* **57**, 2275–2284.

B. Huang, S. Ma and D. Goldfarb (2013), 'Accelerated linearized Bregman method', *J. Sci. Comput.* **54**, 428–453.

J. Huang and T. Zhang (2010), 'The benefit of group sparsity', *Ann. Statist.* **38**, 1978–2004.

M. Jaggi and M. Sulovský (2010), A simple algorithm for nuclear norm regularized problems. In *ICML 2010: Proc. 27th International Conference on Machine Learning, June 2010*, Omnipress, pp. 471–478. www.icml2010.org/papers/196.pdf

A. Juditsky and A. Nemirovski (2011*a*), 'On verifiable sufficient conditions for sparse signal recovery via ℓ_1 minimization', *Math. Program.* B **127**, 57–88.

A. Juditsky and A. Nemirovski (2011*b*), First-order methods for nonsmooth large-scale convex minimization: I General purpose methods; II Utilizing problem's structure. In *Optimization for Machine Learning* (S. Sra, S. Nowozin and S. Wright, eds), The MIT Press, pp. 121–183.

A. Juditsky and A. Nemirovski (2011*c*), 'Accuracy guarantees for ℓ_1 recovery', *IEEE Trans. Inform. Theory* **57**, 7818–7839.

A. Juditsky, F. Kılınç Karzan and A. Nemirovski (2011*a*), 'Verifiable conditions of ℓ_1-recovery of sparse signals with sign restrictions', *Math. Program.* B **127**, 89–122.

A. Juditsky, F. Kılınç Karzan and A. Nemirovski (2013*a*), Randomized first order algorithms with applications to ℓ_1-minimization. *Math. Program.*, to appear.

A. Juditsky, F. Kılınç Karzan, A. S. Nemirovski and B. T. Polyak (2011*b*), On the accuracy of ℓ_1-filtering of signals with block-sparse structure. In *Advances in Neural Information Processing Systems 24* (J. Shawe-Taylor *et al.*, eds), Neural Information Processing Systems Foundation (NIPS), pp. 1260–1268.

A. Juditsky, F. Kılınç Karzan, A. Nemirovski and B. Polyak (2013*b*), Accuracy guarantees for ℓ_1 recovery of block-sparse signals. *Ann. Statist.*, to appear.

A. Juditsky, A. Nemirovski and C. Tauvel (2011*c*), 'Solving variational inequalities with Stochastic Mirror Prox algorithm', *Stochastic Systems* **1**, 17–58.

J. Lee, B. Recht, R. Salakhutdinov, N. Srebro and J. Tropp (2010), Practical large-scale optimization for max-norm regularization. In *Advances in Neural Information Processing Systems 23* (J. Lafferty *et al.*, eds), NIPS, pp. 1297–1305.

C. Lemarechal, A. Nemirovski and Y. Nesterov (1995), 'New variants of bundle methods', *Math. Program.* **69**, 111–148.

H. Liu and J. Zhang (2009), 'Estimation consistency of the group lasso and its applications', *J. Mach. Learning Res. Proc. Track* **5**, 376–383.

H. Liu, J. Shang, X. Jiang and J. Liu (2010), 'The group Dantzig selector', *J. Mach. Learning Res. Proc. Track* **9**, 461–468.

Y.-J. Liu, D. Sun and K.-C. Toh (2012), 'An implementable proximal point algorithmic framework for nuclear norm minimization', *Math Program.* **133**, 399–436.

K. Lounici, M. Pontil, S. van de Geer and A. Tsybakov (2011), 'Oracle inequalities and optimal inference under group sparsity', *Ann. Statist.* **39**, 2164–2204.

M. Lustig, D. Donoho and J. M. Pauly (2007), 'Sparse MRI: The application of compressed sensing for rapid MR imaging', *Magnetic Resonance in Medicine* **56**, 1182–1195.

S. Ma, D. Goldfarb and L. Chen (2011), 'Fixed point and Bregman iterative methods for matrix rank minimization', *Math. Program.* **128**, 321–353.

L. Meier, S. van de Geer and P. Bühlmann (2008), 'The group lasso for logistic regression', *J. Roy. Statist. Soc.* B **70**, 53–71.

N. Meinshausen and B. Yu (2009), 'Lasso-type recovery of sparse representations for high-dimensional data', *Ann. Statist.* **37**, 246–270.

H. Men, N. Nguyen, R. Freund, K. Lim, P. Parrilo and J. Peraire (2011), 'Design of photonic crystals with multiple and combined band gaps', *Phys. Rev.* E **83**, 046703.

Y. Nardi and A. Rinaldo (2008), 'On the asymptotic properties of the group lasso estimator for linear models', *Electron. J. Statist.* **2**, 605–633.

A. Nemirovski (1992), 'Information-based complexity of linear operator equations', *J. Complexity* **8**, 153–175.

A. Nemirovski (2004), 'Prox-method with rate of convergence $O(1/t)$ for variational inequalities with Lipschitz continuous monotone operators and smooth convex–concave saddle point problems', *SIAM J. Optim.* **15**, 229–251.

A. Nemirovski, A. Juditsky, G. Lan and A. Shapiro (2009), 'Stochastic approximation approach to stochastic programming', *SIAM J. Optim.* **19**, 1574–1609.

Y. Nesterov (1983), 'A method for solving a convex programming problem with rate of convergence $O(1/k^2)$', *Soviet Math. Doklady* **27**, 372–376.

Y. Nesterov (2005*a*), 'Smooth minimization of non-smooth functions', CORE Discussion Paper 2003/12 and *Math. Program.* **103**, 127–152.

Y. Nesterov (2005*b*), 'Excessive gap technique in nonsmooth convex minimization', *SIAM J. Optim.* **16**, 235–239.

Y. Nesterov (2007*a*), 'Dual extrapolation and its application for solving variational inequalities and related problems', *Math. Program.* **109**, 319–344.

Y. Nesterov (2007*b*), Gradient methods for minimizing composite objective functions. CORE Discussion Paper 2007/76.

Y. Nesterov (2009), 'Primal–dual subgradient methods for convex problems', *Math. Program.* B **120**, 221–259.

Y. Nesterov and A. Nemirovski (1994), *Interior Point Polynomial Algorithms in Convex Programming.* SIAM Studies in Applied Mathematics.

Y. Nesterov and B. Polyak (2006), 'Cubic regularization of Newton's method and its global performance', *Math. Program.* **108**, 177–205.

S. Osher, Y. Mao, B. Dong and W. Yin (2010), 'Fast linearized Bregman iteration for compressive sensing and sparse denoising', *Commun. Math. Sci.* **8**, 93–111.

F. Parvaresh, H. Vikalo, S. Misra and B. Hassibi (2008), 'Recovering sparse signals using sparse measurement matrices in compressed DNA microarrays', *IEEE J. Selected Topics Signal Process.* **2**, 275–285.

Z. Qin and D. Goldfarb (2012), Structured sparsity via alternating direction methods. *J. Mach. Learning Res.* **13**, 1435–1468.

B. Recht and C. Ré (2011), Parallel stochastic gradient algorithms for large-scale matrix completion.
http://www.optimization-online.org/DB_HTML/2011/04/3012.html

B. Recht, M. Fazel and P. Parrilo (2010), 'Guaranteed minimum-rank solutions of linear matrix equations via nuclear norm minimization', *SIAM Rev.* **52**, 471–501.

B. Recht, C. Ré, S. Wright and F. Niu (2011*a*), Hogwild: A lock-free approach to parallelizing stochastic gradient descent. In *Advances in Neural Information Processing Systems 24* (J. Shawe-Taylor *et al.*, eds), NIPS, pp. 693–701.

B. Recht, W. Xuy and B. Hassibi (2011*b*), 'Null space conditions and thresholds for rank minimization', *Math. Program.* B **127**, 175–211.

L. Rudin, S. Osher and E. Fatemi (1992), 'Nonlinear total variation based noise removal algorithms', *Physica* D **60**, 259–268.

S. Santosa and W. Symes (1986), 'Linear inversion of band-limited reflection seismograms', *SIAM J. Sci. Comput.* **7**, 1307–1330.

K. Scheinberg, S. Ma and D. Goldfarb (2010), Sparse inverse covariance selection via alternating linearization methods. In *Advances in Neural Information Processing Systems 23* (J. Lafferty *et al.*, eds), NIPS, pp. 2101–2109.

J. Shi, W. Yin, S. Osher and P. Sajda (2010), 'A fast hybrid algorithm for large-scale ℓ_1-regularized logistic regression', *J. Mach. Learning Res.* **11**, 713–741.

N. Srebro and A. Shraibman (2005), Rank, trace-norm and max-norm. In *Proc. 18th Annual Conference on Learning Theory* (COLT), Vol. 3559 of *Lecture Notes in Computer Science*, Springer, pp. 545–560.

V. Studer, J. Bobin, M. Chahid, H. Mousavi, E. Candès and M. Dahan (2012), 'Compressive fluorescence microscopy for biological and hyperspectral imaging', *Proc. Natl Acad. Sci. USA* **109**, E1679–E1687.

H. Taylor, S. Banks and J. McCoy (1979), 'Deconvolution with the ℓ_1-norm', *Geophys.* **44**, 39–52.

R. Tibshirani (1996), 'Regression shrinkage and selection via the lasso', *J. Roy. Statist. Soc.* B **58**, 267–288.

J. Tropp (2006), 'Just relax: Convex programming methods for identifying sparse signals', *IEEE Trans. Inform. Theory* **51**, 1030–1051.

P. Tseng (2000), 'A modified forward–backward splitting method for maximal monotone mappings', *SIAM J. Control Optim.* **38**, 431–446.

P. Tseng (2008), On accelerated proximal gradient methods for convex–concave optimization. Technical report. Submitted to *SIAM J. Optim.* www.math.washington.edu/~tseng/papers/apgm.pdf

S. Vasanawala, M. T. Alley, B. A. Hargreaves, R. A. Barth, J. M. Pauly and M. Lustig (2010), 'Improved pediatric MR imaging with compressed sensing', *Radiology* **256**, 607–616.

G. Wagner, P. Schmieder, A. Stern and J. Hoch (1993), 'Application of nonlinear sampling schemes to cosy-type spectra', *J. Biomolecular NMR* **3**, 569–576.

Z. Wen, W. Yin, D. Goldfarb and Y. Zhang (2010), 'A fast algorithm for sparse reconstruction based on shrinkage, subspace optimization, and continuation', *SIAM J. Sci. Comput.* **32**, 1832–1857.

Z. Wen, W. Yin, H. Zhang and D. Goldfarb (2012), 'On the convergence of an active-set method for ℓ_1 minimization', *Optim. Software* **27**, 1127–1146.

J. Yang and X. Yuan (2013), 'Linearized augmented Lagrangian and alternating direction methods for nuclear norm minimization', *Math. Comp.* **82**, 301–329.

W. Yin, S. Osher, D. Goldfarb and J. Darbon (2008), 'Bregman iterative algorithms for L_1-minimization with applications to compressed sensing', *SIAM J. Imaging Sci.* **1**, 143–168.

M. Yuan and Y. Lin (2006), 'Model selection and estimation in regression with grouped variables', *J. Roy. Statist. Soc.* B **68**, 49–67.

For EU product safety concerns, contact us at Calle de José Abascal, 56–1°,
28003 Madrid, Spain or eugpsr@cambridge.org.

www.ingramcontent.com/pod-product-compliance
Lightning Source LLC
LaVergne TN
LVHW062202200926
848484LV00006B/1303